Southeast Asia
North America
East
South America
Southwest Asia & North Africa
Europe
Russia & the Eurasian Republics
Asia
South Asia
Southeast Asia
Central America & the Caribbean
Sub-Saharan Africa
Australia, the Pacific Realm & Antarctica
Europe
South Asia
South Asia
Southeast Asia
Central America & the Caribbean
Australia / The Pacific Realm & Antarctica
East Asia
North America
Sub-Saharan Africa
South America
Southwest Asia & North Africa
South America
North America

Acknowledgments

Grateful acknowledgment is given to the authors, artists, photographers, museums, publishers, and agents for permission to reprint copyrighted material. Every effort has been made to secure the appropriate permission. If any omissions have been made or if corrections are required, please contact the Publisher.

Photographic Credits

Front Cover: © Roger Ressmeyer/Corbis
Back Cover: © Neale Clark/Robert Harding World Imagery/Getty Images

Acknowledgments and credits continued on page R111.

Visit National Geographic Learning online at www.NGSP.com
Visit our corporate website at www.cengage.com

Printed in the USA
RR Donnelley
Jefferson City, MO

ISBN: 978-07362-9000-5

13 14 15 16 17 18 19 20

10 9 8 7 6 5 4 3

Andrew J. Milson

Andrew Milson is a professor of social science education and geography at the University of Texas at Arlington. He taught middle school history and geography near Dallas, Texas. Andy conducts research on geographic education and the use of geospatial technologies in education. He has published more than 30 articles and is an elected member of the Executive Board of the National Council for Geographic Education. He serves as an associate editor of the *Journal of Geography*.

Peggy Altoff

Peggy Altoff's experience includes teaching middle school and high school students, supervising teachers, and serving as adjunct university staff. Peggy served as a state social studies specialist in Baltimore and as a K–12 facilitator in Colorado Springs. She was president of the National Council for the Social Studies (NCSS) in 2006–2007 and was on the task force for the new NCSS National Curriculum Standards.

Mark H. Bockenhauer

Mark Bockenhauer is a professor of geography at St. Norbert College and a former geographer-in-residence at the National Geographic Society. Mark has extensive experience in teacher professional development. He co-wrote *Our Fifty States* and the *World Atlas for Young Explorers, 3rd edition*—both for National Geographic. Mark is coordinator of the Wisconsin Geographic Alliance, and he served as president of the National Council for Geographic Education in 2007.

Janet Smith

Jan Smith is an associate professor of geography at Shippensburg University. Jan began her teaching career as a high school teacher in Virginia where she served as a teacher consultant for the Virginia Geographic Alliance for many years. Her primary research interest focuses on how children develop their spatial thinking skills. Jan served as president of the National Council for Geographic Education in 2008, and she is currently the coordinator for the Pennsylvania Geographic Alliance.

Michael W. Smith

Michael Smith is a professor in the Department of Curriculum, Instruction, and Technology in Education at Temple University. He became a college teacher after 11 years of teaching high school English. His research focuses on how experienced readers read and talk about texts, as well as what motivates adolescents' reading and writing. Michael has written many books and monographs, including the award-winning *"Reading Don't Fix No Chevys": Literacy in the Lives of Young Men.*

David W. Moore

David Moore is a professor of education at Arizona State University. He taught high school social studies and reading before entering college teaching. He currently teaches teacher preparation courses and conducts research in adolescent literacy. David has published numerous professional articles, book chapters, and books, including *Developing Readers and Writers in the Content Areas* and *Principled Practices for Adolescent Literacy.*

CONSULTANTS AND REVIEWERS

Teacher Reviewers

Kayce Forbes
Deerpark Middle School
Austin, Texas

Michael Koren
Maple Dale School
Fox Point, Wisconsin

Patricia Lewis
Humble Middle School
Humble, Texas

Julie Mitchell
Lake Forest Middle School
Cleveland, Tennessee

Linda O'Connor
Northeast Independent
School District
San Antonio, Texas

Leah Perry
Exploris Middle School
Raleigh, North Carolina

Robert Poirier
North Andover Middle School
North Andover, Massachusetts

Heather Rountree
Bedford Heights Elementary
Bedford, Texas

Erin Stevens
Quabbin Regional
Middle/High School
Barre, Massachusetts

Beth Tipper
Crofton Middle School
Crofton, Maryland

Mary Trichel
Atascocita Middle School
Humble, Texas

Andrea Wallenbeck
Exploris Middle School
Raleigh, North Carolina

Reviewers of Religious Content

The following individuals reviewed the treatment of religious content in selected pages of the text.

Charles Haynes
First Amendment Center
Washington, D.C.

Shabbir Mansuri
Institute on Religion and
Civic Values
Fountain Valley, California

Susan Mogull
Institute for Curriculum Reform
San Francisco, California

Raka Ray
Chair, Center for South Asia Studies
University of California
Berkeley, California

National Geographic Society

The National Geographic Society contributed significantly to *World Cultures and Geography*. Our collaboration with each of the following has been a pleasure and a privilege: National Geographic Maps, National Geographic Education Programs, National Geographic Missions Programs, National Geographic Digital Motion, National Geographic Digital Studio, and National Geographic Weekend. We thank the Society for its guidance and support.

Greg Anderson
Linguist
National Geographic Fellow

Greg Anderson records and preserves many different endangered languages. He is also the co-director of the Enduring Voices Project.

Thomas Taha Rassam (TH) Culhane
Urban Planner
National Geographic Emerging Explorer

T.H. Culhane works with residents of Cairo to install solar water heaters.

Katey Walter Anthony
Aquatic Ecologist and Biogeochemist
National Geographic Emerging Explorer

Katey Walter Anthony explores ways to use a greenhouse gas for energy.

Jenny Daltry
Herpetologist
National Geographic Emerging Explorer

Jenny Daltry saves endangered species of reptiles and inspires local people to protect reptiles and their habitats.

Katy Croff Bell
Archaeological Oceanographer
National Geographic Emerging Explorer

Katy Croff Bell uses deep sea technology to explore the depths of the ocean and the Black Sea.

Wade Davis
Anthropologist and Ethnobotanist
National Geographic Explorer-in-Residence

Wade Davis studies plants and people while living within many indigenous cultures around the world.

Alexandra Cousteau
Social Environmental Activist
National Geographic Emerging Explorer

Alexandra Cousteau works to educate people to protect water resources and oceans.

Sylvia Earle
Oceanographer
National Geographic Explorer-in-Residence

Sylvia Earle's research focuses on exploring and preserving marine ecosystems.

Grace Gobbo
Ethnobotanist
*National Geographic
Emerging Explorer*

Grace Gobbo studies traditional medicine practices in Tanzania and plants native to the region.

Beverly Goodman
Geoarchaeologist
*National Geographic
Emerging Explorer*

Beverly Goodman uses her skills to uncover past tsunamis and to help prevent disasters in the future.

David Harrison
Linguist
National Geographic Fellow

David Harrison studies and archives endangered languages and cultures. He is also the co-director of the Enduring Voices Project.

Kristofer Helgen
Zoologist
*National Geographic
Emerging Explorer*

Kristofer Helgen discovers new species of mammals and researches animals from across the world.

Fredrik Hiebert
Archaeologist
National Geographic Fellow

Fredrik Hiebert uncovers mysteries of the past and has traced ancient trade routes, including the Silk Road.

Zeb Hogan
Aquatic Ecologist
National Geographic Fellow

Zeb Hogan studies freshwater fish and educates people on how to save these fish species from extinction.

Shafqat Hussain
Conservationist
*National Geographic
Emerging Explorer*

Shafqat Hussain works with herders in Pakistan to protect the endangered snow leopard.

Beverly Joubert
Filmmaker and Conservationist
*National Geographic
Explorer-in-Residence*

Beverly Joubert has spent years filming and protecting big cats and other wildlife in Africa.

Dereck Joubert
Filmmaker and Conservationist
National Geographic Explorer-in-Residence

Dereck Joubert researches and films big cats and wildlife in Africa, sharing their stories with the world.

Enric Sala
Marine Ecologist
National Geographic Explorer-in-Residence

Enric Sala dedicates his life to finding ways to reverse the damage humans have caused to the seas.

Albert Lin
Research Scientist and Engineer
National Geographic Emerging Explorer

Albert Lin uses computer technologies to search for archeological sites without disturbing the land.

Kira Salak
Writer/Adventurer
National Geographic Emerging Explorer

Kira Salak is an adventure traveler who writes about her explorations in exotic, often dangerous places.

Elizabeth Kapu'uwailani Lindsey
Filmmaker and Anthropologist
National Geographic Fellow

Elizabeth Lindsey strives to preserve the Polynesian culture by using documentary films and education.

Katsufumi Sato
Behavioral Ecologist
National Geographic Emerging Explorer

Katsufumi Soto uses technology to study animal behaviors with the goal of conserving their habitats.

Kakenya Ntaiya
Educator and Activist
National Geographic Emerging Explorer

Kakenya Ntaiya established a school for girls in Kenya, and she continues to improve girls' education there.

Spencer Wells
Population Geneticist
National Geographic Explorer-in-Residence

Spencer Wells studies human migration patterns. He is the director of National Geographic's Genographic Project.

THE ESSENTIALS OF GEOGRAPHY

TECHTREK

myNGconnect.com

Digital Library
Unit 1 GeoVideo
Introduce the Essentials of Geography

Explorer Video Clip
Sylvia Earle, Oceanographer
National Geographic Explorer-in-Residence

NATIONAL GEOGRAPHIC PHOTO GALLERY

Photos of Earth's physical features
and world cultures

Maps and Graphs
Interactive Map Tool

**Interactive Whiteboard
GeoActivities**
• Draw the Stages of an Earthquake
• Build a Climograph
• Map the Spread of Buddhism

Magazine Maker
Create your own presentations

Archaeologists at work

UNIT 2 EUROPE

TECHTREK

myNGconnect.com

Digital Library
Unit 2 GeoVideo
Introduce Europe

Explorer Video Clip
Enric Sala, Marine Ecologist
National Geographic Explorer-in-Residence

NATIONAL GEOGRAPHIC **PHOTO GALLERY**
Regional photos, including Florence, Paris, Amsterdam, and Budapest

Maps and Graphs
Interactive Map Tool

Interactive Whiteboard GeoActivities
• Map Europe's Land Regions
• Compare Greek and Roman Governments
• Analyze Causes and Effects of World War I

Connect to NG
Research links and current events in Europe

Louvre Museum, Paris, France

TECHTREK

myNGconnect.com

Digital Library
Unit 3 GeoVideo
Introduce Russia & the Eurasian Republics

Explorer Video Clip
Katey Walter Anthony, Aquatic Ecologist
National Geographic Emerging Explorer

NATIONAL GEOGRAPHIC PHOTO GALLERY

Regional photos, including St. Basil's Cathedral in Moscow, the Ural Mountains, Siberian tundra, and St. Petersburg

Maps and Graphs
Interactive Map Tool

Interactive Whiteboard GeoActivities
• Analyze Central Asian Economies
• Graph Napoleon's March Through Russia
• Explore the Trans-Siberian Railroad

Magazine Maker
Create your own presentations

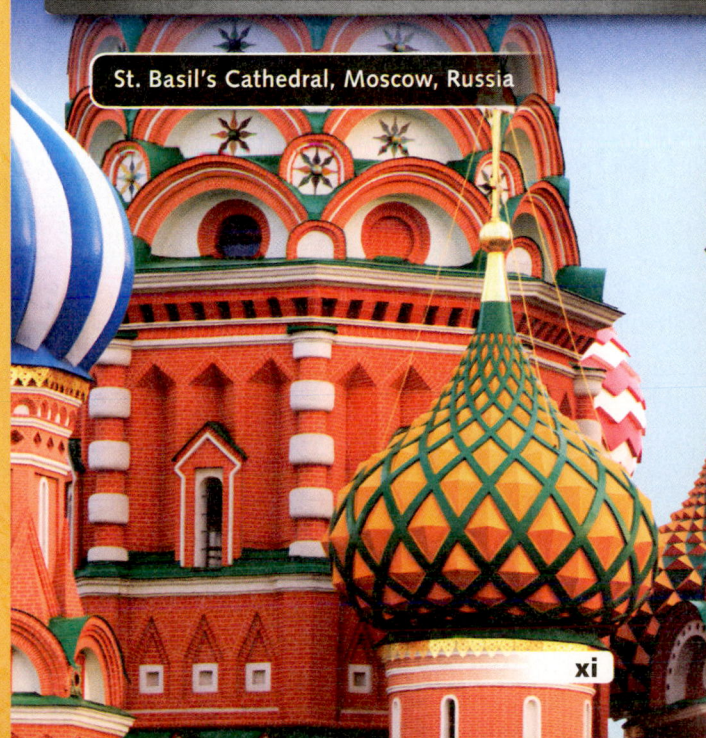

St. Basil's Cathedral, Moscow, Russia

UNIT 4 SUB-SAHARAN AFRICA

TECHTREK

myNGconnect.com

A musician and griot in Senegal

Digital Library
Unit 4 GeoVideo
Introduce Sub-Saharan Africa

Explorer Video Clip
Dereck and Beverly Joubert, Conservationists
National Geographic Explorers-in-Residence

NATIONAL GEOGRAPHIC **PHOTO GALLERY**

Regional photos, including Kenya's
savanna, Victoria Falls, Johannesburg,
and the Great Rift Valley

Music Clips
Audio clips of music from the region

Maps and Graphs
Interactive Map Tool

Interactive Whiteboard GeoActivities
• Compare Precipitation Across Regions
• Locate a Wildlife Reserve
• Research Vanishing Cultures

Connect to NG
Research links and current events in
Sub-Saharan Africa

myNGconnect.com

View from the world's tallest building, Dubai

Digital Library
Unit 5 GeoVideo
Introduce Southwest Asia & North Africa

Explorer Video Clip
Beverly Goodman, Geoarchaeologist
National Geographic Emerging Explorer

NATIONAL GEOGRAPHIC PHOTO GALLERY

Regional photos, including the Blue Mosque, Hagia Sophia, and Topkapi Palace

Maps and Graphs
Outline Maps

Interactive Whiteboard GeoActivities
• Explore an Ancient Irrigation System
• Compare Major Rivers of the World
• Trace the Benefits of Education

Connect to NG
Research links and current events in Southwest Asia & North Africa

TECHTREK

myNGconnect.com

Digital Library
Unit 6 GeoVideo
Introduce South Asia

Explorer Video Clip
Kira Salak, Writer/Adventurer
National Geographic Emerging Explorer

NATIONAL GEOGRAPHIC PHOTO GALLERY

Regional photos, including the Himalayas, the Ganges River, and the Taj Mahal

Music Clips
Audio clips of music from the region

Maps and Graphs
Interactive Map Tool

Interactive Whiteboard GeoActivities
• Draw a Mental Map of South Asia
• Build a Time Line of Colonialism in India
• Graph and Compare Internet Use

Magazine Maker
Create your own presentations

Climbers on Mount Everest

UNIT 7 East Asia

TECHTREK

myNGconnect.com

Digital Library
Unit 7 GeoVideo
Introduce East Asia

Explorer Video Clip
Albert Lin, Research Scientist and Engineer
National Geographic Emerging Explorer

NATIONAL GEOGRAPHIC **PHOTO GALLERY**

Regional photos, including the Great Wall of China, Mount Fuji, bullet trains, and a Buddhist temple in Korea

Maps and Graphs
Interactive Map Tool

Interactive Whiteboard GeoActivities
• Follow the Chang Jiang
• Barter on the Silk Roads
• Explore the Forbidden City

Magazine Maker
Create your own presentations

Mount Fuji, Japan

TECHTREK

myNGconnect.com

A floating market in Thailand

Digital Library
Unit 8 GeoVideo
Introduce Southeast Asia

Explorer Video Clip
Kristofer Helgen, Zoologist
National Geographic Emerging Explorer

NATIONAL GEOGRAPHIC **PHOTO GALLERY**

Regional photos, including the Mekong and Irrawaddy rivers, Angkor Wat, and Hanoi and Jakarta

Maps and Graphs
Interactive Map Tool

Interactive Whiteboard GeoActivities
• Investigate New Species
• Map the Spice Trade
• Analyze Remittances and GDP

Connect to NG
Research links and current events in Southeast Asia

UNIT 9
AUSTRALIA, THE PACIFIC REALM & ANTARCTICA

TECHTREK

myNGconnect.com

Opera House, Sydney, Australia

Digital Library
Unit 9 GeoVideo
Introduce Australia, the Pacific Realm & Antarctica

Explorer Video Clip
David Harrison and Greg Anderson, Linguists
National Geographic Fellows

NATIONAL GEOGRAPHIC PHOTO GALLERY
Regional photos, including the Great Barrier Reef and glaciers in Antarctica

Maps and Graphs
Interactive Map Tool

Interactive Whiteboard GeoActivities
• Research Indigenous Species
• Research and Report on Endangered Languages
• Build a Time Line of Indigenous Rights

Connect to NG
Research links and current events in the region

SPECIAL FEATURES

Infographics

THE GREAT PYRAMID OF KHUFU, 2550 B.C.

NATIONAL GEOGRAPHIC
ATLAS

World Physical

North Pole

Elevation

feet	meters
10,000+	3,050+
5,000	1,524
2,000	610
1,000	305
500	152
0	0
Below sea level	

South Pole

ANTARCTICA

QUEEN MAUD LAND

WEST ANTARCTICA

EAST ANTARCTICA

Ronne Ice Shelf

Ross Ice Shelf

Vinson Massif 16,067 ft (4,897 m)

South Pole

South Geomagnetic Pole

South Magnetic Pole

Weddell Sea

Ross Sea

Antarctic Peninsula

Antarctic Circle

ATLANTIC OCEAN

INDIAN OCEAN

PACIFIC OCEAN

September extent of sea ice

0 400 800 Miles
0 400 800 Kilometers

OCEAN

30°E 60°E 90°E 120°E 150°E

Svalbard (Norway)

Franz Josef Land

Severnaya Zemlya

New Siberian Islands

East Siberian Sea

Wrangel Is.

Laptev Sea

Kara Sea

Barents Sea

Novaya Zemlya

Norwegian Sea

North Sea

SWEDEN FINLAND Kola Pen.

ESTONIA LATVIA LITH. BELARUS

DENMARK POLAND GERMANY CZECH REP. SLOVAKIA

NETH. LUX. FRANCE SWITZ. AUS. SLOV. HUNG. ROM. MOLDOVA UKRAINE

MONACO Sardinia (Italy) ITALY CRO. B. SERB. BULG.

SAN MARINO Sicily (Italy) MAC. ALB. GREECE

MALTA

Mediterranean Sea

CYPRUS LEBANON ISRAEL SYRIA IRAQ

TUNISIA

ALGERIA LIBYA EGYPT

SAHARA

Ahaggar Mts.

NIGER CHAD SUDAN

SAHEL

NIGERIA

BENIN

CAMEROON CENTRAL AFRICAN REP. SOUTH SUDAN ETHIOPIA

Eq. GUINEA SAO TOME & PRINCIPE GABON CONGO UGANDA KENYA

CABINDA (Angola) RWANDA BURUNDI

Congo Basin

DEM. REP. OF THE CONGO TANZANIA

ANGOLA ZAMBIA MALAWI

NAMIBIA ZIMBABWE MOZAMBIQUE

BOTSWANA

KALAHARI DESERT

Namib Desert

SOUTH AFRICA SWAZILAND LESOTHO

Cape of Good Hope

Ural Mountains

Ob R.

Western Siberian Plain

RUSSIA

Central Siberian Plateau

SIBERIA

Lena R.

Kamchatka Peninsula

Sea of Okhotsk

Bering Sea

Aleutian Islands

Lake Baikal

Yenisey R.

Volga R.

Caspian Sea

Aral Sea

KAZAKHSTAN Kazakh Uplands Lake Balkhash

Elbrus 18,510 ft (5,642 m) Caspian Depression

Black Sea

GEORGIA ARM. AZER.

TURKEY

UZBEKISTAN KYRGYZSTAN

TURKMENISTAN TAJIKISTAN

Altay Mountains

MONGOLIA Mongolia Plateau

GOBI

NORTH KOREA SOUTH KOREA

Sea of Japan (East Sea)

JAPAN

Kuril Islands

Sakhalin

Tian Shan

Taklimakan Desert

Kunlun Mountains

Plateau of Tibet

CHINA

Yellow R.

Yellow Sea

East China Sea

JORDAN KUWAIT BAHRAIN QATAR U.A.E.

SAUDI ARABIA

ARABIAN PENINSULA

OMAN

YEMEN

Red Sea

Gulf of Aden

ERITREA DJIBOUTI

Socotra (Yemen)

SOMALIA

Great Rift Valley

Lake Victoria

Kilimanjaro 19,340 ft (5,895 m)

Lake Tanganyika

SEYCHELLES

COMOROS

Mozambique Channel

MADAGASCAR

Réunion (Fr.) MAURITIUS

IRAN

AFGHANISTAN PAKISTAN

Thar Desert

Zagros Mountains

Persian Gulf

Gulf of Oman

Arabian Sea

HIMALAYA

Mt. Everest 29,035 ft (8,850 m)

NEPAL BHUTAN

INDIA

Ganges R.

Deccan Plateau

Western Ghats

Eastern Ghats

Bay of Bengal

BANG.

MYANMAR (BURMA) LAOS

THAILAND CAMB. VIETNAM

Hainan

SRI LANKA

MALDIVES

Andaman Islands (India)

Nicobar Islands (India)

Chagos Archipelago (U.K.)

Diego Garcia (U.K.)

Andaman Sea

South China Sea

MALAYSIA SINGAPORE BRUNEI

Sumatra

INDONESIA

Java Sea

Java

Christmas Island (Aus.)

Cocos (Keeling) Islands (Aus.)

Borneo

Celebes Sea

Celebes

Banda Sea

Timor

TIMOR-LESTE (EAST TIMOR)

Arafura Sea

New Guinea

PAPUA NEW GUINEA

PHILIPPINES

Philippine Sea

TAIWAN

Ryukyu Islands (Japan)

Bonin Islands (Japan)

Volcano Islands (Japan)

Izu Islands (Japan)

Minami Tori Shima (Marcus) (Japan)

Northern Mariana Islands (U.S.)

Guam (U.S.)

FEDERATED STATES OF MICRONESIA

PALAU

NORTH PACIFIC OCEAN

MARSHALL ISLANDS

NAURU KIRIBATI

SOLOMON ISLANDS

TUVALU

VANUATU FIJI

New Caledonia (Fr.)

Norfolk Island (Aus.)

AUSTRALIA

Great Sandy Desert

Western Plateau

Great Victoria Desert

Great Artesian Basin

GREAT DIVIDING RANGE

Great Australian Bight

Darling R.

Mt. Kosciuszko (2,228 m) 7,310 ft

Tasmania

Tasman Sea

Coral Sea

Great Barrier Reef

NEW ZEALAND

North Island

South Island

Auckland Islands (N.Z.)

INDIAN OCEAN

Île Amsterdam (Fr.)

Crozet Islands (Fr.)

Prince Edward Islands (South Africa)

Kerguelen Islands (Fr.)

Heard Island and McDonald Islands (Aus.)

Bouvet (Norway)

September extent of sea ice

South Magnetic Pole

QUEEN MAUD LAND WILKES LAND

South Geomagnetic Pole

ANTARCTICA

0 1,000 2,000 Miles
0 1,000 2,000 Kilometers

30°E 60°E 90°E 120°E 150°E

30°N 30°N 30°S 60°S 0°

A

B

C

D

E

F

G

H

7 8 9 10 11 12 A3

World Political

North Pole (inset)

Anchorage
ALASKA (U.S.)
Yukon
Arctic Circle
Yellowknife
CANADA
ARCTIC OCEAN
North Pole
Franz Josef Land (Russia)
RUSSIA
Svalbard (Norway)
Ob' R.
GREENLAND (Denmark)
Nuuk (Godthåb)
Murmansk
Arkhangel'sk
ICELAND
Reykjavík
FINLAND
SWEDEN
NORWAY
Arctic Circle

0   400   800 Miles
0   400 800 Kilometers

North Pole

Main Map

ARCTIC

Queen Elizabeth Islands
Ellesmere Island
GREENLAND (KALAALLIT NUNAAT) (Denmark)
Jan Mayen (Norway)
Baffin Bay
Banks Is.
Victoria Island
Baffin Island
Beaufort Sea
RUS.
Bering Strait
ALASKA (U.S.)
Gulf of Alaska
Bering Sea
Aleutian Islands
Queen Charlotte Islands
Hudson Bay
CANADA
Labrador Sea
Nuuk (Godthåb)
ICELAND
Reykjavík
Faroe Islands (Denmark)
Arctic Circle
UNITED KINGDOM
IRELAND
Dublin
London
English Channel

NORTH PACIFIC OCEAN

Ottawa
UNITED STATES
Washington, D.C.
St.-Pierre & Miquelon (France)
NORTH ATLANTIC OCEAN
Bermuda (U.K.)
Azores (Port.)
PORTUGAL
Lisbon
ANDORRA
SPAIN
Madrid
Madeira Islands (Port.)
Rabat
MOROCCO
Canary Islands (Sp.)
Laâyoune
WESTERN SAHARA (Morocco)

Tropic of Cancer
30°N
HAWAI'I (U.S.)
MEXICO
Gulf of Mexico
Mexico City
Nassau
BAHAMAS
Havana
CUBA
DOMINICAN REPUBLIC
Kingston
Port-au-Prince
Santo Domingo
PUERTO RICO (U.S.)
MAURITANIA
Nouakchott
MALI
Belmopan
BELIZE
JAMAICA
HAITI
ST. KITTS AND NEVIS
DOMINICA
CAPE VERDE
Praia
Dakar
SENEGAL
GAMBIA
Banjul
Bamako
BURKINA FASO
Guatemala City
GUATEMALA
HONDURAS
Tegucigalpa
ST. LUCIA
BARBADOS
Bissau
GUINEA-BISSAU
Ouagadougou
San Salvador
EL SALVADOR
NICARAGUA
Managua
Caribbean Sea
ST. VINCENT AND THE GRENADINES
TRINIDAD AND TOBAGO
GUINEA
Conakry
SIERRA LEONE
Freetown
GHANA
CÔTE D'IVOIRE (IVORY COAST)
COSTA RICA
San José
Panama City
PANAMA
Caracas
VENEZUELA
Georgetown
Paramaribo
GUYANA
SURINAME
Cayenne
FRENCH GUIANA (France)
Monrovia
LIBERIA
Yamoussoukro
Accra

Bogotá
COLOMBIA
Equator
0°
Galápagos Islands (Ecuador)
Quito
ECUADOR
KIRIBATI
Phoenix Islands
LINE ISLANDS

PERU
Lima
BRAZIL
Brasília
Ascension (U.K.)
Marquesas Islands (Fr.)
La Paz
BOLIVIA
Sucre
São Paulo
St. Helena (U.K.)
SAMOA
Apia
AMERICAN SAMOA (U.S.)
TUAMOTU ARCHIPELAGO
Tahiti
French Polynesia (Fr.)
PARAGUAY
Asunción
Tropic of Capricorn
TONGA
Nuku'alofa
Cook Islands (N.Z.)
Pitcairn Islands (U.K.)
Easter Island (Chile)
URUGUAY
Montevideo
SOUTH ATLANTIC OCEAN
Tristan da Cunha Group (U.K.)
Juan Fernandez Archipelago (Chile)
Santiago
CHILE
Buenos Aires
ARGENTINA
30°S
SOUTH PACIFIC OCEAN
Chatham Islands (N.Z.)
Falkland Islands (U.K.)
South Georgia (U.K.)
Prime Meridian
South Sandwich Islands (U.K.)
Scotia Sea
South Shetland Islands
South Orkney Islands
Antarctic Circle
60°S
Weddell Sea
ANTA

150°W   120°W   90°W   60°W   30°W

NATIONAL GEOGRAPHIC
Endangered Species

CALIFORNIA CONDOR
Condors are threatened by pesticides and habitat loss. A captive breeding program has raised numbers in the wild from 22 in 1987 to more than 160 today.

It's About Habitat

Wild animals need space to live. That means room to roam, room to find food, and room to hide. Quality habitat also means clean water and clean air. However, as the human population grows, the places where people live tend to spread into quality habitat. As the habitat is lost the number of animals decreases. If too much of a species' habitat is lost, the animal is at risk of becoming extinct. The International Union for Conservation of Nature (IUCN) works to save habitat. By identifying the level of endangerment with simple categories, the IUCN helps target the habitats that need to be conserved and the species to be protected.

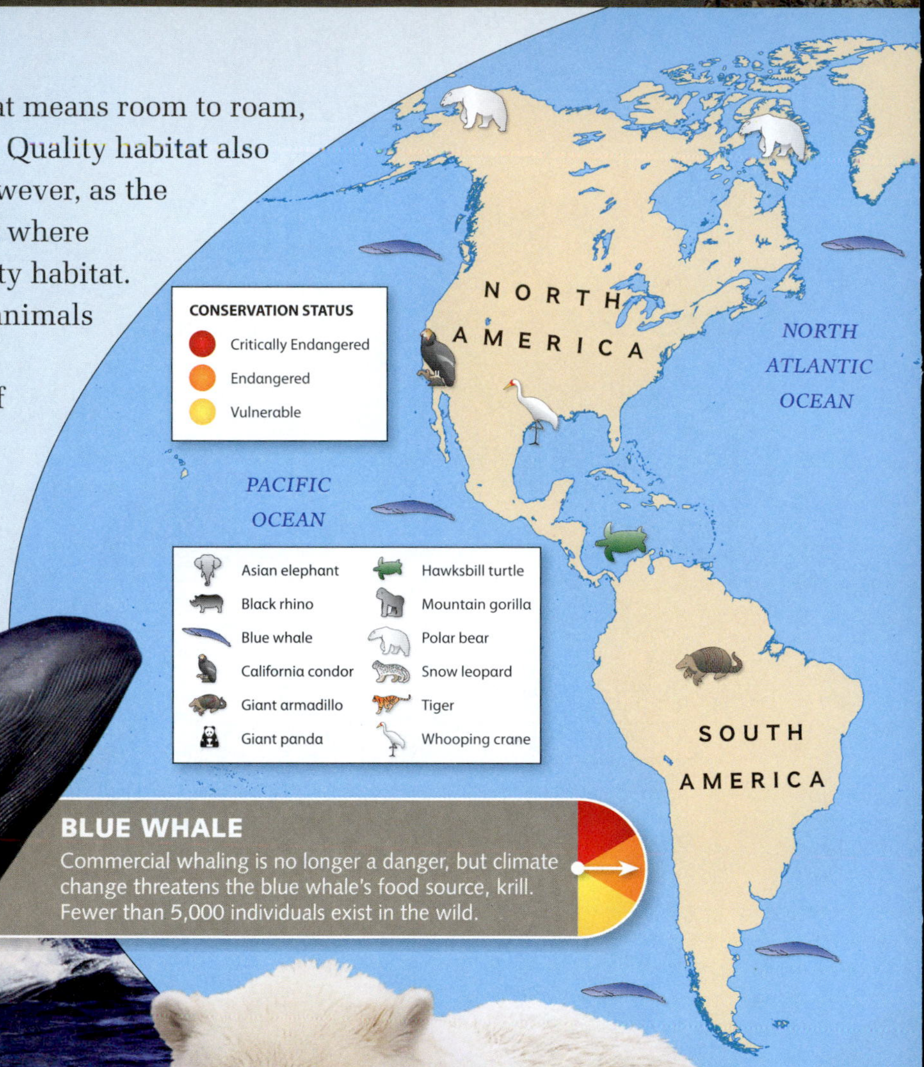

CONSERVATION STATUS
- Critically Endangered
- Endangered
- Vulnerable

NORTH AMERICA

NORTH ATLANTIC OCEAN

PACIFIC OCEAN

- Asian elephant
- Black rhino
- Blue whale
- California condor
- Giant armadillo
- Giant panda
- Hawksbill turtle
- Mountain gorilla
- Polar bear
- Snow leopard
- Tiger
- Whooping crane

SOUTH AMERICA

BLUE WHALE
Commercial whaling is no longer a danger, but climate change threatens the blue whale's food source, krill. Fewer than 5,000 individuals exist in the wild.

HAWKSBILL TURTLE
The Hawksbill's habitat is severely threatened, but a thriving trade in turtle products poses a greater danger.

POLAR BEAR
Climate change is the polar bear's biggest threat. They hunt on Arctic and sub-Arctic ice flows. As the ice melts, they lose access to their main food source, seals.

Species Extinction and Human Population

Populations in billions — 8, 6, 4, 2, 0

Extinctions — 80,000, 60,000, 40,000, 20,000, 0

Year: 1820, 1840, 1860, 1880, 1900, 1920, 1940, 1960, 1980, 2000, 2020

Population

Extinction

Source: www.whole-systems.org/extinctions.html

TIGER

With a territory that once ranged across Asia, tigers remain in only a few pockets of the continent, primarily Southeast Asia. There may be as few as 3,200 left in the wild.

ARCTIC OCEAN

EUROPE

ASIA

AFRICA

SOUTH ATLANTIC OCEAN

INDIAN OCEAN

AUSTRALIA

GIANT PANDA

The giant panda faces an uncertain future. Though fiercely protected, its habitat is threatened by the roads, railroads, and pollution that are part of China's expanding economy.

THE WHOOPING CRANE *Success Story*

The whooping crane was near extinction, with only about 20 birds left in the wild. Thanks to intensive recovery efforts, the species now boasts a population of over 600, with more growth expected. The species is still endangered, but scientists believe the recovery is sustainable.

BLACK RHINOCEROS

Black rhinos suffered a huge decline between 1970 and 1992 because of poaching. Although in severe danger, their numbers have been slowly rising.

North America Physical

North America Political

NATIONAL GEOGRAPHIC

United States Political

Central America & the Caribbean Physical

FLORIDA (U.S.)

Gulf of Mexico

Tropic of Cancer

25°N

Grand Bahama Island
Bimini Islands
Abaco Island
Eleuthera Island
New Providence
Andros Island
Cat Island
Great Exuma
Long Island

B A H A M A S

Straits of Florida

Great Bahama Bank

C U B A

Isle of Youth

G R E A T E R

Yucatán Channel

YUCATÁN PENINSULA

20°N

Cayman Islands (U.K.)
Grand Cayman

MEXICO

JAMAICA

BELIZE

Gulf of Honduras

Usumacinta R.

GUATEMALA

SIERRA MADRE

15°N

Motagua R.

HONDURAS

Coco R.

EL SALVADOR

NICARAGUA

Lake Managua

Rio Grande de Matagalpa

Mosquito Coast

Lake Nicaragua

C A R I B B

PACIFIC

10°N

OCEAN

COSTA RICA

PANAMA CANAL

P A N A M A

Coiba Island

Gulf of Panama

C O L O

N
W   E
S

0   100   200 Miles
0   100   200 Kilometers

A12   1   2   3   4   5   6

70°W   65°W   60°W

A

25°N

B

San Salvador

Rum Cay

ATLANTIC   OCEAN

Tropic of Cancer

Crooked
Island

Mayaguana
Island

Acklins
Island

Turks & Caicos Islands
(U.K.)

Caicos
Islands

Turks
Islands

C

Great
Inagua
Island

20°N

Windward Passage

Île de la Tortue

I

HISPANIOLA

HAITI   DOMINICAN
REPUBLIC

Île de la Gonâve

Virgin
Islands

Tortola

N

Anguilla
(U.K.)

St. Martin (France)

D

Mona Passage

(Puerto Rico) Vieques

St. Thomas

St. Croix

Puerto Rico
(U.S.)

Anegada Passage

St. Maarten (Neth.)

St.-Barthélemy (France)

D

Isla Mona
(Puerto Rico)

Virgin Islands
(U.S.)

(Neth.) Saba

(Neth.) St. Eustatius
St. Kitts

Barbuda

ANTIGUA AND
BARBUDA

Nevis   Antigua

A

N

T

I

L

L

E

S

ST. KITTS AND NEVIS

Montserrat (U.K.)

Grande-Terre
GUADELOUPE (France)

Basse-Terre
Marie-Galante

E

EAN   SEA

Aves (Bird Island)
(Venezuela)

L E S S E R   A N T I L L E S

DOMINICA

15°N

Martinique (France)

ST. LUCIA

BARBADOS

F

ARUBA
(Neth.)

L E S S E R   A N T I L L E S

St. Vincent

Bequia

ST. VINCENT AND
THE GRENADINES

Carriacou

GRENADA

CURAÇAO
(Neth.)

BONAIRE
(Neth.)

Tobago

TRINIDAD AND
TOBAGO

G

Magdalena R.

Trinidad

10°N

Orinoco
River
Delta

Elevation

feet   meters

VENEZUELA

Orinoco R.

10,000+	3,050+
5,000	1,524
2,000	610
1,000	305
500	152
0	0

H

Below
sea level

MBIA

GUYANA

70°W   65°W   60°W

Central America & the Caribbean Political

FLORIDA (U.S.)

BAHAMA

Nassau

Gulf of Mexico

Tropic of Cancer

Straits of Florida

Great Bahama Bank

Havana

CUBA

GREATER

Guantánamo

Yucatan Channel

Santiago de Cuba

MEXICO

Cayman Islands (U.K.)

George Town

JAMAICA

Kingston

Usumacinta R.

BELIZE

Belmopan

Gulf of Honduras

GUATEMALA

Motagua R.

CARIBB

Guatemala City

HONDURAS

San Salvador

EL SALVADOR

Tegucigalpa

Coco R.

NICARAGUA

Lake Managua

Río Grande de Matagalpa

Managua

Lake Nicaragua

PACIFIC

N

W E

S

OCEAN

10°N

San José

COSTA RICA

PANAMA CANAL

PANAMA

Panama City

Gulf of Panama

COLO

0   100   200 Miles

0   100   200 Kilometers

A14   1   2   3   4   5   6

90°W   85°W   80°W

25°N

20°N

15°N

Caribbean Sea

HONDURAS
EL SALVADOR
NICARAGUA
COSTA RICA
PANAMA

10°N

GALÁPAGOS IS.
(ARCHIPIÉLAGO
DE COLÓN)
(Ecuador)

0°

Equator

LLANOS
Orinoco R.
VENEZUELA
GUYANA
SURINAME
GUIANA HIGHLANDS
FRENCH GUIANA
(France)

COLOMBIA
Magdalena R.
Lake Maracaibo

ECUADOR
Negro R.
Amazon R.

A M A Z O N
Marañón R.
Amazon R.
Purus R.
Madeira R.
Tapajós R.
Xingu R.
Araguaia R.
Tocantins R.

B a s i n
Ucayali R.

PERU
Lake Titicaca

BRAZIL

Campos
São Francisco R.

MATO GROSSO
PLATEAU

BRAZILIAN

BOLIVIA

Pantanal
HIGHLANDS

A N D E S

Atacama Desert

PARAGUAY
Paraguay R.
Paraná R.
Gran Chaco

Ojos del Salado
22,615 ft
(6,893 m)

Iguazú
Falls

Entre Ríos
Uruguay R.
URUGUAY

Cerro Aconcagua
22,831 ft
(6,959 m)

CHILE

ARGENTINA

PAMPAS
Paraná R.

Río de
La Plata

Colorado R.

Tropic of Capricorn

PATAGONIA

Valdés
Peninsula

Gulf of
San Jorge

Laguna del Carbón
-344 ft (-105 m)

Strait of
Magellan
TIERRA
DEL FUEGO

FALKLAND ISLANDS
(ISLAS MALVINAS) (U.K.)
Administered by the United Kingdom
(claimed by Argentina)

South Georgia
(U.K.)

Cape Horn
(Cabo de Hornos)

**PACIFIC
OCEAN**

**ATLANTIC
OCEAN**

Scotia Sea

ANTARCTICA

Antarctic Circle

Elevation

feet	meters
10,000+	3,050+
5,000	1,524
2,000	610
1,000	305
500	152
0	0
Below sea level	

0   500   1,000 Miles
0   500   1,000 Kilometers

90°W 80°W 70°W 60°W 50°W 40°W
110°W 100°W 90°W 80°W 70°W 60°W 50°W 40°W 30°W 20°W

10°N 0° 10°S 20°S 30°S 40°S 50°S 60°S

A B C D E F G H

1 2 3 4 5 6

South America Political

NATIONAL GEOGRAPHIC

Caribbean Sea

HONDURAS
EL SALVADOR
NICARAGUA
COSTA RICA
PANAMA

Barranquilla
Maracaibo
Lake Maracaibo
Caracas
Orinoco R.
VENEZUELA
GUYANA
Georgetown
SURINAME
Paramaribo
Cayenne
FRENCH GUIANA (France)
Boundary claimed by Suriname

Medellín
Bogotá
Cali
COLOMBIA

GALÁPAGOS IS. (ARCHIPIÉLAGO DE COLÓN) (Ecuador)

Quito
ECUADOR
Negro R.
Equator
Amazon R.
Manaus
Fortaleza

Amazon R.
Marañón R.
Ucayali R.
Purus R.
Madeira R.
Tapajós R.
Xingu R.
Araguaia R.
Tocantins R.

P E R U

Lima

B R A Z I L

São Francisco R.
Recife

Salvador (Bahia)

Lake Titicaca
La Paz
Santa Cruz
BOLIVIA
Sucre

Brasília

Belo Horizonte

Tropic of Capricorn

PARAGUAY
Asunción
Paraguay R.
Paraná R.

São Paulo
Rio de Janeiro
Curitiba

PACIFIC OCEAN

Córdoba
Santiago
Rosario
Buenos Aires

C H I L E
A R G E N T I N A
Paraná R.
Uruguay R.
URUGUAY
Montevideo
Río de La Plata
Porto Alegre

ATLANTIC OCEAN

Colorado R.

N
W   E
S

Gulf of San Jorge

Strait of Magellan
TIERRA DEL FUEGO

Stanley
FALKLAND ISLANDS (ISLAS MALVINAS) (U.K.)
Administered by the United Kingdom (claimed by Argentina)

South Georgia (U.K.)

Scotia Sea

0   500   1,000 Miles
0   500   1,000 Kilometers

ANTARCTICA

Antarctic Circle

A17

Elevation

feet	meters
10,000+	3,050+
5,000	1,524
2,000	610
1,000	305
500	152
	0
Below sea level	

ICELAND

Norwegian Sea

Arctic Circle

Faroe Islands (Denmark)

Shetland Islands (U.K.)

Hebrides (U.K.)

Orkney Islands (U.K.)

SCOTLAND

NORTHERN IRELAND

IRELAND

UNITED KINGDOM

North Sea

Shannon R.

Irish Sea

DENMARK

JUTLAND

ZEALAND

Gotland (Sweden)

Baltic

NORWAY

SWEDEN

SCANDINAVIA

WALES

ENGLAND

Thames R.

NETHERLANDS

Elbe R.

NORTHER

POL

ATLANTIC OCEAN

Celtic Sea

English Channel

Channel Islands (U.K.)

BELGIUM

GERMANY

Rhine R.

LUXEMBOURG

Seine R.

Oder R.

CZECH REPUBLIC (CZECHIA)

SLO

Bay of Biscay

Loire R.

FRANCE

Black Forest

Danube R.

LIECHTENSTEIN

SWITZERLAND

AUSTRIA

HUNG

Cantabrian Mountains

Douro R.

Iberian Mountains

Ebro R.

PYRENEES

Mt. Blanc 15,781 ft

MASSIF CENTRAL

Rhône R.

4,810 m

MONACO

ALPS

San Marino

SLOVENIA

CROATIA

PORTUGAL

SPAIN

IBERIAN PENINSULA

Tagus R.

Guadiana R.

Sierra Morena

ANDORRA

French Riviera

Ligurian Sea

Corsica (France)

Po R.

A P E N N I N E S

ITALY

Adriatic Sea

BOSNIA AND HERZEGOVINA

MONTENEGRO

Baetic Mountains

Strait of Gibralter

GIBRALTAR (U.K.)

Balearic Sea

Balearic Islands (Spain)

Sardinia (Italy)

VATICAN CITY

Mediterranean Sea

Tyrrhenian Sea

MOROCCO

ALGERIA

TUNISIA

Sicily (Italy)

MALTA

Ionian Sea

0   200   400 Miles

0   200   400 Kilometers

Prime Meridian

FINLAND

Barents Sea

KOLA PENINSULA

White Sea

Pechora R.

Ob R.

Irtysh R.

R U S S I A

E U R O P E A N   P L A I N

Northern Dvina R.

Lake Onega

Gulf of Bothnia

Lake Ladoga

Gulf of Finland

Kama R.

ESTONIA

LATVIA

LITHUANIA

Volga R.

N E

Sea

BELARUS

Vistula R.

Ural R.

KAZAKHSTAN

LAND

UKRAINE

Dniester R.

Dnieper R.

Don R.

Volga R.

Aral Sea

UZBEKISTAN

CARPATHIAN MOUNTAINS

SLOVAKIA

GARY

MOLDOVA

Sea of Azov

CRIMEA

Caspian Sea

TURKMENISTAN

ROMANIA

Danube R.

Black Sea

GEORGIA

AZERBAIJAN

ARMENIA

SERBIA

Balkan Mountains

KOSOVO

BULGARIA

Bosporus

Sea of Marmara

AZERB.

MACEDONIA

ALBANIA

Dardanelles

T U R K E Y

GREECE

Aegean Sea

Euphrates R.

Tigris R.

I R A N

Crete (Greece)

Rhodes (Greece)

CYPRUS

LEBANON

SYRIA

IRAQ

ICELAND

Reykjavik

Norwegian Sea

Arctic Circle

Prime Meridian

N O R W A Y

S W E D E N

Faroe Islands
(Denmark)

Shetland Islands
(U.K.)

Oslo

Stockholm

Hebrides
(U.K.)

Orkney Islands
(U.K.)

SCOTLAND

Edinburgh

NORTHERN
IRELAND

*North
Sea*

Gotland
(Sweden)

IRELAND

Belfast

UNITED
KINGDOM

Dublin

Shannon R.

DENMARK

Copenhagen

Baltic

NETHERLANDS

WALES

ENGLAND

Amsterdam

Hamburg

Elbe R.

Berlin

Cardiff

London

Thames R.

Brussels

GERMANY

P O L

A T L A N T I C

O C E A N

English Channel

BELGIUM

Rhine R.

Channel Islands
(U.K.)

LUXEMBOURG

Luxembourg

Frankfurt

Prague

CZECH REPUBLIC
(CZECHIA)

Oder R.

Vistula R.

Seine R.

Paris

Danube R.

Munich

SLO

Loire R.

F R A N C E

LIECHTENSTEIN

Vienna

Bratislava

*Bay of
Biscay*

SWITZERLAND

Bern

Vaduz

AUSTRIA

Budapest

HUNG

Rhône R.

Milan

Po R.

Ljubljana

SLOVENIA

Zagreb

CROATIA

P O R T U G A L

Douro R.

S P A I N

Ebro R.

ANDORRA

Andorra

SAN
MARINO

BOSNIA AND
HERZEGOVINA

Sarajevo

Lisbon

Madrid

MONACO

I T A L Y

Adriatic Sea

Tagus R.

Barcelona

MONTENEGRO

Podgorica

Guadiana R.

Corsica
(France)

Rome

VATICAN
CITY

Balearic Islands
(Spain)

Sardinia
(Italy)

Tyrrhenian Sea

GIBRALTAR
(U.K.)

M e d i t e r r a n e a n

Sicily
(Italy)

*Ionian
Sea*

MOROCCO

A L G E R I A

TUNISIA

MALTA

Valletta

S e a

0   200   400 Miles

0   200   400 Kilometers

1   2   3   4   5   6

A
B
C
D
E
F
G
H

Barents
Sea

Pechora R.
Ob R.
Irtysh R.
60°N
80°E
70°E
60°E
50°E
40°E
30°E
20°E
70°N

White
Sea

FINLAND

Northern Dvina R.

Lake
Onega

RUSSIA

70°E

Helsinki

Gulf of Bothnia

Gulf of Finland

Lake
Ladoga

Kama R.

50°N

Tallinn
ESTONIA

Riga
LATVIA

Volga R.

Sea

LITHUANIA
Vilnius

Kaliningrad

Minsk
BELARUS

Warsaw

Kiev

Kharkiv

Ural R.

KAZAKHSTAN

Don R.

UKRAINE

Volga R.

Aral
Sea

60°E

UZBEKISTAN

50°N

Kraków

LAND

OVAKIA

Dnieper R.
Dnipropetrovs'k

N
W E
S

Chişinău
MOLDOVA

GARY

ROMANIA

Sea of
Azov

Caspian Sea

TURKMENISTAN

40°N

Belgrade
Bucharest

SERBIA
Danube R.

Black Sea

GEORGIA

AZERBAIJAN

Prishtina
KOSOVO
Sofia
BULGARIA

ARMENIA

AZERB.

Skopje
MACEDONIA

Tirana
ALBANIA

TURKEY

IRAN

GREECE

Aegean Sea

Athens

Euphrates R.

Tigris R.

Crete
(Greece)

Rhodes
(Greece)

CYPRUS

LEBANON

SYRIA

IRAQ

30°N
50°E
40°E
30°E
20°E

ARCTIC

Norwegian Sea

SVALBARD
(Norway)

FRANZ JOSEF LAND

Barents
Sea

NOVAYA ZEMLYA

Kara Sea

UNITED
KINGDOM

North
Sea

DENMARK

GER.

Baltic Sea

POLAND

LITHUANIA
LATVIA
ESTONIA

BELARUS

Dnieper R.

MOLDOVA

UKRAINE

Sea of
Azov

Black Sea

TURKEY

ARMENIA

Euphrates R.
Tigris R.

AZERBAIJAN

IRAQ

KUWAIT

IRAN

N O R W A Y

S W E D E N

FINLAND

Kola
Peninsula

White Sea

Lake
Ladoga

Lake
Onega

Northern Dvina R.

CENTRAL
RUSSIAN
UPLAND

Volga R.

Kama R.

Volga R.

Don R.

Pechora R.

URAL MOUNTAINS

R

U

S

S

Ob R.

WEST

SIBERIAN

PLAIN

Gyda
Peninsula

Yamal Peninsula

Yenisey R.

Pur R.

Taz R.

Ob R.

Irtysh R.

Caspian Depression

Caucasus Mountains
Elbrus
18,510 ft.
(5,642 m)

GEORGIA

Caspian Sea

Aral
Sea

Ural R.

T H E   S T E P P E S

K A Z A K H S T A N

Tobol R.

Esil R.

Ertis R.

Kazakh
Uplands

Lake
Balkhash

Ile R.

Western

TURKMENISTAN

UZBEKISTAN

Syr Darya

Amu Darya

KYRGYZSTAN

TAJIKISTAN

AFGHANISTAN

N
W   E
S

1   2   3   4   5   6

North Pole

OCEAN

Severnaya Zemlya
(North Land)

Taymyr Peninsula

Lake
Taymyr

Laptev
Sea

NEW SIBERIAN ISLANDS

East Siberian
Sea

Wrangel
Island

Chukchi
Sea

Chukchi
Peninsula

Kolyma R.

Indigirka R.

Kolyma Range

Anadyr R.

Koryak Range

Bering
Sea

Central Range

KAMCHATKA
PENINSULA

C E N T R A L

Kotuy R.

Putorana
Plateau

S I B E R I A N

Olenek R.

Lena R.

Yana R.

Cherskiy Range

Verkhoyansk Range

S I B E R I A

Chunya R.

P L A T E A U

Vilyuy R.

Aldan R.

Dzhugdzhur Range

Sea of
Okhotsk

Sakhalin

Angara R.

Lena R.

Olekma R.

Stanovoy Range

Zeya R.

Amur R.

Eastern Sayan Mountains

Vitim R.

Yablonovyy Range

Amur R.

Sikhote Alin' Range

Sayan Mountains

Yenisey R.

Lake
Baikal

Amur R.

Lake
Khanka

Sea of
Japan
(East Sea)

JAPAN

MONGOLIA

0   200   400 Miles
0   200   400 Kilometers

Elevation

feet	meters
10,000+	3,050+
5,000	1,524
2,000	610
1,000	305
500	152
0	0

Below
sea level

NORTH
KOREA

SOUTH
KOREA

Yellow
Sea

PACIFIC
OCEAN

C H I N A

A
B
C
D
E
F
G
H

7   8   9   10   11   12   A23

North Pole
80°N
70°N
60°N
170°W
180°
170°E
160°E
150°E
140°E
130°E
50°N
40°N
30°N
100°E
110°E
120°E

ARCTIC

Norwegian Sea

SVALBARD
(Norway)

FRANZ JOSEF LAND

Barents
Sea

NOVAYA ZEMLYA

Kara Sea

UNITED
KINGDOM

Prime Meridian

Arctic Circle

N O R W A Y

S W E D E N

FINLAND

Murmansk

North
Sea

DENMARK

GER.

Baltic Sea

ESTONIA

St. Petersburg

Archangel

Northern Dvina R.

Pechora R.

Ob R.

Pur R.

Taz R.

Yenisey R.

POLAND

LITHUANIA
Kaliningrad

LATVIA

BELARUS

Dnieper R.

UKRAINE

MOLDOVA

Lake
Ladoga

Lake
Onega

Rostov

Moscow

Volga R.

Nizhniy
Novgorod

Kazan

Kama R.

R   U   S   S

Yekaterinburg

Irtysh R.

Ob R.

Sea of
Azov

Saratov

Volga R.

Samara

Ufa

Chelyabinsk

Tobol R.

Omsk

Novosibirsk

Volgograd

Don R.

Ural R.

Black Sea

Sochi

GEORGIA
Tbilisi

ARMENIA
Yerevan

TURKEY

AZERBAIJAN

Baku

Caspian Sea

K A Z A K H S T A N

Aral
Sea

Esil R.

Ertis R.

Astana

Qaraghandy

Lake
Balkhash

Ile R.

EUPHRATES R.

Tigris R.

IRAQ

IRAN

TURKMENISTAN

Ashgabat

Amu Darya

Syr Darya

UZBEKISTAN

Tashkent

Dushanbe

TAJIKISTAN

AFGHANISTAN

KYRGYZSTAN

Bishkek

Almaty

KUWAIT

N
W E
S

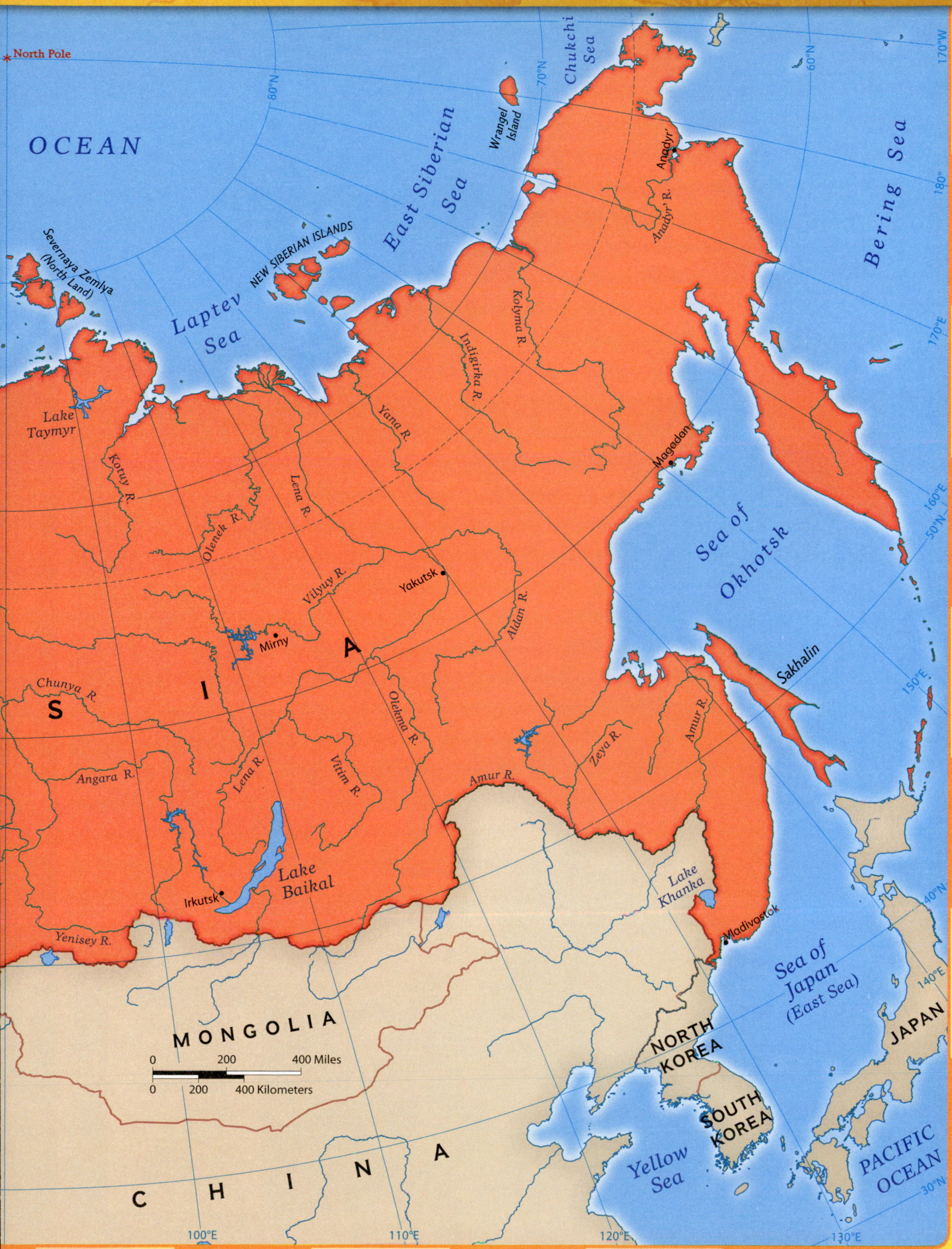

North Pole

OCEAN

Severnaya Zemlya
(North Land)

NEW SIBERIAN ISLANDS

Wrangel
Island

Chukchi
Sea

Bering Sea

Anadyr

Anadyr R.

Laptev
Sea

East Siberian
Sea

Kolyma R.

Lake
Taymyr

Kotuy R.

Indigirka R.

Yana R.

Olenëk R.

Lena R.

Vilyuy R.

Yakutsk

Magadan

Sea of
Okhotsk

Mirny

A

Aldan R.

Chunya R.

S I

Sakhalin

Angara R.

Olëkma R.

Lena R.

Vitim R.

Zeya R.

Amur R.

Amur R.

Lake
Khanka

Lake
Baikal

Irkutsk

Yenisey R.

Vladivostok

Sea of
Japan
(East Sea)

MONGOLIA

0   200   400 Miles

0   200   400 Kilometers

NORTH
KOREA

JAPAN

SOUTH
KOREA

C H I N A

Yellow
Sea

PACIFIC
OCEAN

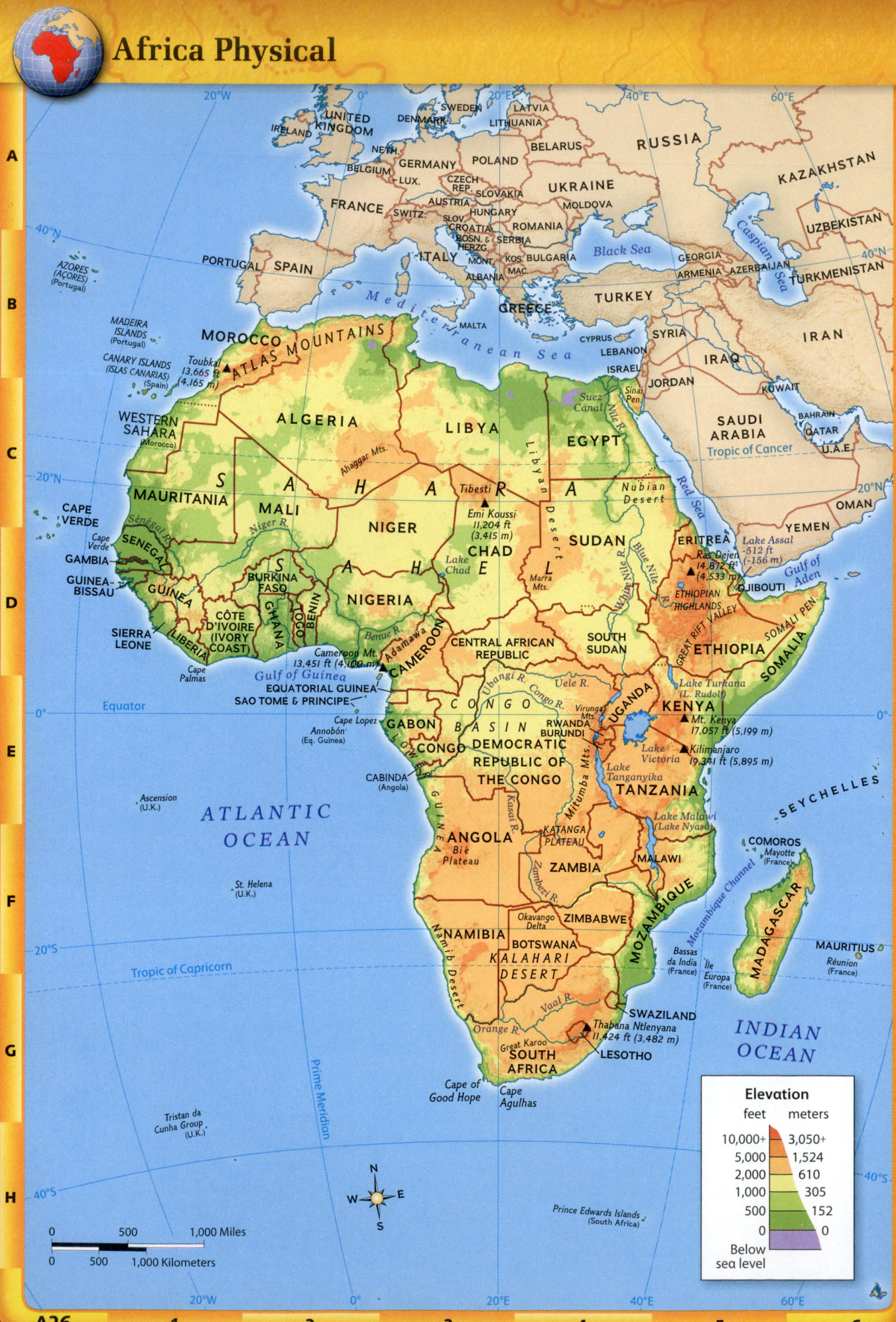

Africa Physical

Elevation

feet	meters
10,000+	3,050+
5,000	1,524
2,000	610
1,000	305
500	152
0	0
Below sea level	

0   500   1,000 Miles

0   500   1,000 Kilometers

Elevation

feet	meters
10,000+	3,050+
5,000	1,524
2,000	610
1,000	305
500	152
0	0
Below sea level	

0   200   400 Miles

0   200   400 Kilometers

POLAND
BELARUS
R U S S I A
SLOVAKIA
HUNG.
UKRAINE
KAZAKHSTAN
ROMANIA
MOLDOVA
Lake Balkhash
BULGARIA
Aral Sea
Syr Darya
UZBEKISTAN
KYRGYZSTAN
Black Sea
Istanbul
Bursa
GEORGIA
Caspian Sea
Amu Darya
TAJIKISTAN
Izmir
Ankara
ARMENIA
AZERBAIJAN
TURKMENISTAN
Kızılırmak R.
Konya
T U R K E Y
Murat R.
Tabriz
Mashhad
Kabul
Jalalabad
Adana
Gaziantep
Mosul
Irbil
Karaj
Tehran
AFGHANISTAN
Harirud R.
CYPRUS
Halab
Nicosia
Euphrates
Tigris R.
Qom
Helmand R.
Mediterranean Sea
Beirut
SYRIA
R.
I R A N
Isfahan
LEBANON
Damascus
Baghdad
ISRAEL
WEST BANK
Ahvāz
PAKISTAN
Jerusalem
Amman
I R A Q
Indus R.
GAZA STRIP
Dead Sea
–1,385 ft (–422 m)
Basra
Shirāz
JORDAN
Kuwait City
EGYPT
Gulf of Suez
Gulf of Aqaba
KUWAIT
Persian Gulf
Strait of Hormuz
Nile R.
S A U D I
Manama
BAHRAIN
Dubai
Gulf of Oman
Tropic of Cancer
INDIA
Medina
Riyadh
Doha
Abu Dhabi
Muscat
Red Sea
QATAR
UNITED ARAB EMIRATES
A R A B I A
Jeddah
Mecca
OMAN
Arabian Sea
SUDAN
ERITREA
Sanaa
YEMEN
Socotra (Yemen)
Aden
Gulf of Aden
DJIBOUTI
ETHIOPIA
SOUTH SUDAN
SOMALIA
KENYA

N
W E
S

| 0 | 200 | 400 Miles |
| 0 | 200 | 400 Kilometers |

A
B
C
D
E
F
G
H

60°E
70°E
80°E
90°E
100°E

40°N
40°N

30°N
30°N

20°N
20°N

Tropic of Cancer

10°N
10°N

0°
0°
Equator

UZBEKISTAN

KAZAKHSTAN

KYRGYZSTAN

TURKMENISTAN

TAJIKISTAN

IRAN

AFGHANISTAN

HINDU KUSH

Karakoram Range

K2 (Godwin Austen, Qogir Feng)
28,251 ft (8,611 m)

CHINA

Hindus R.

Sulaiman Range

Sutlej R.

PAKISTAN

Great Indian Desert

Indus R.

Nanda Devi
(7,817 m) 25,646 ft

Dhaulagiri
(8,167 m) 26,795 ft

NEPAL

Mt. Everest (Sagarmāthā, Qomolangma)
29,035 ft (8,850 m)

Kanchenjunga
28,169 ft
(8,586 m)

HIMALAYA

BHUTAN

Ganges Plain

Brahmaputra R.

Ganges R.

Yamuna R.

BANGLADESH

INDIA

Chota Nagpur
Plateau

Vindhya Range

Narmada R.

Ganges
River
Delta

MYANMAR
(BURMA)

Tapi R.

Godavari R.

DECCAN

Mahanadi R.

EASTERN GHATS

Bay of
Bengal

PLATEAU

Krishna R.

WESTERN GHATS

Arabian
Sea

Andaman
Islands
(India)

Andaman
Sea

Laccadive Sea

SRI LANKA

Nicobar
Islands
(India)

MALDIVES

N
W   E
S

INDONESIA

INDIAN   OCEAN

0   200   400 Miles
0   200   400 Kilometers

70°E
80°E
90°E

South Asia Political

NATIONAL GEOGRAPHIC

East Asia Physical

Elevation

feet	meters
10,000+	3,050+
5,000	1,524
2,000	610
1,000	305
500	152
0	0
Below sea level	

0 400 800 Miles

0 400 800 Kilometers

RUSSIA

KAZAKHSTAN

KYRGYZSTAN

MONGOLIA

Mongolian Plateau

GOBI

ALTAY MOUNTAINS

TIAN SHAN

Taklimakan Desert

KUNLUN MOUNTAINS

PLATEAU OF TIBET

CHINA

North China Plain

NORTH KOREA

SOUTH KOREA

JAPAN

Hokkaido

Honshu

Shikoku

Kyushu

Kuril Is.

Sea of Okhotsk

Sea of Japan (East Sea)

Yellow Sea

East China Sea

PACIFIC OCEAN

RYUKYU ISLANDS

TAIWAN

Philippine Sea

PHILIPPINES

South China Sea

Hainan

NEPAL

BHUTAN

Mt. Everest 29,035 ft (8,850 m)

HIMALAYA

INDIA

BANGLA-DESH

MYANMAR (BURMA)

LAOS

THAILAND

CAMBODIA

VIETNAM

MALAYSIA

BRUNEI

SINGAPORE

INDONESIA

TIMOR-LESTE (EAST TIMOR)

Bay of Bengal

Andaman Sea

Gulf of Thailand

Lake Balkhash

Lake Baikal

Ertis R.

Ob R.

Yenisey R.

Angara R.

Lena R.

Amur R.

Selenge R.

Songhua R.

Liao R.

Huang He (Yellow R.)

Tarim R.

Ganges R.

Brahmaputra R.

Irrawaddy R.

Mekong R.

Chang Jiang (Yangtze R.)

Xi R.

Taiwan Strait

Arctic Circle

Tropic of Cancer

Equator

60°E 70°E 80°E 90°E 100°E 110°E 120°E 130°E 140°E 150°E 160°E

60°N 70°N 50°N 40°N 30°N 20°N 10°N 0° 10°S

East Asia Political

NATIONAL GEOGRAPHIC

A33

Southeast Asia Physical

SOUTH KOREA

JAPAN

CHINA

INDIA

Hkakabo Razi
19,295 ft (5,881 m)

MYANMAR
(BURMA)

LAOS

VIETNAM

TAIWAN

Tropic of Cancer

Red R.

Black R.

Salween R.

Irrawaddy R.

INDOCHINA

THAILAND

PENINSULA

CAMBODIA

Ping R.

Mekong R.

LUZON

Mt. Pinatubo
4,872 ft
(1,485 m)

PHILIPPINES

PACIFIC
OCEAN

Mindoro

*Andaman
Sea*

*Gulf of
Thailand.*

South China Sea

Palawan

*Sulu
Sea*

MINDANAO

Philippine Sea

MALAY PENINSULA

BRUNEI

*Celebes
Sea*

MALAYSIA

SINGAPORE

Kapuas R.

BORNEO

I   N   D   O   N   E   S   I   A

M O L U C C A S

CELEBES

*NEW
GUINEA*

Equator

Maoke Mountains

SUMATRA

Barisan Mountains

Java Sea

Banda Sea

Mt. Merapi
9,738 ft
(2,968 m)

Sumbawa

Flores Sea

JAVA

Bali

Flores

Sumba

Timor

TIMOR-LESTE
(EAST TIMOR)

Arafura Sea

*Timor
Sea*

INDIAN

OCEAN

N
W   E
S

Tropic of Capricorn

AUSTRALIA

Elevation

feet	meters
10,000+	3,050+
5,000	1,524
2,000	610
1,000	305
500	152
0	0

Below
sea level

0   300   600 Miles

0   300   600 Kilometers

Southeast Asia Political

NATIONAL GEOGRAPHIC

Australia, the Pacific Realm & Antarctica Physical

CHINA

TAIWAN

Philippine Sea

PHILIPPINES

NORTHERN MARIANA ISLANDS (U.S.)

Saipan

GUAM (U.S.)

Wake Island (U.S.)

MARSHALL ISLANDS

BRUNEI

MALAYSIA

M I C R O N E S I A

PALAU

FEDERATED STATES OF MICRONESIA

CAROLINE ISLANDS

Ralik Chain

Ratak Chain

NAURU

I N D O N E S I A

M E L A N E S I A

BISMARCK ARCHIPELAGO

Mt. Wilhelm 14,793 ft (4,509 m)

PAPUA NEW GUINEA

SOLOMON ISLANDS

TIMOR-LESTE (EAST TIMOR)

Arafura Sea

Melville Island

Timor Sea

Arnhem Land

Cape York Pen.

Coral Sea

VANUATU

Kimberly Plateau

Berkly Tableland

Gulf of Carpentaria

GREAT DIVIDING RANGE

NEW CALEDONIA (France)

Great Sandy Desert

AUSTRALIA

Great Barrier Reef

20°S

WESTERN

Macdonnell Ranges

GREAT

Fraser Island (Great Sandy Island)

Tropic of Capricorn

Hammersley Range

Northwest Basin

PLATEAU

Gibson Desert

Uluru (Ayers Rock) 2,848 ft (868 m)

Simpson Desert

ARTESIAN

Norfolk Island (Australia)

Great Victoria Desert

Lake Eyre (-16m) -52ft

BASIN

Nullarbor Plain

Darlington Range

Eucla Basin

Darling R.

Lord Howe Island (Australia)

Great Australian Bight

Murray River Basin

GREAT DIVIDING RANGE

Murray R.

North Island

Tasman Sea

Kangaroo Island

Mt. Kosciuszko 7,310 ft (2,228 m)

Australian Alps

Mt. Ruapehu (2,797 m) 9,177

Bass Strait

Flinders Island

NEW ZEALAND

King Island

INDIAN OCEAN

Tasmania

(Mt. Cook) Aoraki (3,754 m) 12,316 ft

Southern Alps

South Is

Stewart Island (Rakiura) (N.Z.)

Elevation

feet	meters
10,000+	3,050+
5,000	1,524
2,000	610
1,000	305
500	152
0	0
Below sea level	

Antip

Auckland Islands (N.Z.)

Campbell Islan (N.Z.)

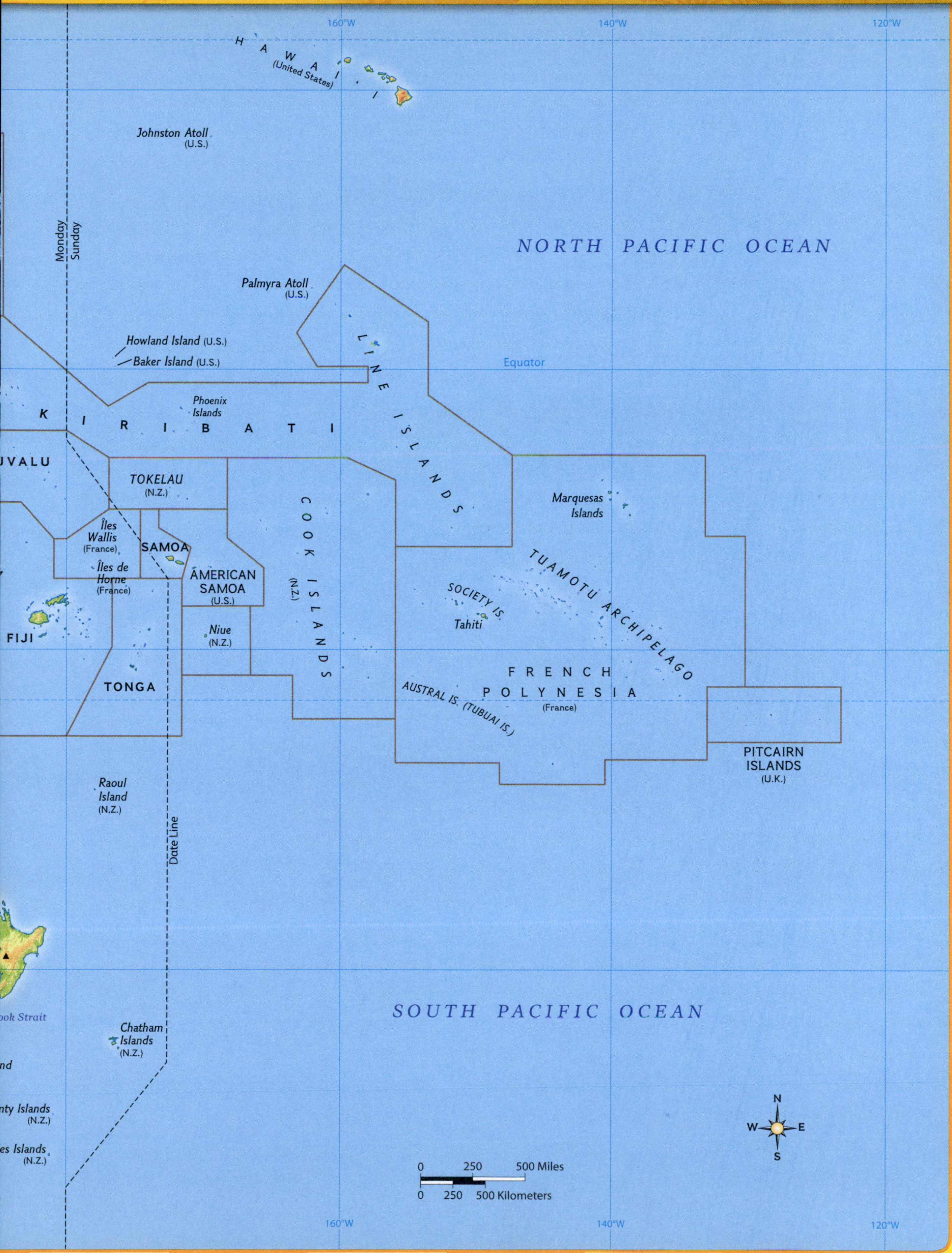

160°W 140°W 120°W

A

NORTH PACIFIC OCEAN

B

H A W A I I
(United States)

Johnston Atoll
(U.S.)

Monday
Sunday

Palmyra Atoll
(U.S.)

C

Howland Island (U.S.)
Baker Island (U.S.)

Equator

Phoenix
Islands

K I R I B A T I

L I N E   I S L A N D S

UVALU

TOKELAU
(N.Z.)

Marquesas
Islands

D

Îles
Wallis
(France)

SAMOA

Îles de
Horne
(France)

AMERICAN
SAMOA
(U.S.)

C O O K

I S L A N D S
(N.Z.)

SOCIETY IS.

Tahiti

T U A M O T U   A R C H I P E L A G O

FIJI

Niue
(N.Z.)

F R E N C H
P O L Y N E S I A
(France)

E

TONGA

AUSTRAL IS. (TUBUAI IS.)

PITCAIRN
ISLANDS
(U.K.)

Raoul
Island
(N.Z.)

Date Line

F

ook Strait

nd

G

SOUTH   PACIFIC   OCEAN

nty Islands
(N.Z.)

Chatham
Islands
(N.Z.)

N
W   E
S

es Islands
(N.Z.)

H

0   250   500 Miles
0   250   500 Kilometers

160°W 140°W 120°W

7   8   9   10   11   12   A37

CHINA

TAIWAN

Tropic of Cancer

A

20°N

Philippine Sea

PHILIPPINES

NORTHERN MARIANA ISLANDS (U.S.)

Saipan ★Capital Hill

Wake Island (U.S.)

MARSHALL ISLANDS

B

BRUNEI

MALAYSIA

(Agana) Hagåtña★ ★GUAM (U.S.)

FEDERATED STATES OF MICRONESIA

Melekeiok★ PALAU

Palikir⊛

★Majuro

C

0°

I N D O N E S I A

Yaren★ NAURU

To (B

Philippine Sea

SOLOMON ISLANDS

T

PAPUA NEW GUINEA

TIMOR-LESTE (EAST TIMOR)

Arafura Sea

Port Moresby★

Honiara★

D

Timor Sea

★Darwin

Gulf of Carpentaria

Great Barrier Reef

Coral Sea

VANUATU

Port Vila★

20°S

NORTHERN TERRITORY

QUEENSLAND

NEW CALEDONIA (France)

Nouméa★

E

Tropic of Capricorn

WESTERN AUSTRALIA

A U S T R A L I A

Lake Eyre

SOUTH AUSTRALIA

Brisbane★

Norfolk Island (Australia)

Perth★

NEW SOUTH WALES

Darling R.

Lord Howe Island (Australia)

F

Great Australian Bight

Adelaide★

Murray R.

Sydney★

Canberra,★ AUSTRALIAN CAPITAL TERRITORY

Tasman Sea

Auckland

VICTORIA

★Melbourne

Bass Strait

40°S

NEW ZEALAND

G

INDIAN OCEAN

TASMANIA

Hobart★

Chri

B

H

Antip

Auckland Islands (N.Z.)

Campbell Isla (N.Z.)

120°E   140°E   160°E

1   2   3   4   5   6

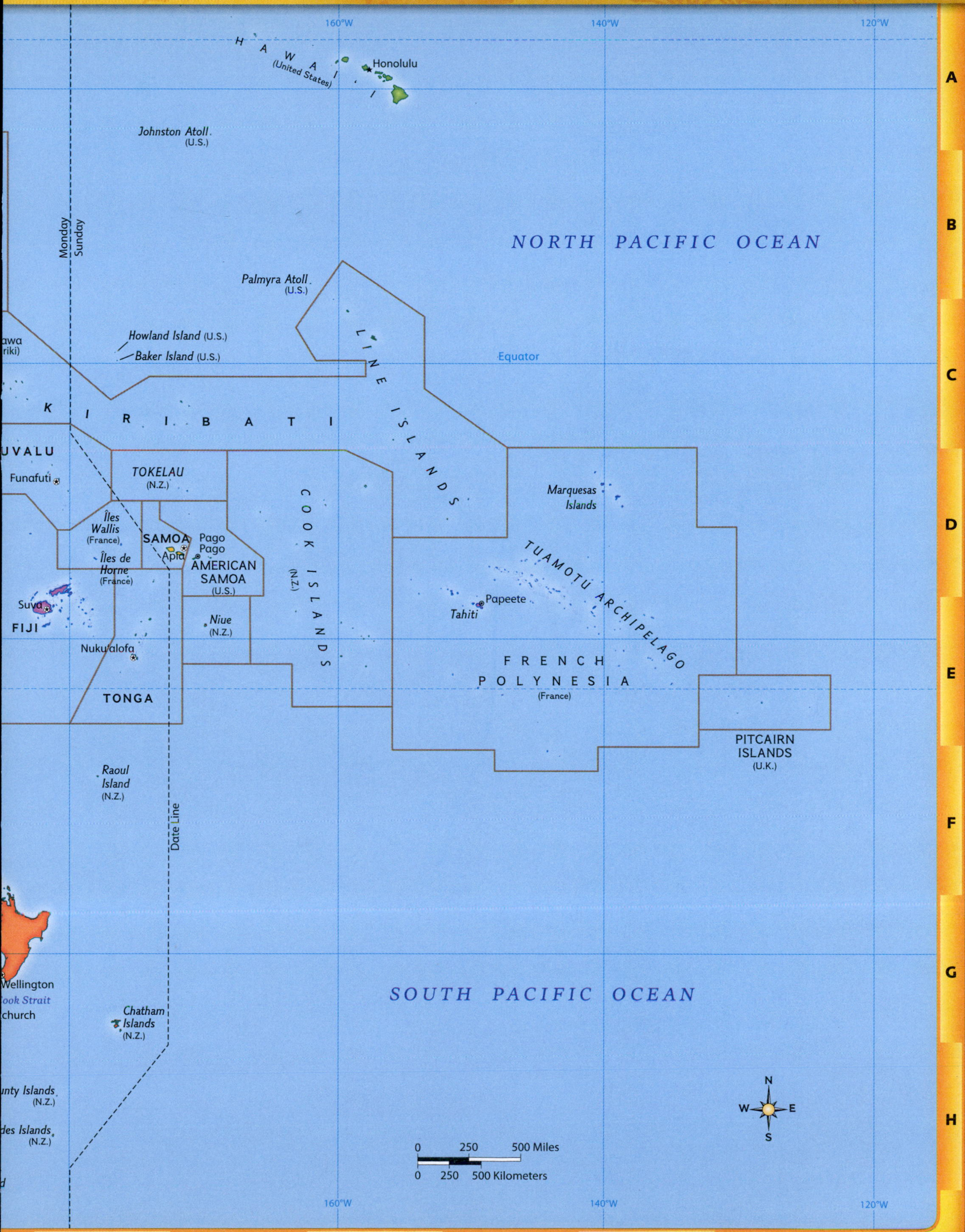

160°W   140°W   120°W

A

H A W A I I
(United States)
★ Honolulu

Johnston Atoll
(U.S.)

B

NORTH PACIFIC OCEAN

Monday
Sunday

Palmyra Atoll
(U.S.)

L
I
N
E

I
S
L
A
N
D
S

Equator

C

awa
riki)

Howland Island (U.S.)

Baker Island (U.S.)

K I R I B A T I

K

UVALU

Funafuti

TOKELAU
(N.Z.)

C
O
O
K

I
S
L
A
N
D
S

(N.Z.)

Marquesas
Islands

D

Îles
Wallis
(France)

SAMOA

Apia

Pago
Pago

AMERICAN
SAMOA
(U.S.)

T
U
A
M
O
T
U

A
R
C
H
I
P
E
L
A
G
O

Îles de
Horne
(France)

Suva

FIJI

Niue
(N.Z.)

Tahiti

Papeete

Nuku'alofa

F R E N C H
P O L Y N E S I A
(France)

E

TONGA

Raoul
Island
(N.Z.)

PITCAIRN
ISLANDS
(U.K.)

Date Line

F

Wellington

ook Strait

church

Chatham
Islands
(N.Z.)

SOUTH PACIFIC OCEAN

G

unty Islands
(N.Z.)

des Islands
(N.Z.)

N
W   E
S

H

0   250   500 Miles

0   250   500 Kilometers

160°W   140°W   120°W

7   8   9   10   11   12   A39

Inspiring people to care about the planet

-National Geographic Society Mission

For more than 100 years, National Geographic Society has sparked our curiosity about the world. NGS supports a network of explorers whose work in the field is vital to the planet. Some of them are shown below. Through education and exploration, the Society works to protect the physical environment and preserve the world's cultures.

Explorers at WORK

Oceanographer
NG Emerging Explorer Katy Croff Bell explores underwater archaeology.

Research Scientist
NG Emerging Explorer Albert Lin uses technology to study artifacts from Asian civilizations.

Archaeologist
NG Emerging Explorer Beverly Goodman researches the ways in which humans affect nature along coastlines.

Conservationist
NG Emerging Explorer Shafqat Hussain protects endangered snow leopards.

Educator
NG Emerging Explorer Kakenya Ntaiya works to improve girls' education in Kenya.

Filmmaker & Anthropologist
NG Fellow Elizabeth Kapu'uwailani Lindsey strives to preserve the Polynesian culture through documentary film.

Across the world, archaeologists, anthropologists, oceanographers, and linguists represent National Geographic Society in their work. Don't let their long titles confuse you—they are all scientists doing exciting work in the field. New information about Earth's physical features is uncovered every day.

These are some of the explorers at work for National Geographic Society.

Oceanographer
NG Explorer-in-Residence Sylvia Earle works to preserve marine ecosystems.

Aquatic Ecologist & Biogeochemist
NG Emerging Explorer Katey Walter Anthony explores ways to use greenhouse gas for energy.

Technology Innovator
NG Emerging Explorer Ken Banks develops mobile technology to connect remote groups.

Anthropologist
NG Explorer-in-Residence Johan Reinhard studies the cultural practices of Andes people.

From the classroom to the world

Tools for EXPLO

National Geographic explorers all depend on tools for exploration. You can use the variety of tools provided with your program at **myNGconnect.com** to explore the world and its cultures.

Connect to NG
Gateway to research links and current events through National Geographic

my eEdition

Interactive Map Tool
A comprehensive online mapmaker at your fingertips

Taj Mahal, India

Digital Library
GeoVideos, Explorer Video Clips, and hundreds of photographs of the world's physical geography and cultures

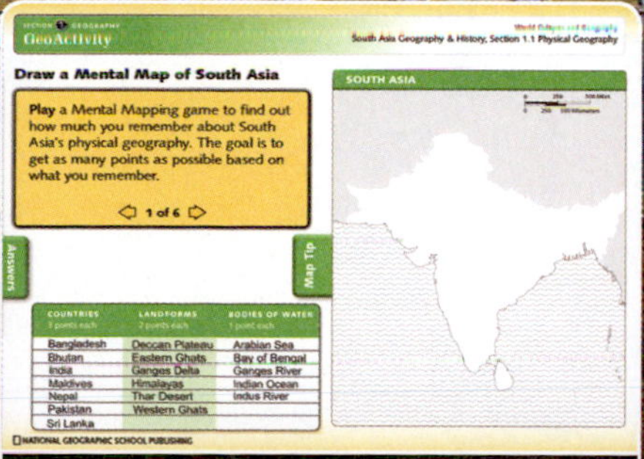

Interactive Whiteboard GeoActivities
Hands-on activities to learn more about how the world works

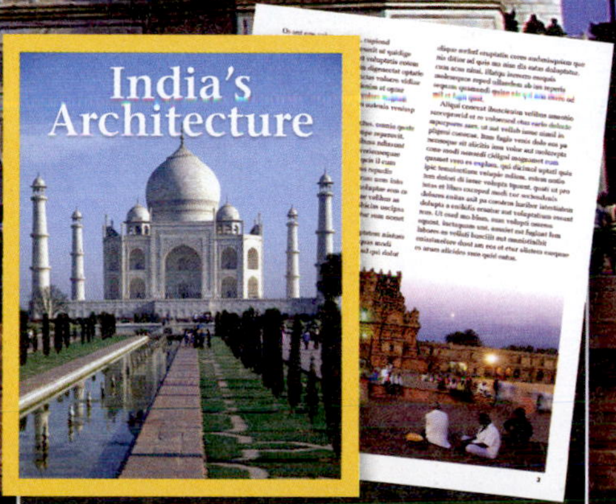

Magazine Maker
Tool for creating student magazines using program resources or by uploading your own photos

India's Architecture

5

From the classroom to the world

The knowledge and skills that explorers need are described in the National Geography Standards, shown here. Keep them in mind as you study *World Cultures and Geography*.

HIGHI

REPORT

1. How to use maps and other geographic representations, tools, and technologies to acquire, process, and **REPORT** information from a spatial perspective

2. How to use mental maps to organize information about people, places, and environments in a spatial context

3. How to **ANALYZE** the spatial organization of people, places, and environments on Earth's surface

ANALYZE

EXPERIENCE

4. The physical and human characteristics of places

5. How people create regions to **INTERPRET** Earth's complexity

6. How culture and **EXPERIENCE** influence people's perceptions of places and regions

INTERPRET

7. The physical processes that shape the patterns of Earth's surface

8. The characteristics and spatial distribution of ecosystems on Earth's surface

9. The characteristics, distribution, and migration of human population on Earth's surface

ER ORDER THINKING

10. The characteristics, distribution, and complexity of Earth's cultural mosaics

11. The **PATTERNS** and networks of economic interdependence on Earth's surface

PATTERNS

12. The processes, patterns, and functions of human settlement

13. How the forces of cooperation and conflict among people influence the division and control of Earth's surface

14. How human actions **MODIFY** the physical environment

15. How physical systems affect human systems

MODIFY

APPLY

16. The changes that occur in the meaning, use, distribution, and importance of resources

17. How to **APPLY** geography to interpret the past

18. How to **APPLY** geography to interpret the present and plan for the future

THE **ESSENTIALS** OF GEOGRAPHY

MEET THE EXPLORER

NATIONAL GEOGRAPHIC

Explorer-in-Residence Spencer Wells is director of the Genographic Project. He analyzes DNA samples to trace humankind's ancient migration patterns and learn where we really come from.

INVESTIGATE GEOGRAPHY

In 2002, NASA released its spectacular "blue marble" image of Earth. It is made from satellite observations gathered over several months, and then carefully combined. The image was inspired by a photo taken by astronauts during a 1972 space mission.

CONNECT WITH THE CULTURE

Culture is how people of a certain region live, behave, and think, but culture is not limited by geography. A busy urban area like New York City (shown here) attracts people from around the world, representing a vast variety of cultures. Those cultures can live together in a single place.

THINK LIKE A GEOGRAPHER

Scientists work to uncover the skeleton of an ancient whale in Wadi Al Hitan, Egypt. Fossils, like those shown here, reveal clues about the earth's past.

NATIONAL GEOGRAPHIC CHAPTER 1

THE GEOGRAPHER'S TOOLBOX

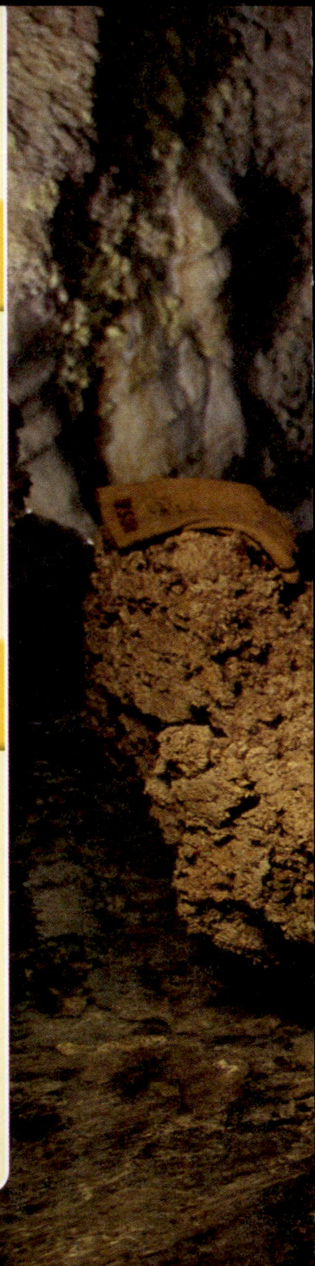

PREVIEW THE CHAPTER

Essential Question How do geographers think about the world?

SECTION 1 • GEOGRAPHIC THINKING

KEY VOCABULARY

- spatial thinking
- geographic pattern
- Geographic Information System (GIS)
- absolute location
- Global Positioning System (GPS)
- relative location
- region
- continent
- terrace

ACADEMIC VOCABULARY

significant, categorize

Essential Question How do people use geography?

SECTION 2 • MAPS

KEY VOCABULARY

- globe
- map
- latitude
- equator
- longitude
- prime meridian
- hemisphere
- scale
- cartographer
- elevation
- relief
- projection

ACADEMIC VOCABULARY

distort, theme

TERMS & NAMES

- North Pole
- South Pole
- Northern Hemisphere
- Southern Hemisphere
- Western Hemisphere
- Eastern Hemisphere

TECHTREK FOR THIS CHAPTER

Student eEdition

Maps and Graphs

Interactive Whiteboard GeoActivities

Digital Library

Connect to NG

Global Positioning System (GPS)

Go to **myNGconnect.com** for information on geographers' tools.

Scientists wear breathing aids while collecting gas samples from a spring in the Cueva (Cave) de Villa Luz in Mexico.

1.1 Thinking Spatially

WESTERN HEMISPHERE

Vancouver, Canada

New York City, United States

Rio de Janeiro, Brazil

Vancouver

Denver

Chicago

New York

Los Angeles

Miami

Veracruz

Lima

Rio de Janeiro

São Paulo

Buenos Aires

PACIFIC OCEAN

ATLANTIC OCEAN

Yukon R.

Mackenzie R.

Peace R.

Missouri R.

Mississippi R.

Colorado R.

Rio Grande

Amazon R.

Tocantins R.

Paraguay R.

Paraná R.

International Date Line

Prime Meridian

Arctic Circle

Tropic of Cancer

Equator

International Date Line

120°E 150°E 180° 150°W 120°W 90°W 60°W 30°W 0° 30°E 60°E

60°N 30°N 0° 30°S

0 1,000 2,000 Miles

0 1,000 2,000 Kilometers

N W E S

A B C D E F G H

1 2 3 4 5 6 7

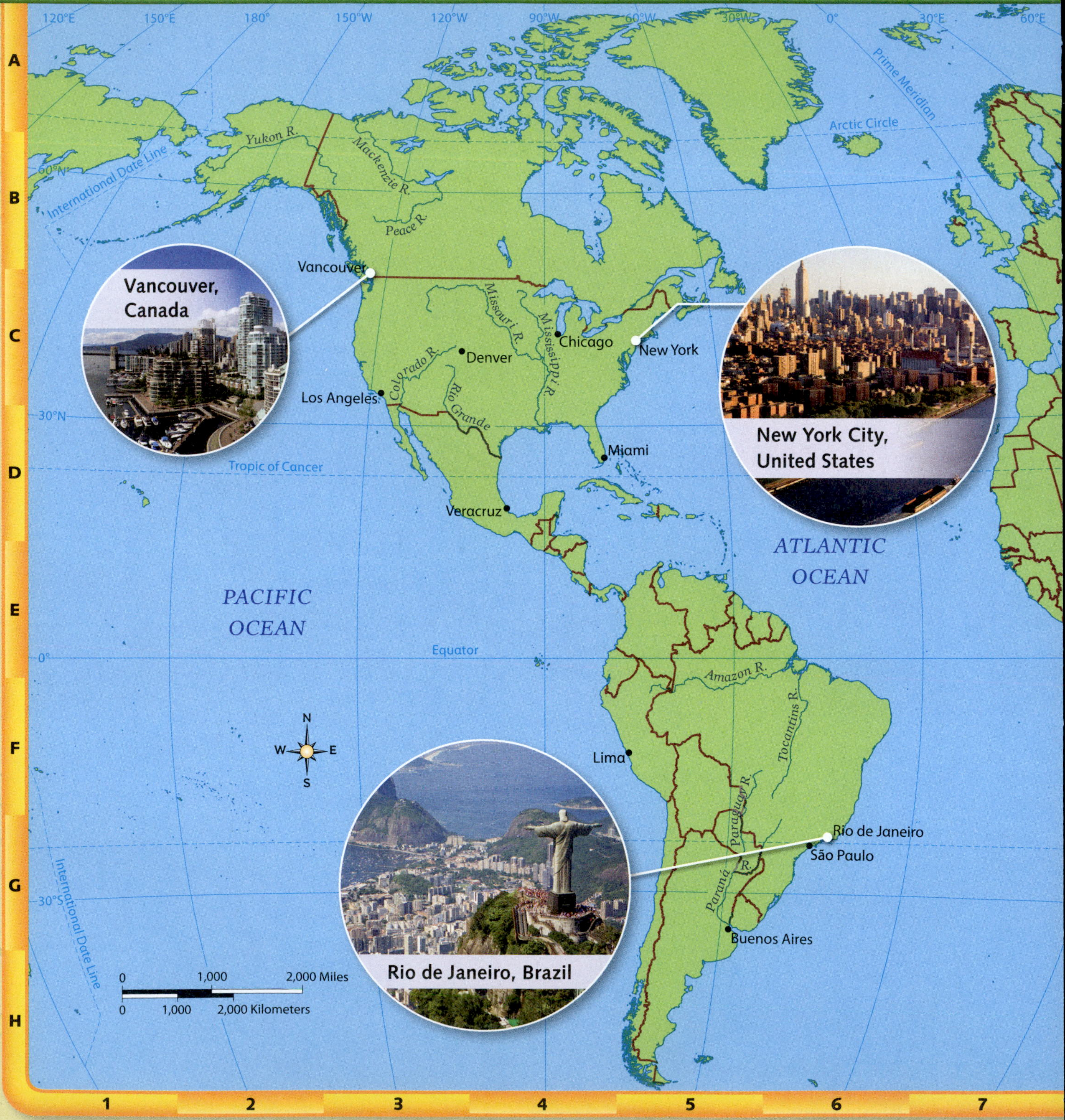

 **Critical Viewing** Denver, Colorado, is located in the foothills of the Rocky Mountains. Chicago, Illinois, is located on the shore of Lake Michigan—one of the five Great Lakes. How might these different locations affect the activities of people who live in these cities?

Main Idea Geographers study the location of places and the people who live there.

Geography is about more than the names of places on a map. It involves **spatial thinking**, or thinking about the space on Earth's surface, including where places are located and why they are there.

Ask Geographic Questions

Geographers use spatial thinking to ask questions, such as "Where is a place located? Why is this location **significant**, or important?"

Look at the photographs of Denver, Colorado, and Chicago, Illinois, above. Denver is near the Rocky Mountains, where gold was once discovered. Chicago is on Lake Michigan, which made it an important shipping center for many years.

Now find New York City on the map. It is on a protected bay. Why is New York's location significant? Like Chicago, it is near water, which is good for trade.

Study Geographic Patterns

By asking and answering many questions, geographers can find patterns. **Geographic patterns** are similarities among places. The location of large cities near water is one example of a geographic pattern.

Many geographers use computer-based **Geographic Information Systems (GIS)**. They create maps and analyze patterns using many layers of data.

Before You Move On

Summarize What do geographers study? How do they study it?

ONGOING ASSESSMENT

MAP LAB
GeoJournal

1. **Location** Where is Rio de Janeiro located on the map? Use directional words in your answer.

2. **Make Inferences** Look at the other cities on the map. Where are most of them located? What pattern do you notice?

3. **Summarize** What are geographic patterns, and how do geographers find them?

1.2 Themes and Elements

TECHTREK
myNGconnect.com For photos and Guided Writing
Digital Library
Student Resources

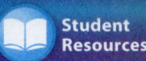

Main Idea Geographers use themes and elements to understand the world.

Geographers ask questions about how people, places, and environments are arranged and connected on Earth's surface. The five themes and six elements will help you **categorize**, or group, information.

The Five Themes of Geography

Geographers use five themes to categorize similar geographic information.

1. **Location** provides a way of locating places. **Absolute location** is the exact point where a place is located. Geographers use a satellite system called the **Global Positioning System (GPS)** to find absolute location. **Relative location** is where a place is in relation to other places. The Great Wall is located near Beijing in northern China.
2. **Place** includes the characteristics of a location. A famous place in the western United States is the Grand Canyon. It has steep rock walls that were carved over centuries by the Colorado River.
3. **Human-Environment Interaction** explains how people affect the environment and how the environment affects people. For example, people build dams to change the flow of rivers.
4. **Movement** explains how people, ideas, and animals move from one place to another. The spread of different religions around the world is an example of movement.
5. **Region** involves a group of places that have common characteristics. North America is a region that includes the United States, Mexico, and Canada.

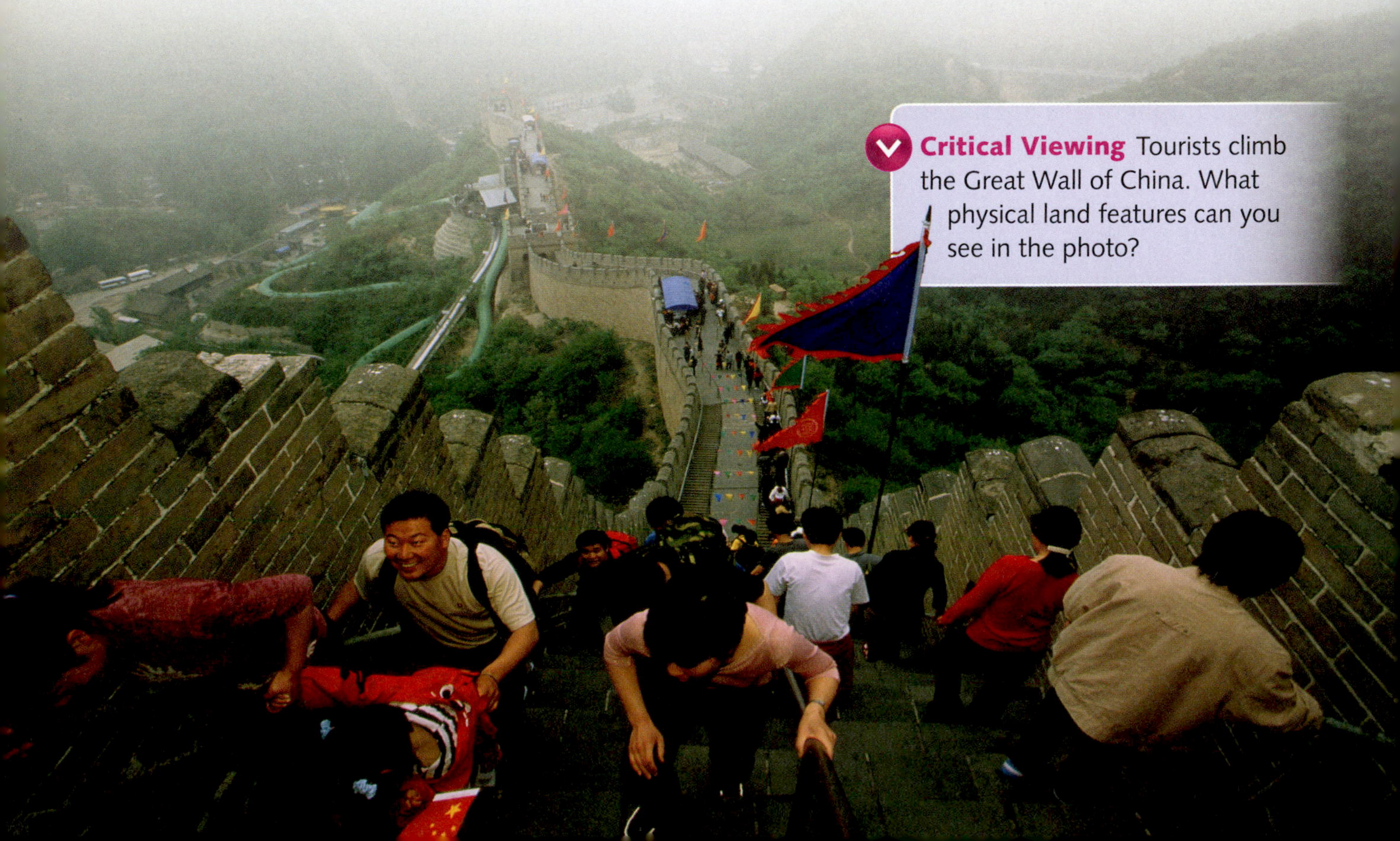

Critical Viewing Tourists climb the Great Wall of China. What physical land features can you see in the photo?

The Great Wall runs through the Chinese countryside.

Six Essential Elements

Some geographers identify essential elements, or key ideas, to study physical processes and human systems.

1. **The World in Spatial Terms** Geographers use tools such as maps to study places on Earth's surface.

2. **Places and Regions** Geographers study the characteristics of places and regions.

3. **Physical Systems** Geographers examine Earth's physical processes, such as earthquakes and volcanoes.

4. **Human Systems** Geographers study how humans live and what systems they create, such as economic systems.

5. **Environment and Society** Geographers explore how humans change the environment and use resources.

6. **The Uses of Geography** Geographers interpret the past, analyze the present, and plan for the future.

Before You Move On

Make Inferences How do geographers use the themes and elements to better understand the world?

WRITING LAB  GeoJournal

1. **Write About Geographic Themes** Write a paragraph in which you use the five themes to describe your community. Explain which theme is the most important in making your community what it is today. Go to **Student Resources** for Guided Writing support.

2. **Categorize** Which theme and which element would you use to categorize information about forms of energy? Explain your answer.

3. **Compare and Contrast** Create a chart like the one below. Use the six essential elements to compare how your town relates to another one that you know about.

ELEMENT	MY TOWN	OTHER TOWN
The World in Spatial Terms	North of the highway	South of the highway
Places and Regions		
Physical Systems		

1.3 World Regions

TECHTREK

myNGconnect.com For an online map and photos of world regions

 Maps and Graphs

 Digital Library

Main Idea Geographers divide the world into regions. Each region is shaped by shared physical and human processes.

In 1413, a Chinese admiral and explorer, Zheng He, sailed from China to Arabia. When he arrived in Arabia, he saw people dressed in ways he had never seen. Yet, like him, these people wanted to trade.

Regions and Continents

Zheng He saw that regions of Earth have similarities and differences. A **region** is a group of places with common traits. The places within a region are linked by trade, culture, and other human activities. They also share similar physical processes and characteristics, such as climate.

A region often includes an entire continent. A **continent** is a large landmass on Earth's surface. Geographers have identified seven continents: Africa, Asia, Australia, Europe, North America, South America, and Antarctica.

Geographers study the world's regions, but they also take a global perspective when they investigate Earth. They might, for instance, study ocean currents around the globe or how one region affects another. Both ways of looking at the world add to our understanding of it.

Before You Move On

Make Inferences Why do geographers study the world by dividing it into regions?

Visual Vocabulary Chinese field workers view the terraced landscape. A **terrace** is a flat surface that is built into a hillside.

Map Labels:

ARCTIC OCEAN

NORTH AMERICA

EUROPE

ASIA

ROCKY MTS

ATLANTIC OCEAN

Tropic of Cancer

URAL MTS

Arctic Circle

HIMALAYA

Yellow R.

Yangtze R.

PACIFIC OCEAN

Equator

Amazon R.

SOUTH AMERICA

ANDES

AFRICA

Nile R.

Congo R.

INDIAN OCEAN

AUSTRALIA

Tropic of Capricorn

0   2,000   4,000 Miles
0   2,000   4,000 Kilometers

Antarctic Circle

ANTARCTICA

Labels: A, B, C, D, E, F, G, H, I, J, K

Coordinates: 180°, 120°W, 60°W, 0°, 60°E, 120°E, 180°, 80°N, 60°N, 40°N, 20°N, 0°, 20°S, 40°S, 60°S, 80°S

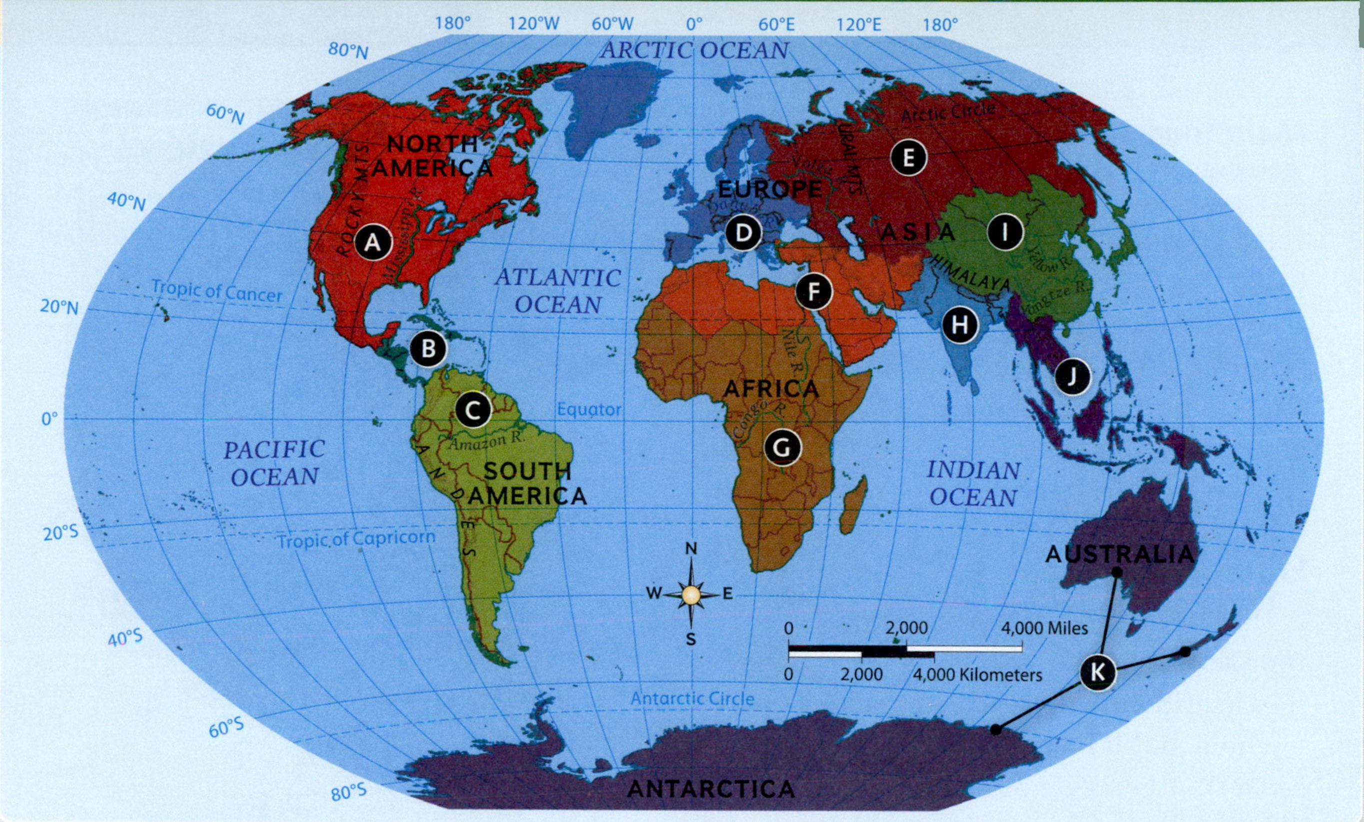

Regions in This Book

(A) North America contains the United States, Canada, and Mexico, along with the Great Lakes—the largest group of freshwater lakes.

(B) Central America and the Caribbean includes the islands of the Caribbean and the countries that connect North and South America.

(C) South America includes Brazil, a growing economic power, and the Amazon Rain Forest, the world's largest tropical rain forest.

(D) Europe includes 29 countries and has nearly 24,000 miles of coastline.

(E) Russia and the Eurasian Republics includes countries that were part of the former Union of Soviet Socialist Republics (U.S.S.R.).

(F) Southwest Asia and North Africa spans two continents—Africa and Asia.

(G) Sub-Saharan Africa includes Africa south of the Sahara, the world's largest desert.

(H) South Asia includes India, one of the world's fastest growing countries.

(I) East Asia includes China, the world's most populous country.

(J) Southeast Asia includes Indonesia, a country made up of over 17,000 islands.

(K) Australia, the Pacific Realm, and Antarctica includes the Pacific island nations north and east of Australia and New Zealand.

ONGOING ASSESSMENT

MAP LAB

 GeoJournal

1. **Interpret Maps** Which regions span more than one continent?

2. **Pose and Answer Questions** Find your region on the map. Write one geographic question about the region and answer the question.

3. **Region** Identify a physical feature in the region in which you live that makes it different from other regions. How does this feature affect you, or those who live near it?

2.1 Elements of a Map

TECHTREK

myNGconnect.com For online maps of geographic regions and photos of antique maps

 Maps and Graphs

 Digital Library

Main Idea Globes and maps are two different tools used to study places on Earth.

Have you ever needed to figure out how to get to a friend's house? Imagine that the only resource you had was a globe. In order to see enough detail to find your friend's house, the globe would have to be enormous—much too big to carry around in your pocket!

Globes and Maps

A three-dimensional, or spherical, representation of Earth is called a **globe**. It is useful when you need to see Earth as a whole, but it is not helpful if you need to see a small section of Earth.

Now imagine taking a part of the globe and flattening it out. This two-dimensional, or flat, representation of Earth is called a **map**. Maps and globes are different representations of Earth, but they have similar features.

A GERMANY'S ECONOMIC ACTIVITY

North Sea
Baltic Sea
Hamburg
Bremen
Elbe R.
Brandenburg
Berlin
Oder R.
Ems R.
Weser R.
Leipzig
Cologne
Zwickau
Dresden
Main R.
GERMANY
Rhine R.
Stuttgart
Danube R.
Munich

Select Industries
- Automobile
- Coal
- Steel

0   50   100 Miles
0   50   100 Kilometers

Map and Globe Elements

A A **title** tells the subject of the map or globe.

B **Symbols** represent information such as natural resources and economic activities.

C **Labels** are the names of places, such as cities, countries, rivers, and mountains.

D **Colors** represent different kinds of information. For example, the color blue usually represents water.

E **Lines of latitude** are imaginary horizontal lines that measure the distance north or south of the equator.

F **Lines of longitude** are imaginary vertical lines that measure the distance east or west of the prime meridian.

G A **scale** shows how much distance on Earth is represented by distance on the map or globe. For example, a half inch on the map above represents 100 miles on Earth.

H A **legend**, or key, explains what the symbols and colors on the map or globe represent.

I A **compass rose** shows the directions north (N), south (S), east (E), and west (W).

J A **locator globe** shows the specific area of the world that is shown on a map. The locator globe on the map above shows where Germany is located.

Latitude

Lines of **latitude** are imaginary lines that run east to west, parallel to the equator. The **equator** is the center line of latitude. Distances north and south of the equator are measured in degrees (°). There are 90 degrees north of the equator and 90 degrees south. The equator is 0°. The latitude of Berlin, Germany, is 52° N, meaning that it is 52 degrees north of the equator.

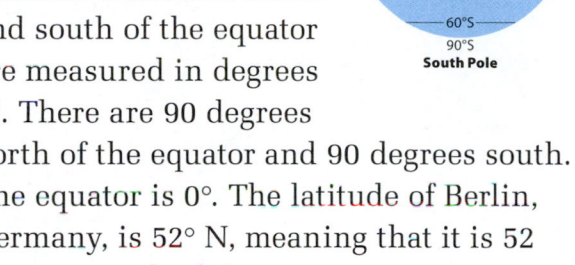

Longitude

Lines of **longitude** are imaginary lines that run north to south from the **North Pole** to the **South Pole**. They measure distance east or west of the prime meridian. The **prime meridian** runs through Greenwich, England. It is 0°. There are 180 degrees east of the prime meridian and 180 degrees west. The longitude of Berlin, Germany, is 13° E, meaning that it is 13 degrees east of the prime meridian.

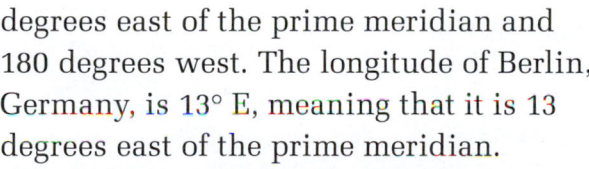

Remember that absolute location is the exact point where a place is located. This point includes a place's latitude and longitude. For example, the absolute location of Berlin, Germany, is 52° N, 13° E. You say this aloud as "fifty-two degrees North, thirteen degrees East."

Hemispheres

A **hemisphere** is half of Earth. The equator divides Earth into the **Northern Hemisphere** and the **Southern Hemisphere**. North America is entirely in the Northern Hemisphere. Most of South America is in the Southern Hemisphere.

The **Western Hemisphere** is west of the prime meridian. The **Eastern Hemisphere** is east of the prime meridian. South America is in the Western Hemisphere. Most of Africa is in the Eastern Hemisphere.

Before You Move On

Monitor Comprehension How are maps and globes different? How is each one used?

ONGOING ASSESSMENT

MAP LAB

 GeoJournal

1. **Interpret Maps** What types of industry are located in Germany? What map elements did you use to find the answers?

2. **Make Inferences** What is the main industry in southern Germany? Why might this industry be located there?

3. **Location** What is the difference between lines of latitude and lines of longitude?

2.2 Map Scale

TECHTREK

myNGconnect.com For online
maps and and photos at different scales

 Maps and
Graphs

 Digital
Library

Main Idea Maps use different scales for different purposes.

On a walk through a city, such as Charlotte, North Carolina, you might use a highly-detailed map that shows only the downtown area. To drive up the Atlantic coast, however, you would use a map that covers a large area, including several states. These maps have different scales.

Interpreting a Scale

A map's scale shows how much distance on Earth is shown on the map. A large-scale map covers a small area but shows many details. A small-scale map covers a large area but includes few details.

A scale is usually shown in both inches and centimeters. One inch or centimeter on the map represents a much larger distance on Earth, such as a number of miles or kilometers.

To use a map scale, mark off the length of the scale several times on the edge of a sheet of paper. Then hold the paper between two points to see how many times the scale falls between them. Add up the distance.

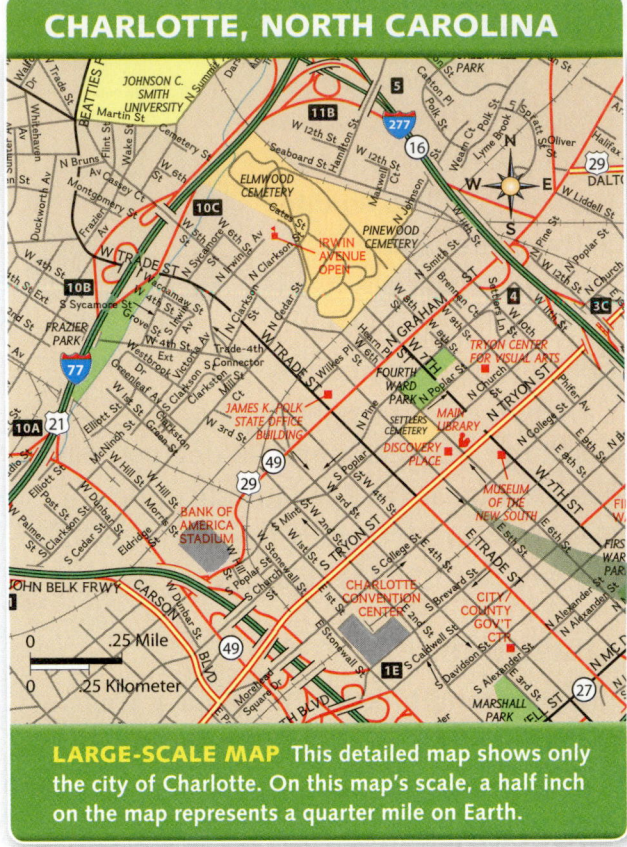

CHARLOTTE, NORTH CAROLINA

LARGE-SCALE MAP This detailed map shows only the city of Charlotte. On this map's scale, a half inch on the map represents a quarter mile on Earth.

Purposes of a Scale

The scale of a map should be appropriate for its purpose. For example, a tourist map of Washington, D.C., should be large-scale, showing every street name, monument, and museum.

Critical Viewing The Atlantic Ocean rolls in along the dunes on the Outer Banks of North Carolina. Which map labels the location of the Outer Banks?

NORTH CAROLINA

MEDIUM-SCALE MAP This map shows the entire state of North Carolina. It includes fewer details and covers a larger area.

EASTERN UNITED STATES

SMALL-SCALE MAP This map shows the Atlantic coast states, from Florida to Maine. It covers a large area and shows even fewer details than a medium-scale map.

Maps of any scale show geographic patterns. The map of Washington, D.C., for instance, would show that many government buildings are in one area.

Before You Move On
Summarize What are the purposes of a small-scale map and a large-scale map?

ONGOING ASSESSMENT

MAP LAB

 **GeoJournal**

1. **Interpret Maps** On the map of North Carolina, approximately how many inches represent 200 miles? On the map of the eastern United States, how many miles does one inch represent?

2. **Location** How far is Washington, D.C., from the southwestern border of Maine? Which map did you use to find the distance? Why?

3. **Synthesize** What geographic pattern do you see in the location of cities in North Carolina? How might you explain this pattern?

2.3 Political and Physical Maps

TECHTREK

myNGconnect.com For online political and physical maps of world regions

 Maps and Graphs

EAST ASIA POLITICAL

60°E · 80°E · 100°E · 120°E · 140°E · 160°E

Ertis R.

Sea of Okhotsk

40°N

Amur R.

Selenge R.

Ulaanbaatar

MONGOLIA

Tarim R.

Beijing

Liao R.

Songhua R.

NORTH KOREA

P'yongyang

Sea of Japan (East Sea)

JAPAN

Tokyo

PACIFIC OCEAN

40°N

Indus R.

Yellow R.

Seoul

SOUTH KOREA

C H I N A

Yellow Sea

Brahmaputra R.

Mt. Everest
29,035 ft
(8,850 m)

Chengdu
Chongqing

Yangtze R.

Shanghai

East China Sea

Philippine Sea

Tropic of Cancer

Ganges R.

20°N

Kunming

Xi R.

South China Sea

Taipei
TAIWAN

20°N

Hong Kong

Bay of Bengal

Irrawaddy R.

Mekong R.

0 · 500 · 1,000 Miles
0 · 500 · 1,000 Kilometers

80°E · 100°E · 120°E · 140°E

EAST ASIA PHYSICAL

60°E · 80°E · 100°E · 120°E · 140°E · 160°E

Ertis R.

Sea of Okhotsk

40°N

ALTAY MOUNTAINS

Amur R.

Selenge R.

M O N G O L I A

Mongolian Plateau

T I A N   S H A N

Tarim R.

G O B I

Liao R.

Songhua R.

NORTH KOREA

Sea of Japan (East Sea)

JAPAN

PACIFIC OCEAN

40°N

Taklimakan Desert

KUNLUN MOUNTAINS

Yellow R.

SOUTH KOREA

Indus R.

PLATEAU

C H I N A

North China Plain

Yellow Sea

OF   TIBET

Elevation

feet	meters
10,000+	3,050+
5,000	1,524
2,000	610
1,000	305
500	152
0	0
Below sea level	

H I M A L A Y A

Brahmaputra R.

Mt. Everest
29,035 ft
(8,850 m)

Yangtze R.

East China Sea

Ganges R.

Philippine Sea

Tropic of Cancer

20°N

Xi R.

20°N

Bay of Bengal

Irrawaddy R.

Mekong R.

TAIWAN

0 · 500 · 1,000 Miles
0 · 500 · 1,000 Kilometers

South China Sea

80°E · 100°E · 120°E · 140°E

Main Idea Political maps show features that humans have created on Earth's surface. Physical maps show natural features.

The governor of a state needs a map that shows counties and cities. A mountain climber needs a map that shows cliffs, canyons, and ice fields. **Cartographers**, or mapmakers, create different kinds of maps for these different purposes.

Political Maps

A political map shows features that humans have created, such as countries, states, provinces, and cities. These features are labeled, and lines show boundaries, such as those between countries.

Physical Maps

A physical map shows natural features of physical geography. It includes landforms, such as mountains, plains, valleys, and deserts. It also includes oceans, lakes, rivers, and other bodies of water.

A physical map can also show elevation and relief. **Elevation** is the height of a physical feature above sea level. **Relief** is the change in elevation from one place to another. Maps show elevation by using color. The physical map at left uses seven colors for seven ranges of elevation.

Before You Move On

Monitor Comprehension How is a political map different from a physical map?

⌃ **Critical Viewing** The Sobaek Mountains cut diagonally across South Korea. Which map best indicates the location of these mountains?

ONGOING ASSESSMENT

MAP LAB
GeoJournal

1. **Interpret Maps** What is the most mountainous country in East Asia? How did you find the answer?

2. **Human-Environment Interaction** Based on elevations shown on the map, what economic activity would you expect to find on the North China Plain?

3. **Draw Conclusions** What do the locations of Hong Kong, Shanghai, and Tokyo have in common? What conclusion can you draw about the location of cities around the world?

2.4 Map Projections

TECHTREK

myNGconnect.com For additional
maps in a variety of projections

Maps and
Graphs

Main Idea Cartographers use various projections to show Earth's curved surface on a flat map.

The world is a sphere, but maps are flat. As a result, maps **distort**, or change, shapes, areas, distances, and directions found in the real world. To reduce distortion, mapmakers use **projections**, or ways of showing Earth's curved surface on a flat map. Five common map projections are the azimuthal, Mercator, homolosine, Robinson, and Winkel Tripel. Each projection has strengths and weaknesses—each distorts in a different way.

When cartographers make maps, they need to choose a map projection. The type of projection depends on the map's purpose. Which elements are acceptable to distort? Which are not acceptable to distort? For example, if a cartographer is creating a navigation map, it is important that directions are not distorted. It may not matter, however, if some areas or shapes are distorted.

Before You Move On

Make Inferences How do cartographers decide which projection to use?

AZIMUTHAL PROJECTION

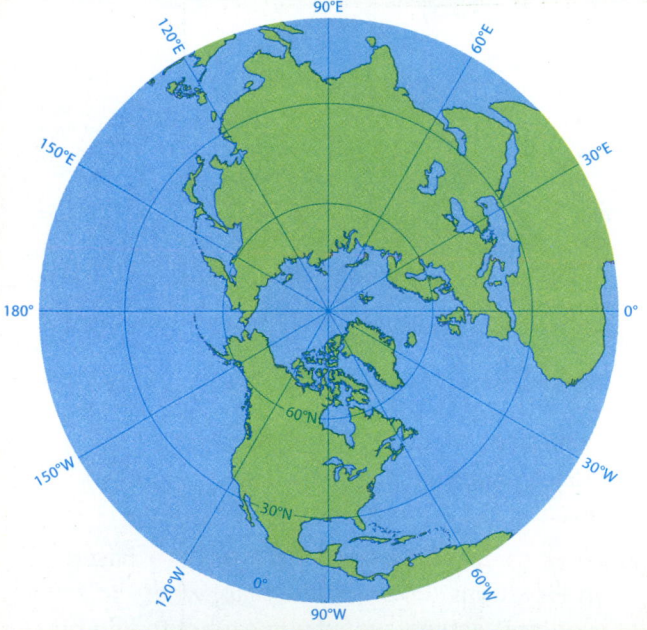

Mapmakers create the **azimuthal projection** by projecting part of the globe onto a flat surface. The projection shows directions accurately but distorts shapes. It is often used for the polar regions.

MERCATOR PROJECTION

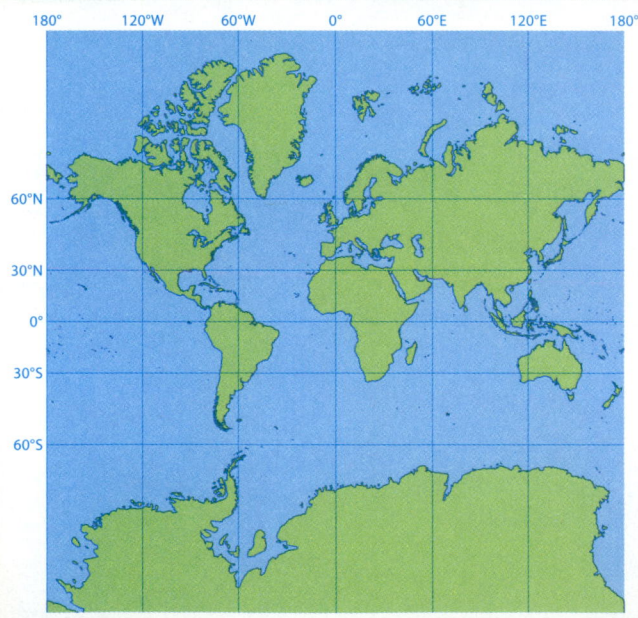

This **Mercator projection** shows much of Earth accurately, but it distorts the shape and area of land near the North and South Poles. This projection shows direction accurately, so it is good for navigation maps.

HOMOLOSINE PROJECTION

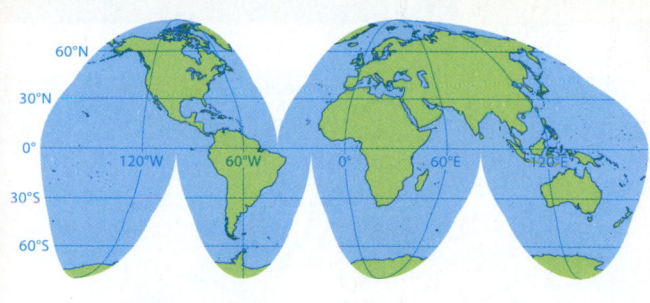

The **homolosine projection** resembles the flattened peel of an orange. It accurately shows the shape and area of landmasses by cutting up the oceans. However, it does not show distances accurately.

ROBINSON PROJECTION

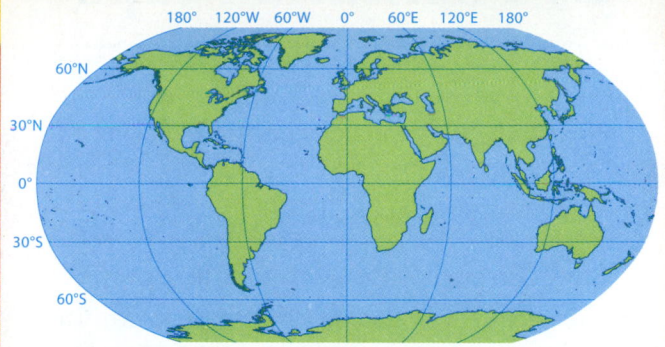

The **Robinson projection** combines the strengths of other projections. It shows the shape and area of the continents and oceans with reasonable accuracy. However, the North and South Poles are distorted.

WINKEL TRIPEL PROJECTION

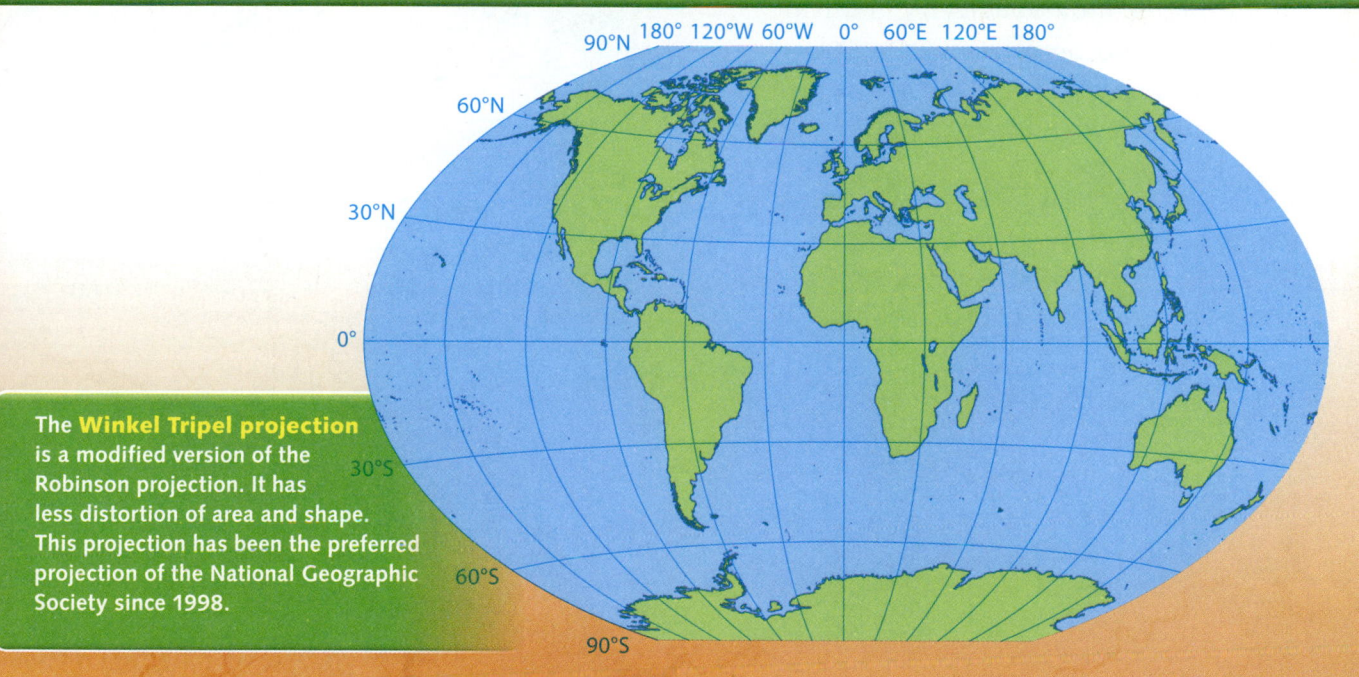

The **Winkel Tripel projection** is a modified version of the Robinson projection. It has less distortion of area and shape. This projection has been the preferred projection of the National Geographic Society since 1998.

ONGOING ASSESSMENT

MAP LAB

 GeoJournal

1. **Compare and Contrast** Locate Greenland on the Mercator projection and on the Robinson projection. What is similar and different in the two maps? Why?

2. **Location** What does the azimuthal projection show about the relative location of Alaska and Russia?

2.5 Thematic Maps

Main Idea Thematic maps focus on specific topics, such as the population density or economic activity in a region or country.

Suppose you wanted to create a map showing the location of sports fields in your community. You would create a thematic map, which is a map about a specific **theme**, or topic.

Types of Thematic Maps

Thematic maps are useful for showing a variety of geographic information, including economic activity, natural resources, and population density. Common types of thematic maps are the point symbol map, the dot density map, and the proportional symbol map.

Before You Move On

Make Inferences Look through this textbook and identify another example of a thematic map. Why did you choose this map?

U.S. ALTERNATIVE ENERGY

U.S. Alternative Energy
- Wind farms
  200 megawatts or greater
- Hydropower
  200 megawatts or greater
- Solar generation
  1 megawatt or greater

POINT SYMBOL MAP This type of map shows the location of activities at different points. For example, this map has symbols that show some sources of wind, water, and solar energy in the United States.

> **Critical Viewing** Solar panels in the Nevada desert absorb light from the sun and turn it into energy. Why might Nevada be a good location for solar fields?

THAILAND'S POPULATION DENSITY

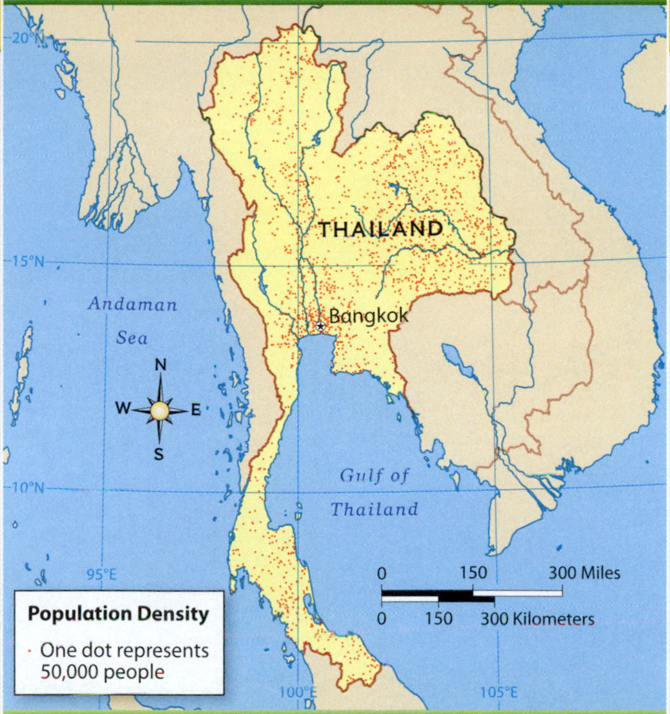

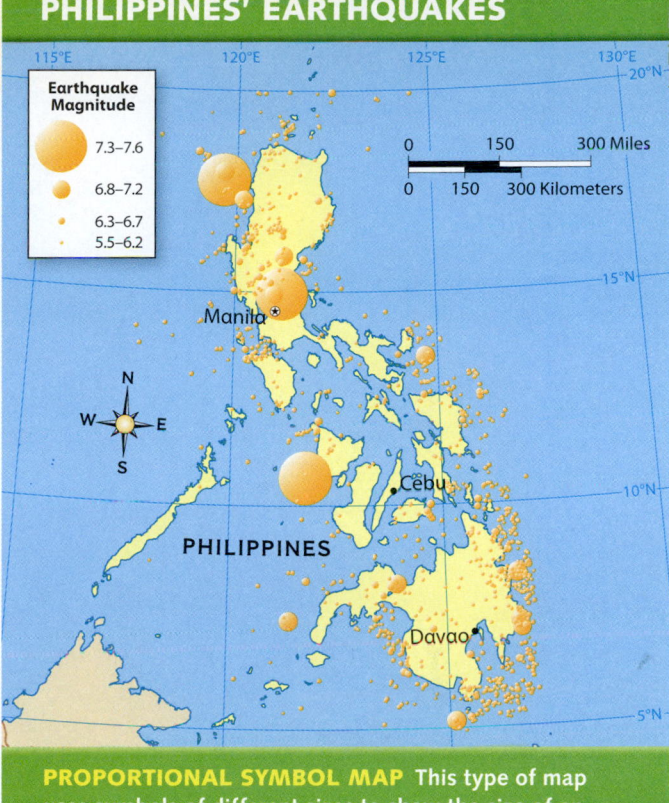

20°N
15°N
10°N
95°E
100°E
105°E
Andaman Sea
Gulf of Thailand
THAILAND
Bangkok
N
S
E
W
0 150 300 Miles
0 150 300 Kilometers
Population Density
One dot represents 50,000 people

DOT DENSITY MAP This type of map uses dots to show how something is distributed in a country or region. Each dot represents an amount. For example, the dots on this map show population density in Thailand.

PHILIPPINES' EARTHQUAKES

115°E
120°E
125°E
130°E
20°N
15°N
10°N
5°N
Earthquake Magnitude
7.3–7.6
6.8–7.2
6.3–6.7
5.5–6.2
0 150 300 Miles
0 150 300 Kilometers
Manila
Cebu
Davao
PHILIPPINES
N
S
E
W

PROPORTIONAL SYMBOL MAP This type of map uses symbols of different sizes to show the size of an event. For example, the size of the circles on this map shows the severity of earthquakes in the Philippines.

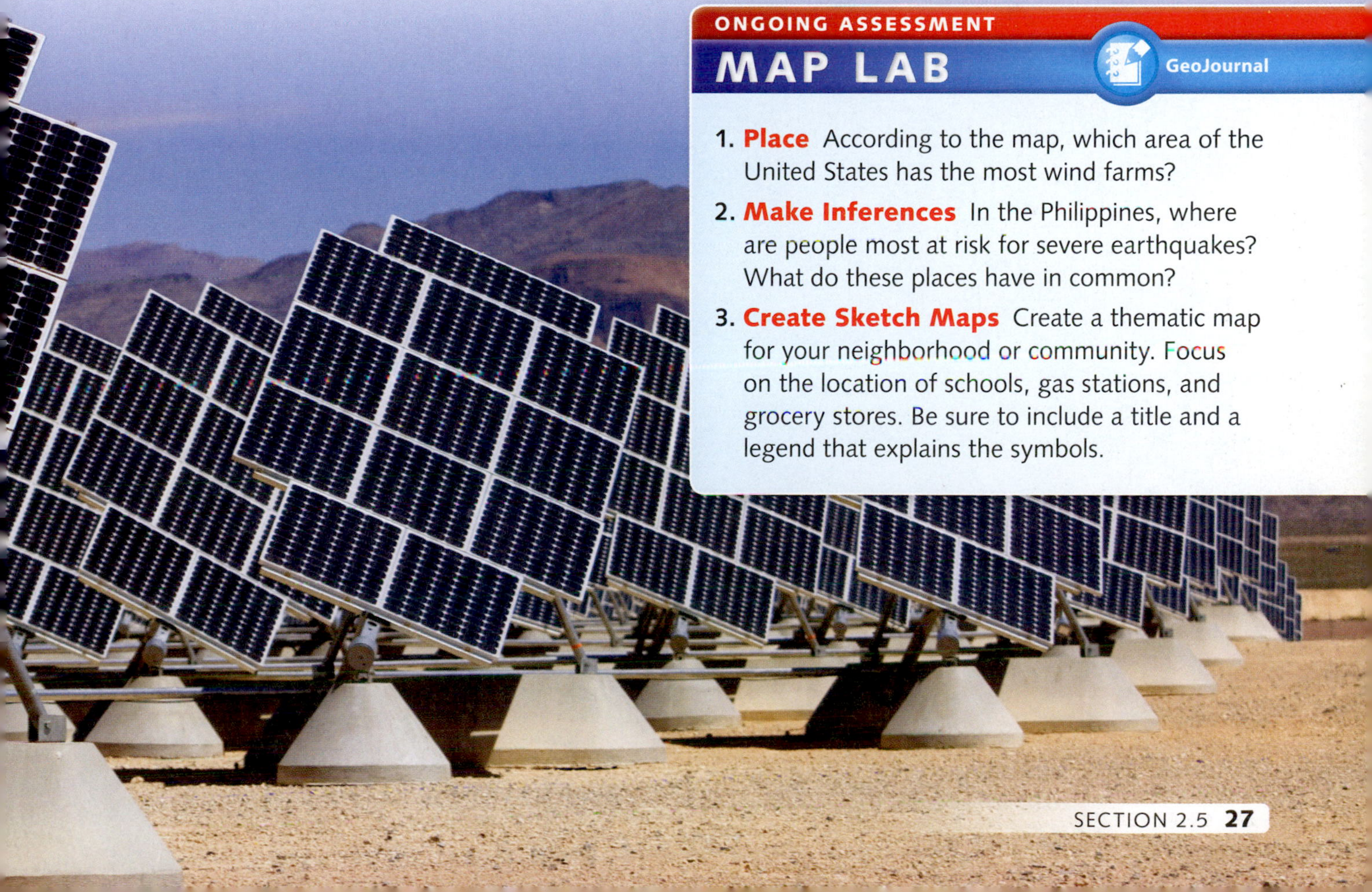

ONGOING ASSESSMENT

MAP LAB

GeoJournal

1. **Place** According to the map, which area of the United States has the most wind farms?
2. **Make Inferences** In the Philippines, where are people most at risk for severe earthquakes? What do these places have in common?
3. **Create Sketch Maps** Create a thematic map for your neighborhood or community. Focus on the location of schools, gas stations, and grocery stores. Be sure to include a title and a legend that explains the symbols.

NATIONAL GEOGRAPHIC

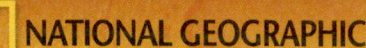

Photo Gallery • The World At Night

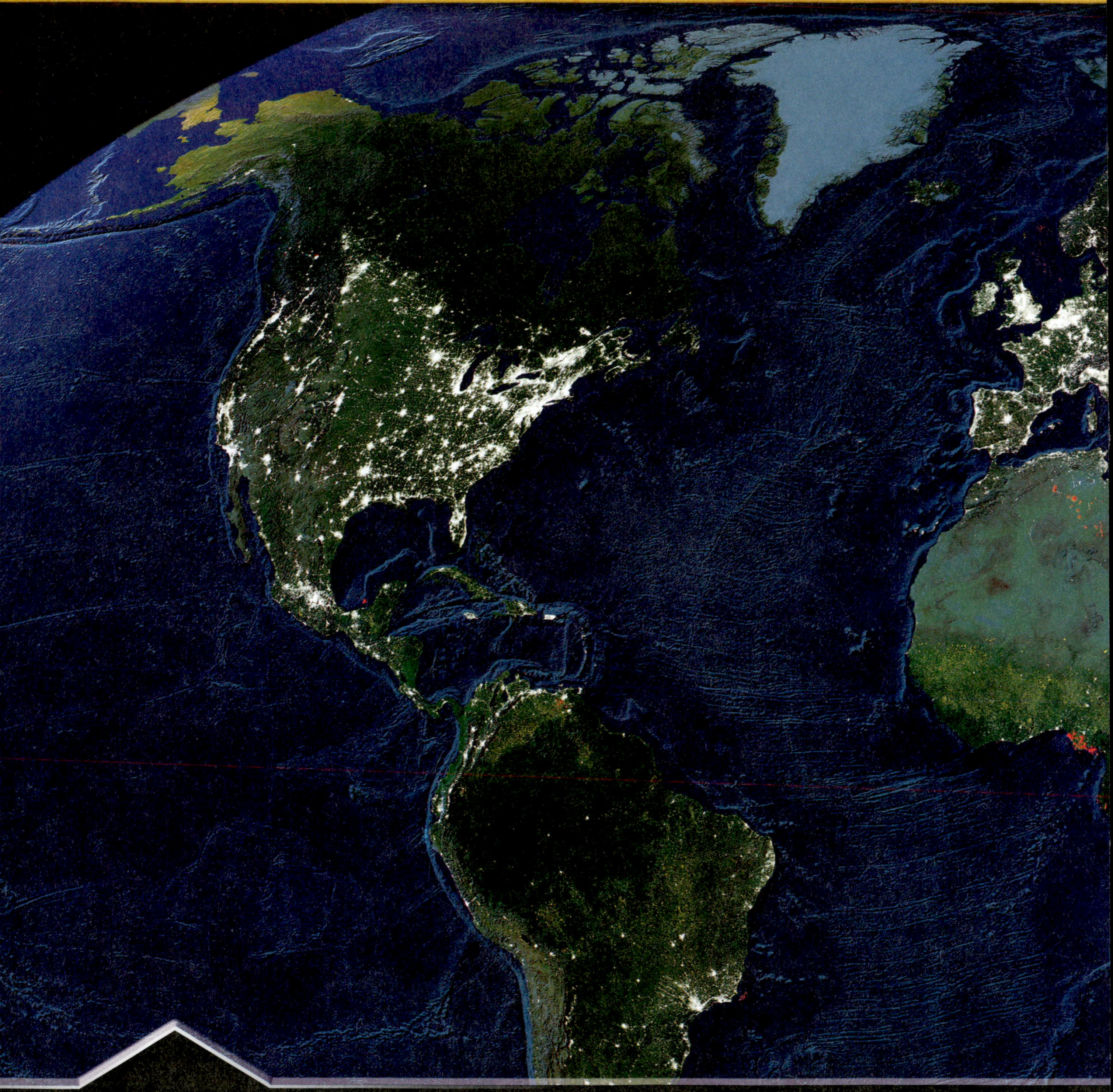

For more photos from the National Geographic Photo Gallery, go to the **Digital Library** at myNGconnect.com.

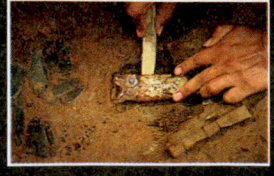

Mayan artifact

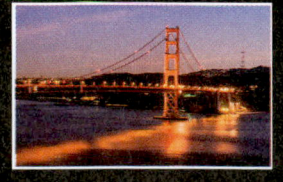

Golden Gate Bridge

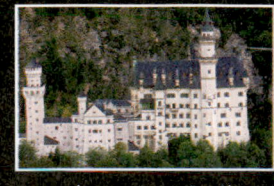

Neuschwanstein Castle

Critical Viewing This view of Earth at night was made by combining images from three satellites over a one-year period. In addition to the lighted areas, fires and natural gas burn-off are shown in orange.

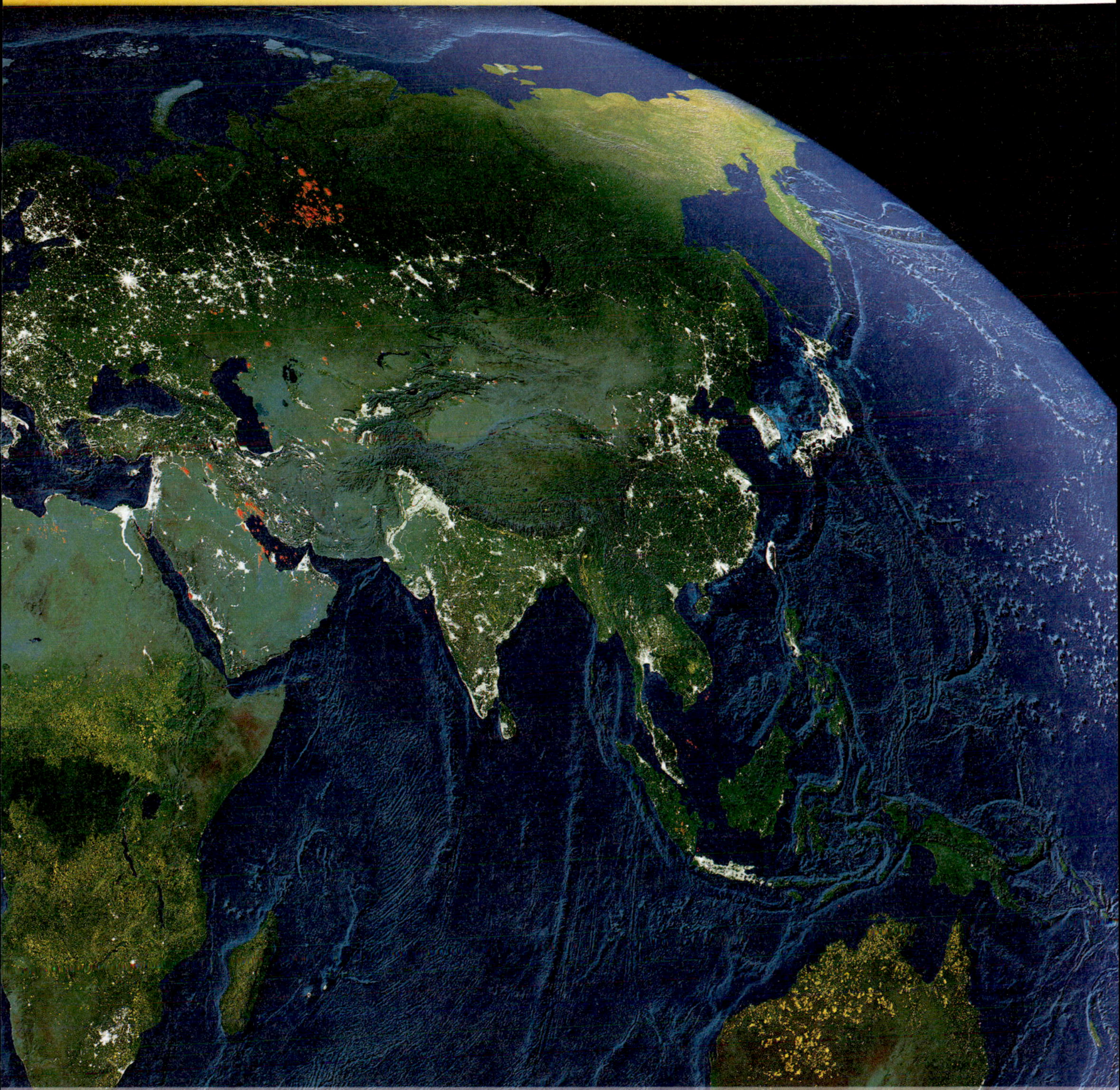

Archaeologist at work

Eastern Hemisphere, 1928

Tokyo, Japan, at night

Lima, Peru

VOCABULARY

For each pair of words, write one sentence that explains the connection between the two words.

1. absolute location; relative location

> The absolute location of Washington, D.C., is 39° N, 77° W; its relative location is on the Potomac River.

2. region; continent

3. latitude; longitude

4. relief; elevation

5. distort; projection

MAIN IDEAS

6. What are Geographic Information Systems? How do geographers use them? (Section 1.1)

7. How are the five themes of geography and the six essential elements similar? How are they different? (Section 1.2)

8. Why is the construction of a highway an example of human-environment interaction? (Section 1.2)

9. What are traits of a region? (Section 1.3)

10. How do latitude and longitude help determine the absolute location of a place? (Section 2.1)

11. Is a map of the world a large-scale map or a small-scale map? Why? (Section 2.2)

12. How do physical maps show elevation and relief? (Section 2.3)

13. How do map projections distort Earth? (Section 2.4)

14. What type of thematic map would a cartographer use to show different types of agriculture in Africa? Explain your answer. (Section 2.5)

GEOGRAPHIC THINKING

ANALYZE THE ESSENTIAL QUESTION

How do geographers think about the world?

Critical Thinking: Describe Geographic Information

15. How do geographers use spatial thinking to make sense of space on Earth's surface?

16. How would you describe New York City using the five themes of geography?

17. What is an example of a physical process, and how can it affect a region?

INTERPRET MAPS

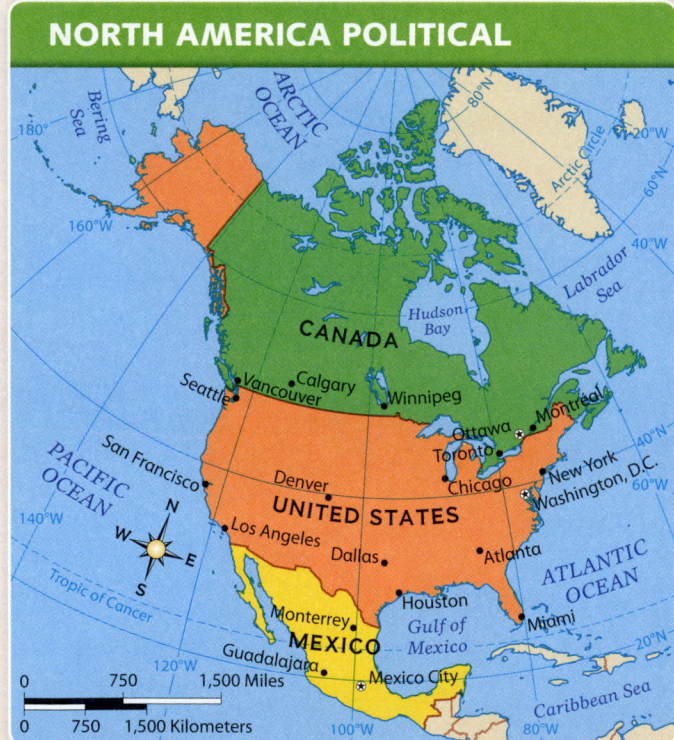

18. **Region** According to the map, what is the relative location of Canada's cities? Why do you think they are located there?

19. **Compare and Contrast** How does the relative location of Mexico's cities compare with the relative location of Canada's cities?

MAPS

ANALYZE THE ESSENTIAL QUESTION

How do people use geography?

Critical Thinking: Make Inferences

20. On a Mercator projection map, Greenland looks larger than South America. However, it is actually much smaller. How could this affect a person's understanding of land areas?

21. A geographer wants to show where economic activities are located. What kind of thematic map should the geographer use? Explain.

22. An explorer is the first to map a region that has hills, canyons, and flat areas. What type of map should the explorer create? Explain.

INTERPRET PRIMARY SOURCES

In 1953, Sir John Hunt led a group of climbers up Mount Everest, the highest mountain in the world. Read Hunt's description of climbing Everest and answer the questions.

> What makes Everest murderous is . . . its cold, its wind, and its climbing difficulties. . . . At 28,000 feet a given volume of air breathed contains only a third as much oxygen as at sea level. On the ground, even if a man were exercising violently, his lungs would need but 50 liters of air per minute. Near Everest's summit he struggles to suck in as much as 200 liters. Since he inhales his air cold and dry and exhales it warm and moist, the stress on his parched lungs and respiratory passages becomes appalling.
>
> –*National Geographic*, July 1954

23. **Find Details** What are the characteristics of the environment on Mount Everest?

24. **Human-Environment Interaction** How is this description an example of human-environment interaction?

25. ACTIVE OPTIONS

Synthesize the Essential Questions by completing the activities below.

26. **Write a Travel Brochure** Write a travel brochure to attract people to a place that you hope to visit. In your brochure, answer geographic questions such as the following: "Where it is located?" "In what region is it located?" "How have people made the environment more livable?" "What are the physical features of the place?" **Share your brochure with the rest of the class.**

> **Writing Tips**
> - Take notes before you begin to write.
> - Write a catchy slogan that grabs the reader's attention.
> - Write a list of reasons that persuades people to travel to the place.

TECHTREK myNGconnect.com For maps and photographs of the five themes of geography

27. **Create a Digital Presentation** Use presentation software to create a digital presentation on the five themes of geography. For each theme, write several bullet points. In addition, explain why each theme is useful. For each theme, select photographs from the **NG Photo Gallery** or other sources, and maps from the **Online World Atlas** that illustrate the theme. Use this format for your slides:

> Region: What is a region?
> - A region shares common physical and/or human characteristics.

PHYSICAL & HUMAN GEOGRAPHY

PREVIEW THE CHAPTER

Essential Question How is the earth continually changing?

KEY VOCABULARY

- solstice
- equinox
- tectonic plate
- continental drift
- plain
- plateau
- continental shelf
- butte
- erosion
- earthquake
- tsunami
- volcano
- evaporation
- condensation
- precipitation

ACADEMIC VOCABULARY
benefit, essential

TERMS & NAMES

- Ring of Fire

Essential Question What shapes the earth's varied environments?

KEY VOCABULARY

- climate
- weather
- vegetation
- hurricane
- cyclone
- tornado
- raw material
- nonrenewable resource
- renewable resource
- habitat
- ecosystem
- marine life

ACADEMIC VOCABULARY
restore, impact

Essential Question How has geography influenced cultures around the world?

KEY VOCABULARY

- culture
- civilization
- communal
- gaucho
- culture region
- kimono
- monotheistic religion
- polytheistic religion
- economy
- capital
- entrepreneurship
- free enterprise economy
- gross domestic product (GDP)
- government
- citizen
- democracy
- human rights

ACADEMIC VOCABULARY
symbol

TERMS & NAMES

- United Nations (UN)
- Universal Declaration of Human Rights

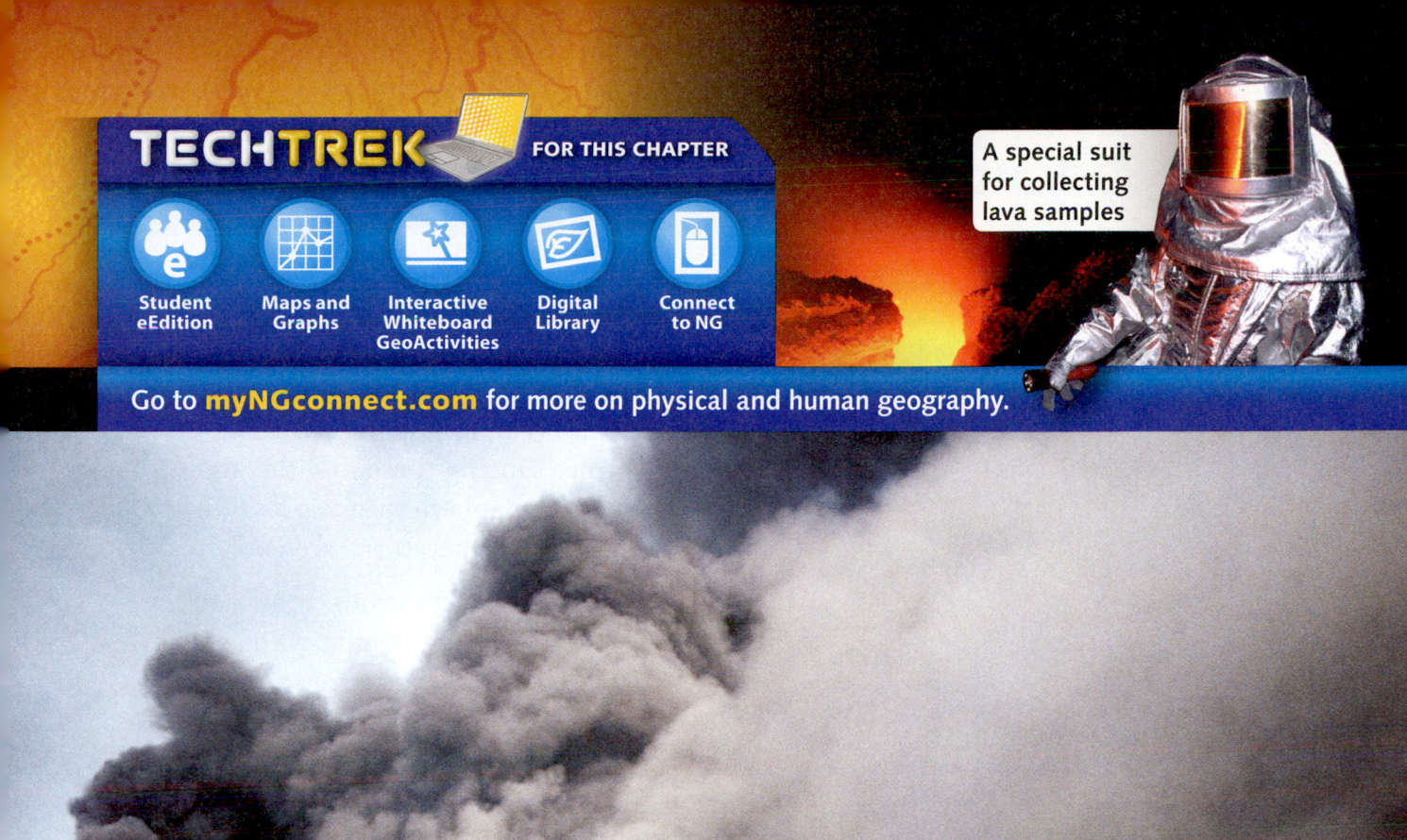

TECHTREK FOR THIS CHAPTER

Student eEdition

Maps and Graphs

Interactive Whiteboard GeoActivities

Digital Library

Connect to NG

Go to **myNGconnect.com** for more on physical and human geography.

A special suit for collecting lava samples

Scientists observe a cloud of ash erupting from Italy's Mount Etna.

1.1 Earth's Rotation and Revolution

TECHTREK

myNGconnect.com For photos and
a model of the four seasons

Digital
Library

Student
Resources

Main Idea Earth's tilt, rotation, and revolution cause weather changes and the four seasons.

In the summer, people who live in the northern parts of Iceland, Norway, Sweden, and Finland have more than 20 hours of daylight. In the winter, these same people have more than 20 hours of darkness. These long days and nights result from Earth's tilt and revolution around the sun.

Revolution and Rotation

The solar system is formed by the sun, Earth, and seven other planets. Earth, which is the third planet from the sun, revolves around the sun at a speed of about 67,000 miles per hour. It takes one year for Earth to make one revolution.

At the same time, Earth rotates on its axis, an imaginary line that runs from the North Pole to the South Pole through the center. Each rotation takes almost one day.

> **Critical Viewing** Bathers in Iceland celebrate the summer solstice with a midnight swim in a hot spring. Based on the photo, what can you tell about the summer solstice in Iceland?

Earth tilts at an angle of about 23.5°. Because of this tilt, the Northern Hemisphere receives more direct sunlight for half the year, and temperatures are warmer. During these months, the Southern Hemisphere receives less direct sunlight, and temperatures are cooler.

As Earth continues around the sun, the Northern Hemisphere faces the sun less directly and temperatures are cooler. Meanwhile the Southern Hemisphere faces the sun more directly, and temperatures are warmer. This process creates the four seasons in both hemispheres.

Summer and Winter Solstices

The exact moment at which summer and winter start is called a **solstice**. June 20 or 21 is the summer solstice in the Northern Hemisphere. It is the longest day of the year. Six months later, on December 21 or 22, the Northern Hemisphere has its winter solstice. This is the shortest day of the year.

The Southern Hemisphere is exactly the opposite. June 20 or 21 is its winter solstice and December 21 or 22 is its summer solstice.

SPRING EQUINOX
(March 21)

WINTER SOLSTICE
(December 21 or 22)

SUN

North Pole

Northern
Hemisphere

24 hours

SUMMER SOLSTICE
(June 20 or 21)

365 days

Southern
Hemisphere

AUTUMN EQUINOX
(September 23)

South Pole

Spring and Autumn Equinoxes

The beginning of spring and autumn is called an **equinox**. Twice a year, the sun's rays hit the equator directly, and day and night are the same length. In the Northern Hemisphere, the spring equinox occurs around March 21, and the autumn equinox occurs around September 23. The Southern Hemisphere is exactly the opposite.

Before You Move On

Monitor Comprehension How do Earth's tilt, rotation, and revolution cause the seasons?

ONGOING ASSESSMENT

VIEWING LAB  GeoJournal

1. **Analyze Models** According to the model above, when do the sun's rays hit the Southern Hemisphere most directly? This is the beginning of which season?

2. **Analyze Visuals** What happens to the sun in Iceland on the day of the summer solstice? Why does this happen?

3. **Compare and Contrast** How are the spring equinox and the autumn equinox alike?

4. **Make Inferences** What happens to the length of days in the Northern Hemisphere after the spring equinox?

1.2 Earth's Complex Structure

TECHTREK

myNGconnect.com For photos and diagrams that illustrate plate tectonics

Digital Library

Student Resources

> **Main Idea** Physical processes within Earth bring about changes on the surface.

If you could dig a tunnel to Earth's center, you would travel through several layers. Each layer would be under tremendous pressure and give off intense heat.

Earth's Layers

On your journey, you would first pass through the crust. This layer includes landmasses and the ocean floor. It is about 30 miles thick.

Next, you would come to the mantle, which consists of molten, or melted, rocks called magma. The mantle is about 1,800 miles thick and has two parts—the upper mantle and the lower mantle.

Descending even deeper, you would find yourself in the outer core, which is about 1,400 miles thick. This layer is mostly liquid, consisting of molten iron and nickel.

At the very center is the inner core. It is about 700 miles thick. It reaches a temperature of 12,000° F—hotter than the surface of the sun. The inner core is made up of iron, which remains solid because the pressure from all the layers above it is so intense.

EARTH'S STRUCTURE

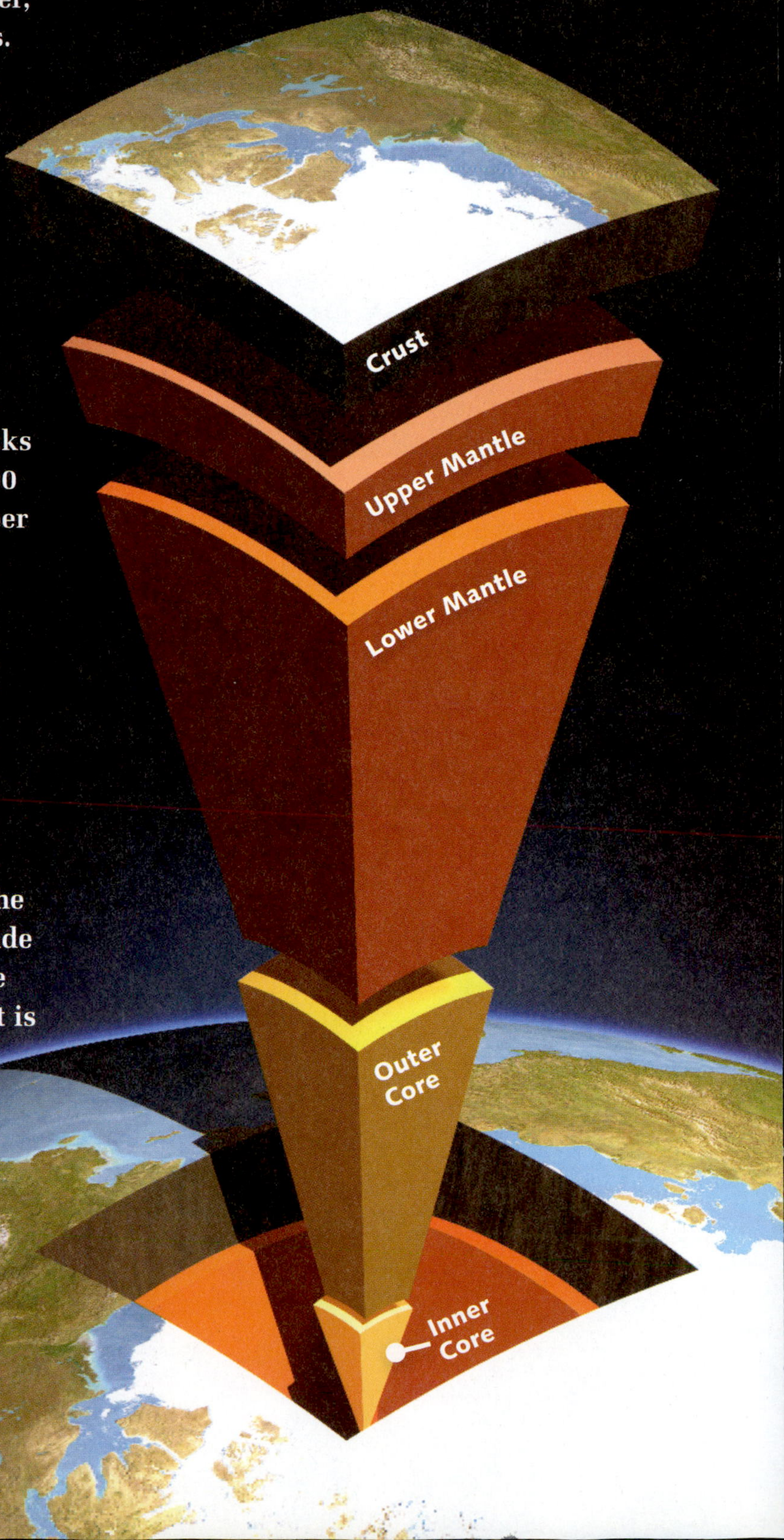

Crust

Upper Mantle

Lower Mantle

Outer Core

Inner Core

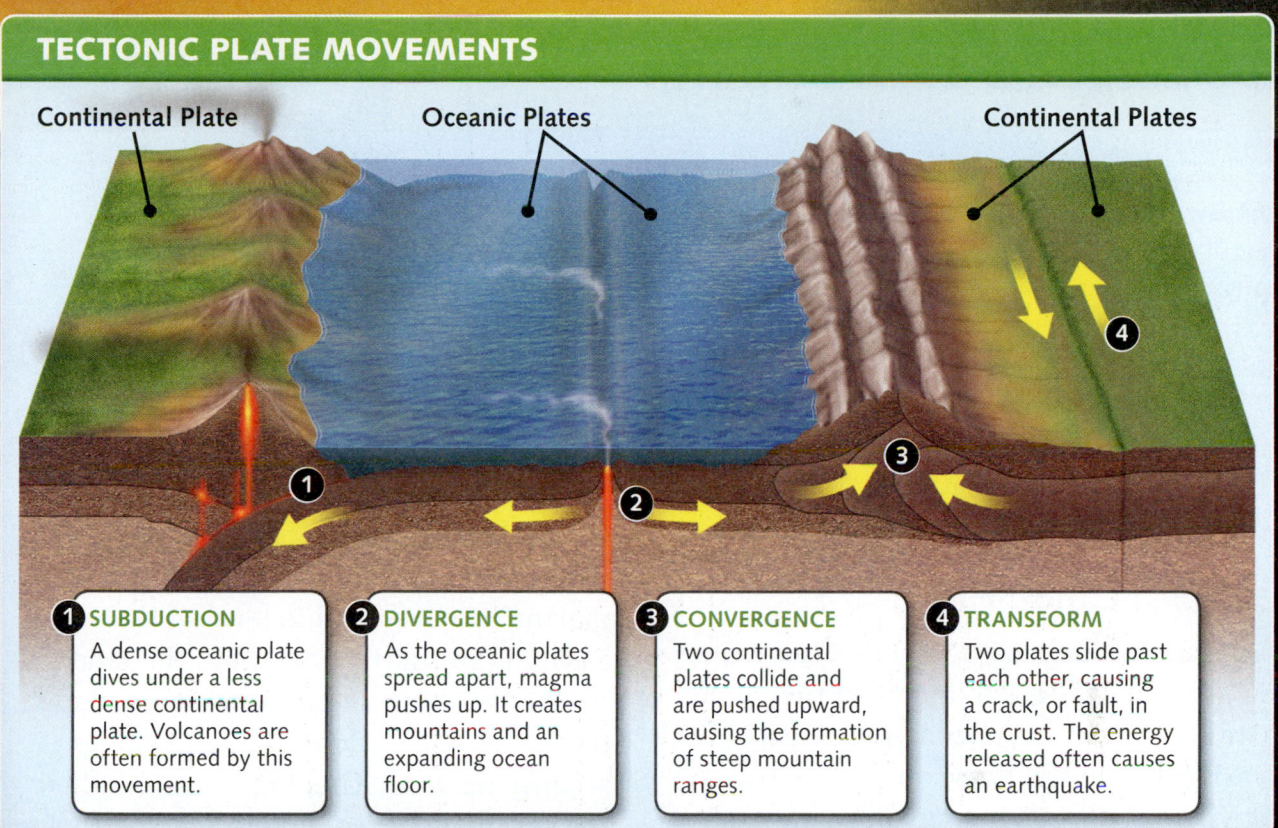

TECTONIC PLATE MOVEMENTS

Continental Plate   Oceanic Plates   Continental Plates

1 SUBDUCTION
A dense oceanic plate dives under a less dense continental plate. Volcanoes are often formed by this movement.

2 DIVERGENCE
As the oceanic plates spread apart, magma pushes up. It creates mountains and an expanding ocean floor.

3 CONVERGENCE
Two continental plates collide and are pushed upward, causing the formation of steep mountain ranges.

4 TRANSFORM
Two plates slide past each other, causing a crack, or fault, in the crust. The energy released often causes an earthquake.

Tectonic Plates

Earth's crust is divided into sections called tectonic plates. The plates float on Earth's mantle. They are constantly shifting and may move up to four inches a year.

The seven continents rest on these tectonic plates. As the plates have shifted over time, the continents have moved into their current positions. This slow movement of the continents is known as continental drift.

Tectonic plates move in four ways, as shown in the diagram above. The enormous force of the movements and collisions creates mountains and causes earthquakes and volcanoes.

Before You Move On
Make Inferences How do the movements of tectonic plates change Earth's surface?

ONGOING ASSESSMENT
VIEWING LAB   GeoJournal

1. **Analyze Visuals** According to the diagram above, which tectonic plate movement often results in volcanoes? Which movement can cause earthquakes?

2. **Place** The Himalaya Mountains formed by convergence when the Indian plate collided with the Eurasian plate. The Indian plate is still moving almost an inch north every year. How do you predict this will affect the Himalayas?

3. **Summarize** What is the main characteristic of each layer of Earth?

1.3 Earth's Landforms

TECHTREK

my NG connect.com For photos and
a diagram of landforms

Digital Library

Student Resources

Main Idea Landforms are physical features on Earth's surface. They are continually reshaped by physical processes.

The Rocky Mountains rise more than 14,000 feet above sea level. The Grand Canyon is more than 5,000 feet deep. Both are landforms, or physical features on Earth's surface.

Surface Landforms

Landforms such as the Rocky Mountains in western North America and the Grand Canyon in Arizona provide a variety of physical environments. These environments support millions of plants and animals.

Several common landforms are found on Earth's surface. A mountain is a high, steep elevation. A hill also slopes upward but is less steep and rugged. In contrast, a **plain** is a level area. The Great Plains, for example, are flat landforms stretching from the Mississippi River to the Rocky Mountains. A **plateau** is a plain that sits high above sea level and usually has a cliff on all sides. A valley is a low-lying area that is surrounded by mountains.

Ocean Landforms

Earth's oceans also have landforms that are underwater. Mountains and valleys rise and fall along the ocean floor. Volcanoes erupt with hot magma, which hardens as it cools to form new crust.

The edge of a continent often extends out under the water. This land is called the **continental shelf**. Most of Earth's marine life lives at this level of the ocean. Beyond the continental shelf, the land develops a steep slope. Beyond the slope, before the ocean floor, the land slopes slightly upward. This landform is called the continental rise. It is formed by rocks and sediment carried by ocean currents. Together, these landforms are known as the continental margin.

> **Visual Vocabulary** A **butte** (BYOOT) is a hill or mountain with steep sides and a flat top. These buttes in Monument Valley, Arizona, are called "the Mittens."

Continental Shelf

Coast

Ocean

Continental Slope

Continental Rise

The Changing Earth

Earth is always changing, and the changes affect plant and animal life. For example, a flood can cause severe erosion, which can ruin farmers' fields. **Erosion** is the process by which rocks and soil slowly break apart and are swept away.

Erosion also results from weathering, which is when air, water, wind, or ice slowly wear away rocks and soil. The buttes in Monument Valley, Arizona, were formed in this way over a span of millions of years.

Before You Move On

Summarize How do physical processes reshape Earth's landforms?

ONGOING ASSESSMENT
SPEAKING LAB GeoJournal

Turn and Talk How does a landform in your community affect your daily life? Make a list of landforms you encounter every day, such as hills or plains. Then turn to a partner and talk about how one of these landforms affects your life. Develop an outline to present to the class. List your topic first. Then list your supporting details.

1.4 The Ring of Fire

TECHTREK

myNGconnect.com For a map of the Ring of Fire and photos of volcanic eruptions

Maps and Graphs

Digital Library

Main Idea Plate boundaries around the Pacific Ocean cause earthquakes and volcanic eruptions.

The **Ring of Fire** is a circle of volcanoes and earthquakes along the rim, or outer edge, of the Pacific Ocean. It exists because a large tectonic plate under the ocean slides against plates in Asia, Australia, South America, and North America. The movements create tremendous pressure, which causes volcanoes and earthquakes.

Earthquakes

An **earthquake** is a violent shaking of Earth's crust. Many earthquakes occur along faults, which are cracks in Earth's surface. Earthquakes are common in the Ring of Fire, but they also occur in other areas on Earth. One area runs from the land around the Mediterranean Sea through East Asia. Other earthquake zones include the middle of the Arctic Ocean and the Atlantic Ocean.

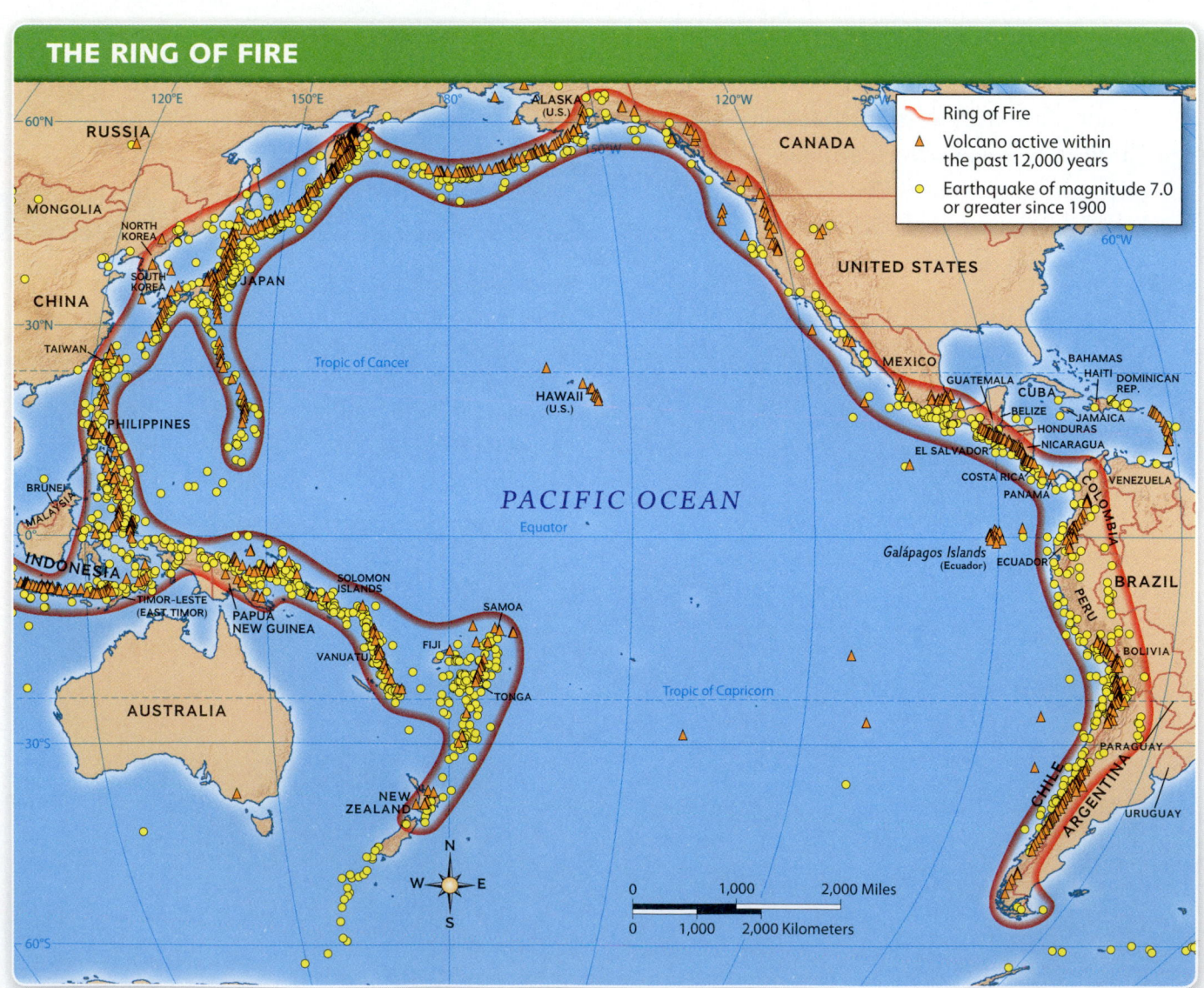

THE RING OF FIRE

Ring of Fire

▲ Volcano active within the past 12,000 years

● Earthquake of magnitude 7.0 or greater since 1900

Mount St. Helens, Washington

On May 18, 1980, Mount St. Helens erupted, blasting away one side of the mountain.

Earthquakes can cause buildings, bridges, and roads to collapse. For example, in 2010, an earthquake in Haiti killed more than 200,000 people. Many people who died were trapped under buildings that collapsed.

Earthquakes beneath the ocean can cause **tsunamis**, which are large, powerful ocean waves that can cause great destruction along the coast.

Volcanoes

The Ring of Fire contains more than 75 percent of the world's volcanoes. A **volcano** is a mountain that erupts in an explosion of molten rock, gases, and ash. Lava, which is molten rock, flows down the side of the mountain.

Volcanoes can cause severe damage. In 1883, Krakatoa in Indonesia spewed ash and rock fragments over an area of 300,000 square miles. It also triggered a tsunami that killed 36,000 people. Yet volcanoes can also **benefit**, or be useful to, plant and animal life. For example, mineral-rich lava turns into fertile soil.

Scientists have learned how to predict volcanic eruptions, and engineers can design buildings that survive earthquakes. As a result, more people can live safely.

Before You Move On
Monitor Comprehension What is the Ring of Fire and why is it significant?

ONGOING ASSESSMENT

WRITING LAB GeoJournal

1. **Make Inferences** Why do so many buildings collapse during an earthquake?

2. **Human-Environment Interaction** How do earthquakes and volcanoes affect people? How have people tried to solve these problems? Copy and complete the chart.

DISASTER	PROBLEM	SOLUTION
earthquake		
volcano		

3. **Write an Action Plan** Imagine that you live along the Ring of Fire. With a partner, write an outline of an action plan to help people survive a serious earthquake.

1.5 **Waters of the Earth**

TECHTREK

myNGconnect.com For photos of water and a diagram of the hydrologic cycle

Digital Library

Student Resources

Main Idea Water is essential for all forms of life on Earth.

The Mississippi River begins as a small stream in northern Minnesota. More than 2,000 miles south, it pours more than 4.7 million gallons of water per second into the Gulf of Mexico. Water flowing in rivers like the Mississippi is **essential**, or necessary, for all forms of life.

Fresh Water

The Mississippi River contains fresh water. People use fresh water to drink, cook, bathe, and irrigate crops. Early civilizations developed along rivers such as the Nile River in Egypt because of the available fresh water.

Different bodies of fresh water exist for different geographic reasons. A river is a path of water that flows from a higher elevation to a lower elevation. Streams, brooks, and creeks are like rivers, but smaller. A lake is a large body of water that is surrounded by land.

Salt Water

Salt water contains salt and other minerals. It is a major source of the world's seafood supply and a means of transportation. Oceans are large bodies of salt water. Earth's four oceans are the Atlantic, the Pacific, the Indian, and the Arctic. Continuously moving flows of water, called currents, circulate through the oceans and affect climates on land.

Seas are smaller bodies of salt water. The Red Sea, for example, lies between the Arabian Peninsula and eastern Africa.

THE HYDROLOGIC CYCLE

The hydrologic cycle is the continual movement of water from Earth's surface into the air and back again.

2 CONDENSATION
During **condensation**, cooler temperatures in the atmosphere cause the water vapor to change into droplets that form clouds.

1 EVAPORATION
During **evaporation**, the sun heats the ocean, and water vapor rises up into the atmosphere.

Before You Move On
Monitor Comprehension How is water essential for all life on Earth?

③ PRECIPITATION

The water droplets grow heavier and fall back to Earth in the form of **precipitation**, which is rain or snow.

④ RUNOFF

Precipitation soaks into the ground and runs into rivers, underground water reservoirs, and, eventually, the ocean.

WORLD'S LONGEST RIVERS

River	Location	Length (miles)
Nile	Africa	4,241
Amazon	South America	4,000
Chang Jiang (Yangtze)	Asia	3,964
Mississippi-Missouri	North America	3,710
Yenisey-Angara	Asia	3,440

Source: *National Geographic Atlas of the World*, 8th ed.

DATA LAB

 GeoJournal

1. **Interpret Charts** According to the chart, which continent has two of the longest world rivers, and what are they? How do you think the two rivers have affected that continent?

2. **Interpret Models** How does the hydrologic cycle explain why rivers and lakes do not run out of water?

3. **Location** St. Louis, Missouri, is located just south of where the Missouri River flows into the Mississippi River. Why is this a good location for a major city?

2.1 Climate and Weather

TECHTREK
myNGconnect.com
For maps and photos of climate and weather
Maps and Graphs
Digital Library

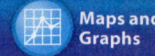

Main Idea Climate and weather are different, but they both influence life on Earth.

People who live in Sacramento, California, have mild winters. When they go skiing in the nearby Sierra Nevada Mountains, they wear parkas to protect themselves from the colder temperatures. They have adapted to a different climate.

Climate Elements

Climate is the average condition of the atmosphere over a long period of time. It includes average temperature, average precipitation, and the amount of change from one season to another. For example, Fairbanks, Alaska, has a cold climate. In the winter, the temperature can reach -8°F. Yet the temperature can rise to 90°F in the summer. The city goes through changes from one season to another.

Four factors that affect a region's climate are latitude, elevation, prevailing winds, and ocean currents. Places at high latitudes, such as Fairbanks, experience more change between winter and summer. Places close to the equator have nearly the same temperature throughout the year. Places at higher elevations have generally colder temperatures than places closer to sea level.

Prevailing winds are winds coming from one direction that blow most of the time. In Florida in the summer, the prevailing winds come from the south, making a warm climate even hotter.

Ocean currents also affect climate. The Gulf Stream is a current that carries warm water from the Caribbean Sea toward Europe. Air passing over the water becomes warm and helps create a mild winter climate in England and Ireland.

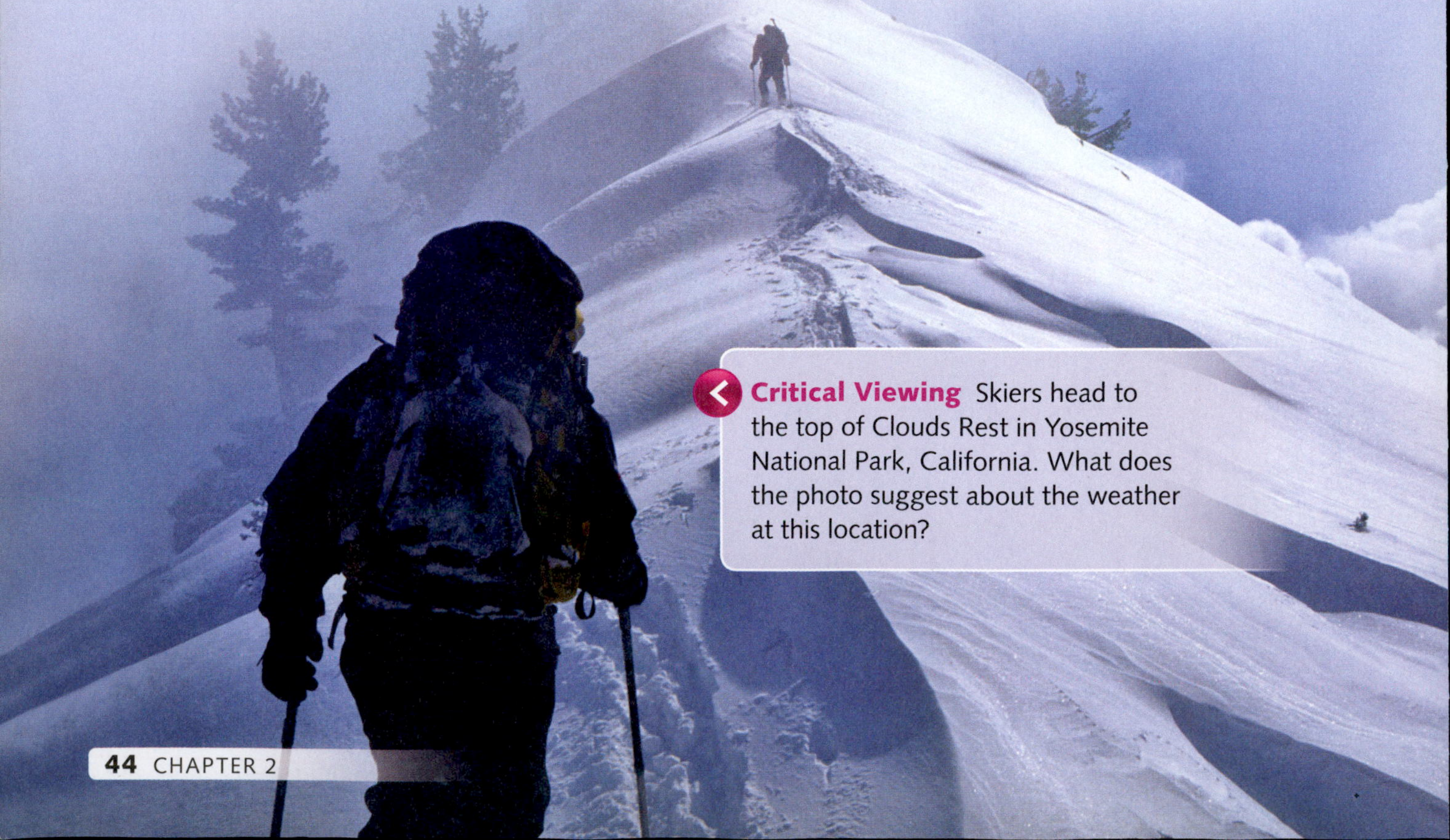

Critical Viewing Skiers head to the top of Clouds Rest in Yosemite National Park, California. What does the photo suggest about the weather at this location?

WESTERN UNITED STATES: CLIMATE

Climate Regions
- Humid Temperate–No dry season
- Humid Temperate–Dry summer
- Unclassified highlands
- Dry–Semiarid
- Dry–Arid
- Humid Cold–No dry season

CANADA
PACIFIC OCEAN
UNITED STATES
MEXICO

WESTERN UNITED STATES: WEATHER

- Wet, stormy weather
- Sunny, dry weather
- Cold front
- Warm front

CANADA
PACIFIC OCEAN
UNITED STATES
MEXICO

Weather Conditions

Weather is the condition of the atmosphere at a particular time. It includes the temperature, precipitation, and humidity for a particular day or week. Humidity is the amount of water vapor in the air. If a weather forecaster says the humidity is at 95 percent, he or she means that the air is holding a large amount of water vapor.

Weather changes because of air masses. An air mass is a large area of air that has the same temperature and humidity. The boundary between two air masses is called a front. If a forecaster talks about a warm, humid front, he or she usually means that thunderstorms are headed toward the area.

Before You Move On

Monitor Comprehension What is the difference between climate and weather?

MAP LAB

 GeoJournal

1. **Interpret Maps** Compare the weather map and the climate map of the western United States. How do you think ocean currents and mountains affect the climate and the weather?

2. **Make Inferences** How might the climate and weather of the western United States influence everyday life there?

3. **Place** The chart shows average temperatures and rainfall for Los Angeles, California. Ask and answer questions about the data.

Los Angeles, CA	January	July	November
Temperature (°F)	57.1	69.3	61.6
Rainfall (inches)	2.98	0.03	1.13

Source: National Drought Mitigation Center

2.2 World Climate Regions

myNGconnect.com For an online map and photos of world climate regions

 Maps and Graphs

 Digital Library

Main Idea Geographers identify climate regions to help them understand and categorize life on Earth.

A climate region is a group of places that have similar temperatures, precipitation levels, and changes in weather. Geographers have identified 5 climate regions that are broken down into 12 subcategories. Places that are located in the same subcategory often have similar **vegetation**, or plant life.

Before You Move On

Make Inferences How might climate regions help geographers analyze life in a particular place?

WORLD CLIMATE REGIONS

ARCTIC OCEAN

NORTH AMERICA

PACIFIC OCEAN

ATLANTIC OCEAN

SOUTH AMERICA

Ⓐ **Dry Climates** have little to no rain or snow and both hot and cold temperatures. Plant life includes shrubs and cacti.

Saguaro Cactus, Sonoran Desert, Arizona

Ⓑ **Humid Temperate Climates** have cool winters, warm summers, and ample rainfall. Plant life includes mixed forests with evergreens and leafy trees.

Mixed Forest, Great Smoky Mountains, North Carolina

Ⓒ **Humid Equatorial Climates** are found near the equator. They have high temperatures and rainfall all or most of the year. Plant life includes tropical plants and rain forests or grasslands with trees.

Bromeliads, Amazon rain forest, Peru

D **Tundra** or **Ice Climates** are north of the Arctic Circle and south of the Antarctic Circle. They have long, cold winters and short summers. Plant life includes mosses or no vegetation.

Mosses, Disko Bay, Greenland

E **Humid Cold Climates** have cold winters, warm summers, rain, and snow. Plant life includes evergreen or deciduous (leafy) forests.

Natural Park, Eastern Siberia, Russia

EUROPE

ASIA

AFRICA

PACIFIC OCEAN

INDIAN OCEAN

AUSTRALIA

ANTARCTICA

N
W E
S

TECHTREK

Go to **myNGconnect.com** to explore this map with the Interactive Map Tool.

Humid Equatorial
- 🟥 No dry season
- 🟧 Short dry season
- 🟨 Long dry season

Dry
- 🟨 Semiarid
- 🟡 Arid

Humid Temperate
- 🟩 No dry season
- 🟩 Dry summer
- 🟩 Dry winter

Humid Cold
- 🟪 Dry winter
- 🟪 No dry season

- 🟦 Tundra or ice
- ⬜ Unclassified highlands

ONGOING ASSESSMENT

MAP LAB

GeoJournal

1. **Interpret Maps** What is the most common climate in northern Africa? How might this climate affect population?

2. **Compare and Contrast** How does the climate of western Europe differ from that of eastern Europe?

3. **Human-Environment Interaction** What advantages would humid temperate climates have for farming? For logging?

TECHTREK

myNGconnect.com For a map and photos of extreme weather and a diagram of tornado formation

Maps and Graphs

Digital Library

Student Resources

Main Idea Extreme weather can cause great destruction, but scientists are lessening its effect.

On August 29, 2005, Hurricane Katrina raced toward New Orleans, Louisiana. The water level in the Gulf of Mexico rose 34 feet and flooded 80 percent of the city. Thousands of people lost their homes and many businesses were destroyed.

Wild Weather

Katrina is an example of extreme weather, which is weather so powerful it can deeply affect human lives. A **hurricane** such as Katrina is a strong storm with swirling winds and heavy rainfall. Winds rotate fiercely and can reach 200 miles per hour. A hurricane is a type of **cyclone**, which is a storm with rotating winds. In the Eastern Hemisphere, a cyclone is called a typhoon.

A **tornado** is a smaller storm than a cyclone, but it has even more powerful winds that can reach 300 miles per hour. A tornado follows an unpredictable path and can rip buildings from their foundations. Tornadoes occur all over the world, but most of them form in the United States, east of the Rocky Mountains.

Other types of extreme weather are not as dangerous as cyclones or tornadoes, but they still put people at risk. A flood occurs when water covers an area of land that is usually dry. Floods often occur after a cyclone. A blizzard is a heavy snowstorm with strong winds and very cold temperatures. A drought results when the amount of rainfall drops far below the average amount. It is sometimes accompanied by a heat wave, or unusually high temperatures over a period of time.

HOW A TORNADO FORMS

Air rises from the ground into the bottom of a thunderstorm cloud.

The air begins to rotate and extends to the ground in a funnel shape.

Extreme Weather in the Continental United States

Tornado Alley

Coastal areas most prone to hurricane landfall

Areas most prone to blizzards

Areas most prone to severe drought

0   250   500 Miles

0   250   500 Kilometers

Scientific Solutions

Scientists are working to lessen the effects of extreme weather on humans. For example, they can often predict the path of a hurricane. They have also worked with engineers to design levees, or walls to hold back floodwaters. Many residents of New Orleans believe that better levees might have limited the damage caused by flooding after Hurricane Katrina.

The ability to predict tornadoes has also improved. Today, the National Weather Service uses radar and satellite imagery, as well as a network of "spotters," to track big storms. Today's instant communications technology makes it possible to broadcast warnings before a storm strikes.

Before You Move On

Summarize How are scientists helping to lessen the impact of extreme weather?

VIEWING LAB  GeoJournal

1. **Analyze Visuals** What happens in each stage of tornado formation?

2. **Interpret Maps** According to the map, which state is at risk for all four types of extreme weather?

3. **Draw Conclusions** Even though scientists can predict when and where tornadoes may occur, why do these storms still catch people by surprise?

4. **Human-Environment Interaction** How might a severe drought affect people who live in the Great Plains area of the United States?

2.4 Natural Resources

TECHTREK

myNGconnect.com For an online resource map and photos of energy resources

Maps and Graphs

Digital Library

Main Idea Natural resources are central to economic development and basic human needs.

What materials make up a pencil? Wood comes from trees. The material that you write with is a mineral called graphite. The pencil is made from natural resources, which are materials on Earth that people use to live and to meet their needs.

Earth's Resources

There are two kinds of natural resources. Biological resources are living things, such as livestock, plants, and trees. These resources are important to humans because they provide us with food, shelter, and clothing.

Mineral resources are nonliving resources buried within Earth, such as oil and coal. Some mineral resources are **raw materials**, or materials used to make products. Iron ore, for example, is a raw material used in making steel. The steel, in turn, is used to make skyscrapers and automobiles.

Categories of Resources

Geographers classify resources in two categories. **Nonrenewable resources** are resources that are limited and cannot be replaced. For example, oil comes from wells that are drilled into Earth's crust. Once a well runs dry, the oil is gone. Coal and natural gas are other examples of nonrenewable resources.

Renewable resources never run out, or a new supply develops over time. Wind, water, and solar power are all renewable. So are trees because a new supply can grow to replace those that have been cut down.

> **Critical Viewing** Pumpjacks pump oil at a field in California. Based on the photo, how does this action affect the land?

SELECTED NATURAL RESOURCES OF THE WORLD

ARCTIC OCEAN

ATLANTIC OCEAN

PACIFIC OCEAN

PACIFIC OCEAN

INDIAN OCEAN

Energy Resources
- Coal
- Oil
- Natural gas
- Uranium

Mineral Resources
- Gold
- Silver
- Copper
- Iron
- Diamonds
- Aluminum

Natural resources are an important part of everyday life, yet countries with a large supply are not always wealthy. Nigeria, for example, is a major supplier of oil, but seven out of every ten Nigerians live in poverty. Japan is one of the wealthiest countries in the world—yet it must import oil from other countries.

Before You Move On
Monitor Comprehension Why are natural resources important?

ONGOING ASSESSMENT
MAP LAB
GeoJournal

1. **Interpret Maps** Copper is in demand for electric wiring and other uses. What part of the world do you think benefits from the demand for copper? Why?

2. **Location** Where are supplies of oil found in the world? How do these supplies impact people living in these regions?

3. **Describe Geographic Information** What are examples of a biological and a mineral resource? How are these examples different?

2.5 Habitat Preservation

TECHTREK

myNGconnect.com For photos of animal habitats

 Digital Library

Main Idea Plants and animals depend on their natural habitats to survive.

At the beginning of the 20th century, millions of elephants roamed across Africa. Today the African elephant population is fewer than a half million. These elephants are an endangered species, which is a plant or an animal in danger of becoming extinct.

Natural Habitats

The African elephant is endangered for several reasons. One is the demand for their ivory tusks. Poachers, or people who hunt animals illegally, slaughtered elephants at a rapid rate in the early 1970s.

Another reason elephants are endangered is the loss of their **habitat**. A habitat is a plant or an animal's natural environment. African elephants' habitats are grasslands and forests. Unfortunately, much of this land is being turned into farms and villages to feed and house Africa's growing human population. Thousands of other plants and animals have lost their natural habitats in this way.

Another threat to habitats is pollution, or human activity that harms the environment. During the 1960s, for example, Lake Erie in the United States was a polluted habitat, and fish nearly disappeared from its waters.

Critical Viewing Elephants roam the Samburu National Reserve in Kenya. What can you tell about their natural habitat?

Habitat Loss and Restoration

The loss of habitats can destroy an entire ecosystem. An **ecosystem** is a community of plants and animals and their habitat. Earth has many different ecosystems that interact with each other. The destruction of one ecosystem affects all the others. For example, many scientists believe the destruction of rain forest habitats has led to global climate change.

People around the world have taken steps to save ecosystems and preserve natural habitats. In 1973, for example, the United States passed the Endangered Species Act, which protects the habitats of endangered species. People have also **restored**, or brought back, habitats such as forests by planting trees.

Before You Move On
Monitor Comprehension How do plants and animals lose their natural habitats?

Tiny spruce trees grow in a cut forest near Olympic National Park in the state of Washington.

ONGOING ASSESSMENT

PHOTO LAB GeoJournal

1. **Analyze Visuals** Based on the photo of the reserve in Kenya, why might an elephant's habitat be desirable to people for farming?

2. **Describe** Based on the photo above, what steps are taken to restore forest habitats?

3. **Make Inferences** In a forest ecosystem, wolves eat deer, and deer carry the parasite that causes Lyme disease in humans. If the wolf population declines, how might it affect the number of cases of Lyme disease?

4. **Human-Environment Interaction** What is your natural habitat? With a partner, think of resources and interactions that are part of your daily life. Write a short paragraph describing your habitat.

2.6

SECTION **2** PHYSICAL GEOGRAPHY

NATIONAL GEOGRAPHIC

TECHTREK

myNGconnect.com For photos of oceans and an Explorer Video Clip

Maps and Graphs

Digital Library

Exploring the
World's Oceans
with Sylvia Earle

Main Idea The oceans, a natural habitat for thousands of plant and animal species, face many challenges.

Life in the Ocean

As a child, Sylvia Earle loved oceans. This love continued into her adult life. In over 50 years, Explorer-in-Residence Sylvia Earle has led roughly 70 diving expeditions to explore **marine life**, or the plants and animals of the ocean. During these dives, she has seen the incredible variety of ocean life—more than 30 major divisions of animals.

However, Earle has also seen the different ways people have harmed the oceans. "Taking too much wildlife out of the sea is one way," she claims. "Putting garbage, toxic chemicals, and other wastes in is another." Earle has witnessed a huge drop in the number of fish in the ocean. She has also noted pollution's destructive **impact**, or effect, on the oceans' coral reefs. These "rain forests of the sea" house one-fourth of all marine life.

myNGconnect.com

For more on Sylvia Earle in the field today

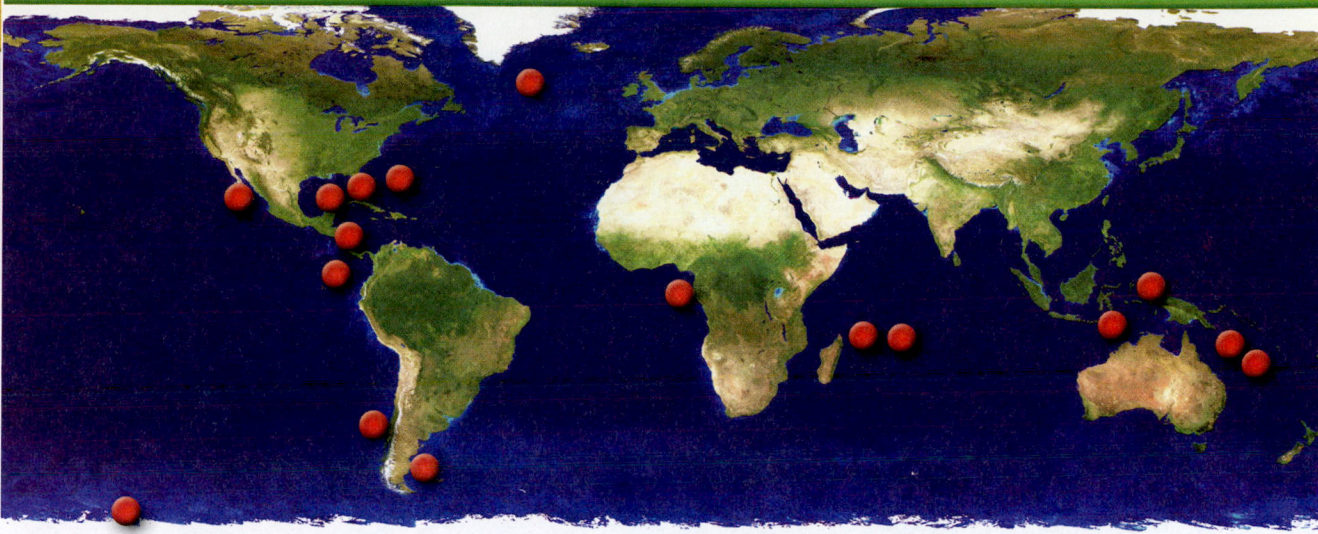

Critical Viewing This satellite map of the world shows the location of 17 "hope spots," places that are important to the overall health of Earth's oceans. What patterns, if any, do you notice about the location of these spots?

Mission Blue

Many people believe human activity has no effect on the vast oceans. They also do not understand the impact oceans have on all forms of life. "The ocean is the cornerstone of our life support system," says Earle. "Take away life in the ocean and we don't have a planet that works."

Earle is trying to educate the public. In 2009, she launched Mission Blue, a program that seeks to heal and protect Earth's oceans. One of the program's goals is to establish marine protected areas (MPAs) in endangered hot spots, or "hope spots" as she calls them. These spots are ocean habitats that can recover and grow if human impact is limited.

Voice of the Ocean

Earle's efforts to save Earth's oceans have earned her many awards and honors, including the title "Hero for the Planet." She continues to work tirelessly to protect the world's oceans.

In 2010, the Gulf of Mexico was hit with the largest oil spill in U.S. history. Earle went before Congress to testify about the impact of the spill on the Gulf's natural resources. "I really come to speak for the ocean," she began.

Before You Move On
Summarize According to Sylvia Earle, what challenges do our oceans face?

ONGOING ASSESSMENT

READING LAB GeoJournal

1. **Monitor Comprehension** What does Sylvia Earle hope to accomplish through the Mission Blue program?
2. **Analyze Cause and Effect** How has human activity affected the world's oceans? Use a chart like the one below to list some of the causes and effects of these activities.

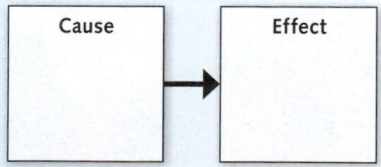

Cause → Effect

NATIONAL GEOGRAPHIC

Photo Gallery • Slip Point Tidal Pool

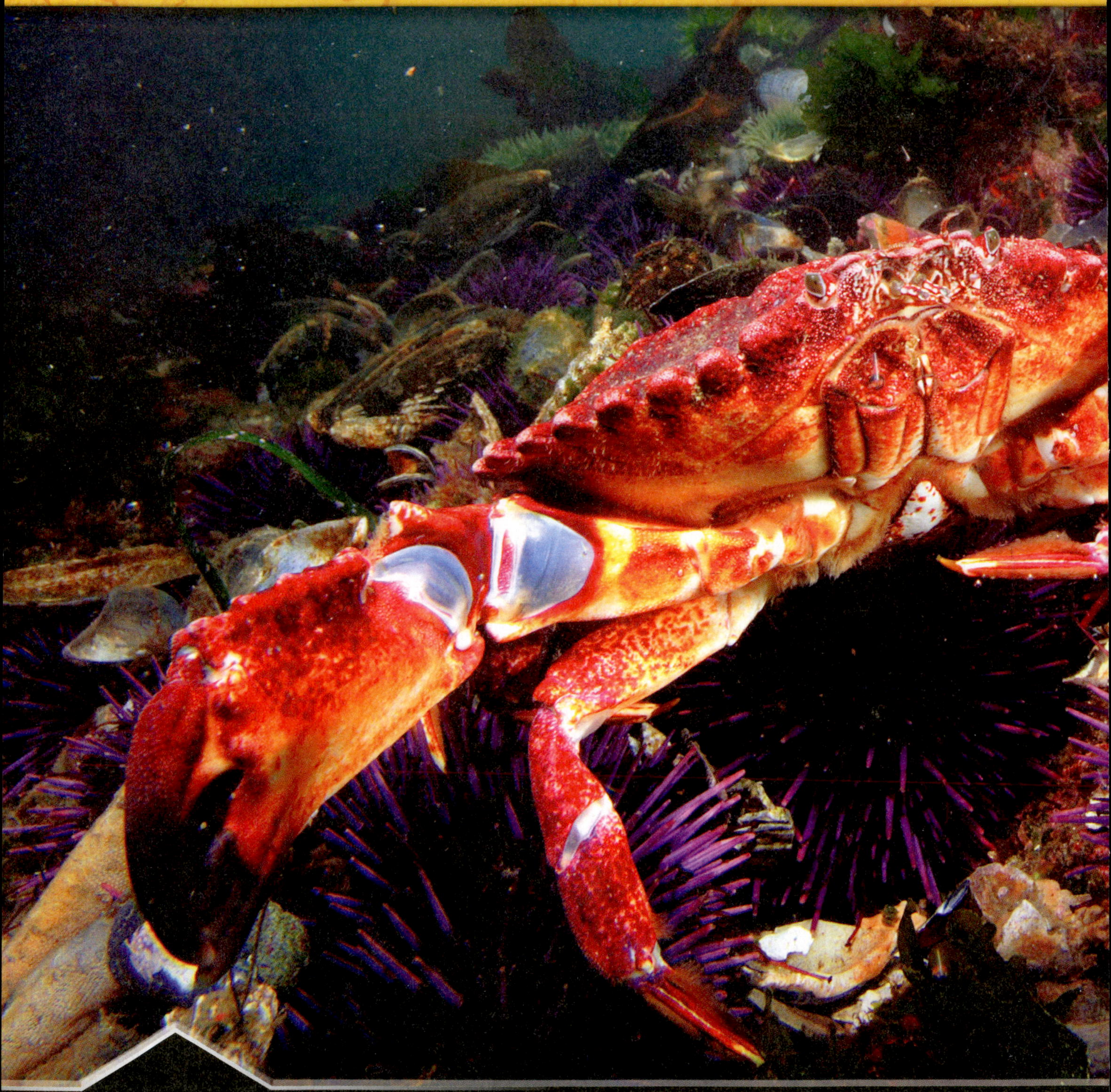

For more photos from the National Geographic Photo Gallery, go to the **Digital Library** at myNGconnect.com.

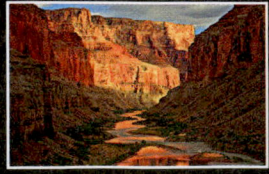

Grand Canyon

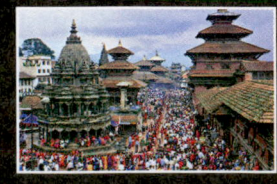

Kathmandu, Nepal

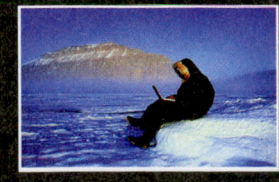

Japanese man with laptop

Critical Viewing A crab walks over sea urchins in a tidal pool at Slip Point in Clallam Bay, Washington. River outlets and tides create this plentiful and diverse coldwater habitat.

Maori carving

Polar bears

Woman and baby, Peru

Tornado

3.1 World Cultures

> **Main Idea** The ways people speak, eat, work, play, and worship are all part of culture.

A Japanese tea ceremony is an old tradition. The host greets the guests and prepares the tea. The guests remain silent. Once the tea is poured, though, a lively conversation follows.

Expressions of Culture

The tea ceremony is an important part of Japan's culture. **Culture** is how people in a region live, behave, and think. Expressions of culture include language, religion, beliefs, and customs. Culture also includes the arts, such as music, dance, literature, theater, and film.

Culture is reflected in symbols that people recognize and respect. A **symbol** is an object that stands for something else. For example, the stars on the American flag are symbols for the 50 states.

Civilization and Culture

Culture is a main trait of civilizations. A **civilization** is a society that has a highly developed culture and technology. People in a civilization are not born knowing their culture. They learn it by watching and imitating others.

A civilization's culture affects people's lives. It guides how people meet their basic needs of food, clothing, and shelter. It also influences people's values and beliefs.

Visual Vocabulary Senegalese children gather around a **communal**, or shared, plate of food. Communal dishes are common in African cuisine.

Visual Vocabulary A **gaucho** (GOW cho), or cowboy, herds sheep in South America.

Culture Regions

Geographers study culture regions, areas that are unified by common cultural traits, or characteristics. For instance, some geographers group Mexico, Central America, the islands of the Caribbean, and South America in a culture region called Latin America. Many of the people in this region speak Spanish or Portuguese and practice the Roman Catholic religion. Many also share a history of Spanish colonization.

Before You Move On

Monitor Comprehension What are some expressions of culture?

Visual Vocabulary Women dressed in kimonos, traditional Japanese clothing, bow at a fashion show. In Japan, it is customary to bow when meeting someone.

ONGOING ASSESSMENT
PHOTO LAB
GeoJournal

1. **Place** Each photo and caption shows an expression of culture in a different country. How do these expressions compare to various expressions of U.S. culture?

2. **Turn and Talk** What is one activity, such as playing music or dancing, that is important in your culture? Discuss with a partner and prepare a response to share with the class.

3.2 Religions and Belief Systems

TECHTREK

myNGconnect.com For an online map and photos of world religions

 Maps and Graphs   Digital Library

Main Idea Religions and belief systems are important parts of cultures around the world.

A religion is a set of beliefs and practices that is often focused on one or more deities, or gods. Major world religions include Christianity, Hinduism, Islam, Buddhism, Judaism, and Sikhism.

Elements of Religion

Religion is a powerful influence that helps people answer questions such as, "What is the purpose of life?" At the center of many religions is the belief in a deity. **Monotheistic religions** are those with a belief in one deity. Christianity, Islam, and Judaism are monotheistic religions. **Polytheistic religions**, such as Hinduism, have many deities.

Every religion has a doctrine, or a set of basic beliefs. For example, Christians believe that Jesus was the Son of God.

Scriptures are sacred, or highly respected, texts that communicate the beliefs of a religion. The Bible is the sacred text for Christianity, the Koran is the sacred text for Islam, and the Torah is the sacred text for Judaism.

One of the important ways in which religion affects culture is through a code of conduct, or beliefs about right and wrong behavior. For example, the Ten Commandments are a code of conduct for Jews and Christians.

Origin and Spread of Religions

Several of the major world religions are based on the teachings of an individual. For example, Christianity grew from the teachings of Jesus Christ nearly 2,000 years ago. Other religions, such as Hinduism, grew from the beliefs of ancient peoples.

BUDDHISM
Founder Siddhartha Gautama, the Buddha
Followers 400 million
Basic Beliefs People reach enlightenment, or wisdom, by following the Eightfold Path and understanding the Four Noble Truths.

CHRISTIANITY
Founder Jesus of Nazareth
Followers 2.3 billion
Basic Beliefs There is one God, and Jesus is the only Son of God. Jesus was crucified but was resurrected. Followers reach salvation by following the teachings of Jesus.

HINDUISM
Founder Unknown
Followers 860 million
Basic Beliefs Souls continue to be reborn. The cycle of rebirth ends only when the soul achieves enlightenment, or freedom from earthly desires.

ISLAM
Founder The Prophet Muhammad
Followers 1.6 billion
Basic Beliefs There is one God. Followers must follow the Five Pillars of Islam in order to achieve salvation.

ARCTIC OCEAN

NORTH AMERICA

EUROPE

ASIA

PACIFIC OCEAN

PACIFIC OCEAN

AFRICA

ATLANTIC OCEAN

SOUTH AMERICA

INDIAN OCEAN

AUSTRALIA

Major Religions

	Buddhism
	Christianity
	Hinduism
	Indigenous
	Islam
✡	Judaism
☬	Sikhism
	Uninhabited

ANTARCTICA

Religions have grown and spread around the world. For example, Buddhism began in India, but it spread to Japan, China, Korea, and Southeast Asia through migration and trade. Religions have also spread due to the work of missionaries, people who convert others to follow their religion.

Before You Move On

Make Inferences How is religion an important part of culture?

JUDAISM

Founder Abraham
Followers 15 million
Basic Beliefs There is one God. People serve God by living according to his teachings. God handed down the Ten Commandments to guide human behavior.

ONGOING ASSESSMENT

READING LAB GeoJournal

1. **Monitor Comprehension** What is the difference between a monotheistic religion and a polytheistic religion?

2. **Compare and Contrast** How are Hinduism and Buddhism similar?

3. **Region** What major religions are found in the area around the eastern Mediterranean Sea? Why do you think they are found in this region?

SIKHISM

Founder Guru Nanak
Followers 25 million
Basic Beliefs There is one God. Souls are reborn. The goal is to achieve union with God, which a person does by acting selflessly, meditating, and helping others.

3.3 Economic Geography

TECHTREK

my**NG**connect.com For an online
economic indicator chart

**Student
Resources**

Main Idea People produce, buy, and sell goods in a variety of ways.

Singapore became a trading colony of the British Empire in 1824. Today, it is an independent country, but trade is still an important part of its economy. An economy is a system in which people produce, sell, and buy things.

Economic Activity

The production of goods and services is known as economic activity. Geographers divide this activity into different sectors. The primary sector involves taking raw materials from the soil or water. It includes mining, farming, fishing, and forestry. The secondary sector involves using raw materials to manufacture products, such as cars. The tertiary, or third, sector includes services, such as banking and health care.

Factors of Production

Geographers study where economic activity occurs and how this activity is connected around the world. A country is more likely to have a strong economy if it has all four factors of production—land, labor, capital, and entrepreneurship. Land includes all the natural resources used to produce goods and services. Labor involves the size and education level of the workforce. Capital is a country's wealth and infrastructure. The fourth factor, entrepreneurship, involves the creativity and risk needed to develop new goods and services.

Economic Systems

Economic systems are ways in which countries organize the production of goods and services. Four main systems are found around the world:

- In a traditional economy, people trade goods and services without money.

- In a free enterprise economy, privately owned businesses create goods that people buy in markets. This is also called a market economy or capitalism.

- In a command economy, the government owns most parts of the economy and decides what will be produced and sold.

- A mixed economy has elements of a free enterprise and command economy.

Economic Indicators

The strength of a country's economy can be measured by several indicators, or signs. One is gross domestic product (GDP). It is the total value of the goods and services that a country produces. The GDP per capita is the value of products that a country produces per person. Other indicators include income, literacy rate, and life expectancy.

Economies fall in one of two categories. Countries with high GDPs are more developed countries. Most of their economic activity is in the tertiary sector. Countries with low GDPs are less developed countries. Most of their activity is in the primary or secondary sector.

Before You Move On

Summarize What are four ways in which countries organize the production of goods and services?

ECONOMIC INDICATORS OF SELECTED COUNTRIES*

Country	Population	GDP (in U.S. dollars)	GDP Per Capita (in U.S. dollars)	Life Expectancy	Literacy Rate (percent)
Afghanistan	29.1 million	10.6 billion	366	44	28.0
Brazil	191.9 million	1.6 trillion	8,536	72	90.0
China	1.3 billion	4.5 trillion	3,422	73	93.3
Ethiopia	80.7 million	25.9 billion	321	55	35.9
Germany	82.1 million	3.7 trillion	44,525	80	99.0
Haiti	9.8 million	6.4 billion	649	61	62.1
Mexico	106.3 million	1.1 trillion	10,249	75	92.8
Singapore	4.8 million	193.3 billion	39,950	81	94.4
United States	304.3 million	14.4 trillion	47,210	78	99.0

Sources: The World Bank and the United Nations
*All figures are for 2008 with the exception of literacy rate, which is for 2007.

< Critical Viewing This shipping terminal in Singapore is busy all day and all night. Based on the photo and the chart, what type of country is Singapore—a more developed or a less developed country?

ONGOING ASSESSMENT

DATA LAB

 GeoJournal

1. **Interpret Charts** Which country has the largest population? the largest GDP per capita? What can you conclude about the relationship between the two?

2. **Synthesize** What sector is probably the main source of Haiti's economic activity? Explain.

3. **Region** In the region in which you live, what is an example of an economic activity from each of the three sectors?

3.4 Political Geography

TECHTREK

myNGconnect.com For photos of governments in action

Digital Library

Main Idea Countries around the world have different forms of government.

How is trash collection related to the protection of free speech? Both are responsibilities of government. A **government** is an organization that keeps order, sets rules, and provides services for a society. Political geographers study boundaries between different places, where different types of government exist, and how geography affects government.

Government and Citizens

Governments govern citizens. A **citizen** is a person who lives within the territory of a government and is granted certain rights and responsibilities by that government.

Governments are either limited or unlimited. A limited government does not have complete control over its citizens. The citizens have some individual rights and responsibilities. An unlimited government has complete control over every aspect of its citizens' lives.

Critical Viewing Voters in South Africa line up for miles to vote. How important to them is the right to vote? How can you tell?

Types of Government

In the modern world, five types of government are common. The major differences among them are in the power and rights that citizens have.

- In a **democracy**, citizens elect representatives to govern them. A legislature creates laws, an executive branch carries out laws, and a judicial branch interprets laws. Citizens have many rights, such as those in the Bill of Rights of the U.S. Constitution. The United States was the first modern country to establish a representative democracy.

- In a monarchy, a king, queen, or emperor rules society. The ruler usually inherits, or is born into, the office. Citizens in an absolute monarchy, such as Saudi Arabia, have few or no rights. In a constitutional monarchy, such as the United Kingdom, the queen or king shares power with a government organized by a constitution.

- In a dictatorship, one person, the dictator, rises to power and rules society. The dictator controls all aspects of life, including education and the arts. Citizens have few or no rights. North Korea has been ruled by a dictator for more than a half century.

- In an oligarchy, a group of a few people rules society. The ruling group usually is wealthy or has military power, and citizens have few or no rights. The government of Myanmar (Burma) has been an oligarchy since 1988.

 Critical Viewing Female soldiers march in a military parade in North Korea to celebrate the country's 60th anniversary. Based on the photo, what qualities are valued by North Korea's government?

- Communism is a type of command economy in which the government, controlled by the Communist Party, owns all the property. Citizens have few or no rights. Cuba has been a communist country since 1959.

Before You Move On

Summarize What are common types of government in the modern world?

ONGOING ASSESSMENT
READING LAB   GeoJournal

1. **Monitor Comprehension** Which of the five types of government are limited governments? Which are unlimited governments?

2. **Create Charts** Create a chart in which you compare the five forms of government. Use the following format:

Type of Government	Source of Power	Leader or Ruler	Citizens' Rights
democracy			

3. **Make Inferences** Look at the map of Russia in the National Geographic Atlas in the front of your textbook. How might Russia's geography affect the government's ability to rule?

3.5 Protecting Human Rights

TECHTREK

myNGconnect.com For more on the Universal Declaration of Human Rights

Digital Library

Global Issues

Main Idea The United Nations adopted the Universal Declaration of Human Rights to state how all people deserve to be treated.

KEY VOCABULARY

human rights, n., political, economic, and cultural rights that all people should have

Nesse Godin was 13 years old during World War II when the Nazis occupied her town in Lithuania. Because she and her family were Jewish, they were transferred to a concentration camp. Then Nesse was sent to different labor camps, where she worked digging ditches. In January 1945, she and other prisoners were forced to march in the cold weather with little food. Many prisoners died.

The United Nations

Nesse Godin survived the events of the Holocaust, but 6 million other people did not. At the end of World War II, people vowed to never let it happen again.

In 1945, 51 countries from around the world formed the **United Nations (UN)**. The main goals of this organization were to keep peace, to develop friendly relationships among countries, and to protect people's human rights.

The UN established the Commission on Human Rights to decide which rights all people should have. The members of this commission came from different cultural backgrounds, but they worked together to create a "common standard of achievement for all people and all nations." On December 10, 1948, the UN General Assembly approved the commission's **Universal Declaration of Human Rights**.

The declaration has 30 articles, or sections. Article 1 states that all people should be treated with respect:

All human beings are born free and equal in dignity and rights. They are endowed with reason and conscience and should act towards one another in a spirit of brotherhood.

Twenty-one of the articles explain political rights, such as the right to equality before the law, the right to freedom from torture, and the right to take part in government. Six of the articles address people's economic and cultural rights, such as the right to work, the right to education, and the right to participate in the cultural life of the community.

The Impact of Human Rights

The declaration has had an impact, or effect, on people and governments. For example, during the 1960s–1980s, countries around the world pressured the Republic of South Africa to grant human rights to its non-white population. Many countries refused to trade with South Africa, and the country was barred from participating in the Olympic Games from 1964–1990.

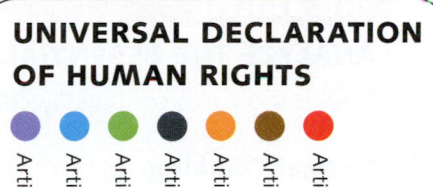

UNIVERSAL DECLARATION OF HUMAN RIGHTS

Article 26.1 · Article 25.2 · Article 24 · Article 18 · Article 5 · Article 4 · Article 1

No one shall be held in slavery or servitude ...

Motherhood and childhood are entitled to special care and assistance.

All human beings are born free ...

Everyone has the right to freedom of thought, conscience, and religion ...

No one shall be subjected to torture or to cruel, inhuman, or degrading treatment or punishment.

Everyone has the right to education.

Everyone has the right to rest and leisure ...

In 1994, South Africa finally gave in to the pressure and held elections in which all people could vote. The people elected Nelson Mandela, a leader of the African population, as president. This action showed the power of human rights.

Before You Move On

Monitor Comprehension According to the Universal Declaration of Human Rights, how do all people deserve to be treated?

ONGOING ASSESSMENT

WRITING LAB   GeoJournal

1. **Form and Support Opinions** Choose one of the rights from the Universal Declaration of Human Rights that are listed above. In a paragraph, explain what this right means and why it is important.

2. **Write Reports** Select a news story in the newspaper or on television. In a short report, explain how the story shows the importance of protecting human rights.

VOCABULARY

For each pair of vocabulary words, write one sentence that explains the connection between the two words.

1. solstice; equinox

Summer begins at the solstice while spring begins at the equinox.

2. plateau; continental shelf

3. renewable resource; nonrenewable resource

4. habitat; restore

5. free enterprise economy; democracy

MAIN IDEAS

6. How does Earth's tilt affect the seasons? (Section 1.1)

7. What layers would you pass through on a journey to Earth's center? (Section 1.2)

8. How does erosion change Earth's surface? (Section 1.3)

9. How have scientists tried to lessen the impact of earthquakes and volcanoes? (Section 1.4)

10. In what ways is climate different from weather? (Section 2.1)

11. How is a humid equatorial climate different from a humid temperate climate? (Section 2.2)

12. What is an example of a renewable resource and a nonrenewable resource? (Section 2.4)

13. What are endangered species, and what factors threaten them? (Section 2.5)

14. What are four ways in which culture affects people's lives? (Section 3.1)

15. How is a free enterprise economy different from a command economy? (Section 3.3)

16. What are the common types of government in the modern world? (Section 3.4)

THE EARTH

ANALYZE THE ESSENTIAL QUESTION

How is the earth continually changing?

Critical Thinking: Make Inferences

17. What impact do the changing seasons have on how farmers grow food?

18. Why did early civilizations develop along rivers?

19. How does the hydrologic cycle return water to Earth?

PHYSICAL GEOGRAPHY

ANALYZE THE ESSENTIAL QUESTION

What shapes the earth's varied environments?

Critical Thinking: Draw Conclusions

20. How would plants in humid cold climates and humid equatorial climates differ?

21. What factors might prevent a country rich in natural resources from using them effectively?

INTERPRET CHARTS

U.S. AND WORLD ENDANGERED SPECIES			
Group	United States	Other Countries	Total Number
Mammals	71 species	255 species	326 species
Reptiles	13 species	66 species	79 species
Fish	74 species	11 species	85 species
Birds	76 species	184 species	260 species

Source: U.S. Fish and Wildlife Service

22. **Analyze Data** What percentage of endangered animals in the United States are mammals? What percentage in other countries are mammals?

23. **Human-Environment Interaction** What steps might wildlife conservationists take to reduce the number of endangered species?

ANALYZE THE ESSENTIAL QUESTION

How has geography influenced cultures around the world?

Critical Thinking: Find Main Ideas and Details

24. What are some characteristics of American culture?

25. In what ways do different religions spread?

26. What is an example of each sector of economic activity?

27. What are two examples of different ways in which governments govern?

INTERPRET MAPS

CANADA'S BIOLOGICAL RESOURCES

Legend:
- Barley
- Cattle
- Corn
- Fish
- Forest products
- Potatoes
- Swine
- Wheat

28. **Place** Are Canada's resources mostly in the north or the south? Why do you think this is the case?

ACTIVE OPTIONS

Synthesize the Essential Questions by completing the activities below.

29. **Write a Public Service Announcement (PSA)** Write a PSA about what to do during an extreme weather emergency in your community. Focus on one type of weather event that is common in your area. Explain the supplies that a household should have. Provide an escape route or hiding place. Also, explain how people should behave during the emergency. **Share your PSA with the class.**

> **Writing Tips**
> - Do research to find the advice of weather experts.
> - Take notes and organize your information to make it clear to readers.
> - Use a tone that is calm and shows that you are informed.

Go to **Student Resources** for Guided Writing support.

TECHTREK myNGconnect.com For photographs of elements of culture

30. **Create a Visual Overview** Choose a part of culture that interests you, such as music, dance, language, or sports. Use photos from the **Digital Library** or from other online or print sources to create a visual introduction to that aspect of culture around the world. Show similarities and differences among countries around the world. Copy and complete the following organizer to organize your ideas:

Element of Culture: _____

Examples: _____

Similarities: _____

Differences: _____

Climate

In this unit, you learned that climate is the average condition of the atmosphere in a region over a long period of time. Climate is determined by the latitude of a location, as well as by ocean currents, air currents, and elevation.

Two important factors of climate are temperature and precipitation. These factors determine the length and timing of the growing season and influence the types of economic activities in a region. For instance, a country with year-round warm temperatures and precipitation probably has a better agricultural industry than a country with low temperatures and a short rainy season.

Compare

- Brazil
- Russia
- United States

CLIMATE MAPS AND CLIMOGRAPHS

A climate map provides an overview of the climate in a region. However, sometimes more specific information about a city or country is needed. Geographers use a special tool called a **climograph** (climate + graph) to graphically show a range of temperature and precipitation in a place over a period of time. A climograph includes a bar graph that shows the amount of precipitation for a location. Average monthly temperature is shown by a line connecting 12 points, one for each month of the year.

Climographs are useful in comparing the climates of two different locations. They help geographers and others better understand the effects of climate on human activities in those locations.

INTERPRET CLIMOGRAPHS

Look at the climographs for the cities of Belem, Brazil, and Omsk, Russia, on the opposite page. The months of the year are listed across the bottom of each climograph. A scale on the left vertical axis measures precipitation, or rainfall, in inches. A scale on the right vertical axis measures temperature in degrees Fahrenheit (°F).

The bars on the climographs show the rainfall for Belem, Brazil, and Omsk, Russia. The lines connecting the dots show the range of temperatures for each location.

Study the data in the climographs to analyze the climate in Belem and Omsk. Then answer the questions at right.

AVERAGE MONTHLY TEMPERATURE AND PRECIPITATION

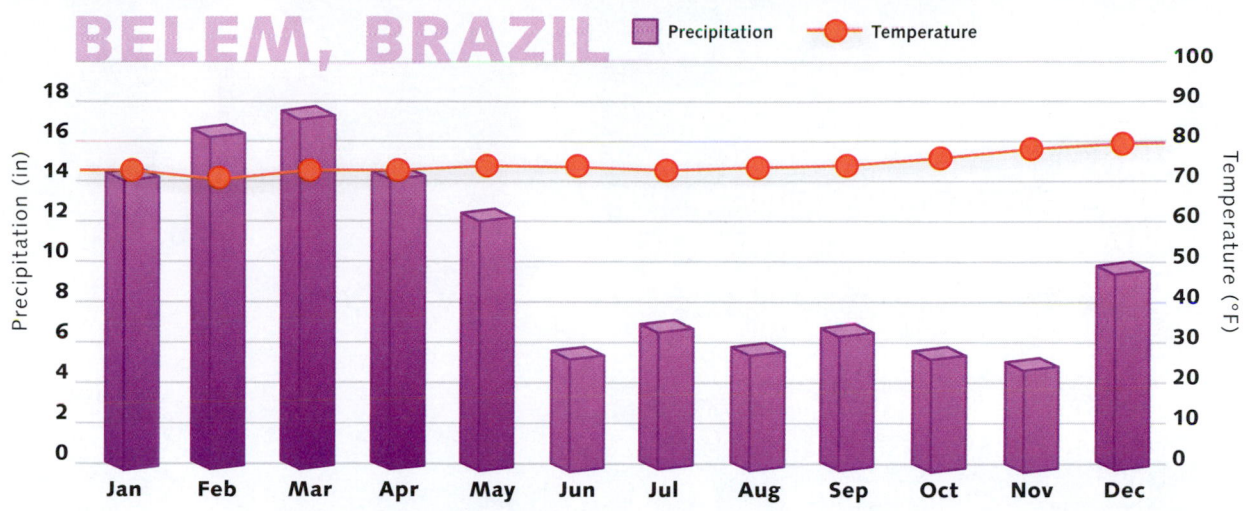

BELEM, BRAZIL ● Precipitation ● Temperature

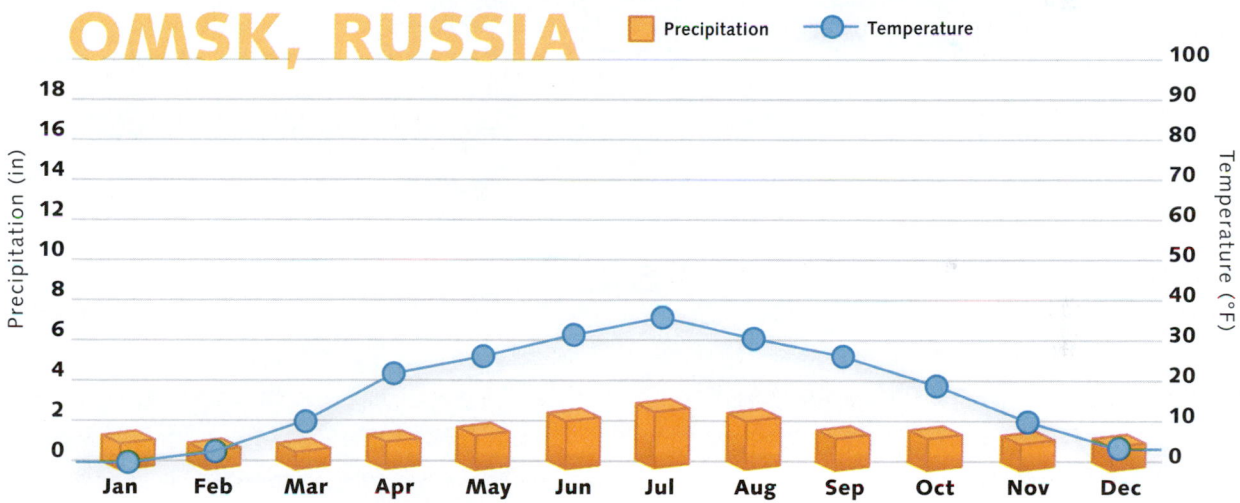

OMSK, RUSSIA ● Precipitation ● Temperature

Source: National Drought Mitigation Center, University of Nebraska–Lincoln

RESEARCH LAB GeoJournal

1. **Interpret Graphs** How would you describe the range of temperatures in Belem? What is the average temperature in Omsk in January? In July?

2. **Make Inferences** Most crops need water and warm temperatures in order to grow. Based on this information, what can you tell about the growing season in each city?

Research and Create a Climograph Choose a city in the United States and do research to find the average temperature and rainfall for each month of the year. Record the data in a chart. Use the data to create a climograph for that city. Then, with a partner, write three questions to help someone analyze and compare data on your two climographs.

Active Options

TECHTREK
myNGconnect.com
For writing templates

 Student Resources

 Magazine Maker

ACTIVITY 1

Goal: Extend your understanding of the environment.

Compose a Top Ten List

For over 40 years, countries around the world have celebrated Earth Day. The purpose of the day is to appreciate our planet and to raise awareness of the need to protect the environment. With your classmates, create a Top Ten list of actions students in your school can plan to do to show appreciation for Earth on the next Earth Day. See if you can get permission to post copies of your list around the school to share with other students.

A reforestation event in the Philippines

ACTIVITY 2

Goal: Learn about culture through religious architecture.

Create a Catalog of Religious Architecture

Some of the world's greatest architecture has been inspired by religious beliefs. Use the **Magazine Maker** to create a visual catalog of examples of architecture representing several religions around the world. Include important information in your catalog, such as the name of the architect, the name of the structure, and the year the structure was created.

ACTIVITY 3

Goal: Research the history of human migration.

Write a Migration History

Explorer Spencer Wells is researching the history of human migration. Interview a friend to find out more about his or her migration history. Include these questions:

- When did your ancestors arrive in the United States?
- Why did they come?
- What did they do when they arrived?

Then write an account of your friend's migration history.

Explore EUROPE WITH NATIONAL GEOGRAPHIC

MEET THE EXPLORER

NATIONAL GEOGRAPHIC

Some archaeological sites are underwater. Emerging Explorer Katy Croff Bell works with archaeologists in the Mediterranean and Black seas to help them figure out where to look for submerged secrets.

INVESTIGATE GEOGRAPHY

The Alps are the highest and most extensive mountain range in Europe. They stretch across central Europe and are concentrated in France, Germany, Italy, Switzerland, and Austria. This is the Lauterbrunnen Valley in Oberland, Switzerland.

STEP INTO HISTORY

The Colosseum in Rome is one of the Roman Empire's greatest architectural and engineering achievements. The arena, completed in A.D. 80, seated nearly 50,000 spectators who watched gladiator games, performances, and even mock naval battles.

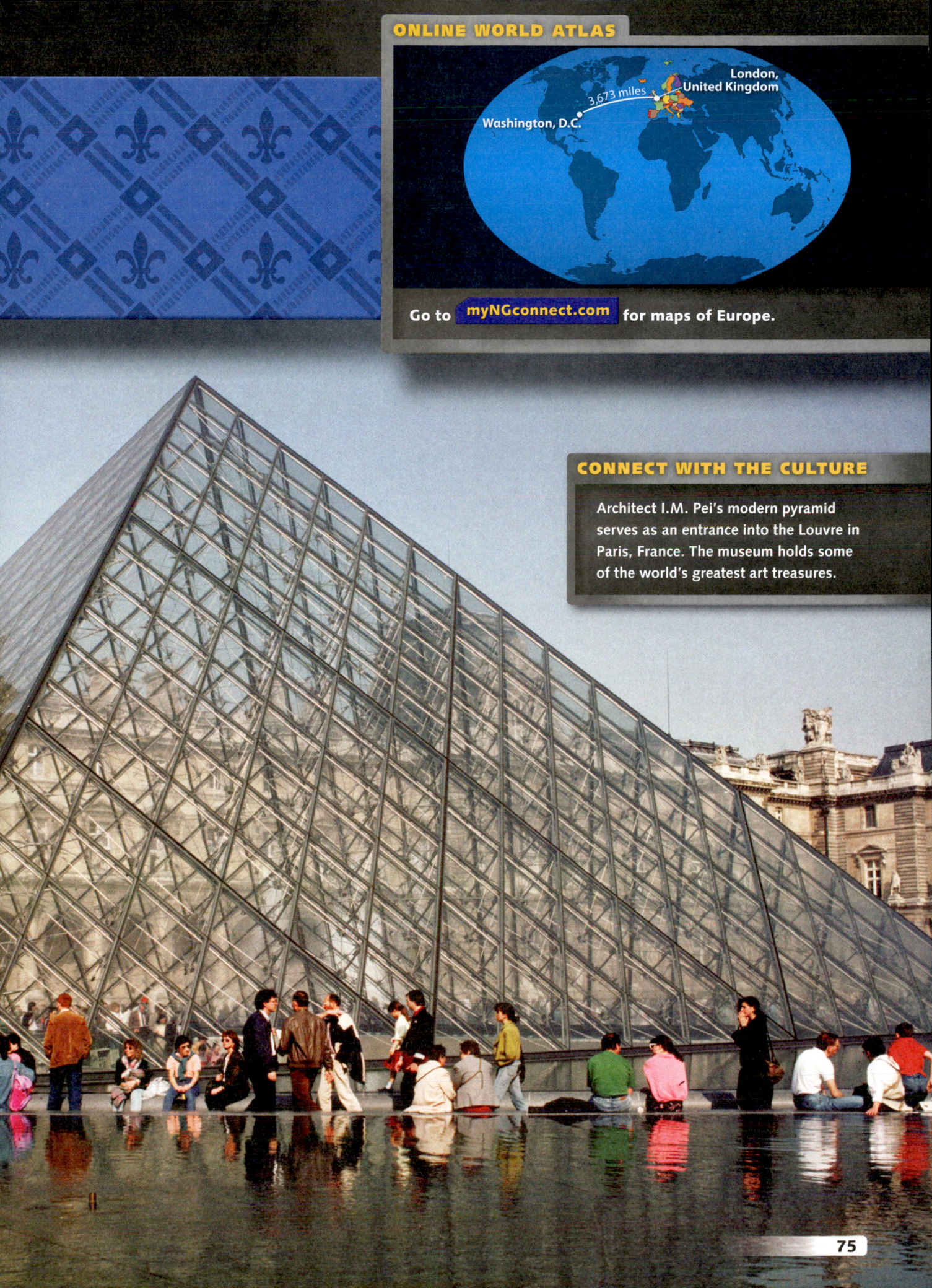

London, United Kingdom

3,673 miles

Washington, D.C.

Go to **myNGconnect.com** for maps of Europe.

CONNECT WITH THE CULTURE

Architect I.M. Pei's modern pyramid serves as an entrance into the Louvre in Paris, France. The museum holds some of the world's greatest art treasures.

CHAPTER 3

EUROPE GEOGRAPHY & HISTORY

PREVIEW THE CHAPTER

Essential Question How did Europe's physical geography encourage interaction with other regions?

SECTION 1 • GEOGRAPHY

KEY VOCABULARY

- peninsula
- uplands
- polder
- bay
- fjord
- canal
- waterway
- ecosystem
- marine reserve

ACADEMIC VOCABULARY
navigable, erosion

TERMS & NAMES

- Northern European Plain
- Alps
- Danube River
- Rhine River

Essential Question How did European thought shape Western civilization?

SECTION 2 • EARLY HISTORY

KEY VOCABULARY

- democracy
- city-state
- golden age
- philosopher
- republic
- patrician
- plebeian
- barbarian
- aqueduct
- feudal system
- serf
- perspective
- indulgence

ACADEMIC VOCABULARY
aristocrat, veto

TERMS & NAMES

- Acropolis
- Alexander the Great
- Julius Caesar
- Augustus
- Christianity
- Middle Ages
- Crusades
- Renaissance
- Johannes Gutenberg
- Martin Luther
- Reformation
- Counter-Reformation

Essential Question How did Europe develop and extend its influence around the world?

SECTION 3 • EMERGING EUROPE

KEY VOCABULARY

- navigation
- colony
- textile
- factory system
- radical
- guillotine
- natural rights
- apartheid
- nationalism
- trench
- reparations
- concentration camp

ACADEMIC VOCABULARY
convert, alliance

TERMS & NAMES

- Industrial Revolution
- Enlightenment
- John Locke
- Reign of Terror
- Napoleon Bonaparte
- Treaty of Versailles
- Great Depression
- Adolf Hitler
- Holocaust
- Iron Curtain
- Cold War
- Berlin Wall

TECHTREK FOR THIS CHAPTER

Student eEdition

Maps and Graphs

Interactive Whiteboard GeoActivities

Digital Library

Connect to NG

Go to **myNGconnect.com** for more on Europe.

Chestnut horses

30°W 20°W 10°W Jan Mayen 0° 10°E 20°E 30°E 40°E 50°E 60°E
(Norway)
70°N

Reykjavik ICELAND

Arctic Circle

Norwegian Sea

Barents Sea

70°N

60°N

Volga R.

B

ATLANTIC OCEAN

Faroe Islands (Denmark)

Shetland Islands (U.K.)

Prime Meridian

N O R W A Y

S W E D E N

F I N L A N D

60°N

60°N

Hebrides (U.K.)

Orkney Islands (U.K.)

Gulf of Bothnia

Helsinki

Oslo

Stockholm

Tallinn

ESTONIA

C

NORTHERN IRELAND

SCOTLAND

Edinburgh

North Sea

Gotland (Sweden)

Baltic Sea

Riga

LATVIA

LITHUANIA

Vilnius

50°N

IRELAND

Belfast

UNITED KINGDOM

Dublin

Shannon R.

WALES

Cardiff

ENGLAND

London

Thames R.

NETHERLANDS

Amsterdam

DENMARK

Copenhagen

Elbe R.

Berlin

Oder R.

Kaliningrad

Minsk

BELARUS

Warsaw

Vistula R.

POLAND

Kiev

Dnieper R.

Don R.

50°N

D

English Channel

Channel Islands (U.K.)

BELGIUM

Brussels

GERMANY

Frankfurt

Rhine R.

Prague

CZECH REPUBLIC (CZECHIA)

Kraków

UKRAINE

LUXEMBOURG

Paris

Seine R.

Loire R.

FRANCE

Danube R.

LIECHTENSTEIN

Vienna

SLOVAKIA

Bratislava

CARPATHIAN MOUNTAINS

MOLDOVA

Chișinău

Sea of Azov

E

Bay of Biscay

Mt. Blanc (4,810 m) 15,781 ft.

Rhône R.

SWITZERLAND

Bern

AUSTRIA

SLOVENIA

Ljubljana

Budapest

HUNGARY

Zagreb

ROMANIA

Belgrade

Bucharest

Danube R.

Black Sea

40°N

PYRENEES

Douro R.

ANDORRA

SPAIN

Madrid

Tagus R.

PORTUGAL

Lisbon

GIBRALTAR (U.K.)

Po R.

SAN MARINO

MONACO

Corsica (France)

ITALY

Rome

VATICAN CITY

Sardinia (Italy)

Balearic Islands (Spain)

Tyrrhenian Sea

Adriatic Sea

CROATIA

BOSNIA AND HERZEGOVINA

Sarajevo

MONTENEGRO

Podgorica

SERBIA

Prishtina

KOSOVO

Skopje

MACEDONIA

Tirana

ALBANIA

Sofia

BULGARIA

GREECE

Ionian Sea

Aegean Sea

Athens

Rhodes (Greece)

40°N

G

Sicily (Italy)

MALTA

Valletta

Mediterranean Sea

Crete (Greece)

30°N

H

0 200 400 Miles

0 200 400 Kilometers

0° 10°E 20°E 30°E

1 2 3 4 5

SECTION 1 GEOGRAPHY

1.1 Physical Geography

TECHTREK

myNGconnect.com For online maps
of Europe and Visual Vocabulary

Maps and
Graphs

Digital
Library

EUROPE PHYSICAL

Visual Vocabulary
uplands

Visual Vocabulary
peninsula

Elevation

feet	meters
10,000+	3,050+
5,000	1,524
2,000	610
1,000	305
500	152
0	0
Below sea level	

30°W   20°W   10°W   Jan Mayen   0°   10°E   20°E   30°E   40°E   70°N   60°E
(Norway)

ICELAND

Norwegian Sea

Arctic Circle

Barents Sea

A

60°N

Faroe Islands
(Denmark)

Shetland Islands
(U.K.)

Prime Meridian

Orkney Islands
(U.K.)

NORWAY   SWEDEN   FINLAND

Gulf of Bothnia

B

IRELAND   Shannon R.   UNITED KINGDOM

North Sea

DENMARK

Gotland
(Sweden)

Baltic Sea

ESTONIA

LATVIA

LITHUANIA

EUROPEAN PLAIN

C

50°N

ATLANTIC OCEAN

Thames R.

English Channel

Channel Islands
(U.K.)

NETHERLANDS

BELGIUM

LUXEMBOURG

Rhine R.   NORTHERN   GERMANY   Elbe R.   Oder R.   POLAND   Vistula R.

BELARUS

Don R.

50°N

D

Dnieper R.

UKRAINE

N
W   E
S

Seine R.

Loire R.

FRANCE

LIECHTENSTEIN

CZECH REPUBLIC
(CZECHIA)

SLOVAKIA

CARPATHIAN MOUNTAINS

MOLDOVA

Sea of Azov

E

Bay of Biscay

Mt. Blanc
(4,810 m) 15,781 ft

SWITZERLAND

Danube R.

AUSTRIA

HUNGARY

ALPS

ROMANIA

Po R.

MASSIF CENTRAL

Rhône R.

SAN MARINO

MONACO

SLOVENIA

CROATIA

Danube R.

SERBIA

Black Sea

40°N

F

Cantabrian Mountains

Douro R.

Iberian Mountains

PYRENEES

ANDORRA

Corsica
(France)

APENNINES

BOSNIA AND HERZEGOVINA

MONTENEGRO

KOSOVO

BULGARIA

MACEDONIA

ALBANIA

40°N

PORTUGAL

IBERIAN   Tagus R.   PENINSULA

SPAIN

Balearic Islands
(Spain)

Sardinia
(Italy)

VATICAN CITY

Adriatic Sea

ITALY

Aegean Sea

GREECE

G

Strait of Gibraltar

GIBRALTAR
(U.K.)

Tyrrhenian Sea

Ionian Sea

Mediterranean

Sicily
(Italy)

Crete
(Greece)

30°N

H

MALTA

Sea

30°N

0   200   400 Miles

0   200   400 Kilometers

0°   10°E   20°E   30°E

1   2   3   4   5   6   7

Main Idea Europe is made up of several peninsulas with varied land regions and climates.

Europe is a "peninsula of peninsulas." A **peninsula** is a body of land surrounded on three sides by water.

A Peninsula of Peninsulas

Europe forms the western peninsula of Eurasia, the landmass that includes Europe and Asia. In addition, Europe contains several smaller peninsulas, including the Italian, Scandinavian, and Iberian. Europe also consists of significant islands, including Great Britain, Ireland, Greenland, Iceland, Sicily, and Corsica.

Four land regions form Europe. The Western Uplands are made up of **uplands**, or hills, mountains, and plateaus, that stretch from the Scandinavian Peninsula to Spain and Portugal. The **Northern European Plain** is made up of lowlands that reach across northern Europe. The Central Uplands are hills, mountains, and plateaus at the center of Europe. The Alpine region consists of the **Alps** and several other mountain ranges.

Varied Climates

Most of Europe lies within the humid temperate climate region. The North Atlantic Drift, an ocean current of warm water, keeps temperatures relatively mild. Winds also affect climate. The sirocco (shuh RAH koh) sometimes blows over the Mediterranean Sea and brings wet weather to southern Europe at different seasons. The mistral is a cold wind that sometimes blows through France and brings cold, dry weather to the country.

CLIMATE

Climate Regions
- Semiarid
- Humid Temperate– No dry season
- Dry Summer
- Humid Cold– No dry season
- Tundra & ice
- Unclassified highlands

In general, a Mediterranean climate brings mild, rainy winters and hot, dry summers and supports a long growing season. Hardy plants grow best in this climate. In contrast, Eastern Europe has a humid continental climate with long, cold winters. Iceland, Greenland, and northern Scandinavia have a polar climate and a limited growing season.

Before You Move On
Monitor Comprehension What are the main land regions and climates in Europe?

ONGOING ASSESSMENT

MAP LAB

 GeoJournal

1. **Interpret Maps** Study both maps in this lesson. Which climate regions are found on the Scandinavian Peninsula? Based on the climate, where do you think most of the peninsula's population is concentrated?

2. **Compare and Contrast** Use both maps to determine what places in Europe have the coldest climates. What geographic characteristics do these places have in common?

1.2 A Long Coastline

TECHTREK

myNGconnect.com For an online map and photos of Europe's coastal features and ports

 Maps and Graphs

 Digital Library

Main Idea Europe's long coastline helped to promote trade, industry, exploration, and settlement on the continent.

Europe has more than 24,000 miles of coastline. If you walked 25 miles a day along the continent's coasts, it would take more than four years to walk the entire distance. These extensive coastlines provided early Europeans with great access to oceans and seas.

Trade and Industry

Europe's water access has benefited the continent in many ways. These benefits include the growth of trade and the development of industry.

Trade has been central to Europe's growth. The civilizations of ancient Greece and Rome flourished largely because of trade. Early sailors traveled to nearly every port on the roughly 2,500-mile-long Mediterranean Sea. They brought back from other lands goods and ideas, such as grains, olive oil, and new religions, that greatly influenced European culture.

Europe also developed several industries that depend on oceans and seas, including a fishing industry. In fact, Europeans have fished along their coastlines for thousands of years.

In lowland areas such as the Netherlands, the people created a way to drain water from the sea in order to increase their farming industry. They built dikes, or giant walls, to hold back the sea in order to create **polders**. Most of the low-lying land of a polder, which once was part of the seabed, was transformed into farms. Today, the Netherlands has about 3,000 polders.

Critical Viewing A boat docks at a harbor in Gdansk (guh DANTSK), Poland, on the Baltic Sea. What do you notice about the harbor?

MAP TIP
This map shows Europe's landforms and rivers, but it also includes country borders within the continent. You can use the map in Section 1.1 to identify the countries.

Map of Europe showing major landforms and rivers with labels including:

Reykjavik, Norwegian Sea, Arctic Circle, Scandinavian Peninsula, Bergen, Stockholm, Helsinki, Göteborg, North Sea, Copenhagen, Baltic Sea, Hartlepool, Grimsby, Bremerhaven, Gdansk, London, Rotterdam, Amsterdam, Hamburg, Vistula R., Antwerp, Oder R., Dunkirk, Elbe R., Le Havre, Rhine R., Seine R., Loire R., Danube R., Carpathian Mountains, Dnieper R., Bordeaux, Rhône R., ALPS, Po R., Trieste, Odessa, Crimea, Bilbao, Pyrenees, Marseille, Genoa, Danube R., Constanța, Black Sea, Iberian Peninsula, Douro R., Tagus R., Corsica, Apennines, Adriatic Sea, Balkan Peninsula, Lisbon, Barcelona, Italian Peninsula, Naples, Valencia, Sardinia, Tyrrhenian Sea, Athens, Aegean Sea, Strait of Gibraltar, Algeciras, Mediterranean Sea, Sicily, Malta, Crete, Volga R., Don R., Sea of Azov, Atlantic Ocean, Jutland, Prime Meridian

0   300   600 Miles
0   300   600 Kilometers

Exploration and Settlement

The location of the continent near large bodies of water also encouraged exploration. In the 1400s, explorers helped European rulers obtain raw materials, spread religious beliefs, and build empires.

Over time, people settled around the ports where the ships docked. Towns often grew up near **bays**, which are bodies of water surrounded on three sides by land. Some of the towns, including Hamburg, Germany, became large cities as trade and industry expanded. In contrast, the deep and narrow bays of Norway, called **fjords** (fee ORDZ), did not encourage settlement.

Before You Move On

Summarize In what ways has Europe benefited from its long coastline?

ONGOING ASSESSMENT

MAP LAB

GeoJournal

1. **Compare and Contrast** Use the map above and those in Section 1.1 to compare and contrast the Italian Peninsula with the Scandinavian Peninsula. What do the two have in common? What differences do you see?

2. **Analyze Cause and Effect** Complete the chart by writing one effect for each cause.

CAUSE	EFFECT
Trade was conducted on Europe's oceans and seas.	Goods and ideas spread.
Industry developed on the coasts.	
Explorers traveled to new lands.	
Cities grew along the coasts.	

SECTION **1** GEOGRAPHY

TECHTREK

myNGconnect.com For an online map and
photos of Europe's landforms and natural resources

 Maps and
Graphs

 Digital
Library

1.3 Mountains, Rivers, and Plains

Main Idea The landforms and resources in Europe support many economic activities.

As you have already learned, Europe consists of four main land regions. A great variety of landforms lie within these regions, including mountains and a vast plain. Many important rivers also cross the continent.

Mountain Chains

Europe's Alpine region contains several mountain chains. The Alps stretch from Austria and Italy to Switzerland, Germany, and France. The Pyrenees are located to the west of the Alps and separate Spain and France. South of the Alps lie the Apennines, which run along the Italian Peninsula. The Carpathians extend through Poland, Romania, and Ukraine.

All of these mountain chains provide natural resources for industries, including forests, which supply wood, and mineral resources, such as iron ore. The valleys between the mountains contain fertile land for growing crops.

Rivers and Plains

Europe has a wealth of rivers. Many are **navigable**, which means that boats and ships can travel easily on them. The **Danube River** is an important transportation route. The river starts in Germany and passes along or through ten countries before emptying into the Black Sea. The **Rhine River** is another vital body of water used to transport goods deep inland. The river originates in Switzerland, winds through Germany, and flows into the North Sea.

Visual Vocabulary A **canal** is a human-made water passage used for travel, shipping, and irrigation. The boats in this canal in Venice, Italy, are being used to move goods within the city.

Land Use
- Forest
- Woodland
- Grassland
- Mixed-use, including crops
- Cropland
- Wetland
- Desert, barren land
- Ice, cold desert, tundra

Natural Resources
-  Coal
- Copper
- Fish
- Iron ore
- Natural gas
- Oil
- Uranium

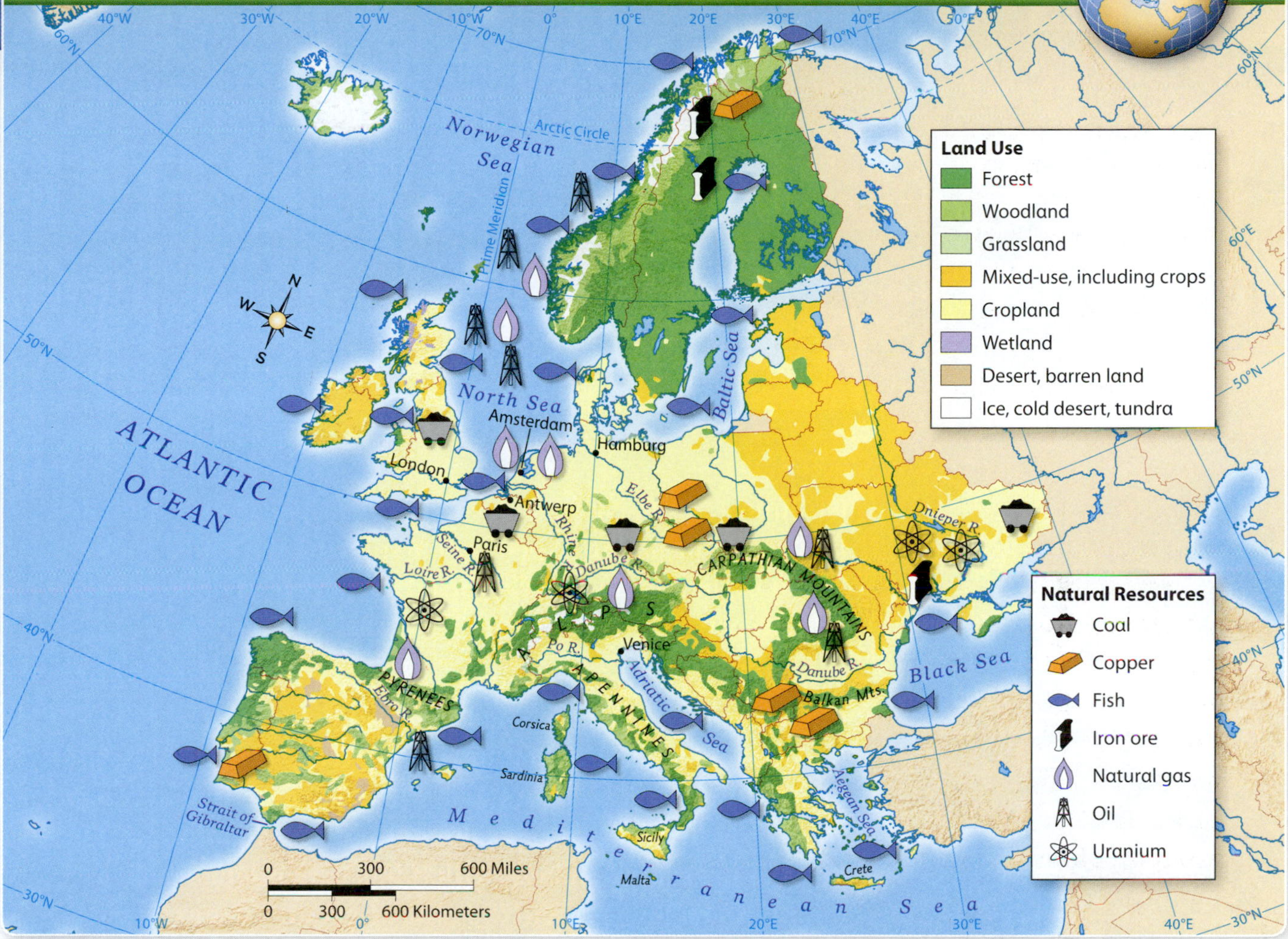

For centuries, Europeans have built canals. When linked together, canals and rivers form **waterways**, or navigable routes of travel and transport. The small country of the Netherlands has more than 3,000 miles of rivers and canals.

Many of the rivers in Europe cross the Northern European Plain. This vast lowland region stretches across France, Belgium, Germany, and Poland all the way to Russia. The fertile soil on the plain makes it ideal for growing crops, and thousands of farms are sprinkled throughout the region. The plain also contains some of the largest and most heavily populated cities, or urban centers, in Europe, including Paris, France.

Before You Move On

Summarize What economic activities are supported by Europe's landforms and resources?

ONGOING ASSESSMENT

MAP LAB
GeoJournal

1. **Interpret Maps** What natural resources are found in the Carpathian Mountains? What geographic challenges might workers deal with when they extract these resources?

2. **Draw Conclusions** Find the Danube River on the map. What natural resources are found along the river? What role might the Danube play in handling these resources?

3. **Make Inferences** What natural resources are found in the North Sea? What impact might they have on the countries bordering the sea?

SECTION **1** GEOGRAPHY

NATIONAL GEOGRAPHIC

1.4

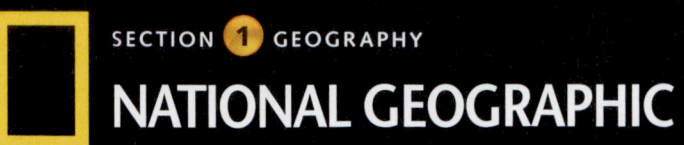

TECHTREK
myNGconnect.com For a map, photos, and an Explorer Video Clip

 Maps and Graphs   Digital Library

Protecting the
Mediterranean
with Enric Sala

> **Main Idea** Human activities have harmed the Mediterranean Sea's natural environment.

Under the Sea

On June 4, 2010, National Geographic Explorer-in-Residence Enric Sala began an expedition: exploring the underwater world of the Mediterranean Sea. He wanted to find out how human activities have affected the sea's ecosystem.

An **ecosystem** is a community of living organisms and their natural environment. Three human activities have had an impact on the Mediterranean's ecosystem. One is overfishing, which occurs when people catch fish at a faster rate than the fish can reproduce. Another is pollution. The third is overdevelopment, which has occurred as coastal populations have grown. (See the chart on the next page.) Growing populations have added to the Mediterranean's pollution and the **erosion**, or wearing away, of its coastline.

myNGconnect.com

For more on Enric Sala in the field today

◀ **Critical Viewing** Sala swims with a sea turtle. What equipment is he using during this underwater exploration?

Sala's work was inspired by Jacques Cousteau, the French underwater explorer. When he became an explorer himself, Sala sailed with Cousteau's son, Pierre-Yves Cousteau. They compared the condition of the Mediterranean Sea now with its condition 65 years earlier and found that the sea has been damaged. Sala concluded, "We have lost most of the large fish and the red coral because of centuries of exploitation [misuse]."

Marine Reserves

In spite of the harm that has been done, Sala sees signs of hope. During their Mediterranean expedition, he and Cousteau visited the Scandola Natural Reserve, near Italy. Scandola is a **marine reserve**, or protected area where people are prohibited from fishing, swimming, or anchoring their boats. As a result, marine life is thriving at the reserve. "This marine reserve," Sala said, "has restored the richness that Jacques Cousteau showed us 65 years ago."

Before You Move On
Monitor Comprehension What can be done to protect the Mediterranean Sea from the human activities that have harmed its environment?

POPULATION OF MEDITERRANEAN CITIES (IN MILLIONS)		
City	1960	2015 (projected)
Athens, Greece	2.2	3.1
Barcelona, Spain	1.9	2.73
Istanbul, Turkey	1.74	11.72
Marseille, France	0.8	1.36
Rome, Italy	2.33	2.65

Source: UN, 2002

ONGOING ASSESSMENT
DATA LAB  GeoJournal

1. **Interpret Charts** Based on the chart, which Mediterranean city will have undergone the greatest growth by 2015? What impact will this growth have on the city's coastline?

2. **Make Inferences** According to the chart, which city will have undergone the least growth by 2015? What does this projected number mean for the city's coastline?

3. **Turn and Talk** What could be done today to preserve fish populations for the future? Get together with a partner and come up with one or two specific suggestions. Share your ideas with the rest of the class.

NATIONAL GEOGRAPHIC

Photo Gallery • Scotland's Rugged Coast

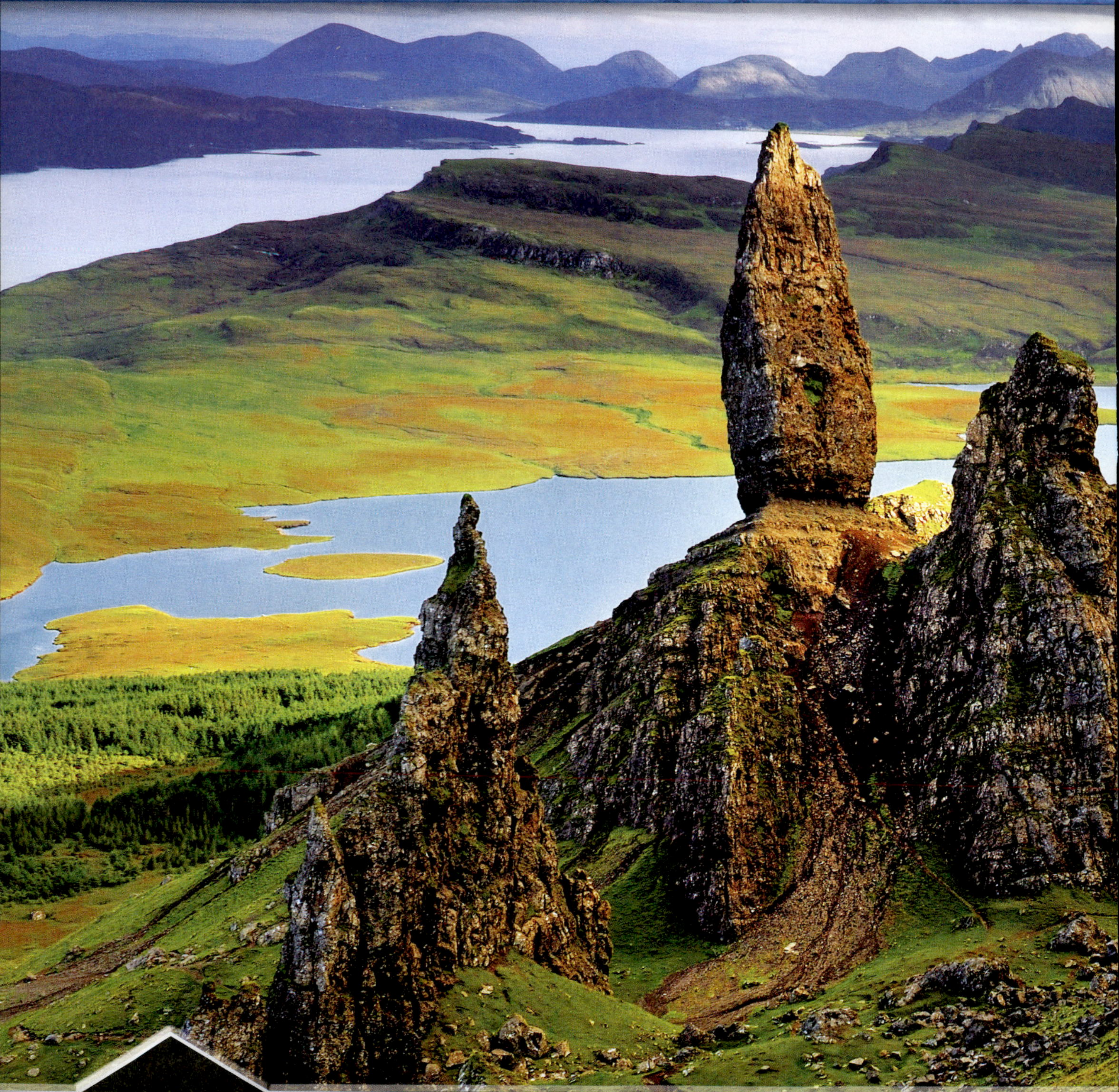

For more photos from the National Geographic Photo Gallery, go to the **Digital Library** at myNGconnect.com.

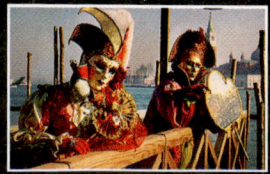

The Carnival in Venice

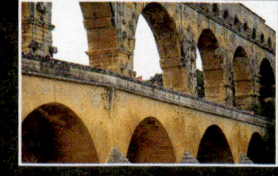

Ancient Roman aqueduct

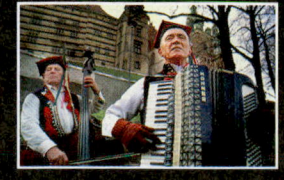

Musicians in Krakow, Poland

Critical Viewing These rocky pinnacles, or pointed formations, took shape thousands of years ago on a hill in Scotland called the Storr. The largest of the formations shown here is known as the Old Man of Storr.

Italy's Amalfi Coast

French pastries on display

The Parthenon in Athens

Armor from the 1600s

2.1 Roots of Democracy

Main Idea The ideas out of which democracy grew first took root in ancient Athens.

The rulers of Athens laid the groundwork for **democracy**, a government in which people can influence law and vote for representatives. Democracy was one of the great achievements of Greek civilization.

The Greek City-States

Greece lies on both the Balkan and Peloponnesus (pehl uh puh NEE suhs) peninsulas. The two are connected by an isthmus, or narrow strip of land. People first arrived in Greece around 50,000 B.C. Early civilizations developed between 1900 B.C. and 1400 B.C.

Around 800 B.C., several Greek city-states started to thrive. A **city-state** is an independent community that includes a city and its surrounding territory. The mountains on the peninsulas made transportation and communication difficult. As a result, each city-state developed independently. The two largest and most important Greek city-states were Athens and Sparta.

Each city-state established its own community and government. The earliest form of government in the city-states was a monarchy, in which a king or queen rules. Over time, a group of upper-class noblemen called **aristocrats** began to act as advisors to the king. In some city-states, the aristocrats set up a ruling council that served as the government. This council was a form of oligarchy (AHL ih gahr kee), in which a small group rules.

Around 650 B.C., tyrants in many of the city-states seized power away from the councils, took control of the government, and re-established one-person rule. Today, any harsh ruler may be called a tyrant. However, not all tyrants in ancient Greece were bad leaders. Some were fair and had the support of the Greek people.

Democracy in Athens

Around 600 B.C. in Athens, a statesman named Solon controlled the government of the city-state. He established assemblies in which all the wealthy people of Athens—not just the aristocrats—made the laws.

Then in 508 B.C., a leader named Cleisthenes (KLIHS thuh neez) increased the people's power even more. He established a direct democracy. Under this government, all citizens voted directly for laws. However, only Athenian adult males were citizens and had the right to vote.

Athens and Sparta

Democracy developed in Athens but not in all city-states. Sparta, Athens' rival, had an oligarchy ruled by a small group of warriors. They supervised a military training system for Spartan boys.

In 490 B.C., Athens and Sparta joined together to defeat the invading army of the Persian Empire under King Darius I. After that, however, the two city-states became fierce enemies.

Before You Move On

Summarize What ancient Greek ideas served as the roots of modern democracy?

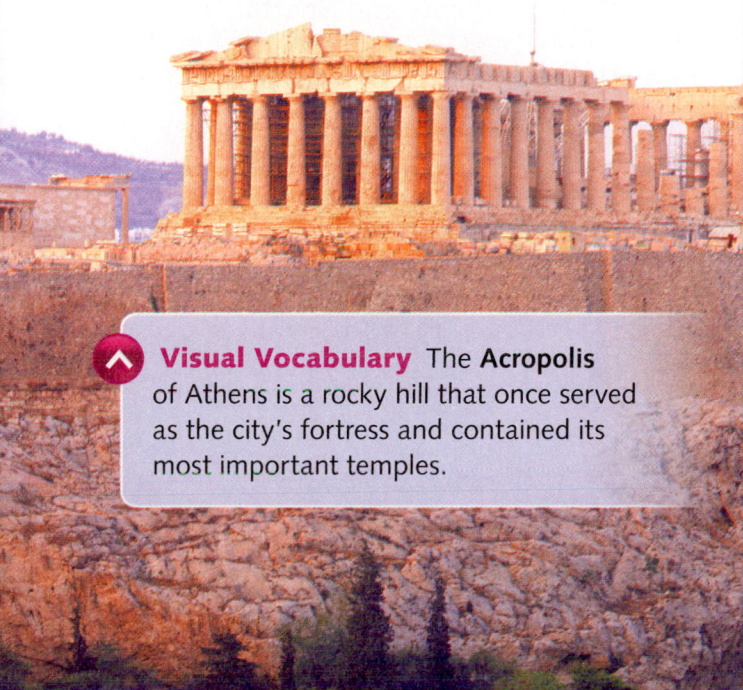

Visual Vocabulary The **Acropolis** of Athens is a rocky hill that once served as the city's fortress and contained its most important temples.

ANCIENT GREEK CITY-STATES

Government in Ancient Greece
510–323 B.C.

- Limited democracy
- Oligarchy
- Monarchy
- Tyranny
- Mixed government

ONGOING ASSESSMENT

READING LAB GeoJournal

Synthesize Use the Greek roots in the chart to form a word in English that completes each of the following sentences:

a. The form of government that represents the people is called _____ (use your own paper) _____.

b. People empowered to enforce a city's laws are the _____ (use your own paper) _____.

c. The ruler of a kingdom is also called a _____ (use your own paper) _____.

SELECTED ENGLISH WORDS FORMED FROM GREEK ROOTS		
Greek Root	**Meaning**	**English Word**
demos	people	democracy
polis	city-state	policy
aristo	best	aristocracy
monos	one	monarchy
oligo	few	oligarchy

TECHTREK

myNGconnect.com For an online map and photos of Classical Greece

 Maps and Graphs

 Digital Library

Main Idea Greek ideas about democracy, architecture, philosophy, and science have had a lasting influence on Western culture.

As you have learned, democracy began in ancient Greece. In 461 B.C., Pericles became the leader of Athens. His rule began a **golden age**, a period of wealth and power during which democracy developed further and Greek culture flourished.

Golden Age of Greece

Pericles had three goals for Greece. The first was to strengthen democracy. He accomplished this goal by paying citizens who held public office. This meant that even people who were not wealthy could afford to serve in government.

The leader's second goal was to expand the empire. Pericles built a strong navy and used it to increase Athens' power over the other Greek city-states.

Pericles' third goal was to make Athens more beautiful. He began rebuilding the city, including the Acropolis. Many of Athens' temples had been destroyed during the war with Persia. Pericles constructed a new temple called the Parthenon, dedicated to the goddess Athena for whom Athens was named.

Greek Achievements

The golden age of Greece was a period of extraordinary achievements. Greek architects designed temples and theaters with graceful columns. **Philosophers**, people who examine questions about the universe, searched for the truth. Socrates (SAHK ruh teez) and his student, Plato (PLAY toh), were leading philosophers.

In the sciences, the mathematician Euclid (YOO klihd) developed the principles of geometry. The physician Hippocrates (heh PAH kruh teez) changed the practice of medicine by insisting that illnesses originated in the human body and were not caused by evil spirits.

Greek Culture Spreads

Greece's golden age ended around 431 B.C., when war broke out between Athens and Sparta. The conflict, known as the Peloponnesian War, lasted 27 years and weakened both Athens and Sparta.

Statue of the goddess Athena

1900–1400 B.C.
Early Greek civilizations develop.

1900 B.C.

1000 B.C.

Early Greek gold lion's head

800 B.C.
Greek city-states begin to thrive.

750 B.C.

ALEXANDER'S EMPIRE, 330 B.C.

0 200 400 Miles
0 200 400 Kilometers
Empire of Alexander the Great at its height, c. 330 B.C.
EUROPE
MACEDONIA
Athens
Black Sea
Caucasus Mountains
Caspian Sea
Aral Sea
ASIA
Alexandria Eschate (Kokand)
Mediterranean Sea
Alexandria
MESOPOTAMIA
Babylon
Alexandria Areion (Herat)
Hindu Kush
Alexandria Arachoton (Kandahar)
PERSIA
Persepolis
AFRICA
EGYPT
Red Sea
ARABIA
Persian Gulf
INDIA
Arabian Sea
Tropic of Cancer
20°E
30°E
40°E
50°E
60°E
70°E
80°E
30°N
40°N
50°N

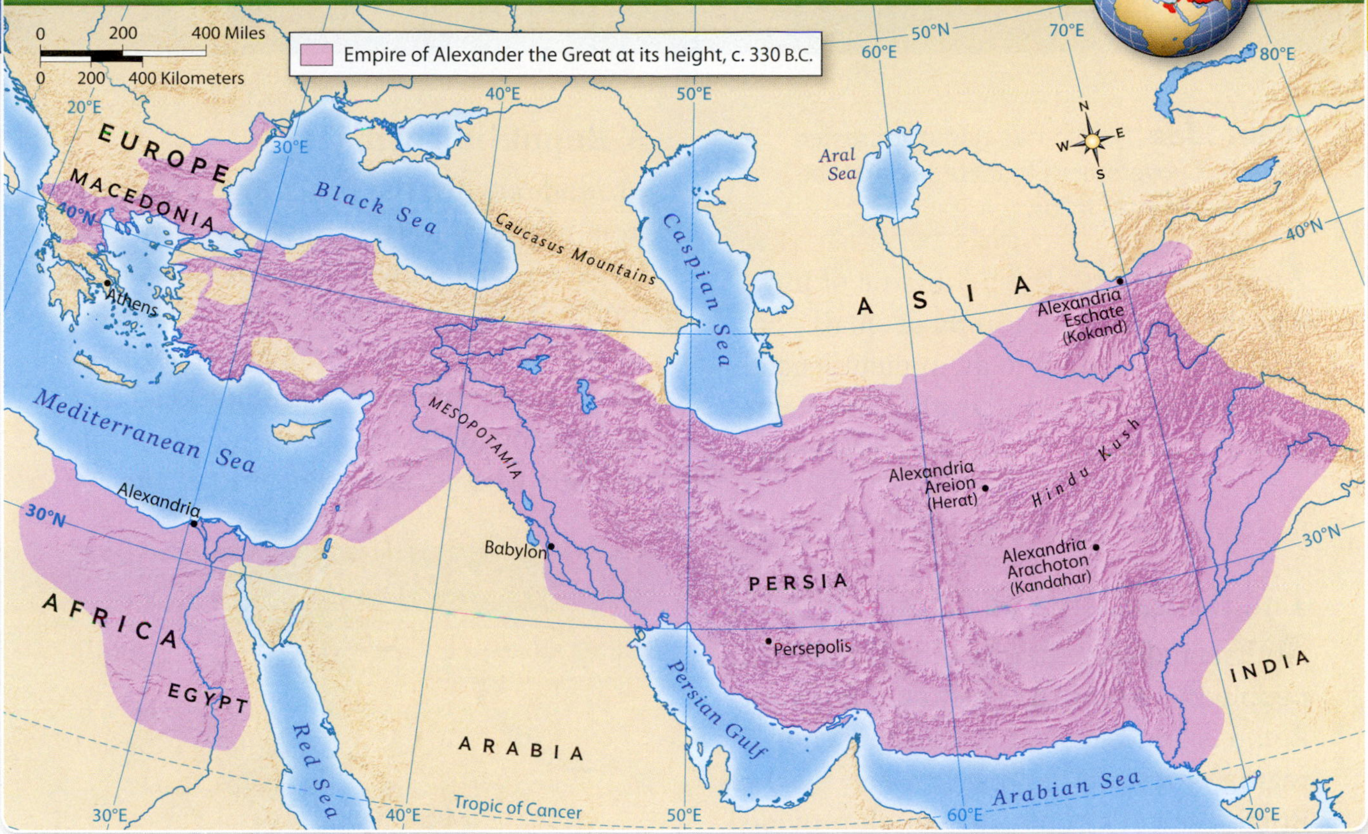

Around 340 B.C., King Philip II of Macedonia took advantage of the weakened city-states and conquered Greece. In 334 B.C., Philip's son, **Alexander the Great**, became king and began to extend his father's empire. Alexander loved Greek culture and spread its ideas throughout the lands he conquered. Alexander died in 323 B.C. at the age of 33. The Greek ideas about democracy, science, and philosophy that he helped spread shaped the modern world.

Before You Move On

Monitor Comprehension What Greek ideas have had a lasting influence on Western culture?

ONGOING ASSESSMENT

MAP LAB

GeoJournal

1. **Draw Conclusions** Study the map. Across which continents did Alexander's empire spread? What helped him unite his vast empire?
2. **Make Inferences** Based on the map, what great empires did Alexander conquer? What do these conquests suggest about Alexander?

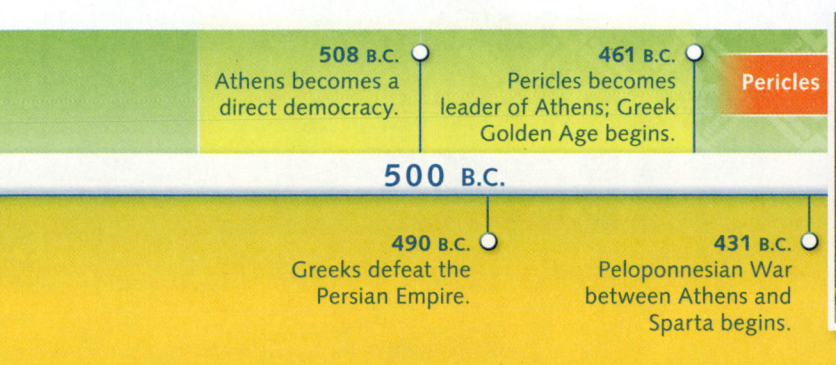

508 B.C. Athens becomes a direct democracy.

490 B.C. Greeks defeat the Persian Empire.

461 B.C. Pericles becomes leader of Athens; Greek Golden Age begins.

431 B.C. Peloponnesian War between Athens and Sparta begins.

Pericles

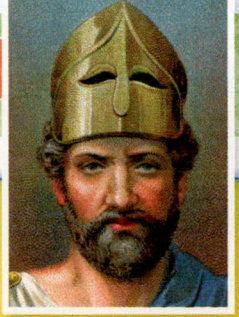

340 B.C. Philip II of Macedonia conquers Greece.

334 B.C. Alexander the Great begins to extend his father's empire.

500 B.C.

250 B.C.

2.3 The Republic of Rome

TECHTREK

myNGconnect.com For an online map of ancient Rome and photos of Roman ruins

 Maps and Graphs

 Digital Library

Main Idea The Roman Republic created a form of government that Europe and the West would later follow.

Around 1000 B.C., the peninsula of Italy was dotted with hundreds of small villages. According to an ancient legend, two brothers named Romulus and Remus founded Rome in 753 B.C. The brothers were said to be the children of a god and to have been raised by a wolf.

The Beginnings of Rome

Archaeologists actually believe that people known as the Latins founded Rome around 800 B.C. They came from a region of Italy called Latium and lived on Rome's seven steep hills, which provided protection from enemy attack. The Tiber River, which flows through Rome, provided water for farming and a route for trade. Over time, Rome developed into a wealthy city-state.

Critical Viewing The Roman Forum contained the ancient city's most important buildings, including the Senate. In what ways does this photo reflect Rome's former glory?

A Republic Forms

Around 600 B.C., the Etruscans, a people from northern Italy, conquered Rome. One Etruscan king named Tarquin was a brutal tyrant. In 509 B.C., the Romans rebelled against him, and Roman leaders began to create a republic. A **republic** is a form of government in which the people elect officials who govern according to law.

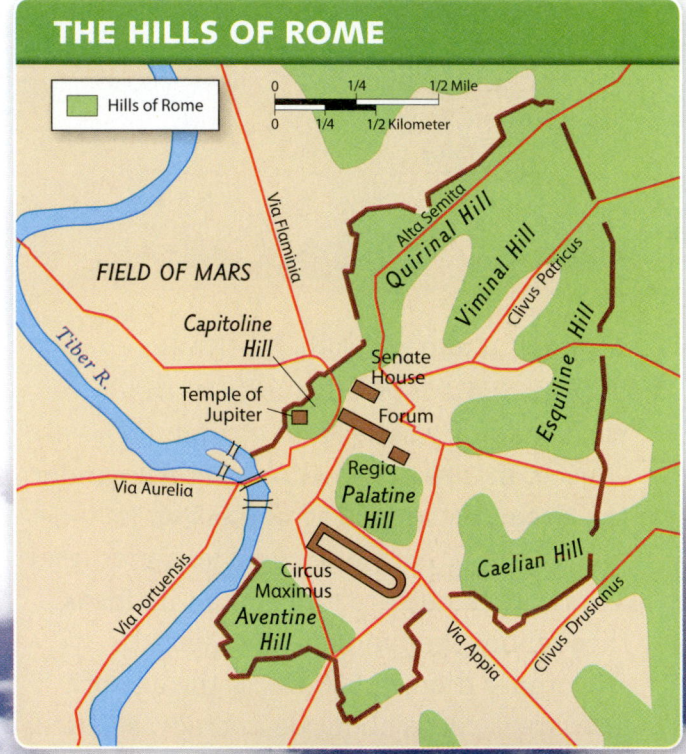

THE HILLS OF ROME

Hills of Rome

0   1/4   1/2 Mile
0   1/4   1/2 Kilometer

FIELD OF MARS

Tiber R.

Via Flaminia

Alta Semita

Quirinal Hill

Viminal Hill

Clivus Patricus

Esquiline Hill

Capitoline Hill

Senate House

Temple of Jupiter

Forum

Regia

Palatine Hill

Via Aurelia

Caelian Hill

Circus Maximus

Aventine Hill

Via Portuensis

Via Appia

Clivus Drusianus

GOVERNMENT OF THE ROMAN REPUBLIC

EXECUTIVE	LEGISLATIVE		JUDICIAL
• Two consuls • Elected to one-year term • Led the government and controlled the army	**Senate** • 300 members made up of patricians • Not elected; selected by the consuls to serve for life • Made the laws and advised consuls	**Assembly** • Made up of plebeians • Elected tribunes as representatives • Made the laws and selected consuls	• Eight judges • Governed provinces • Served for one year **Legal Code** • Twelve Tables • Established rights and responsibilities of Roman citizens

Two main classes of people lived in Rome at this time. The **patricians** were mostly wealthy landowners. The **plebeians** were mostly farmers. At first, only the patricians could take part in government. They controlled the Senate and made laws.

In 490 B.C., plebeians gained the right to form an assembly and elect legislative representatives called tribunes. The assembly made laws and elected the consuls, the two executive officials who led the government for a year at a time. One consul could **veto**, or reject, a decision made by the other consul.

The judicial branch was made up of eight judges who served for one year. These judges oversaw the lower courts and governed the provinces.

Around 450 B.C., the government published the Twelve Tables. These were bronze tablets that set down the rights and responsibilities of Roman citizens. At this time, only adult male landowners born in Rome were citizens. Roman women were citizens but could not vote or hold office.

The Roman Way

Citizens of Rome believed in values that were known as the Roman Way. These values included showing self-control, working hard, doing one's duty, and pledging loyalty to Rome. The Roman Way helped to unite all Roman citizens.

The Romans applied these values as the Republic began to conquer new lands and expand. During the second century B.C., Rome defeated the empire of Carthage in northern Africa. By 100 B.C., Rome controlled most of the lands around the Mediterranean. About this time, tensions began to grow between patricians and plebeians. These tensions triggered a war between the two groups. The war set the stage for the end of the Roman Republic—and the birth of the Roman Empire.

Before You Move On

Summarize What structures, laws, and values made up the government of the Roman Republic?

ONGOING ASSESSMENT

DATA LAB

 GeoJournal

1. **Interpret Charts** According to the chart, members of the Senate were selected by the consuls. However, the assembly elected the consuls. In what way did this arrangement help to control the Senate's power?

2. **Turn and Talk** Study the chart and consider what you know about the U.S. government. Then, with a partner, compare and contrast the two systems of government.

2.4 The Roman Empire

TECHTREK

myNGconnect.com For an online map of the Roman Empire and photos of Roman architecture

 Maps and Graphs

 Digital Library

Main Idea The Roman Empire was one of the largest in history and left a legacy in technology and language.

As you have read, tensions grew between the patricians and plebeians not long after Rome defeated Carthage. The Roman soldiers had brought back great wealth from the conquered territory. They used their wealth to buy large plots of farmland. Small farmers could not compete with them, and the gap between rich and poor widened. In 88 B.C., war erupted between the patricians and plebeians.

Creation of the Empire

After years of war, a general named **Julius Caesar** rose to power and became the sole ruler of Rome in 46 B.C. Caesar started projects to help the poor and tried to re-establish order in Rome, but he developed powerful enemies. In 44 B.C., a small group of senators stabbed him to death.

Octavian, Caesar's nephew, fought in the long, internal war for power that followed Caesar's death. Octavian won, but the war put an end to the Roman Republic. Calling himself **Augustus**, which means "honored one," Octavian became the ruler of the Roman Empire in 27 B.C. His rule began a period called the *Pax Romana*, which is Latin for "Roman peace."

Rome's Decline

For about 500 years, the Roman Empire was the most powerful in the world and extended over three continents. Around A.D. 235, however, Rome had a series of poor rulers. In addition, German tribes, whom the Romans called **barbarians**, began invading the empire from the north.

In 330, Emperor Constantine moved the capital of the weakened empire from Rome to Byzantium, in present-day Turkey, and renamed the city Constantinople. He also made Christianity lawful throughout the empire. **Christianity**, which is based on Jesus' life and teachings as described in the Bible, began in the Roman Empire.

In 395, the empire was divided into an Eastern Empire and a Western Empire with two different emperors. In 476, invaders overthrew the last Roman emperor and ended the Western Empire.

Rome's Legacy

The Western Empire fell, but Rome left the world a great legacy, or heritage. For example, Roman engineers built a network of roads that connected the empire. Many of the roads are still in use.

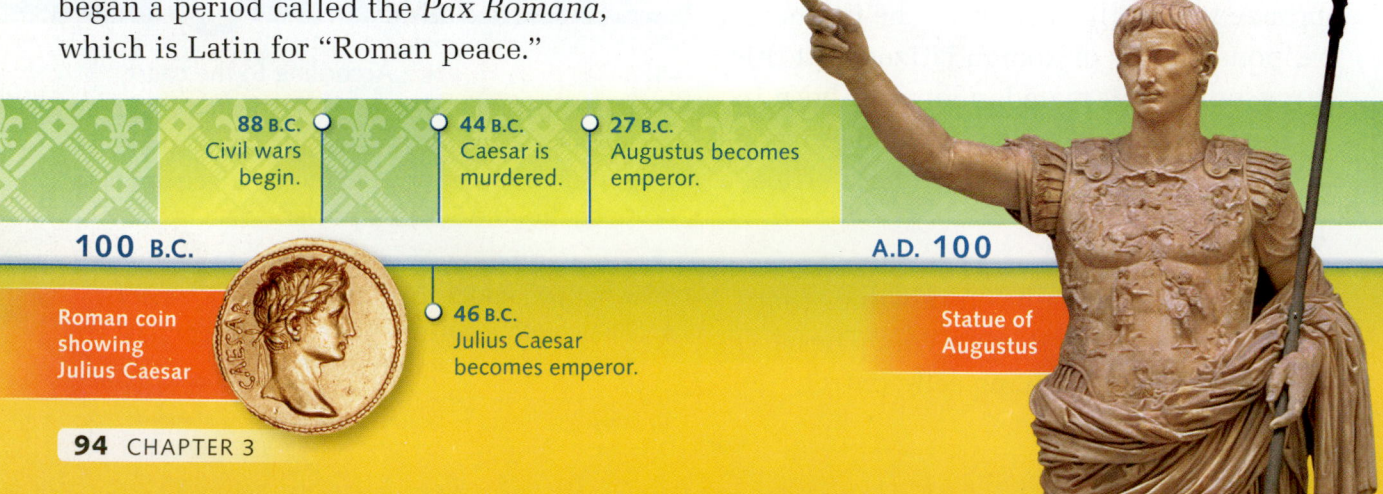

88 B.C. Civil wars begin.

44 B.C. Caesar is murdered.

27 B.C. Augustus becomes emperor.

100 B.C.

A.D. 100

Roman coin showing Julius Caesar

46 B.C. Julius Caesar becomes emperor.

Statue of Augustus

JUTLAND

BRITAIN
Londinium

BELGIUM
GERMANY

ATLANTIC
OCEAN

E U R O P E

GAUL
Lugdunum

Rhine R.

Danube R.

Don R.

Caspian Sea

50°N

40°N

A
L
P
S

Potaissa

DACIA

CRIMEA
Panticapaeum

CAUCASUS MTS.

Narbo

Pyrenees

Ebro R.

Po R.

Apennines

Ravenna
Salonae

ILLYRIA

Danube R.

Balkan Mts.

Black Sea

Sinope

Amisus

ARMENIA

Massilia

Corsica

Rome

ITALY

THRACE

Byzantium

CAPPADOCIA

Tigris R.

A S I A

SPAIN

Tarraco

Sardinia

Brundisium

MACEDONIA

Thessalonica

Pergamum
Ephesus

Tarsus

Antioch

SYRIA

Euphrates R.

Gades

New Carthage

Carthage

Sicily
Syracuse

Malta

Athens

ACHAEA

Cyprus

Damascus

30°N

MAURETANIA

NUMIDIA

Crete

Mediterranean Sea

Cyrene

JUDAEA

Aelana

A F R I C A

LIBYA

Alexandria

EGYPT

Nile R.

Red Sea

Tropic of Cancer

40°E

Expansion of the Roman Empire

■ Roman Republic in 264 B.C.
■ Roman Empire at its height, c. A.D. 200

0   200   400 Miles

0   200   400 Kilometers

N E S W

0°   10°E   20°E   30°E

The engineers also developed the arch and used it to construct buildings and **aqueducts**, which carried water to parts of the empire. Latin, the language of Rome, became the basis for Romance languages, such as Spanish and Italian. Many English words have Latin roots.

Before You Move On

Summarize Describe the Roman Empire's rise, fall, and legacy.

ONGOING ASSESSMENT

MAP LAB

 GeoJournal

1. **Interpret Maps** According to the map, over which continents did the Roman Empire extend? What challenges might the size of the empire have presented to its rulers?

2. **Location** Find Byzantium on the map. Why do you think Constantine chose this location to become the capital of the Eastern Empire?

A.D. **395**
Empire is divided into an Eastern Empire and a Western Empire.

Painting of German invader

A.D. **300**

A.D. **500**

A.D. **330**
Constantine moves the capital from Rome to Byzantium.

A.D. **476**
Rome falls to invaders.

2.5 Middle Ages and Christianity

TECHTREK

myNGconnect.com For portrayals
of the Crusades in fine art

Digital
Library

Main Idea The Roman Catholic Church and the feudal system influenced Western Europe during the Middle Ages.

After the fall of the Roman Empire, Western Europe entered a period known as the **Middle Ages**, which lasted from about 500 to 1500. During this period, Western Europe consisted of numerous kingdoms. Castles, like those on the Rhine River in Germany, served as defensive fortresses. The Roman Catholic Church helped unite people during the Middle Ages, and the feudal system provided a social structure.

The Roman Catholic Church

In 1054, Christianity officially divided into two parts: the Roman Catholic Church in Western Europe and the Eastern Orthodox Church in Eastern Europe. The Roman Catholic Church was the center of life for most people in Western Europe. It cared for the sick, provided education, and helped preserve books and learning.

The Church also played a leading role in government. It collected taxes, made its own laws, and waged wars. In 1096, the Church began a series of **Crusades**. These were military expeditions undertaken to take back holy lands in Southwest Asia from Muslim control. The Crusades cost many lives and ended in 1291.

The Feudal System

The many kingdoms of Western Europe were often at war. From about 400 to 800, a German group called the Franks stopped the fighting and unified most of Western Europe. Their most important leader was Charlemagne (SHAHR luh mayn).

When Charlemagne died in 814, warfare between the kingdoms returned and Western Europe again became divided. To provide security for each kingdom, the feudal system developed. The **feudal system** was a social structure that was organized like a pyramid. At the top was a king who owned vast territory. Beneath the king were lords, powerful noblemen who owned land. The lords gave pieces of their land to vassals, who pledged their loyalty and service to the lord. Some vassals also served as knights, who were warriors on horseback.

Each lord lived on an estate called a manor, which functioned as a small village. **Serfs**, who farmed the lord's land in return for shelter and protection, were at the bottom of the pyramid. Serf families dwelt in small huts on the manor and gave most of the crops they grew to their lord.

The Growth of Towns

In time, the growth of towns helped end the feudal system. Trade and businesses developed, and people began to leave the manors. A deadly disease called the bubonic plague, which swept through Europe in 1347, also weakened the feudal system. The plague killed millions and greatly reduced the workforce in the towns. Desperate for workers, employers offered higher wages. As a result, farmers left the country to seek the higher-paying jobs in the towns.

Before You Move On

Summarize In what ways did the Roman Catholic Church and the feudal system influence Western Europe during the Middle Ages?

MANOR IN THE
MIDDLE AGES

This illustration shows a simplified view of a feudal manor. Rolling fields and farmland lay outside its walls.

The church was the center of life on the manor.

The lord lived in relative safety and ease in his castle.

Serfs lived in tiny huts with dirt floors.

Guards protected the manor from rival lords.

ONGOING ASSESSMENT

VIEWING LAB GeoJournal

1. **Interpret Visuals** What details in the illustration suggest the measures taken to protect those who lived in the manor?

2. **Make Inferences** Notice the position of the church in the manor. Why might it have been positioned near the lord's castle?

3. **Compare and Contrast** Based on the illustration and what you have read, in what ways did life probably differ for lords and serfs?

Main Idea Both the Renaissance and the Reformation brought great change to Europe.

You have learned that the growth of towns in Western Europe helped put an end to the feudal system. The Roman Catholic Church also began to lose some of its power at this time. As these key structures of the Middle Ages weakened, the **Renaissance** began to take hold. The Renaissance was a rebirth of art and learning that started in Italy in the 1300s and had spread through Europe by 1500.

The Renaissance

Several other factors led to the Renaissance. Increased trade in the growing towns brought some Italian merchants great wealth. This wealth allowed them to buy the artists' work.

As you have learned, the Western part of the Roman Empire fell in 476. The Eastern part continued and became known as the Byzantine Empire. The fall of this empire in 1453 also advanced the Renaissance. Scholars from the empire came to Italy, bringing with them ancient writings of the Greeks and Romans. Studies of these works encouraged humanism, which focuses on human rather than religious values. In addition,

a German inventor named **Johannes Gutenberg** developed a printing press in 1450 that printed many books in a short amount of time. Soon, more people in Europe had access to knowledge—including the new humanist ideas.

The result was an explosion in art, architecture, and literature. Renaissance artists such as Leonardo da Vinci, Michelangelo (my kuhl AN juh loh), and Raphael (rah fee ELL) used **perspective** to make a painting look as if it had three dimensions. Architects used elements of ancient Greek and Roman design to create churches and buildings. Writers wrote in the vernacular, the language spoken in a particular region. For example, Dante Alighieri (ah lah GYER ee), who bridged the Middle Ages and Renaissance, wrote his work, *The Divine Comedy*, in Italian, not Latin.

The Reformation

Meanwhile, some people began looking more critically at the Church. **Martin Luther**, a monk in Germany, was shocked by the corrupt practices of some priests. To raise funds, they often sold **indulgences**, which relaxed the penalty for sin.

Gutenberg's printing press

1300s
Renaissance begins in Italy.

1300

1308
Dante starts writing *The Divine Comedy* in Italian.

Portrait of Dante

1400

1450s
Johannes Gutenberg develops the printing press.

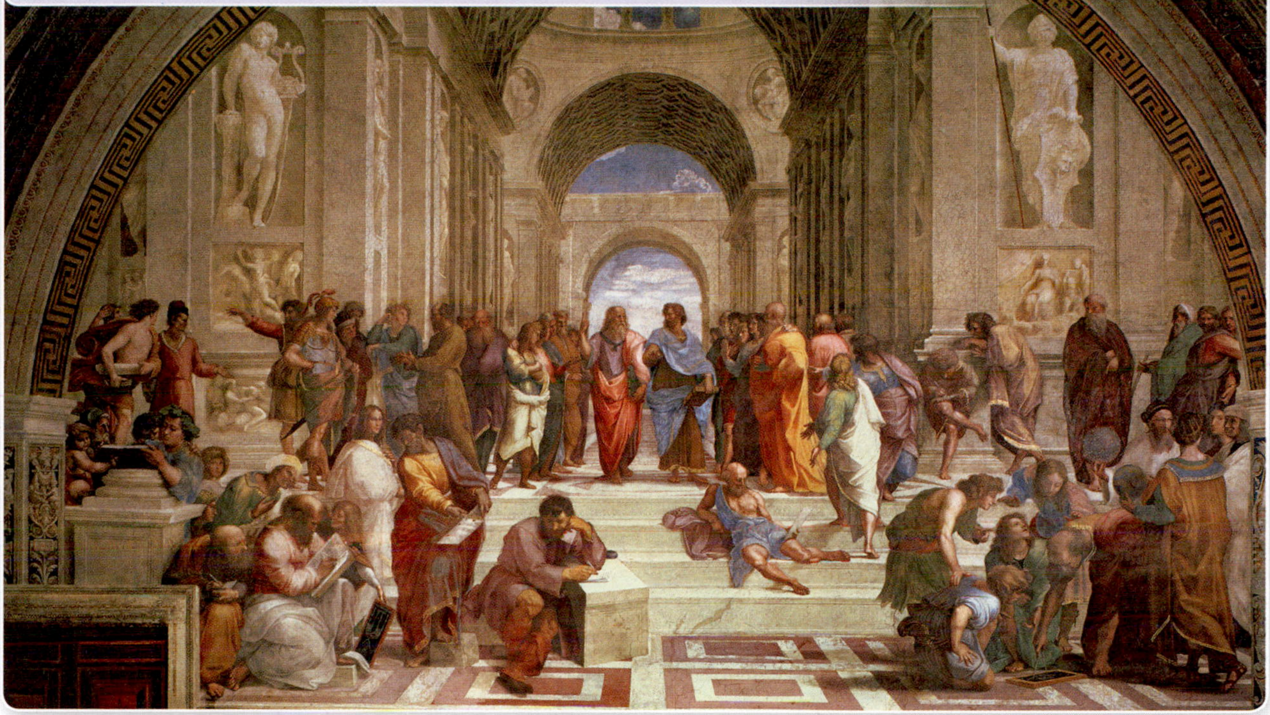

In 1517, Luther wrote the 95 Theses, in which he objected to such practices, and nailed them to a church door. Luther's actions started the **Reformation**, the movement to reform Christianity. Over time, people founded Protestant churches. The term comes from the word *protest*.

In response, the Church began a reform movement called the **Counter-Reformation** and placed more emphasis on faith and religious behavior. Nonetheless, the conflict between Catholics and Protestants would continue for the next 300 years.

Before You Move On

Monitor Comprehension What changes did the Renaissance and Reformation bring to European culture and society?

ONGOING ASSESSMENT

READING LAB  GeoJournal

1. **Summarize** What was the Renaissance?
2. **Analyze Cause and Effect** In what ways did the development of Gutenberg's printing press help spread humanist ideas?
3. **Draw Conclusions** What humanist ideas might have led people to look at the Roman Catholic Church more critically?

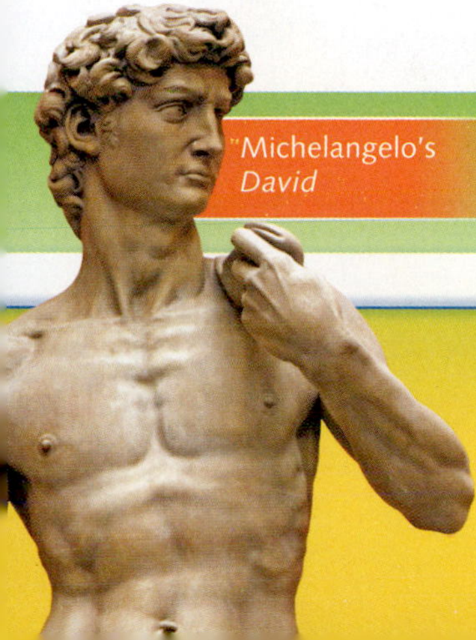

"Michelangelo's *David*

1504 Michelangelo completes the sculpture *David*.

1517 Martin Luther nails his 95 Theses to the door of a church in Wittenberg, Germany.

1500

1600

1497 Leonardo da Vinci finishes painting *The Last Supper*.

In this illustration, Martin Luther posts the 95 Theses.

3.1 Exploration and Colonization

TECHTREK

myNGconnect.com For an online map of European colonization in Africa, Asia, and the Americas

Maps and Graphs

Main Idea To expand trade, Europeans explored Africa, Asia, and the Americas and established colonies on all three continents.

Around 1415, Prince Henry of Portugal decided that he would send explorers to Africa to establish new trade routes. Henry, who became known as Prince Henry the Navigator, founded a **navigation** school. The school taught sailors about mapmaking and shipbuilding and marked the beginning of the Age of Exploration.

European Exploration

Portugal was the first of many European countries to sponsor voyages of exploration. Europeans wanted to find gold and establish trade with Asia to obtain spices, silk, and gems. They also wanted people in other lands to **convert**, or change their religion, to Christianity.

The voyages were filled with danger. Explorers often sailed for months in ships that were small and not always able to withstand strong storms at sea. The men also faced disease and attacks by native peoples. Furthermore, the explorers were traveling to unknown lands. Mapmakers often marked unexplored places with the phrase "Here be dragons."

Nonetheless, Portuguese explorers such as Bartolomeu Dias and Vasco da Gama sailed along the coast of Africa in the late 1400s to open up trade with Asia. Italian explorer Christopher Columbus uncovered a "new world"—the continents of North America and South America—in 1492. In the 1530s, Jacques Cartier (kahr TYAY) explored parts of North America for France. An Englishman, Sir Francis Drake, sailed around the world in 1577.

> **Critical Viewing** In this painting, Columbus and his crew land in North America as Native Americans arrive to meet the explorers in their canoes. What qualities must explorers have had to undertake their voyages?

European Colonies c. 1750

Britain and possessions	France and possessions
Spain and possessions	Netherlands and possessions
Portugal and possessions	Denmark and possessions
	Russia and possessions

Establishing Colonies

In addition to trade, Europeans used the voyages of exploration to claim lands for their own countries. When explorers landed in a new place, they declared it a colony. A **colony** is an area controlled by a distant country. As you have learned, Spanish explorers claimed colonies in Mexico and South America. The French and the English also established colonies in North America. By 1650, European countries controlled parts of Africa and Asia as well.

European exploration and colonization resulted in a sharing of goods and ideas known as the Columbian Exchange. From the Americas, Europeans obtained new foods, such as potatoes, corn, and tomatoes. Europeans introduced wheat and barley to the Americas. They also introduced diseases like smallpox. The diseases killed millions of native peoples.

Before You Move On

Monitor Comprehension What inspired Europeans to undertake voyages of exploration, and what did they gain as a result?

ONGOING ASSESSMENT

MAP LAB

 GeoJournal

1. **Interpret Maps** According to the map, where in Asia did France establish a large colony? Why was this location beneficial geographically?

2. **Identify Problems and Solutions** Study the map. Who was Spain's main rival for colonies in South America? What problems might have arisen from their rivalry?

TECHTREK

myNGconnect.com For an online map and images of the Industrial Revolution

Maps and Graphs

Digital Library

Main Idea The Industrial Revolution was an age of great developments in technology that changed how people worked and lived.

The Age of Exploration opened up trade around the world and brought great wealth to many western European countries. To increase this wealth, businesses looked for new ways to expand production. The result was the **Industrial Revolution**, a period when industry grew rapidly, and the production of machine-made goods greatly increased.

The Revolution Begins

The Industrial Revolution started in Great Britain in the 1700s as a result of new inventions and technologies. The **textile** industry, which deals with the manufacturing of cloth, was the first to be transformed by the revolution. In 1769, textile manufacturers began using machines that were run using water from a stream. Then, around 1770, James Hargreaves invented the spinning jenny. This machine allowed workers to make cotton and wool yarn at a much faster rate.

Before these inventions, most people made cloth by hand in their homes. However, the new machines were too large and expensive to use in small houses. Instead, the machines were placed in factories, and workers manufactured the goods there. In these early factories, each person worked on a small part of the product. This way of producing goods is called the **factory system**.

At first, factories were powered by water. Then around 1776, James Watt developed the steam engine, which was powered by coal. As a result, coal became an important raw material, and Britain benefited from its rich deposits of the fuel.

In the late 1700s, the Industrial Revolution spread to the rest of Europe. France and Belgium became leading manufacturers of textiles. Germany built factories for processing iron. Railroad systems developed in the 1800s. In 1825, George Stephenson built the first railroad in England. By 1850, thousands of miles of tracks crossed Europe.

> **Critical Viewing** England's Iron Bridge, built in 1779, was the world's first arch bridge made of iron. Based on what you have read, what made this bridge possible?

INDUSTRIES IN EUROPE, 1840–1890

Legend:
- Coal
- Iron ore
- Textiles
- Railroad
- International boundary

Map labels: SWEDEN, NORWAY, St. Petersburg, DENMARK, RUSSIAN EMPIRE, Glasgow, North Sea, Baltic Sea, GREAT BRITAIN, Birmingham, NETHERLANDS, Hamburg, Berlin, Warsaw, Amsterdam, London, Brussels, BELGIUM, GERMAN EMPIRE, Frankfurt, Prague, ATLANTIC OCEAN, Paris, FRANCE, LUXEMBOURG, Munich, LIECH., Vienna, Budapest, Nantes, SWITZERLAND, AUSTRO-HUNGARIAN EMPIRE, ROMANIA, Lyon, Milan, PORTUGAL, ANDORRA, Marseille, SAN MARINO, SERBIA, Black Sea, Madrid, ITALY, Rome, MONTENEGRO, OTTOMAN EMPIRE, Barcelona, SPAIN, Mediterranean Sea, Sicily, GREECE

0 200 400 Miles
0 200 400 Kilometers

Impact of the Revolution

The Industrial Revolution had a tremendous impact on how people worked and lived. Cities grew rapidly because people migrated there for factory jobs. Standards of living rose, and a prosperous middle class grew.

However, factory workers often faced harsh conditions. Laborers worked as many as 16 hours a day. Child labor was common. Boys and girls as young as five years of age worked in factories and mines. Some were chained to their machines.

Many workers lived in small, crowded houses in neighborhoods where open sewers were common. Diseases spread quickly in these cramped buildings.

Over time, the workers' quality of life improved as sewer systems were created and other public health acts were passed.

Before You Move On

Summarize In what ways did the Industrial Revolution change how people lived and worked?

ONGOING ASSESSMENT

MAP LAB
GeoJournal

1. **Interpret Maps** Where in Europe were most of the industries concentrated? What does this suggest about the economies of countries in other parts of Europe?

2. **Human-Environment Interaction** Which industry was the most widespread in Europe? Why was this such an important industry?

3. **Evaluate** Based on the map, which countries probably imported the fewest raw materials?

3.3 The French Revolution

TECHTREK

myNGconnect.com For images
of the French Revolution

Digital
Library

> **Main Idea** The late 1700s in France was a period of economic and political unrest, which led to the French Revolution and the rise of Napoleon.

By the summer of 1789, the French people had not yet benefited from the Industrial Revolution. Harvests were poor, and prices skyrocketed. On July 14, mobs attacked the Bastille, Paris's ancient prison. This action sparked the French Revolution.

Roots of the Revolution

For years, France's lower and middle classes had suffered injustices. French society was composed of three large groups, called the Three Estates. The First Estate was made up of clergy. The Second Estate was made up of the nobility, or aristocrats. The Third Estate included everyone else, from merchants to peasants. The Third Estate paid most of the taxes but had no voice in government.

The people of the Third Estate began to call for change. Many of them were influenced by the **Enlightenment**. This movement stressed the rights of the individual. The ideas of Enlightenment thinkers like Voltaire and **John Locke** had helped inspire the American Revolution in 1776. The American Revolution, in part, inspired the revolution in France.

The Revolution Begins

In May 1789, the Third Estate demanded reforms, but the king of France, Louis XVI, refused. In response, the Third Estate formed the National Assembly. On August 26, 1789, the assembly issued the *Declaration of the Rights of Man and of the Citizen.* This document guaranteed liberty, equality, and property to citizens. The assembly tried to form a new government in which Louis would share power with an elected legislature. However, he again refused to cooperate.

French citizens stormed the Bastille because they thought it held guns and gunpowder.

The guillotine was considered an efficient and painless method of execution.

1793
King Louis XVI and Marie Antoinette are executed; Reign of Terror begins.

1785

1790

1795

1789
Mobs attack the Bastille.

1792
Jacobins seize power.

1794
Robespierre is executed, and the Reign of Terror ends.

The Radicals Take Over

Finally, in 1792, the Jacobins, a group of ==radicals==, or extremists, seized power and formed the National Convention. The following year, the group executed Louis XVI and Marie Antoinette, his queen.

The violence soon got worse. Jacobin leader Maximilien Robespierre led a **Reign of Terror**. The Jacobins used a machine called the ==guillotine== (GHEE uh teen) to cut off the heads of an estimated 40,000 people. In July 1794, the French finally turned on Robespierre and executed him.

Napoleon's Rise

After five years of violence, the French were exhausted. France was at war with Prussia, Austria, and Britain, and the government was not ruling effectively.

A young general, **Napoleon Bonaparte**, saw his chance and overthrew the government. Over the next five years, Napoleon increased his powers. He then declared himself Emperor Napoleon I and set about conquering other European powers and building an empire. Britain and Prussia finally defeated him in 1815.

Before You Move On

Summarize What led to the French Revolution and the rise of Napoleon?

 Critical Viewing Marie Antoinette, shown here, was often accused of reckless spending. What details in this painting support this accusation?

ONGOING ASSESSMENT
SPEAKING LAB
GeoJournal

Express Ideas Through Speech Get together in a group and do research to prepare a panel discussion in which you will present the viewpoints of various figures from this section.

Step 1 Decide who each person in your group will be. You might choose from King Louis XVI, Marie Antoinette, Maximilien Robespierre, and Napoleon or be a member of the Third Estate.

Step 2 Come up with a few questions that your panel will discuss. The questions should focus on the French Revolution, the Reign of Terror, and Napoleon's rise.

Step 3 Present the panel discussion before the class. At its conclusion, invite questions and answer them in character.

1804 Napoleon names himself Emperor.

1800

1805

1810

1799 Napoleon overthrows the French government.

Statue of Napoleon on horseback

1815 Napoleon is defeated.

3.4 Declarations of Rights

myNGconnect.com For photos of the documents and Guided Writing

 Digital Library   **Student Resources**

As you have learned, thinkers like John Locke and Voltaire led the Enlightenment. They asserted that people have natural rights, or rights that people possess at birth, such as life, liberty, and property. Two key documents describe these rights: the American Declaration of Independence and the French *Declaration of the Rights of Man and of the Citizen*. In 1993, Nelson Mandela of South Africa received the Nobel Peace Prize. In his speech at the ceremony, he explained that the rights detailed in the declarations are still important.

DOCUMENT 1

from the **Declaration of Independence** (July 4, 1776)

> We hold these truths to be self-evident, that all men are created equal, that they are endowed [provided] by their Creator with certain unalienable [guaranteed] Rights, that among these are Life, Liberty, and the pursuit of Happiness; that, to secure these rights, Governments are instituted among Men, deriving their just powers from the consent of the governed.

This painting illustrates the signing of the Declaration of Independence.

CONSTRUCTED RESPONSE

1. What rights are citizens guaranteed?

DOCUMENT 2

from the **Declaration of the Rights of Man and of the Citizen** (August 26, 1789)

> The representatives of the French people, organized as a National Assembly, . . . have determined to set forth in a solemn declaration the natural, unalienable, and sacred rights of man. Articles:
>
> 1. Men are born and remain free and equal in rights. Social distinctions [classes] may be founded only upon the general good.
>
> 2. The aim of all political association is the preservation of the natural . . . rights of man. These rights are liberty, property, security, and resistance to oppression.

CONSTRUCTED RESPONSE

2. Think about what you learned in Section 3.3 about the roots of the French Revolution. In what ways might the ideas in this document have inspired the French people to revolt?

Mandela and fellow Nobel recipient, F. W. de Klerk, were elected co-presidents of South Africa in 1994.

DOCUMENT 3

from **Nobel Lecture** by Nelson Mandela (December 10, 1993)

Nelson Mandela helped lead the struggle to end apartheid (uh PAHRT hyt) in South Africa. This system had denied black South Africans their rights. In recognition of his efforts, Mandela received the Nobel Peace Prize. The following excerpt is from his acceptance speech.

> The value of our shared reward will and must be measured by the joyful peace which will triumph, because [of] the humanity that bonds both black and white into one human race. . . .
>
> Thus shall we live, because we will have created a society which recognizes that all people are born equal, with each entitled in equal measure to life, liberty, prosperity, human rights, and good governance.

CONSTRUCTED RESPONSE

3. How do the rights Mandela discusses reflect those described in Documents 1 and 2?

ONGOING ASSESSMENT

WRITING LAB GeoJournal

DBQ Practice Think about the ideas in the Declaration of Independence and the *Declaration of the Rights of Man and of the Citizen*. How did these ideas influence Nelson Mandela?

Step 1. Review your answers to Constructed Response questions 1, 2, and 3.

Step 2. On your own paper, jot down notes about the main ideas expressed in each document.

> Document 1: Declaration of Independence
>
> Main Idea(s) _____
>
> Document 2: Declaration of the Rights of Man and of the Citizen
>
> Main Idea(s) _____
>
> Document 3: Nobel Lecture
>
> Main Idea(s) _____

Step 3. Use your notes to construct a topic sentence that answers this question: How did the Declaration of Independence and the *Declaration of the Rights of Man and of the Citizen* influence Nelson Mandela?

Step 4. Write a paragraph that explains specific phrases and ideas in the Declaration of Independence and the *Declaration of the Rights of Man and of the Citizen*. Go to **Student Resources** for Guided Writing support.

3.5 Nationalism and World War I

myNGconnect.com For an online map of Europe before World War I

Maps and Graphs

> **Main Idea** Nationalism, new alliances, and growing tensions in Europe led to World War I.

After the French Revolution, the French people developed powerful feelings of nationalism. **Nationalism** is a strong sense of loyalty to one's country. During the 1800s, nationalism swept through Europe.

Italy and Germany Unify

Nationalism led to unification efforts in Italy and Germany. In 1800, the Italian Peninsula was made up of separate city-states. In 1870, the states came together to form a unified Italy. Germany was also composed of many different states in the early 1800s. Beginning in 1865, Prussia, the most powerful German state, led the way to unification. Driven by nationalist feelings, Prussia fought to take control of other German states away from their non-German rulers. In 1871, the states came together as a united German Empire.

Growing Tensions in Europe

By 1900, tensions had begun to grow among European powers. Nationalism had united some countries from within. However, nationalism also created fierce competition among rival countries.

Mainly, the countries competed for raw materials and colonies in Africa and Asia. To strengthen their position, Britain, France, and Russia formed an **alliance**, or agreement to work toward a common goal, called the Triple Entente. The German Empire and Austria-Hungary formed an alliance known as the Central Powers.

These alliances were tested in June 1914, when Archduke Franz Ferdinand of Austria-Hungary was assassinated in Serbia by a nationalist from Bosnia-Herzegovina. The assassin belonged to a group that was unhappy with Austrian rule of Bosnia-Herzegovina and wanted to unite with Serbia. Immediately after the assassination, Austria-Hungary declared war on Serbia. Then, because Serbia was a Russian ally, Russia declared war on Austria-Hungary. Within weeks, much of Europe had been drawn into war.

A Brutal War

The Great War, as it was called, dragged on for four brutal years. Both sides fought from **trenches**, or long ditches that protected soldiers from the enemy's gunfire. Both sides also used deadly technology, including machine guns, airplanes, tanks, and poison gas. German U-boats, or submarines, sank British ships.

1870
Italy unifies.

Prussian prime minister Otto von Bismarck oversaw German unification.

Illustration of Archduke Ferdinand's assassination

1870

1885

1900

1871
German states unite to form the German Empire.

1914
Archduke Ferdinand is assassinated; World War I begins.

In 1917, Germany seemed to gain an advantage when the Communist Party seized control of Russia's government and economy and made peace with Germany. That same year, the United States entered the war on the side of France and Britain. The fresh American troops helped turn the tide against Germany. In 1918, Germany surrendered to France, Britain, and the United States. By the time the war ended, ten million soldiers had died. About seven million civilians also lost their lives.

Impact of the War

In 1919, Germany signed the **Treaty of Versailles**. Under this peace treaty, Germany was forced to pay several billion dollars in damages and accept full blame for the war. Many of Germany's territories were taken away, and new countries were formed, including Austria, Hungary, Czechoslovakia, Yugoslavia, and Turkey. The treaty angered and humiliated the German people and did little to ease tensions in Europe. These tensions would help lead the way to another world war in a little more than 20 years.

EUROPE BEFORE WORLD WAR I 1914

Triple Entente
Neutral countries that joined the Triple Entente
Central Powers
Neutral countries that joined the Central Powers
Countries that remained neutral

Before You Move On

Monitor Comprehension In what ways did nationalism, new alliances, and growing tensions in Europe lead to World War I?

ONGOING ASSESSMENT

MAP LAB
GeoJournal

1. **Region** According to the map, which empires ruled much of Europe in 1914?

2. **Make Inferences** Note the countries that remained neutral in the war. What geographic factors might have encouraged their neutrality?

Poster illustrating the alliance of Britain, France, and Russia in 1915

Leaders who signed the treaty

1917 United States enters the war.

1919 Treaty of Versailles is signed.

1915

1930

1918 World War I ends.

French prime minister Georges Clemenceau

American president Woodrow Wilson

British prime minister David Lloyd George

3.6 World War II and the Cold War

TECHTREK
myNGconnect.com For an online map of post-war Europe

Maps and Graphs

Main Idea After World War II was fought to defeat the Axis Powers, the Cold War developed between the democratic United States and the Communist Soviet Union.

At the end of World War I, Germany lost its military power. As you have read, the Treaty of Versailles also placed full blame for the war on Germany and forced it to pay **reparations**, or money to cover the losses suffered by the victors. The Great Depression, which began in 1929, further damaged Germany's economy. The **Great Depression** was a severe downturn in the world's economy. During this crisis, **Adolf Hitler** rose to power in Germany.

World War II

Hitler became the leader of the National Socialist German Workers' Party, or the Nazis. In 1936, Hitler made an alliance with Italy. Germany also formed an alliance with Japan, where the military had seized power. Germany, Italy, and Japan formed the Axis Powers.

Germany's invasion of Poland in 1939 started World War II. Two of Poland's allies, Great Britain and France, declared war on Germany soon after the invasion. Germany responded by conquering Poland and then quickly took over most of Europe, including France.

In 1941, Japan attacked the United States at Pearl Harbor, Hawaii. As a result, the United States abandoned its neutrality and entered the war on the side of Britain and the Soviet Union. Together, they were known as the Allies. Over time, many other countries took sides and joined either the Allies or the Axis Powers.

After more than five years of war, Germany surrendered on May 8, 1945. Allied troops were stunned to find the Nazi **concentration camps** where six million Jews and other victims had been murdered. This mass slaughter was called the **Holocaust**. Japan continued to fight until the United States dropped atomic bombs on Hiroshima and Nagasaki. Japan surrendered on September 2, 1945.

The Cold War

After World War II, the Soviet Union established Communist governments in Eastern Europe. Germany was divided into Communist East Germany and democratic West Germany. The imaginary boundary that separated Eastern and Western Europe was called the **Iron Curtain**. The division marked the beginning of the **Cold War**, a period of great tension between the United States and the Soviet Union.

To defend against possible attack, both sides forged military alliances. Western Europe and the United States formed NATO (North Atlantic Treaty Organization), while Communist Eastern Europe formed the Warsaw Pact. The two never directly waged war against each other during the course of the Cold War.

In the 1980s, many eastern European countries overthrew their Communist governments. In 1991, the Soviet Union itself collapsed. The Cold War ended, and democracy replaced communism throughout Eastern Europe.

Before You Move On
Make Inferences In what ways did World War II help lead to the Cold War?

POST-WORLD WAR II EUROPE, 1950

The Iron Curtain
- NATO member countries
- Warsaw Pact member countries
- Neutral countries, non-Communist
- Neutral countries, Communist
- Iron Curtain

Visual Vocabulary

The **Berlin Wall** divided Communist East Berlin from democratic West Berlin. It was torn down in 1989.

PHOTO LAB

GeoJournal

1. **Analyze Visuals** Study the photo. In what way was the Berlin Wall a solid symbol of the Iron Curtain?

2. **Evaluate** Soldiers patrolled the wall on East Berlin's side. How can you tell that this photo was taken on the West Berlin side?

VOCABULARY

Match each word in the first column with its definition in the second column.

WORD	DEFINITION
1. ecosystem	a. sold to relax penalty for sin
2. democracy	b. strong sense of loyalty to one's country
3. plebeians	c. community of living organisms and their environment
4. indulgence	d. area controlled by a distant country
5. colony	e. common people
6. nationalism	f. government of the people

MAIN IDEAS

7. What are some of the significant islands in Europe? (Section 1.2)

8. Where is much of Europe's farming industry located? (Section 1.3)

9. What events led to the development of democracy in ancient Greece? (Section 2.1)

10. How were plebeians represented in the Roman Republic? (Section 2.3)

11. In what way did the Roman Empire influence language? Why? (Section 2.4)

12. In what ways did the Roman Catholic Church serve as a unifying force in Western Europe? (Section 2.5)

13. What achievements in arts and literature did the Renaissance inspire? (Section 2.6)

14. How did the Industrial Revolution change Europe? (Section 3.2)

15. Why did the French people welcome Napoleon's rise to power? (Section 3.3)

16. Why did Russia declare war on Austria-Hungary in 1914? (Section 3.5)

17. What was the Cold War? (Section 3.6)

GEOGRAPHY

ANALYZE THE ESSENTIAL QUESTION

How did Europe's physical geography encourage interaction with other regions?

Critical Thinking: Evaluate

18. In what ways do rivers like the Danube make trade easier within Europe?

19. Why has trade been central to Europe's growth throughout its history?

EARLY HISTORY

ANALYZE THE ESSENTIAL QUESTION

How did European thought shape Western civilization?

Critical Thinking: Draw Conclusions

20. What elements of the democracy practiced in ancient Greece did the United States adopt?

INTERPRET MAPS

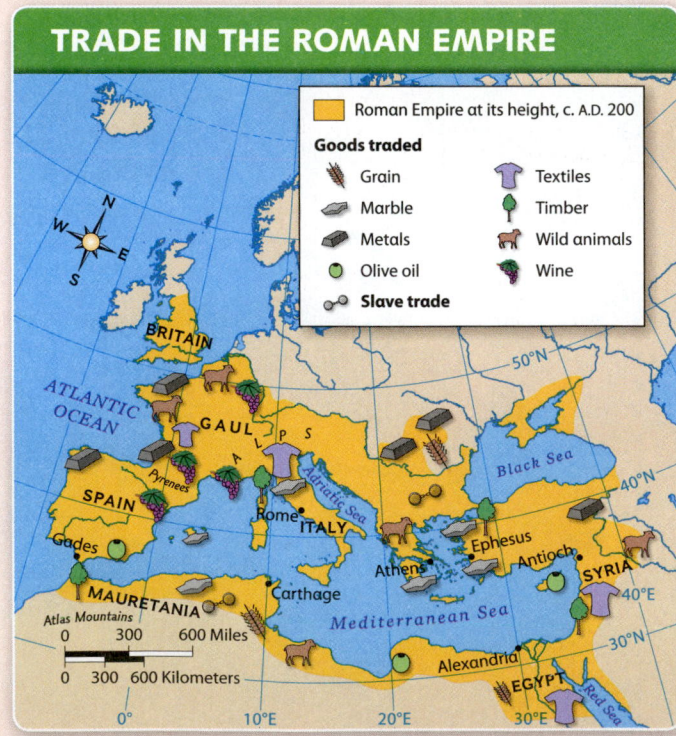

TRADE IN THE ROMAN EMPIRE

Roman Empire at its height, c. A.D. 200

Goods traded
- Grain
- Marble
- Metals
- Olive oil
- Slave trade
- Textiles
- Timber
- Wild animals
- Wine

21. **Movement** From what part of the Empire did Rome obtain its grains? its textiles?

ANALYZE THE ESSENTIAL QUESTION

How did Europe develop and extend its influence around the world?

Critical Thinking: Make Inferences

22. In what way did improvements in navigation and shipbuilding lead Europe to establish colonies in other parts of the world?

23. Why did the Industrial Revolution make European leaders eager to establish colonies in the Americas and Asia?

24. What are some of the positive effects of nationalism? What are some negative effects?

INTERPRET TABLES

MILES OF RAILWAY TRACK IN SELECTED EUROPEAN COUNTRIES (1840–1880)			
	1840	1860	1880
Austria-Hungary	144	4,543	18,507
Belgium	334	1,730	4,112
France	496	9,167	23,089
Germany	469	11,089	33,838
Great Britain	2,390	14,603	25,060
Italy	20	2,404	9,290
Netherlands	17	335	1,846
Spain	0	1,917	7,490

Source: Modern History Sourcebook

25. **Analyze Data** How much railway track did the Germans build between 1840 and 1880? What might account for this increase?

26. **Draw Conclusions** France and Spain are almost the same size. Note the difference in the extent of the railway system in each one in 1880. What does this suggest about the level of industrialization in each country?

ACTIVE OPTIONS

Synthesize the Essential Questions by completing the activities below.

27. **Write Tour Notes** Suppose that you are going to lead a group of tourists on a trip on one of Europe's rivers. Select and research the river. You might choose the Danube, Rhine, Tiber, Rhone, Thames, or Seine. Then write notes for a guided tour of the river. Start by describing its location, size, and appearance. Next, point out important sites along the river and explain their historical significance. Finally, discuss the river's uses today. **Gather photos of the river and conduct the tour with your group of "tourists."**

> **Writing Tips**
> - Use language that appeals to the senses to help your tourists see and experience the river.
> - Include stories about the sites and historical events to hold your audience's interest.
> - Involve your audience by comparing the river to one that is familiar to them.

TECHTREK   myNGconnect.com For research links on European history

28. **Create a Slide Show** Prepare a slide show of famous European buildings, using **Connect to NG** or other online sources. Research and identify five buildings, such as the Pantheon in Rome, Italy. Write two sentences that explain the importance of each building to European history. Copy the chart below to help you organize your information.

BUILDING	IMPORTANCE TO EUROPE
1.	
2.	
3.	

PREVIEW THE CHAPTER

Essential Question How is the diversity of Europe reflected in its cultural achievements?

KEY VOCABULARY

- dialect
- heritage
- perspective
- abstract
- troubadour
- opera
- genre
- epic poem
- novel
- staple
- cuisine

ACADEMIC VOCABULARY
cosmopolitan

TERMS & NAMES

- Romantic period
- impressionism
- Baroque period
- Classical period

Essential Question What are the costs and benefits of European unification?

KEY VOCABULARY

- tariff
- currency
- euro
- sovereignty
- eurozone
- consumer
- democratization
- privatization
- demographics
- aging population

ACADEMIC VOCABULARY
exchange, assimilate

TERMS & NAMES

- Common Market
- European Union (EU)
- Orange Revolution

TECHTREK FOR THIS CHAPTER

Student eEdition

Maps and Graphs

Interactive Whiteboard GeoActivities

Digital Library

Connect to NG

Go to **myNGconnect.com** for more on Europe.

Buildings with traditional red-tiled roofs line this square in Prague in the Czech Republic.

1.1 Languages and Cultures

TECHTREK

myNGconnect.com For photos reflecting European culture

Digital Library

Main Idea Europe has a great variety of languages, cultures, and cities.

Europe has more than a half billion people, yet they live in an area that is one-half the size of the United States. In addition, the continent of Europe contains more than 40 countries. The result is a great diversity, or wide variety, of languages and cultures.

European Languages

Many of the languages spoken in Europe today fall into three language groups: Romance, Germanic, and Slavic. The Romance languages include French, Spanish, and Italian. The Germanic languages are spoken mostly in northern Europe and include German, Dutch, and English. Most people in Eastern Europe speak Slavic languages, such as Russian, Polish, and Bulgarian.

Some countries in Europe have more than one official language. Belgium, for example, has three: Dutch, French, and German. Even in countries with only one official language, people may speak different dialects. A **dialect** is a regional variety of a language. In Italy, for instance, people in Rome speak a dialect of Italian that differs from that in other cities.

Cultural Traditions

Because Europe is composed of many different countries and ethnic groups, it has a rich cultural **heritage**, or tradition. Europe's cultural diversity is reflected in its religions and celebrations.

Christianity is the dominant religion in Europe. Today, about 45 percent of the continent's total population is Catholic. Protestantism is most common in Northern Europe.

> **Critical Viewing** In this photo of the Palio, a centuries-old cultural tradition is honored as these horses race in Italy. What details in the photo convey the excitement of the race?

In recent years, Islam has become the fastest-growing religion in Europe. Immigrants from Turkey, North Africa, and Southwest Asia move to Europe and bring their Muslim faith with them.

Many of the holidays celebrated in Europe are rooted in religion. However, Europeans enjoy other kinds of festivals as well. One of the most colorful is the Palio, a horse race held each summer in Siena, Italy. In this race, which dates back to the Middle Ages, ten riders from ten of the city's neighborhoods compete.

City Life

More than 70 percent of Europeans live in urban areas. In Belgium, over 95 percent of the people live in or near its cities. Most of Europe's cities are **cosmopolitan**, which means that they bring together many different cultures and influences. London is an example of a cosmopolitan city. Its restaurants and shops reflect the South Asian, Caribbean, and East Asian origins of some of its newer citizens.

Many European cities date back hundreds of years. As a result, they developed in ways very different from American cities. These cities are often smaller in area than those in the United States and have narrow, winding streets. Most people live in apartments rather than individual houses. For recreation, city dwellers visit their many parks. They also tend to use public transportation more often than most Americans.

Before You Move On

Make Inferences What might be some of the advantages and disadvantages of Europe's great variety of languages and cultures?

ONGOING ASSESSMENT
PHOTO LAB
 GeoJournal

1. **Analyze Visuals** Note the way the riders are dressed in the photo. What do you suppose their clothes represent?

2. **Draw Conclusions** Study the photo and recall what you have read about the Palio. How is this race an example of a cultural tradition?

3. **Turn and Talk** Get together with a partner and discuss the holidays and festivals that you celebrate. Take notes on your discussion and be prepared to share with the rest of the class.

1.2 Art and Music

Main Idea European art and music have developed over thousands of years.

Throughout the centuries, European art and music have changed to reflect different styles and beliefs.

European Art

European art grew out of the artistic achievements of ancient Greece and Rome. Greek and Roman gods and goddesses were frequent subjects of the artists from these cultures, but they were portrayed to represent realistic human forms.

Much of the art of the Middle Ages reflected the influence of Christianity. The religious subjects were often presented as two-dimensional figures. During the Renaissance, artists used **perspective** to give their work greater depth. Although religious subjects were common, artists also painted portraits of people.

In the **Romantic period** of the early 1800s, artists moved away from religious themes to paint landscapes and other natural scenes that would convey emotion. **Impressionism** emerged in the late 1800s. Impressionist artists, such as Claude Monet, used light and color to capture a moment. By 1900, artists wanted to create a new form of art. These modern artists often worked in an **abstract** style, which emphasized form and color over realism.

⌃ Critical Viewing The *Mona Lisa*, by Italian Renaissance artist Leonardo da Vinci, is probably the most famous portrait of all time. What about this painting might account for its popularity?

⌃ Critical Viewing This painting, *Impression, Sunrise,* by French artist Claude Monet, gave the impressionist movement its name. What kind of mood is conveyed in this painting?

Most opera houses contain a stage, an orchestra pit, and several levels of balconies. The ceiling of the Paris Opéra, shown here, was painted by Russian-born artist Marc Chagall.

European Music

Like art, European music began in ancient Greece and Rome. Musicians played on a few simple instruments and were often accompanied by singers.

During the Middle Ages, music was used in religious ceremonies. Singers called **troubadours** performed songs about knights and love. These songs influenced Renaissance music, when instruments such as the violin were introduced.

The new instruments helped inspire the complex rhythms in the music of the **Baroque period**, which lasted from about 1600 to 1750. **Opera**, which tells a story through words and music, was born then.

The **Classical** and Romantic periods followed the Baroque period and continued until about 1910. Composers from these two periods, such as Ludwig van Beethoven (BAY toh vuhn) of Germany, wrote works using instruments and techniques that are still used today.

Before You Move On

Monitor Comprehension What styles and beliefs have influenced European art and music?

ONGOING ASSESSMENT

LISTENING LAB GeoJournal

1. **Analyze Audios** Listen to the music clip of Beethoven's Fifth Symphony in the **Digital Library**. Describe the music's mood. What instruments help to convey this mood?

2. **Form and Support Opinions** What do you think of the opening? Support your opinion by referring to specific details in the music.

1.3 Europe's Literary Heritage

TECHTREK
myNGconnect.com For photos of European writers and Guided Writing

 Digital Library

 Student Resources

Main Idea European literature has reflected new ways of thinking over the centuries.

Plays by the English playwright William Shakespeare (1564–1616) are performed almost every day. European writers such as Shakespeare have influenced literature for centuries. They wrote in many different **genres**, or forms of literature, including poetry, plays, and novels.

Literary Origins

European literature began with the ancient Greeks and Romans. Around 800 B.C., the Greek poet Homer wrote the epic poems *The Iliad* and *The Odyssey*. An **epic poem** is a long poem that tells the adventures of a hero who is important to a particular nation or culture. Around 20 B.C., the Roman poet Virgil wrote *The Aeneid*, an epic poem about the founding of Rome.

One of the greatest writers of the late Middle Ages and early Renaissance was the Italian poet Dante (1265–1321). As you have learned, Dante wrote *The Divine Comedy* in Italian, not in Latin. The epic poem deals with the religious beliefs and politics of his time.

Many later works of the Renaissance focused on human behavior. Shakespeare explored this theme in plays such as *Hamlet*. Spanish writer Miguel de Cervantes (1547–1616) wrote what is considered the first modern novel, *Don Quixote* (kee HO tee). A **novel** is a long work of fiction, containing characters and a plot. The printing press, which Johannes Gutenberg developed in the 1450s, helped spread the popularity of these books.

The 1700s and 1800s

In the mid-1700s, Enlightenment ideas about reason and government inspired the movement toward democracy. These ideas, in turn, led French and English writers of the time, such as Voltaire and John Locke, to explore the rights of the individual.

In the 1800s, writers of the Romantic period continued this exploration, with an emphasis on emotion and nature. For example, German author Johann Wolfgang von Goethe (GHER tuh) (1749–1832) wrote *The Sorrows of Young Werther*, a novel about a sensitive young artist.

Other writers of the 1800s took a much more realistic look at life. In novels such as *Sense and Sensibility*, British writer Jane Austen (1775–1817) used humor to examine women's role in society.

Critical Viewing Inspired by tales of knights, Don Quixote (right) goes to battle evil with his servant Sancho Panza (left). What details in this painting suggest that Cervantes' novel is a comedy?

 Critical Viewing Austen's novels typically end with marriage, such as the one shown in this scene from a film adaptation of *Sense and Sensibility*. Based on the photo, how would you describe an English wedding in the 1800s?

Another British writer, Charles Dickens (1812–1870), commented on social issues, including poverty, in such novels as *Oliver Twist*. Norwegian playwright Henrik Ibsen (1828–1906) wrote plays, such as *A Doll's House*, which criticized the traditional role of husbands and wives at that time.

Modern Literature

The two world wars had a great impact on the modern literature of the 20th century. Writers at this time reflected the sense that life was uncertain and unpredictable. Some rejected traditional genres and experimented with writing new forms of plays, poems, and novels.

Many modern writers examined the inner workings of the mind. In the novel *Ulysses*, for example, Irish writer James Joyce (1882–1941) focused on the thought processes of the main character over the course of a single day. Romanian playwright Eugene Ionesco (1909–1994) used ridiculous situations to comment on

what he saw as the emptiness of life. Many European and other writers today have been influenced by these authors.

Before You Move On

Summarize What new ways of thinking has European literature reflected over the centuries?

ONGOING ASSESSMENT

WRITING LAB   GeoJournal

Write Reports Think about what you have learned about Europe's literary heritage. Then consider the following question: In what ways do beliefs and events influence literature? Select a writer mentioned in this lesson and write a report in which you answer this question.

Step 1 Research to learn more about the writer and the period in which he or she lived.

Step 2 Find out how the beliefs and events of the time influenced the writer.

Step 3 Write a brief report in which you explain these influences. Support your ideas with specific references to one or two of the writer's works. Go to **Student Resources** for Guided Writing support.

SECTION 1 CULTURE

1.4 Cuisines of Europe

TECHTREK
myNGconnect.com For photos of traditional European foods
Digital Library

Main Idea Landforms and climate have influenced the cooking traditions of Europe.

Throughout most of Europe, foods such as meat, bread, and cheese are **staples**, or basic parts of people's diets. However, the **cuisine**, or cooking traditions, of most European countries is largely determined by the landforms and climate of a particular region.

Foods of Western Europe

The hot, dry climate in the Mediterranean countries of Spain, France, Italy, and Greece is ideal for growing olives, tomatoes, and garlic. As a result, these are key ingredients in their cuisines. Fish from the bodies of water surrounding the countries is also an important menu item.

The cuisines of France and Italy have influenced cooking throughout the world and are especially known for their sauces. French sauces are typically made of milk or cheese, while those of Italy are often tomato based.

People in western European countries with cooler climates often eat more filling fare. In Germany, Great Britain, and Ireland, potatoes grow well and are popular side dishes. In contrast with the light sauces of France and Italy, German cooks often serve heavier gravies.

In Scandinavian countries such as Sweden and Norway, the people often eat herring and other fish. Scandinavians also enjoy deer meat provided by the herding culture of these northern countries.

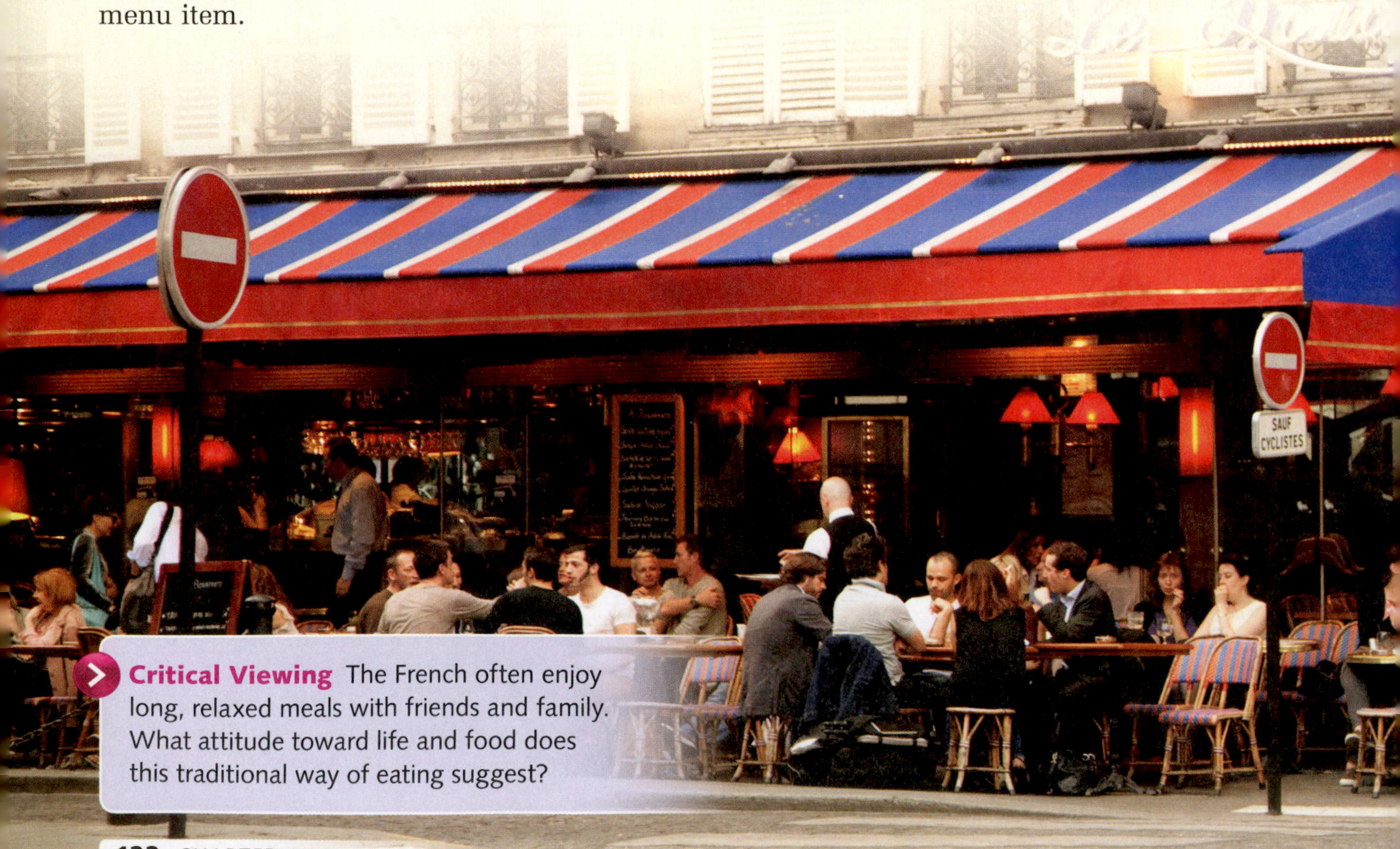

Critical Viewing The French often enjoy long, relaxed meals with friends and family. What attitude toward life and food does this traditional way of eating suggest?

Foods of Eastern Europe

Eastern Europe's cold climate results in a shorter growing season than that of Western Europe. In Russia, root vegetables such as turnips and beets are well adapted to the country's climate. A soup called *borscht*, which is made from beets, is a traditional offering on cold winter nights.

The fertile soil of Hungary allows Hungarian farmers to grow grains and potatoes. These crops are used to make a variety of breads and dumplings. A meat stew called *goulash* is Hungary's national dish. It is made with beef, potatoes, and vegetables, and seasoned with paprika. Paprika is a red spice that was brought to Hungary by the Turks in the 1500s.

Bread like this round loaf is the traditional centerpiece at a Ukrainian wedding.

Like Hungary, Ukraine has fertile soil and fields of wheat and other grains. The country is known for its bread. On special occasions, cooks prepare breads decorated with ornaments made of dough.

Before You Move On

Summarize How have landforms and climate influenced European cooking traditions?

RUE
GUILLAUME
APOLLINAIRE

RUE
BONAPARTE

SPEAKING LAB GeoJournal

Turn and Talk What is the traditional cuisine of the United States? What impact have dishes brought by immigrants from other countries had on how and what Americans eat? Get together in a small group and discuss these questions. Be prepared to share your ideas with other groups.

2.1 The European Union

Main Idea The European Union was formed to unite Europe and benefit it economically.

In 1948, the United States established a program called the Marshall Plan to help Europe rebuild after World War II. To manage the U.S. aid money, European countries formed the Organization for European Economic Cooperation in 1948. As a result, European countries discovered that they could rebuild their countries faster when they worked together.

The Common Market

In 1957, some European countries sought even closer economic ties. They formed the European Economic Community (EEC), which became known as the **Common Market**. The first countries to join were Belgium, France, Italy, Luxembourg, the Netherlands, and West Germany. Several more countries joined during the 1980s.

The Common Market pledged to create "an ever closer union among the European peoples." However, it was primarily formed to create a single market among the member nations. A single market is one in which a group of countries trades across its borders without restrictions or tariffs. A **tariff** is a tax paid on imports and exports.

A United Europe

In 1992, the countries of the Common Market sought to extend their economic organization throughout Europe. They met in Maastricht in the Netherlands and signed the Treaty of Maastricht, which created the **European Union (EU)**. By 2010, the EU had 27 member nations with a total of more than 500 million people. When considered as a single economy, the EU is the largest in the world.

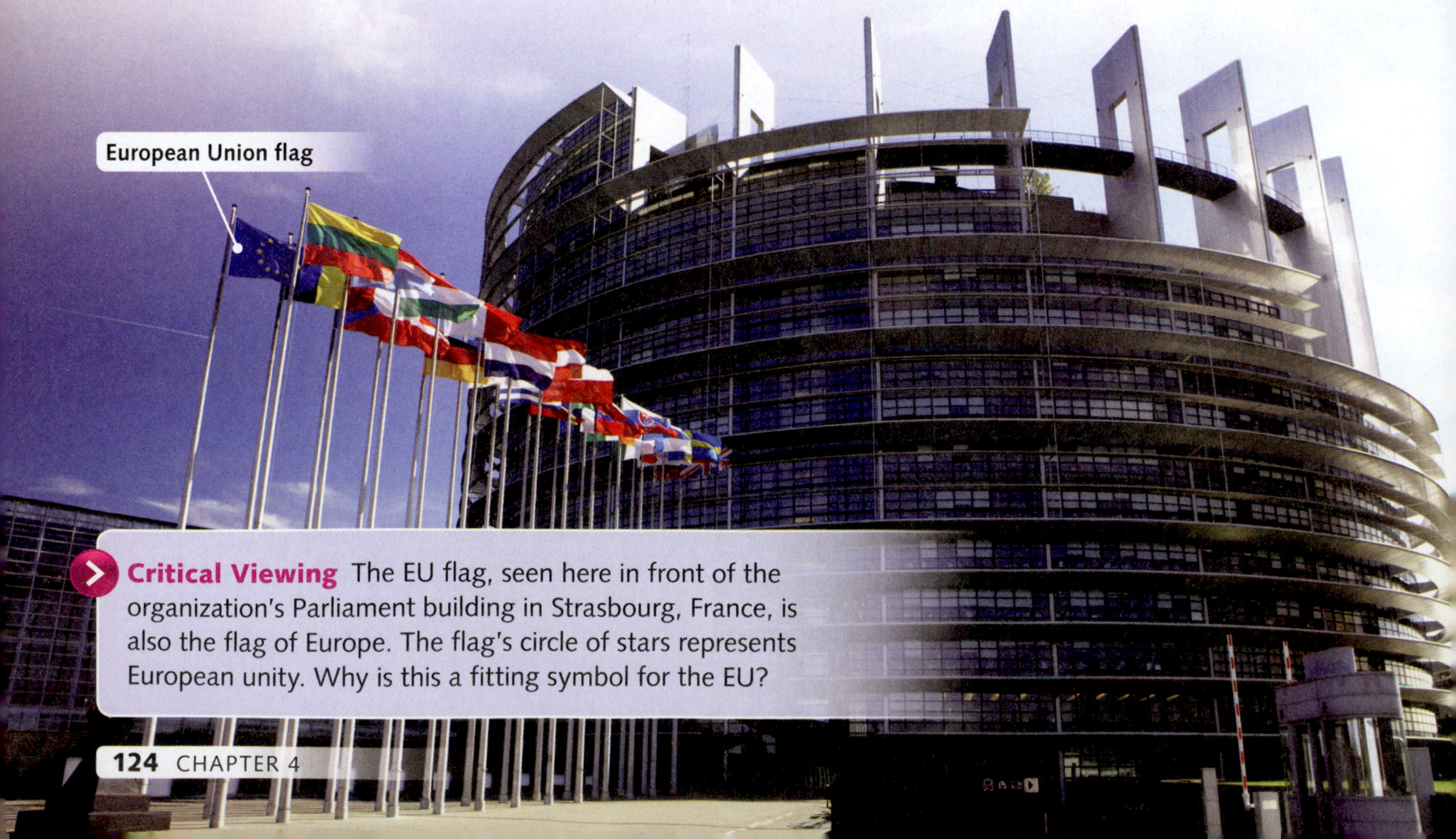

European Union flag

> **Critical Viewing** The EU flag, seen here in front of the organization's Parliament building in Strasbourg, France, is also the flag of Europe. The flag's circle of stars represents European unity. Why is this a fitting symbol for the EU?

MAP TIP On this map, the year beneath a country's name indicates when it joined either the EEC (between 1957 and 1991) or the EU (beginning in 1992).

Legend:
- Member
- Candidate
- Potential candidate

The EU has a government with an executive, a legislative, and a judicial branch. These branches propose, pass, and enforce the organization's policies and legislation. The EU also has agencies that direct economic policies. Through these agencies, the EU has eliminated tariffs among most member nations and founded a European bank. In 1999, the EU created a common **currency**, or form of money, called the **euro**. By 2011, 17 member nations had adopted this currency.

One of the requirements for joining the EU is having a stable democracy that respects human rights. Some countries that have applied for membership, such as Turkey, are still under review. However, other European countries, including

Norway, have chosen not to join. Norway does not want to give up its **sovereignty**, or control over its own affairs.

Before You Move On

Monitor Comprehension How has the European Union helped unite Europe?

ONGOING ASSESSMENT

MAP LAB

GeoJournal

1. **Interpret Maps** Study the map and recall what you have learned about Europe's geography. In what ways does Europe's geography encourage unity?

2. **Draw Conclusions** Based on the map, in what area are most of the members who joined in later years located? What political situation might have prevented them from joining earlier?

2.2 The Impact of the Euro

TECHTREK

myNGconnect.com For photos of the euro

Digital Library

Main Idea The euro has helped to unify Europe both economically and politically.

As you have learned, the European Union (EU) created the euro in 1999. Since then, many of the member nations have adopted the euro, which has had a significant impact on Europe.

The Euro Arrives

The euro was launched in 1999, but paper money and coins of the currency were not issued until 2002. The 17 countries that use the euro are known as the <mark>eurozone</mark>. Some countries that belong to the EU, including Romania, hope to join the eurozone soon. Other countries, including Great Britain and Denmark, have not adopted the euro. They believe that giving up their own currency might result in a loss of control over their economies.

The symbol for the euro is €. Euro paper money, or banknotes, is the same throughout the eurozone. Euro coins, however, differ from country to country. The front, or common, side of each coin has the same image and a number indicating its value. The back, or national, side shows a design that was chosen by the member nation.

Economic Benefits of the Euro

The euro allows people, money, and goods to move freely within the eurozone. Before the creation of the euro, when a French citizen traveled to Germany, for example, he or she had to pay a fee to <mark>exchange</mark>, or convert, francs—the French currency—into marks—the German currency. Because the common currency has made travel easier and less expensive, tourism has increased within much of Europe.

The prices of fruit in this Italian market are in euros.

A single currency means lower fees for conducting business. As a result, trade has increased among European nations by an estimated 10 percent since 2002. The currency has also made costs easier to compare for companies within the eurozone. As a result, companies can import the least expensive products and then pass along the savings to **consumers**, the people who buy the goods.

Political Benefits of the Euro

In 2010, the unity of the eurozone was tested. Greece and Ireland—countries in the eurozone—were deeply in debt. To help them manage their debt, the other eurozone nations loaned the two countries money. In return, Greece and Ireland had to raise taxes and reduce spending. By cooperating, the eurozone was able to help two of its member nations.

1-EURO COIN OF THREE EUROZONE COUNTRIES

Front	Back	
		**AUSTRIA** Mozart, Austrian composer
		**GERMANY** Eagle, German symbol
		**IRELAND** Harp, Irish symbol

Before You Move On

Summarize In what ways has the euro helped Europe unify economically and politically?

ONGOING ASSESSMENT

VIEWING LAB GeoJournal

1. **Compare and Contrast** Study the euros in the diagram. Note that the front of the coin is the same for all three countries shown. What elements appear on the front of each euro? What does each element represent?

2. **Make Inferences** Why might the countries have wanted their own design on the euro?

3. **Conduct Internet Research** Go online to find out what impact helping Greece and Ireland has had on the euro and the eurozone. Share your findings with the class.

2.3 Democracy in Eastern Europe

TECHTREK

myNGconnect.com For photos reflecting democratic progress in Eastern Europe

Digital Library

Main Idea Eastern European countries have faced many challenges in their transition to democracy.

After World War II, many eastern European countries came under the control of the Soviet Union. The citizens of these countries lacked democratic freedoms and had a low standard of living. In 1981, Poland rebelled peacefully against its Communist government. By the late 1980s, similar rebellions had spread throughout Eastern Europe. Finally, in 1991, Russia and several other republics declared their independence, and the Soviet Union collapsed.

The Road to Democracy

Since gaining their independence, Poland, Hungary, and the Czech Republic developed stable democratic governments In other countries, **democratization**, or the process of becoming a democracy, has been more difficult to achieve. In 1991, civil war broke out among ethnic groups in Yugoslavia. Over time, the country divided into several new democratic countries, including Serbia and Croatia.

Ukraine has also had setbacks. In 2004, the Ukrainian people staged the **Orange Revolution** and peacefully removed their prime minister, Viktor Yanukovych.

Critical Viewing Polish citizens shop and relax in a spacious mall in Warsaw. What does the shopping complex in this photo suggest about Poland's economy?

Many believed that he was corrupt and was being controlled by Russia. However, their new leader, Viktor Yushchenko, disappointed the Ukrainians. Some believed he had become anti-democratic and blamed him for their weakened economy. In 2010, the voters brought Yanukovych back to power.

Rebuilding Economies

The former Communist countries of Eastern Europe also began to rebuild their economies. They changed from government-controlled economies to market economies. They accomplished this goal through **privatization**. That means that government-owned businesses became privately owned.

Eastern European countries have had mixed results since making the adjustment to a market economy. Poland has had the greatest success. It has a fast-growing economy and exports goods throughout Europe. Other countries have been slower to establish new businesses and become competitive. They have also experienced rises in prices and unemployment.

The leaders of many eastern European countries wish to integrate with the rest of Europe. They want to join the European Union and NATO, a military alliance of democratic states in Europe and North America. While some citizens of eastern European countries believe that they were more secure under Communist leaders, others—particularly young people— disagree. They favor democracy and feel that this form of government can better help them solve their countries' problems.

Before You Move On

Summarize What challenges have eastern European countries faced in their transition to democracy and a market economy?

ONGOING ASSESSMENT

READING LAB   GeoJournal

1. **Identify Problems and Solutions** What problem did Ukraine face in its transition to democracy? In what way did the solution to the problem reflect democratization?

2. **Make Inferences** Why do you think younger eastern Europeans might be more willing than older people to support the democratic movements in their countries?

2.4 Changing Demographics

TECHTREK

myNGconnect.com For an online graph of Europe's changing demographics

Maps and Graphs   Global Issues

Main Idea New immigrants are changing Europe.

Every May, people from Germany, Denmark, Hungary, Bulgaria, and other European countries come together to celebrate Europe's diversity on Europe Day. The celebrations reflect Europe's changing **demographics**, the characteristics or the profile of a human population. The population has become more diverse as people from Africa and Asia have immigrated to Europe.

KEY VOCABULARY

demographics, n., the characteristics of a human population, such as age, income, and education

aging population, n., a trend that occurs as the average age of a population rises

ACADEMIC VOCABULARY

assimilate, v., to be absorbed into a society's culture

An Aging Population

For years, Europe has had an **aging population**. In other words, the average age of people on the continent has been rising. This trend has had a number of causes. For one thing, Europeans have been living longer because of better medical care. For another, most families are having fewer children. So today, senior citizens form a higher percentage of Europe's total population.

The trend created a need for more workers to replace the many senior citizens who were retiring. Workers were also needed to keep the economy strong and pay taxes to support such social services as education and health care. As a result, immigrants came to Europe to take the newly available jobs. People immigrated to Great Britain from former colonies, such as India, Pakistan, and Bangladesh. France also attracted immigrants from former colonies, especially Algeria and Morocco. In the 1970s, people from Turkey began coming to Germany for jobs. Germany now has approximately 2 million people of Turkish heritage. The fall of communism in Eastern Europe in the 1980s and 1990s also resulted in increased migration within Europe.

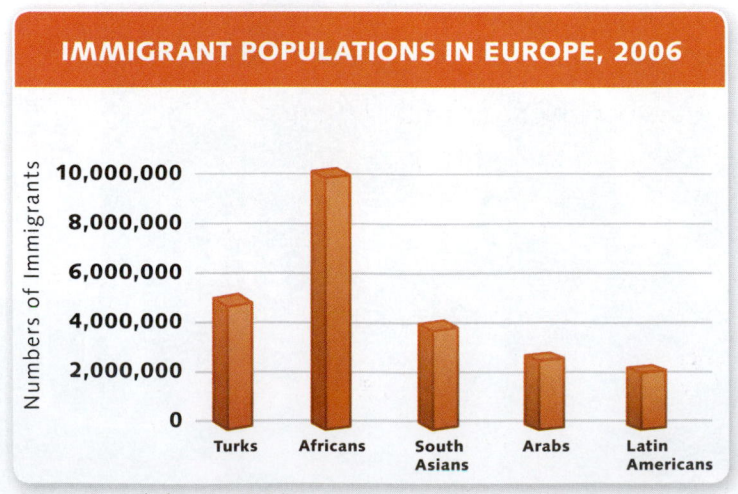

IMMIGRANT POPULATIONS IN EUROPE, 2006

Numbers of Immigrants: 0, 2,000,000, 4,000,000, 6,000,000, 8,000,000, 10,000,000

Turks, Africans, South Asians, Arabs, Latin Americans

Source: Council of Europe, 2006

Before You Move On

Summarize In what way did Europe's aging population create a need for immigrants?

Reasons for Immigration

Most immigrants have come to Europe to seek a better life. Some left their home countries for economic reasons—usually to find jobs. Others left for political reasons—to escape conflict in their countries or an unjust government. Once they have found work in Europe, many immigrants send money to relatives back home. Eastern Europeans have migrated to Western Europe for similar reasons. Membership within the European Union (EU) has made their migration easier.

Challenges of Immigration

Tensions have sometimes developed between immigrants and native citizens. These tensions often arise when the two groups compete for the same jobs. Also, some immigrants enter European countries illegally—according to some EU estimates, about a half million each year. Housing, educating, and caring for these illegal immigrants can create economic strains on the host countries.

The mix of different cultures has also created problems. Many immigrants are Muslims, with cultural and religious practices that differ from those of their Christian neighbors. Some Europeans would like to see the Muslim immigrants **assimilate**, or be absorbed into their society's culture. They believe the immigrants should adopt European traditions and values. Others believe a more multicultural approach is better. This approach encourages tolerance and embraces all cultures.

COMPARE ACROSS REGIONS

Australia's Skilled Immigrants

Like Europe, Australia has an aging population. Many in the Australian government believe that immigration can help change that trend. As a result, each year the government identifies gaps in the country's workforce. It then determines the number of skilled immigrants that can come to Australia. Between 2008 and 2009, more than 110,000 immigrants arrived on this skilled migration program. Most of these immigrants came from Great Britain, India, and China.

Of course, Europe receives many more immigrants than Australia each year. Italy alone took in more than 400,000 immigrants in 2008. Many Europeans appreciate the cultural enrichment that immigrants bring to their countries. However, some believe that Europe, like Australia, should set immigration limits.

Before You Move On
Monitor Comprehension What are the ways in which new immigrants are changing Europe?

ONGOING ASSESSMENT
READING LAB  GeoJournal

1. **Interpret Graphs** According to the bar graph, from which two continents has the immigrant population in Europe primarily come?

2. **Make Inferences** Why might some immigrants resist assimilation?

3. **Identify Problems and Solutions** What has Australia done to solve some of the problems posed by immigration?

VOCABULARY

Match each word in the first column with its meaning in the second column.

WORD	DEFINITION
1. abstract	a. tax on imports and exports
2. dialect	b. control over one's own affairs
3. sovereignty	c. art style emphasizing form and color
4. tariff	d. absorb into another culture
5. privatization	e. privately owned businesses
6. assimilate	f. regional language

MAIN IDEAS

7. Why are there so many different dialects in Europe? (Section 1.1)

8. What is the fastest-growing religion in Europe? Why might that be so? (Section 1.1)

9. During the Romantic period, what themes did artists use in their paintings? (Section 1.2)

10. Why do you think the ancient Greeks and Romans celebrated historic events in epic poems? (Section 1.3)

11. How does the soup called *borscht* reflect Russia's geography? (Section 1.4)

12. What is one of the requirements for joining the European Union? (Section 2.1)

13. In what way has the euro helped increase trade among European nations? (Section 2.2)

14. Why did the Ukrainian people stage the Orange Revolution? (Section 2.3)

15. What factors have led people from other parts of the world to immigrate to Europe? (Section 2.4)

CULTURE

ANALYZE THE ESSENTIAL QUESTION

How is the diversity of Europe reflected in its cultural achievements?

Critical Thinking: Draw Conclusions

16. In what ways do the many dialects in Europe show its great diversity?

17. What aspects of ancient Greek and Roman art inspired Renaissance artists?

18. Why did pasta with tomato sauce develop in Italy rather than Russia?

INTERPRET MAPS

CHRISTIANITY IN EUROPE

- Protestant
- Roman Catholic
- Eastern Orthodox
- Other religions

19. Region Where do most Protestants live in Europe? Where do most Catholics live?

20. Make Inferences Find the countries on the map in which a relatively small part of the population belongs to a different Christian denomination. What challenges might the people in the minority religion face?

GOVERNMENT & ECONOMICS

ANALYZE THE ESSENTIAL QUESTION

What are the costs and benefits of European unification?

Critical Thinking: Analyze Cause and Effect

21. In what way did the Marshall Plan help bring about the formation of the European Union?

22. What was the response of the eurozone countries when Greece and Ireland became deeply in debt in 2010?

23. What was the impact of the fall of communism on Eastern Europe? What was the impact on Western Europe?

24. What problems are caused by illegal immigration to Europe?

INTERPRET CHARTS

COST OF A TEN-MINUTE PHONE CALL TO THE U.S. IN EUROS (€)*		
Country	1997	2006
Belgium	7.50	1.98
Czech Republic	3.09	2.02
Denmark	7.41	2.38
Ireland	4.61	1.91
Spain	6.17	1.53
France	6.78	2.32
United Kingdom	3.50	2.23

* 1997 prices have been converted to euros
Source: Eurostat

25. **Analyze Data** According to the chart, how did the cost of making a telephone call change between 1997 and 2006?

26. **Analyze Cause and Effect** What move made by the European Union might have brought about the change in the cost of a telephone call?

ACTIVE OPTIONS

Synthesize the Essential Questions by completing the activities below.

27. **Write a Speech** Suppose that you are the leader of a European country that has been invited to join the European Union. Write a speech that will persuade the citizens of your country to vote in favor of joining. **Deliver your speech to the class and ask the members of your audience to vote on whether they are in favor of joining the European Union.**

> **Writing Tips**
> - Take notes on three benefits that would result from joining the European Union.
> - Support each benefit using facts and statistics.
> - Address any concerns your audience might have about joining and explain why the advantages outweigh any disadvantages.

TECHTREK myNGconnect.com For photos of European art

28. **Create an Art Chart** Select three European works of art. You can choose Raphael's *School of Athens* or Leonardo da Vinci's *Mona Lisa* from the **Digital Library**, or you can search for other works online. Then research each piece to find out what period it is from and what theme, or subject matter, it represents. Copy the chart below to help you organize your information. Display the artwork and your findings on a poster. Be prepared to explain the relationship between each work's period and theme.

WORK OF ART	PERIOD	THEME

World Languages

TECHTREK
myNGconnect.com For an online graph and research links on world languages

 Maps and Graphs

 Connect to NG

In this unit, you learned about the diversity of languages in Europe. Every culture in the world uses language to communicate. Scholars estimate that there are about 7,000 languages spoken today.

Most countries name one or more languages as official languages. An official language is the one used by a country's government. For example, French is the official language of France, and English and French are the official languages of Canada. Most countries have groups of people whose first language differs from the official language. It is estimated that at least half of the people in the world speak one or more languages in addition to their first language.

Compare

- Africa
- Americas
- Asia
- Europe
- Oceania

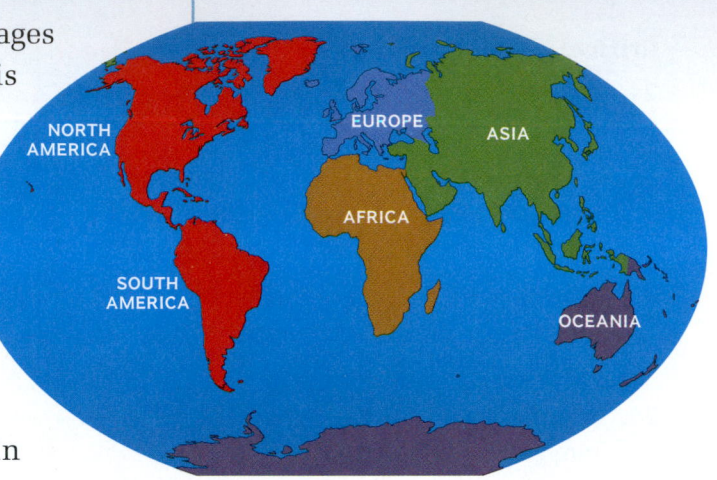

LIVING LANGUAGES

Although there are thousands of world languages, many are not widely spoken. In addition, sometimes the distinction, or difference, between a language and a dialect is not clear.

The following are the ten most widely spoken languages in the world in order of their ranking. Each is spoken as a native language by at least 100 million people. Some are official languages in widely different regions of the world. Some of the languages appear in only one region of the world.

1. Mandarin Chinese
2. Spanish
3. English
4. Hindi/Urdu
5. Arabic
6. Bengali
7. Portuguese
8. Russian
9. Japanese
10. German

DYING LANGUAGES

Some languages are spoken by so few people that they are in danger of dying out. In fact, linguists, or people who study languages, estimate that one language dies every two weeks.

Languages die for different reasons. Some simply disappear with the death of the last speaker. Others fade more slowly as a dominant language replaces it. A few linguists are trying to document some of these languages to preserve the history and culture of the people who spoke them.

The graph at right shows the number of languages spoken on the world's continents, as well the names of a few of the languages spoken. Note that "Americas" consists of the number of languages spoken on the North American and South American continents. Compare the data in the graph and use it to answer the questions.

NUMBER OF LANGUAGES SPOKEN BY CONTINENT

Source: *Ethnologue*, 16th Edition, 2009

EUROPE
AFRICA
ASIA
AMERICAS
OCEANIA

2,322

2,110

1,250

993

234

▷Hola
▷Guten tag

Hello

▷Kia ora   ▷Hujambo

▷Ohayou

RESEARCH LAB   GeoJournal

1. **Compare and Contrast** On which continent are the most languages spoken? the fewest? What do these numbers suggest about the cultural unity of each continent?

2. **Analyze Data** Study the graph and the list of languages with the most native speakers. How many of the total languages in Europe are among those spoken most in the world?

Research and Create Charts Research to find out more about the people who speak Hindi and Portuguese. Create a chart for each language showing approximately how many people speak it and where it is spoken. Which language has more native speakers? Which language is an official language in more places? What might account for this?

Active Options

TECHTREK

myNGconnect.com For photos of Renaissance art, nuclear power plants in Europe, and European cuisine

 Digital Library   **Connect to NG**   **Magazine Maker**

ACTIVITY 1

Goal: Extend your understanding of Renaissance art.

Write a Renaissance Arts Magazine

The Renaissance was a period of great artistic activity in Europe. Choose a city in Europe that was influenced by the Renaissance between 1400 and 1600. With a group, plan and publish a magazine showcasing that city's artistic achievements. Use the Magazine Maker to find photos and information on the following:

- art
- architecture
- literature
- fashion

Brunelleschi's dome atop the Cathedral of Florence in Italy

ACTIVITY 2

Goal: Research the use of nuclear power in Europe.

Create a Pro-and-Con Chart

Some European countries are planning to build new plants, while others have chosen to close existing plants. Use the research links at **Connect to NG** to create a pro-and-con chart that explains some of the advantages and disadvantages of nuclear power. Be prepared to present your chart and explain the issues.

ACTIVITY 3

Goal: Learn about European culture through its food.

Plan a Dinner Menu

Get together in a group and plan a dinner menu featuring typical European dishes. Discuss what the courses for the meal will be. Each group member should be in charge of one course, each of which should come from a different European country. Design a poster presentation of the menu.

Explore
Russia & THE EURASIAN REPUBLICS
WITH NATIONAL GEOGRAPHIC

MEET THE EXPLORER

NATIONAL GEOGRAPHIC

Tracing ancient trade routes, NG Fellow Fredrik Hiebert excavated a 4,000-year-old Silk Road city in Turkmenistan. He also searches for underwater settlements in the Black Sea.

INVESTIGATE GEOGRAPHY

North of the Arctic Circle in Russia, reindeer herded by a nomadic clan charge across the tundra. Many groups in the Arctic rely heavily on the reindeer. The animals, also known as caribou, provide food, clothing, and shelter, as well as transportation.

CONNECT WITH THE CULTURE

A young Mongolian Kazak family stands in front of a yurt, a shelter used by nomads. Kazaks are a nomadic, animal-herding people. Yurts allow them to live in harsh climates and give them the freedom to move and graze their animals in traditional tribal communities.

4,859 miles

Moscow, Russia

Washington, D.C.

Go to **myNGconnect.com** for maps of Russia and the Eurasian Republics.

STEP INTO HISTORY

Saint Basil's Cathedral in Red Square, Moscow, Russia, was built between 1554 and 1560. The cathedral has nine chapels, each topped with an onion-shaped dome.

CHAPTER 5

Russia & the Eurasian Republics
GEOGRAPHY & HISTORY

PREVIEW THE CHAPTER

Essential Question How have size and extreme climates shaped Russia and the Eurasian republics?

SECTION 1 • GEOGRAPHY

KEY VOCABULARY

- landmass
- steppe
- permafrost
- tundra
- taiga
- nonrenewable fossil fuel
- peat
- hydroelectric power
- methane
- greenhouse gas
- semiarid
- arid
- pesticide

ACADEMIC VOCABULARY
isolate, remote

TERMS & NAMES

- Ural Mountains
- North Atlantic Drift
- Siberia
- Black Sea
- Caspian Sea
- Aral Sea

Essential Question How has geographic isolation influenced the region's history?

SECTION 2 • HISTORY

KEY VOCABULARY

- state
- tribute
- czar
- reign
- secular
- invader
- scorched earth policy
- serf
- strike
- communism
- socialism
- collective farm
- propaganda

ACADEMIC VOCABULARY
expand, promote

TERMS & NAMES

- Slav
- Kievan Rus
- Genghis Khan
- Mongol Empire
- Silk Roads
- Peter the Great
- Catherine the Great
- Nazi Germany
- Industrial Revolution
- V. I. Lenin
- Bolshevik
- Russian Revolution
- Soviet Union
- Cold War
- Mikhail Gorbachev

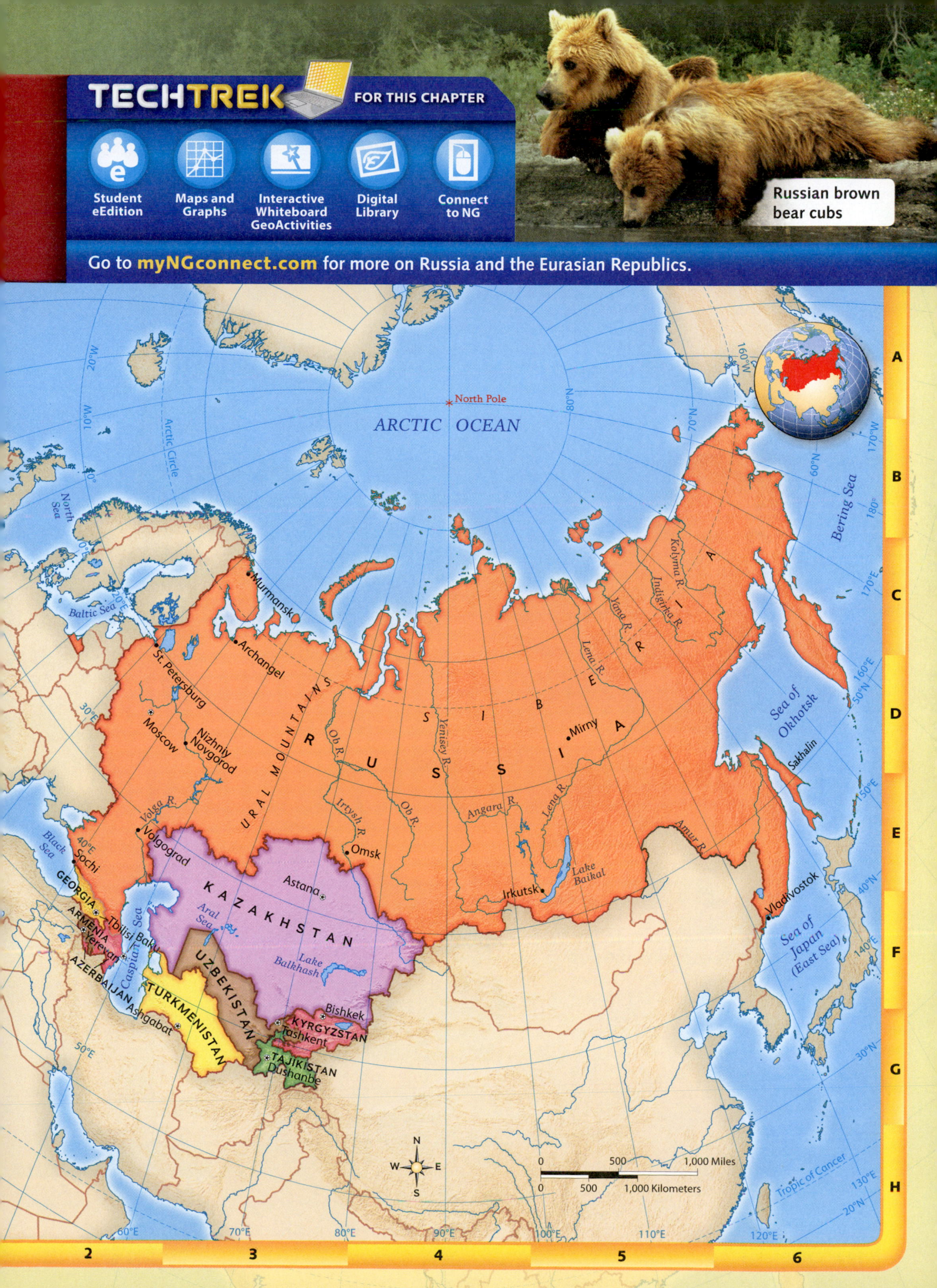

TECHTREK FOR THIS CHAPTER

Student eEdition

Maps and Graphs

Interactive Whiteboard GeoActivities

Digital Library

Connect to NG

Go to **myNGconnect.com** for more on Russia and the Eurasian Republics.

Russian brown bear cubs

ARCTIC OCEAN

North Pole

Arctic Circle

Bering Sea

North Sea

Baltic Sea

Murmansk

Archangel

St. Petersburg

Moscow

Nizhniy Novgorod

URAL MOUNTAINS

R   U   S   S   I   A

S I B E R I A

Ob R.

Ob R.

Irtysh R.

Yenisey R.

Angara R.

Lena R.

Lena R.

Kolyma R.

Indigirka R.

Yana R.

Mirny

Omsk

Irkutsk

Lake Baikal

Amur R.

Sea of Okhotsk

Sakhalin

Sea of Japan (East Sea)

Vladivostok

Volga R.

Volgograd

Astana

KAZAKHSTAN

Aral Sea

Lake Balkhash

Caspian Sea

Black Sea

Sochi

GEORGIA

Tbilisi

ARMENIA

Yerevan

AZERBAIJAN

Baku

TURKMENISTAN

Ashgabat

UZBEKISTAN

Tashkent

Bishkek

KYRGYZSTAN

TAJIKISTAN

Dushanbe

Tropic of Cancer

N W E S

0   500   1,000 Miles

0   500   1,000 Kilometers

20°W
10°W
30°E
40°E
50°E
60°E
70°E
80°E
90°E
100°E
110°E
120°E
130°E
140°E
150°E
160°E
170°E
180°
170°W
160°W
20°N
30°N
40°N
50°N
60°N
70°N
80°N

A B C D E F G H

2 3 4 5 6

SECTION 1 GEOGRAPHY

1.1 Physical Geography

TECHTREK

myNGconnect.com For maps of Russia and the Eurasian republics and Visual Vocabulary

Maps and Graphs

Digital Library

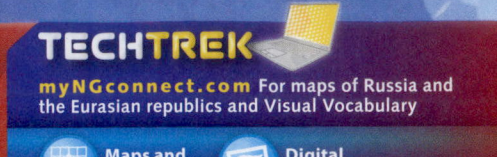

RUSSIA AND THE EURASIAN REPUBLICS PHYSICAL

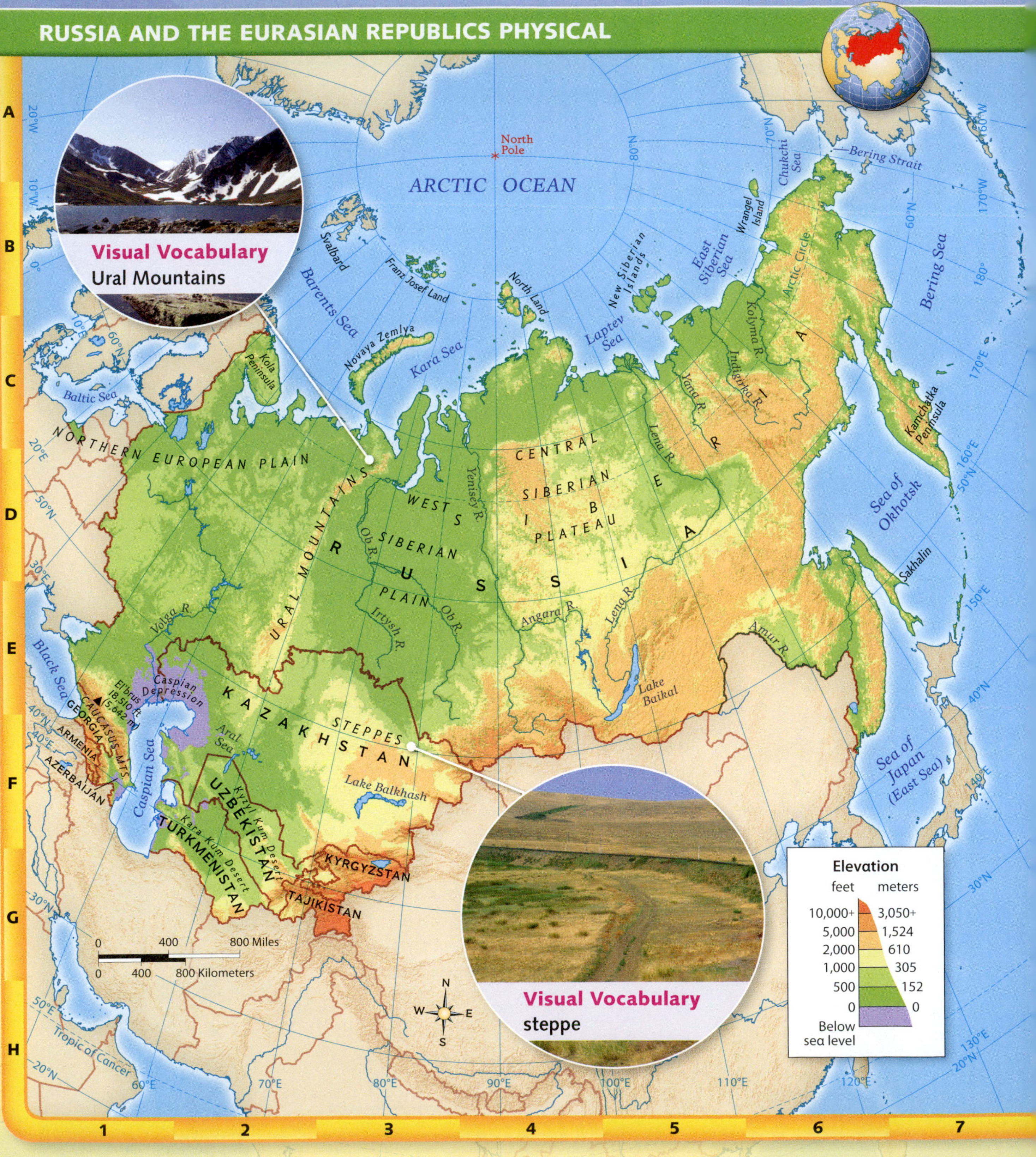

Visual Vocabulary
Ural Mountains
Visual Vocabulary
steppe
North Pole
ARCTIC OCEAN
Svalbard
Franz Josef Land
North Land
New Siberian Islands
Wrangel Island
Chukchi Sea
Bering Strait
East Siberian Sea
Laptev Sea
Kara Sea
Barents Sea
Novaya Zemlya
Bering Sea
Arctic Circle
Kolyma R.
Indigirka R.
Yana R.
Lena R.
Kamchatka Peninsula
Sea of Okhotsk
Sakhalin
Amur R.
Sea of Japan (East Sea)
Kola Peninsula
Baltic Sea
NORTHERN EUROPEAN PLAIN
URAL MOUNTAINS
WEST SIBERIAN PLAIN
CENTRAL SIBERIAN PLATEAU
RUSSIA
Yenisey R.
Ob R.
Irtysh R.
Angara R.
Lake Baikal
Volga R.
Black Sea
Caspian Depression
El'brus (18,510 ft.) (5,642 m)
CAUCASUS MTS.
GEORGIA
ARMENIA
AZERBAIJAN
Caspian Sea
Aral Sea
KAZAKHSTAN
STEPPES
Lake Balkhash
UZBEKISTAN
Kyzyl Kum Desert
Kara-Kum Desert
TURKMENISTAN
KYRGYZSTAN
TAJIKISTAN
Tropic of Cancer
0 400 800 Miles
0 400 800 Kilometers
N S E W
Elevation
feet meters
10,000+ 3,050+
5,000 1,524
2,000 610
1,000 305
500 152
0 0
Below sea level
A B C D E F G H
1 2 3 4 5 6 7

Main Idea Russia and the Eurasian republics cover a huge area and contain a variety of geographic features.

Russia and the Eurasian republics take up about one-sixth of the land surface of the entire earth. The region's geographic features have limited its population.

Huge Landmass

Russia is the largest country in the world in total area. Its **landmass**, or continuous extent of land, stretches almost 6,000 miles from east to west. The Eurasian republics lie south of Russia. The republics in the Caucasus Mountains are Armenia, Azerbaijan (ahz ur by JAHN), and Georgia. Those in Central Asia include Kazakhstan (kah zahk STAHN), Kyrgyzstan (kihr gih STAN), Tajikistan (tah jik ih STAN), Turkmenistan (turk MEN ih stan), and Uzbekistan (ooz BEK ih stan).

Plains, or large areas of level ground, cover much of the region. The relatively low **Ural Mountains** separate the Northern European Plain from the West Siberian Plain. In much of southwestern Russia and northern Kazakhstan, the very large plains are called **steppes**. Much of this land is good for agriculture and grazing.

Natural Barriers

Russia has oceans on its northern and eastern borders and mountainous areas along much of its southern border. These natural barriers separate Russia from its neighbors. The deserts and mountains of Central Asia keep republics such as Kyrgyzstan and Tajikistan **isolated**, or cut off from other countries.

POPULATION DENSITY

One dot represents 50,000 people

Because so much of Russia's coastline lies to the north of the Arctic Circle, few ports stay open to ships and trade all year long. Murmansk, located far to the north on Kola Peninsula, is one such port. An ocean current called the **North Atlantic Drift** warms the waters around Murmansk and keeps them ice-free most of the time.

Before You Move On

Monitor Comprehension What are some of the key geographic features of this region?

1. **Place** Find the Ural Mountains on the physical map. What is the elevation of the mountains? Locate the steppes. How does its elevation compare with that of the Ural Mountains?

2. **Interpret Maps** Use the population density map to determine how density differs west and far east of the Urals. Then study the physical map. What physical features might contribute to the population distribution?

1.2 Land of Extreme Climates

TECHTREK
myNGconnect.com For a climate map and photos of extreme climates
Maps and Graphs
Digital Library

Main Idea The extreme climates of this region have an impact on where and how people live.

About half the land in Russia is so cold that it has **permafrost**, or permanently frozen ground, beneath it. Yet parts of Russia can also reach 100°F in summer, and large areas of Central Asia are desert. Because of these extremes, most of the population lives in the western part of the region, where the climate is not as harsh.

Cold, Dark Winters

Latitude is an important factor in the climate of a region. The northern boundary of Russia is a coastal plain along the Arctic Ocean, with no natural barriers to keep out arctic winds. The high northern latitudes of Moscow and areas to its north help bring this region long, dark, snowy winters. St. Petersburg, for example, has a latitude of almost 60° N. For about one month each winter, there is has hardly any daylight in the city.

Climate and Vegetation

Climate affects the types of vegetation that grow in different areas. **Tundra**, or flat land found in arctic and subarctic regions, exists in **Siberia**, which lies in central and eastern Russia. Here, only small plants can grow. Permafrost prevents most trees from growing because their roots can't spread deep under the ground.

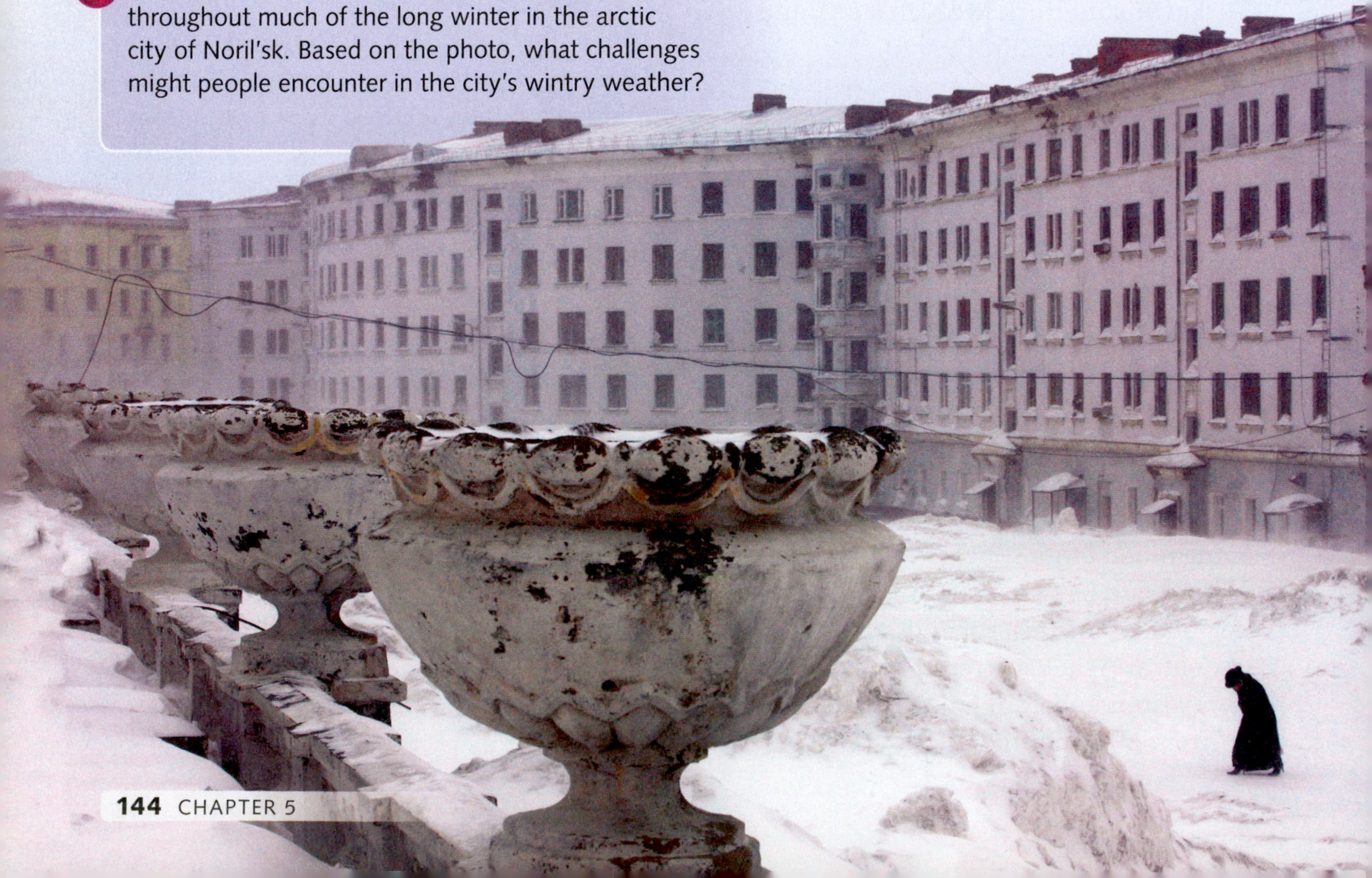

Critical Viewing Snow and ice cover the ground throughout much of the long winter in the arctic city of Noril'sk. Based on the photo, what challenges might people encounter in the city's wintry weather?

ARCTIC OCEAN

North Pole

PACIFIC OCEAN

Bering Sea

Sea of Okhotsk

Sea of Japan (East Sea)

Lake Baikal

S I B E R I A

R U S S I A

Moscow

St. Petersburg

Noril'sk

Black Sea

Caspian Sea

GEORGIA

ARMENIA

AZERBAIJAN

TURKMENISTAN

UZBEKISTAN

KYRGYZSTAN

TAJIKISTAN

K A Z A K H S T A N

Aral Sea

Lake Balkhash

Arctic Circle

Climate Regions	
Dry–Semiarid	Humid Cold–No dry season
Dry–Arid	Humid Cold–Dry winter
Humid Temperate–No dry season	Cold Polar–Tundra & ice
	Unclassified Highlands

0   400   800 Miles

0   400   800 Kilometers

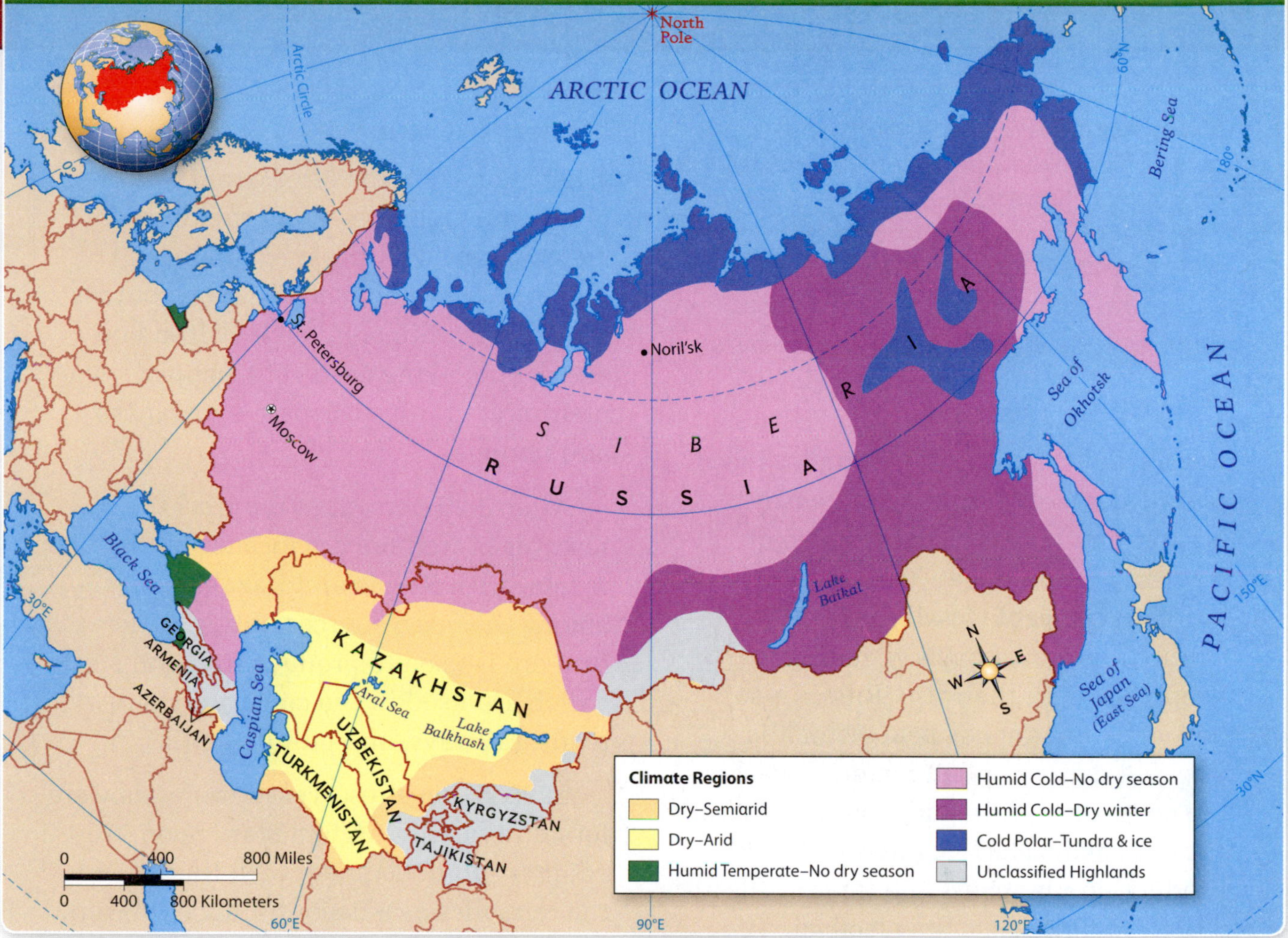

Just south of the tundra is the **taiga** (TY guh), or forest area. Mostly small evergreens like pines grow here. This area provides valuable timber resources.

Extremes in temperature and moisture make it hard for some areas to be used for agriculture. Much of the northern territory has short summers and, as a result, short growing seasons. The semiarid and desert areas are limited to herding and grazing. Farming is concentrated in the fertile soils of the western plains and steppes, along the **Black Sea**, the **Caspian Sea**, and in some river valleys.

Before You Move On

Summarize In what ways do this region's extreme climates affect the people who live there?

ONGOING ASSESSMENT
MAP LAB
 GeoJournal

1. **Location** What is the climate like in the northeastern part of Russia? How does latitude affect the climate in this region?

2. **Interpret Maps** What areas in the region have arid or semiarid climates? What areas have a humid cold climate with dry winters?

3. **Pose and Answer Questions** Study the map and then create another climate question to ask a partner.

4. **Human-Environment Interaction** Look at the population map in Section 1.1 and then compare it with the climate map on this page. In what ways might climate determine where people live in the region?

1.3 Natural Resources

Main Idea Russia and the Eurasian republics have plentiful natural resources, but many of them are in remote locations.

This region is among the world's richest in natural resources. These resources are important for the countries' economies.

Energy Resources

Russia and the Eurasian republics have plentiful energy resources, especially oil and natural gas. Russia is also a leading coal producer. These resources are **nonrenewable fossil fuels**. They cannot reproduce quickly enough to keep pace with their use. Russia also has large amounts of **peat**, which is very old decayed plant material. Peat is burned like coal. In addition, some rivers provide **hydroelectric power**. Power production plants use the force of the rivers' water to generate electricity.

Mineral Resources

The region contains large quantities of mineral resources that provide raw materials for factories and support industrial development. These resources include metallic ores, such as iron and aluminum, along with gold, copper, platinum, uranium, cobalt, manganese, and chrome.

Almost 20 percent of the world's reserves of iron ore are located in the region, with Russia and Kazakhstan among the main sources of this mineral. Iron ore is used to produce iron and steel, which are used in the construction of roads, railways, and buildings.

In 2010, huge reserves of minerals were found in nearby Afghanistan. Soon, this country might compete with Russia and the republics as a major producer of iron, copper, and other metals.

> **Critical Viewing** Open-pit mines, like this one in Siberia, are dug when the diamonds are near the surface. What benefits might this mine bring to the town?

The mine is in the often snowy town of Mirny. **1**

Stepped walls help prevent landslides.

A ramp allows trucks to carry away rocks and diamonds.

Water often pools at the bottom of the mine.

Map legend:

Fish		Iron		Aluminum	
Coal		Copper		Hydropower	
Oil		Gold		Diamonds	
Natural gas		Lead		Forest products	
Zinc		Nickel			

The Challenge of Location

Much of the region's resources are in **remote**, or hard to reach, locations. For example, Siberia contains oil fields, hydroelectric power sources, and minerals, such as nickel and gold. Many of these resources are located in the far-eastern, and coldest, parts of Siberia. The permafrost there makes it difficult to drill or mine for the natural resources and transport them to market. As a result, the resources in these areas of Siberia remain largely untouched.

Before You Move On

Summarize What are some important natural resources of Russia and the Eurasian republics, and why are some of them hard to reach?

ONGOING ASSESSMENT

DATA LAB — GeoJournal

Create Charts You have learned that Russia and the Eurasian republics contain plentiful mineral resources that provide raw materials for factories. Identify the region's mineral resources in the map above. Then research to find out what consumer goods are made from these minerals. Create a chart like the one below to record your findings.

MINERALS	GOODS
Iron	steel, medicine, magnets, auto parts, paper clips
Aluminum	kitchen utensils, drink cans, foil, airplane and car parts

1.4

SECTION **1** GEOGRAPHY

NATIONAL GEOGRAPHIC

TECHTREK

myNGconnect.com For photos of the explorer's work and an Explorer Video Clip

 Digital Library

Exploring Siberian Lakes

with Katey Walter Anthony

Main Idea As permafrost thaws in Siberia, it releases methane gas into the atmosphere.

Expedition to Siberia

Emerging Explorer Katey Walter Anthony first went to Siberia as a high school exchange student. Now she works with other scientists from Alaska and Russia at the Northeast Science Station in Cherskiy, Siberia, **1** ▶ to study how climate change is affecting the area—and possibly the entire world. Siberia's frigid climate makes the work difficult.

Anthony and some other scientists are concerned about Siberia's permafrost. They believe that because of global warming, the permafrost below its lakes is thawing, releasing carbon that was locked inside the frozen ground. Carbon is formed from dead prehistoric animals and the plants they ate. The carbon is then turned into **methane**, a colorless, odorless natural gas that can have a negative impact on the environment.

Anthony checks for methane on a frozen lake.

ANTHONY'S EXPEDITIONS

East Siberian Sea

Cape Chukochiy

Cape Baranov

• Ambarchik

Kolymskoye

1 ◀

Cherskiy

Nizhnekolymsk

NORTHEAST SCIENCE STATION

Bilibino

Kobma R.

Omolon R.

Greater Anyuy R.

Lesser Anyuy R.

N W E S

0   50   100 Miles

0   50   100 Kilometers

PROCESS OF METHANE RELEASE

3 Methane is released into the atmosphere.

2 Methane is released into the lake as the permafrost thaws.

1 Organic matter is trapped in the permafrost.

soil

frozen lake

sediments

permafrost

METHANE RELEASE Warming temperatures thaw the permafrost and release methane gas. The methane heats the air as it is released into the atmosphere as a greenhouse gas.

Methane and Climate Change

Methane is a greenhouse gas, a gas that traps the sun's heat over the earth. Anthony explains, "It's 25 times more powerful than carbon dioxide on 100-year time scales" and could have the most powerful effect of all on global warming. Siberian lakes could release about ten times as much methane as is in the atmosphere now. Some experts believe this could cause temperatures across the world to rise higher and faster.

To check for methane, Anthony chops holes into the ice on the lakes to collect gas samples that she brings back to the lab. Sometimes she wants to know right away what gases might be present, so she lights a match. If a giant flame shoots up, she knows she's found methane.

Before You Move On
Monitor Comprehension What impact does methane have on the atmosphere?

ONGOING ASSESSMENT

VIEWING LAB GeoJournal

1. **Analyze Visuals** Go to the **Digital Library** to view the video clip on Emerging Explorer Katey Walter Anthony. How might her work in Siberia be applied to other places in the world?

2. **Interpret Models** Study the model. What does the methane have to pass through to enter the atmosphere? Why does the gas penetrate this material so easily?

3. **Analyze Cause and Effect** List some causes and effects of methane release using a chart like the one below. Why do some scientists believe that increasing methane in Siberian lakes is both a cause and an effect of global warming?

CAUSES	EFFECTS

1.5 Central Asian Landscapes

TECHTREK
myNGconnect.com For online maps
and photos of Central Asia

Maps and Graphs

Digital Library

Main Idea Human activities have led to the shrinking of the Aral Sea, which has damaged the surrounding landscape.

As you have learned, methane gas is threatening Siberia's environment. Central Asia's landscape has also been damaged. Human activities have nearly destroyed one of its most important bodies of water.

Adapting to Dry Conditions

Central Asia includes landforms such as deserts, mountains, forests, and steppes. Though the area doesn't experience the extreme cold of northern Russia, there are places in northern Kazakhstan that can reach 0°F in winter. In general, summers are hot and longer than in the northern parts of the region. Central Asia's temperatures vary so widely in part because it is not protected by a large body of water, which would help to keep temperatures moderate.

Large parts of Central Asia are **semiarid** or **arid**, meaning there is little or no rainfall. These dry lands are best suited for livestock grazing. Because of irrigation in some river valleys, however, farmers have also been able to grow crops such as cotton.

The Shrinking Aral Sea

Major efforts to grow cotton led to the shrinking of the **Aral Sea** in Kazakhstan and Uzbekistan. The body of water is actually a salt-water lake. The rivers that fed the lake were redirected into canals for irrigation. The Aral Sea was once the fourth largest lake in the world, but now it is only a fraction of the size it was in 1960.

Pollution contributed to problems in the Aral Sea. Fertilizer and **pesticides**, which are chemicals that kill harmful insects and weeds, ran into it. As the lake shrank, salt and pesticides destroyed the habitat of many plants and animals and threatened human health. The lake's once-thriving fishing industry was also damaged. One resident of the area said, "My father and grandfather were fishermen in this town, but as you can see, the boats are now sitting in the middle of a desert."

In 2005, Kazakhstan, with the help of the World Bank, built a dam to save the North Aral Sea. That part has increased in size, and fishing has returned to the area. However, the southern part of the lake, in Uzbekistan, is almost completely gone.

Before You Move On
Monitor Comprehension What human activities caused the Aral Sea to shrink, and what damage to the landscape has occurred as a result?

> **Critical Viewing** Camels walk past a stranded ship on land where the Aral Sea used to be. What does this picture suggest about the Aral Sea?

North Aral Sea, 1973

Aral Sea, 1973

This satellite image shows what the Aral Sea looked like in 1973. At this time, the North and South Aral seas were full of water.

North Aral Sea, 2009

Aral Sea, 2009

This satellite image shows what the Aral Sea looked like in 2009. The photo reveals that efforts to save the North Aral Sea have worked.

ONGOING ASSESSMENT

PHOTO LAB · GeoJournal

1. **Analyze Visuals** Study the photos above of the Aral Sea. How much time passed between the photo on the left and the photo on the right? What happened during that time period?

2. **Identify** What details in the photo below tell you that this part of the Aral Sea has been devastated for a long time?

3. **Human-Environment Interaction** Why did people redirect the Aral Sea? What steps have been taken to fix what happened as a result?

2.1 Early History

TECH TREK

myNGconnect.com For an online map of the Silk Roads and photos of artifacts

Maps and Graphs

Digital Library

Main Idea Settlers and conquerors from Europe and Asia shaped the early history of Russia and the Eurasian republics.

Settlement of Russia by different groups dates back to around 1200 B.C. At that time, a people called the Cimmerians lived north of the Black Sea in what is now southern Ukraine. Over hundreds of years, many other groups ruled this region.

Kievan Rus

The people whose culture had the most lasting influence on early Russia were the **Slavs** (slahvz). Some historians think they were farmers near the Black Sea around 700 B.C. or earlier. Others think they came from Poland around the A.D. 400s. By the A.D. 800s, the Slavs had built towns near rivers in Ukraine and western Russia.

In 862, Vikings from Scandinavia called the Varangian Russes (vahr ANG ee ehn ROOS ehz) took control of the Slavic town of Novgorod. Russia's name may have come from this tribe. About 20 years later, a Varangian prince captured Kiev (KEE ehf). He established a **state**, or defined territory with its own government, that came to be known as **Kievan Rus**.

Mongol Rule and Trade

By the late 1100s, Kiev's power had declined. Struggles for control within the ruling family weakened the state.

In the early 1200s, **Genghis Khan** (JEHNG gihs KAHN) established the **Mongol Empire** in Central Asia. In 1240, his grandson, Batu Khan, extended the empire by taking over Kievan Rus and much of Russia.

Russian princes had to declare their loyalty and pay **tribute**, or taxes, to the Mongol ruler, the khan. Mongols selected one of the princes to serve as grand prince and represent Russian interests.

The empire isolated Russia from European influence for more than 200 years. However, Mongol rule kept Russia and Central Asia open to the East. The **Silk Roads**, ancient trade routes, carried goods and new ideas throughout the Mongol Empire. As you can see on the map on the next page, the Silk Roads connected Southwest Asia and Central Asia with China. Goods traded included gold, jade, and silk. Merv, in Turkmenistan, and Samarkand, in Uzbekistan, were important stops on the Silk Roads.

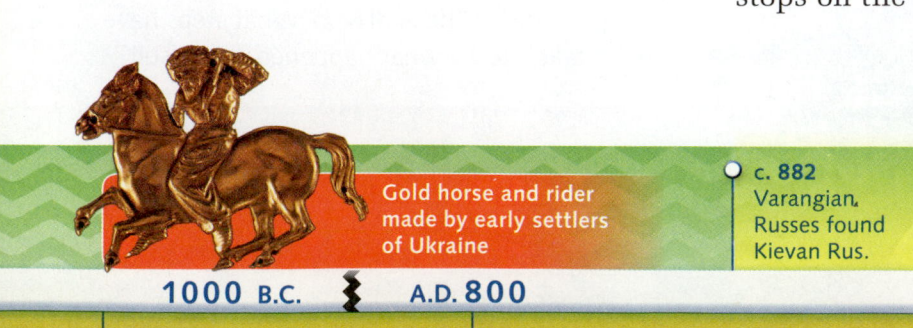

Gold horse and rider made by early settlers of Ukraine

c. 882 Varangian Russes found Kievan Rus.

1000 B.C. | **A.D. 800** | **1000**

1200 B.C. Cimmerians settle in southern Ukraine.

A.D. 800s Slavs build towns on the Dnieper River.

Ceiling of St. Sophia, Kievan Rus cathedral

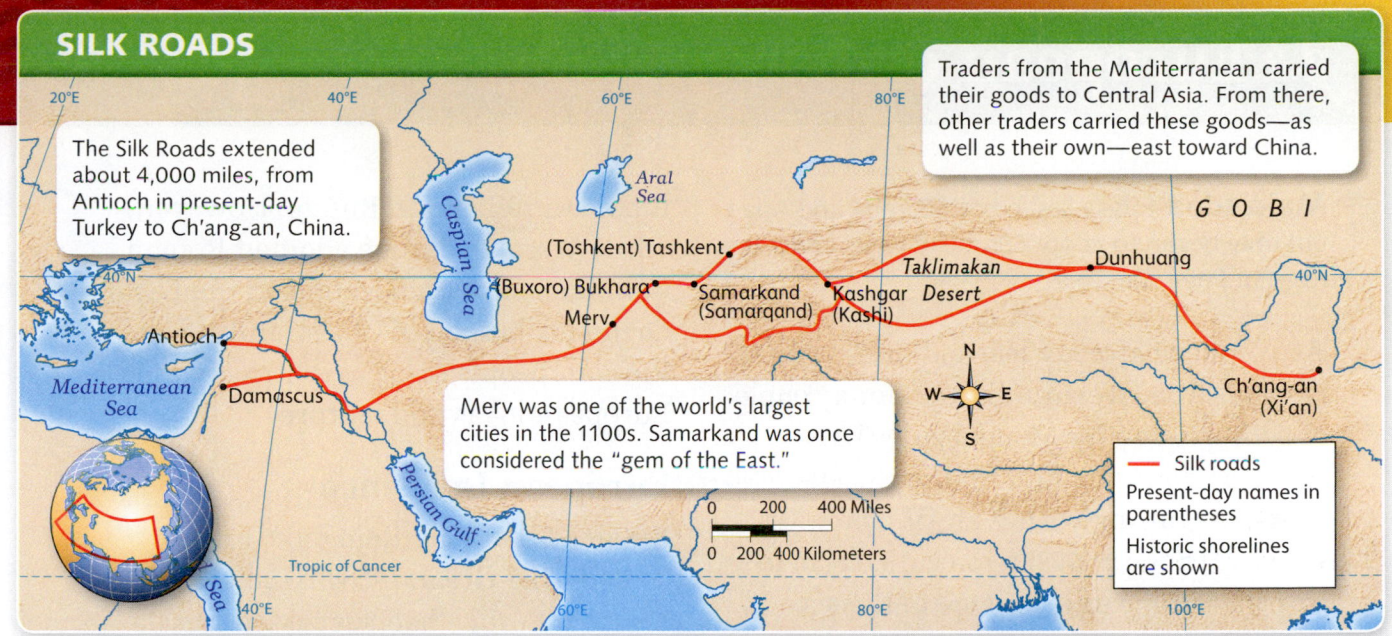

SILK ROADS

The Silk Roads extended about 4,000 miles, from Antioch in present-day Turkey to Ch'ang-an, China.

Traders from the Mediterranean carried their goods to Central Asia. From there, other traders carried these goods—as well as their own—east toward China.

Merv was one of the world's largest cities in the 1100s. Samarkand was once considered the "gem of the East."

Silk roads
Present-day names in parentheses
Historic shorelines are shown

The Rise of Moscow

Around 1330, the Mongols allowed Grand Prince Ivan I of Moscow to collect tribute for them. Ivan kept some of the money for himself and used it to buy land and **expand** his territory. Moscow became stronger as Mongol rule weakened.

In 1380, Grand Prince Dmitry defeated the Mongols in battle. In 1480, Ivan III—also called Ivan the Great—refused to pay tribute to the Mongols, and they never demanded it again. This action ended Mongol rule in Russia. Ivan the Great's grandson, Ivan IV, expanded Russia and made it into an empire. He became Russia's first **czar** (zahr), or emperor.

Before You Move On
Summarize How did settlers and conquerors shape the early history of Russia?

ONGOING ASSESSMENT
MAP LAB
GeoJournal

1. **Place** Why do you think the main route of the Silk Roads split in two between the cities of Kashgar and Dunhuang?

2. **Interpret Maps** Study the map. Why was Samarkand an important stop on the routes?

3. **Analyze Cause and Effect** What actions strengthened Moscow and what resulted from its rise? Use a chart like the one below to list the causes and effects.

CAUSES	EFFECTS

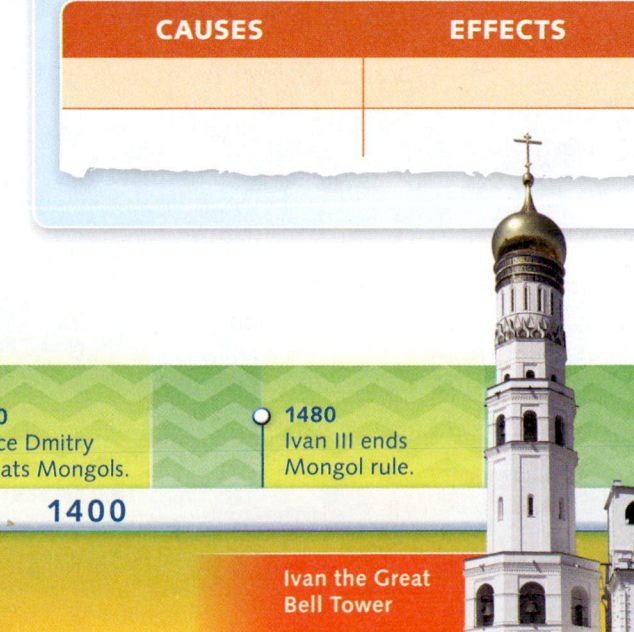

1237–1240	1380	1480
Mongols conquer Russia.	Prince Dmitry defeats Mongols.	Ivan III ends Mongol rule.

1200

1400

Mongol emperor Genghis Khan

Ivan the Great Bell Tower

2.2 European or Asian?

TECHTREK

myNGconnect.com For an online map and photos of historic buildings

 Maps and Graphs

 Digital Library

Main Idea Russia lies in both Europe and Asia but only began to adopt European ideas under the reign of Peter the Great.

As you have already learned, Russia was isolated from Western Europe for a couple of centuries under Mongol rule. This isolation continued under the Russian czars, beginning with Ivan IV. Then, starting in the 1600s, two czars began to bring a European influence to Russia.

Spanning Europe and Asia

Geographically, Russia extends across the landmass of Europe and Asia, which is often called Eurasia. The Ural Mountains separate the continents. Many of the Russian people live in the European area west of the mountains.

In spite of lying partly in Europe, Russia did not develop a European culture. In the late 900s, Christianity became the main religion in Russia. However, while Western Europe adopted Roman Catholicism, Russia embraced the Eastern Orthodox branch of Christianity. The Russian people came to distrust western European ideas and culture.

European Influence

Finally, Czar Peter Romanov, known as **Peter the Great**, recognized that Western Europe had surpassed Russia both economically and militarily. Peter, who ruled from 1682 to 1725, decided to modernize Russia. As a result, he became the first Russian czar to travel to Western Europe. From his travels, he brought back European ideas about how government and business should be run. He also introduced a Western style of architecture. Many buildings in St. Petersburg, including the Peterhof Palace shown below, reflect this style.

> **Critical Viewing** The Peterhof Palace is often called "the Russian Versailles," after the French palace of King Louis XIV. What words would you use to describe the palace?

Catherine the Great, empress from 1762 to 1796, also introduced European ideas to Russia, focusing on the arts and education. Western forms of entertainment, such as opera, became popular during her <mark>reign</mark>, or rule. In addition, Catherine founded many new schools and supported the idea of educating women. She also built hospitals and <mark>promoted</mark>, or encouraged, vaccinating against smallpox. Planners redesigned cities, and European architecture replaced older Russian styles.

The empress attempted to create a more <mark>secular</mark> country, one in which the church was less powerful. In fact, Catherine wanted to pass laws that would allow citizens to practice the religion of their choice. However, her attempt at reform was unsuccessful.

Before You Move On

Summarize What European ideas did Russia adopt under Peter the Great?

EUROPEAN AND ASIAN RUSSIA

ARCTIC OCEAN

URAL MTS.

RUSSIA

- Asian Russia
- European Russia
- Ural Mountains

0   800   1,600 Miles
0   800   1,600 Kilometers

 MAP TIP Map projections sometimes distort, or bend, the area shown. The Mollweide projection here makes Europe look somewhat mashed together but it is helpful in showing vast expanses of polar regions in small spaces.

ONGOING ASSESSMENT

READING LAB GeoJournal

1. **Make Inferences** Based on the map, in what ways do you think Russia's location encouraged its isolationism?

2. **Synthesize** How was Russia under Peter and Catherine different than under Mongol rule?

3. **Interpret Maps** In which continent does most of Russia lie?

4. **Analyze Visuals** What impression of Peter the Great is this statue meant to convey? Explain your answer with specific references to the statue.

2.3 Defending Against Invaders

TECHTREK
myNGconnect.com For an online
map and photos of historical figures

 Maps and Graphs

 Digital Library

Main Idea For centuries, Russia used its geographic strengths—location, size, isolation, and climate—to defend against invaders.

Throughout history, Russians adapted to the challenges of distance and climate in their country. **Invaders**, or enemies entering by force, however, often had a hard time overcoming these challenges.

Russia Builds an Empire

You have learned that Ivan IV became the first czar of Russia in 1547. Ivan began building Russia's empire by conquering territory from the Mongols. Eventually, Russia expanded across Siberia to the Pacific Ocean.

As Russia continued to grow under Peter the Great and Catherine the Great, the country was often forced to defend its empire against invaders. Enemies from Europe had to cope with Russia's vast expanse and harsh climate. They were rarely successful at invading the country. Two famous examples are the failed invasions of Napoleon and Hitler.

Napoleon Invades Russia

Napoleon I became emperor of France in 1804, and his empire included much of Europe. Angered that Britain had defeated him in battle, he came up with a plan to limit Britain's trade with European countries. Czar Alexander I refused to go along with the plan. Alexander thought reducing trade with Britain would be bad for Russia's economy.

In revenge, Napoleon invaded Russia in the summer of 1812. The Russians followed a **scorched earth policy**, where troops retreat in front of the advancing army and destroy crops and other resources that might supply the enemy. When Napoleon arrived in Moscow, the city was nearly deserted. The Russians had burned it to destroy any supplies that might have helped the French troops.

By mid-October, Napoleon started his retreat. He knew the harsh Russian winter was coming. Soon, snow and extreme cold began to weaken the French army. Russian troops attacked the army as it fled. Of the roughly 420,000 troops that set out for Moscow, only about 10,000 survived.

Ivan IV stands before St. Basil's, which he had constructed.

1500

1600

1700

1604–1613
Civil war and invasions

1703
St. Petersburg is founded.

1547
Ivan IV is crowned first czar.

1613
Romanov rule begins. The family, which included Peter the Great and Catherine the Great, ruled until 1917.

Picture of Nicholas II, the last of the Romanovs

NAPOLEON'S 1812 INVASION OF RUSSIA

Napoleon's March to Moscow, 1812
→ Napoleon's advance
← Napoleon's retreat

JUNE, KOVNO
Troops: 422,000
Napolean and his army cross the Neman River into Russia.

AUGUST, SMOLENSK
Troops: 145,000
Almost one-third die from famine, heat, and exhaustion.

SEPTEMBER, MOSCOW
Troops: 100,000
Napolean enters empty city; citizens had set fire to it as they fled.

OCTOBER, MOSCOW
Troops: 100,000
Napolean leaves the city in a pouring rain.

DECEMBER, KOVNO
Troops: 10,000
Only about 10,000 French soldiers survive.

NOVEMBER, SMOLENSK
Troops: 37,000
Cold, starvation, attacks, and disease further reduce army.

−26.5°F
−35.5°F
−15.25°F

Western Dvina R.
Volga R.
Moscow
Moscow R.
Borodino
Vitebsk
Smolensk
Kovno
Vilna
Molodechno
Minsk
Dnieper R.
Bug R.

PRUSSIA
GRAND DUCHY OF WARSAW

N
W E
S

0   50   100 Miles
0   50   100 Kilometers

Hitler Invades Russia

History repeated itself when Adolf Hitler sent troops from **Nazi Germany** to invade Russia during World War II. Frigid winter weather resulted in the deaths of many Nazi soldiers. Hitler was unable to take Moscow in 1941, or Stalingrad in 1943. Of the 300,000 Nazi troops that fought at Stalingrad, only about 5,000 came home.

Before You Move On

Monitor Comprehension What role did Russia's geography play in defeating Napoleon and Hitler?

ONGOING ASSESSMENT

MAP LAB

 GeoJournal

1. **Movement** Using the map, figure out how far Napoleon's army had to travel to get to Moscow from Kovno.

2. **Interpret Maps** What does the map tell you about the weather as the French troops retreated from Moscow?

3. **Compare and Contrast** In what ways were Nazi Germany's experiences in Russia similar to those of Napoleon?

1800

1812
Napoleon defeated in Russia

1900

1943
Nazi forces defeated at Stalingrad in World War II

PRESENT

In this painting, Moscow burns as the French army battles Russian soldiers.

2.4 Serfdom to Industrialization

TECHTREK

myNGconnect.com For an online map of Russia's resources and industries

Maps and Graphs

Main Idea Both peasants and industrial workers led difficult lives, but they played important roles in Russia's history.

For centuries, most Russian workers were peasants. They worked the land for wealthy landlords. The peasants were free to leave the estate as long as they paid their debts after the harvest.

The Beginning of Serfdom

Ivan IV, also known as Ivan the Terrible, changed this system in the 1500s. He murdered many nobles and gave their lands to people who supported him. These people needed the peasants to farm their estates, so Ivan passed laws that tied the peasants to the land as **serfs**. Serfs had their own houses and a small plot of land to farm, but they had to pay the landlord rent. Serfs could not leave the estate without permission and had few rights.

In 1861, Czar Alexander II freed the serfs. He wanted these workers to be free to work in industry and help modernize Russia. As a result, he ended their legal ties to their landlords. Serfs could then become industrial workers for wages.

Industrial Revolution in Russia

Under Peter the Great, early industries, such as shipbuilding and metalworking, had begun. Yet the **Industrial Revolution** didn't really begin in Russia until the 1890s. By 1913, it had become the fifth largest industrial nation in the world. As in other countries, however, the change from rural work to factory work was difficult. Most industrial workers and peasants were very poor.

Critical Viewing Serfs clear the land of large stones in this illustration. Based on the illustration, what can you conclude about the type of work serfs performed?

Many peasants starved because of poor harvests, and city factory workers were unhappy with their working conditions. Frequent worker **strikes**, or work stoppages, and protests gave way to political unrest. Revolutionary activist and politician **V. I. Lenin** led a political group called the **Bolsheviks**. They wanted workers to take over industry and the government. In February 1917 the Bolsheviks began the **Russian Revolution** and overthrew the czar. Lenin became leader of the new government.

Before You Move On
Summarize What roles did serfs and industrial workers play in Russian history?

RESOURCES AND INDUSTRIES IN WESTERN RUSSIA, 1900

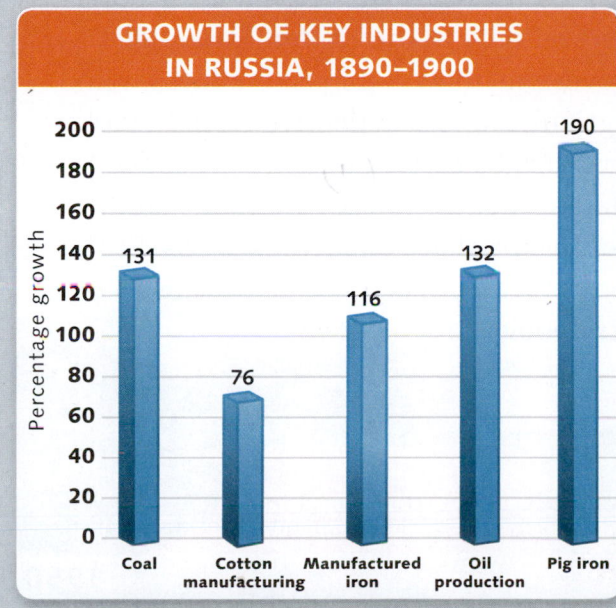

Legend:
- Iron ore
- Coal
- Oil
- Textiles
- Metalworking
- Sugar production
- — Railroads

GROWTH OF KEY INDUSTRIES IN RUSSIA, 1890–1900

Percentage growth:
- Coal: 131
- Cotton manufacturing: 76
- Manufactured iron: 116
- Oil production: 132
- Pig iron: 190

Source: Peter Stearns, *The Industrial Revolution in World History*, 1998

ONGOING ASSESSMENT
DATA LAB

GeoJournal

1. **Interpret Graphs** According to the graph, which industry grew the least between 1890 and 1900? The most? Why do you think these industries grew at these rates?

2. **Synthesize** Based on the graph, the map, and the text, what conclusions can you draw about the Industrial Revolution in Russia?

3. **Movement** Examine the map shown above. Why do you think several railroad lines ran through Moscow?

2.5 The Soviet Union

TECHTREK

myNGconnect.com For photos of Soviet propaganda

Digital Library

Main Idea The Communist Soviet Union had a powerful central government that controlled the region from 1922 until 1991.

The Bolsheviks, who led the Russian Revolution, believed that a Communist form of government and a socialist economic system were the answers to the problems of the Industrial Revolution. Under **communism**, a single political party controls the government and economy. **Socialism** is a system in which the government controls economic resources. The Bolsheviks wanted to end private ownership of land and resources and establish a classless society.

A Communist State

In 1922, Russia, Ukraine, Belarus, and the Transcaucasian republics of Armenia, Azerbaijan, and Georgia formed the Union of Soviet Socialist Republics (U.S.S.R.), also known as the **Soviet Union**. Later, other republics came under the control of the Soviet Union and the central Communist government in Moscow.

From 1927 to 1953, the Soviet people lived under the total command of Josef Stalin. Stalin's government isolated its citizens from contact with the West.

The Cold War

After World War II, the Soviet Union and the United States were the two most powerful countries in the world. Tension and conflict arose between the two because of their very different political and economic systems. The conflict came to be known as the **Cold War** because the countries did not fight each other directly.

The Cold War led the United States and Soviet Union to develop nuclear weapons. The United States went to war in Korea and Vietnam during the 1950s and 1960s to try to prevent communism from spreading to these countries. The Cold War also resulted in a "space race." The Soviet Union won the race in 1957 when it launched its Sputnik satellite into space.

A Controlled Economy

The Soviet Union also became an industrial leader and a world power, second only to the United States. The government owned most businesses and agriculture. On **collective farms**, workers produced a certain amount of food— determined by the government—and received a share of surplus crops. Still, the Soviet Union had trouble feeding all of its people.

A banner of Lenin overlooks Red Square as Soviet Russia celebrates the anniversary of the 1917 Russian Revolution.

1922
Soviet Union forms.

1941–1945
Soviet Union fights Germany in World War II.

1925

1950

1924
Stalin becomes Soviet leader.

1945
Cold War begins.

Quality of life was poor, too. People had guaranteed jobs, but their standard of living was much lower than that in Western countries. For example, they had little access to consumer goods. President **Mikhail Gorbachev** (mih KYL GAWR buh chawf) tried to reform and improve the economy. However, a movement toward adopting democratic forms of government was spreading across Eastern Europe. In 1991, the Soviet Union collapsed, and its republics gained their independence.

Before You Move On

Monitor Comprehension In what ways did the Communist government of the Soviet Union control its republics?

ONGOING ASSESSMENT

VIEWING LAB GeoJournal

1. **Analyze Visuals** In the propaganda poster, the figure in the back is holding a flag with Stalin's picture on it. What are the other figures holding? What might each figure represent?

2. **Make Inferences** Notice the attitude and gestures of the figures in the poster. How do you think the poster was supposed to make the Russian people feel?

3. **Draw Conclusions** How do you think the American people felt when the Soviet Union sent the first satellite into space? How might the space race have intensified the Cold War?

1957 Soviets launch Sputnik satellite.

U.S. president Ronald Reagan and Gorbachev

1985 Gorbachev begins reform programs and works with President Reagan to end Cold War.

1975

PRESENT

1961 Soviets send first person into space.

1991 Soviet Union collapses.

VOCABULARY

On your paper, write the vocabulary word that completes each of the following sentences.

1. _____ under the tundra keeps trees from growing there.

2. Some rivers provide _____, using the power of water to generate electricity.

3. _____ is information designed to influence people's opinions.

4. Czar Alexander I freed the _____ in 1861.

5. _____ is a form of socialism in which a single party controls the government and economy.

MAIN IDEAS

6. What natural barriers separate Russia and the republics from their neighbors? What impact do the barriers have on these countries? (Section 1.1)

7. Why are Russian winters cold and dark in some places? (Section 1.2)

8. What are the two main types of natural resources in this region? (Section 1.3)

9. What does Katey Walter Anthony hope to learn by studying the presence of methane in Siberian lakes? (Section 1.4)

10. What type of agriculture is suited to the dry climate of Central Asia? (Section 1.5)

11. How was Kievan Rus founded? (Section 2.1)

12. What western European influences did Peter the Great and Catherine the Great introduce to Russia? (Section 2.2)

13. What factors helped defeat the Nazis in Russia? (Section 2.3)

14. In what ways were serfs different from peasants in Russia? (Section 2.4)

15. What methods did the government of the Soviet Union use to control the country's economy? (Section 2.5)

GEOGRAPHY

ANALYZE THE ESSENTIAL QUESTION

How have size and extreme climates shaped Russia and the Eurasian republics?

Critical Thinking: Analyze Cause and Effect

16. Why does Russia have few ice-free ports?

17. What impact do Russia's size and extreme climate have on its use of natural resources?

18. In what way did the climate in Central Asia contribute to the shrinking of the Aral Sea?

INTERPRET MAPS

KAZAKHSTAN POLITICAL

19. **Location** Which two cities are best located to take advantage of energy resources in the Caspian Sea?

20. **Make Inferences** Why do you think Kazakhstan moved its capital from Almaty to Astana in 1997?

HISTORY

ANALYZE THE ESSENTIAL QUESTION

How has geographic isolation influenced the region's history?

Critical Thinking: Make Inferences

21. What effect did the Mongols have on Russia's relationship with Europe?

22. During Napoleon's invasion of Russia, what impact might distance have had on the army's horses and supplies?

23. In what way might physical geography have prevented the Industrial Revolution from starting in Russia as early as it started in western European countries?

INTERPRET TABLES

TIME INDUSTRIAL LABORERS NEEDED TO WORK TO BUY SELECTED GOODS, 1986		
	Moscow	United States
Loaf of bread	11 min*	18 min
Liter of milk	20 min	4 min
Grapefruit	112 min	6 min
Chicken	189 min	18 min
Bus fare (2 miles)	3 min*	7 min
Postage stamp	3 min	2 min
Pair of jeans	56 hrs	4 hrs
Washing machine	177 hrs	48 hrs

* price supports by government

Source: Radio Free Europe, published in the *New York Times*, June 28, 1987

24. **Compare and Contrast** What type of food must have had the biggest price difference? What food was probably more expensive in the United States than in Moscow?

25. **Make Generalizations** Based on the chart, what do you think life was like for industrial laborers in the Soviet Union in 1986?

ACTIVE OPTIONS

Synthesize the Essential Questions by completing the activities below.

26. **Write an Email** Schools sometimes conduct exchange programs with schools in other countries. Given what you know about the geography, history, and culture of Russia, decide which area of Russia you would like to visit. Write an email to a classmate recommending a particular location in Russia using the tips below. **Send the email to a classmate and ask for feedback on the information provided.**

> **Writing Tips**
> - Organize your ideas under two or three main headings before you begin writing.
> - Use a clear, straightforward style to present specific, useful, and interesting details in your email.
> - Make sure you explain why you offer the advice that you do.

Go to **Student Resources** for Guided Writing support.

TECHTREK myNGconnect.com For research links on Russia and the Eurasian republics

27. **Gather and Share Information** Work in a group to gather two new facts about Russia and each of the republics. Use the research links at **Connect to NG** and other online sources to help you find the facts. Then create a chart to record your facts and share them in an oral presentation. Compare your findings with those of other groups.

Russia	Fact 1: _____ Fact 2: _____
Armenia	Fact 1: _____ Fact 2: _____
Azerbaijan	Fact 1: _____ Fact 2: _____

CHAPTER 6

Russia & THE EURASIAN REPUBLICS TODAY

PREVIEW THE CHAPTER

Essential Question What features, such as size and climate, have influenced Russian culture?

SECTION 1 • CULTURE

KEY VOCABULARY

- culture
- nomad
- yurt
- terrain
- gauge
- port
- diplomacy

ACADEMIC VOCABULARY

enlist

TERMS & NAMES

- Trans-Siberian Railroad
- Hermitage Museum

Essential Question How have Russia and the Eurasian republics dealt with recent political, economic, and environmental challenges?

SECTION 2 • GOVERNMENT & ECONOMICS

KEY VOCABULARY

- perestroika
- glasnost
- coup
- federal system
- proportional representation
- revenue
- pipeline
- radioactive
- fallout
- half-life
- contaminate

ACADEMIC VOCABULARY

autonomy, vulnerable

TERMS & NAMES

- Russification
- Kremlin
- Chernobyl

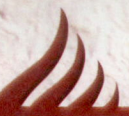

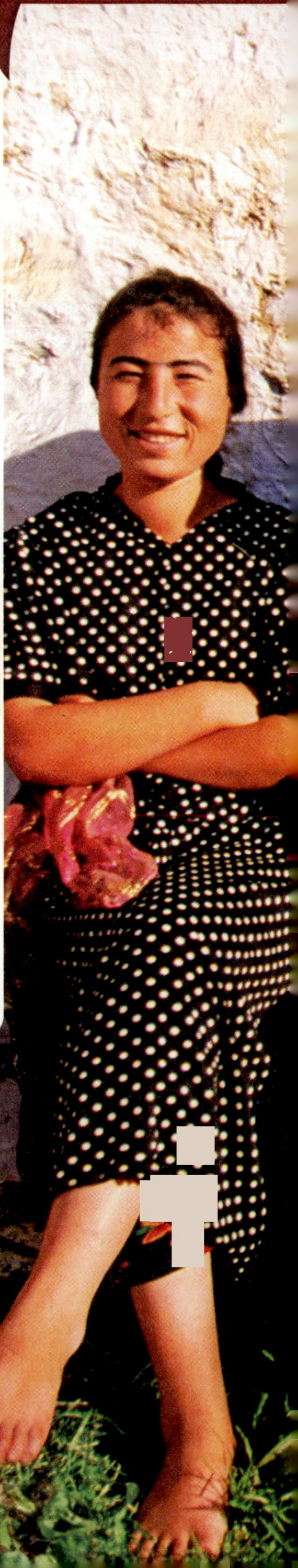

TECHTREK FOR THIS CHAPTER

Student eEdition

Maps and Graphs

Interactive Whiteboard GeoActivities

Digital Library

Connect to NG

Go to **myNGconnect.com** for more on Russia and the Eurasian Republics.

Uzbek women and girls from a mountain village wear a *rumol*, a traditional head scarf.

1.1 Climate and Culture

TECHTREK

myNGconnect.com For photos reflecting climate and culture in Russia and the republics

 **Digital Library**

Main Idea The variety of climates in Russia and the Eurasian republics has a major impact on the region's cultures.

Culture refers to a group's unique way of life. Cultural traits include food—what people eat and how they obtain it—shelter, clothing, religion, and language. The first three traits are influenced by climate.

Enduring the Cold in Siberia

About 30 different native peoples live in the tundra and taiga of Siberia in Russia. These include the Yakut, Nenets, Evenki, Chukchi, and Inuit (IHN yoo iht). All of these people have similar cultures but different languages.

People living in these cold climates, with their long winters and short, hot summers, traditionally dress in clothes made from reindeer fur. Some people, such as the Nenets, still live as **nomads**, moving from place to place according to the seasons in search of food. This way of life is changing, however, and today many native people live in towns or cities.

Herding in Central Asia

Like the Nenets in Siberia, some herders in Kazakhstan and other parts of Central Asia also continue to lead nomadic lives. In the region's dry climate, this traditional culture focuses on herding sheep, camels, cattle, and horses. Animals provide food and milk as well as hides and wool for clothing and tents. Although herding continues to be important in this region, many herders now live in rural villages in houses made of mud bricks that have been baked by the sun.

Farming in the Steppes

On the steppes of Russia and parts of Central Asia, the rich soil and moderate climate—cold winters and warm summers—are suited to agriculture. A peasant farming culture developed in this climate. In this culture, people grew grain crops, especially wheat, and raised livestock. They lived in settled villages. Later they worked for landlords on large estates and became serfs.

> **Critical Viewing** A Nenet boy studies for school in the frozen tundra of northwest Siberia. What impact might the intense cold reflected in this photo have on everyday activities?

ADAPTING TO CLIMATE

Cold

Dry

Moderate

SIBERIA Many native people of Siberia center their culture around reindeer. The warm, durable clothing worn by this family is made from reindeer skin. Reindeer meat and fat also provide nutritious food.

CENTRAL ASIA The animals the Kazakhs of Central Asia raise provide almost everything the herders need to live. Felt, or wool fibers pressed together, is used to make hats and traditional tents called **yurts**.

STEPPES Life for farmers on the steppes revolves around the seasonal growing cycle. Farmers often use simple tools to grow their crops. Animals help with the work and provide meat and milk.

Today, large corporations or individuals own the farms. Farmers still grow wheat as in the past, and they grow other grain crops as well, including maize, or corn, and barley. Many of these grain products are exported to other countries. Beets and potatoes are also common crops, along with sunflowers, which are grown to make oil used in cooking.

Before You Move On

Summarize What impact have the variety of climates in Russia and the Eurasian republics had on the region's cultures?

ONGOING ASSESSMENT

PHOTO LAB  GeoJournal

1. **Analyze Visuals** The Central Asian yurt shown above is movable. Why might this be an important consideration for nomadic herders?

2. **Compare and Contrast** Study all three photos above. In what ways does the women's clothing reflect the climate in each region?

3. **Human-Environment Interaction** In what ways has climate influenced traditional culture in Siberia, Central Asia, and the steppes?

1.2 Trans-Siberian Railroad

TECHTREK

myNGconnect.com For an online map of the Trans-Siberian Railroad and photos of sites on its routes

 Maps and Graphs

 Digital Library

Main Idea The Trans-Siberian Railroad links the western and eastern parts of Russia.

The **Trans-Siberian Railroad** is the world's longest continuous railroad, spanning about 6,000 miles and crossing eight time zones. *Trans-Siberian* means "across Siberia." The railroad links Moscow with eastern Russia. It also carries people and goods to East Asia and Europe.

Building the Railroad

Russia began to build the railroad in 1891. It was designed to connect Moscow with Vladivostok (vlah duh VAWH stock), a busy port on the Pacific Ocean. At that time, no reliable form of transportation linked this far eastern port with the European part of Russia.

Laborers began at both ends of the route and worked their way toward the center, but the climate and **terrain**, or physical features of the land, made progress very difficult. Workers had to lay tracks across long stretches of permafrost, and the route went through mountains, forests, rivers, and lakes. Materials—including explosives used to blast through rocks and cliffs— had to be brought thousands of miles to the work sites.

The construction project required many workers. As a result, thousands of Russian peasants, convicts, and soldiers were **enlisted**, or selected, to labor on the railroad. They were given only picks and shovels to do their difficult work. Human workers and horses hauled the heavy materials. The laborers worked long hours in extreme heat in the summer and extreme cold in the winter. They also had to deal with attacks by thieves and, occasionally, tigers. The main route, running from Moscow to Vladivostok, was finally completed in 1916.

Critical Viewing A Trans-Siberian Railroad train runs along the banks of Lake Baikal in Siberia. In this photo, the conductor looks out of the train as it curves around the lake. What challenges did the workers probably face while building this part of the railroad?

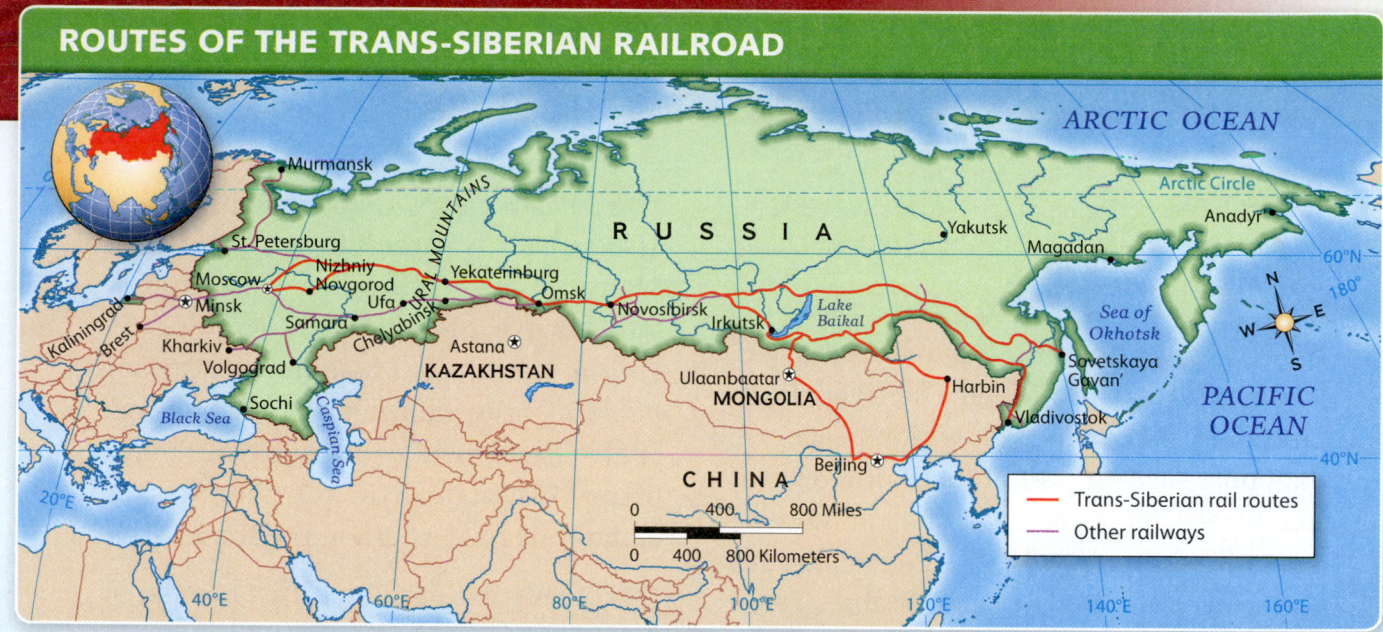

ROUTES OF THE TRANS-SIBERIAN RAILROAD

Map legend:
— Trans-Siberian rail routes
— Other railways

Effects of the Railroad

The railroad transformed Siberia and its traditional culture. Between 1891 and 1914, more than five million people immigrated to Siberia. New towns and cities grew up along the train's route. Soviet leaders began to industrialize Siberia and mine its plentiful raw materials. During World War II, the railroad moved Soviet troops and materials across Russia.

The Railroad Today

Today, the railroad operates several more routes and has replaced all of the old steam engines with electric trains that carry passengers and freight. The railroad also plays an important role in the world economy. Container cargo, or goods packed in large steel boxes, travels from China and other parts of East Asia to Europe. One challenge is that the **gauge**, or width of the tracks, in Russia is wider than in Europe or China. As a result, containers need to be transferred to different trains at the borders. Still, shipping containers by land across Russia is much faster—and cheaper—than shipping by sea.

Before You Move On

Monitor Comprehension In what ways has the Trans-Siberian Railroad helped to link Russia?

ONGOING ASSESSMENT

MAP LAB

 GeoJournal

1. **Interpret Maps** Use the map scale to determine the distance from Novosibirsk to Irkutsk on the Trans-Siberian Railroad.

2. **Movement** Trace the Trans-Siberian Railroad route to China. What benefit does it offer to manufacturers of consumer goods in China?

3. **Make Inferences** Why are there many tunnels and bridges along the route of the Trans-Siberian Railroad?

1.3 St. Petersburg Today

TECHTREK

myNGconnect.com For photos of St. Petersburg

Digital Library

Main Idea St. Petersburg is Russia's second largest city and a center of culture, industry, and trade.

St. Petersburg is located in northwest Russia on the Neva River. The river flows into the Gulf of Finland, which is at the most eastern part of the Baltic Sea. Because of its far northern location—at a latitude of about 60°N—St. Petersburg has long winter nights. In fact, for about one month a year, there's barely any daylight. For about three weeks in summer, the sky never gets completely dark. Special music and dance events are held during these "White Nights" to take advantage of the long days.

Window to the West

St. Petersburg originally sat on isolated swampland, making it a poor place on which to build. Nevertheless, Peter the Great chose the site in 1703 to gain a **port**, or harbor, on the Baltic Sea for trade. He also wanted to create a modern city—for the times—that would resemble those in Western Europe and become Russia's "window to the West." Peter got his wish. He had St. Petersburg filled with islands, canals, and bridges like those in the western European cities of Amsterdam and Venice. He also built wide boulevards like those in Paris and London.

FAST FACTS ON ST. PETERSBURG	
Population	4.6 million people
Land Area	550 square miles
Location	Average temperature in January: 21°F Average temperature in July: 65°F Latitude: 59° 57'N; longitude: 30° 19'E
Date Founded	1703

St. Petersburg's historic architecture also reflects western European influences. Many of the buildings from the 1700s still survive, including the Winter Palace, **Hermitage Museum**, and summer palaces of several czars. St. Petersburg is also filled with many beautiful and historic Russian Orthodox churches and cathedrals in Western and Eastern style.

The city is Russia's cultural center. St. Petersburg boasts world-famous museums and ballet companies, such as the Kirov. It also contains many universities and theaters and the country's oldest music academy. In addition, St. Petersburg offers a variety of contemporary music, including jazz and rock.

A Vibrant Economy

Manufacturing and construction are important industries in St. Petersburg. The city is also a center for trade and distribution of goods to and from Europe.

Above all, the local economy depends on tourism. More than three million visitors came to see the city's attractions in 2003 on its 300th anniversary. Many historic buildings that had been damaged during World War II were restored in time for the celebration. Vladimir Putin, Russia's president in 2003, is from St. Petersburg. He wanted the city to become a center of **diplomacy**, a place where international affairs could be conducted. Like Peter the Great, Putin wanted Russia to be more connected to the West.

Before You Move On

Summarize In what ways is St. Petersburg a center of culture, industry, and trade?

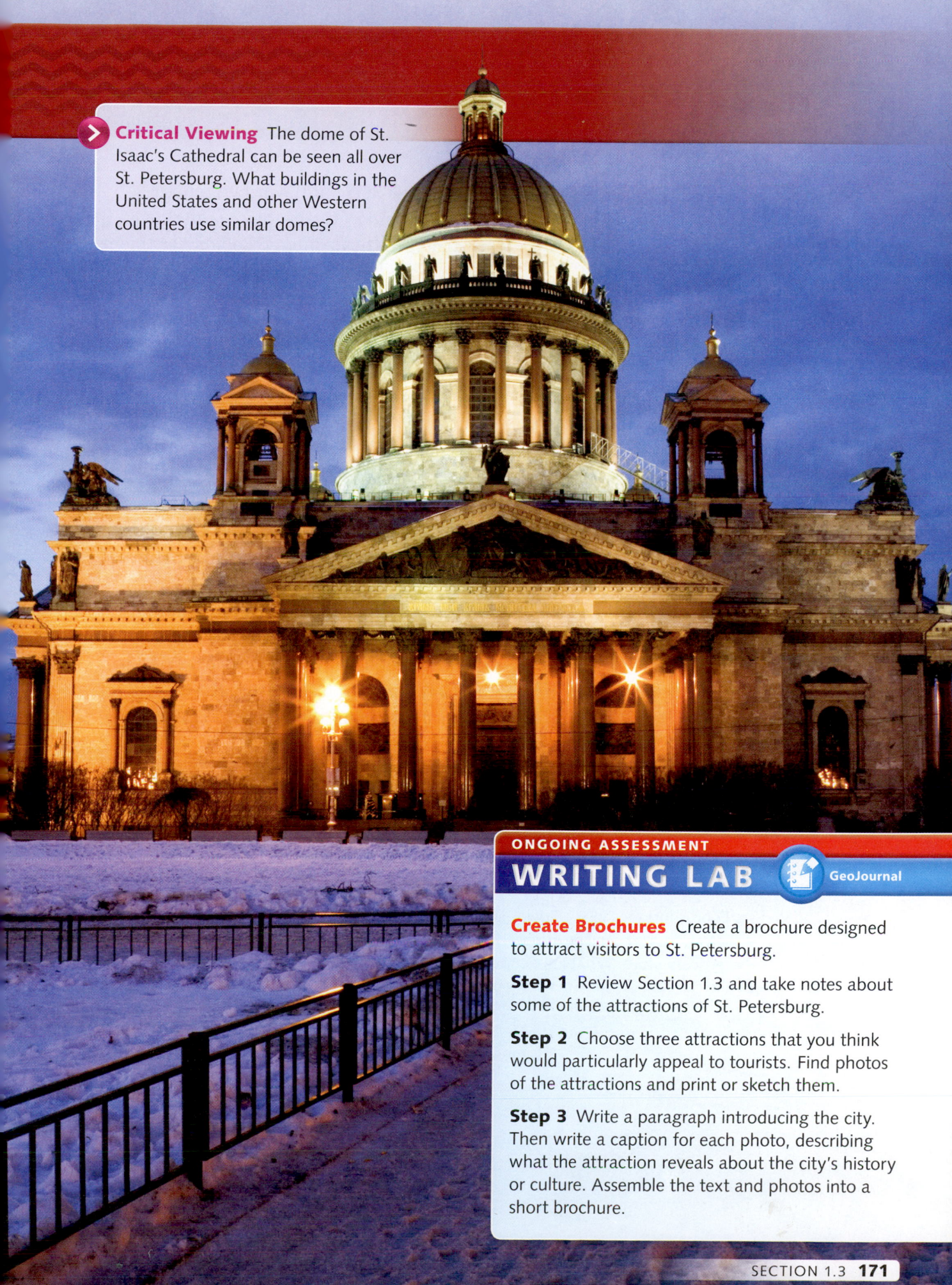

> **Critical Viewing** The dome of St. Isaac's Cathedral can be seen all over St. Petersburg. What buildings in the United States and other Western countries use similar domes?

ONGOING ASSESSMENT

WRITING LAB   GeoJournal

Create Brochures Create a brochure designed to attract visitors to St. Petersburg.

Step 1 Review Section 1.3 and take notes about some of the attractions of St. Petersburg.

Step 2 Choose three attractions that you think would particularly appeal to tourists. Find photos of the attractions and print or sketch them.

Step 3 Write a paragraph introducing the city. Then write a caption for each photo, describing what the attraction reveals about the city's history or culture. Assemble the text and photos into a short brochure.

2.1 The Soviet Collapse

TECHTREK

myNGconnect.com For photos of the Soviet Union

Digital Library

Main Idea Economic problems and people's desire for independence caused the Soviet Union to collapse in 1991.

As you have learned, the Soviet Union formed in 1922. During the 1970s and 1980s, the economy of the huge region slowed. Soviet money had almost no value outside the country. Stores had little food on their shelves and few goods to buy. In addition, people in the Eurasian republics had long resented the Soviet policy of **Russification**. Under this policy, the Soviet Union moved Russians into the republics and put them in charge. They also forced local people to learn the Russian language.

Gorbachev Brings Reform

In 1985, Mikhail Gorbachev became head of the Soviet Communist Party and began to introduce reforms. He promoted a movement known as **perestroika** (pehr ih STROY kuh), which means "restructuring." Gorbachev wanted to restructure the economy. Less government control, he believed, would make the economy more effective.

After the economy had failed to improve by 1990, the government presented a new plan to change the economy in 500 days. According to this plan, the republics would have greater control over their economies, and state ownership of businesses would decline.

Gorbachev also introduced the policy of **glasnost** (GLAHS nuhst), or "openness," which encouraged people to speak openly about the government. This freedom of expression went beyond what Gorbachev had intended, however. People started to criticize and protest against the central government, and they began to demand even more freedom.

> **Critical Viewing** Food shortages in the Soviet Union resulted in empty shelves, as you can see in this 1990 photo of a Moscow grocery store. What impact do you think seeing stores like this had on the Soviet people?

The Soviet Union Falls

After the collapse of the Berlin Wall in Germany in November 1989, the Communists no longer had control over Eastern Europe. People in many of the Soviet republics also wanted freedom from the central Communist government. Weakened by the economy, the government could no longer manage its vast empire. By the fall of 1990, all of the republics had declared their **autonomy**, or determination to govern themselves.

In August 1991, conservative Communists who opposed Gorbachev's political and economic reforms attempted a coup against him. A **coup** (KOO) is a sudden overthrow of the government by force. The coup failed, but Gorbachev knew he had lost power. On December 25, 1991, he resigned as president, and the Soviet Union dissolved. All of the republics became independent countries.

Life After Independence

Some groups within the newly independent republics and Russia's own borders sought independence for themselves. In Azerbaijan, Armenians continue to fight over a part of the country where most of the people are ethnic Armenians. Within Russia, the Muslim republic of Chechnya (CHEHCH nee uh) has struggled to gain full independence.

The economic transition has been difficult as well. Less government control in Russia has led to a large gap between rich and poor, higher prices, unemployment, and an increase in corruption and organized crime. Some of the Eurasian republics still have centrally controlled economies and governments.

Before You Move On

Make Inferences In what ways did attempts to reform the Soviet economy and political system contribute to the collapse of the Soviet Union?

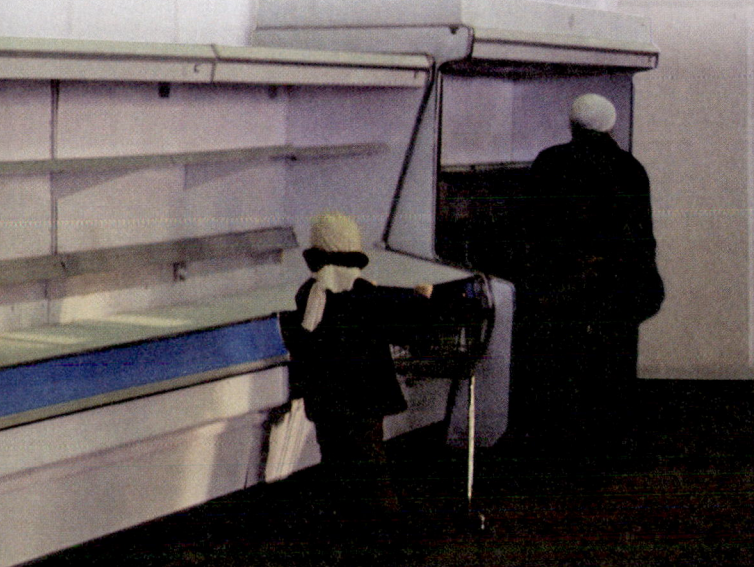

ONGOING ASSESSMENT

SPEAKING LAB GeoJournal

Turn and Talk With a partner, discuss the causes and effects of the Soviet Union's collapse. Use a cause-and-effect chart like the one below to list and organize your ideas.

CAUSES	EFFECTS
1.	
2.	

2.2 Russia's Government

TECHTREK

myNGconnect.com For current
events on Russia and photos of its leaders

🖱 **Connect
to NG**   ✉ **Digital
Library**

Main Idea Russia's central government consists of three branches and, together with the president, holds most of the power.

Following the collapse of the Soviet Union, Russia adopted a new constitution. This set up a ==federal system==, with a strong central government and local government units. The United States also has a federal system. Russia's government is more democratic than that of the former Soviet Union. Everyone 18 years and older can vote, and there are several political parties. However, the central government and the president hold most of the power.

Presidential Power

In Russia, the president is the head of the executive branch and the most powerful government leader. In 2008, the presidential term was extended from four to six years. That year, Dmitry Medvedev (med VYED if) was elected president.

The president appoints the prime minister. Medvedev appointed Vladimir Putin, who had been president from 2000 to 2008 (two four-year terms). Unlike earlier Russian prime ministers, Putin exercises a great deal of power. Some observers believe that he is the real leader of Russia, not Medvedev.

Legislators and Judges

There are two houses in the legislative branch. The lower house is the State Duma. Members are elected based on ==proportional representation==. Under this system, a political party gets the same percentage of seats as the percentage of votes it received. A party must receive at least seven percent of the vote to get seats. Putin leads the United Russia Party, which holds about 64 percent of the seats.

The upper house of the legislative branch is the Federation Council. Executive and legislative leaders in each of the local government units appoint the members of this house. The State Duma is the more powerful legislative house. All bills must first be considered in the State Duma, even those proposed by the upper house.

> **Critical Viewing** The **Kremlin** is a historic complex of palaces and churches in the heart of Moscow. What feature in the photo suggests that the Kremlin once served as the city's fortress?

COMPARE RUSSIAN AND U.S. GOVERNMENTS

Russia	Government Branches	United States
• President, elected to a six-year term Prime Minister, appointed • Government Ministers, appointed	Executive	• President and Vice President, both elected to a four-year term • Cabinet, appointed
• Federal Assembly: – Federation Council (166 members), appointed to a four-year term – State Duma (450 members), elected to a five-year term	Legislative	• Congress: – Senate (100 members), elected to a six-year term – House of Representatives (435 members), elected to a two-year term
• Constitutional Court Judges, appointed for life	Judicial	• Supreme Court Judges, appointed for life

The highest court in Russia is the Constitutional Court. The Federation Council appoints these judges based on the president's recommendations. The judges are appointed for life. In general, the judicial branch in Russia is more **vulnerable**, or open, to political pressure than in most Western democracies. That means that officials in the executive and legislative branches are sometimes able to influence the judges.

Central Control

The central government in Moscow still tries to control most levels of government. Those in power generally choose the people they want to be elected.

In 2000, then-president Putin reduced the number of local government units in Russia from 89 to 7 to increase central control of them. The president already exercised some power over these units because he nominated their governors.

Before You Move On

Summarize What branches make up the central government, and in what ways do the government and the president exercise their power?

ONGOING ASSESSMENT

DATA LAB

GeoJournal

1. **Identify** Based on the chart, which officials in the Russian executive branch probably serve a role similar to that of the Cabinet members in the U.S. government?

2. **Compare and Contrast** Study the chart. In what way is the legislative branch in Russia similar to that in the United States? In what way is it different?

3. **Express Ideas Through Speech** Is the government in Russia more or less democratic than that in the United States? Think about which officials are elected or appointed and how power is handled. Discuss with a partner.

2.3 Unlocking Energy Riches

TECHTREK
my NGconnect.com For an online map and photos of the region's energy pipelines

Maps and Graphs

Digital Library

Main Idea Oil and natural gas enrich the economies of Russia and several countries around the Caspian Sea.

In addition to reforming its government, Russia has worked to develop its economy. Oil and natural gas are its greatest sources of wealth. In fact, the country is the largest exporter of oil and natural gas in the world. In 2008, these resources accounted for about two-thirds of the value of all Russia's exports and one-third of its **revenue**, or income.

Siberian Boom Towns

About 70 percent of Russian oil comes from western Siberia. Since the end of the Soviet Union, the region has been booming, or growing rapidly. Workers from western Russia and Central Asia come to Siberia for good-paying jobs. They earn enough to buy apartments in cities such as Surgut. New suburbs have developed, and many people enjoy a higher standard of living. Oil wealth has also helped fund new airports, museums, and schools.

Caspian Sea Riches

Some experts believe that the Caspian Sea may actually have more energy reserves than the Persian Gulf. Russia would like to control the energy from the Caspian. In 2007, Russia, Turkmenistan, and Kazakhstan signed an agreement to build a pipeline that would transport natural gas from the Caspian through Kazakhstan into Russia. From there, Russia would export the gas to Europe and make a large profit.

In 2009 and 2010, new gas pipelines were opened to carry gas from Turkmenistan to China and Iran. Pipelines in Kazakhstan also bring oil from the Caspian Sea to China.

> **Visual Vocabulary** Oil workers seal a pipeline in Kazakhstan. A **pipeline** is a series of connected pipes used to transport liquids or gases.

OIL AND NATURAL GAS PIPELINES

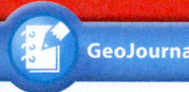

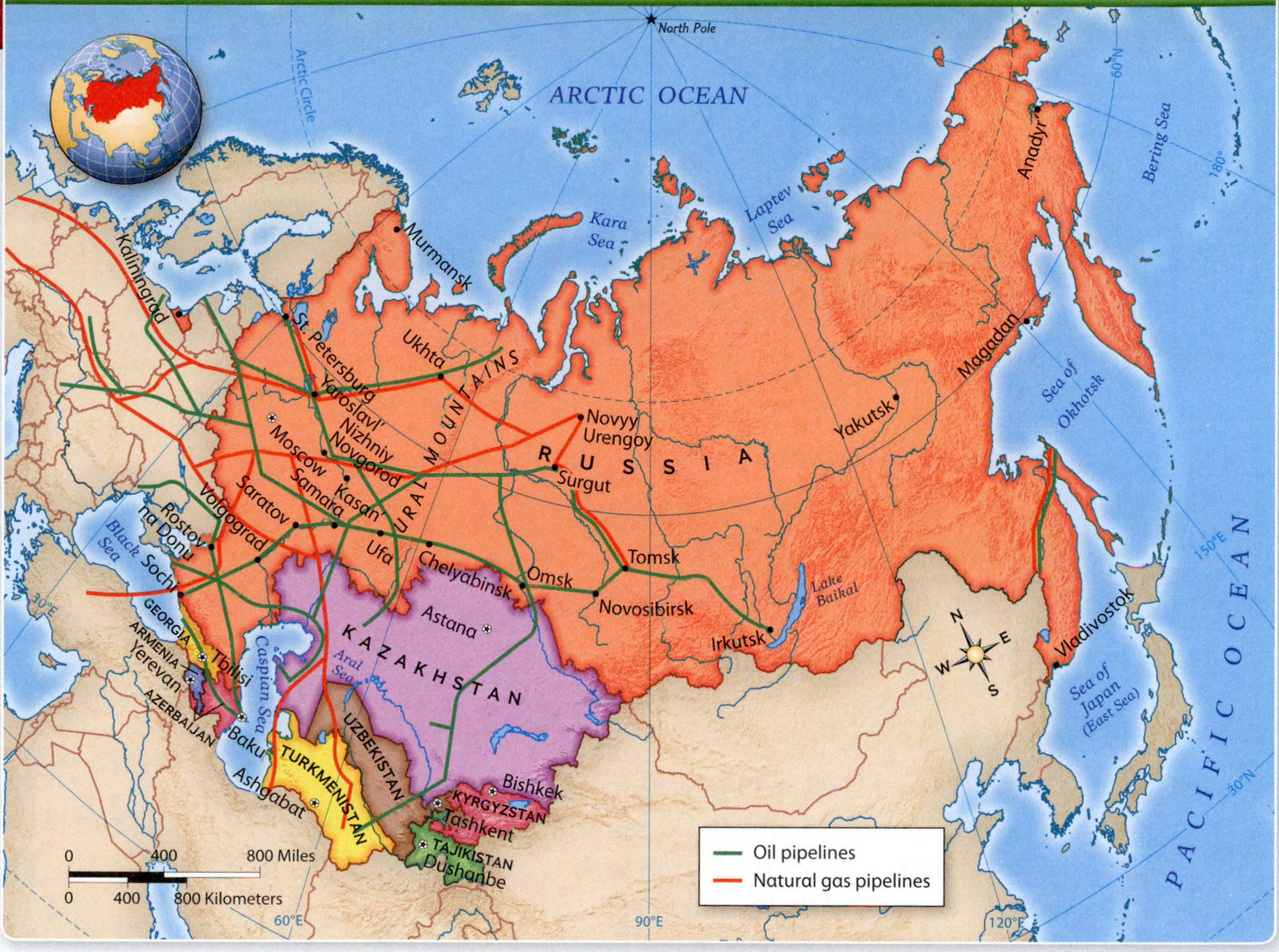
North Pole
ARCTIC OCEAN
Arctic Circle
Kara Sea
Laptev Sea
Bering Sea
Sea of Okhotsk
Sea of Japan (East Sea)
PACIFIC OCEAN
Black Sea
Caspian Sea
Aral Sea
Lake Baikal
URAL MOUNTAINS
RUSSIA
KAZAKHSTAN
UZBEKISTAN
TURKMENISTAN
KYRGYZSTAN
TAJIKISTAN
GEORGIA
ARMENIA
AZERBAIJAN
Murmansk
Kaliningrad
St. Petersburg
Ukhta
Yaroslavl'
Nizhniy Novgorod
Moscow
Samara
Kasan
Saratov
Volgograd
Rostov na Donu
Sochi
Ufa
Chelyabinsk
Novyy Urengoy
Surgut
Omsk
Tomsk
Novosibirsk
Irkutsk
Yakutsk
Magadan
Anadyr'
Vladivostok
Astana
Bishkek
Tashkent
Dushanbe
Ashgabat
Baku
Tbilisi
Yerevan
60°N
180°
150°E
30°N
30°E
60°E
90°E
120°E
N
S
E
W
0
400
800 Miles
0
400
800 Kilometers
Oil pipelines
Natural gas pipelines

Azerbaijan has large reserves of both oil and natural gas. It used to send the fuel through pipelines to Russia. Now it makes more money by trading mainly with the United States and by piping the fuel directly through Georgia and Turkey. By so doing, Azerbaijan does not have to share export profits with Russia.

For now, Russia gains a great deal of wealth from its exports of natural gas and oil. However, by relying too heavily on energy exports, the country risks using up these natural resources.

Before You Move On

Monitor Comprehension In what ways have oil and natural gas enriched Russia and countries around the Caspian Sea?

ONGOING ASSESSMENT

MAP LAB

GeoJournal

1. **Movement** About how far does natural gas travel by pipeline from Surgut in western Siberia to St. Petersburg in Russia? Why do you think pipelines might be a good way to transport the gas over long distances?
2. **Pose and Answer Questions** Come up with your own question about the information shown on the map. Then challenge a partner to answer your question.
3. **Draw Conclusions** What would Russia gain by controlling the energy from the Caspian Sea?
4. **Make Inferences** Why are countries in Central Asia developing pipelines to transport oil and natural gas directly to China and Iran?

2.4 After Chernobyl

Main Idea The Chernobyl nuclear disaster severely damaged the environment in parts of Belarus, Ukraine, and western Russia.

On April 26, 1986, a nuclear reactor at a power plant in Chernobyl exploded and caught fire, resulting in the worst nuclear disaster in history. **Chernobyl** is in the country of Ukraine. At the time of the disaster, Ukraine was part of the Soviet Union. A **radioactive** cloud about 3,280 feet high spread over parts of Ukraine, Belarus, and Russia. Winds carried the **fallout** into parts of northern and central Europe. These radioactive materials have caused harm to humans, animals, and plants.

Health Effects

Scientists know that it takes some radioactive materials a long time to break down and disappear. Each material has a particular **half-life**, the time needed for half of its atoms to decay and decrease. For example, the highly radioactive element cesium has a half-life of 30 years. That means that after 30 years, half of its atoms will still be radioactive.

Thirty people died within three months of the accident, most from being exposed to huge amounts of radioactive material. Over time, thousands of the people who helped clean up the disaster have developed health problems. Some of the children born to exposed parents carry the effects in their genes. In addition, radioactive iodine got into the milk of cows that grazed on **contaminated**, or infected, grass after the accident. This caused an increase in thyroid cancer, especially in children. Millions of people are still living on contaminated land. Officials cannot yet predict the long-term effects on people's health.

Before You Move On

Summarize What are some of the health problems that occurred as a result of the Chernobyl nuclear disaster?

KEY VOCABULARY

radioactive, adj., giving off energy caused by the breakdown of atoms

fallout, n., radioactive particles from a nuclear explosion that fall through the atmosphere

half-life, n., the time needed for half the atoms in a radioactive substance to decay and decrease

contaminated, adj., unfit for use because of the presence of unsafe elements

CHERNOBYL DISASTER BY THE NUMBERS

400
Estimated number of atomic bombs it would take to equal the accident

4,000
Estimated number of people who may die from cancers caused by radiation exposure from the accident

600,000
Estimated number of people who received significant radiation exposure, including evacuees, residents, and those who helped clean up after the accident

5 million
Estimated number of people still living on contaminated areas in Ukraine, Belarus, and Russia

Source: UN Chernobyl Forum, 2006

Environmental Damage

Belarus was hardest hit because of the direction the wind blew just after the accident. About 23 percent of the country was contaminated, including agricultural and forest land. In Ukraine, 7 percent of the land and 40 percent of its forests were contaminated. In Russia, the area bordering with Belarus was most affected.

Radiation is expected to remain in the soil for many years. Plants that grow in the region's forests are contaminated and so are the animals that feed on them. Many reindeer herds had to be killed shortly after the accident because they were so infected by the radiation.

The fallout has especially contaminated the fish in the rivers and lakes of Ukraine because these waters flow down from the site of the disaster. Some radiation that has gotten into groundwater sources will last for hundreds of years.

Few humans live in the 18-mile fenced area around the Chernobyl reactor now.

As a result, the people know that the plants are technologically more advanced and safer than the one in Ukraine. The French scientists and engineers who built them learned the lesson of Chernobyl.

Before You Move On

Monitor Comprehension What has been learned from the Chernobyl disaster?

COMPARE ACROSS REGIONS

France's Nuclear Program

In spite of what happened at Chernobyl, many countries have continued to develop nuclear energy. France, for example, has more than 55 nuclear plants that generate more than 75 percent of its electricity. Many of the French people welcome a new nuclear plant in their town because it brings jobs and prosperity to the area.

The French are aware of the dangers of nuclear energy plants, but they have little fear. Many of the plants offer tours, and advertisements help reinforce the idea that nuclear energy is a fact of life in France.

ONGOING ASSESSMENT

READING LAB GeoJournal

1. **Summarize** What damage did the Chernobyl disaster cause to the environment?

2. **Make Predictions** In what ways do you think the disaster will continue to affect people in Ukraine, Belarus, and Russia?

3. **Write Comparisons** Imagine that two new nuclear plants were being proposed: one in Russia and one in France. What do you think would be the reactions of the townspeople in each country? Write two sentences comparing their reactions. Make sure that you explain what led you to believe that the people in each country would feel that way. Go to **Student Resources** for Guided Writing support.

VOCABULARY

For each pair of vocabulary words, write one sentence that explains the connection between the two words.

1. nomad; yurt

> For centuries, nomads in Central Asia have traditionally lived in felt tents called yurts.

2. perestroika; glasnost

3. federal system; proportional representation

4. radioactive; contaminated

5. fallout; half-life

MAIN IDEAS

6. Describe some of the different types of climate in Russia and the Eurasian republics. (Section 1.1)

7. In what ways has Russia benefited from the Trans-Siberian Railroad? (Section 1.2)

8. Why is St. Petersburg called Russia's "window to the West"? (Section 1.3)

9. How did installing Russian officials in the republics help bring about the collapse of the Soviet Union? (Section 2.1)

10. In what way did Gorbachev's changes to the government contribute to the fall of the Soviet Union? (Section 2.1)

11. Which branch of Russia's central government has the most power? Which branch has the least power? (Section 2.2)

12. Why is western Siberia said to be "booming"? (Section 2.3)

13. What was the impact of the Chernobyl nuclear disaster on Ukraine, Belarus, and Russia? (Section 2.4)

CULTURE

ANALYZE THE ESSENTIAL QUESTION

What factors, such as size and climate, have influenced Russian culture?

Critical Thinking: Make Inferences

14. What impact does climate have on nomadic herders in Siberia and Central Asia?

15. Why was it difficult to build the Trans-Siberian Railroad across Russia?

16. At what time of year might tourists prefer to visit St. Petersburg? Why?

INTERPRET MAPS

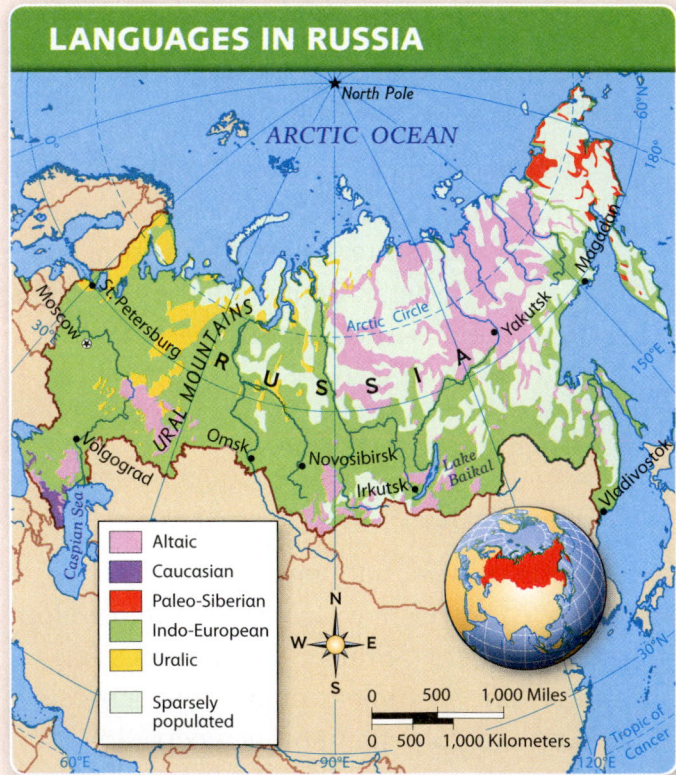

17. **Region** What are the two major language families in Russia?

18. **Compare and Contrast** What do the languages spoken west and east of the Ural Mountains have in common? In what ways do they differ?

GOVERNMENT & ECONOMICS

ANALYZE THE ESSENTIAL QUESTION

How have Russia and the Eurasian republics dealt with recent political, economic, and environmental challenges?

Critical Thinking: Make Generalizations

19. What impact did the collapse of the Soviet Union have on Russia's role in the region?

20. Why did Russia's central government reduce the number of local government units?

21. In what ways do Russia's location and power help the country control energy resources in the region?

22. Why was Russia less affected by the Chernobyl nuclear disaster than Ukraine and Belarus?

INTERPRET TABLES

NATURAL GAS RESERVES (2009)		
Country	**Amount (cu m)***	**World Rank**
Russia	47.6 trillion	1
Turkmenistan	7.5 trillion	4
United States	6.9 trillion	6
Kazakhstan	2.4 trillion	15
Uzbekistan	1.8 trillion	19
Ukraine	1.1 trillion	25
Azerbaijan	850.0 billion	27

Source: CIA Factbook / *cu m = cubic meters

23. Analyze Data According to the table, about how much more natural gas reserves does Russia have than the United States?

24. Evaluate What does this table suggest about where most of the world's natural gas reserves can be found?

25. Draw Conclusions Why do you think Ukraine imports natural gas from Russia?

ACTIVE OPTIONS

Synthesize the Essential Questions by completing the activities below.

26. Write Journal Entries Write several journal entries from the point of view of a worker on the Trans-Siberian Railroad. Describe the difficulties of the work and the obstacles you encounter. Use the writing tips below to help you write your entries. **When you have finished, trade your entries with a partner and compare them.**

> **Writing Tips**
> • Include details on where you worked, the tools you used, and the conditions you endured.
> • Describe any unexpected experiences, such as an encounter with thieves or wild animals along the tracks.
> • Express your feelings about having been enlisted to carry out the work.

TECHTREK myNGconnect.com For research links on Russia and the Eurasian republics

27. Create Charts Make a chart in which you compare the size, population, and major religions of Russia and the Eurasian republics. Include Kazakhstan, Kyrgyzstan, Tajikistan, Turkmenistan, and Uzbekistan in your chart. Use the research links at **Connect to NG** and other sources to gather the data. With a partner, discuss what the biggest similarities and differences are between these countries.

COMPARE COUNTRIES			
Country	**Size**	**Population**	**Major Religions**
Russia			
Armenia			
Azerbaijan			
Georgia			

Natural Resources and Energy

TECHTREK

myNGconnect.com For an online graph and research links on energy sources

 Maps and Graphs

Connect to NG

In this unit, you learned that Russia and many of the Eurasian republics have plentiful supplies of natural gas and oil. In fact, Russia has the world's largest supply of natural gas. Russia's Asian neighbor, China, is the world's biggest producer of coal. These energy sources are in great demand from countries all over the world.

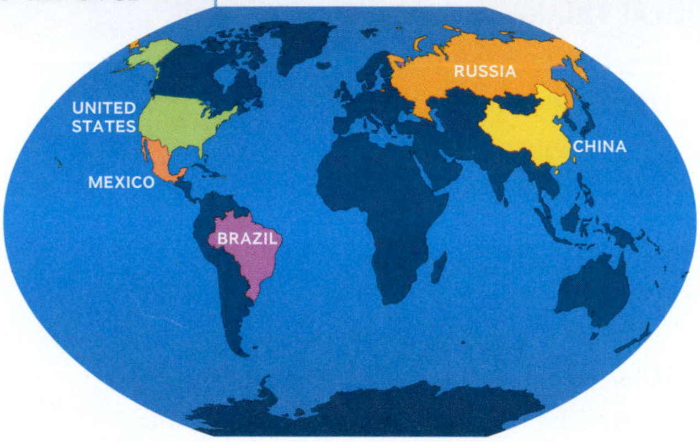

The choices people make about energy are critical to a healthy planet. As you have learned, some energy sources pollute the environment as they burn. Many scientists believe that the carbon dioxide those sources give off contributes to an overall global temperature increase. To ease these problems, they would like to reduce the use of fossil fuels and develop nonfossil energy sources.

Compare

- Brazil
- China
- Mexico
- Russia
- United States

FOSSIL FUELS

Fossil fuels are formed by buried plants and animals that have been dead for millions of years. This type of fuel is found in deposits beneath the earth's surface and must be burned to release its energy. These fuels supply about 85 percent of the world's energy. Fossil fuels include coal, oil, and natural gas.

These energy sources are classified as nonrenewable because they take millions of years to form and supplies are being used up faster than new ones can redevelop. Although natural gas is relatively clean—meaning that it doesn't create much pollution—fossil fuels, especially coal, tend to be dirty and release a great deal of harmful carbon dioxide into the air.

NONFOSSIL FUELS

Nonfossil fuels are alternative sources of energy and include the following: hydroelectric power from water, nuclear power, wind power, solar energy from the sun, and biofuels made from vegetable oil.

Many of these energy sources are renewable because they can be quickly replenished. Some countries use significant amounts of nuclear energy and hydroelectric power. However, the other nonfossil fuels make up only two percent of the world's energy.

Russia and China use both fossil and nonfossil fuels. The graphs on the next page show the energy consumption in both countries. Compare the data in the graphs and use it to answer the questions.

ENERGY CONSUMPTION PER COUNTRY

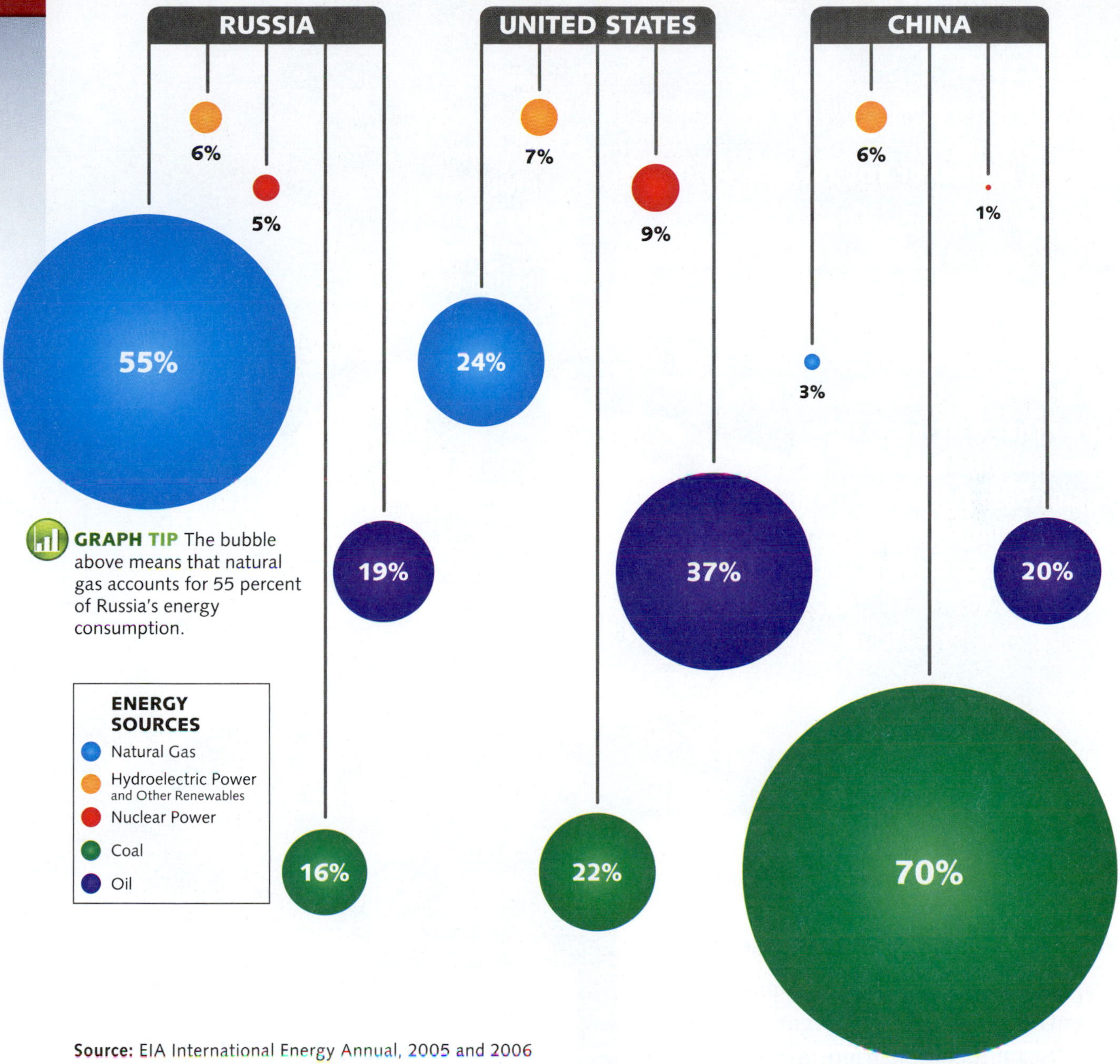

RUSSIA

6%

5%

55%

GRAPH TIP The bubble above means that natural gas accounts for 55 percent of Russia's energy consumption.

19%

16%

UNITED STATES

7%

9%

24%

37%

22%

CHINA

6%

1%

3%

20%

70%

ENERGY SOURCES
- Natural Gas
- Hydroelectric Power and Other Renewables
- Nuclear Power
- Coal
- Oil

Source: EIA International Energy Annual, 2005 and 2006

RESEARCH LAB GeoJournal

1. **Explain** What is the main source of energy in each country? Why might that be the case?

2. **Analyze Data** What is the energy consumption of nonfossil fuels in each country? What does this suggest about the development of alternative energy in Russia and China?

Research and Make Comparisons Research energy consumption in Brazil and Mexico and then compare with that in Russia and China. Which country appears to be investing more in nonfossil and renewable energy sources? What might this statistic suggest about its pollution levels?

Active Options

TECHTREK

myNGconnect.com For photos of natural and cultural sites in Russia and the Eurasian republics and other UNESCO World Heritage sites

Digital Library

Magazine Maker

ACTIVITY 1

Goal: Extend your understanding of animals native to Russia and the Eurasian republics.

Design a Poster for Native Animals

Russia is home to both land and sea animals. Some of these animals, including those listed on the right, have been over-hunted or are considered undesirable neighbors. Research one of these animals and use the information to design a poster that will help viewers appreciate it. Consider the animal's habitat, surviving population, and efforts made to protect it.

- Blue whale
- Brown bear
- Gray wolf
- Polar bear
- Siberian tiger

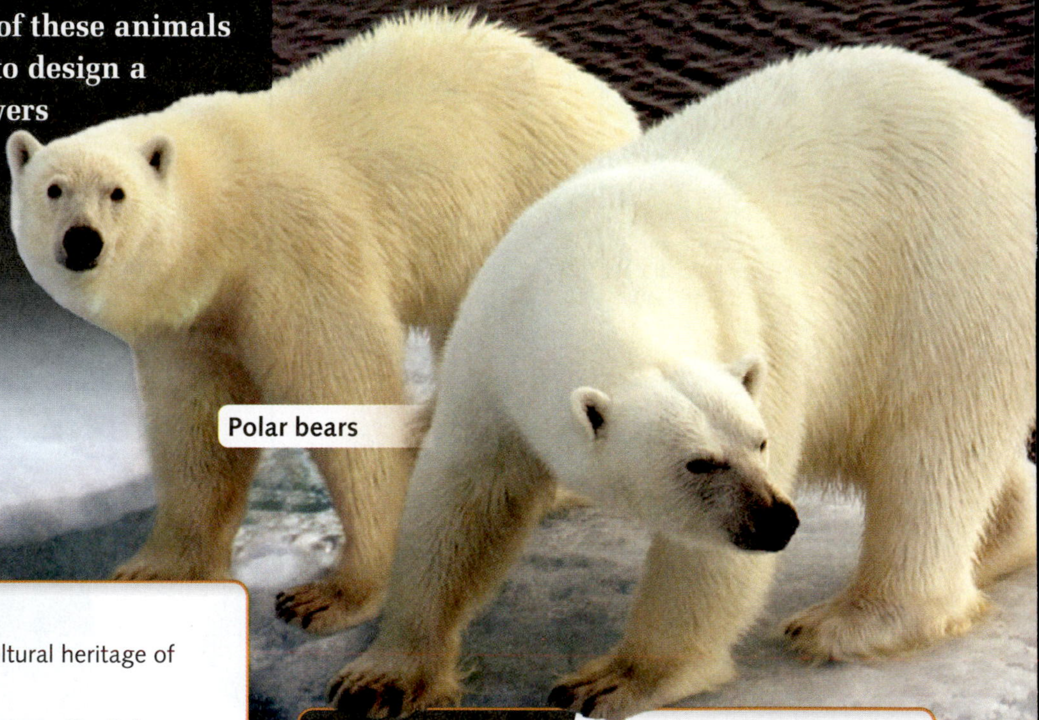

Polar bears

ACTIVITY 2

Goal: Research the natural and cultural heritage of Russia and the Eurasian republics.

Write a World Heritage Guide

The United Nations Educational, Scientific and Cultural Organization (UNESCO) has identified hundreds of sites that form part of the world's natural and cultural heritage. A variety of these World Heritage sites are located in the Russia Federation and the Eurasian republics. Prepare a guide for students visiting the area. Choose five to ten UNESCO World Heritage sites in the region. Use the **Magazine Maker** to create the guide and provide brief background information about each site.

ACTIVITY 3

Goal: Learn about Russian culture through the country's folk tales.

Hold a Russian Folk Tale Festival

Folk tales, such as "The Old Man and the Bear" and "Baba Yaga," are an age-old part of the Russian culture. With a group, find some Russian folk tales online and read them. Decide which ones you will present in the festival. Then decide how you will present the tales. You may each choose to retell a story, or you may decide as a group to act out one or more of the tales.

Explore Sub-Saharan Africa

with NATIONAL GEOGRAPHIC

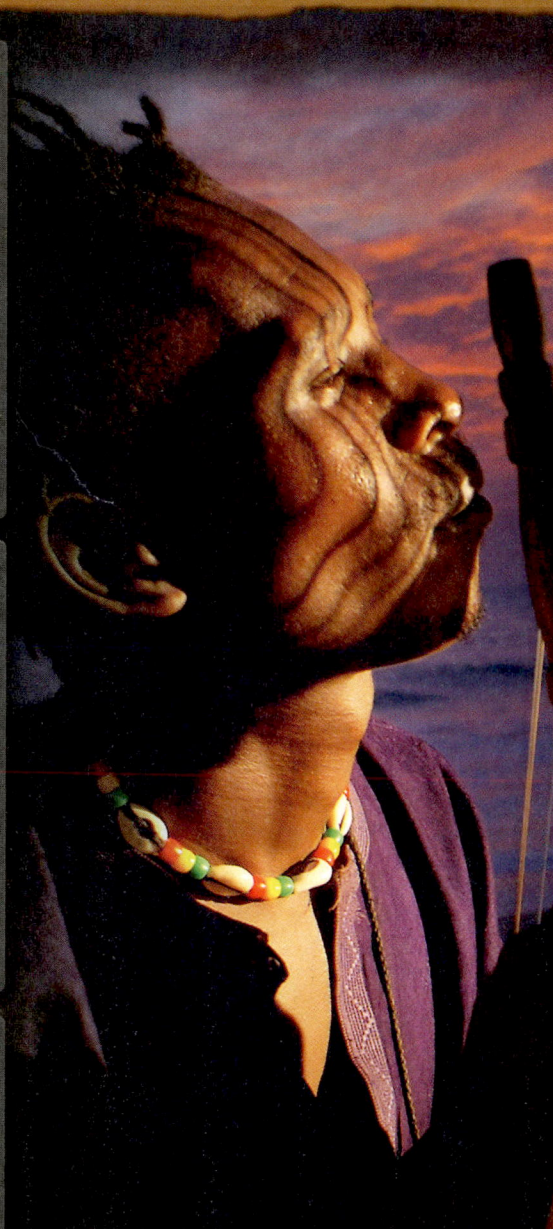

MEET THE EXPLORER

NATIONAL GEOGRAPHIC

Emerging Explorer Kakenya Ntaiya, educator and activist (shown at center), founded the first primary school for girls in her village. Her work provides African girls with an education and an opportunity to develop leadership skills.

INVESTIGATE GEOGRAPHY

Giraffe mothers and their young run in Serengeti National Park, Tanzania. The Serengeti, connected to the Masai Mara National Park in neighboring Kenya, is the only place in Africa where large animal herds, including gnu, zebras, and gazelles, still make vast land migrations.

STEP INTO HISTORY

This ancient Dogon village was built into the sandstone cliffs of Mali, West Africa. Today, most Dogon people are farmers, and practice the tradtional religion of their ancestors. Much of their social organization and cultural practices relates to this belief system.

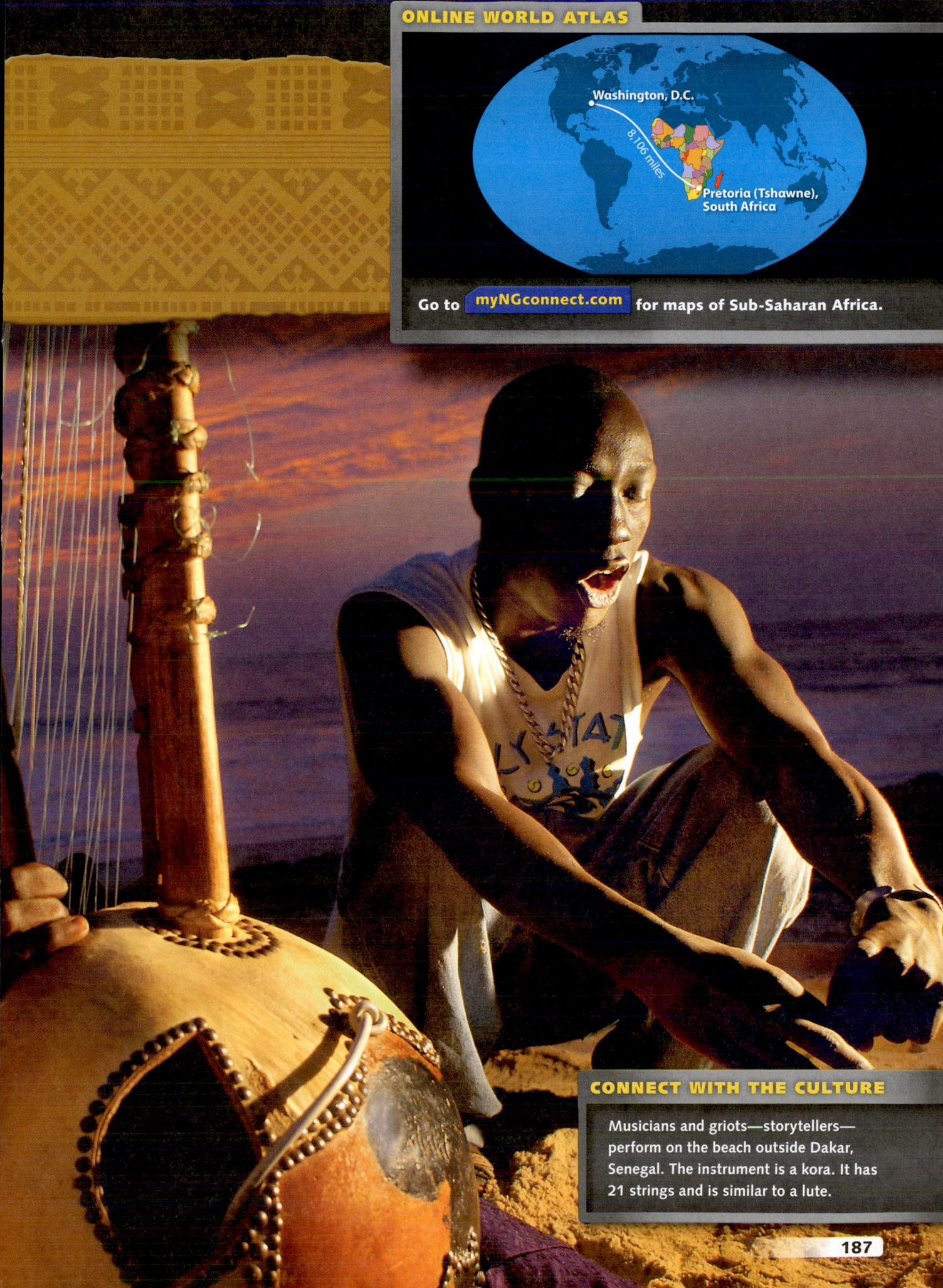

Washington, D.C.

8,106 miles

Pretoria (Tshawne),
South Africa

Go to **myNGconnect.com** for maps of Sub-Saharan Africa.

CONNECT WITH THE CULTURE

Musicians and griots—storytellers—
perform on the beach outside Dakar,
Senegal. The instrument is a kora. It has
21 strings and is similar to a lute.

SUB-SAHARAN AFRICA
GEOGRAPHY & HISTORY

PREVIEW THE CHAPTER

Essential Question How has the varied geography of Sub-Saharan Africa affected people's lives?

SECTION 1 • GEOGRAPHY

KEY VOCABULARY
- basin
- savanna
- desertification
- rift valley
- pride
- interior
- deforestation
- transition zone
- highlands
- rain forest
- hydroelectric power
- landlocked
- escarpment
- habitat
- poaching
- nocturnal
- ecotourism

ACADEMIC VOCABULARY
dormant

TERMS & NAMES
- Sahel
- Kalahari
- Great Rift Valley
- Kilimanjaro
- Congo River
- Great Escarpment
- Zambezi River
- Okavango Delta

Essential Question How did trade networks and migration influence the development of African civilization?

SECTION 2 • HISTORY

KEY VOCABULARY
- agricultural revolution
- first language
- caravan
- *lingua franca*
- trans-Saharan
- alluvial
- city-state
- trans-Atlantic slave trade
- malnutrition
- imperialism
- colonialism
- missionary

ACADEMIC VOCABULARY
incentive

TERMS & NAMES
- Bantu
- Swahili
- Timbuktu
- Kongo
- Great Zimbabwe
- Middle Passage
- Berlin Conference
- Pan-Africanism
- Jomo Kenyatta
- Kwame Nkrumah

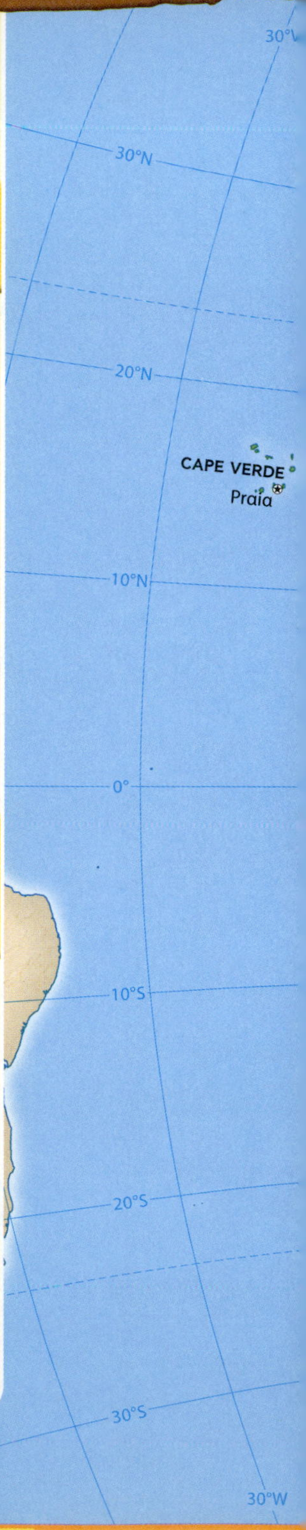

CAPE VERDE
Praia

TECHTREK
FOR THIS CHAPTER

Student eEdition

Maps and Graphs

Interactive Whiteboard GeoActivities

Digital Library

Connect to NG

Go to **myNGconnect.com** for more on Sub-Saharan Africa.

African elephants

WESTERN SAHARA (Morocco)

Laayoune

MAURITANIA
Nouakchott

MALI
Bamako

SENEGAL
Dakar
Banjul
GAMBIA
Bissau
GUINEA-BISSAU
Conakry
Freetown
SIERRA LEONE
Monrovia
LIBERIA
Yamoussoukro

GUINEA

BURKINA FASO
Ouagadougou

CÔTE D'IVOIRE (IVORY COAST)
GHANA
Accra
Abidjan

Niamey
NIGER

NIGERIA
Abuja
Porto-Novo
TOGO
BENIN
Lomé
Cotonou

SAHARA

SAHEL

CHAD
N'Djamena

SUDAN
Khartoum

ERITREA
Asmara

DJIBOUTI
Djibouti

Gulf of Aden

Tropic of Cancer

Nile R.
Red Sea

White Nile R.
Blue Nile R.

Addis Ababa
ETHIOPIA

SOUTH SUDAN
Juba

SOMALIA
Mogadishu

CAMEROON
Yaoundé

CENTRAL AFRICAN REPUBLIC
Bangui

BIOKO
Malabo
EQUATORIAL GUINEA
RÍO MUNI
SAO TOME & PRINCIPE
São Tomé
Libreville
Annobón (Eq. Guinea)

GABON

CONGO
Brazzaville
Kinshasa
CABINDA (Angola)
Luanda

DEMOCRATIC REPUBLIC OF THE CONGO

Congo R.
Uele R.

UGANDA
Kampala

RWANDA
Kigali
BURUNDI
Bujumbura

KENYA
Nairobi

Kilimanjaro
19,340 ft (5,895 m)

TANZANIA
Dodoma
Dar es Salaam

Equator

Victoria
SEYCHELLES

ANGOLA

ZAMBIA
Lusaka

Zambezi R.

MALAWI
Lilongwe

Harare
ZIMBABWE

MOZAMBIQUE

COMOROS
Moroni
Mayotte (France)

Mozambique Channel

MADAGASCAR
Antananarivo

Port Louis
MAURITIUS
Réunion (France)

ATLANTIC OCEAN

Ascension (U.K.)

St. Helena (U.K.)

NAMIBIA
Windhoek

BOTSWANA
Gaborone

Mbabane
Lobamba
SWAZILAND
Maputo

(Tshwane) Pretoria
Bloemfontein

Maseru
LESOTHO

SOUTH AFRICA

Cape Town

Bassas da India (France)
Île Europa (France)

INDIAN OCEAN

Tropic of Capricorn

Prime Meridian

Tristan da Cunha Group (U.K.)

N W E S

0 500 1,000 Miles
0 500 1,000 Kilometers

20°W 10°W 0° 10°E 20°E 30°E 40°E 50°E 60°E

2 3 4 5 6

A B C D E F G H

TECHTREK

myNGconnect.com For an online map of Sub-Saharan Africa and Visual Vocabulary

Maps and Graphs

Digital Library

SUB-SAHARAN AFRICA PHYSICAL

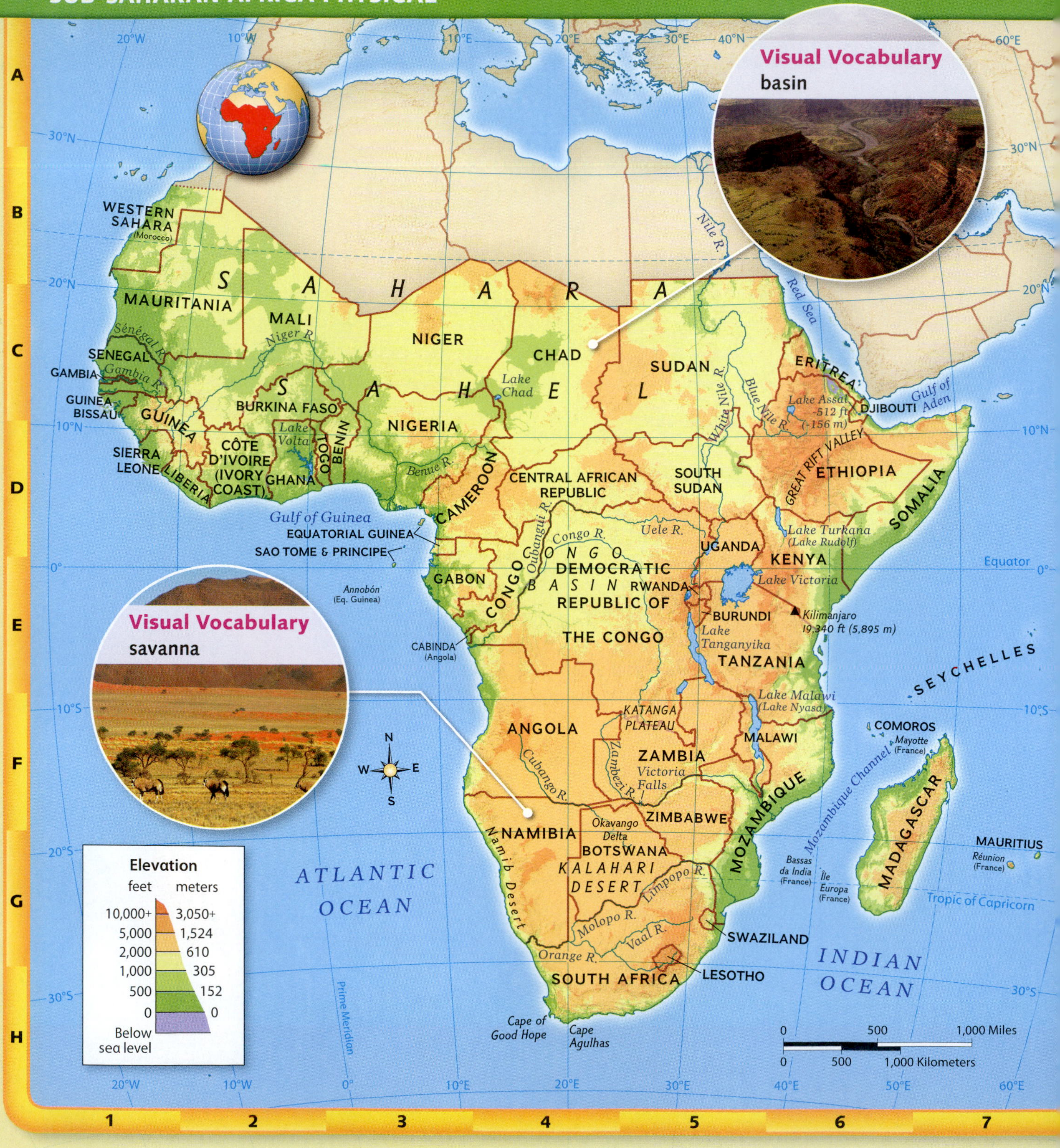

Visual Vocabulary
basin

Visual Vocabulary
savanna

WESTERN SAHARA (Morocco)

S A H A R A

MAURITANIA
MALI
NIGER
CHAD
SUDAN
ERITREA

Lake Assal
−512 ft
(−156 m)
DJIBOUTI
Gulf of Aden

S A H E L

SENEGAL
GAMBIA
GUINEA BISSAU
GUINEA
SIERRA LEONE
LIBERIA
BURKINA FASO
CÔTE D'IVOIRE (IVORY COAST)
GHANA
TOGO
BENIN
NIGERIA
CAMEROON
CENTRAL AFRICAN REPUBLIC
SOUTH SUDAN
ETHIOPIA
SOMALIA

Niger R.
Sénégal R.
Gambia R.
Lake Volta
Benue R.
Lake Chad

GREAT RIFT VALLEY

EQUATORIAL GUINEA
SAO TOME & PRINCIPE
Gulf of Guinea

Oubangui R.
Congo R.
Uele R.
UGANDA
KENYA
Equator

GABON
CONGO
C O N G O   B A S I N
DEMOCRATIC REPUBLIC OF THE CONGO
RWANDA
BURUNDI
Lake Turkana (Lake Rudolf)
Lake Victoria
▲ Kilimanjaro 19,340 ft (5,895 m)

Annobón (Eq. Guinea)
CABINDA (Angola)

Lake Tanganyika
TANZANIA

SEYCHELLES

ANGOLA
KATANGA PLATEAU
ZAMBIA
MALAWI
Lake Malawi (Lake Nyasa)
COMOROS
Mayotte (France)

Cubango R.
Zambezi R.
Victoria Falls
ZIMBABWE
MOZAMBIQUE
Mozambique Channel
MADAGASCAR
MAURITIUS
Réunion (France)

Okavango Delta
NAMIBIA
BOTSWANA
KALAHARI DESERT
Bassas da India (France)
Île Europa (France)

Namib Desert
Limpopo R.
Molopo R.
Orange R.
Vaal R.
SWAZILAND
SOUTH AFRICA
LESOTHO

Tropic of Capricorn

ATLANTIC OCEAN
INDIAN OCEAN

Prime Meridian

Cape of Good Hope
Cape Agulhas

N
W   E
S

Elevation

feet	meters
10,000+	3,050+
5,000	1,524
2,000	610
1,000	305
500	152
0	0
Below sea level	

0   500   1,000 Miles
0   500   1,000 Kilometers

Nile R.
Red Sea
White Nile R.
Blue Nile R.

Main Idea Sub-Saharan Africa is divided into four parts, each with varied geographic features.

Sub-Saharan Africa lies south of the Sahara desert. The region extends from Senegal in western Africa to Ethiopia and Somalia in eastern Africa, and southward to the tip of the continent. Sub-Saharan Africa includes the island nation of Madagascar (ma duh GAS kahr).

Africa's Continental Drift

As you have learned, the continents once belonged to one supercontinent called Pangaea. The movement of Earth's tectonic plates caused the continents to drift apart, and Africa became a separate continent. Plate tectonics also caused the formation of certain physical features on the African continent. Lakes, basins, and valleys were all formed as the plates moved. A **basin** is a region drained by a river system.

Landforms and Water

Sub-Saharan Africa is divided into four parts: west, central, east, and south. West Africa includes savannas and much of the Sahel. The **Sahel** (saa HEL) is a semiarid grassland that separates the Sahara in the north from the tropical grasslands, or **savannas**, in the south. In the Sahel, one climate gradually changes to another. Parts of the Sahel, for example, are changing to desert. This process, called **desertification**, means there is less fertile land to grow food. Desertification is caused by many factors including climate change and overpopulation.

Central Africa's primary landform is rain forest, especially in the Congo Basin. In East Africa, **rift valleys**—deep valleys that formed when Earth's crust separated and broke apart—stretch from the Red Sea southward through Mozambique. Southern Africa features great plateaus and another major desert, the **Kalahari**. The Kalahari has limited surface water, but supports a variety of plants and wildlife.

Access to fresh water is an issue for sub-Saharan Africans due to limited water resources and poor sanitation. The situation is improving, however. For example, between 1990 and 2006, the percentage of people in Namibia with access to clean water increased from 57 percent to 93 percent as the government and individual communities began working together to solve the problem.

Before You Move On

Monitor Comprehension What are the main geographic features of sub-Saharan Africa's four parts?

ONGOING ASSESSMENT

MAP LAB

 GeoJournal

1. **Human-Environment Interaction**
   According to the map, the approximate width of the Sahara is 3,000 miles from east to west. Why is the Sahara considered a natural boundary?

2. **Location** Use the map to point out two sources of fresh water in sub-Saharan Africa.

3. **Describe Geographic Information** Describe the causes of desertification and the problem it poses for sub-Saharan farming.

1.2 East Africa and the Rift Valley

TECHTREK

myNGconnect.com For online maps of East Africa and photos of the Great Rift Valley

 Maps and Graphs

 Digital Library

Main Idea East Africa is best known for the Great Rift Valley and the deep lakes found there.

East Africa is easy to spot on a map—it is shaped like a rhinoceros horn. In fact, the region is known as the Horn of Africa. The "horn" is formed by the countries of Somalia, Djibouti (ji BOO tee), Eritrea, and Ethiopia. Other countries in East Africa are Sudan, Kenya, Uganda, Rwanda, Burundi, and Tanzania.

The Great Rift Valley

The most important physical feature of East Africa is the **Great Rift Valley**. This valley is part of a chain of valleys that stretch from southwest Asia to southern Africa. In some places the valley is 60 miles wide. Valley walls often rise more than 6,000 feet in height.

This chain of valleys was formed by tectonic plates that separated and created deep cracks, or rifts, in the earth's crust. The rift valleys have been forming for about 20 million years and continue to develop today.

Plate movements also created the freshwater lakes in the Great Rift Valley. As low spots and rifts developed, they slowly filled with rainwater to create lakes. West of Tanzania is Lake Tanganyika (TANG guhn YEE kuh). At 4,700 feet deep, it is the second deepest freshwater lake in the world.

Plateaus and Savannas

East Africa sits mostly on plateaus. As the climate map shows, the two most substantial elevated areas are in Ethiopia and Kenya. The higher elevations of these areas means the temperatures are cooler, even though much of the area is on or near the equator.

In addition to the rift valleys and lakes, plate movements created volcanoes. Both **Kilimanjaro** (19,340 feet) in Tanzania and Mount Kenya (17,058 feet) in Kenya are dormant, or inactive, volcanoes. The volcanic soil around these mountains is fertile, meaning crops grow well.

> **Visual Vocabulary** A pride, or group, of lions moves through tall savanna grass in Kenya. A pride averages about 15 members.

EAST AFRICA POLITICAL

EAST AFRICA CLIMATE

Climate Regions
- Humid Equatorial–Long dry season
- Dry–Semiarid
- Dry–Arid
- Humid Temperate–Dry winter
- Unclassified Highlands

Both Tanzania and Kenya have vast savannas where wildlife such as lions, giraffes, and elephants roam freely or in protected reserves.

Before You Move On

Summarize Describe the physical characteristics of the Great Rift Valley.

ONGOING ASSESSMENT

MAP LAB

 GeoJournal

1. **Location** Using the maps, locate three sources of freshwater in East Africa.

2. **Interpret Maps** Based on the climate map, which countries have the largest variety of climate regions in East Africa? How does that information add to your understanding of this region?

3. **Make Inferences** How might climate, elevation, and access to water determine where people live in East Africa?

1.3 West Africa's Steppes

TECHTREK

myNGconnect.com For an online map and photos of desertification

 Maps and Graphs

 Digital Library

Main Idea West Africa's physical geography includes steppes and highlands as well as tropical coast and dry desert.

West Africa runs from the Atlantic coast south of the Sahara, eastward to the continent's **interior**, or area away from the coast. This part of Africa supports several different ways of life, depending on the landforms and climates found in each area.

Steppes and Highlands

Semiarid steppes, or grasslands, define part of West Africa. As you know, the grassland between the Sahara desert and the tropical savannas is the Sahel. The Sahel runs through the middle of West Africa. The area has a short rainy season and, as a result, is very dry.

The growing population in West Africa has increased the demand for food crops. To create more cropland, West Africans have had to cut down forests, a practice called **deforestation**. Deforestation in West Africa has left the arid soil unprotected.

Soil has eroded, or washed away, because of overuse. These conditions, along with climate change, contribute to desertification, an ongoing problem in sub-Saharan Africa.

Highlands, which are areas of higher mountainous land, are also found in West Africa. The Adamawa Highlands are on Nigeria's eastern border. The Futa Jallon Highlands rise in Senegal on its border with Guinea. This series of sandstone plateaus is marked by rugged canyons.

Tropical Coast, Dry Desert

Coastal countries differ from interior countries in West Africa. Countries on the coast have more people and more cities than those in the interior. Coastal countries often have tropical climates. Some cities on the Gulf of Guinea get more than 80 inches of rain per year, and the coast of Côte d'Ivoire can get more than 10 feet per year. In contrast, the northern desert areas of Niger, an interior country, get less than 10 inches of rain per year.

> **Visual Vocabulary** A **transition zone** is an area between two geographic regions that has characteristics of both. The Sahel (shown here in Mali) is a transition zone between the Sahara in the north and the savannas in the south.

The largest cities in West Africa are located near or on the Atlantic coast. The coastal city of Lagos, Nigeria, is the largest city in West Africa and one of the largest in the world. Such coastal cities have many advantages. Adequate rainfall, fishing, and trade all make it easier for West African coastal cities to support growing populations.

The interior countries of West Africa are mostly covered by desert. Deforestation and desertification, along with the lack of water, make farming in the poor soil a constant struggle. About 80 percent of people in Chad make their living from subsistence farming and raising livestock. In 2003, Chad began exporting oil. This new economic resource may improve Chad's economy and make money available for improved farming technology.

Before You Move On
Monitor Comprehension In what ways are West Africa's steppes different from the coasts?

WEST AFRICA POLITICAL

CITIES AND WATER Nearly all of the West African capitals are either near the coast or near a river. Water is essential for cities to grow. Areas with good access to water have better farming, more industry, and the ability to support larger populations. Large cities around the world tend to build up around good sources of water.

PHOTO LAB  GeoJournal

1. **Analyze Visuals** Based on the photo, what characteristics of the Sahel in Mali make it a transition zone?

2. **Location** What does the photo suggest about the challenges of living in this transition zone?

3. **Interpret Maps** What makes the capital of Nigeria different from other capitals in West Africa?

1.4 Rain Forests and Resources

TECHTREK
myNGconnect.com For a land use map and photos of wildlife
Maps and Graphs
Digital Library

Main Idea Central Africa is defined by the rain forests of the Congo River Basin and a variety of natural resources.

Central Africa is bordered by the Adamawa Highlands of West Africa and the Great Rift Valley of East Africa. On the north and south, Central Africa lies between the plateaus of the Sahel and Southern Africa.

Rain Forest in the Congo Basin

The Congo Basin is the main geographic feature of Central Africa. The basin is located on the equator and surrounded by higher elevations. Within the basin is a **rain forest**, which is a forest with warm temperatures, plentiful rain, high humidity, and thick vegetation.

The rain forest in Central Africa is second in the world in size only to the South American rain forest. Vegetation grows so thick that sometimes sunlight does not reach the ground. The plants also make ground movement difficult, so most people live on the edges of the rain forest.

Spectacular wildlife also thrives in the rain forest. The okapi, related to the giraffe, lives in the Congo Basin. Some other rain forest animals include gorillas similar to the one shown here, leopards, and rhinoceroses.

The **Congo River** is a major waterway in Central Africa. Like the Amazon River of South America, the Congo is located in an equatorial area and flows into the Atlantic Ocean. Several large rivers in Central Africa feed into the Congo River, including the Ubangi, the Aruwimi, and the Lomami rivers.

Resources of Central Africa

Central Africa includes the Central African Republic, Congo, Cameroon, São Tomé (sow too MAY) and Príncipe, Equatorial Guinea, and Gabon.

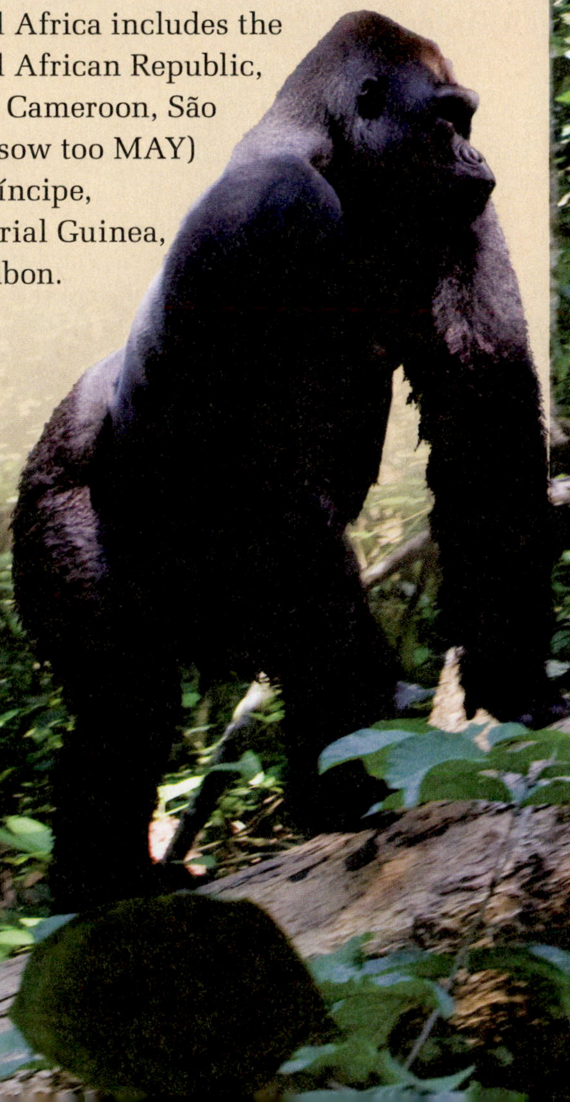

Critical Viewing A western lowland gorilla walks in the forest in Democratic Republic of the Congo. Why might the rain forest be able to support so many plants and animals?

CENTRAL AFRICA'S LAND USE

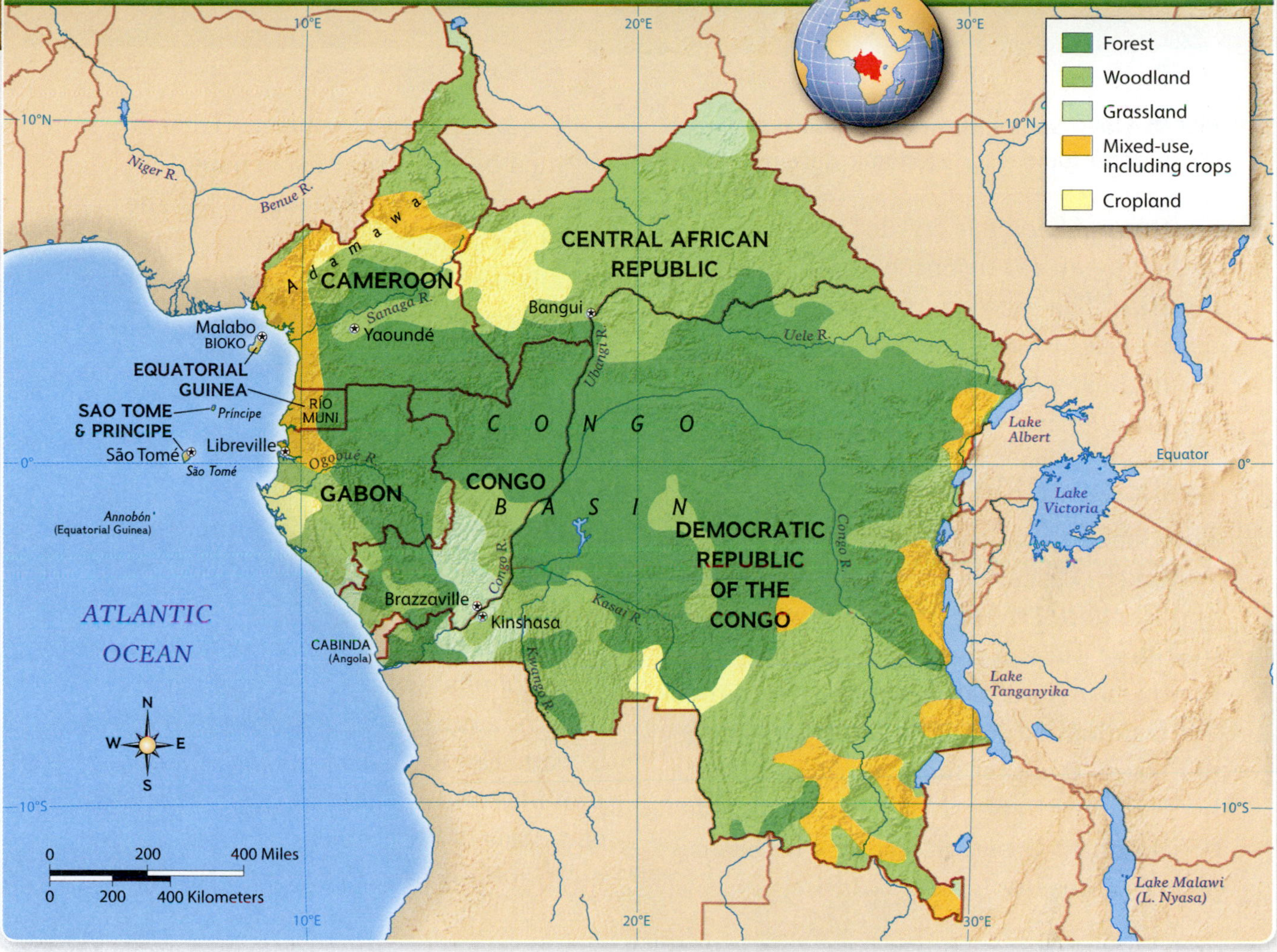

Forest
Woodland
Grassland
Mixed-use, including crops
Cropland

Niger R.
Benue R.
10°N
10°N
CAMEROON
CENTRAL AFRICAN REPUBLIC
Adamawa
Sanaga R.
Malabo
BIOKO
Bangui
Uele R.
Yaoundé
EQUATORIAL GUINEA
SAO TOME & PRINCIPE
Príncipe
RÍO MUNI
C O N G O
Lake Albert
Equator
Ogooué R.
São Tomé
Libreville
0°
0°
São Tomé
GABON
CONGO
B A S I N
DEMOCRATIC REPUBLIC OF THE CONGO
Lake Victoria
Annobón (Equatorial Guinea)
Congo R.
Congo R.
ATLANTIC OCEAN
Brazzaville
Kasai R.
Kinshasa
Lake Tanganyika
CABINDA (Angola)
Kwango R.
10°S
10°S
0   200   400 Miles
0   200   400 Kilometers
Lake Malawi (L. Nyasa)
10°E
20°E
30°E

However, the largest, most populous country in Central Africa is the Democratic Republic of the Congo (DRC). The capital of the DRC is Kinshasa.

The DRC has a wealth of natural resources, including copper, forests, diamonds, and the Congo River itself. The river provides **hydroelectric power**, or electricity produced by a water source such as a river. The river is forced through turbines, or engines, and produces electricity for the DRC and other countries.

Before You Move On
Monitor Comprehension Describe the rain forest and natural resources found in Central Africa.

ONGOING ASSESSMENT
MAP LAB

GeoJournal

1. **Interpret Maps** How does the latitude of the Congo Basin help explain its climate and vegetation?

2. **Conduct Internet Research** Use the Internet to research the Amazon River Basin of South America. On a chart like the one below, compare and contrast the Congo River Basin and Amazon River Basin.

	LOCATION	SIZE	RAINFALL
Congo River Basin			
Amazon River Basin			

1.5 Southern Plateaus and Basins

TECHTREK
myNGconnect.com For online maps and photos of Southern Africa
Maps and Graphs
Digital Library

Main Idea Southern Africa's physical geography offers opportunities for economic development.

Southern Africa has fertile farmland, valuable natural resources, and abundant wildlife. The income from exporting natural resources gives Southern Africa the highest standard of living in sub-Saharan Africa.

Basins and Plateaus

The Congo Basin extends into the Southern African countries of Angola and Zambia. From there, the land rises to a large plateau that spans most of Southern Africa. Six of the area's countries are **landlocked**, or have no direct access to a coast. Another major African basin, the Kalahari in Southern Africa, includes the Kalahari Desert. However, the basin also has areas rich with wildlife.

The plateau of Southern Africa is defined by the **Great Escarpment**. An **escarpment** is a steep slope. The Great Escarpment is the steep slope from the plateau down to the coastal plains of Southern Africa. The Great Escarpment is most dramatic in the countries of South Africa and Lesotho. It also extends northeast into Zimbabwe and northwest into Namibia and Angola.

The **Zambezi River** in Southern Africa collects water from the entire south-central part of Africa. The Zambezi flows through Angola, Zambia, along Zimbabwe's border, and through Mozambique (mo zam BEEK) to the Indian Ocean. The Kariba Dam on the Zambezi provides hydroelectric power. In fact, the countries of Zambia and Zimbabwe get most of their electricity from the dam.

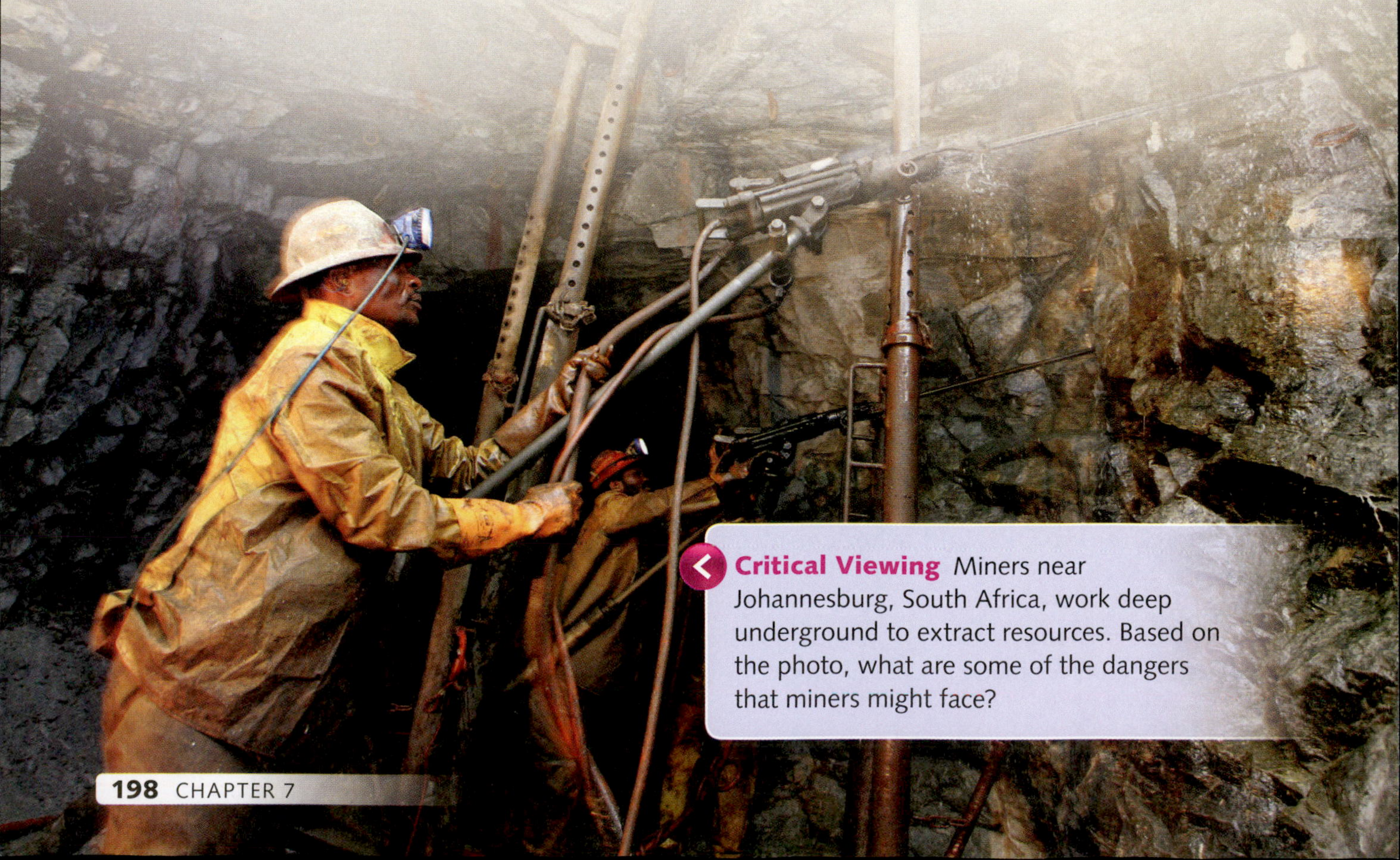

Critical Viewing Miners near Johannesburg, South Africa, work deep underground to extract resources. Based on the photo, what are some of the dangers that miners might face?

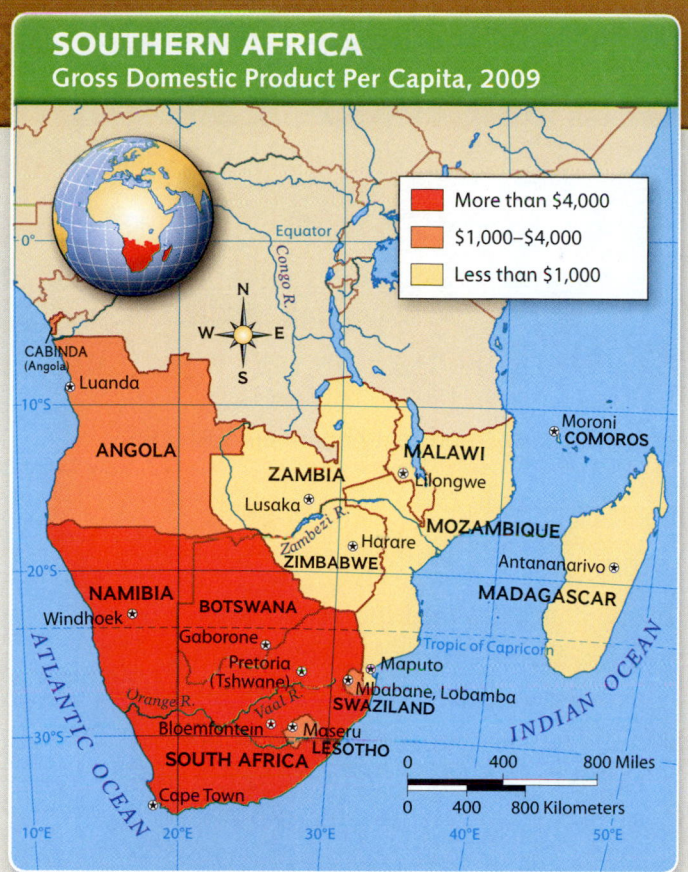

SOUTHERN AFRICA
Gross Domestic Product Per Capita, 2009

More than $4,000
$1,000–$4,000
Less than $1,000

SOUTHERN AFRICA
Natural Resources

Coal
Gold
Copper
Uranium
Diamonds
Tea
Corn
Wheat
Tobacco
Sugarcane
Sheep
Fish

Mining and Farming

A zone of mineral deposits winds through Zambia, Zimbabwe, and South Africa. Copper, gold, and diamonds are mined in this zone, and destined for jewelry and industrial uses. In fact, South Africa is one of the world's largest gold producers. Many people migrate there to work in the mines even though the work can be dangerous.

Southern Africa's temperate climate supports a variety of crops. For example, South Africa has many vineyards that thrive on its plateau, and Zimbabwe has tea plantations on its eastern escarpment. Angola produced nearly 20 percent of the world's coffee until 1975 when a civil war began. When the war ended in 2002, coffee production started again. Fruits, such as bananas, pineapples, and apples, are grown throughout Southern Africa. Corn, wheat, and other grains can also be found across the area.

South Africa, Botswana, and Namibia have the highest Gross Domestic Product in Southern Africa. Other countries are working to overcome factors that have weakened their economies, such as civil war and disease. For them, economic success is a goal for the future.

Before You Move On

Monitor Comprehension In what ways do the area's resources and crops support its economy?

ONGOING ASSESSMENT

MAP LAB

 GeoJournal

1. **Compare and Contrast** According to the Gross Domestic Product map, which countries have the highest GDP? The lowest GDP?

2. **Interpret Maps** In what ways might South Africa's natural resources contribute to differences on the GDP map?

3. **Movement** What circumstances have led many workers to migrate to Southern Africa?

TECHTREK

myNGconnect.com For an online map, photos, and an Explorer Video Clip

Maps and Graphs

Digital Library

Exploring
Africa's Wildlife
with Dereck and Beverly Joubert

> **Main Idea** Humans are working to protect endangered African big cats and their homes.

Big Cats in Africa

Big cats, such as lions, cheetahs and leopards, are a vital part of African wildlife. Since the 1940s, the number of lions in Africa has been reduced from about 450,000 to 20,000—and humans caused most of this reduction. Hunting, movement into big cat **habitats** (natural homes), and **poaching** (illegal hunting) all contribute to fewer big cats in the wild.

Working to Protect Habitats

"It seems like we were explorers from birth, wandering the wild Earth with a passion," said Dereck Joubert. The Jouberts, who are National Geographic Explorers-in-Residence, learned about wildlife on game reserves in Southern Africa. Today, they live in Botswana. "On our first trip to Botswana and the **Okavango** (oh kuh VAANG oh) **Delta**, in 1981, we felt we had come home," the Jouberts noted.

That same year, the Jouberts joined the Chobe Lion Research Institute in Botswana. They began an intensive study of lions,

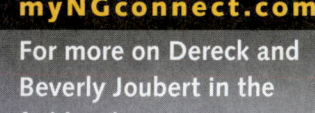

myNGconnect.com

For more on Dereck and Beverly Joubert in the field today

AFRICAN WILDLIFE RESERVES

Legend:
- National Park, Wildlife Reserve
- Chobe Lion Research Institute

Tropic of Cancer

Banc d'Arguin Nat. Park
Aïr and Ténéré Nat. Res.
Nafka Wildlife Res.
Faro Res.
Babile Elephant Sanctuary
Salonga Nat. Park
Léfini Faunal Res.
Masai Mara Nat. Res.
Serengeti Nat. Park
Rungwa Game Res.

Equator

ATLANTIC OCEAN

Okavango Delta
Skeleton Coast Park
Central Kalahari Game Res.

Tropic of Capricorn

0   600   1,200 Miles
0   600   1,200 Kilometers

20°W   40°S   0°   20°E   40°E   60°E

INDIAN OCEAN

Addo Elephant Nat. Park

BIG CATS BY THE NUMBERS

5

Distance in miles from which a lion's roar can be heard

12

Average life span of a male lion in the wild

23

Distance in feet a cheetah can cover in one stride

70

A cheetah's top speed in miles per hour

550

Weight in pounds of a fully grown male lion

Source: Smithsonian National Zoo, National Geographic Society

---

which included adopting a **nocturnal**, or night-based, lifestyle. They worked during the night in the African wilderness for months at a time. Since leaving the Chobe Institute, the Jouberts have worked on their own in many different filmmaking, photography, and conservation projects.

The Jouberts have a mission: conserving big cat habitats. Protecting habitats also protects the biodiversity, or variety of life, in the habitat. The Jouberts' projects draw attention to the plight of big cats in Africa. Their knowledge of how big cats live helps other conservationists to develop habitat protection programs.

The Jouberts also support **ecotourism**, which is tourism that is focused on wildlife protection and responsible use of land and resources. Ecotourism teaches visitors about conservation issues. For the Jouberts, teaching is an essential part of their work. Ultimately, they believe that "We are part of a global community of lions and leopards, buffalo, dung beetles, snakes, trees, and ice caps, not somehow apart from it all."

Before You Move On
Summarize What are some ways of protecting African big cats?

ONGOING ASSESSMENT
READING LAB   GeoJournal

1. **Make Inferences** Using what you know about sub-Saharan geography, what might explain the lack of wildlife reserves in the north?

2. **Interpret Data** What do the cheetah's speed and stride distance imply about its ability as a hunter?

2.1 Bantu Migrations

TECHTREK

myNGconnect.com For a map of the Bantu migrations and photos of Bantu culture

Maps and Graphs

Digital Library

Main Idea Bantu-speaking people migrated from West Africa and influenced the language and culture of the African continent.

As you have read, sub-Saharan Africa's varied physical features support a variety of crops today. About 10,000 years ago an agricultural revolution began in Central Africa. During the **agricultural revolution**, humans began to grow crops instead of gathering plants. Sometime after the agricultural revolution, around 2000 B.C., one of the greatest migrations in human history began—the **Bantu** migration. By A.D. 1000, Bantu peoples had spread from their Central African homeland south and east across sub-Saharan Africa.

The Bantu People

People migrate for economic, political, religious, social, and environmental reasons. When groups of people move, cultures blend, new languages form, technology spreads—and sometimes conflict occurs.

Historians and anthropologists are not certain why the Bantu chose to migrate when and where they did. Whatever the reason, the Bantu people began to move and carried their cultural traits and skills along with them.

The Bantu had knowledge of iron working, and their use of iron weapons gave them an advantage over other tribal groups. As the Bantu people moved, they forced other groups to move or become absorbed into Bantu culture.

The many different groups that became part of Bantu culture kept much of their own culture as well. As a result, the nearly 85 million people who trace their history to the Bantu migrations share a very diverse culture today.

> **Critical Viewing** Bantu descendants pick tea in South Africa. One possible reason for the Bantu migration was to find better farmland. What does this photo suggest about farming methods in South Africa today?

Bantu Languages

Today, people whose ancestors were Bantu exist in more than 400 ethnic groups, including the Zulu, the Swahili, and the Kikuyu. Original Bantu languages have evolved into more than 450 languages.

Swahili (also known as Kiswahwali) is one of the best-known of the surviving Bantu languages. For more than 5 million people, Swahili is the language they learn as children, or their ==first language==. For 30 million other people, Swahili is their second language. Swahili is spoken mainly in East Africa and can be heard in several different forms, or dialects.

Swahili is heavily influenced by the Arabic language. Arab traders from northern Africa and Bantu-speaking people of eastern Africa met and exchanged goods over many centuries. Over time, Swahili became the language used to conduct trade.

Beginning in the early 19th century, Arab trade ==caravans==—groups of merchants traveling together for safety—traveled farther into the interior of Africa, spreading Swahili to more people. Eventually, Swahili would be used by some Europeans who colonized parts of Africa where the language was spoken.

Today, many areas in Africa have two groups of people with different first languages. In these areas, Swahili is frequently spoken by both groups as a way to communicate. As a result, Swahili is the ==lingua franca== (LEEN gwa FRAWN kah), or common language between multiple groups of people.

BANTU MIGRATIONS

Paths of migrations
- Bantu homeland, c. 2000 B.C.
- Northwestern Bantu region, by A.D. 500
- Eastern Bantu region, by A.D. 500
- Western Bantu region, by A.D. 500

Before You Move On
Summarize How did the Bantu influence the culture and language of the African continent?

ONGOING ASSESSMENT
MAP LAB
GeoJournal

1. **Interpret Maps** What landform in Africa might have influenced the Bantu people to migrate south instead of north?

2. **Movement** Based on the map, over what period of time did the Bantu migration take place?

3. **Draw Conclusions** How was the migration of the Bantu people supported by their use of metal-working technology?

4. **Make Inferences** Why might it be important for an area to have a *lingua franca*, or common language?

2.2 Early States and Trade

TECHTREK

myNGconnect.com For an online map of the African empires and and photos of Great Zimbabwe

 Maps and Graphs

 Digital Library

Main Idea Trade helped develop powerful states and empires in sub-Saharan Africa.

Trade was important in the development of sub-Saharan Africa. The promise of valuable trade goods brought Arab traders to a transportation corridor from North Africa to West Africa. This **trans-Saharan** trade, or trade across the Sahara, introduced Africans to the Islamic religion, which spread from the Arabian peninsula beginning in the 8th century A.D.

West African Empires

In West Africa, a number of empires arose over the centuries and all thrived because of the gold and salt trade. **Alluvial** (a LOO vee ahl) gold, or gold deposited by a river, was found in forests, and salt was found in deserts. These goods were traded in the savanna between the forest and desert, where the empire of Ghana began.

Ghana gained wealth and power by taxing the gold and salt trade, controlling West Africa from A.D. 700 to the 1200s. Ghana declined, and the kingdom of Mali, led by King Sundiata, overtook Ghana. His great-nephew, Mansa Musa, continued to control trade, spread Islam, and make the city of **Timbuktu** a center of education.

Another empire that traded gold and salt was Songhai. Brought to its height by Askia Muhammad, Songhai prospered from the 900s to the 1400s. South of Songhai, the kingdom of Benin lasted from the 1200s to the 1800s. Benin actively traded with European countries such as Portugal and the Netherlands.

East African Empires and States

The powerful East African empire of Aksum was located in present-day Ethiopia. Aksum flourished between A.D. 300 and 600. Adulis, its main port, was located on the Red Sea and served as a center of trade. Many **city-states** (independent states made up of a city and the territories depending on it), such as Mogadishu, formed on the coast of East Africa as trading grew.

Other African States

Powerful states also arose in Central and Southern Africa. In Central Africa, the state of **Kongo** (different from the modern country of Congo) was founded in 1390. Kongo became known for its highly-organized government. Soon after its founding, the Portuguese arrived and became involved in many aspects of the state, including politics, trade, and religion.

Aksum Coins

300

600

900

A.D. **300**
East African empire of Aksum begins growing to its height in present-day Ethiopia.

A.D. **700**
Ghana becomes center of gold and salt trade in West Africa.

Modern vendor cuts a slab of salt to sell in a Mali market.

SUB-SAHARAN AFRICAN EMPIRES

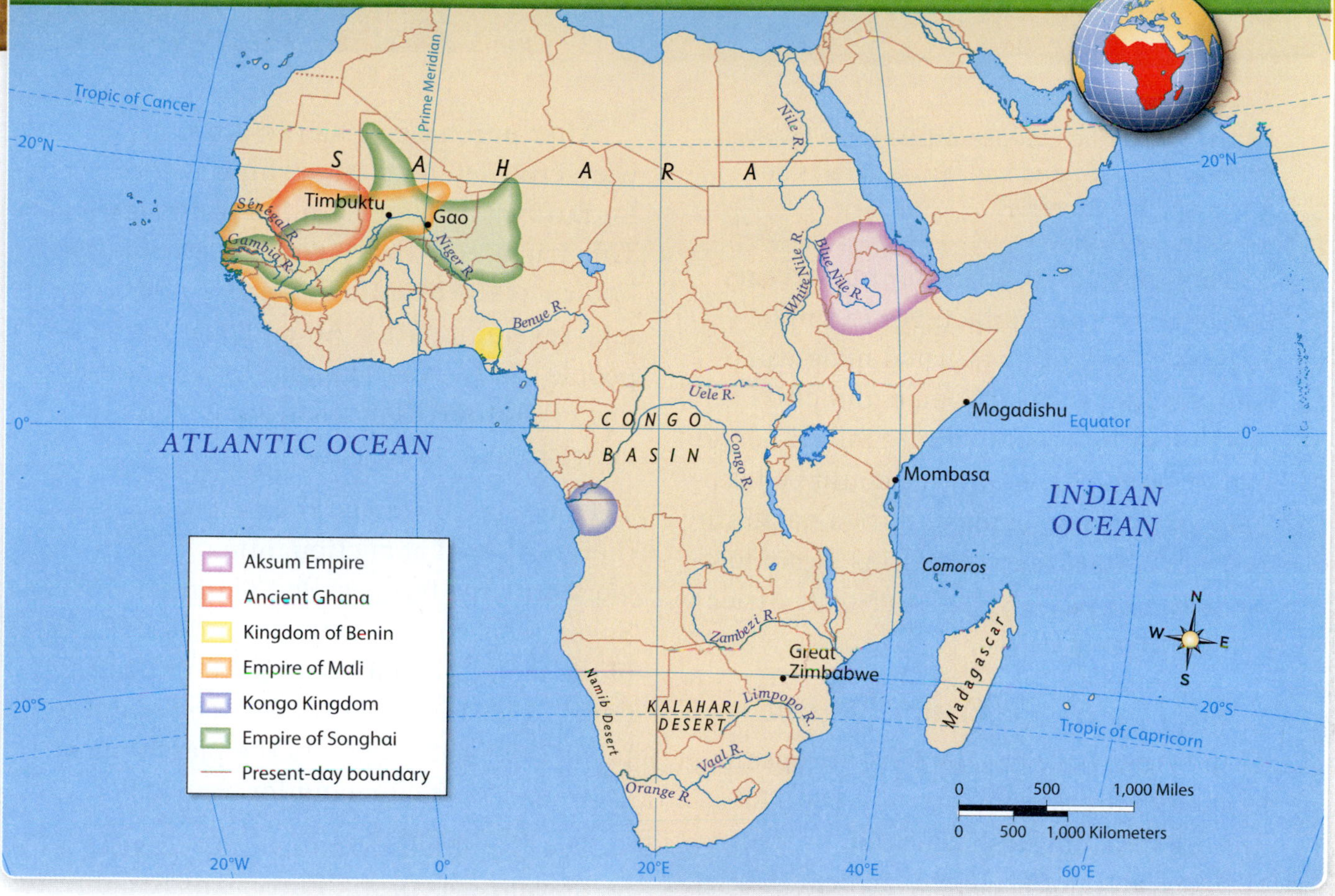
Tropic of Cancer
20°N
S A H A R A
Timbuktu
Gao
Sénégal R.
Gambia R.
Niger R.
Prime Meridian
Nile R.
White Nile R.
Blue Nile R.
Benue R.
Uele R.
C O N G O
B A S I N
Congo R.
ATLANTIC OCEAN
INDIAN OCEAN
Mogadishu
Equator
Mombasa
Comoros
Madagascar
Zambezi R.
Great Zimbabwe
Limpopo R.
KALAHARI DESERT
Namib Desert
Vaal R.
Orange R.
Tropic of Capricorn
0°
20°S
20°W
20°E
40°E
60°E
N
S
E
W
0 500 1,000 Miles
0 500 1,000 Kilometers
Aksum Empire
Ancient Ghana
Kingdom of Benin
Empire of Mali
Kongo Kingdom
Empire of Songhai
Present-day boundary

Between 1200 and 1450, the Shona people in Southern Africa built a walled city out of stone, called **Great Zimbabwe** (zim BAH bwe). *Zimbabwe* is a Shona word meaning "stone houses." The Shona traded gold, copper, and iron with places as far away as China and India.

Before You Move On

Summarize In what ways did trade help develop the early states and kingdoms in Africa?

ONGOING ASSESSMENT

MAP LAB

GeoJournal

1. **Place** Locate the city-state of Mogadishu on the map. Why would its location be beneficial for trade?
2. **Interpret Maps** Based on the map and the time line, which empire was oldest?
3. **Make Inferences** What might explain why the West African empires of Ghana, Mali, and Songhai overlapped?

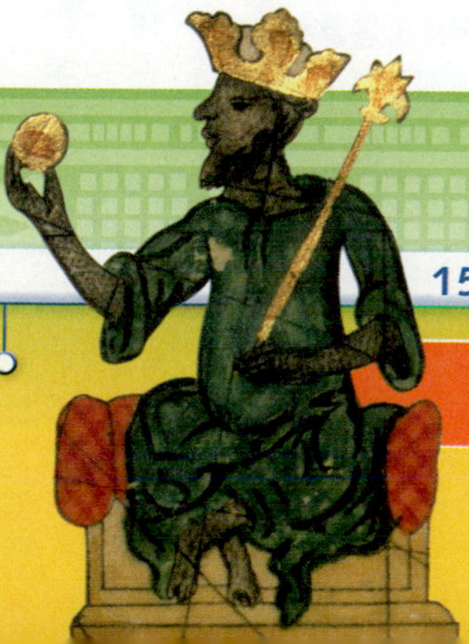

1200 1500 1800

1300s Mali empire in West Africa at its height

1591 Songhai empire comes to an end.

Mansa Musa ruled Mali for about 25 years.

2.3 Impact of the Slave Trade

TECHTREK

myNGconnect.com For research links on the slave trade and Guided Writing

Connect to NG

Student Resources

Main Idea The European slave trade involved millions of people and had lasting effects on Africa and the Americas.

Slavery existed in Africa for many years before European contact. For example, African tribal groups turned male war captives into slaves. Women and children were often incorporated into families, and the children of some slaves could be born free. When Islam came to Africa beginning in the A.D. 700s, some Muslims began to capture and sell Africans to North Africa and Southwest Asia.

European Slave Trade Begins

The Portuguese were the first Europeans to explore the African coast in the 1400s. The **trans-Atlantic slave trade**, or trading of slaves across the Atlantic Ocean, started around 1500. Enslaved people were brought to African coastal cities and held captive until sold. The Portuguese, Spanish, Dutch, French, and English all purchased slaves at African coastal ports.

After purchase, enslaved Africans were crowded onto large ships headed for European colonies in the Americas. This trip across the Atlantic Ocean, known as the **Middle Passage**, could take several months. About 2 million people died in the Middle Passage, many due to **malnutrition** (inadequate food or nourishment) or disease.

Once slaves arrived in the Americas, they were sold at auction, often to go to work on large farms called plantations. Sugar, tobacco, and cotton were some of the major plantation crops. European demand for these crops increased and the plantations got bigger. As the plantations grew, so did the demand for slave labor.

The **incentive**, or motivating reason, for slavery was profit. Europeans bought enslaved people in order to have a cheap and captive labor source. The plantation owners made more money because they did not have to pay the slaves.

> **Critical Viewing** This is a European ship used to transport enslaved Africans. What does this illustration suggest about conditions on the ships for enslaved Africans?

Consequences of the Slave Trade

The trans-Atlantic slave trade lasted from the 1500s to the mid-1800s. Historians estimate that more than 12 million Africans were enslaved and shipped to the Western Hemisphere. The majority of slaves were sent to Brazil and the Caribbean.

People forced into slavery were generally young because they had a better chance of surviving the Middle Passage. Also, when they arrived at their destination, it was expected that young people would be able to work longer and harder in the fields.

THE MIDDLE PASSAGE		
	LEFT FROM AFRICA	**ARRIVED IN THE AMERICAS**
1500–1600	277,506	199,285
1601–1700	1,875,631	1,522,677
1701–1800	6,494,619	5,609,869
1801–1867	3,873,580	3,370,825
TOTAL	12,521,336	10,702,656

Source: http://slavevoyages.org/tast/assessment/estimates.faces

Many Africans taken were male, and many were potential leaders in their community. Families were often torn apart. These losses weakened many African communities and completely destroyed others.

Millions of people in North America, the Caribbean, and South America are descendents of enslaved Africans. These people have shaped cultures in those regions by sharing their languages, customs, and traditions. The impact of the slave trade has lasted for centuries.

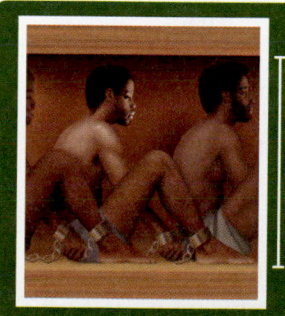

3.25 feet

The journey across the Atlantic could take up to 90 days depending on the weather. Enslaved Africans were packed onto ships and chained below deck. They were only taken above deck for brief periods. Unable to stand up or move, many Africans died where they sat.

Before You Move On

Summarize How did the European slave trade develop and how did it change cultures?

ONGOING ASSESSMENT

DATA LAB GeoJournal

1. **Interpret Charts** Based on the chart, how many people died during the Middle Passage?
2. **Analyze Data** During which time period did the greatest number of enslaved Africans die?
3. **Movement** How did the forced movement of Africans affect communities around the world?
4. **Write Reports** Research one of the following topics and then write a short report on how each affected sub-Saharan Africa: sugar plantations, the triangular trade, or slave ports. Go to **Student Resources** for Guided Writing support.

2.4 Colonization to Independence

Main Idea European powers colonized and ruled large parts of Africa until Africans began independence movements in the mid-1900s.

As you have learned, the Portuguese began exploring the African coast in the 1400s. By the 1500s, many European nations were seeking to control large parts of Africa.

Imperialism and Colonialism

Imperialism is the practice of extending a nation's influence by controlling other territories. European imperialism in Africa began when Europeans started trading with slave merchants on the coast. Little by little, Europeans moved into African lands in search of profitable resources. Eventually, several European countries conquered African lands and established colonies. The practice of directly controlling and settling foreign territories is known as **colonialism**.

By the mid-1800s, European powers began to fight over their African colonies. They wanted more natural resources to fuel industrialization, or the transition to large-scale industries, in Europe. Because of advanced European weapons, there was little the Africans could do to stop them.

Scramble for Africa

In 1884, Europeans held the **Berlin Conference** to settle their disputes about colonial claims in Africa. No Africans were invited to attend. Europeans at the conference divided Africa among themselves. By 1910, France, Germany, Belgium, Portugal, Italy, Spain, and Great Britain had established themselves as colonial powers in all parts of Africa.

Many Europeans believed that African culture and religion were inferior to those of Europe. They wanted to change African traditions. Europeans sent **missionaries** —people sent by a church to spread their religion among native populations—to convert Africans to Christianity.

African Independence

In the early 1900s, **Pan-Africanism**, a movement to unify African people, grew among African leaders in London and other cities around the world. By the 1950s and 1960s, this nationalist movement had brought together many African leaders. **Jomo Kenyatta** (JOH moh ken YAA taa) of Kenya and **Kwame Nkrumah** (KWAA may en KROO mah) of Ghana helped to gain independence for their people.

A caravel ship used in trade voyages

1500	1600	1700

1500s
European imperialism in Africa begins.

1642
The Dutch take possession of Portuguese forts in West Africa.

EUROPEAN COLONIES, 1938

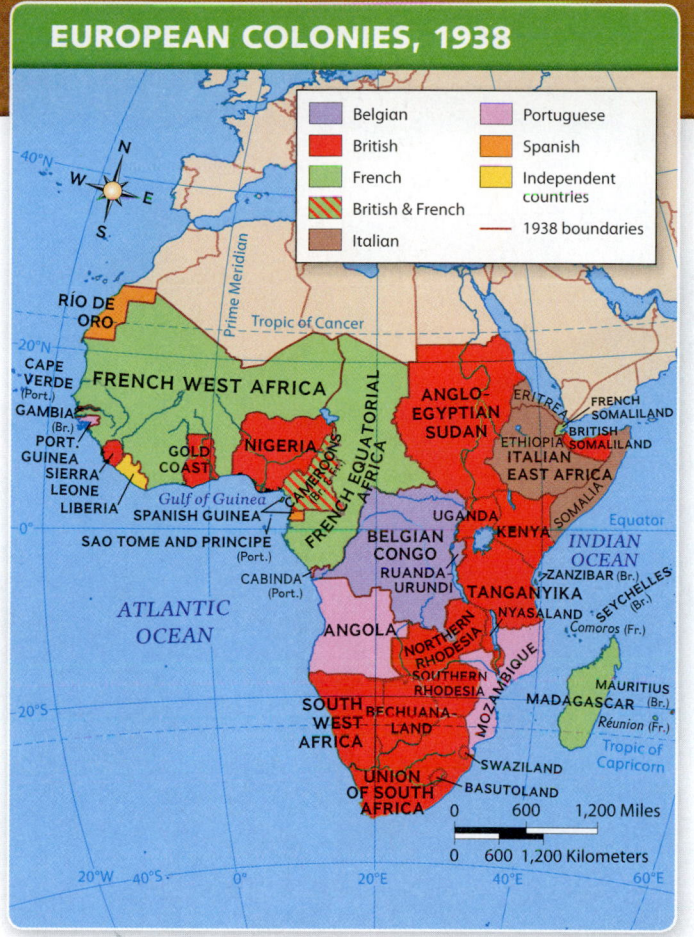

Legend:
- Belgian
- British
- French
- British & French
- Italian
- Portuguese
- Spanish
- Independent countries
- 1938 boundaries

RÍO DE ORO, CAPE VERDE (Port.), FRENCH WEST AFRICA, GAMBIA (Br.), PORT. GUINEA, SIERRA LEONE, LIBERIA, GOLD COAST, NIGERIA, SPANISH GUINEA, SAO TOME AND PRINCIPE (Port.), CABINDA (Port.), Gulf of Guinea, CAMEROONS (Br.), FRENCH EQUATORIAL AFRICA, ANGLO-EGYPTIAN SUDAN, ERITREA, FRENCH SOMALILAND, BRITISH SOMALILAND, ETHIOPIA, ITALIAN EAST AFRICA, UGANDA, BELGIAN CONGO, RUANDA URUNDI, KENYA, TANGANYIKA, ZANZIBAR (Br.), NYASALAND, SEYCHELLES (Br.), Comoros (Fr.), ANGOLA, NORTHERN RHODESIA, SOUTHERN RHODESIA, MOZAMBIQUE, MADAGASCAR, MAURITIUS (Br.), Réunion (Fr.), SOUTH WEST AFRICA, BECHUANA-LAND, SWAZILAND, UNION OF SOUTH AFRICA, BASUTOLAND

ATLANTIC OCEAN, INDIAN OCEAN

0   600   1,200 Miles
0   600   1,200 Kilometers

AFRICAN NATIONS, 2011

WESTERN SAHARA (Morocco), CAPE VERDE, MAURITANIA, MALI, NIGER, CHAD, SUDAN, ERITREA, DJIBOUTI, SENEGAL, GAMBIA, GUINEA BISSAU, GUINEA, BURKINA FASO, BENIN, TOGO, NIGERIA, CENTRAL AFRICAN REP., SOUTH SUDAN, ETHIOPIA, SIERRA LEONE, LIBERIA, CÔTE D'IVOIRE, GHANA, EQUATORIAL GUINEA, SAO TOME & PRINCIPE, CAMEROON, GABON, CONGO, DEM. REP. OF THE CONGO, RWANDA, BURUNDI, UGANDA, KENYA, SOMALIA, TANZANIA, SEYCHELLES, ANGOLA, MALAWI, COMOROS, ZAMBIA, ZIMBABWE, MOZAMBIQUE, MADAGASCAR, Réunion (France), MAURITIUS, NAMIBIA, BOTSWANA, SWAZILAND, SOUTH AFRICA, LESOTHO

ATLANTIC OCEAN, INDIAN OCEAN

0   600   1,200 Miles
0   600   1,200 Kilometers

In 1963 the Organization of African Unity (OAU) was founded to promote Pan-Africanism. The OAU is known today as the African Union (AU). It continues to promote African unity and cooperation, but also functions as an economic group similar to the European Union.

Before You Move On

Make Inferences What action by Africans helped bring colonialism to an end?

ONGOING ASSESSMENT

MAP LAB

GeoJournal

1. **Location** Based on the maps, which modern countries made up the colony of French Equatorial Africa?

2. **Interpret Maps** Looking at the maps, how would you describe the changes to Africa's internal borders from 1938 to today?

3. **Analyze Cause and Effect** Why did European countries colonize Africa? How were Africans affected by colonization?

This political cartoon (1892) suggests that Europe now controls the African continent.

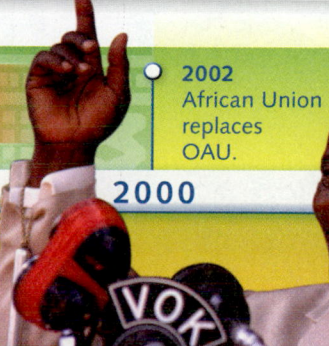

1800

1900

2000

1884
Berlin Conference divides Africa.

1963
Organization of African Unity (OAU) is founded.

2002
African Union replaces OAU.

Jomo Kenyatta

NATIONAL GEOGRAPHIC

Photo Gallery • Nigerian Salt Ponds

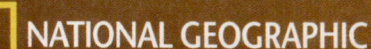

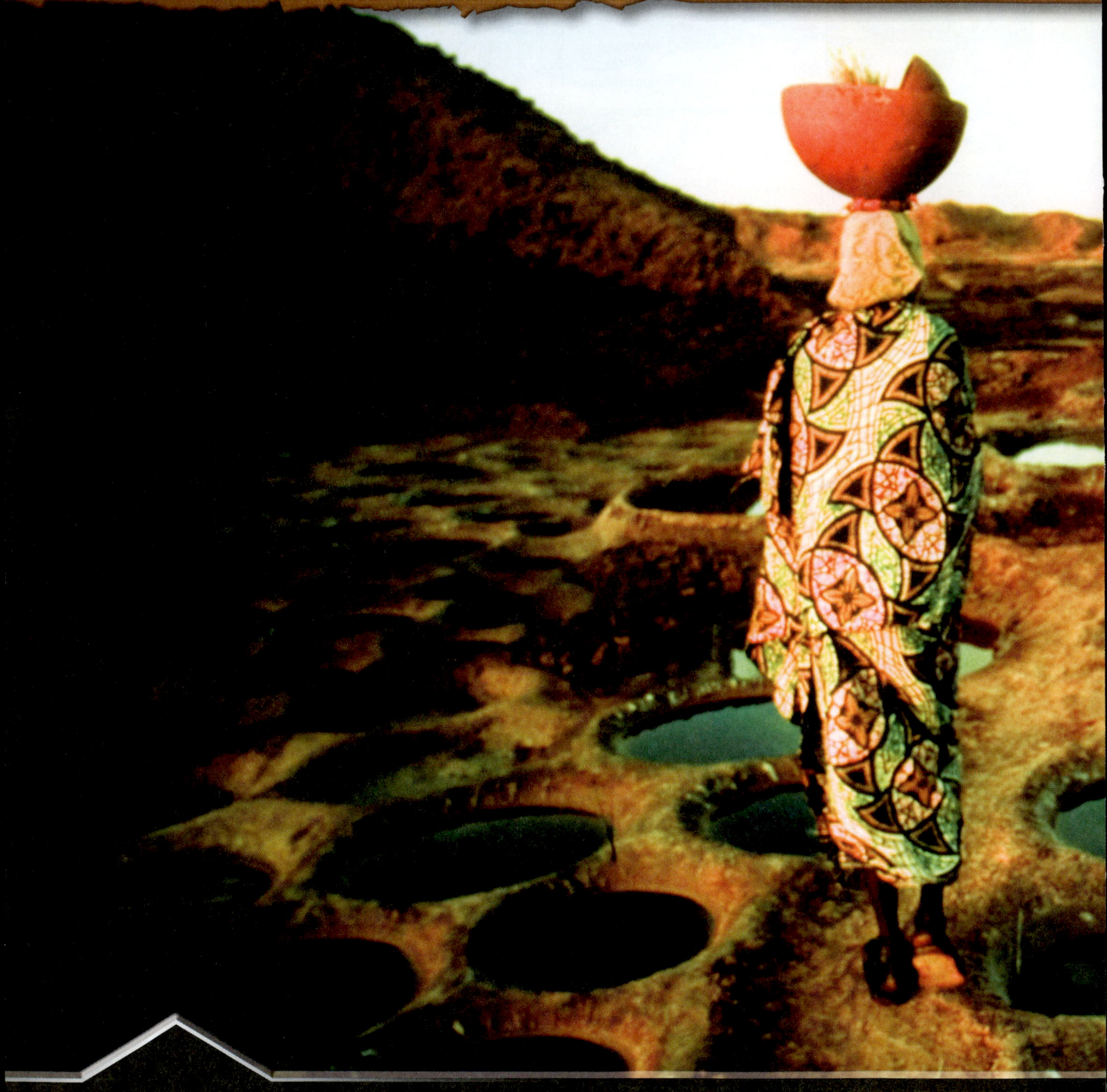

For more photos from the National Geographic Photo Gallery, go to the **Digital Library** at myNGconnect.com.

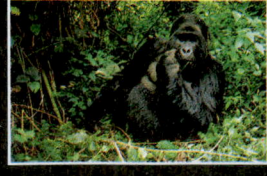

Mountain gorilla, Africa

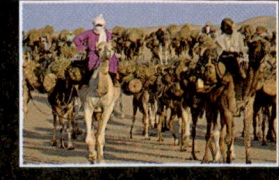

Nigerian camel caravan

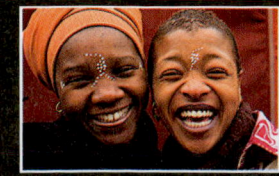

South African women

∨ Critical Viewing These circular ponds contain a mixture of water and salty earth. The mud is moved to evaporation areas where salt is left behind. The color of the salt depends on where the mud came from.

Students in Botswana

African lions

Victoria Falls, Zimbabwe

Zulu village, South Africa

Review

VOCABULARY

For each pair of vocabulary words, write one sentence that explains the connection between the two words.

1. savanna; Sahel

> The Sahel runs south of the Sahara and north of the savannas, forming a border between the two.

2. ecotourism; habitats
3. Swahili; *lingua franca*
4. Middle Passage; trans-Atlantic slave trade
5. imperialism; Pan-Africanism

MAIN IDEAS

6. What are the four regions of sub-Saharan Africa? (Section 1.1)
7. Which large landform in East Africa was created by the separation of tectonic plates? (Section 1.2)
8. Which region of West Africa supports more population, and why? (Section 1.3)
9. What is the major river in Central Africa, and what effect does it have on the region? (Section 1.4)
10. What are the two main types of landforms that make up Southern Africa? (Section 1.5)
11. What factors threaten the big cats of Africa? (Section 1.6)
12. From what African region did the Bantu people begin their migration? (Section 2.1)
13. How did gold and salt help Ghana develop as a kingdom? (Section 2.2)
14. What were some effects of slavery on communities in Africa? (Section 2.3)
15. What reasons did Europeans have for wanting to colonize Africa? (Section 2.4)

GEOGRAPHY

ANALYZE THE ESSENTIAL QUESTION

How has the varied geography of sub-Saharan Africa affected people's lives?

Critical Thinking: Draw Conclusions

16. What might be some ways for Africans to combat the challenges of desertification?
17. How does not having access to a coastal port impact a country?
18. What advantages or challenges does physical geography create for economic development in Central Africa?

INTERPRET TABLES

IMPACT OF CHANGES IN LAND				
Change	Environment	Human Health	Safety Issues	Politics & Economics
Desertification	Loss of habitat; decline in variety of plant and animal life; increased soil erosion	Malnutrition, hunger	Wars over arable land and limited water resources	Poverty; decreased political and economic influence; population movement
Deforestation	Decline in variety of plant and animal life; loss of habitat; reduced resources	Loss of potential new medical products	Increased landslides and flooding	Loss of forest products; loss of indigenous communities; loss of tourism opportunities
Soil Erosion	Loss of soil and habitat; loss of farmland	Loss of food and water; hunger, malnutrition	Risk of flooding and landslides	Loss of property; reduced farm development

Source: http://www.eoearth.org/article/Global_Environment_Outlook_ (GEO-4):_Chapter_3#Introduction

19. **Summarize** How does desertification lead to loss of habitat?
20. **Form and Support Opinions** Think about the causes of desertification, deforestation, and soil erosion. Which one do you think would be easiest to solve? Explain.

HISTORY

ANALYZE THE ESSENTIAL QUESTION

How did trade networks and migration influence the development of African civilization?

Critical Thinking: Analyze Cause and Effect

21. How can the effects of the Bantu migration be seen in Africa today?

22. In what ways did trade networks help build African kingdoms?

INTERPRET MAPS

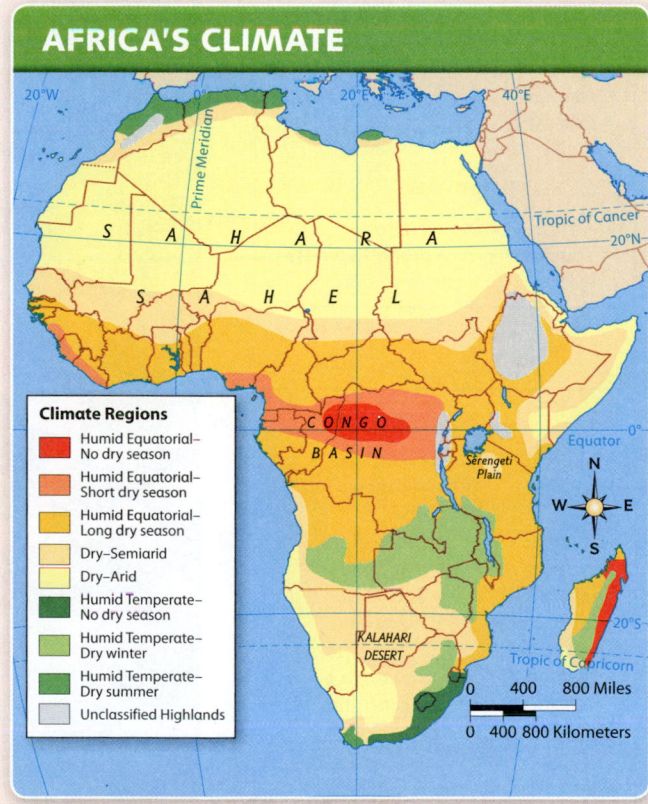

AFRICA'S CLIMATE

Climate Regions
- Humid Equatorial– No dry season
- Humid Equatorial– Short dry season
- Humid Equatorial– Long dry season
- Dry–Semiarid
- Dry–Arid
- Humid Temperate– No dry season
- Humid Temperate– Dry winter
- Humid Temperate– Dry summer
- Unclassified Highlands

23. **Location** What climate zone covers most of the Sahara?

24. **Region** Think about the four areas of sub-Saharan Africa: West, East, Central, and Southern. What area has the largest number of different climate zones?

25. **Make Inferences** What is the relationship between the climate zones and the rain forests of the Congo Basin?

ACTIVE OPTIONS

Synthesize the Essential Questions by completing the activities below.

26. **Create Time Lines** Review this chapter for information on the history of sub-Saharan African countries. Then use a computer or pen and paper to create a time line similar to those found in this book. **Post the time line in your classroom so others can add more entries.**

Time Line Tips
- Take notes before you begin.
- Include important dates for a time line.
- Include important rulers and events.
- Find at least one photo that represents your country.

TECHTREK myNGconnect.com For research links on Sub-Saharan Africa

27. **Create Sketch Maps** Create a natural resources sketch map of sub-Saharan Africa. Use maps in this chapter and research links at **Connect to NG** and other online sources to gather information about natural resources in sub-Saharan Africa. Add them to an outline map similar to the one below. Use a map key to show the resources.

AFRICA

CHAPTER 8

SUB-SAHARAN AFRICA TODAY

PREVIEW THE CHAPTER

Essential Question What historical and geographic factors have influenced the cultures of sub-Saharan Africa?

SECTION 1 • CULTURE

KEY VOCABULARY

- ethnic group
- transportation corridor
- griot
- oral tradition
- reserve
- domestic policy
- modernization
- literacy rate
- ethnobotanist
- medicinal plant

ACADEMIC VOCABULARY

evident

TERMS & NAMES

- African Union
- Youssou N'Dour
- Nairobi
- Jomo Kenyatta

Essential Question How have conflict and government instability slowed economic development in sub-Saharan Africa?

SECTION 2 • GOVERNMENT & ECONOMICS

KEY VOCABULARY

- mineral
- commodity
- coup
- famine
- erosion
- legume
- microcredit
- epidemic
- pandemic
- vaccine
- infectious
- refugee
- clan
- failed state
- segregation
- apartheid
- homeland

ACADEMIC VOCABULARY

concentrated

TERMS & NAMES

- Lost Boys of Sudan
- African National Congress
- Nelson Mandela
- Steve Biko

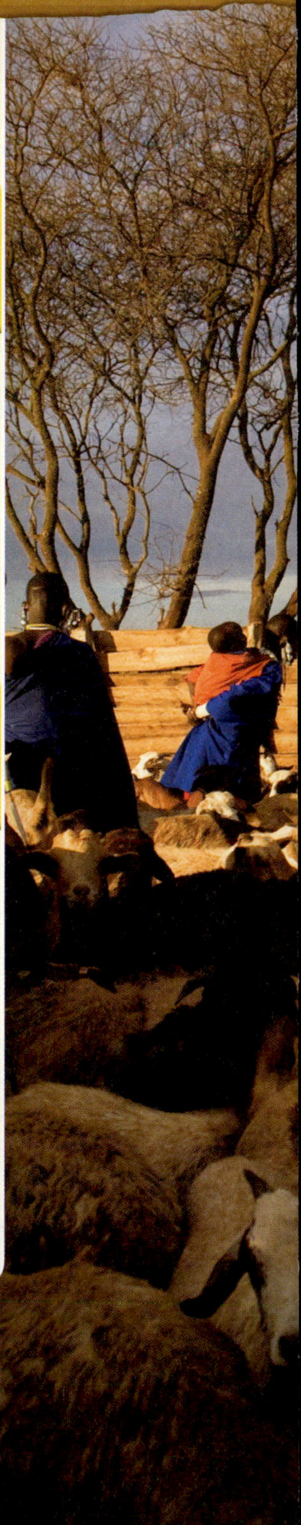

TECHTREK

FOR THIS CHAPTER

Student eEdition

Maps and Graphs

Interactive Whiteboard GeoActivities

Digital Library

Connect to NG

Go to **myNGconnect.com** for more on sub-Saharan Africa.

Traditional Masai in Tanzania
herd a flock of goats into a pen.

1.1 Africa's Borders and Cultures

TECHTREK

myNGconnect.com For a language map and photos of African culture

Maps and Graphs

Digital Library

Main Idea European powers created new colonial borders that ignored the existing borders of traditional African cultures.

Africa is a continent of diverse cultures and countries. Today, about two thirds of Africans live in rural villages that have varied customs and languages. About 1,000 languages are spoken in Africa by different **ethnic groups**, which are groups of people who share a common culture, language, and sometimes racial heritage. Many Africans identify themselves first as members of a tribe rather than a country.

The Impact of Colonialism

Before Europeans arrived, borders between African cultural groups developed by agreement or conflict. Natural features such as oceans, lakes, mountain ranges, and rivers often defined borders. Natural borders were easier to control than borders set by groups of people. Borders along bodies of water also provided **transportation corridors**, or routes to easily move people and goods from place to place.

When Europeans colonized the continent, they routinely ignored the borders of African cultural groups. Instead, colonial borders were set to address the European need for resources. To do so, some sub-Saharan African cultural groups were divided, forced to share territory with rivals, or both.

As African countries began to gain independence in the 1900s, they did not change their colonial borders. As a result, few sub-Saharan African countries share one common culture.

Conflict and Cooperation

Cultural differences across Africa have often led to civil wars over political control, territory, and resources. Somalia's ethnic groups, for example, have never united to form a single nation. Military dictatorships often have been necessary to impose order.

> **Critical Viewing** Fishermen sun-dry fish caught in a lake near their village in Malawi. What can you infer about this fishing culture from the photo?

COOPERATIVE INTERNATIONAL ORGANIZATIONS IN AFRICA

Organization	Purpose
African Union	Peace-keeping; oppose colonization; promote unity; cooperate for economic development
Economic Community of West African States (ECOWAS)	Promote economy, industry, transportation, energy resources, agriculture, and natural resources
Southern African Development Community	Support local economy, transportation networks, and political interaction
Economic Community of Central African States (ECCAS)	Promote industry, transportation, communication, energy, natural resources, economy, tourism, education
East African Community (EAC)	Improve cooperation in transport and communication, industry, security, immigration, and economic matters

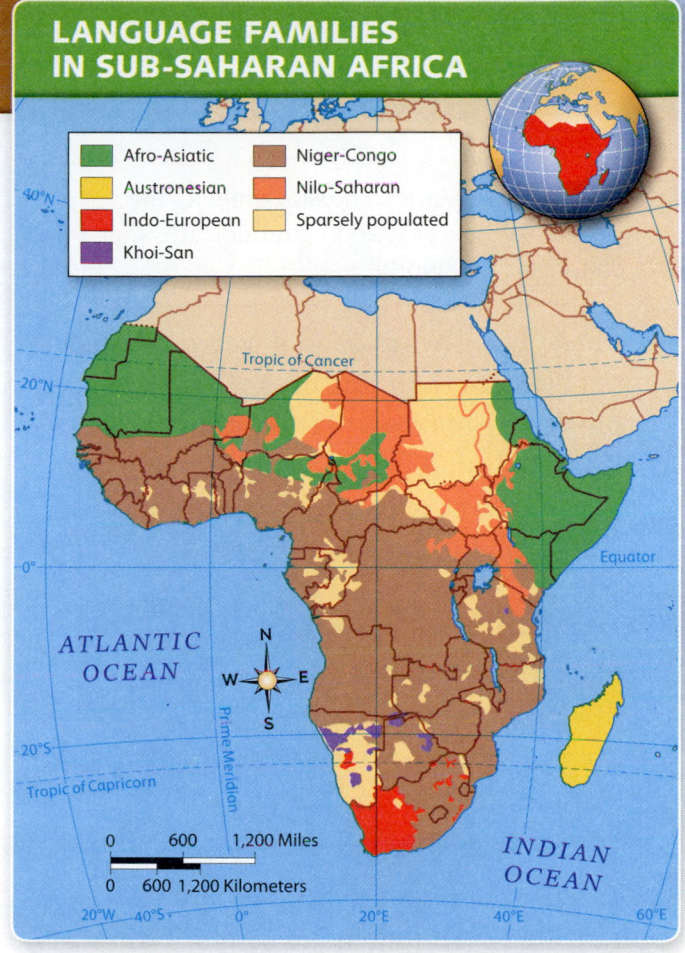

LANGUAGE FAMILIES IN SUB-SAHARAN AFRICA

- Afro-Asiatic
- Austronesian
- Indo-European
- Khoi-San
- Niger-Congo
- Nilo-Saharan
- Sparsely populated

To address these challenges, African countries have developed organizations (see chart above) such as the **African Union** and the East African Community (EAC). These organizations promote cooperation and economic progress within and between countries.

Before You Move On

Summarize How were colonial borders different from sub-Saharan Africa's traditional borders?

MAP LAB

GeoJournal

1. **Interpret Maps** Locate country borders near areas with the Nilo-Saharan language group. What can you infer about the location of language groups and the borders that were created during colonization?

2. **Human-Environment Interaction** On the map, identify the oceans that surround Africa. How do natural features such as rivers, lakes, and oceans help countries control territory and methods of transportation?

3. **Make Generalizations** How did colonization lead to conflict between present-day cultures in Africa?

4. **Synthesize** What is one goal that is shared by the organizations shown in the chart?

1.2 African Music Goes Global

TECHTREK

myNGconnect.com For audio clips of African music and photos of musicians

 Digital Library

 Magazine Maker

Main Idea African music connects the people to their past and communicates their cultures to the world.

Music has always been a way for Africans to celebrate their cultures. Today, African music has become part of many cultures. Experts link African music to the development of jazz, blues, rock-and-roll, and gospel in the United States.

A Wealth of Music

Each area of sub-Saharan Africa has distinctive music that reaches a wide range of audiences. Advancements in communication and transportation have helped to spread African music.

West African music is influenced by the stories and music of griots. **Griots** (GREE oh) are traditional storytellers. For centuries, they orally passed on the histories of West African cultures. Because they were not written down, these histories are part of the **oral tradition**, the practice of orally passing stories from one generation to the next.

Griots accompany their songs with harps, lutes, and drums. Their influence is **evident**, or clearly present, in West African music today.

Much West African music is a fusion, or blend, of griot music and other African music with global music. Mbalax (uhm BALAKS), for example, is a blend of griot percussion and songs with Afro-Cuban influences. **Youssou N'Dour** is a contemporary griot who plays mbalax music. His popularity has helped spread the music out of West Africa to audiences around the world.

In South Africa, music sometimes focused on politics. One example is protest music. Miriam Makeba and a number of other South African musicians left South Africa to protest government policies in the 1970s and 1980s—gaining global audiences for their music.

Before You Move On

Summarize What are some examples of how African music connects the people to their past and communicates their culture to the world?

South African singer Lira performs in Soweto, South Africa.

Critical Viewing A griot playing a drum and blowing a whistle helps to welcome a newborn child in this ceremony in Senegal. What might music contribute to this sort of ceremony?

ONGOING ASSESSMENT

PHOTO LAB

GeoJournal

1. **Draw Conclusions** Both of these photos show African music being played or sung. What do the photos suggest about the significance of music in African cultures?

2. **Create a Photo Essay** You've seen two photos of musical traditions in Africa. Now go to the **NG Photo Gallery** to find more photos of musical traditions. Choose several photos and paste them on a sheet of paper or use the **Magazine Maker** to create a photo essay of musical traditions in Africa. Write a caption for each photo that describes the details shown in the photo and explains the tradition.

1.3 Kenya Modernizes

Main Idea Kenya is modernizing its economy and improving its standard of living.

Kenya's geographic features range from beaches, savannas, deserts, and farmland to snow-capped mountains such as Mount Kenya, for which the country is named. Large mammals such as the African elephant, giraffes, and lions roam Kenya's **reserves**, land set aside for the protection of animals. The diverse geography and wildlife are key to Kenya's economic future.

Cultural Diversity

You can hear many languages and see an array of traditional clothing on a busy street in **Nairobi** (ny ROE bee), Kenya's capital. More than 40 ethnic groups live in Kenya. The largest are the Kikuyu (ki KOO yoo), Luhya, Luo, Kalenjin, and Kamba. This diversity makes Kenya a multicultural society.

As you've learned, cultural diversity has sometimes led to rivalry and conflict in Africa. Even so, diverse groups have been able to work toward common goals.

For example, the people of Kenya united in their struggle for independence from British rule. After independence in 1963, **Jomo Kenyatta**, a Kikuyu, became Kenya's first elected leader. To help all Kenyans pull together, he appointed members from different ethnic groups as his advisors.

Modernization

Kenyatta and other leaders encouraged **domestic policies**, or government plans within a country, that led to economic growth and modernization. **Modernization** is the development of policies and actions designed to bring a country up to world standards in technology and other areas.

Education is an important factor in modernization for Kenya. Most children there attend free elementary schools. The adult **literacy rate**, or percentage of people who can read, soared from 32 percent in 1970 to 85 percent in 2003.

Critical Viewing Students work at Kibagare Good News Centre School in Nairobi, Kenya. What details in the photo remind you of a day in your classroom?

Nairobi, Kenya

KENYA'S PARKS AND RESERVES

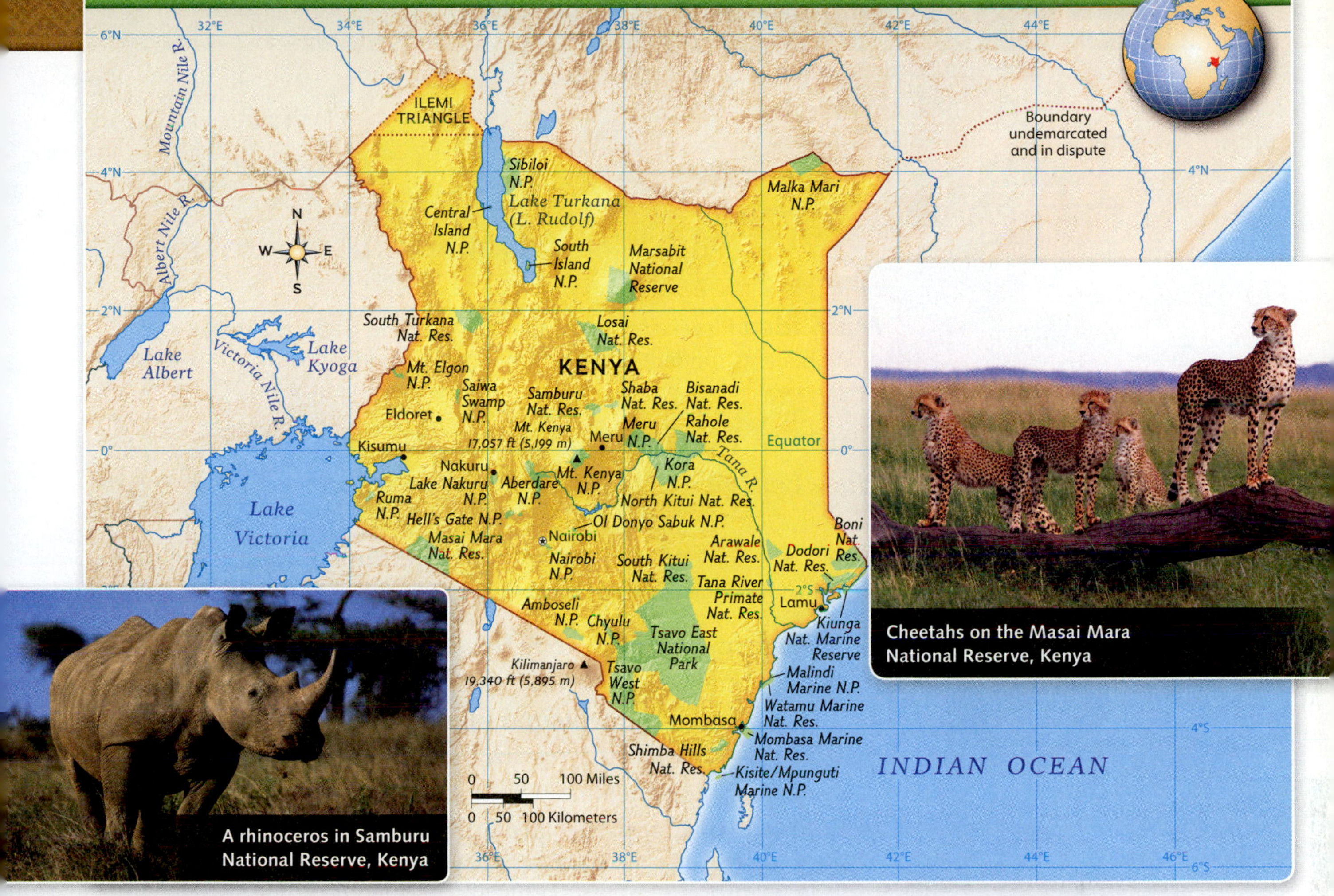

Map labels:

ILEMI TRIANGLE

Boundary undemarcated and in dispute

Mountain Nile R.
Albert Nile R.
Victoria Nile R.
Lake Albert
Lake Kyoga

Sibiloi N.P.
Lake Turkana (L. Rudolf)
Central Island N.P.
South Island N.P.

Malka Mari N.P.

Marsabit National Reserve

South Turkana Nat. Res.
Losai Nat. Res.

KENYA

Lake Victoria
Kisumu
Mt. Elgon N.P.
Saiwa Swamp N.P.
Eldoret
Samburu Nat. Res.
Mt. Kenya 17,057 ft (5,199 m)
Meru N.P.
Shaba Nat. Res.
Bisanadi Nat. Res.
Rahole Nat. Res.

Equator

Nakuru
Lake Nakuru N.P.
Aberdare N.P.
Mt. Kenya N.P.
Kora N.P.
North Kitui Nat. Res.
Ruma N.P.
Hell's Gate N.P.
Masai Mara Nat. Res.
Ol Donyo Sabuk N.P.
Nairobi
Nairobi N.P.
South Kitui Nat. Res.
Arawale Nat. Res.
Boni Nat. Res.
Dodori Nat. Res.
Tana River Primate Nat. Res.
Lamu
Tana R.

Amboseli N.P.
Chyulu N.P.
Tsavo East National Park
Kilimanjaro 19,340 ft (5,895 m)
Tsavo West N.P.
Mombasa
Shimba Hills Nat. Res.
Kiunga Nat. Marine Reserve
Malindi Marine N.P.
Watamu Marine Nat. Res.
Mombasa Marine Nat. Res.
Kisite/Mpunguti Marine N.P.

INDIAN OCEAN

0   50   100 Miles
0   50   100 Kilometers

A rhinoceros in Samburu National Reserve, Kenya

Cheetahs on the Masai Mara National Reserve, Kenya

Kenya's modernization is also evident in its growing tourist industry. The scenic beauty and unique wildlife found in Kenya's national parks and reserves attracts tourists. The government is determined to protect the parks and reserves because of the jobs and money they bring into the country.

In 2000, Kenya, Uganda and Tanzania formed the East African Community in order to improve trade and industry in the region. These efforts to expand the economy should further improve the standard of living for Kenya's citizens.

Before You Move On

Summarize What steps has Kenya taken to modernize?

MAP LAB
GeoJournal

1. **Interpret Maps** What does the number of national parks and reserves shown on the map indicate about their importance in Kenya?

2. **Place** In what ways are the national parks and reserves shown on the map part of the economic modernization of Kenya?

3. **Form and Support Opinions** What makes Kenya diverse? Explain an advantage or disadvantage of diversity in Kenya. Support your opinion with details from the map and from what you have read about Kenya.

4. **Turn and Talk** Think about Jomo Kenyatta's decision to include different ethnic groups in his council. Talk to a partner and discuss effective ways of bringing people together.

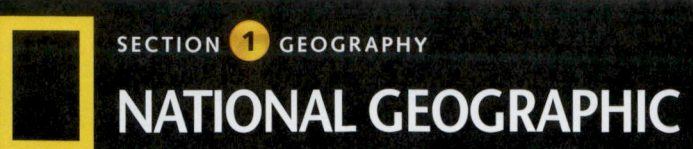

SECTION **1** GEOGRAPHY

NATIONAL GEOGRAPHIC

1.4

TECHTREK

myNGconnect.com For photos of Explorers at work and an Explorer Video Clip

Digital Library

Exploring
Traditional Cultures
with Grace Gobbo and Wade Davis

Main Idea Preserving indigenous cultures and traditional ways of life benefits modern societies around the world.

Cures in the Forest

When Emerging Explorer Grace Gobbo walks through a Tanzanian rain forest, she sees more than trees, flowers, and vines. She sees possible cures. As an **ethnobotanist**, she studies the relationship between cultures and plants.

Tanzanians have relied on traditional healers for centuries. Gobbo interviewed more than 80 traditional healers who rely on **medicinal plants**, or plants used to treat illnesses. Scientific research has confirmed that many of the plants these healers use do help treat illnesses. In fact, large drug companies are now using the same plants in many medicines.

However, many of the forests where these plants grow are at risk from logging and clear-cut farming. By teaching the value of medicinal plants, Gobbo hopes to inspire people to preserve their forests and maintain their traditional healing methods. "Nothing was written down," she explains. "The knowledge is literally dying out with the elders."

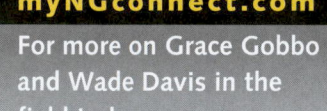

myNGconnect.com

For more on Grace Gobbo and Wade Davis in the field today

AFRICAN PLANTS AND USES

Plant Latin name	Location	Medicinal Use
Tulbaghia violacea	Southern Africa	lowers blood pressure
Catharanthus roseus	Madagascar	treats rheumatism (joint or muscle disorder)
Peltophorum africanum	Southern Africa	relieves stomach problems
Strychnos madagascariensis	Southern Africa	relieves stomach problems
Harpagophytum procumbens	Southern Africa and Madagascar	reduces swelling
Sutherlandia frutescens	Southern Africa	treats cancer
Agapanthus praecox	Southern Africa	treats heart disease and chest pain

Preserving Culture

Like Gobbo, National Geographic Explorer-in-Residence Wade Davis is an ethnobotanist who has studied the use of native plants around the world. His current focus, however, is indigenous cultures around the globe—the customs and practices of people in their daily lives. Davis has lived among 15 cultural groups in the Americas, Asia, Africa, and the Arctic.

Davis has proposed that there is a cultural web of life, just as there is a biological connection between all forms of life. He believes that every culture is connected and that all indigenous cultures have contributed to the cultural web. Davis hopes to preserve this cultural diversity, just as Gobbo works to protect medicinal plants and knowledge of traditional healing in Tanzania.

Davis believes that the development of modern industry and technology has contributed to a decline in the number of indigenous cultures. As a result, the entire cultural web suffers from the loss of each culture. Davis further believes that, like the world's plants and animals, human cultures should be protected so that they can continue to contribute to humanity.

Before You Move On

Monitor Comprehension Why are Gobbo and Davis working to preserve traditional cultures?

ONGOING ASSESSMENT

VIEWING LAB GeoJournal

1. **Analyze Visuals** Go to the **Digital Library** to watch the Explorer Video Clip of Wade Davis. How are his contributions similar to and different from those of Grace Gobbo?

2. **Summarize** According to Wade Davis, why is preserving indigenous cultures important for humanity?

3. **Make Predictions** What image or idea from the video clip was most surprising to you? Write down two questions you still have.

National Geographic Explorer-in-Residence Wade Davis

2.1 Prized Mineral Resources

TECHTREK

myNGconnect.com For an online map and photos of mining in sub-Saharan Africa

Maps and Graphs

Digital Library

Main Idea Sub-Saharan Africa has mineral resources that could improve life for its people.

The mineral resources of sub-Saharan Africa have the potential to lift the economies of sub-Saharan countries. However, careful management is needed to maximize benefits for the African people.

Mineral Riches

Sub-Saharan Africa has large deposits of gold, diamonds, and other **minerals**, which are inorganic solid substances formed through geological processes. Many of these minerals are exported to Europe, North America, and Asia where they are used to make automotive and electronic products.

Many sub-Saharan countries use mineral resources as **commodities**, which are materials or goods that can be bought, sold, or traded. These commodities can bring in enormous wealth, but economic progress has been slow. Over time, government corruption has taken much of the profits from mineral mining.

SELECTED OIL AND DIAMOND EXPORTS SUB-SAHARAN AFRICA, 2005–2006

Oil		
Country	World Rank	Percentage of Country's Total Exports
Nigeria	6	91.9
Angola	12	96.6
Equatorial Guinea	21	92.7
Democratic Republic of the Congo	22	89.6
Sudan	29	88.0
Chad	37	94.6
Diamonds		
Country	World Rank	Percentage of Country's Total Exports
South Africa	5	6.9
Botswana	9	83.5
Namibia	14	43.5
Angola	16	2.4
Democratic Republic of the Congo	17	41.5
Central African Republic	32	36.8

Source: The World Bank

> **Critical Viewing** Gold miners dig in the Chudja mine near the village of Kobu in northeastern Congo. What does the photo suggest about working conditions in the mines?

SUB-SAHARAN AFRICA'S RESOURCES

Legend:
- Aluminum
- Coal
- Copper
- Diamonds
- Gold
- Natural gas
- Oil
- Other mineral resources

Economic Improvement

Today, unstable governments continue to challenge many sub-Saharan countries. For example, the Central African Republic has significant diamond resources as the chart shows, but a lack of infrastructure, smuggling, and political unrest have limited the development of its diamond mining industry. As a result, the country remains one of the poorest in the world.

Some countries such as South Africa, Namibia, and Tanzania have successfully used their mineral profits to build their economies. In Botswana, the government is a partner in the country's largest diamond mining company. It has invested the profits in education, infrastructure, and health care. Countries with resources that follow Botswana's example can improve the stability of their economies and improve the lives of their citizens.

Before You Move On

Monitor Comprehension In what ways do sub-Saharan countries' mineral resources improve life for their people?

2.2 Nigeria and Oil

TECHTREK

myNGconnect.com For an online map of oil production and oil industry photos

 Maps and Graphs

 Digital Library

Main Idea Nigeria's people have received little benefit from the country's oil wealth.

Nigeria has a population of more than 150 million people, the largest in Africa. Its coastline has major ports, and its river system and delta are among the world's largest. Nigeria also produces more oil than any other African country.

Ethnic Conflict

When oil was discovered in Nigeria in the 1950s, it became the focus of the country's economy. However, Nigeria has faced many obstacles in attempting to use its oil riches. The most difficult challenge has been ethnic conflict.

British colonialism left Nigeria divided among three major ethnic groups—the Hausa-Fulani, the Igbo, and the Yoruba. There were also 250 smaller groups and at least two major religions. When Nigeria gained independence in 1960, these groups struggled for control.

Not long after independence, the new government was overthrown in a **coup**, or an illegal takeover by force. In a series of coups, rival ethnic groups replaced existing governments—often putting military leaders in charge. After Muslims in the north adopted *sharia*, or Islamic law, Christians in the area began moving south. This movement further divided the country, this time along religious lines.

Challenges of Oil Wealth

In the early 1970s, world oil prices increased and the Nigerian economy grew. Cities expanded and farmers left their rural land for better paying jobs in the cities. As a result, agricultural production declined and Nigeria was forced to import food to feed its people.

The oil industry has added to the ethnic tension. Different groups struggle to control the oil and its profits. Conflict and corruption have prevented Nigerians from

> **Critical Viewing** Fishermen cast a net near an oil refinery in the Niger River delta. Think about the oil industry and the simple fishing shown in the photo. In what ways are they connected?

sharing in the oil wealth. Most people living in the Niger River delta, where oil production is **concentrated**, or centered, live in poverty.

The delta region is also heavily polluted from oil spills. People in the delta have protested against the pollution and demanded that more of the oil profits be used to improve the area.

Nigeria's Progress

Between 2007 and 2009, Nigeria made economic gains. Its leaders received favorable loans to repay the country's debt to other countries. Officials began putting in place economic reforms designed to improve Nigeria's infrastructure. Maintaining these improvements while resolving ethnic conflicts will be a challenge for Nigeria's future.

Before You Move On

Summarize Why have Nigeria's people benefited so little from the country's oil wealth?

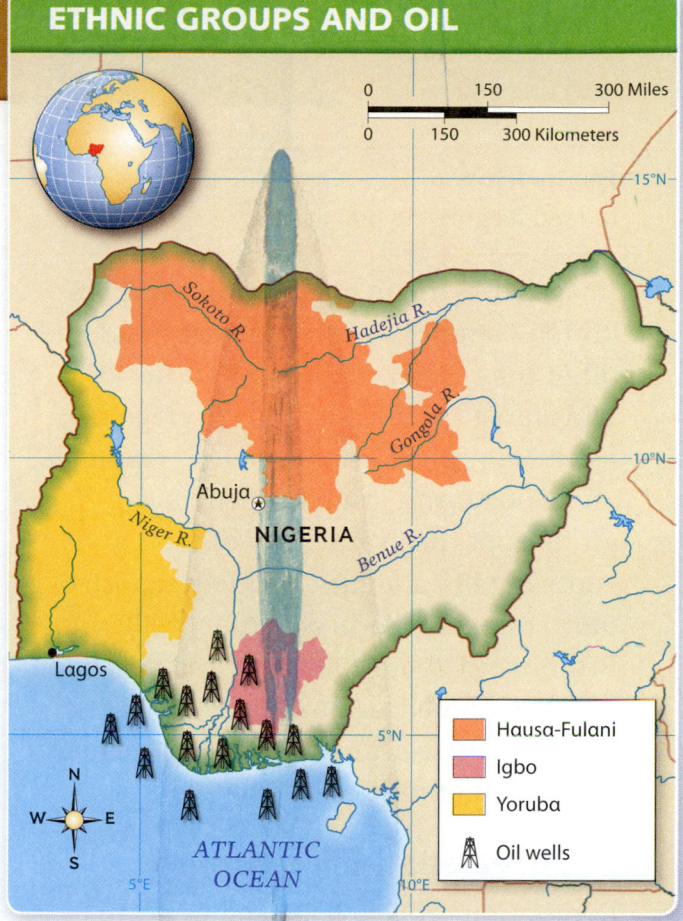

ETHNIC GROUPS AND OIL

Legend:
- Hausa-Fulani
- Igbo
- Yoruba
- Oil wells

ONGOING ASSESSMENT

MAP LAB

GeoJournal

1. **Location** What ethnic group appears to have the most access to the country's oil industry?

2. **Interpret Maps** Identify the delta region on the map. What are some possible consequences of polluting this region?

3. **Analyze Cause and Effect** Use a chart similar to the one below to list effects related to Nigeria's oil industry.

CAUSE	EFFECTS
Nigeria's oil industry	1.
	2.
	3.

2.3 Agriculture and Food Supply

TECHTREK

myNGconnect.com For photos of agriculture and Guided Writing

 Student Resources

 Digital Library

Main Idea Africa is improving its ability to feed its growing population.

Images of African children in need are all too common. Their hunger is real and often the result of famine. **Famine** is a widespread and sustained shortage of food. Natural disasters, such as droughts and floods, and armed conflicts can all disrupt the availability of food and contribute to a famine. Widely reported famines occurred in Sudan in 1998, Ethiopia in 2000, 2002, and 2003, Malawi in 2002, and Niger in 2005. Famine is a constant danger in Africa and threatens the health of millions of people.

A Threatened Food Supply

Population growth has been rapid in some parts of Africa, but food production has not increased as rapidly. The average African eats 10 percent fewer calories today than he or she did 20 years ago. The number of undernourished people in sub-Saharan Africa increased from about 90 million in 1970 to 225 million in 2008. Many African children suffer from malnutrition, which is a lack of nutrients essential to good health. Because of hunger and malnutrition, African children typically have shorter life expectancies than other children around the world.

Critical Viewing Women harvest cotton in Mali. Based on the photo, how would you describe this Mali cotton farm?

Visual Vocabulary Erosion is the wearing away of Earth's surface by natural forces. The grid-like structures in the photo are fences made from branches. The fences keep sand from eroding onto land used for crops and grazing.

African food crops include corn, yams, and sorghum, which is a type of grain. Before colonization, Africans grew these crops with the goal of feeding everyone. After colonization, European settlers took the most fertile land, and farming shifted from food crops to cash crops for export. Cash crops such as coffee and cotton are grown because they can be sold for more money than food crops. The income from cash crops allows farmers to purchase more, but less food is grown to feed the African people.

Since independence, some land has been returned to Africans. However, African land can be easily overworked and overgrazed, exhausting the soil. Farmers also struggle with droughts, soil erosion, and desertification.

New and Better Farming

Africans are working to improve agricultural practices. For example, farmers are moving animals from place to place to avoid overgrazing. They are also growing different crops on the same plot to avoid exhausting the soil. More **legumes**, which are peas or beans, are being planted. These add to the food supply and release helpful nutrients into the soil. Farmers are also enriching the soil by using fertilizer with animal and plant waste instead of chemicals.

Microcredit, or small loans, has helped poor farmers invest in land, tools, and seeds. Relief agencies have begun to provide free seeds and tools to farmers. Scientists hope to develop seeds that are more productive and drought resistant. These seeds could help farmers grow more food crops for their own use as well as cash crops for export. If successful, these efforts will feed Africans, extend life expectancies, and improve economies.

Before You Move On

Summarize What is Africa doing to improve its agriculture?

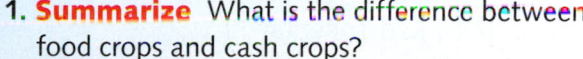

ONGOING ASSESSMENT

WRITING LAB GeoJournal

1. **Summarize** What is the difference between food crops and cash crops?

2. **Analyze Cause and Effect** What is the relationship between reduced food production and life expectancy?

3. **Write Comparisons** Write a paragraph comparing African food crops and cash crops. Describe why is it important for African farmers to balance growing food crops with cash crops. Then work in groups and share your paragraphs. Go to **Student Resources** for Guided Writing support.

2.4 Improving Public Health

> **Main Idea** National governments and international agencies are committed to improving health care in Africa by reducing the impact of diseases.

Nearly one million people die of malaria each year, and 9 out of 10 of them are in Africa. Eighty-five percent of the victims are children under the age of five. Malaria is one of many chronic diseases that plague the continent.

Climate, Poverty, and Disease

In 2008, the average life expectancy for a woman in sub-Saharan Africa was 49 years. That is 32 years shorter than the average in North America. One cause is the deadly toll taken by diseases such as malaria, sleeping sickness, and yellow fever.

Only 15 percent of the world's people live in Africa, but 90 percent of tropical disease cases occur there. The organisms that cause disease thrive in its tropical climates, as do the insects that carry the diseases.

For example, malaria is carried by a mosquito and causes flu-like symptoms. Severe forms can cause brain and nerve damage and death. Malaria probably began in ancient times as an **epidemic**—an outbreak limited to a particular community. Later, it became a **pandemic**, meaning it spread over a large area, in this case, across much of the world. Malaria can be prevented by draining or treating water where insects breed and by sleeping under mosquito nets treated with chemicals to kill insects.

In addition to the tropical climate, poverty contributes to the spread of disease. Poor Africans may lack mosquito nets, screened windows, and technology for draining insect breeding grounds. Poverty also raises the rates of other infections because of overcrowding, poor sanitation, and contaminated water.

KEY VOCABULARY

epidemic, n., an outbreak of disease affecting a large number of people within a community at the same time

pandemic, n., an outbreak of disease occurring over a wide geographic area

vaccine, n., treatments to increase immunity to a particular disease

infectious, adj., capable of spreading rapidly to others

Students in Ethiopia read before school under the protection of a mosquito net.

Before You Move On

Summarize What are some factors that contribute to the spread of disease in Africa?

The Fight Against Disease

The fight against diseases such as malaria, HIV/AIDs, and yellow fever has become a global commitment. Many agencies offer medication to treat diseases and vaccines to prevent them. **Vaccines** are treatments designed to increase immunity, or resistance, to a particular disease. Scientists are trying to find new vaccines and cures for **infectious** diseases, or diseases that can spread rapidly.

Eliminating malaria is a global goal for the 21st century. Worldwide, there are an estimated 250 million cases every year. A non-governmental organization (NGO) called Malaria No More works to fight malaria in Africa. An NGO is a nonprofit volunteer group founded by citizens. Malaria No More hands out mosquito nets to protect people.

The World Health Organization (WHO) battles malaria around the world. The group's Mekong Malaria Program in Southeast Asia works with governments and NGOs to monitor disease outbreaks and treatment programs in the region.

A major source of funding for malaria treatment programs is the Bill and Melinda Gates Foundation. The foundation funds research for a vaccine and more effective anti-malarial drugs for people who have already been infected.

All of these programs reflect a renewed commitment on the part of national governments, private foundations, and NGOs to fight malaria and other diseases around the world.

MALARIA BY THE NUMBERS

38
Countries worldwide that reduced cases of malaria by over 50 percent between 2000 and 2008

45
Number of seconds between each death of an African child due to malaria

109
Countries worldwide reporting malaria cases in 2009

2,414
Estimated number of people worldwide who die every day from malaria

801,000
Estimated number of Africans who die every year from malaria

Source: World Health Organization, United Nations Children's Fund

Before You Move On

Monitor Comprehension What are the main tools in the global fight against disease?

READING LAB GeoJournal

1. **Analyze Cause and Effect** What is one cause of the short life expectancy among sub-Saharan Africans?

2. **Make Predictions** How might development of a vaccine against malaria affect the life expectancy in Africa?

3. **Make Inferences** How might you expect malaria to impact the economies of countries with large numbers of cases?

2.5 Sudan and Somalia

TECHTREK

myNGconnect.com For online maps and current events in Sudan and Somalia

 Maps and Graphs

 Connect to NG

Main Idea Civil wars involving ethnic and religious groups in Sudan and Somalia have limited progress in these countries.

For years, Sudan and Somalia have been plagued by war, famine, and disease. Many people survived conflict only to die of starvation. In both countries, climate, history, and culture all combined to help create this difficult situation.

As you've learned, colonial rule in Africa left countries divided among different ethnic and religious groups. Sudan and Somalia also face a geographic division: The Sahel runs through both of these countries, dividing them between northern Africa and sub-Saharan Africa.

Conflict in Sudan

Sudan is mostly desert in the north. Nomadic Muslim herders of Arabic descent live there. The south is swamp and savanna that was settled by farmers of African descent. Northerners are mainly Muslim, while southerners practice mainly Christianity or indigenous religions. Both north and south include a mix of ethnic groups.

These religious and ethnic differences have led to conflict. Since Sudan's independence in 1956, the government has oppressed Christians in the country. In Darfur, the government created a largely Arab militia in 2003. The militia attacked government protestors, but also attacked militia members' personal enemies. Almost 400,000 people were killed, and about 2.5 million people became **refugees**, or people who flee a place to find safety.

Neighboring countries such as Chad set up refugee camps for people fleeing Sudan. One camp housed more than 250,000 people in 2010. The camps have no permanent shelter or sewer system, often suffer food shortages, and rely on the host countries and relief agencies for help.

The refugees included a group known as the **Lost Boys of Sudan**. These young men were orphaned by the civil war and stuck together to escape the violence. They traveled to Ethiopia, then back to Sudan, and then to camps in Kenya. In 2001, the United States took in more than 3,500 of these young men. Many attended high school and college in the United States.

> **Critical Viewing**
> Eyl (AY uhl), shown here, is a town in Somalia's Puntland state. What might you infer from this photo about economic conditions in Eyl?

SUDAN POLITICAL

Predominant Religion
- Islam
- Christianity
- Indigenous
- Islam/Indigenous
- Christianity/Indigenous
- Uninhabited
- Darfur
- South Sudan

SOMALIA POLITICAL

- Somaliland
- Puntland

Somalia

Somalians have suffered through conflict among five major **clans**, or large, family-based units with loyalty to the group. Even within the clan, sub-clans may clash. As a result, Somalians have never united to form a single nation.

In 1991, clan-based groups overthrew a crumbling military government and battles raged between rival clans. This conflict disrupted farms that were already threatened by flood and drought. With no central government, some clans turned to piracy, which continues today. They attack and rob foreign ships or hold the ships and their crews for ransom.

Somalia is considered by many to be a **failed state**, a country in which government, economic institutions, and civil order have broken down. The regions of Somaliland and Puntland have claimed

independence but have not received international recognition. In the 21st century, forces from the United Nations and the African Union have been trying to keep peace in the region.

Before You Move On
Summarize How have conflicts limited progress in Sudan and Somalia?

ONGOING ASSESSMENT
MAP LAB
GeoJournal

1. **Interpret Maps** Using the map of Sudan, describe the geographic distribution of the main religions.

2. **Region** Using the map of Somalia, describe how it is divided. Based on your description, what challenges might Somalia face?

3. **Draw Conclusions** What is a failed state, and why might Sudan and Somalia both be considered failed states?

2.6 Ending Apartheid

TECHTREK
myNGconnect.com For photos of South Africa
Digital Library

Main Idea In 1994, South Africa moved from a minority, racist government to a democratically elected government.

In 1994, long lines of South Africans of all races stood for hours under a blazing hot sun. The majority of them waited to vote for the first time in their lives. This day marked the end of racial **segregation**, or separation by race, and the birth of a new democracy.

Rich Resources Lure Colonists

South Africa stretches from warm tropics in the north to chilly waters in the south. Its geography and climate support a variety of crops, and it has large deposits of diamonds, gold, and other minerals.

In the 19th century, both Dutch and British colonists laid claim to these lands. The Dutch settlers, known as Boers or Afrikaners, formed their own republics. Boers had enslaved Africans and imported other people as laborers from Asia. Both the British and the Boers seized lands from Zulus, Xhosa, and other Africans. Even more Africans were forced to work for Boer and British colonists after the discovery of large gold and diamond deposits around 1870.

The Beginning of Apartheid

In 1902, after a series of wars, the Boer territories became British colonies known as the Union of South Africa. The British colonial government divided South Africa into white and black areas. Only white Africans could vote. Most of the land, and all of the best land, was reserved for the white minority. Black Africans were left with the land that was less useful or productive. Cities were declared white, and black Africans could enter them only to work.

Critical Viewing A crowd in South Africa celebrates at a soccer match. What can you infer from the photo about how a sporting event might help bring people together?

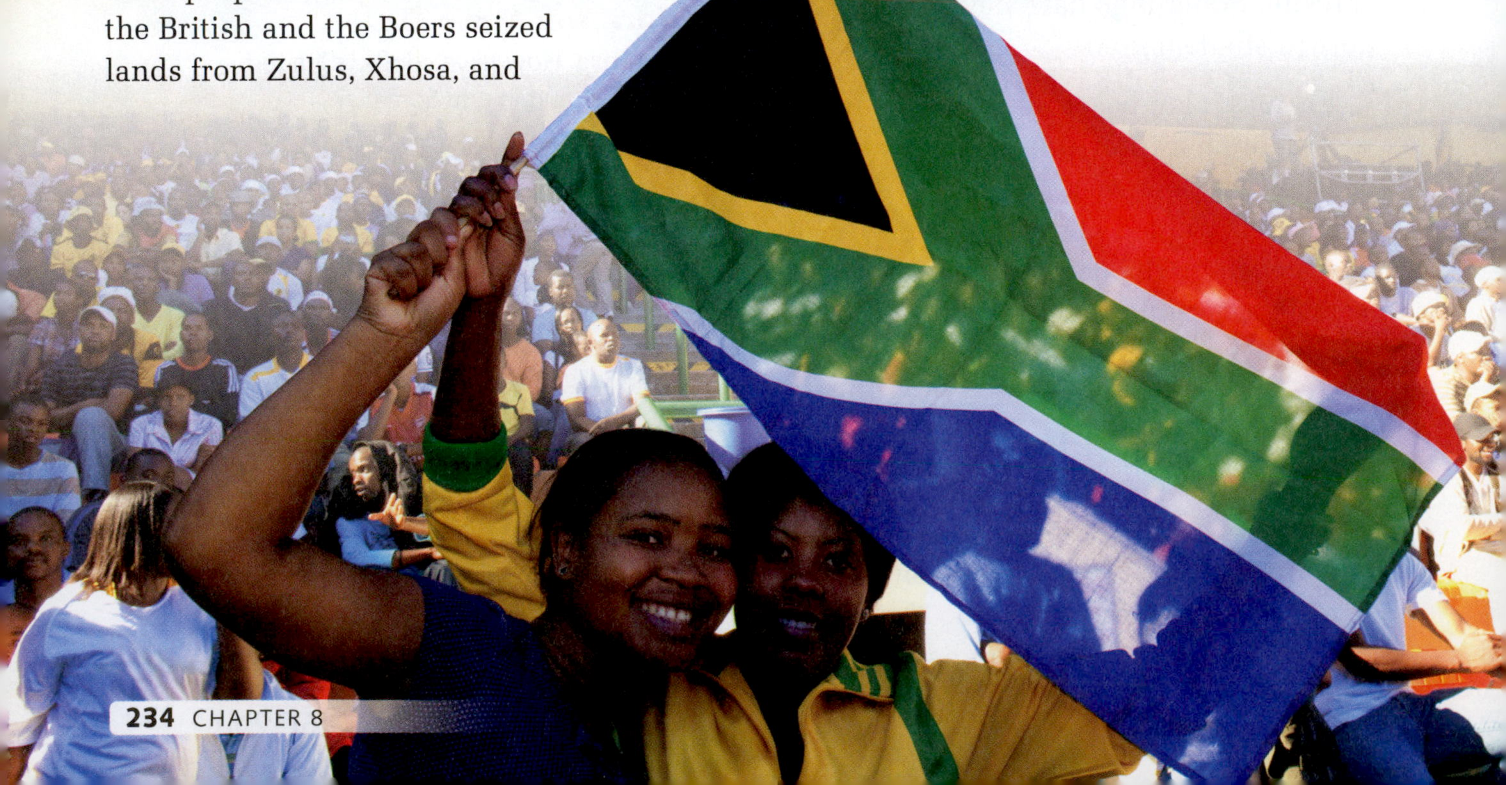

In 1948, new laws created **apartheid** (uh PART hite), or the legal separation of races. In addition to being divided by race, South Africans could only own land in areas assigned to them. Black South Africans were required to carry identification papers at all times.

In 1970, the government made all black South Africans citizens of a homeland instead of citizens of South Africa. **Homelands** were supposed to be self-governing areas, but the national government actually controlled them. The homelands separated the races even more.

The End of Apartheid

In the early 20th century, black Africans formed the **African National Congress (ANC)** to protest their treatment. Later, South Africa outlawed the ANC and imprisoned its leaders, including **Nelson Mandela**. In 1977, **Stephen Biko**, president of a student protest organization, was arrested and beaten to death. He and Mandela became symbols for the protest movement in South Africa.

In 1989, a new white president, F. W. de Klerk, began to change apartheid laws. He legalized the ANC and released its leaders, including Mandela. In 1994, voting rights were extended to all South Africans, and the people elected Mandela president.

Decades of apartheid left South Africa far from racial equality. The government has worked to give black South Africans access to better jobs and farmland, but a large economic gap remains between most blacks and whites in the country. In 2002, the government took control of

Nelson Mandela gestures to the crowd at a rally shortly after his release from prison.

the country's mineral resources, in part to make sure that black South Africans profited fairly. Progress is slow, but South Africa's ability to harness its resources has allowed it to develop the most prosperous economy in Africa.

Before You Move On

Summarize How did South Africa change from minority rule to a more democratically elected government?

VOCABULARY

For each pair of vocabulary words, write one sentence that explains the connection between the two words.

1. ethnic group; transportation corridor

> Bodies of waters can act as transportation corridors between different African ethnic groups.

2. griot; oral tradition

3. ethnobotanist; medicinal plant

4. commodity; mineral

5. famine; erosion

6. epidemic; pandemic

MAIN IDEAS

7. What factors were ignored when the colonial boundaries of African countries were drawn? (Section 1.1)

8. How has Africa's music influenced music in the United States? (Section 1.2)

9. What challenges has Kenya faced in modernizing its economy? (Section 1.3)

10. How are Grace Gobbo and Wade Davis helping to preserve traditional cultures? (Section 1.4)

11. What circumstances have kept Africans from profiting from mineral resources? (Section 2.1)

12. How has population growth made it harder for Africa to feed its people? (Section 2.3)

13. What is being done to reduce the spread of malaria in Africa? (Section 2.4)

14. What are some of the causes of the conflict in Sudan and Somalia? (Section 2.5)

15. How has life improved in South Africa since the end of apartheid? (Section 2.6)

CULTURE

ANALYZE THE ESSENTIAL QUESTION

What historical and geographic factors have influenced the cultures of sub-Saharan Africa?

Critical Thinking: Draw Conclusions

16. Without the influence of colonialism, would African countries be as ethnically diverse as they are today? Why or why not?

17. In what ways have sub-Saharan Africa's geography, climate, and resources helped to shape the lifestyles of its people?

INTERPRET MAPS

BOTSWANA POLITICAL

18. **Location** Which city can be found near 22°S latitude and 28°E longitude?

19. **Human-Environment Interaction** What type of geographic feature defines the irregular shape of Botswana's southern border?

GOVERNMENT & ECONOMICS

ANALYZE THE ESSENTIAL QUESTION

How have conflict and government instability slowed economic development in sub-Saharan Africa?

Critical Thinking: Analyze Cause and Effect

20. Describe some specific natural resources that have helped sub-Saharan African countries develop economically.

21. How did colonialism slow down Africa's economic development?

22. What effects did apartheid have on racial equality in South Africa?

INTERPRET GRAPHS

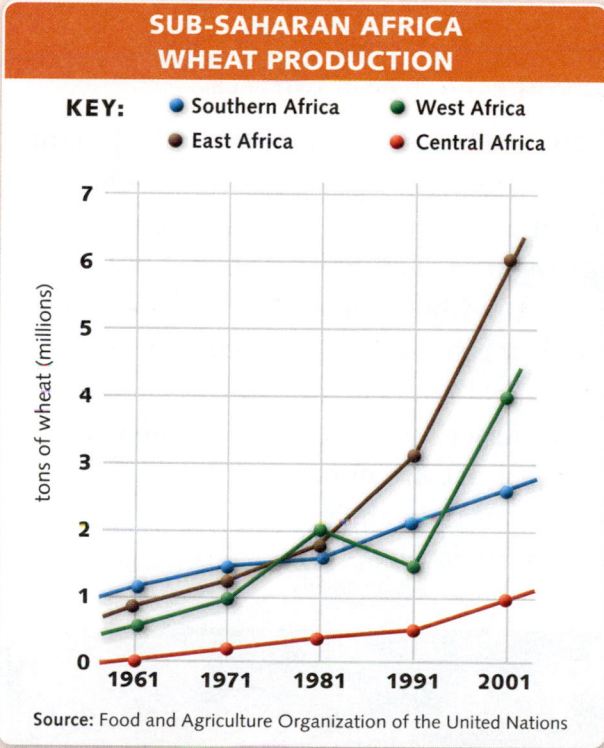

SUB-SAHARAN AFRICA WHEAT PRODUCTION

KEY: ● Southern Africa ● West Africa
     ● East Africa ● Central Africa

Source: Food and Agriculture Organization of the United Nations

23. **Analyze Data** In which region of Africa did food production increase the most?

24. **Evaluate** What additional information is needed in order to determine growth in food production per capita?

ACTIVE OPTIONS

Synthesize the Essential Questions by completing the activities below.

25. **Write a Letter** Imagine that you are a member of an aid team working on improving public health in Africa. Pick a specific country and write a letter to donors explaining why your organization and team need money to support your work. **Read your letter aloud to a partner to practice presenting the information.**

> **Writing Tips**
> • Take notes from the chapter about the health concerns faced by many people in Africa.
> • Present clear, strong evidence to support your proposal.
> • Make specific suggestions about how the money can best be spent to improve the quality of life for all the country's people.

TECHTREK myNGconnect.com For research links on sub-Saharan Africa

26. **Create a Chart** Make a three-column chart showing comparisons among three countries in sub-Saharan Africa. Be sure each country comes from a different area (West, East, Central, or Southern). Use the research links at **Connect to NG** and other online sources to gather the data for the categories shown below. Based on the data, which country has the most people? The fewest?

	Ghana (West Africa)	Uganda (East Africa)	Lesotho (South Africa)
Year Country Gained Independence			
Population			
Square Miles of Land			
Main Economic Resources			

Deserts of the World

TECHTREK

myNGconnect.com For an online chart and research links about deserts

Student Resources

Connect to NG

Deserts make up about one fifth of the land on Earth. Geologically speaking, deserts as we know them are relatively young—forming within the last 65 million years. Deserts are dry lands that can lose more water through evaporation than they get from precipitation. Rainfall in deserts is usually less than 10 inches a year. Limited rainfall, low humidity, often high daytime temperatures, and winds all contribute to the desert's dryness.

Deserts sometimes cover large areas and extend beyond country borders. For example, the Sonoran Desert crosses the border between Mexico and the United States. The Kalahari Desert covers parts of three countries—Botswana, Namibia, and South Africa.

Compare

- Botswana
- China
- Mexico
- Mongolia
- Namibia
- South Africa
- United States

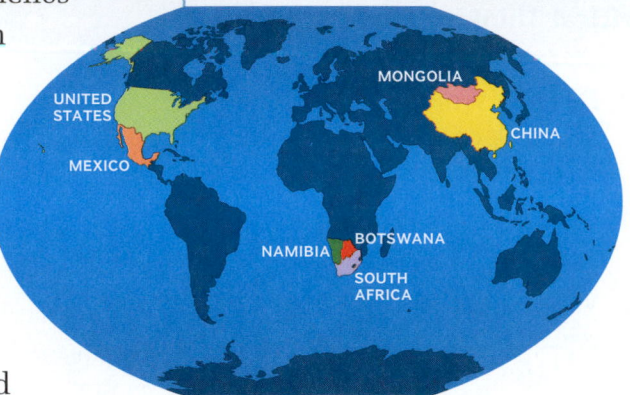

PHYSICAL CHARACTERISTICS

Many people think of a desert as a hot, sandy wasteland with no water and no life. In fact, there is an amazing diversity among deserts. Some are hot and sandy, but some have rainy seasons and can get very cold. While about one fourth of deserts are made of sand, the rest are composed of dirt, clay, rock, ice, and other materials. For example, Antarctica is a desert. Deserts can be flat or hilly, below sea level or in mountains.

DESERT PLANTS AND ANIMALS

Deserts are home to a wide array of plants and animals that have adapted to the harsh climates. Desert plants, such as the saguaro cactus (below), can go long periods without water. Some plants have shallow root systems that cover a large area to gather as much water as possible. Many desert animals, such as the elf owl (below), are nocturnal, hunting for food after the sun goes down and the temperatures drop.

Namib Desert, Namibia

Elf owl in a saguaro cactus, Sonoran Desert, United States

SELECTED DESERTS OF THE WORLD

Gobi Desert

Sonoran Desert

Kalahari Desert

SONORAN

Location:
United States, Mexico

Approximate size:
100,000 square miles

Average yearly rainfall:
4–12 inches

Interesting animal species:
Gila monster

Examples of temperature ranges:
120°F (summer)
32°F (winter)

KALAHARI

Location:
Botswana, South Africa, Namibia

Approximate size:
220,000 square miles

Average yearly rainfall:
3–7.5 inches

Interesting animal species:
Meerkat

Examples of temperature ranges:
113°F (summer)
7°F (winter)

GOBI

Location:
China, Mongolia

Approximate size:
500,000 square miles

Average yearly rainfall:
2–8 inches

Interesting animal species:
Bactrian Camels

Examples of temperature ranges:
113°F (summer)
-40°F (winter)

Source: World Wildlife Fund; worldatlas.com; gobidesert.org; Arizona-Sonoran Desert Museum

---

ONGOING ASSESSMENT

RESEARCH LAB GeoJournal

1. **Analyze Data** Which desert has the widest range in temperatures?

2. **Compare and Contrast** How do the deserts compare in size?

Research and Create Charts Locate the Patagonian and Sahara deserts on a map. Do research to compare the deserts and use a chart like the one above to present your findings. Adjust the categories depending on the information you want to compare.

Active Options

ACTIVITY 1

Goal: Extend your understanding of endangered animals.

Write a Briefing

Sub-Saharan Africa is famous for the variety of animals that live there. Unfortunately, many of these animals are endangered. Some of these animals are listed below. Choose one or a different threatened animal in the region. Do research and prepare a written briefing about the animal. A briefing is a short summary of important information about a topic. Give your audience facts about the animal and about efforts to protect it.

- African elephant
- addax
- black rhinoceros
- mandrill (shown at right)
- pygmy hippopotamus
- lowland gorilla

mandrill

ACTIVITY 2

Goal: Research sub-Saharan Africa's culture.

Keep a Travel Journal

Congratulations! You have won a seven-day, all-expenses-paid tour to one of these countries: Kenya, Tanzania, South Africa, Botswana, Zimbabwe, or Zambia. The only condition is that you keep a daily journal describing what you see and do. Choose your country and begin your tour. Use the **Magazine Maker** to share your journal with friends back home.

ACTIVITY 3

Goal: Review sub-Saharan Africa with a game.

Play a Capital Game

With a group, look at a map and choose 25 countries in sub-Saharan Africa. Write the name of each country on a small card. On another small card, write the capital of each country. Now you're ready to play. Each player draws a country card and tries to identify the capital. If the player misses, the card is put at the bottom of the pile, and it is the next player's turn. If the player names the capital correctly, he or she keeps the card and it is the next player's turn. The one with the most cards wins. Vary the game by choosing a capital card and trying to name the country.

Explore SOUTHWEST ASIA & NORTH AFRICA
with NATIONAL GEOGRAPHIC

MEET THE EXPLORER

NATIONAL GEOGRAPHIC

Emerging Explorer and urban planner Thomas Taha Rassam (TH) Culhane works in Cairo's poorest neighborhoods. He installs rooftop solar water heaters and household biogas systems using environmentally friendly technology.

INVESTIGATE GEOGRAPHY

The Ramses temple, on the Nile River, was carved thousands of years ago in Egypt. Ancient Egypt was an oasis in the desert of northeastern Africa. Egypt depended on annual flooding of the river to support its society.

STEP INTO HISTORY

Jerusalem is home to the Dome of the Rock (shown here), an Islamic and Jewish holy place. It also contains many sites of importance to Christianity, including the Church of the Holy Sepulchre.

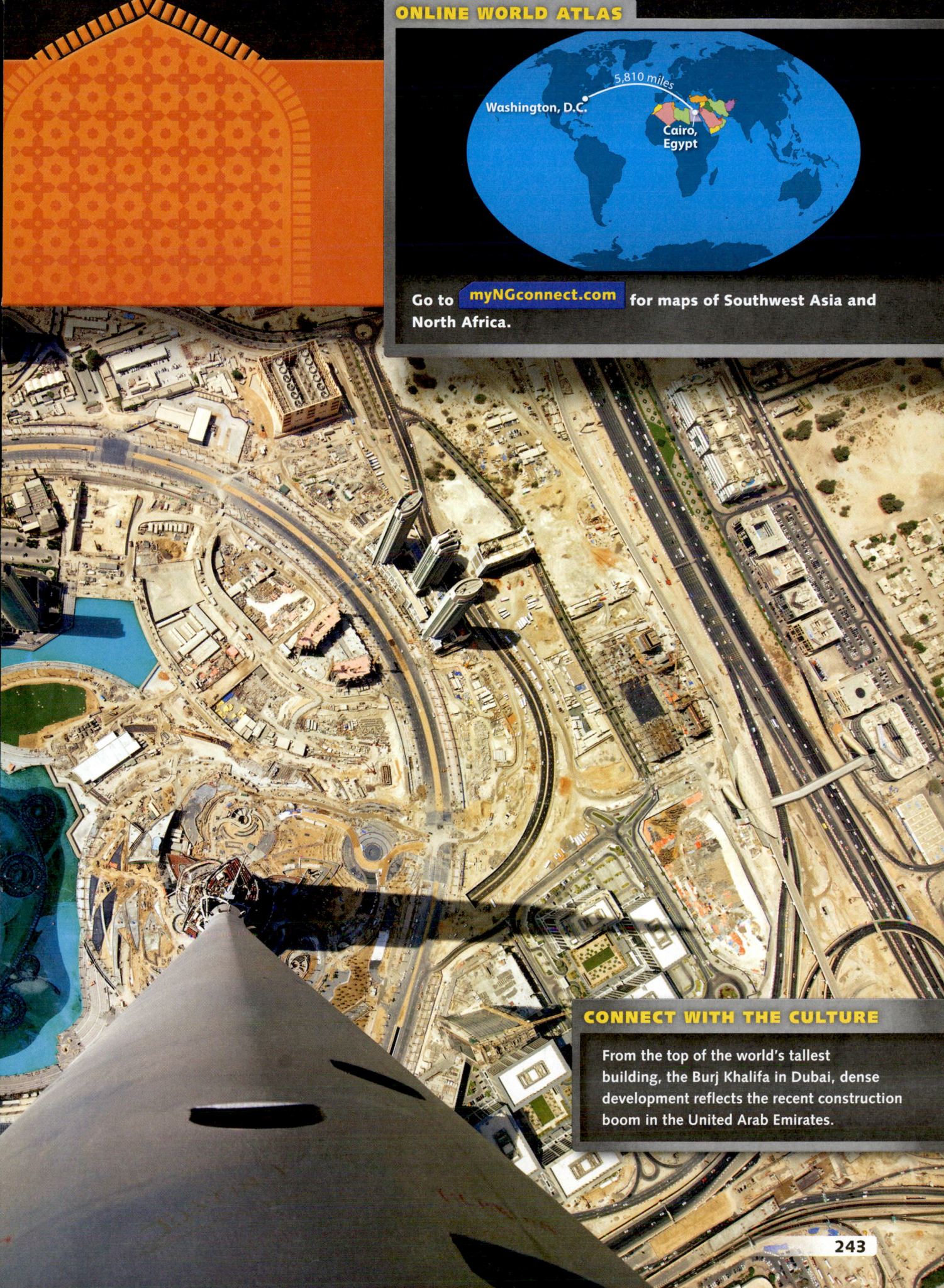

ONLINE WORLD ATLAS

5,810 miles

Washington, D.C.

Cairo, Egypt

Go to **myNGconnect.com** for maps of Southwest Asia and North Africa.

CONNECT WITH THE CULTURE

From the top of the world's tallest building, the Burj Khalifa in Dubai, dense development reflects the recent construction boom in the United Arab Emirates.

PREVIEW THE CHAPTER

Essential Question How have climate and location influenced the region in the past and today?

KEY VOCABULARY
- arid
- desertification
- alluvial plain
- silt
- irrigation
- petroleum
- nonrenewable
- fault
- oasis
- qanat

ACADEMIC VOCABULARY
reliable

TERMS & NAMES
- Sahara Desert
- Nile River
- Fertile Crescent
- Rub Al Khali
- North Anatolian Fault

Essential Question How did civilizations develop in Southwest Asia and North Africa?

KEY VOCABULARY
- cultural hearth
- agricultural revolution
- domesticate
- city-state
- cuneiform
- monotheistic
- messiah
- pilgrimage
- diffusion
- adherent
- sultan
- religious tolerance

ACADEMIC VOCABULARY
extent

TERMS & NAMES
- Ur
- Hammurabi
- Nebuchadnezzar
- Judaism
- Christianity
- Islam
- Hebrew Bible
- Christian Bible
- Qur'an
- Diaspora
- Constantine
- Ottoman Empire
- Byzantine Empire
- Osman
- Suleyman I

TECHTREK FOR THIS CHAPTER

Student eEdition

Maps and Graphs

Interactive Whiteboard GeoActivities

Digital Library

Connect to NG

Arabian camel

Go to **myNGconnect.com** for more on Southwest Asia and North Africa.

Essential Question How did an advanced civilization develop in Egypt?

SECTION 3 • FOCUS ON EGYPT

KEY VOCABULARY
- floodplain
- hydroelectric power
- hieroglyphics
- dynasty
- pyramid
- pharaoh
- deity
- papyrus
- tomb
- sarcophagus

ACADEMIC VOCABULARY
regulate

TERMS & NAMES
- Aswan High Dam
- Hatshepsut
- Ramses II
- Re
- Isis
- Giza
- Great Pyramid of Khufu

SOUTHWEST ASIA AND NORTH AFRICA PHYSICAL

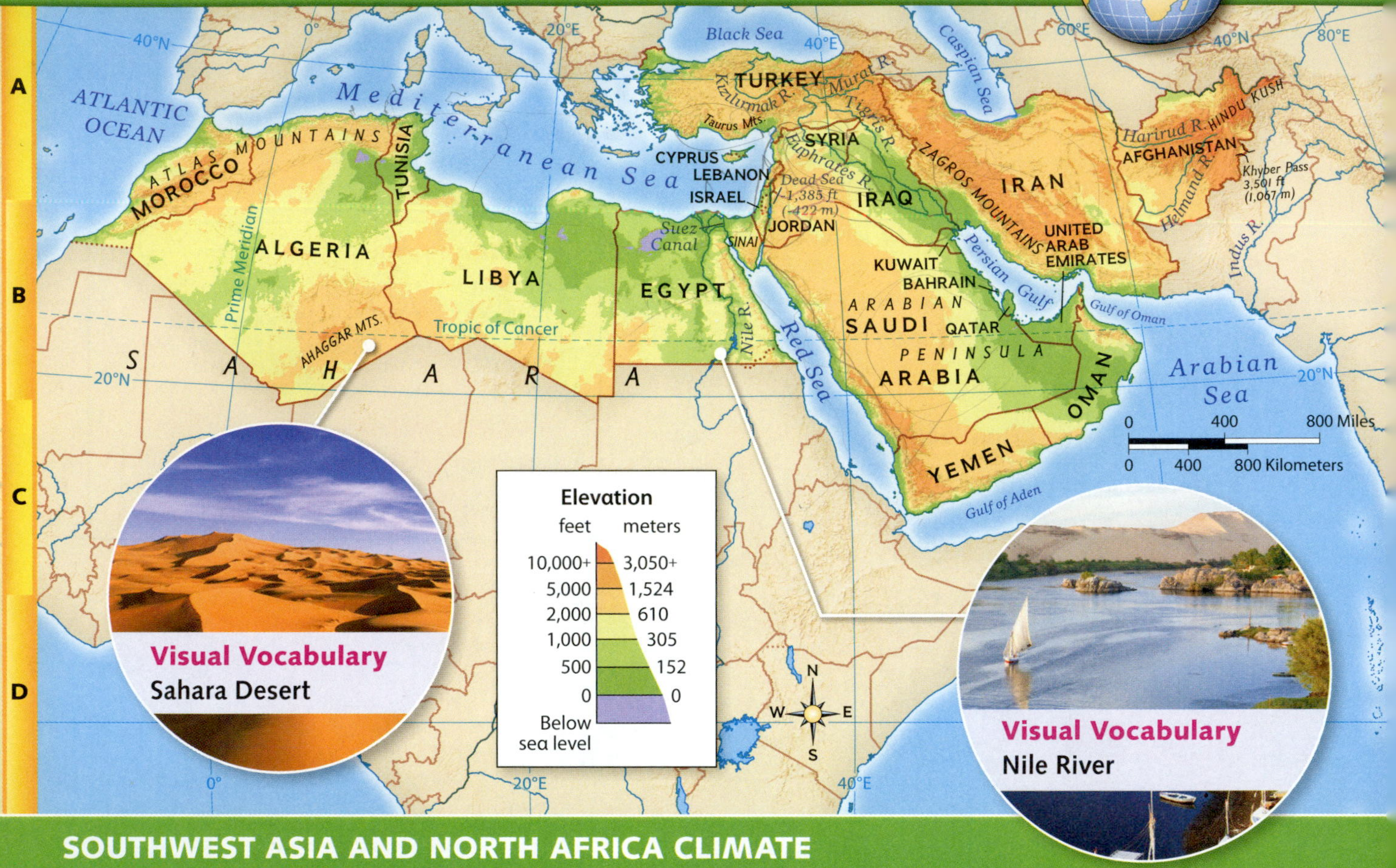

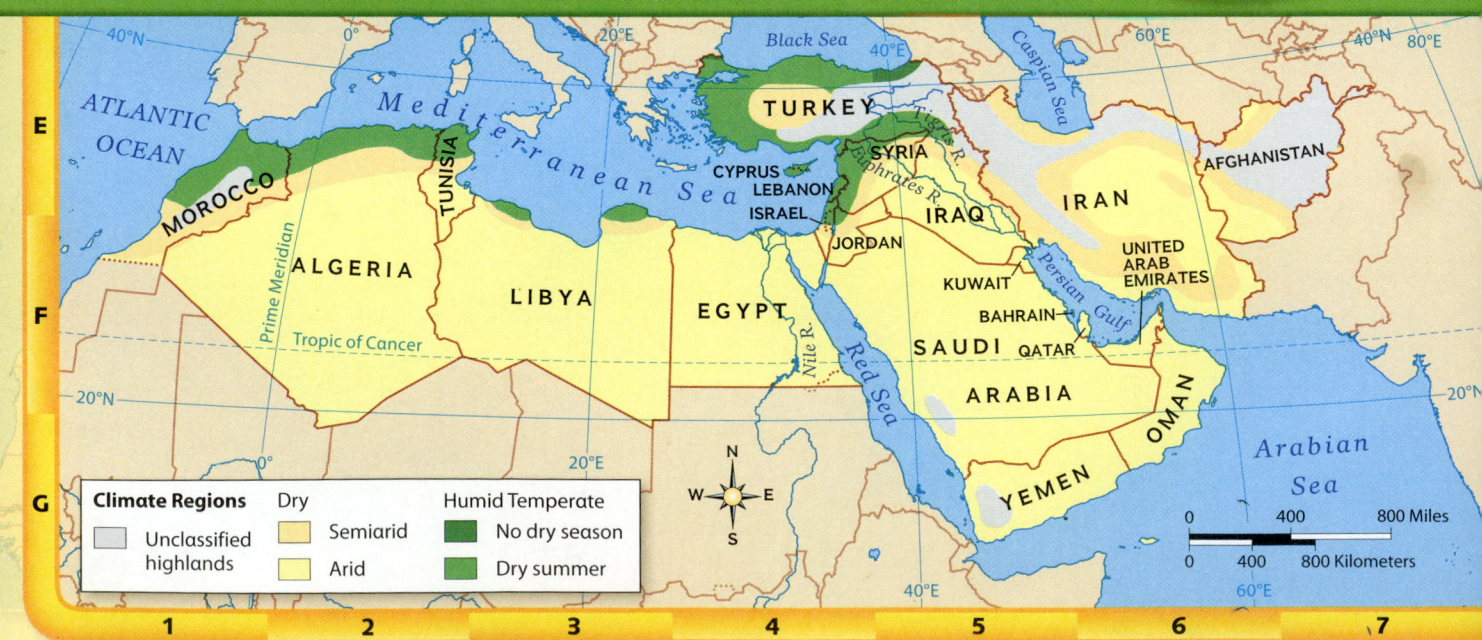

ATLANTIC OCEAN

Mediterranean Sea

Black Sea

Caspian Sea

Kızılırmak R.
Murat R.
Taurus Mts.
TURKEY
CYPRUS
LEBANON
ISRAEL
SYRIA
Tigris R.
Euphrates R.
Dead Sea -1,385 ft (-422 m)
IRAQ
JORDAN
ZAGROS MOUNTAINS
IRAN
AFGHANISTAN
Harirud R.
HINDU KUSH
Helmand R.
Khyber Pass 3,501 ft (1,067 m)
Indus R.

ATLAS MOUNTAINS
MOROCCO
TUNISIA
ALGERIA
Prime Meridian
LIBYA
EGYPT
AHAGGAR MTS.
Tropic of Cancer
S A H A R A
SINAI
Suez Canal
Nile R.
Red Sea

KUWAIT
BAHRAIN
ARABIAN SAUDI
QATAR
UNITED ARAB EMIRATES
Persian Gulf
Gulf of Oman
PENINSULA
ARABIA
OMAN

Arabian Sea

YEMEN
Gulf of Aden

Elevation

feet	meters
10,000+	3,050+
5,000	1,524
2,000	610
1,000	305
500	152
0	0
Below sea level	

0 400 800 Miles
0 400 800 Kilometers

N W E S

Visual Vocabulary
Sahara Desert

Visual Vocabulary
Nile River

SOUTHWEST ASIA AND NORTH AFRICA CLIMATE

ATLANTIC OCEAN

Mediterranean Sea

Black Sea

Caspian Sea

TURKEY
Tigris R.
CYPRUS
LEBANON
ISRAEL
SYRIA
Euphrates R.
JORDAN
IRAQ
IRAN
AFGHANISTAN

MOROCCO
TUNISIA
ALGERIA
Prime Meridian
LIBYA
EGYPT
Tropic of Cancer
Nile R.
Red Sea

KUWAIT
BAHRAIN
SAUDI
QATAR
UNITED ARAB EMIRATES
Persian Gulf
ARABIA
OMAN

Arabian Sea

YEMEN

Climate Regions

	Dry	Humid Temperate	
Unclassified highlands	Semiarid	No dry season	
	Arid	Dry summer	

0 400 800 Miles
0 400 800 Kilometers

N W E S

> **Main Idea** The expansive region of Southwest Asia and North Africa is hot, and water in the region is sometimes scarce.

Southwest Asia and North Africa span parts of the Asian and African continents. Two notable physical features located in this region are the world's longest river and its largest desert. Changes in climate in some places led to new patterns of migration and settlement.

Physical Features and Climate

Much of Southwest Asia and North Africa is made up of vast, barren deserts. This desert area is <mark>arid</mark>, or very dry, with extremely high daytime temperatures. For example, most of the **Sahara Desert** and Arabian Peninsula receive fewer than four inches of rain per year. Daytime temperatures can climb to 130°F.

The region includes several different mountain ranges. The Atlas Mountains are located at the tip of northwest Africa. The Zagros Mountains in Iran stretch along the Persian Gulf, and the Taurus Mountains in Turkey border the Mediterranean Sea. The climate in these mountains is semi-arid and temperatures can fall below 0°F.

Three important rivers have historically supported life in this region, and they continue to do so today. The **Nile River** flows through Egypt, and the Tigris and Euphrates rivers flow through Syria and Iraq. Water is scarce in most of this region, except near rivers and coasts. Cultures have developed and grown along these waterways. Many of the world's oldest cities are located here.

Desert Transformation

Today, the Sahara stretches 3,500 miles across North Africa from the Red Sea to the Atlantic Ocean. However, this broad area of the continent has not always been desert. The Sahara's climate has changed dramatically over time. Nearly 10,000 years ago, the world's largest desert was tropical grassland.

Around 5300 B.C., seasonal rains that watered the Sahara began shifting southward. Over time, the Sahara became a desert. The gradual transition from fertile to less productive land is called <mark>desertification</mark>. The drier climate prompted people living in the Sahara to migrate out of the desert. They settled in the Nile River Valley, where they found a <mark>reliable</mark>, or dependable, water source.

Before You Move On

Monitor Comprehension In what ways have heat and water resources defined this region?

ONGOING ASSESSMENT

MAP LAB GeoJournal

1. **Location** On the physical map, locate the Nile, the Tigris, and the Euphrates rivers. Into which bodies of water do these rivers flow? What other major bodies of water can you infer might be important in the region?

2. **Make Inferences** Even though most of the region is hot and dry, variations exist. Based on the climate map, compare the climates of Lebanon and Yemen. Which country can you infer receives more rainfall?

3. **Describe Geographic Information** How did the climate of the Sahara change over time, and what caused the change?

TECHTREK

myNGconnect.com For an online map and photos of the Tigris and Euphrates rivers

 Maps and Graphs

 Digital Library

Main Idea The Tigris and Euphrates rivers have supported life for thousands of years.

One of the world's earliest civilizations began between two rivers in a desert. The ancient name for this area is Mesopotamia, which means "land between the rivers."

Two Rivers

As you have read, the climate of this region is hot and dry. However, the area surrounding and extending from the Tigris and Euphrates rivers is known as the **Fertile Crescent.** This fertile area arcs from the Mediterranean Sea to the Persian Gulf. Its rivers and floodplains were an ideal location for early agriculture.

The source of both the Tigris and Euphrates rivers is in the mountains of Turkey. The rivers flow through parts of Turkey, Syria, and Iraq. They join at Al Qurnah, Iraq, before emptying into the Persian Gulf.

Though the Tigris and the Euphrates flow in similar directions, each river has distinct features. As the longest river in southwestern Asia, the Euphrates River is about 1,740 miles long—about 600 miles shorter than the Mississippi River in the United States. After leaving eastern Turkey, the Euphrates flows southeast through Syria and across Iraq, and is fed by two major tributaries. Over its journey, the flow of the river slows down. As the Euphrates winds through hot desert land, much of its water evaporates.

The Tigris River is also a long river, about 1,180 miles in length. After leaving Turkey, it flows southeast through Iraq. Fed by several fast-moving tributaries, the Tigris carries more water than the Euphrates. The speed and unpredictable flow of the Tigris sometimes cause major floods. Though floods can be destructive, they have also enabled development of agriculture along the rivers.

◄ Critical Viewing This man tends sheep along the Euphrates River in Syria. From the photo, how would you describe the climate and landscape of this river valley?

Fertile Land

The land between the rivers is an **alluvial plain**, which is a flat area of land located next to a stream or river that floods. Regular flooding deposits **silt**, or fine particles of soil, along the riverbanks. This silt makes the soil fertile.

Historically, farmers have relied on the natural flooding of the rivers to water their crops. However, farmers have also used irrigation. **Irrigation** is the process of redirecting water to crops using channels and ditches. When irrigated, crops in dry areas can grow over a wider area.

The Tigris and Euphrates rivers continue to support life. For example, most of Iraq's population lives between the two rivers and Iraq's capital, Baghdad, lies on the Tigris River. Frequent flooding used to overrun the city, but today Baghdad uses dams and embankments, or walls, to help control floods and regulate irrigation.

Before You Move On

Summarize In what ways are the Tigris and Euphrates rivers important to this region?

TIGRIS AND EUPHRATES RIVERS

ONGOING ASSESSMENT

MAP LAB
GeoJournal

1. **Interpret Maps** According to the map, where do the Tigris and Euphrates rivers join? Where is the mouth of these rivers?

2. **Location** Look at the map and identify where most cities in Iraq are located. What pattern do you notice?

3. **Draw Conclusions** In what ways has flooding of the Tigris and Euphrates rivers been both productive and destructive? Create a chart like the one below to help you answer the question.

TIGRIS AND EUPHRATES RIVER FLOODS

Productive	Destructive

1.3 The Arabian Peninsula

TECHTREK

myNGconnect.com For an online map and photos of the Arabian deserts

 Maps and Graphs

 Digital Library

Main Idea The Arabian Peninsula is primarily desert and provides a large percentage of the world's petroleum.

Though the Arabian Peninsula is nearly surrounded by water, it is mostly desert. This peninsula's geologic history, or the way Earth developed over millions of years, provides the basis of its economic activity today.

Sand and Heat

The Arabian Peninsula covers more than one million square miles. It is bordered on the west by the Red Sea, on the south by the Arabian Sea, and in the northeast by the Persian Gulf. Saudi Arabia is the largest country on the peninsula. Kuwait, Oman, Qatar, the United Arab Emirates, Yemen, Bahrain, and parts of Jordan and Iraq are also on the peninsula. Temperatures in this area often rise to 130°F, and very little rain falls.

Several deserts lie on the Arabian Peninsula. The Syrian Desert is located in the northern and central part of the peninsula. Another desert, the **Rub al Khali,** covers 250,000 square miles of southern Saudi Arabia—almost the size of the state of Texas. Its name, Rub al Khali, means "empty quarter." Except for small groups of nomads, almost no one lives in the Rub al Khali.

The coasts of the peninsula contrast with its desert interior. The western part of the peninsula along the Red Sea features mountain peaks, some as high as 9,000 feet. Fertile soil along the coasts even allows for some farming. For example, date palms, which can tolerate salty soils, grow abundantly along the Persian Gulf's coastal salt flats. Salt flats are soils with high concentrations of salt. Sand and gravel cover most of this area but a valuable resource lies under it.

> **Critical Viewing** The Matrah district in the Muscat Sultanate of Oman sits on the Gulf of Oman. Based on the photo, how would you describe the district?

Oil

Large **petroleum**, or unrefined oil, reserves lie underneath the Arabian Peninsula. Petroleum develops from tiny animals and plants that died millions of years ago. Over time, heat and pressure change these organic materials into a new substance that moves through rock layers and collects in large deposits. Once extracted from the ground, petroleum can be refined into gasoline, diesel fuel, and other products.

Because it takes so long to form, petroleum is a **nonrenewable** natural resource. As you have read, a nonrenewable resource is a resource that cannot reproduce quickly enough to keep pace with its use.

Twenty-five percent of the world's known petroleum reserves are on the Arabian Peninsula. In fact, petroleum production is the most important industry on the peninsula. Most of the countries in the world depend on petroleum in one

THE ARABIAN PENINSULA

Elevation

feet	meters
10,000+	3,050+
5,000	1,524
2,000	610
1,000	305
500	152
0	0
Below sea level	

form or another for their energy needs. This dependence links the countries on the Arabian Peninsula to the rest of the world in critical ways.

Before You Move On

Summarize What role does the Arabian Peninsula play in the production of the world's petroleum?

ONGOING ASSESSMENT

PHOTO LAB  GeoJournal

1. **Analyze Visuals** Look at the photo and read the caption. Then locate approximately where this district lies on the map. Based on the photo and the map, how would you describe the physical features of this area? Consider elevation in your response.

2. **Make Inferences** Based on the photo and the text, what can you infer about patterns of settlement on the Arabian Peninsula?

3. **Human-Environment Interaction** In what ways does climate impact agriculture on the Arabian Peninsula?

1.4 Anatolian and Iranian Plateaus

TECHTREK
myNGconnect.com For an online map and photos of Turkey and Iran

Maps and Graphs · Digital Library

Main Idea The Anatolian and Iranian plateaus have a long history as a crossroads of trade.

Humans have lived on the Anatolian and Iranian plateaus for thousands of years. Though these plateaus are located in different parts of this region, they share characteristics and patterns of settlement.

Plateaus and Mountains

Most of Turkey sits on a peninsula called Anatolia. The Anatolian Plateau **1** is located in the central part of this peninsula. The Iranian Plateau **2** is located in the center of present-day Iran.

Seismic activity formed both plateaus. The **North Anatolian Fault** runs east to west just south of the Black Sea. Activity along this **fault**, or fracture in Earth's crust, has caused many earthquakes. Scientists believe the North Anatolian Fault may have triggered a huge, long-lasting flood about 7,500 years ago, creating the Black Sea. Tectonic shifts also shaped the Iranian Plateau. When the

Arabian and Eurasian plates pushed into each other, the Zagros Mountains formed.

The high mountains on each plateau help create a rain shadow. A rain shadow is a dry area on one side of a mountain range. Moisture from surrounding seas rises and condenses, but rain and snow fall only as the air rises up the side facing the moist winds. Land on the other side, sheltered from winds by the mountains, receives little precipitation.

Trade and Settlement

The dryness of the plateaus has meant sparse settlement. However, the climate did not discourage travelers and traders. **Oases**, or fertile places in dry areas where water is found, dot the plateaus. Across the centuries, these oases were stops for trade caravans.

Throughout history, traders have crossed the Anatolian peninsula on their way between Europe and East Asia. Traders to and from ancient Greece

> **Critical Viewing** The Zagros Mountains border the Iranian Plateau. Based on what you see in the photo, what might be some challenges of living on this plateau?

MAP TIP The dark blue lines on this map show the boundaries of the tectonic plates on which these continents lie. The plates themselves are labeled in purple. The red lines indicate country boundaries. The North Anatolian Fault lies on the boundary between the Eurasian and the Anatolian plates.

Black Sea

Bosporus

Caucasus Mountains

Caspian Sea

30°E   40°E   50°E

North Anatolian Fault

40°N

Sea of Marmara

ANATOLIA

TURKEY

Amu Darya

ANATOLIAN PLATE

Mt. Ararat 16,854 ft (5,137 m)

Taurus Mts.

Elburz Mountains

EURASIAN PLATE

CYPRUS

Mt. Damavand 18,606 ft (5,671 m)

Dasht-e Kavir

AFGHANISTAN

Indus R.

Mediterranean Sea

LEBANON

SYRIA

Tigris R.

ZAGROS

Zard Küh 14,921 ft (4,548 m)

I   R   A   N   I   A   N

DASHT-E LUT

ISRAEL

30°N

JORDAN

IRAQ

M O U N T A I N S

Euphrates R.

Khuzestan Plain

IRAN

P L A T E A U

30°N

ARABIAN PLATE

KUWAIT

Nile R.

EGYPT

AFRICAN PLATE

Red Sea

S A U D I
A R A B I A

BAHRAIN

QATAR

Persian Gulf

UNITED ARAB EMIRATES

OMAN

Gulf of Oman

INDIAN PLATE

Tropic of Cancer

OMAN

40°E   50°E   60°E

— Tectonic plate boundary
— Country boundary

0   200   400 Miles

0   200   400 Kilometers

traveled along the Mediterranean coast. Some people moved inland, across the Taurus Mountains, and settled on the Anatolian Plateau. These settlements became stops on important trade routes.

Early human settlement took place on the Iranian Plateau, too. About 2,500 years ago, people invented a system to bring water into their arid lands. They built **qanats** (kuh NOTZ), or underground tunnels, to carry mountain waters to dry plains. This technology is still used today in Iran's capital, Tehran.

Before You Move On

Monitor Comprehension In what way does the location of the Anatolian and Iranian plateaus make them an important crossroads?

ONGOING ASSESSMENT

MAP LAB

GeoJournal

1. **Interpret Maps** Locate the tectonic plate boundaries on the map. Which plates does the North Anatolian Fault border? On which plates do Turkey and Iran lie?

2. **Draw Conclusions** Locate the Zagros Mountains on the map. What conclusions can you draw about how these mountains formed?

3. **Place** What climate effect do the surrounding mountain ranges have on the Anatolian and Iranian plateaus?

4. **Explain** Over time, how did trade influence settlement on the plateaus?

2.1 Mesopotamia

Main Idea Mesopotamia's early civilization contributed much to other cultures.

Mesopotamia is known as an early **cultural hearth**, or center of civilization from which ideas and technology spread to other cultures. The emergence of farming there more than 10,000 years ago allowed for advanced societies to grow.

Agriculture Develops

As you have learned, the Fertile Crescent extends from the Persian Gulf to the eastern shore of the Mediterranean Sea. About 9500 B.C., people in this fertile land began to shift from gathering food to growing food. This shift is called the **agricultural revolution**. This revolution enabled groups of people to settle in one place and eventually develop advanced civilizations.

In addition to farming, Mesopotamians began to **domesticate** animals, or keep them as a source of animal labor and food. Farming villages grew into bigger settlements and then cities. Eventually, these cities unified into the world's first **city-states**, or independent political units.

MESOPOTAMIA

Black Sea · Caspian Sea · Taurus Mountains · Tigris R. · Zagros Mountains · Ninevah · Assur · Euphrates R. · MESOPOTAMIA · Mediterranean Sea · Syrian Desert · Babylon · Umma · Lagash · Uruk · Ur · Arabian Desert · Persian Gulf

Fertile Crescent
- - - Ancient coastline (about 5000 B.C.)

0   150   300 Miles
0   150   300 Kilometers

Sumer

The first Sumerian city-states formed around 3500 B.C. Sumer was an ancient Mesopotamian region in what is now southeastern Iraq. Located in the lower valley of the Tigris and Euphrates rivers, **Ur** was one of the most important Sumerian city-states. Between 2800 and 1850 B.C., Ur was a center of trade.

9500 B.C.
Agricultural revolution begins in Fertile Crescent.

Ancient Mesopotamian map with cuneiform text

1900 B.C.
Amorites conquer Mesopotamia.

9500 B.C.   **3500 B.C.**   **2500 B.C.**

3500 B.C.
City-states begin to develop in Sumer.

Golden helmet from Ur, a Mesopotamian city-state

Its location on the Euphrates River helped it become a major port. Sea-traders from Ur connected Mesopotamia with people as far away as the Indus Valley in South Asia.

Sumerians made many significant cultural contributions. They used advanced mathematics and invented the first wheeled vehicles and the first codes of law. Sumerians also used written language to record their knowledge. Sumerian writing, called **cuneiform** (kyoo NEE uh form), is the earliest known form of writing. Clay tablets carved with cuneiform provide scientists with many details about daily life and culture in Sumer.

Babylonia

Around 1900 B.C., the Amorites, a nomadic group from Arabia, conquered Mesopotamia. The Amorites adopted much of Sumerian culture and continued using cuneiform. Eventually, the conquered lands became known as Babylonia. The Babylonian Empire included all of southern Mesopotamia. During the 1700s B.C., King **Hammurabi** developed a code of law known as Hammurabi's Code.

After Hammurabi's death, outside invaders weakened the Babylonian Empire. Nearly 1,000 years after Hammurabi's rule, another strong ruler emerged. **Nebuchadnezzar** (nehb buh kuhd NEHZ uhr) was Babylonia's king from 605 to 562 B.C. His ambitious projects included rebuilding the port at Ur on the Persian Gulf and the creation of the Ishtar Gate in Babylon, the empire's capital. The Persians captured Babylonia in 539 B.C. Then, in 331 B.C., Alexander the Great from Macedonia conquered the Persians and Babylonia was never independent again.

Before You Move On

Summarize In what ways did Mesopotamia's early civilizations contribute to other cultures?

ONGOING ASSESSMENT

SPEAKING LAB GeoJournal

1. **Express Ideas Through Speech** What was the impact of the agricultural revolution? Work with a partner to develop an oral presentation that describes the agricultural revolution and how it impacted human development.

2. **Conduct Internet Research** Go to **Connect to NG** to research the cultural contributions of Mesopotamian civilizations. Discuss your findings with a partner.

1750 B.C.
Babylonian Empire begins decline.

Detail from Ishtar Gate of ancient Babylonia

331 B.C.
Alexander the Great takes over Babylonia.

1500 B.C.

500 B.C.

1792 B.C.
Hammurabi becomes king of Babylonia.

Sculpture of Hammurabi

539 B.C.
Persians conquer Babylonia.

2.2 Birthplace of Three Religions

TECHTREK

myNGconnect.com For a map of the birthplace of three religions

Maps and Graphs

Main Idea Three of the world's most influential religions began in Southwest Asia.

Judaism, **Christianity**, and **Islam** all began in Southwest Asia. All three religions are monotheistic, meaning that they worship one god.

Judaism

The Jewish people trace their ancestry to Abraham, who lived in southern Mesopotamia around 1800 B.C. According to the Hebrew Bible, God told Abraham to move his people to Canaan—where Lebanon and Israel are located today. Over hundreds of years, the descendants of Abraham developed a distinct religion, Judaism. In 1000 B.C., King David made Jerusalem ▶ his capital. Jews built and rebuilt sacred temples there. Jerusalem remains the holiest city for Jews today.

The Jewish belief in one god was new. At the time, most people believed in more than one god. The establishment of monotheism marked a shift in religious practices. The **Hebrew Bible** is the sacred text of Judaism. It contains Jewish history and teaches how to lead a moral life.

Christianity

Christianity also developed as a monotheistic religion. Among Jewish teachings was the expectation of a messiah, a leader or savior. Jews believe the messiah is yet to come. Christians, however, believe the messiah was a Jew named Jesus. Born in Nazareth ▶, Jesus began to preach in Galilee sometime around A.D. 30. Jesus drew many followers as he preached. Roman leaders who resented his popularity sentenced Jesus to death and he died in Jerusalem.

According to Christian literature, Jesus rose from the dead and told his followers to spread his message. The sacred book of Christianity is the **Christian Bible**, which includes the Old Testament, containing the Hebrew Bible, and the New Testament. The New Testament is about the life and teachings of Jesus.

The Dome of the Rock (far left) in Jerusalem was built on the site where Muslims believe Muhammad ascended into heaven. Just below the Dome, where people are standing and praying, is the Western Wall, Judaism's holiest site.

Islam

Another monotheistic religion that developed in this region was Islam. According to its followers, called Muslims, in the early 600s, a man from Mecca  named Muhammad received revelations from Allah, the Arabic name for God. Muslims recognize Abraham, Jesus, and others as God's messengers, or prophets. For Muslims, Muhammad is the last prophet.

The holy book of Islam is the **Qur'an**, which contains the revelations Muhammad received until his death. Muslims fulfill religious duties known as the Five Pillars of Islam. One of these duties includes a **pilgrimage**, or journey, to Mecca, in Saudi Arabia, at least once in a lifetime.

Before You Move On

Monitor Comprehension What characteristics do these three religions share?

BIRTHPLACE OF THREE RELIGIONS
Judaism, Christianity, and Islam

Critical Viewing The Church of the Holy Sepulchre in Jerusalem was built on the site where Christians believe Jesus was buried. What details in the photo convey the religious nature of this setting?

2.3 Diffusion of Religions

> **Main Idea** Religions spread around the world through migration, missionaries, and trade.

In the years since becoming established as major religions, Judaism, Christianity, and Islam have spread thousands of miles from their birthplace in Southwest Asia. The **diffusion**, or spread, of each religion happened in different ways.

Migration, Missionaries, Trade

Judaism spread through the migration of the Jewish people. In A.D. 135, in response to harsh Roman rule, Jews began to move to avoid persecution, or discrimination because of their beliefs. By the 1200s, they had migrated as far as Eastern Europe. The spread of Jews around the world is called the **Diaspora**. As Jews in Europe faced increasing persecution, some migrated back to Southwest Asia. In 1948, the Jewish people established modern Israel. Today, about 15 million Jews live throughout the world.

After the Roman Emperor **Constantine** legalized Christianity in 313, it became the official religion of the Roman Empire. As the empire expanded, so did Christianity. During the Middle Ages, European kings promoted Christianity in conquered lands. As European countries became more powerful, their influence spread around the world. Beginning in the 1500s, predominantly Christian and European countries began to colonize the Americas, Africa, and other regions. Christian missionaries spread their faith on behalf of their countries. Today, Christianity has about 2.3 billion **adherents**, or followers.

Islam spread through the expansion of Muslim rule and through trade. By the 1500s, Islam had moved from the Arabian Peninsula throughout North Africa, Southwest Asia, Southeast Europe, and parts of India. Muslim traders spread Islam as they traveled and exchanged goods. Today, Islam has about 1.6 billion followers, more than any religion except for Christianity.

Before You Move On
Summarize Compare and contrast the ways in which Judaism, Christianity, and Islam spread.

1 **Trade** The Great Mosque (above) in Kairouan, Tunisia, was founded in 670. Kairouan was an important stop for desert trade caravans. The Great Mosque continues to be an important holy site for Muslims today.

2 Migration In 1845, 37 Jewish immigrants from Germany founded the Temple Emanu-el synagogue in New York City. As the congregation grew, it constructed new buildings, including this one, built in 1928. Temple Emanu-el is the largest synagogue in the world.

SELECTED EXAMPLES OF RELIGIOUS DIFFUSION

3 Missionaries Catholic missionaries arrived in present-day Bolivia in the 1540s to convert the Inca to Christianity. This cathedral in Sucre, Bolivia, was built in 1559. Christianity is the main religion in Bolivia and most of South America today.

ONGOING ASSESSMENT
VIEWING LAB
 GeoJournal

1. **Movement** Identify each location indicated on the map. What can you infer about how and where each religion spread?

2. **Evaluate** How does the religious community at Temple Emanu-el reflect the Diaspora?

3. **Explain** How did Christianity reach the Americas? How might colonization account for the number of Christian adherents today?

2.4 The Ottoman Empire

TECHTREK

myNGconnect.com For an online map of the Ottoman Empire and photos of Ottoman architecture

Maps and Graphs

Digital Library

Main Idea The Ottoman Empire was a powerful empire in Southwest Asia and North Africa that lasted more than five centuries.

The heart of the **Ottoman Empire** was located in what is present-day Turkey. As you have learned, Turkey is on the Anatolian Plateau, through which several trade routes crossed. Many groups had tried to control this area, including the Hittites, the Greeks, the Persians, and the Romans, who took control in A.D. 30.

The Birth of the Ottoman Empire

When the Roman Empire split in 395, this area became part of the **Byzantine Empire**, the eastern part of the Roman Empire. The Byzantines ruled parts of Anatolia and Southeast Europe for 1,000 years. Turks from Central Asia began to invade and conquer parts of the Byzantine Empire in

the 1300s. These Turks became known as Ottomans, after **Osman,** the name of their first leader.

In 1453, the Ottomans defeated the Byzantine Empire. They captured the city of Constantinople, renamed it Istanbul, and made it the capital of the empire. Istanbul became an important center of trade and wealth under Ottoman rule.

The Empire at Its Height

The Ottoman Empire expanded its reach of power in the mid-1500s. Under the rule of **Suleyman I**, the Ottoman Empire stretched from present-day Hungary in Europe to the Persian Gulf and Red Sea in Asia. After Suleyman's reign, the empire continued to grow. It reached its farthest **extent**, or degree of spread, in the late 1600s.

> **Critical Viewing** Topkapi Palace in Istanbul served as a palace for Ottoman sultans from the mid-1400s to the early 1900s. Based on the photo, how might you describe this palace?

THE OTTOMAN EMPIRE, 1683

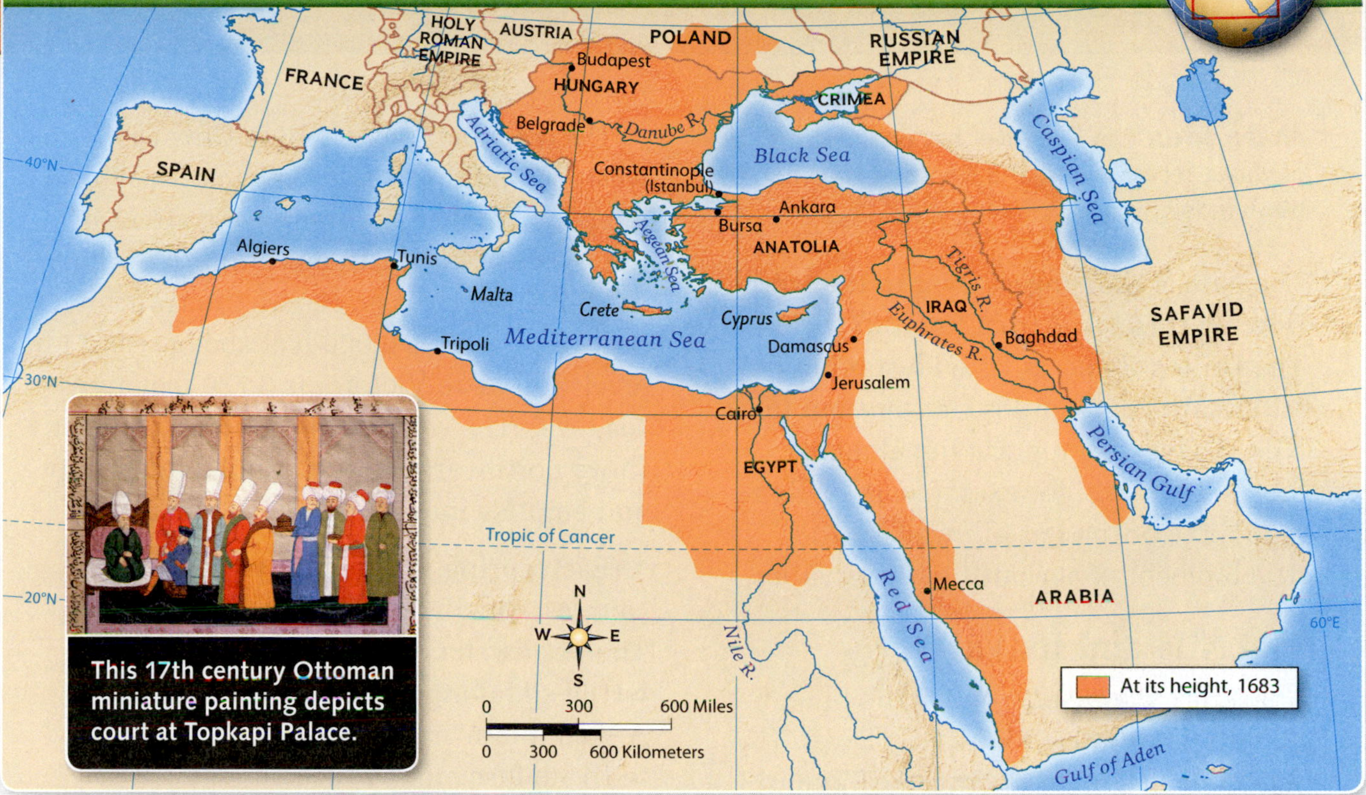

FRANCE • **HOLY ROMAN EMPIRE** • **AUSTRIA** • **POLAND** • **RUSSIAN EMPIRE**

SPAIN • HUNGARY • Budapest • CRIMEA • Belgrade • *Danube R.* • Black Sea • Constantinople (Istanbul) • *Adriatic Sea* • Ankara • Bursa • ANATOLIA • *Caspian Sea* • Algiers • Tunis • *Aegean Sea* • *Tigris R.* • IRAQ • SAFAVID EMPIRE • Malta • Crete • Cyprus • Damascus • Baghdad • *Euphrates R.* • Tripoli • *Mediterranean Sea* • Jerusalem • Cairo • EGYPT • *Persian Gulf* • Tropic of Cancer • *Nile R.* • Mecca • ARABIA • *Red Sea* • Gulf of Aden

40°N • 30°N • 20°N • 60°E

N W E S

0 300 600 Miles
0 300 600 Kilometers

At its height, 1683

This 17th century Ottoman miniature painting depicts court at Topkapi Palace.

Most Ottoman wealth was gained through trade and taxation. The busiest trade routes in the region ran through the empire. The Ottomans controlled trade on rivers, as well as ports on important seas such as the Black Sea and the Mediterranean Sea. When the empire conquered other lands, Ottoman **sultans**, or leaders, appointed officials to collect taxes from their new subjects. Taxes were a major source of wealth in the Ottoman empire.

Because it covered such a vast area, the Ottoman Empire was composed of many different ethnic groups, including Turks, Greeks, Slavs, Arabs, and Armenians. The Ottomans were Muslim and they spread Islam throughout the empire. They were known for their **religious tolerance**. Existing religious groups maintained their own practices and communities within the empire.

Internal conflicts and wars with European countries began to weaken the empire by the late 1600s. The empire lasted into the 1900s, but after World War I, it lost most of its remaining territory. In 1923, Turkey, the last remnant of the Ottoman Empire, became a republic.

Before You Move On

Make Inferences How did the Ottoman Empire grow so powerful and last so long?

ONGOING ASSESSMENT

MAP LAB

 GeoJournal

1. **Region** According to the map, which bodies of water did the Ottoman Empire control at its height? How did such control benefit the empire economically?

2. **Describe Geographic Information** Look at the map. Where is the capital of the Ottoman Empire? Why might that be a good location for the capital?

SECTION 3 FOCUS ON EGYPT

3.1 The Nile River Valley

TECHTREK

myNGconnect.com For an online map and photos of the Nile River Valley

Maps and Graphs

Digital Library

Main Idea The Nile River has provided a source of water, fertile plains, and transportation for thousands of years.

The Nile is the longest river in the world, flowing through Africa for 4,132 miles. The United States at its widest point is still nearly 1,500 miles shorter than the Nile. The river supports life along its banks. In Egypt, for example, the Nile's regular flooding has allowed for the development of agriculture in its valley.

The Nile and Its Valley

The Nile's main sources are the Blue Nile River and the White Nile River. The Blue Nile originates from Lake Tana in the Ethiopian highlands. The White Nile originates from Lake Victoria, which lies in Tanzania, Uganda, and part of Kenya.

Unlike most rivers, which flow east, west, or south, the Nile flows north. This happens because the southern sources of the river are higher in elevation than the mouth of the river on the Mediterranean Sea. Lake Tana is 6,000 feet above sea level, and Lake Victoria is 3,720 feet above sea level. The river drops in elevation as it flows northward through Sudan and Egypt and empties into the Mediterranean.

Each spring, snows melt at the river's sources, causing predictable flooding. The floodwaters deposit rich and fertile silt along the river's floodplain. A **floodplain** is the low-lying land next to rivers formed by sediment deposited by flooding. Historically, these favorable farming conditions supported permanent settlements and the development of an advanced civilization. The ebb and flow of the river allowed farmers to plan around its flood cycles.

Critical Viewing Palm trees grow along the Nile at Armana. How does this photo fit with the written description of the Nile River Valley?

NILE RIVER VALLEY

Mediterranean Sea

Alexandria

Avaris
(Tell el Daba)

LOWER EGYPT

El Giza · Cairo

Memphis
(Mit Rahina)

Beni Suef

Nile R.

Eastern

SINAI

Gulf of Suez

Gulf of Aqaba

Beni Hasan
el Shuruq

Akhetaten
(Amarna)

Western

Desert

UPPER EGYPT

Desert

Red
Sea

Girga

(Luxor) Thebes

Floodplain area

City
(modern name in parentheses)

Aswan High
Dam · Syene (Aswan)

Tropic of Cancer

Lake
Nasser

0   100   200 Miles
0   100   200 Kilometers

This satellite image captures the Nile River in Egypt at night. The majority of Egypt's population lives along the Nile.

The Nile Valley Today

Use of the Nile River has changed since the 1960s, when construction of a major dam began. In 1970, the **Aswan High Dam** was completed. One reason for its creation was to generate **hydroelectric power**, a form of energy created using flowing water. The Aswan High Dam provides about 15 percent of Egypt's electric power. Another reason the dam was built was to **regulate**, or control, the flooding of the Nile. Man-made irrigation allows for crops to be grown year-round because farmers are no longer dependent on the natural flooding cycle.

Today, the Nile River Valley is a densely populated area. In fact, about 95 percent of Egyptians live along the Nile. Surrounded by desert, the river continues to support life and agriculture along its banks.

Before You Move On
Summarize In what ways is the Nile River important to life in the region?

ONGOING ASSESSMENT

PHOTO LAB  GeoJournal

1. **Analyze Visuals** Based on the photo of the Nile River, what details demonstrate characteristics of a floodplain?

2. **Make Generalizations** Look at the satellite image above. What generalization can you make about settlement in the Nile Valley?

3. **Interpret Maps** What advantage might cities located along the Nile River have?

4. **Human-Environment Interaction** In what way did the human use of the Nile River change in 1970?

3.2 Egypt's Ancient Civilization

TECHTREK

myNGconnect.com For photos
of ancient Egyptian artifacts

Digital
Library

Main Idea Egypt's ancient civilization developed
significantly during four main time periods.

As you have read, as the Sahara became
a desert, people who lived there began
to migrate to the Nile River Valley in
search of a reliable source of water. They
brought with them skills in pottery,
metalworking, and agriculture. By around
3000 B.C., a writing system based on
pictures called **hieroglyphics** (HY ruh
GLIHF ihks) had developed. For the next
several thousand years, ancient Egyptian
civilization thrived along the Nile River.

Old and Middle Kingdoms

Ancient Egyptian civilization is divided
into four main periods, shown on the
time line below. Around 3000 B.C., the
kingdoms in Egypt unified under King
Menes. His rule began the first **dynasty**,
or series of rulers in the same family.
Many dynasties followed. In the Early
Dynastic period, agriculture and trade
developed along the Nile.
Egyptians began to
build stone monuments
called **pyramids**.

The Old Kingdom continued the growth
of dynasties. Egypt expanded from the
Nile Delta southward along the banks
of the Nile. More advanced farming and
trade practices enriched the Old Kingdom.
Between 2150 and 2040 B.C., however,
the annual floods of the Nile were not
as strong, which led to a decrease in
crops. This decrease created a long-term
economic and political crisis in Egypt.

Menuhotep II's reign restored Egypt to
stability and power in the 1900s B.C. and
is considered the beginning of the Middle
Kingdom. During this period, classic
Egyptian arts, architecture, and literature
flourished. In the 1600s B.C., however,
Hyksos invaders from the East weakened
and took over the Middle Kingdom.

New Kingdom

Egyptians rebelled against the Hyksos
and the defeat of the Hyksos ushered
in the New Kingdom—ancient Egypt's
greatest period of power and wealth. Its
boundaries extended into the desert, to the
Red Sea and along the eastern coast of the
Mediterranean Sea.

During this period, Egyptians began
to call their kings **pharaohs**. One New
Kingdom pharaoh was **Hatshepsut**
(hat SHEHP soot), a female pharaoh.

3200 B.C. Migration to Nile River Valley	The Great Sphinx at Giza lies near King Khafre's pyramid.	2575–2150 B.C. Old Kingdom	1975–1640 B.C. Middle Kingdom

3000 B.C.   **2500** B.C.   **2000** B.C.

2950–2575 B.C.
Early Dynastic Period

Egyptian
hieroglyphics

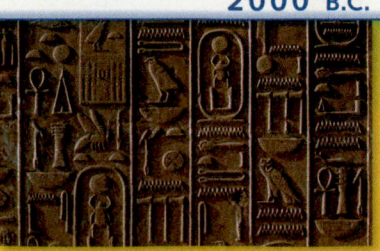

During her reign, Egypt's economy was strengthened through trade. Another New Kingdom pharaoh, **Ramses II,** expanded Egypt's empire through conquest.

The New Kingdom began to decline about 1075 B.C. In 525 B.C., the Persian Empire conquered Egypt. In 332 B.C., Alexander the Great liberated Egypt from Persian rule. Alexander was a Macedonian king who overthrew many empires, including the Persians. Eventually, Egypt became a province of the Roman Empire.

Egyptian Legacies

Ancient Egyptians made important contributions to culture, knowledge, and technology. Ancient Egyptian religion emphasized the belief in an afterlife, and was built around worship of <mark>deities</mark>, or gods. Two important deities were the sun god, **Re**, and the goddess, **Isis**.

Ancient Egyptians invented a paper-like material called <mark>papyrus</mark>. They used engineering skills to build cities and pyramids, and their mathematical and scientific observations endure today.

Before You Move On
Summarize In what ways did ancient Egypt's civilization develop over its four main time periods?

 **Critical Viewing** Artifacts like this statue from the tomb of Tutankhamen show how ancient Egyptians buried their royalty. What details from the photo indicate that King Tut is royalty?

ONGOING ASSESSMENT
READING LAB   GeoJournal

1. **Summarize** Ancient Egypt is frequently described as "the gift of the Nile." Based on your reading, summarize the details that might lead to this description.

2. **Movement** When and why did migration to the Nile River Valley occur?

3. **Interpret Time Lines** Based on the text and the time line, what explains the gaps in years between the Old, Middle, and New Kingdoms?

1539–1075 B.C.
New Kingdom

Painting of the Egyptian goddess Isis

30 B.C.
Egypt becomes province of Roman Empire.

1500 B.C.   **1000** B.C.   **500** B.C.

332 B.C.
Alexander the Great conquers Egypt.

3.3 The Great Pyramids

TECHTREK

myNGconnect.com For online photos of the Egyptian pyramids

Digital Library

Main Idea The ancient Egyptians built great tombs for their pharaohs.

The pyramids are among the enduring legacies of ancient Egypt. Archaeologists and engineers still marvel at how the Egyptians built the pyramids without modern machines and technology.

Royal Tombs

Egyptian pyramids were built for two reasons. One purpose of the pyramids was to provide individual <mark>tombs</mark>, or burial places, for royalty. Secondly, kings and pharaohs demonstrated their power by having huge pyramids built for them. Ten pyramids were built at **Giza** alone.

Pyramids were holy sites, meant to transport the pharaohs to the afterlife. Ancient Egyptians believed the dead could enjoy earthly possessions, so they filled burial chambers with clothes, food, and furniture for that purpose.

The Great Pyramid of Khufu

Built around 2550 B.C., the largest pyramid is the **Great Pyramid of Khufu,** which took 20 years to build. Initially, the pyramid stood 481 feet high. Over time, erosion has worn away the limestone and granite surface that had covered the outside of the pyramid. As a result, today the Great Pyramid is only 449 feet tall.

Critical Viewing Khufu's Great Pyramid (far right) is the oldest and tallest. It stands with the pyramids for Khafre (middle) and Menkaure (far left). What details in the photo convey the sizes of the pyramids?

The interior of the Great Pyramid has three separate burial chambers and several passageways. Because the Great Pyramid was looted, or burglarized, by thieves seeking valuable artifacts, only Khufu's **sarcophagus**, or coffin, remains in his burial chamber.

Before You Move On

Summarize For what purposes did ancient Egyptians build the pyramids?

1. **Analyze Visuals** Based on the illustration, what does the Great Pyramid's construction reveal about ancient Egyptian engineering skills?

2. **Explain** Why might the King's Chamber be located in the center of the pyramid?

3. **Draw Conclusions** In what way did pyramid-builders try to prevent robberies? What conclusions can you draw about why guarding against theft was important?

THE GREAT PYRAMID OF KHUFU, 2550 B.C.

3 KING'S CHAMBER
Khufu was buried in this chamber, which lies nearly at the exact center of the pyramid.

2 GRAND GALLERY
The Grand Gallery leads to the King's chamber.

4 QUEEN'S CHAMBER
Though it is called the Queen's Chamber, none of Khufu's wives were buried there. Egyptologists believe Khufu wanted more than one burial option.

5 UNFINISHED CHAMBER
The Unfinished Chamber lies underground, beneath the Great Pyramid.

1 ENTRANCE
Pyramid-builders tried to guard against robberies by constructing heavy walls to seal off entrances. However, most pyramids were looted by other Egyptians and many treasures disappeared.

VOCABULARY

For each pair of vocabulary words, write one sentence that explains the connection between the two words.

1. pharaoh; tomb

Pyramids were built as tombs for the pharaohs.

2. hieroglyphics; papyrus
3. petroleum; nonrenewable
4. agricultural revolution; domesticate
5. desertification; arid

MAIN IDEAS

6. Describe the climate in Southwest Asia and North Africa. (Section 1.1)

7. In what ways are the Tigris and Euphrates rivers important to the region? (Section 1.2)

8. What natural resource on the Arabian Peninsula contributes most to the economy? (Section 1.3)

9. What characteristics do the Anatolian and Iranian plateaus share? (Section 1.4)

10. Why did Mesopotamia become a cultural hearth? (Section 2.1)

11. What core belief do Judaism, Christianity, and Islam share? (Section 2.2)

12. How did Judaism, Christianity, and Islam spread? (Section 2.3)

13. In what ways did the Ottoman Empire gain wealth and power? (Section 2.4)

14. How did the Nile River help early civilizations form? (Section 3.1)

15. What contributions did ancient Egyptians make to human knowledge? (Section 3.2)

16. For what purpose did ancient Egyptians build the pyramids? (Section 3.3)

GEOGRAPHY

ANALYZE THE ESSENTIAL QUESTION

How have climate and location influenced the region in the past and today?

Critical Thinking: Analyze Cause and Effect

17. When the Sahara changed from grassland to desert, how did people living there respond?

18. In what ways did the flooding of rivers influence development in the region?

HISTORY

ANALYZE THE ESSENTIAL QUESTION

How did civilizations develop in Southwest Asia and North Africa?

Critical Thinking: Summarize

19. Summarize the ways in which European countries spread Christianity.

20. What circumstances and events led to the end of the Ottoman Empire?

INTERPRET TABLES

RELIGION IN SOUTHWEST ASIA AND NORTH AFRICA				
	Christianity	Islam	Judaism	Other
Egypt	10.0%	90.0%	—	—
Israel	2.1%	16.8%	75.5%	5.6%
Jordan	6.0%	92.0%	—	2.0%
Lebanon	39.0%	59.7%	—	1.3%
Morocco	1.1%	98.7%	2.0%	—
Saudi Arabia	—	100.0%	—	—
Turkey	—	99.8%	—	0.2%

Source: CIA World Factbook, 2010

21. **Compare and Contrast** What country has the largest percentage of Christians?

22. **Make Inferences** Why does Israel have the largest Jewish population?

ANALYZE THE ESSENTIAL QUESTION

How did an advanced civilization develop in Egypt?

Critical Thinking: Draw Conclusions

23. In what ways does the Nile River support life along its banks today?

24. Why is ancient Egypt referred to as "the gift of the Nile"?

25. In what ways do the pyramids demonstrate the advanced architectural and engineering skills of the ancient Egyptians?

INTERPRET MAPS

ANCIENT EGYPT

Egyptian Kingdoms
- Old Kingdom (c. 2575–2150 B.C.)
- Middle Kingdom territorial expansion (c. 1975–1640 B.C.)
- New Kingdom territorial expansion (c. 1539–1075 B.C.)

26. **Location** At its height, how far did ancient Egyptian kingdoms extend? Use the map scale to determine the north-south extent of the kingdoms.

27. **Explain** Which of Egypt's ancient kingdoms reached the furthest south? Based on what you know, what explains this expansion?

ACTIVE OPTIONS

Synthesize the Essential Questions by completing the activities below.

28. **Write a Travel Article** Imagine you are a travel writer who has just returned from a trip to the region. Write a travel article that describes the history, physical features, and important cities, and which city has the most interesting history. Use the following tips to help you write your article. **Present your article to a small group of classmates.**

> **Writing Tips**
> - Before you write, make an outline that organizes your major points.
> - Include important dates and locations for specific events and points of interest.

Go to **Student Resources** for Guided Writing support.

TECHTREK myNGconnect.com For research links on Southwest Asia and North Africa

29. **Research Archaeological Sites** Many important artifacts from ancient civilizations have been found in Southwest Asia and North Africa. Conduct research, and prepare a chart that highlights two important archaeological sites in the region. Use the research links at **Connect to NG** and other online sources to gather your data. Organize your research with a chart similar to the one below.

	SITE #1	SITE #2
name and location of site		
date of discovery		
ancient civilization		
important artifacts found there		
what artifacts reveal about their civilization		

CHAPTER 10

SOUTHWEST ASIA & NORTH AFRICA TODAY

PREVIEW THE CHAPTER

Essential Question How have resources and migration shaped culture in Southwest Asia and North Africa?

KEY VOCABULARY

- migration
- guest worker
- strait
- mosque
- breakwater
- tsunami
- sheikh
- emirate

ACADEMIC VOCABULARY
sprawl, interact

TERMS & NAMES

- Bedouin
- Tuareg
- Bosporus Strait
- Grand Bazaar
- Hagia Sophia
- Blue Mosque
- Dubai
- United Arab Emirates

Essential Question What forces have affected the development of modern countries in the region?

KEY VOCABULARY

- hereditary
- suffrage
- reserve
- petrochemical
- self-rule
- intifada
- coalition
- totalitarian
- extremist
- terrorist
- literate
- vocational

ACADEMIC VOCABULARY
distribute, disregard

TERMS & NAMES

- Knesset
- Jerusalem
- Palestine Liberation Organization (PLO)
- Shi'ite
- Sunni
- Kurd
- Saddam Hussein
- Taliban
- al-Qaeda

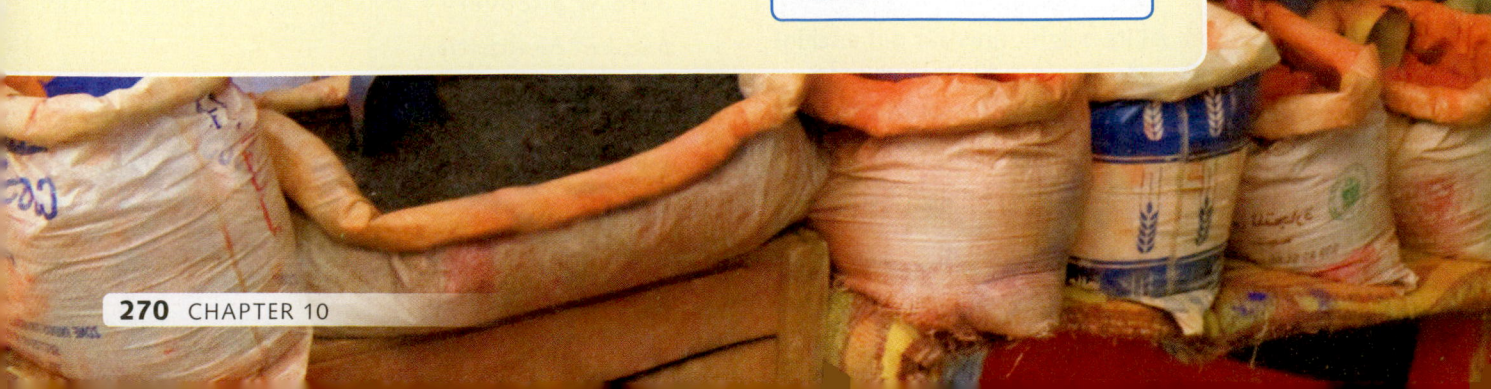

Women visit a souk, or market, in Morocco.

1.1 Migration and Trade

TECHTREK

myNGconnect.com For photos of the Tuareg and Bedouin

Digital Library

Main Idea Trade and migration continue to define Southwest Asia and North Africa.

Migration, or the movement from one place to another, is a familiar process in Southwest Asia and North Africa. Three kinds of migration mark this region: movement for herding, movement for trade, and movement for job opportunities.

Nomadic Herders

The **Bedouin** (BEHD u ihn) are a nomadic, Arabic-speaking people who live in the deserts of Saudi Arabia, Iraq, Syria, and Jordan. Most Bedouin trace their ancestry to the Arabian Peninsula. Strongly independent, Bedouin often identify themselves first as Bedouin rather than as citizens of a country.

The Bedouin move from place to place as they herd camels, sheep, goats, or cattle. Their migration patterns depend on the season and the needs of their herds.

Salt Traders

As you have learned, the Sahara is a vast, hot desert and a difficult physical barrier to cross. However, for those who overcame the difficulties, trade across the Sahara could be very profitable. One group of traders, the **Tuareg** (TWAH rehg), historically dominated the trans-Saharan caravan trade—especially the salt trade. The Tuareg are a semi-nomadic people who live in various North African countries, including Algeria and Libya.

> **Critical Viewing** A Tuareg nomad leads a camel caravan across the Sahara. What details in this photo illustrate what the climate is like?

Each winter, Tuareg traders travel in small caravans across a wide expanse of barren sand dunes in the Sahara. Caravans stop at oases along the way for rest and water. At these oases, the Tuareg trade goats for salt and millet, a grain, for dates, which are dried fruits of the date palm. After returning home, the Tuareg sell the salt and dates at markets for a profit.

This desert highway is located in the western Sahara.

Guest Workers

Today's migration is prompted by a variety of different factors. Resource-rich countries such as Saudi Arabia draw millions of **guest workers**, or temporary laborers who migrate to work in another country. Companies employ guest workers to fill labor needs and shortages. Guest workers, seeking better wages than they might make in their home countries, work in industries such as oil, mining, and construction. In some cities, such as Dubai in the United Arab Emirates, guest workers represent a large percentage of the total population.

Before You Move On

Summarize How would you describe the different types of migration in the region?

ONGOING ASSESSMENT

PHOTO LAB

 GeoJournal

1. **Analyze Visuals** Based on the photo, how would you describe a caravan? What do people use to protect themselves in this environment?

2. **Compare and Contrast** Based on the photos of the caravan and the highway, compare and contrast modern and traditional methods of movement in this region. What might be benefits and drawbacks of both methods?

3. **Movement** What factors draw guest workers to this region today?

SECTION **1** CULTURE

1.2 Istanbul: Bridging East and West

TECHTREK

myNGconnect.com For a map of the Bosporus Strait and photos of Istanbul

 Maps and Graphs

 Digital Library

Main Idea For centuries, Istanbul has been a thriving city at the crossroads of Europe and Asia.

As the only city in the world located on two continents, Istanbul ▶ is a cultural bridge between Europe and Asia. Istanbul has had several names. It began as the ancient city of Byzantium and became Constantinople in 330. When the Ottomans claimed the city as their new capital in 1453, Constantinople became known as Istanbul. Though it is no longer a capital, Istanbul is the cultural and industrial center of Turkey.

Center of Trade and Culture

Istanbul sits on both sides of the **Bosporus Strait.** A **strait** is a narrow passage of water that connects larger bodies of water. The Bosporus Strait connects the Black Sea with the Sea of Marmara, the Aegean Sea, and, ultimately, the Mediterranean Sea. The Bosporus Strait is an important waterway that links Asia with Europe and North Africa.

Istanbul's location made this city an important center of trade for thousands of years. The **Grand Bazaar** stands as a reminder of Istanbul's history as a commerical hub. It has 5,000 shops and has been important for trade since the mid-1400s.

Istanbul is home to world-famous architecture. The Byzantines built the **Hagia Sophia** in the 500s as a Christian cathedral. In 1453, the building was converted into a **mosque**, or a Muslim place of worship. Another historical building is the **Blue Mosque**. Built in the early 1600s, it features six minarets, or tall, slender towers. The high ceilings of the Blue Mosque are decorated with more than 20,000 blue tiles that give the mosque its name.

People and cultures from all over the region converge in Istanbul. Most residents are Muslim Turks, but Istanbul is also home to many other ethnic and religious groups. Historically, diverse groups have practiced their own religions in this Muslim city.

Rapid Growth

Today, more than 12 million people live in Istanbul, a hilly city that **sprawls**, or spreads out, for more than 90 square miles. Since the 1950s, rapid population growth has created a housing shortage. Many homes were built quickly and are not earthquake-safe, a real danger for a city that sits on the very active North Anatolian Fault.

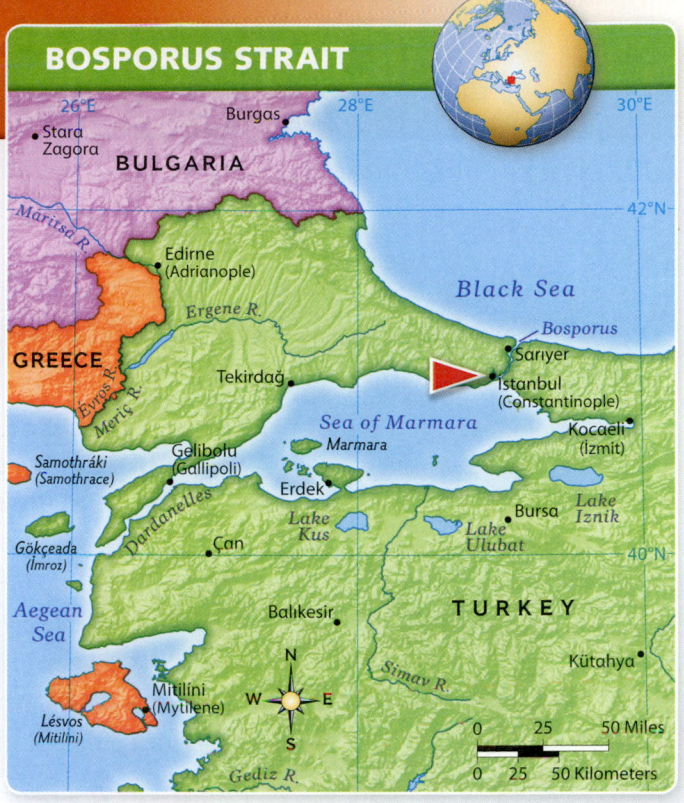

BOSPORUS STRAIT

Critical Viewing This café is located near the Grand Bazaar. What details do you notice about how this tea is prepared and served?

Determined to keep pace with growth, city leaders are modernizing infrastructure and services. For example, one major building project is the construction of a tunnel under the Bosporus Strait for high-speed commuter trains. Once completed, this modern—and earthquake-safe—tunnel will help transport 1.5 million people back and forth across the Bosporus every day.

Before You Move On

Make Inferences In what ways has Istanbul connected Europe and Asia?

ONGOING ASSESSMENT

WRITING LAB   GeoJournal

Write Reports Work with a partner to research and write a short report on Istanbul.

Step 1 Make a three-column chart with the labels Location, Trade, and Culture.

Step 2 Together, fill in the columns with details from the reading. Go to **Connect to NG** to add details and information to your chart.

Step 3 Use the chart to help you write a two-paragraph report that describes the city as a hub of trade and culture, both historically and today.

Critical Viewing The Hagia Sophia (left) and the Blue Mosque (right) are located in Istanbul. How would you describe the city's architecture?

1.3

SECTION **1** GEOGRAPHY

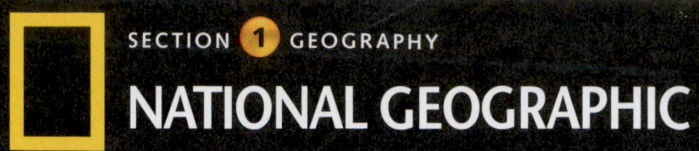
NATIONAL GEOGRAPHIC

TECHTREK
myNGconnect.com For photos of
Goodman at work and an Explorer Video Clip

**Digital
Library**

Exploring
Ancient Israel
with Beverly Goodman

Main Idea Underwater exploration reveals clues about historical events in ancient Israel.

"Coastlines are the most dynamic natural environments on earth," says National Geographic Emerging Explorer Beverly Goodman. Goodman uses archaeology, geology, and anthropology to study how nature and people **interact**, or affect one another, on constantly changing coastlines.

Herod's Harbor

In 2003, Beverly Goodman and a team of archaeologists were in Caesarea (say suh REE ah), Israel, to explore the ruins of the harbor built by Herod the Great, King of Judea, at the end of the first century B.C. Caesarea was a major port of trade between the Roman Empire and Asia. It was located on the coast of the Mediterranean Sea, in present-day Israel.

The harbor at Caesarea was one of the first harbors constructed on the open sea and had no peninsula or other natural protection. Instead, Herod's builders used concrete blocks to construct two huge **breakwaters**, or barriers, to protect the harbor. It was a gigantic structure, but for some reason it was heavily damaged and today lies in ruins on the seafloor.

myNGconnect.com

For more on Beverly Goodman in the field today

> **Critical Viewing** Caesarea was an important port city on the Mediterranean coast. How might the surrounding physical features make it difficult to preserve the ruins?

An Ancient Tsunami

Initially, Goodman focused her work on understanding the construction of the ancient harbor. However, while diving, Goodman's team found layers of pottery, stone, and shells. This discovery would not have been unusual, except that the layer of shells was more than three feet thick. Goodman and her team determined that all the shells were deposited by a single, rapid, violent event. The discovery proved that a tsunami struck and destroyed the harbor sometime in the first or second century. A **tsunami** is a giant wave caused by an earthquake, volcano, or landslide.

Goodman is making similar excavations throughout the Mediterranean. She hopes to determine a pattern of tsunami activity across the region. Her findings may help the increasing number of people living on coasts anticipate tsunamis in the future.

Now an archaeological park open to visitors, Caesarea is located about 60 miles northwest of Jerusalem.

Before You Move On

Monitor Comprehension In what way has underwater exploration led to an increased understanding of events in ancient Israel?

ONGOING ASSESSMENT
VIEWING LAB  GeoJournal

1. **Analyze Visuals** Watch the Explorer Video Clip and consider this question: What did you find interesting about Beverly Goodman's work?

2. **Pose and Answer Questions** After viewing the video clip, write two questions about what you saw. Ask a partner to answer your questions, and answer your partner's questions.

3. **Human-Environment Interaction** In what way does the construction of Herod's harbor demonstrate how people interacted with the environment to meet their needs?

1.4 Dubai: Desert City

TECHTREK
myNGconnect.com For photos of Dubai's sand islands

Digital Library

Main Idea Dubai is a multicultural and rapidly developing city on the Persian Gulf.

In 1959, the <mark>sheikh</mark>, or Arab leader, of **Dubai** decided to turn a small fishing village into a modern city in the desert. Today, Dubai is one of the fastest growing cities in the world.

Built on Sand

As both a city and a state, Dubai is one of seven <mark>emirates</mark>, or states, that make up the **United Arab Emirates** (UAE). The emirates are located on the Persian Gulf between Qatar, Saudi Arabia, and Oman. The UAE's location on the Persian Gulf makes it a prime trading center in this oil-rich part of the world.

Compared to other cities in the region, Dubai is a new city. Islands created out of sand from the Persian Gulf provide homes for wealthy residents and hotels for tourists. The Palm Jumeirah, the island featured below, is shaped like a palm tree. Just as Dubai extended its shoreline, it has also grown its skyline. In 2009, the tallest building in the world (shown at right) opened to tourists and businesses.

Multicultural City

Dubai attracts tourists, investors, and workers from more than 150 countries. International business and tourism drive its economic growth. Businesses from around the world have moved here because they do not have to pay corporate taxes or income taxes. Dubai also has a well-developed international banking system. The city has created enormous wealth. In 2007, its economic growth rate was 16 percent, an unusually high rate.

Only about one in eight residents of Dubai are citizens of the United Arab Emirates. Guest workers from South Asian countries such as India represent more than 60 percent of the population. The official language of the emirate is Arabic, but people also speak Hindi, Urdu, English, and Bengali.

Because of its multicultural population, Dubai also shows religious tolerance. Located on the Arabian Peninsula, where the population is mostly Muslim, this city's Islamic mosques, Christian churches, and Hindu temples accommodate people of different faiths. Religious and ethnic conflicts among its residents are rare.

Critical Viewing The Palm Jumeirah is an island built out of sand from the Persian Gulf. In what ways might Dubai's sand islands extend its shoreline?

Before You Move On

Make Inferences What factors contribute to Dubai's rapid growth?

POPULATION OF DUBAI*

	Male	Female	Total
Employed Citizens	30,725	11,053	41,778
Employed Non-Citizens	1,183,914	126,556	1,310,470
Total Employed Persons	1,214,639	137,609	1,352,248

*Aged 15 and older

Source: Dubai Statistics Centre—Labor Force Survey, 2009

Critical Viewing The Burj Khalifa stands 2,716.5 feet high and is the tallest building in the world. Based on what you can see in this photo, in what ways does the city contrast with its desert surroundings?

DATA LAB
GeoJournal

1. **Analyze Data** Look at the chart. Compare the total numbers of employed non-citizens with employed citizens in Dubai. What do you think explains the difference in numbers?

2. **Draw Conclusions** The total population of Dubai, employed and non-employed, is 1,570,923. What conclusions can you draw when you compare this number to the total number of employed citizens?

2.1 Comparing Governments

TECHTREK

myNGconnect.com For photos of governments in the region

Digital Library

Connect to NG

> **Main Idea** Countries in Southwest Asia and North Africa have varied types of governments.

Many governments in Southwest Asia and North Africa are monarchies. Other countries have established representative democracies. These varying governments must coexist as neighbors in the same region.

Monarchies

Saudi Arabia is a monarchy ruled by a king. The position of king is **hereditary**, or passed on through family. The Council of Ministers helps the king govern but power rests with the monarch. Ministers, many of whom are from the royal family, are appointed by the king and can be dismissed at any time. Saudi kings govern by Islamic law and the country has no formal constitution. Opposition to those in power is not tolerated and political parties are banned. Only males at least 21 years of age have **suffrage**, or the right to vote. However, as of 2011, no national elections had yet been held.

Like Saudi Arabia, a hereditary monarch rules Jordan. Unlike Saudi Arabia, however, Jordan is a constitutional monarchy. In this form of government, monarchs must follow the constitution of the country and their power is not absolute. Jordan's legislative branch includes two parts: the National Assembly and the Chamber of Deputies. Monarchs appoint the members of the National Assembly but citizens elect the members of the Chamber of Deputies. All Jordanians 18 years of age and older can vote.

Democracies

While countries such as Saudi Arabia and Jordan are monarchies, other countries in the region are democracies. Israel and Turkey are two examples.

Israel is a parliamentary democracy. In this type of democracy, instead of voting for individuals, citizens cast ballots for a particular party. Based on election results, each party is assigned a number of seats in the **Knesset** (kuh NEH set), or the Israeli parliament. The Knesset then elects a president and a prime minister. Israel does not have a formal constitution, but instead has a set of basic laws passed by the Knesset as its foundation.

After the end of Ottoman rule in 1923, Turkey became a republic. A republic is a form of democratic government in which representatives elected by the people hold power. Although it was established as a republic, only one party existed in Turkey for about 25 years. Since the 1950s, though, Turkey has been a multiparty republic. In 2007, voters in Turkey approved a constitutional amendment to establish direct presidential elections.

Unrest in North Africa

Some governments in the region are technically democracies but have had a history of suppressing fair elections and limiting citizens' rights. Beginning in 2010, economic crises in these countries fueled political revolt.

In January 2011, several uprisings led to shifts in leadership. In Tunisia, President Ben Ali stepped down after days of protest by Tunisians. Encouraged by Tunisia's success, protesters in Egypt gathered in Cairo for 18 days, demanding that President Hosni Mubarak (moo BAR uhk) step down. Initially, Mubarak refused, but on February 11, he resigned the office he had held for 30 years. Libyans watching the events in Egypt also began to demand—at great risk—that their leader, Colonel Mu'ammar al-Qadhafi (kuh DAH fee), surrender power.

Before You Move On

Summarize Describe at least two forms of government in the region.

> **Critical Viewing** Many Egyptians celebrated the resignation of Hosni Mubarak in Cairo's Tahir Square (shown below) on February 11, 2011. What details in the photo convey the importance of the moment?

ONGOING ASSESSMENT

SPEAKING LAB GeoJournal

1. **Turn and Talk** In a small group, compare and contrast the ways in which monarchies and democracies in the region govern. As you discuss, take notes on your group's comparisons and contrasts.

2. **Conduct Internet Research** Go to **Connect to NG** and research the form of government of another country in this region. Compare that country's government with those in the lesson. Be prepared to present your research and comparison to the class.

SECTION 2 GOVERNMENT & ECONOMICS
2.2 Oil and Wealth

TECHTREK

myNGconnect.com For an online map and photos of oil-rich countries in the region

Maps and Graphs

Digital Library

Main Idea Oil-producing countries in Southwest Asia and North Africa possess important sources of energy and wealth.

Petroleum, or oil in its unrefined form, was discovered in the region in the 1900s. In fact, most of the world's known petroleum deposits are concentrated in Southwest Asia and North Africa. Much of the world, including the United States, depends on the region for this important energy source.

An Oil-Rich Region

By extracting and exporting oil, several countries have grown wealthy. Among those countries, Saudi Arabia, Iran, Iraq, Kuwait, and the United Arab Emirates possess more than half of the world's known petroleum **reserves**, or future supply. About 20 percent of the world's known oil reserves are located in Saudi Arabia alone.

Even when oil prices drop, the major oil-exporters rank among the world's wealthiest countries. Some countries, such as Saudi Arabia, have used this wealth to build modern ports, airports, highways, and industrial plants. A few countries, including Kuwait and the UAE, have built industries that manufacture modern plastics and **petrochemicals**, or products made from petroleum.

Jobs in oil-rich countries have attracted guest workers. Some guest workers fill oil-producing and construction jobs. Others work in service industries or as domestic workers. Guest workers fill jobs that tend to be low-paying and temporary. In Saudi Arabia, guest workers comprise about 80 percent of the labor force; in Kuwait, they represent about 60 percent.

Critical Viewing This oil tanker is anchored in the Persian Gulf. What can you infer from the photo about the size of ships used to export oil?

OIL IN SOUTHWEST ASIA AND NORTH AFRICA

Daily Oil Production
Thousand barrels per day
- Less than 10
- 10 to 100
- 100 to 1,500
- 1,500 to 3,000
- 3,000 to 5,000
- More than 5,000
- ◆ Oil fields

0   300   600 Miles

0   300   600 Kilometers

Oil Wealth and the Income Gap

Like oil itself, oil wealth is not evenly **distributed**, or spread out, within the region, or even within individual countries. Oil-rich countries include large areas of desert and remote villages. People who live in these underdeveloped rural areas tend not to benefit from oil wealth. Many rural residents have migrated to cities in the region seeking better economic opportunities.

However, the income gap is widening even within large, modern cities. For example, Riyadh, Saudia Arabia, is a rapidly-expanding city of more than four million people. Although it is the capital of one of the wealthiest countries in the world, housing is expensive. Many young Saudis are unemployed, and poverty is an increasing problem.

Before You Move On

Monitor Comprehension In what ways has wealth gained from oil benefited and changed countries in the region?

OIL RESERVES / PRODUCTION, TOP FIVE OIL-PRODUCING COUNTRIES					
	Saudi Arabia	Iran	Iraq	Kuwait	UAE
Barrels of Reserves	266.7 billion	137.6 billion	115 billion	104 billion	97.8 billion
Barrels Produced Per Day	9.764 million	4.172 million	2.399 million	2.494 million	2.798 million

Source: CIA World Factbook, 2009 estimates

ONGOING ASSESSMENT

MAP LAB

 GeoJournal

1. **Location** Look at the map. Near what geographic feature are most oil fields located? How might this feature benefit the countries as exporters?

2. **Interpret Maps** Of the top five oil-producing countries, which are located on the Arabian Peninsula? What other oil-producing countries are located there?

3. **Analyze Data** Which country in this region is second to Saudi Arabia in terms of oil production and reserves? According to the map, which countries produce less than 100,000 barrels per day?

2.3 Tensions in Southwest Asia

Main Idea Israelis and Palestinians have struggled over issues of land, self-rule, and security for many years.

Current tensions in Southwest Asia have a long history. Complicated matters of land, security, and **self-rule**, or the government of a country by its own people, are critical issues for both Israelis and Palestinians.

Founding Modern Israel

After World War I, many territories once ruled by the Ottomans became spheres of European rule. The area of present-day Jordan, Israel, the West Bank, and the Gaza Strip was placed under British control and named the British Mandate. Many Jews immigrated there, joining already established Jewish communities. An increased Jewish presence caused resentment among Palestinian Arabs.

The experience of the Holocaust during World War II prompted the United Nations (UN) to create a state for the Jewish people. In 1947, the UN voted to divide the British Mandate into two parts: Arab and Jewish. Surrounding Arab countries and the Palestinians themselves rejected the UN's decision. The Jews accepted the UN plan and in 1948 declared Israel an independent state.

Immediately, six Arab countries— Egypt, Iraq, Jordan, Syria, Saudi Arabia, and Lebanon—declared war against Israel. Before the war, many Palestinians fled to neighboring countries or to Arab towns in the West Bank. Israel won the war, and a Palestinian state never formed.

Israeli and Palestinian leaders meet to negotiate peace in 1993.

A series of Arab-Israeli wars followed for the next several decades. During these wars, Israel and its Arab neighbors fought over territory and security. During the 1960s, Palestinian leaders created the **Palestine Liberation Organization (PLO).** At the time, the PLO wanted to create a Palestinian state in place of Israel.

Israelis and Palestinians Today

In 1987, Palestinians launched an ==intifada==, or mass uprising. Palestinians protested—sometimes violently—against Israeli control of the Gaza Strip and the West Bank. In the 1990s, leaders of Israel, Arab countries, and the Palestinians began peace talks. Israel agreed to give the Palestinians self-rule in the Gaza Strip and the West Bank. Palestinians agreed to acknowledge Israel's right to exist. Later, Israel withdrew completely from the Gaza Strip and turned over control of much of the West Bank to the Palestinians.

Efforts toward peace stalled, however. Israelis and Palestinians could not agree on several issues—especially **Jerusalem**. Israel's capital is Jerusalem, but Palestinians also want to establish East Jerusalem as their capital. Tensions over the collapse of the peace process led to a second intifada in 2000. In response to the violence, in 2002, Israel began building a security barrier along the boundary between Israel and the West Bank. Permits are required to pass through the checkpoints along the barrier, which divides many workers from their workplaces and people from basic services. However, violence against Israeli civilians has decreased by 90 percent.

ISRAEL POLITICAL

MAP TIP
Palestinians have self-rule in the Gaza Strip and limited rule in the West Bank. The permanent status of these areas is undecided.

Palestinians still desire a state. Israelis still desire security. World leaders are helping work toward a peaceful solution but progress is slow.

Before You Move On
Make Inferences What issues divide Israelis and Palestinians?

2.4 Iraq's Problems and Promise

TECHTREK
myNGconnect.com For a map of Iraq and current events

 Maps and Graphs

 Connect to NG

Main Idea Internal division and wars have caused great problems for Iraq, but a move toward democracy promises a better future.

For much of its modern history, Iraq has been torn by war and ruled by foreign rulers or dictators. In recent years, however, democracy—and with it, hope for the future—has begun to take root.

Religious and Ethnic Divisions

Internal divisions have a long history in Iraq. When the Ottoman Empire ended after World War I, Great Britain established a monarchy in Iraq and defined the country's borders. The new territory brought together two distinct Arabic groups, **Shi'ite** (SHEE eyt) and **Sunni** (SOO nee) Muslims.

These two Muslim groups have been divided since the death of Muhammad, Islam's founder, in 632. The Shi'ites believe that the leaders of Islam should be descendants of Muhammad. The Sunnis believe that Islam's leaders should be chosen from those most qualified. About 75 percent of the world's Muslims are Sunni. However, Shi'ites represent about 60 percent of Iraq's population.

Iraq's newly established borders also included members of an ethnic group called the Kurds. The **Kurds** are Sunni Muslims but have a history, language, and culture that differs from their Arabic neighbors. Kurds represent 15 to 20 percent of Iraq's population. Because they are a minority ethnic group, the Kurds have at different times in their history experienced serious discrimination.

War-Torn Nation

Iraq waged war in the 1980s with neighboring Iran partly as a result of the Sunni-Shi'ite divide. The largely Shi'ite Iran had overthrown its monarch in 1979 and established an Islamic government. Iraqi president **Saddam Hussein** was a Sunni Muslim who rose to power that same year. He feared that the Iranian Shi'ites would persuade the Shi'ites in Iraq to overthrow his government. The two countries fought for eight years, and the war ended with no clear winner.

In 1990, Iraq invaded Kuwait. Hussein claimed that Kuwait had been stealing Iraqi oil. In late 1990 and early 1991, the United States formed a **coalition**, or alliance, with other countries. Coalition countries drove Iraqi forces from Kuwait.

Move Toward Democracy

In 2003, another coalition of countries invaded Iraq. The coalition believed that Iraq was concealing weapons of mass destruction. Although none were found, the coalition removed Hussein from power. Many Iraqis were relieved. As a **totalitarian** dictator, Hussein had demanded complete obedience and ruled through terror.

In 2005, free elections were held in Iraq and legislators drafted a democratic constitution. Violence still occurs, but Iraqis are working to stabilize and rebuild their country.

Before You Move On
Summarize What factors do you think make Iraq's future look promising?

IRAQ'S RELIGIOUS AND ETHNIC DIVISIONS

TURKEY

Mosul

SYRIA

Tigris R.

Samarra'

Euphrates R.

Baghdad

I R A Q

Al Kut

Najaf

Al Qurnah

Nasiriyah

Basra

SAUDI ARABIA

KUWAIT

IRAN

36°N

32°N

28°N

40°E

44°E

48°E

Legend:
- Kurd
- Sunni
- Sunni and Kurd
- Shi'ite
- Shi'ite and Sunni
- Sparsely populated

0   100   200 Miles

0   100   200 Kilometers

MAP LAB

GeoJournal

1. **Interpret Maps** Locate the Tigris River. Which groups live along the river north of Baghdad? Which groups live along the river south of Baghdad? Which cities are located on the Tigris River?

2. **Location** Several different groups live in Iraq. According to the map, where do most Sunnis in Iraq live? Where do most Kurds live?

3. **Analyze Cause and Effect** Locate Iraq's Shi'ite population and Iraq's neighbor, Iran. How did the establishment of a Shi'ite government in Iran lead to the Iran-Iraq war?

Critical Viewing Iraqi women hold up ink-stained fingers indicating participation in provincial elections in 2009. What details in the photo suggest their feelings of pride?

2.5 Afghanistan: Moving Forward

TECHTREK
myNGconnect.com For a
map of Afghanistan

Maps and Graphs

> **Main Idea** Afghanistan struggles to define itself and move forward in the modern political world.

Afghanistan is one of the world's poorest countries. After decades of war and upheaval, the country is trying to build a stable central government.

Afghanistan Divided

The physical features and ethnic diversity of Afghanistan complicate efforts to unify the country. First, Afghanistan is landlocked and its mountains and deserts make travel in the country difficult. Second, about 80 percent of Afghanistan's population is rural. Widely scattered, isolated villages focus on local, rather than national, affairs. Third, Afghanistan's population includes several ethnic groups. Rural Afghans identify with their ethnic groups rather than as citizens of Afghanistan. Of Afghanistan's ethnic groups, Pashtun are most numerous, followed by Tajik, Hazara, and Uzbek.

In 1978, a Communist political party took power. This new government **disregarded**, or ignored, local ethnic and religious customs and forced policies of modernization. The speed and extent of change triggered rebellions throughout the country. Afghan leaders asked another communist government, the Soviet Union, for help. In 1979, the Soviets invaded Afghanistan. Islamic forces known as mujahedeen (moo ja heh DEEN) fought the Soviets for ten years.

After the Soviet Union withdrew from Afghanistan in 1989, the Afghan government fell, civil war tore the country apart, and locally armed leaders took control of much of the country. A group of Pashtuns known as the **Taliban** brought these local leaders under control and in 1996 seized the capital, Kabul.

The Impact of Extremism

The Taliban are **extremists**, or people who hold rigid religious or political views. The Taliban immediately imposed their own brand of Islamic law. They destroyed non-Islamic works of art and enforced strict dress and behavior codes. The Taliban forbade women to work, go to school, or leave the house without a male relative.

Because of Taliban rule, Afghanistan became attractive to other extremist groups, including a group called **al-Qaeda**. On September 11, 2001, 19 al-Qaeda **terrorists**, or people who use violence to achieve political results, hijacked four U.S.

AFGHANISTAN

UZBEKISTAN
40°N
KYRGYZSTAN
60°E
65°E
70°E
75°E

TAJIKISTAN
Pamirs

TURKMENISTAN

Amu Darya
• Feyzabad
• Mazar-e Sharif
• Baghlan

Torkestan Mountains

Hindu Kush

35°N
Paropamisus Range
• Qal'eh-ye Now
Charikar • Kabul
Boundary claimed by India

Herat •
Harirud R.
• Jalalabad
Khyber Pass
3,501 ft
(1,067 m)

AFGHANISTAN

Naomid Plain
Kuh-e Sangan ▲
12,274 ft
(3,741 m)

Helmand R.
Kafar Jar Ghar Range

Indus R.

Khash Desert

• Kandahar

IRAN

Rigestan

PAKISTAN

30°N

N
W E
S

0   100   200 Miles
0   100   200 Kilometers

Chagai Hills

airplanes. They flew the planes into the towers of the World Trade Center in New York City and the Pentagon in Washington D.C. A plane headed for the White House crashed in a Pennsylvania field. In response, the United States and its allies bombed terrorist camps in Afghanistan. The Taliban were removed from power and a new government was established, but violent conflict continued.

Today, Afghanistan remains divided, but efforts to unify the country through elections continue. Most Afghan leaders believe that with political stability, economic development will follow.

Before You Move On

Make Inferences What difficulties does Afghanistan face as it tries to move forward?

ONGOING ASSESSMENT

READING LAB  GeoJournal

1. **Summarize** What factors have complicated efforts to unify Afghanistan?

2. **Analyze Cause and Effect** Why did the Soviet Union invade Afghanistan? What were the effects of the Soviet Union's withdrawal from Afghanistan ten years later?

3. **Movement** On the map, notice how much of Afghanistan is covered by mountains. How might Afghanistan's physical features influence settlement patterns?

Critical Viewing Afghan girls collect water outside Kabul, Afghanistan's capital. From what you can infer from the photo, in what ways is Kabul a modern city, and in what ways is it still developing?

2.6 Building Schools

Main Idea Building schools in Afghanistan and other developing countries is critical to the future.

KEY VOCABULARY

literate, adj., able to read and write

vocational, adj., related to a job or profession

For more than 30 years, Afghanistan has endured armed conflict, invasion, and instability. As a result, more than half of Afghan children do not attend school. Building schools in this country and in other parts of the developing world is a key part of improving children's lives and futures.

Building Schools

According to the United Nations Children's Emergency Fund (UNICEF), only 28 percent of all Afghans are literate, or able to read and write. Lack of schools is one reason for low literacy rates in the war-torn country.

As you have read, when Taliban rulers took power in Afghanistan in the 1990s, they imposed their own form of Islamic law. Under Taliban rule, Afghan children, especially girls, had little or no access to education. In early 2002, just after the Taliban was pushed out, fewer than one million children in Afghanistan attended school. None of these students were girls.

Recognizing an immediate humanitarian need, multiple agencies worked with tribal leaders, military officials, and teachers in Afghanistan to build and re-open schools for Afghan children. By 2010, seven million students in Afghanistan attended school. Of these students, 37 percent were girls. New or re-built schools have been established in rural areas of Afghanistan and neighboring Pakistan, and literacy levels are climbing. In addition to teaching children to read and write, these schools offer health education and provide vocational, or job skills, training.

Before You Move On

Summarize What factors motivated agencies to build or re-open schools in Afghanistan?

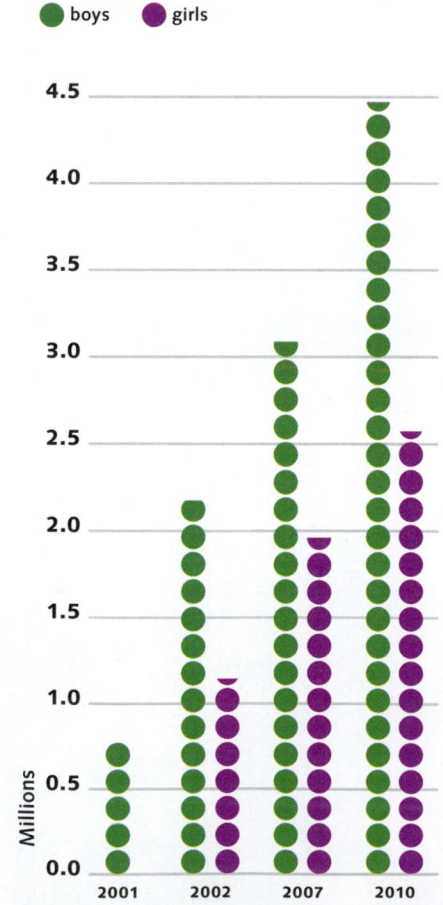

SCHOOL ATTENDANCE* IN AFGHANISTAN

● boys ● girls

*Equivalent of Grades 1–6

Source: Ministry of Education, Afghanistan, 2010; UNESCO Institute for Statistics, 2010

Educating Girls

One challenge facing developing countries such as Afghanistan is that many girls do not attend school or they leave school at an early age. For example, on average, girls in Afghanistan attend school four fewer years than boys. Limited attendance at school results in lower rates of literacy.

Research shows that girls and women equipped with education can bring about positive change in their families and in their communities. Worldwide, a higher level of education for mothers is linked to better infant and child health. In addition, the number of years a girl spends in school is linked to her level of earnings as an adult. When women and girls earn income in the developing world, they usually invest 90 percent of that income back into their families.

COMPARE ACROSS REGIONS

Schools for Africa

As one of the agencies working in Afghanistan, UNICEF recognizes the importance of building schools in war-torn and developing countries. Through its Schools for Africa program, UNICEF also partners with communities and local governments to build schools, improve classrooms, and increase access to education and opportunities for more than eight million children in at least 11 African countries.

One of these countries is Niger. Like Afghanistan, Niger is an extremely poor country. Only 38 percent of school-aged children attend school. Also like Afghanistan, school attendance rates

Critical Viewing Afghan girls and boys study together in this school in Bamyan Province. What details in the photo convey a positive learning environment?

for girls are lower than for boys. Because of a lack of education, many girls marry and have children at a young age and tend to remain in poverty.

Schools for Africa is active in other sub-Saharan countries, including Malawi, Angola, and Rwanda. Through partnership with local residents, Schools for Africa aims to improve children's lives.

Before You Move On
Monitor Comprehension In what ways does educating girls help strengthen families and communities?

ONGOING ASSESSMENT

READING LAB GeoJournal

1. **Summarize** What factors might explain the lack of educational opportunity in Afghanistan? How does this lack impact girls specifically?

2. **Interpret Graphs** Based on the graph and the text, what prediction might be made about future literacy levels in Afghanistan?

3. **Make Generalizations** In what ways are war and poverty linked to low literacy rates?

VOCABULARY

For each pair of vocabulary words, write one sentence that explains the connection between the two words.

1. guest worker; migration

> Modern migration in the region includes guest workers from many countries.

2. breakwater; tsunami

3. petrochemical; reserve

4. sheikh; emirate

5. terrorist; extremist

6. literate; vocational

MAIN IDEAS

7. In what ways have migration and trade shaped this region? (Section 1.1)

8. What factors make Istanbul a center of commerce? (Section 1.2)

9. How might underwater exploration of ancient sites help people living on coasts today? (Section 1.3)

10. What factors have encouraged the rapid economic growth of Dubai? (Section 1.4)

11. What are characterstics of two systems of government in the region? (Section 2.1)

12. What important resources do countries in this region possess? (Section 2.2)

13. What issues have led to violence among Israelis and Palestinians? (Section 2.3)

14. What challenges does Iraq face at home and internationally? (Section 2.4)

15. In what ways do ethnic and geographic factors divide Afghanistan? (Section 2.5)

16. Why is building schools in Afghanistan an important goal? (Section 2.6)

CULTURE

ANALYZE THE ESSENTIAL QUESTION

How have resources and migration shaped culture in Southwest Asia and North Africa?

Critical Thinking: Make Generalizations

17. In what ways has migration defined the population and culture of Dubai?

18. What factors have made Istanbul the cultural center of Turkey?

19. What characteristics do different patterns of migration in the region share?

INTERPRET GRAPHS

INTERNATIONAL MIGRANTS IN SELECTED COUNTRIES, 2010

Percentage of total population, countries with one million or more residents

Country	Percentage
QATAR	86.5
UNITED ARAB EMIRATES	70.0
KUWAIT	68.8
JORDAN	45.9
SINGAPORE	40.7
ISRAEL	40.4
OMAN	28.4
SAUDI ARABIA	27.8
SWITZERLAND	23.2
NEW ZEALAND	22.4

GRAPH TIP This number means 86.5 percent of Qatar's population is composed of people from other countries.

Source: United Nations, Migration Policy Institute, 2010

20. **Explain** Of the countries selected for this graph, which of them are located on the Arabian Peninsula? What factors draw international migrants to countries located on the Arabian Peninsula?

21. **Make Inferences** Based on what you know, why might the United Arab Emirates have more international migrants than Israel?

ANALYZE THE ESSENTIAL QUESTION

What forces have affected the development of modern countries in the region?

Critical Thinking: Evaluate

22. In what ways has conflict over land caused problems in Israel?

23. What are some ways in which Afghanistan is moving forward in the modern world?

24. What impact did the discovery of petroleum have on the culture of oil-rich countries?

INTERPRET MAPS

GDP PER CAPITA, ARABIAN PENINSULA

Gross Domestic Product Per Capita, 2008
- More than $30,000
- $3,000–$30,000
- Less than $3,000

25. **Make Inferences** What can you infer from the map about the economic significance of the oil industry?

26. **Compare** Compare the GDP per capita of Kuwait and Yemen. Based on what you know, what might explain the difference?

ACTIVE OPTIONS

Synthesize the Essential Questions by completing the activities below.

27. **Write Discussion Questions** Explore the geographic, cultural, and historical factors that have shaped the region by writing discussion questions for a roundtable. As a group, choose three countries in this region and write discussion questions about each country. Elect a roundtable host, and decide which questions the host will ask. **Present your roundtable in class. Be prepared to answer questions from your audience.**

> **Writing Tips**
> - Research your country selection thoroughly.
> - Write questions that require more than a "yes" or "no" or one-word answer.

Go to **Student Resources** for Guided Writing support.

TECHTREK myNGconnect.com For research links on Southwest Asia and North Africa today

28. **Create Diagrams** Countries in this region have rich cultures. Working with a partner, choose two countries and do research to answer this question: Which cultural traits do these countries share, and which are unique to each country? Use the research links at **Connect to NG** and other online sources to gather your information. Use a Venn diagram to organize the information you find. Share your diagrams with other groups.

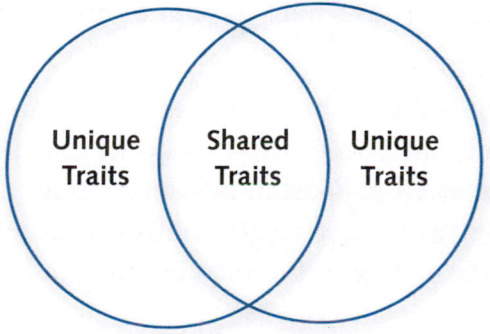

Unique Traits — Shared Traits — Unique Traits

The World's Bread

In Egypt, *baladi* and *pita* are breads eaten almost every day. All over the world, bread is a staple, or main food. Most cultures have one or more traditional forms of bread. Bread comes in a variety of shapes and sizes. It can be made from the ground flour of grains such as wheat, barley, rye, oats, or flaxseed.

Two main categories of bread are leavened and unleavened. Leavened breads are made with yeast, a microorganism used to make bread dough rise. Sandwich bread is an example of leavened bread. Unleavened breads—like many flatbreads eaten around the world—are made without yeast.

Compare

- Egypt
- France
- India
- Mexico
- Peru
- Russia
- United States

BREAD IN HISTORY

People have eaten forms of bread for thousands of years. Wheat grown in ancient Mesopotamia and Egypt was ground and mixed with water to make a paste. The paste was hardened over a fire into bread that kept for several days. When yeast was added to the paste—which may have happened accidentally—leavened bread was developed.

The Greeks learned bread making from the Egyptians and shared their knowledge with the Romans. The ancient Mayans made tortillas, their version of bread, from ground maize, or corn.

For many cultures, bread has featured prominently in daily life and in rituals and festivals. Yeasted breads such as baguettes in France and flatbreads such as naan bread in India have fed familes for many generations.

REGIONAL GRAINS, LOCAL BREADS

Types of bread vary around the world. Bread can be made from locally grown grains, or even vegetables. Flour made from barley is used to make Egyptian flatbreads. In Peru, potatoes provide a base for potato rolls. Both cornmeal and flour are used to make two different kinds of tortillas in Mexico and other Latin American countries.

How much processing goes into flours for bread varies by region. In highly industrialized countries such as the United States, wheat is processed to remove the rough outer layer of the wheat grain in order to produce a light, white flour. Because the bran has been removed, processed flours contain less fiber and are considered less nutritious. Look at the illustration on the opposite page to compare regional breads.

BREAD AROUND THE WORLD

The price of bread to consumers varies around the world. Some governments subsidize the price, or pay to reduce the cost, in order to make bread affordable to its poorest citizens. Elsewhere, grain prices directly sway the market. The prices per loaf shown here are estimates.

35¢
Potato rolls
(Peru)

When wheat prices soared in 2008, the Peruvian government promoted bread made from locally grown potatoes.

$1.37
White bread
(United States)

Many Americans have switched to whole wheat loaves for health reasons, but heavily processed white bread is still popular with consumers.

89¢
Brown bread
(Russia)

The Russian governnment hopes increased wheat plantings will continue to lower the cost of this staple.

1¢
Pita (Egypt)

A hollow, leavened flatbread, pita is the foundation of many meals. The Egyptian government subsidizes the price.

$1.50
Baguette (France)

An icon of French cuisine, the price of baguettes has increased steadily. In 2002, a loaf cost about 70 cents.

Source: *National Geographic* October, 2008

ONGOING ASSESSMENT

RESEARCH LAB GeoJournal

1. **Describe** What is the difference between leavened and unleavened bread?

2. **Make Generalizations** Which of the breads featured above is most expensive? Which is least expensive and why?

3. **Make Inferences** How might less-processed forms of bread make more sense nutritionally?

Research and Create Charts Research the breads of Mexico and India. Learn about different flours used to make breads in those countries, and determine whether their governments subsidize the price. Identify breads made both for everyday use and for special holidays or festivals. Create a chart organizing your information.

TECHTREK

myNGconnect.com For photos of items from the Silk Road and research links

Digital Library | Magazine Maker | Connect to NG

ACTIVITY 1

Goal: Extend your understanding of trade in the region.

Draw a Map

For 2,000 years, the Silk Roads connected Asia, Europe, and Africa was a major route for trade and cultural exchange. Do research and draw a map that shows the path of the Silk Roads as they crossed Southwest Asia and North Africa. Include major cities and any other relevant detail.

Caravan crossing the Silk Roads, as detailed on a 14th century Spanish map

ACTIVITY 2

Goal: Research ancient Egyptian culture.

Write Field Notes

Use the research links at **Connect to NG** to take a virtual tour of the Valley of the Kings. Take notes on what you observe, and keep a written log of all the stops you make. Use the **Magazine Maker** to record in each log entry what you see there. Include any visuals that might let others see what you are experiencing.

ACTIVITY 3

Goal: Learn more about recycling.

Host a Panel Discussion

National Geographic Emerging Explorer Thomas Taha (TH) Rassam-Culhane helps people in Egypt and around the world think creatively about how to recycle used materials. With a small group, use the research links at **Connect to NG** to learn about new ways to reuse materials around your home or school. Present a panel discussion to share your ideas with the class.

Explore South Asia
with NATIONAL GEOGRAPHIC

MEET THE EXPLORER

NATIONAL GEOGRAPHIC

Emerging Explorer Shafqat Hussain created Project Snow Leopard to save the endangered snow leopard species and boost the local economy in the mountain regions of Pakistan.

INVESTIGATE GEOGRAPHY

Mount Everest, on the border of Nepal and China, is the ultimate climbing challenge. In 2006, 70-year-old Takao Arayama became the oldest to climb it. In 2010, 13-year-old Jordan Romero became the youngest. At 29,035 feet, Everest is the tallest mountain in the world.

STEP INTO HISTORY

The Taj Mahal, a UNESCO World Heritage site, has attracted visitors to India since it was built in the 1600s. The white marble masterpiece was built by Emperor Shah Jahan as a memorial to his wife. The project took 20,000 workers 22 years to complete.

7,486 miles

Washington, D.C.

New Delhi, India

Go to **myNGconnect.com** for maps of South Asia.

CONNECT WITH THE CULTURE

These women are participating in a wedding in Rajasthan, India. Weddings in South Asia are elaborate celebrations that may last for several days and include hundreds of guests.

CHAPTER 11

South Asia
GEOGRAPHY & HISTORY

PREVIEW THE CHAPTER

Essential Question How do South Asia's water systems affect how people in the region live?

SECTION 1 • GEOGRAPHY

KEY VOCABULARY

- subcontinent
- plate
- delta
- elevation
- subsistence farmer
- monsoon
- evaporation
- drought
- arable
- famine
- sustainable
- conservation
- ecosystem
- pollution
- sanitation
- aquifer

ACADEMIC VOCABULARY
collide, reverse

TERMS & NAMES

- Himalaya Mountains
- Deccan Plateau
- Indus River
- Ganges Delta

Essential Question How have physical features, religion, and empires shaped South Asia's borders?

SECTION 2 • HISTORY

KEY VOCABULARY

- isolation
- cultural hearth
- caste system
- empire
- tolerance
- deity
- reincarnation
- colonialism
- civil disobedience
- mythology
- symbol

ACADEMIC VOCABULARY
isolation, displace

TERMS & NAMES

- Harappan
- Aryans
- Sanskrit
- Asoka
- Taj Mahal
- Hinduism
- Buddhism
- Jainism
- Sikhism
- Islam
- Vedas
- Mohandas Gandhi
- Partition
- Bhagavad Gita

TECHTREK FOR THIS CHAPTER

Student eEdition

Maps and Graphs

Interactive Whiteboard GeoActivities

Digital Library

Connect to NG

Go to **myNGconnect.com** for more on South Asia.

snow leopard

60°E

80°E

90°E

100°E

70°E

Boundary claimed by India

HINDU KUSH

Boundary claimed by India

Boundary claimed by Pakistan

Boundary claimed by China

Boundary claimed by China

Islamabad

H

30°N

Lahore

Indus R.

PAKISTAN

Delhi
New Delhi

NEPAL

Kathmandu

L

A

Y

A

Thimphu
BHUTAN

Brahmaputra R.

Cherrapunji

Turbat

Jodhpur

Yamuna R.

Ganges R.

Varanasi

BANGLADESH
Dhaka

Karachi

Tropic of Cancer

Narmada R.

Kolkata

20°N

Arabian
Sea

INDIA

Godavari R.

Mumbai

WESTERN GHATS

EASTERN GHATS

Hyderabad

Krishna R.

Bay of
Bengal

Andaman
Islands
(India)

Bangalore

Chennai

N
W E
S

F

10°N

SRI LANKA

Colombo
Sri Jayewardenepura
Kotte

Nicobar
Islands
(India)

M
A
L
D
I
V
E
S

Male

INDIAN   OCEAN

0   250   500 Miles
0   250   500 Kilometers

70°E

80°E

90°E

A

B

C

D

E

F

G

H

2   3   4   5   6

SECTION 1 GEOGRAPHY

1.1 Physical Geography

TECHTREK
myNGconnect.com For online maps of South Asia and Visual Vocabulary
Maps and Graphs
Digital Library

SOUTH ASIA PHYSICAL

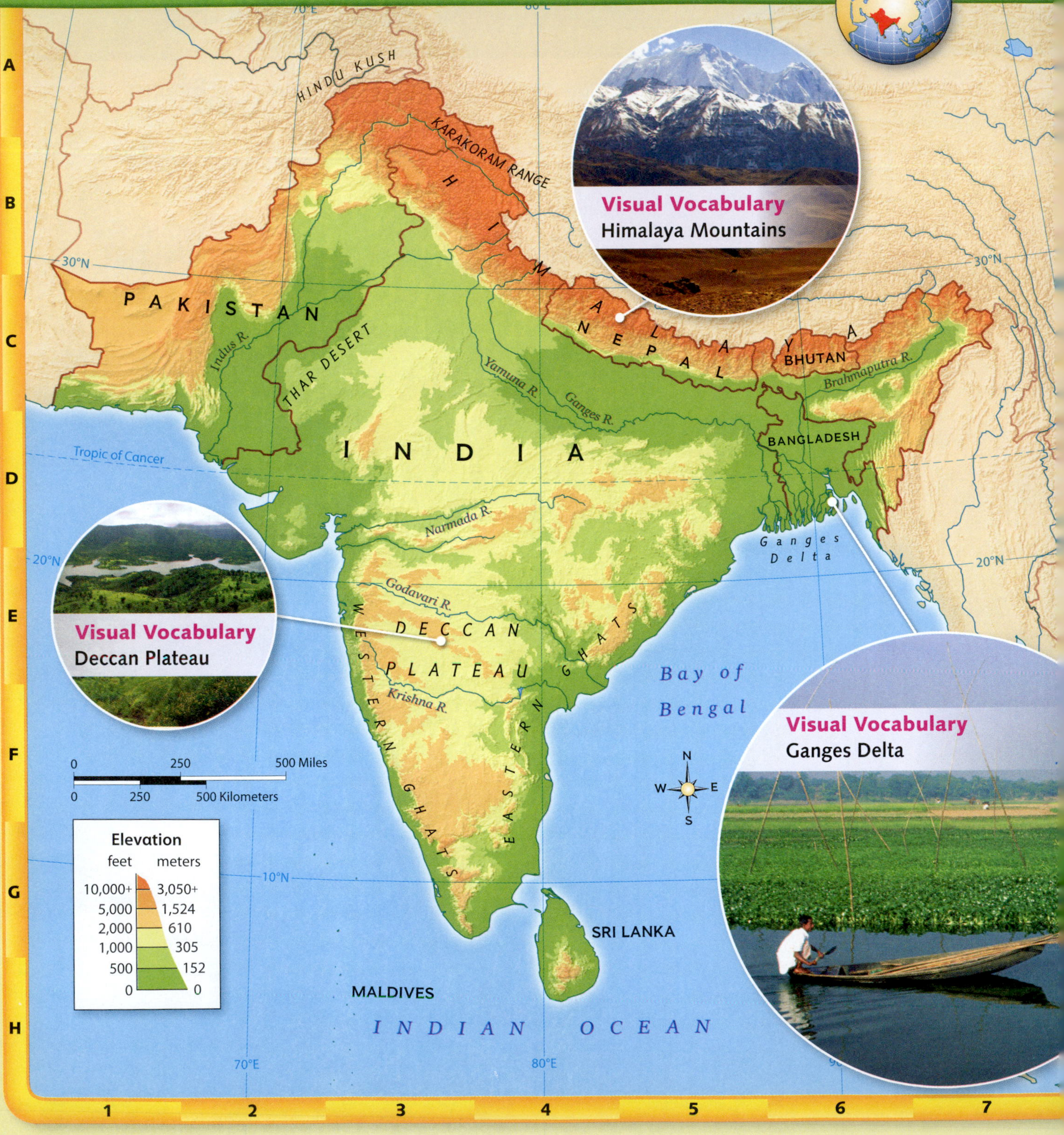

HINDU KUSH
KARAKORAM RANGE
HIMALAYA
PAKISTAN
Indus R.
THAR DESERT
INDIA
NEPAL
BHUTAN
Brahmaputra R.
Yamuna R.
Ganges R.
BANGLADESH
Tropic of Cancer
Narmada R.
Ganges Delta
Godavari R.
DECCAN PLATEAU
WESTERN GHATS
EASTERN GHATS
Krishna R.
Bay of Bengal
SRI LANKA
MALDIVES
INDIAN OCEAN
30°N
20°N
10°N
70°E
80°E
Visual Vocabulary
Himalaya Mountains
Visual Vocabulary
Deccan Plateau
Visual Vocabulary
Ganges Delta
0 250 500 Miles
0 250 500 Kilometers
Elevation
feet meters
10,000+ 3,050+
5,000 1,524
2,000 610
1,000 305
500 152
0 0
N
S
E
W
A
B
C
D
E
F
G
H
1
2
3
4
5
6
7

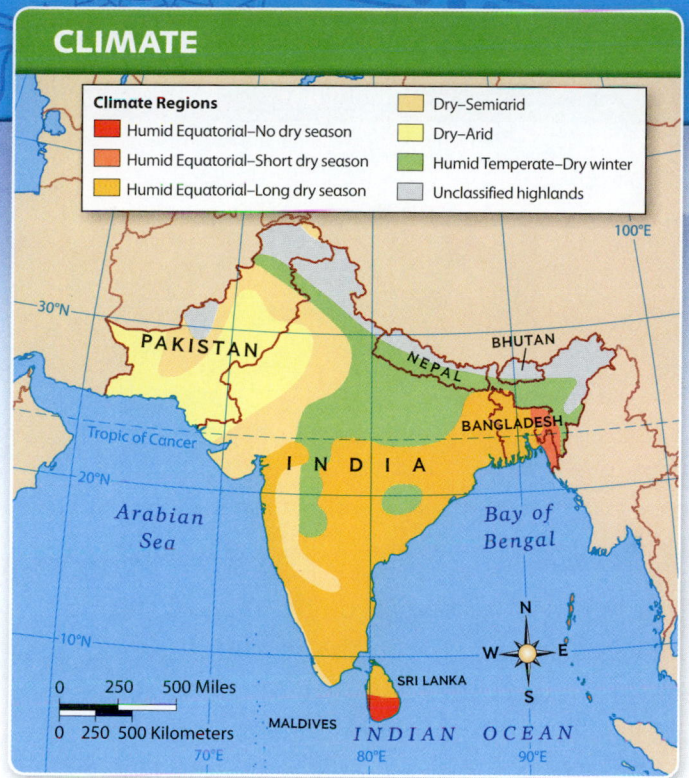

CLIMATE

Climate Regions
- Humid Equatorial–No dry season
- Humid Equatorial–Short dry season
- Humid Equatorial–Long dry season
- Dry–Semiarid
- Dry–Arid
- Humid Temperate–Dry winter
- Unclassified highlands

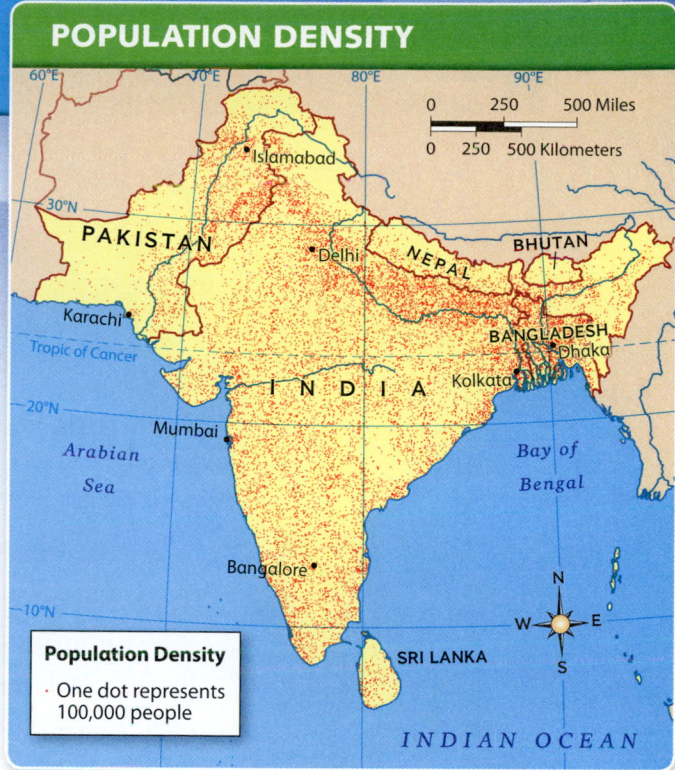

POPULATION DENSITY

Population Density
· One dot represents 100,000 people

Main Idea The key physical features of South Asia include mountains, rivers, and a delta.

On a map, South Asia looks like a diamond-shaped chunk of land that was shoved into Asia. At the center of this diamond is the Indian **subcontinent**, a separate region of the Asian continent.

Continental Collision

South Asia is a separate **plate**, a rigid section of the earth's crust that can move independently. The plate is still moving. As a result, the **Himalaya Mountains** are being pushed higher. They are the highest mountain range in the world. The Himalayas were formed about 50 million years ago when the Indian plate **collided** with the Eurasian plate. This movement produced the world's tallest mountain, Mount Everest in Nepal. The Himalayas separate South Asia from the rest of Asia. Rainwater and snowmelt from these mountains form South Asia's rivers.

Major River Systems

Much of South Asia's water drains into the Ganges River and the **Indus River**. These river systems provide drinking water and nourish farmland, and are considered holy by Hindus. In Bangladesh, the Brahmaputra River joins the Ganges, forming the low, fertile **Ganges Delta**. A **delta** is an area where a river deposits sediment as it empties into a body of water.

Before You Move On
Summarize What are some key physical features of South Asia?

ONGOING ASSESSMENT

MAP LAB

 GeoJournal

1. **Location** Where is the Ganges River located?
2. **Identify** What is the climate like in northeast India near Dhaka and Kolkata?
3. **Draw Conclusions** The population in northeastern South Asia is very dense. What physical features might account for this?

1.2

SECTION **1** GEOGRAPHY

NATIONAL GEOGRAPHIC

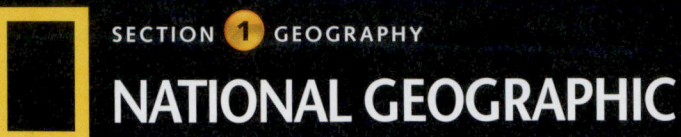

TECHTREK

myNGconnect.com For a map of the Snowman Trek and an Explorer Video Clip

 Maps and Graphs   Digital Library

Exploring the Himalayas
with Kira Salak

Main Idea South Asia's mountain systems create unique challenges for its people.

The Snowman Trek

The Snowman Trek involves hiking through some of the tallest and most isolated mountains in the world. Most of the passes between the mountain peaks are over 16,000 feet high. I eased comfortably into the first few days of the Trek, absorbing the beauty of the environment. On the third day we were introduced to the extreme **elevation**—the height above sea level—that we would experience the rest of the climb. It was nine hours, all uphill, to 11,800 feet.

We came to our first high mountain pass at Nyile La (roughly 16,000 feet). **1** This marked the first test for altitude sickness, a condition caused by a reduced level of oxygen in the air. As we climbed, the clouds parted to reveal many snow-topped peaks—the "rooftop of the world."

Life in the Mountains

Later, we hiked to one of the most isolated villages, Thanza, **2** with an elevation of 13,700 feet. Only about 300 people live there. Most are **subsistence farmers**. They grow food for their families, but nothing to sell to anyone else. Food consists of a high-altitude rice, with meat and cheese from yaks, which are long-haired Tibetan oxen. Homes are built from what can be found. Walls are made from stone, and shingles from wood. Clay is the mortar that holds it together.

Map Legend

- ⊛ National capital
- • Town
- ～ River
- ⟩𝒸 Pass
- — International boundary
- ▬ Snowman trek

Elevation

feet	meters
10,000+	3,050+
5,000	1,524
2,000	610
1,000	305
500	152
0	0

CHINA

HIMALAYA

Thanza ▷2

Nyile La ▷1

Nikka Chhu

Bumtang Tang ▷3

• Gasa

• Lhuntshi

Dur Village

• Tashi Yangtse

• Thimphu ⊛

• Punakha

B L A C K   M T S

• Tongsa

B   H   U   T   A   N

• Paro

Sankosh R.

Bumthang R.

Kuru R.

Wong Chu R.

Mangde R.

• Tashigang

Amo R.

• Chukha

Mongar •

Manas R.

• Phuntsholing

• Sarbhang

• Geylegphug

• Dewangiri

INDIA

| 0 | 25 | 50 Miles |
| 0 | 25 | 50 Kilometers |

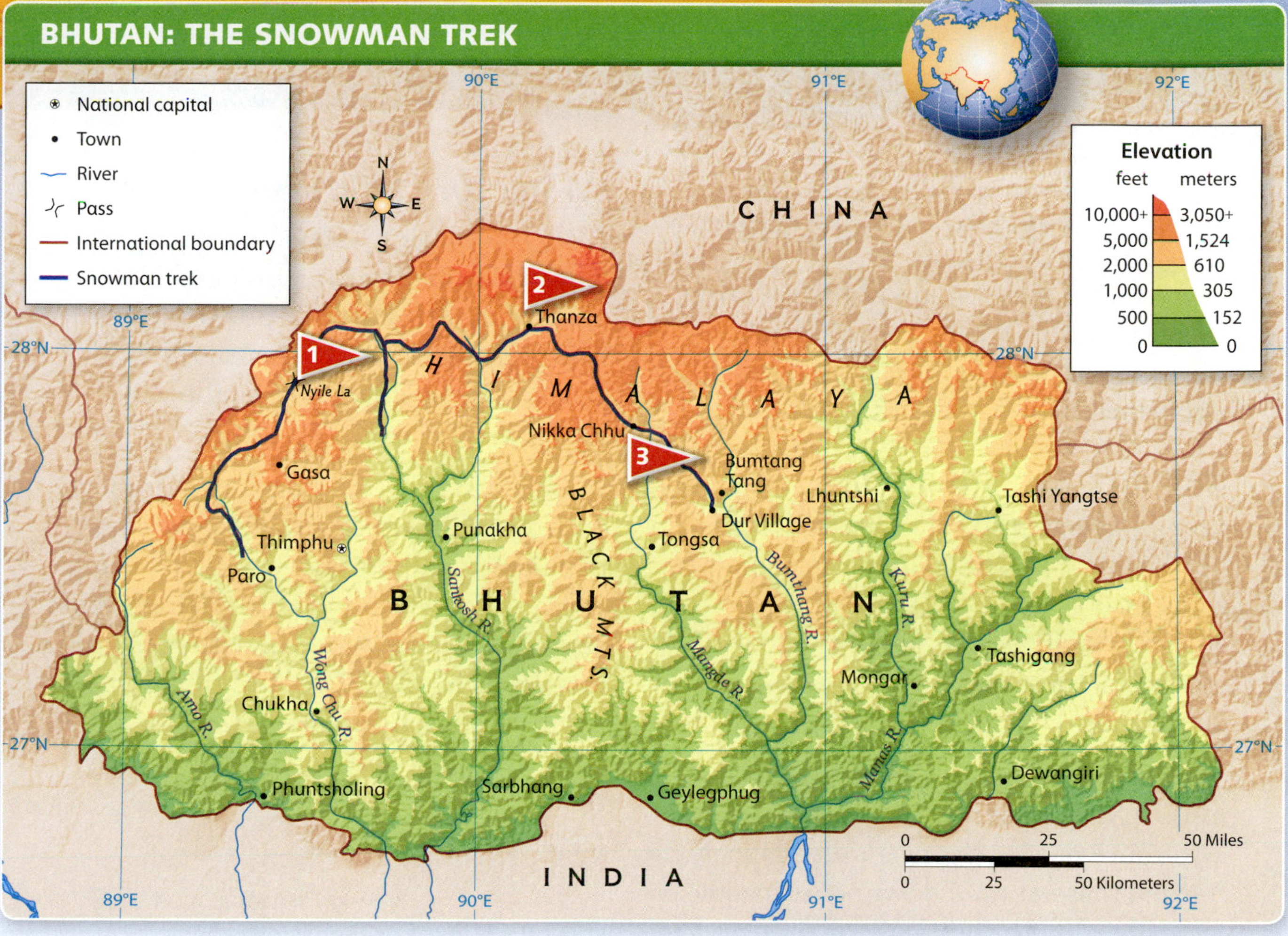

As we left Thanza, all signs of human presence vanished. Even our trail disappeared from underfoot. When we were done with the Trek and arrived in Nikka Chhu village ▷3 we calculated that we'd each taken half a million steps and walked at least 216 miles.

Before You Move On

Monitor Comprehension What key challenges do people living in South Asia's mountain systems face?

ONGOING ASSESSMENT

MAP LAB

 **GeoJournal**

1. **Place** Based on the map and description of Bhutan, why do you think people built most of their towns where they did?

2. **Create Charts** Make a chart like this one. What geographic factors explain the differences between your life and life in Bhutan?

NECESSITIES	MY LIFE	BHUTAN
shelter		
access to water		
access to food		
level of fitness required		

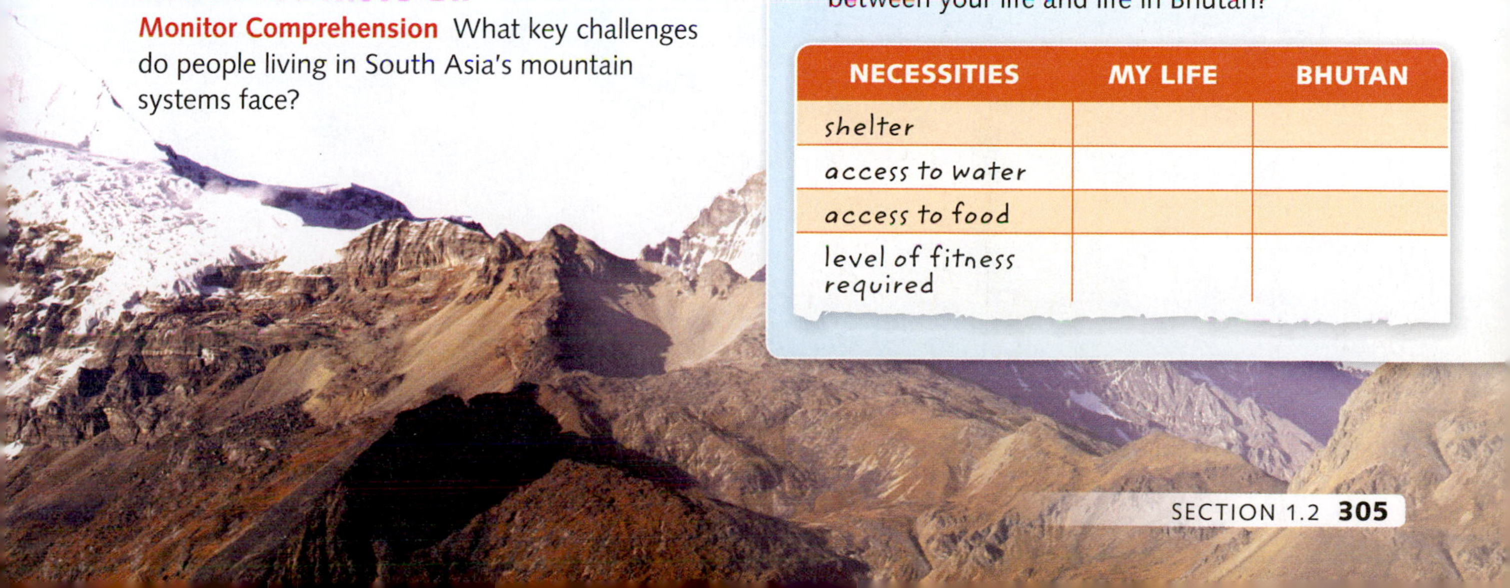

1.3 Living with Monsoons

TECHTREK
myNGconnect.com For online monsoon maps and photos of monsoons
Maps and Graphs
Digital Library

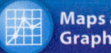

Main Idea Seasonal monsoons provide water for crops and bring fresh soil to farmland.

Monsoons are seasonal winds that bring intense rainfall during part of the year. These powerful wind patterns are an important feature of South Asia's climate.

Summer: Wet Monsoons

From May to early October, winds blow northward from the Indian Ocean and bring heavy rain. Certain areas receive 100 inches of rain per year. During very wet years, these areas might receive over 300 inches.

The summer monsoon rains irrigate crops and fill reservoirs. The rains can also cause deadly floods and landslides. Despite these annual downpours, people forge ahead with their daily activities. They adapt by finding ways to save crops and navigate flooded streets.

Winter: Dry Monsoons

From November to April, the winter monsoons **reverse** direction and blow southward. The air from this direction is usually dry. The dry monsoons do not produce as much rain as the wet monsoons do.

In fact, a very dry monsoon season can destroy crops, which threatens farmers' livelihoods and the economic well-being of the region. Because of this, people must carefully manage water they stored during the wet monsoons. They use this water to drink and to irrigate their crops. By the following June, after months of hot, dry weather, the heavy rain is welcome.

Before You Move On

Make Inferences How do seasonal monsoons affect the region's economic development?

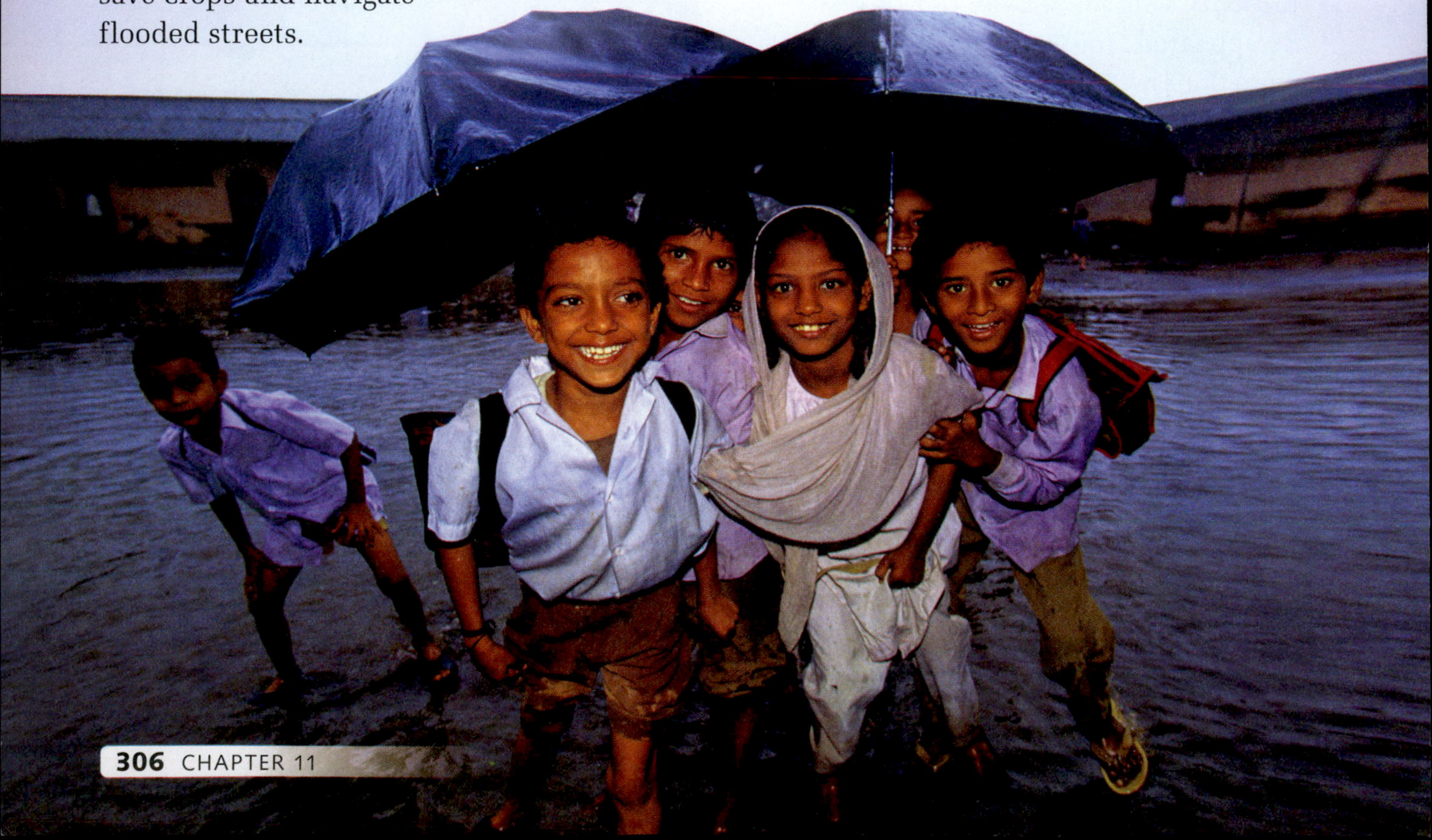

SUMMER MONSOON
Average Summer Precipitation

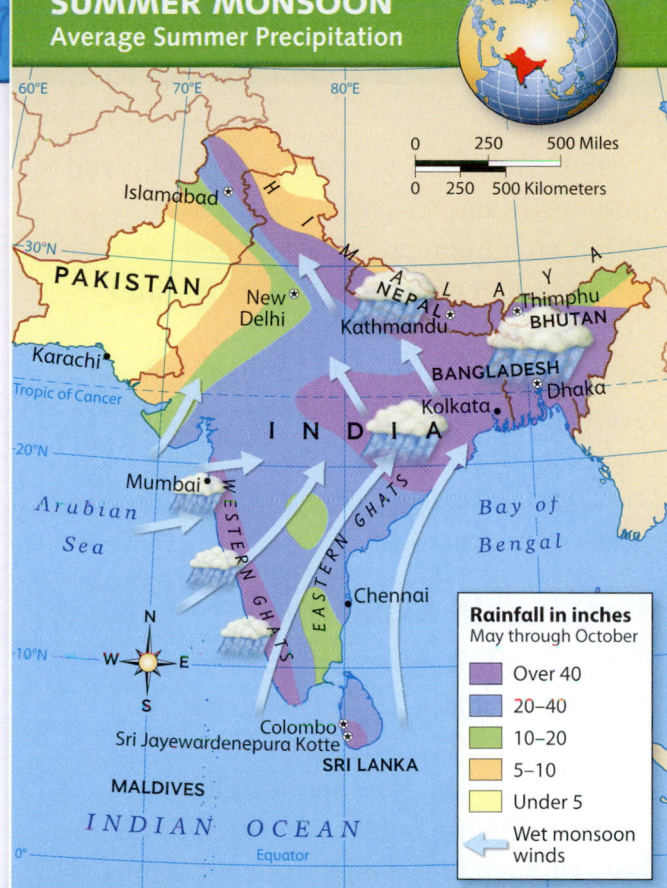

Rainfall in inches
May through October

- Over 40
- 20–40
- 10–20
- 5–10
- Under 5
- ← Wet monsoon winds

WET MONSOONS The wet monsoon winds pick up moisture from the Indian Ocean as they flow northwest. This moisture turns into heavy rain when it crashes into the mountains of the Indian subcontinent. The rain saturates the land and swells the rivers. This provides nourishment for crops, animals, and people alike.

WINTER MONSOON
Average Winter Precipitation

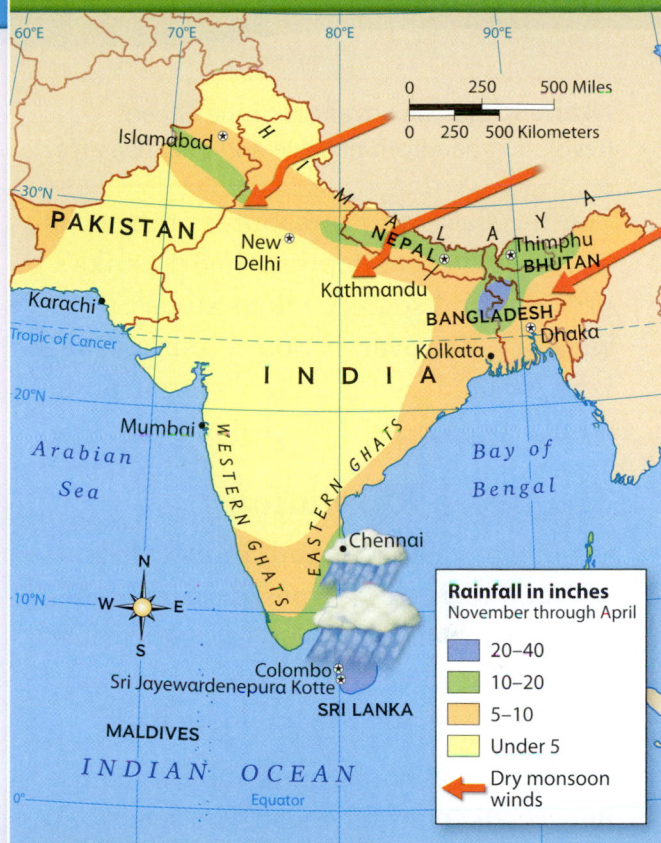

Rainfall in inches
November through April

- 20–40
- 10–20
- 5–10
- Under 5
- ← Dry monsoon winds

DRY MONSOONS The dry monsoons begin in the northeast and pick up moisture from the land as they move southwest. This is **evaporation**, the process of a liquid becoming a gas. If the wet monsoons do not produce enough rainfall, the dry monsoons can lead to a **drought**, or prolonged period without rain.

‹ Critical Viewing These children don't seem bothered by their flooded schoolyard. What does this photo suggest about the children's life during monsoon season?

ONGOING ASSESSMENT

MAP LAB   GeoJournal

1. **Location** Locate Kolkata on the summer monsoon map. Why do you think Kolkata receives so much rain in the summer?

2. **Compare** Find Kolkata on the winter monsoon map. Does the city receive more or less rainfall in the winter? Why is there a difference in rainfall from summer to winter?

3. **Human-Environment Interaction** What might life be like for the people in Kolkata in summer? In winter?

Main Idea The advances of the Green Revolution had both positive and negative effects.

South Asia's natural resources range from India's huge coal deposits to Sri Lanka's precious and semi-precious stones. The big story, however, is South Asia's **arable** land, or land that is suitable for farming, which feeds nearly 1.5 billion people.

The Green Revolution

In spite of vast farmland, India often suffered terrible **famines**, or times when people starved because of a lack of food. During the 1950s and 1960s, India made slow progress in growing more food.

Then, in 1966, farmers started using new seeds to grow wheat. The seeds dramatically increased crop yield, or the amount of food grown on a unit of land. India's wheat production shot up from 10.3 million metric tons in 1960 to 20 million metric tons in 1970. Farmers also increased production of rice, fruits, sugarcane, and vegetables. This rapid and significant rise in food production was known as the Green Revolution.

However, the high-yield seeds required more fertilizer, irrigation, and pesticides, which are chemicals that kill diseases and insects. Fertilizers, pesticides, and irrigation cost money. As a result, the Green Revolution benefited mostly wealthy farmers. The Green Revolution also had negative effects on the environment. Rain washed fertilizers and pesticides into rivers, causing pollution.

Sustainable Agriculture

Today, South Asian governments and farmers are adapting to the physical environment. They use new technologies and methods that are **sustainable**, or capable of being continued without long-term damage to the environment. More farmers are using natural fertilizers, such as manure, to enrich soils. They are also using crop rotation, in which they change by season the crops grown on a plot of land. These methods help to grow more food and also protect the environment.

Before You Move On

Summarize What have been the positive and negative effects of the farming methods used during the Green Revolution?

Visual Vocabulary **Arable** refers to land that is suitable for farming. This field is designed to let water soak into the ground. In India, more than 50 percent of the land is arable.

Map legend:

- Copper
- Iron ore
- Coal
- Natural gas
- Oil
- Fish
- Forest
- Woodland
- Grassland
- Mixed-use, including crops
- Cropland
- Intensive cropland
- Wetland
- Desert, barren land
- Ice, cold desert, tundra

Map labels: PAKISTAN, Indus R., NEPAL, Brahmaputra R., BHUTAN, INDIA, Ganges R., BANGLADESH, Narmada R., Godavari R., Arabian Sea, Bay of Bengal, MALDIVES, SRI LANKA, INDIAN OCEAN, Tropic of Cancer

Scale: 0 250 500 Miles / 0 250 500 Kilometers

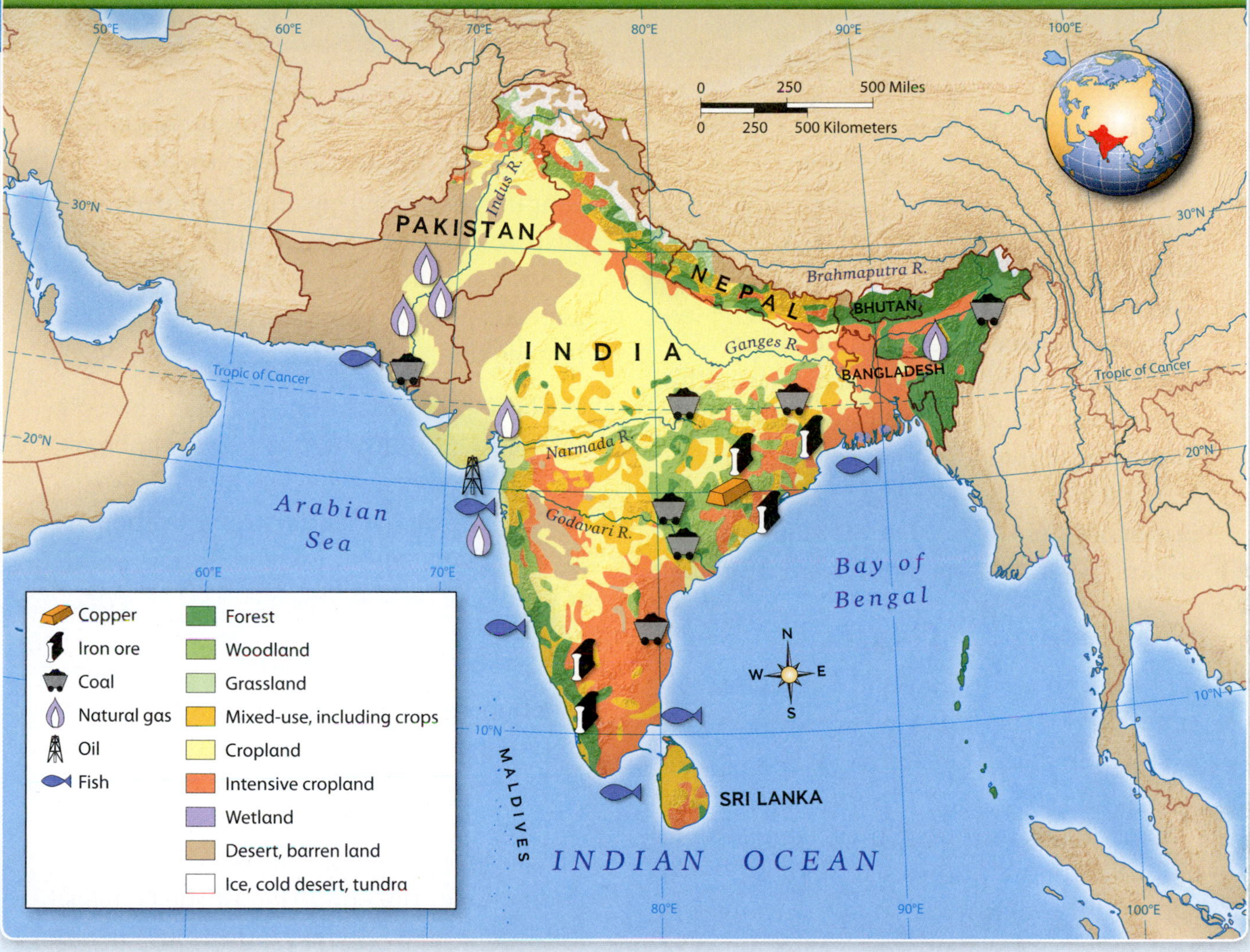

GRAIN PRODUCTION AND POPULATION (1950–2008)			
	Food Grain Production (millions of metric tons)	Wheat Production (millions of metric tons)	Population (millions)
1950	50.8	6.4	361
1960	82.0	10.3	439
1970	108.4	20.0	548
1980	129.6	31.8	683
1990	176.4	49.8	846
2000	201.6	76.3	1,000
2008	227.8	78.0	1,148

Sources: CIA Factbook; Indian National Science Academy

ONGOING ASSESSMENT

MAP LAB  GeoJournal

1. **Location** What types of land use are most common around the main rivers in South Asia?

2. **Compare** Look at the physical map in Section 1.1 and compare it with the land use map on this page. Why do you think there are fewer natural resources in the northeast?

3. **Interpret Charts** How much wheat per person did India produce in 1950? In 2008? What happened to the amount of wheat grown?

4. **Human-Environment Interaction** How do physical features, climate, and land use in South Asia affect humans in the region?

1.5 **Conservation**

TECHTREK

myNGconnect.com For an online population map and an Explorer Video Clip

Maps and Graphs   Digital Library

> **Main Idea** People have contributed to the problems affecting the Ganges River.

Humans have a powerful impact on the land and the way it is used. One way they can protect the earth is through **conservation**, which is the protection of the environment. In South Asia, this includes protecting tigers in India and Bangladesh and preserving marine ecosystems in the Maldives. An **ecosystem** is a group of interacting organisms and their natural environment.

Conserving the Ganges

In India, conservation of the rivers is a national issue. The Ganges River is a valuable natural resource. It is also considered sacred by Hindus, India's main religious group. However, a growing population and poor waste disposal have led to the contamination of the Ganges.

One source of pollution in the Ganges is raw sewage, including household wastes. In 2008, Varanasi had three sewage plants that could process 100 million liters of sewage a day. However, the city produces up to 300 million liters a day.

Another source of pollution comes from industry. In industrial centers such as Kanpur, big leather-producing tanneries flush waste containing arsenic, chromium, and mercury into the Ganges. Chromium can cause cancer, and mercury can damage the nervous system.

These pollutants also damage the river habitat, which threatens animals such as the river dolphin. Parts of the Ganges are so polluted that scientists refer to them as "dead." They can no longer support plant or animal life.

Cleaning Up Pollution

Local non-governmental organizations (NGOs) are helping to reverse the damage. NGOs are nonprofit volunteer groups founded by citizens. For example, Rakesh Jaiswal founded EcoFriends to raise awareness of the condition of the Ganges at Kanpur. Jaiswal is hopeful that one day the Ganges will be clean again.

Before You Move On

Monitor Comprehension What are the main sources of pollution in the Ganges River?

Critical Viewing Pollution clogs the Ganges River. In what ways might pollution of the Ganges affect human life along its banks?

POPULATION OF THE GANGES RIVER BASIN

MAP TIP This is a proportional symbol map. Each circle represents a different population size along the rivers. The larger the circle, the larger the population estimate.

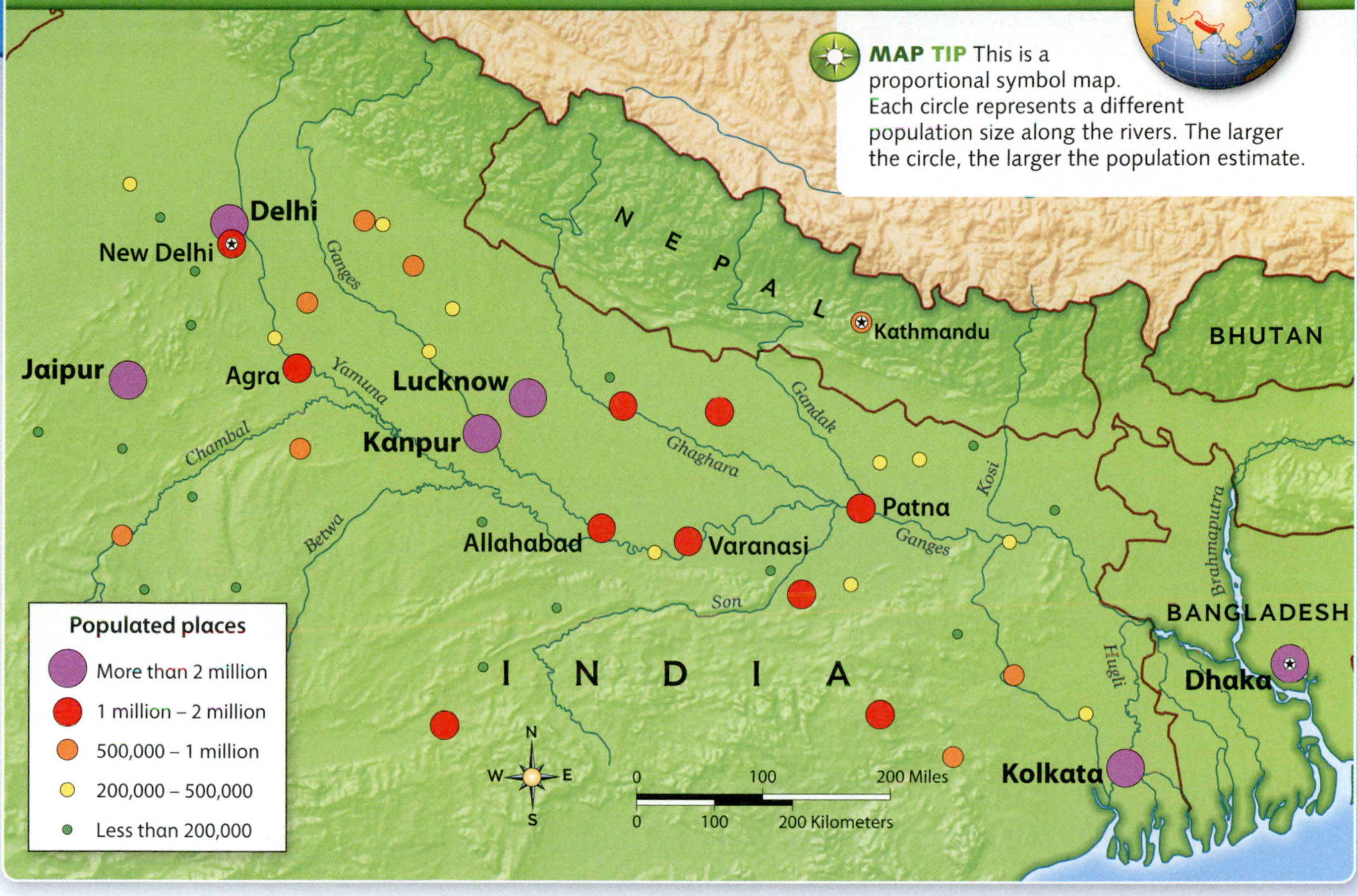

Delhi
New Delhi

NEPAL
Kathmandu
BHUTAN

Jaipur
Agra
Lucknow
Kanpur

Ganges
Yamuna
Chambal
Betwa
Ghaghara
Gandak
Kosi
Brahmaputra

Allahabad
Varanasi
Patna
Ganges
Son

INDIA

BANGLADESH
Hugli
Dhaka

Kolkata

Populated places

- ● More than 2 million
- ● 1 million – 2 million
- ● 500,000 – 1 million
- ● 200,000 – 500,000
- ● Less than 200,000

0   100   200 Miles
0   100   200 Kilometers

VIEWING LAB

GeoJournal

1. **Analyze Visuals** Go to the **Digital Library** to watch the video clip about pollution. What most interested you?

2. **Pose Questions** Draw this chart on your own paper. Write down three things you learned from the video clip and pose three questions you have after watching the clip.

National Geographic Emerging Explorer Alexandra Cousteau with Rakesh Jaiswal

WHAT I LEARNED	QUESTIONS I STILL HAVE
1. Kanpur is the 7th most polluted city in the world	
2.	

3. **Make Inferences** What connections can you make between the condition of the Ganges and population?

1.6 South Asia's Water Crisis

Main Idea South Asia faces a water crisis because of pollution, water scarcity, and flooding.

Several times a week, 12-year-old Somnath Dantoso drops a magnet attached to a fishing line into the Yamuna River. The river flows through New Delhi, India's capital city. He hauls rupees (India's coin money) out of the water, but he also hauls out garbage. The Yamuna is choking on **pollution**. It is loaded with food, tires, chemicals, sewage, and other waste. In fact, more than half of New Delhi's garbage ends up in the river. South Asia is facing a major water crisis.

Polluted Water and Disease

Many other rivers in India are polluted. Water pollution is just as bad in other parts of South Asia. In Pakistan, 38.5 million people do not have safe drinking water. More than 50 million people lack proper **sanitation**, such as sewers to carry away dirty water. Rivers in many of Nepal's cities are so polluted that people cannot use them for drinking water.

In Bangladesh, millions of wells have arsenic, a poison, in the water. Experts believe that 20,000 Bangladeshis may die every year from drinking poisoned water. One United Nations scientist said that it could be "the largest mass poisoning of a population in history."

The pollution has severe consequences for the health of South Asians. Contaminated water kills 500,000 infants a year in South Asia and Southeast Asia. It can cause cancer and other deadly diseases.

Before You Move On

Monitor Comprehension How does water pollution affect South Asia?

KEY VOCABULARY

pollution, n., making the environment dirty, foul, unclean, or contaminated

sanitation, n., measures such as sewers to protect public health

aquifer, n., a bed or layer beneath the surface of the earth that contains water

POLLUTION BY THE NUMBERS (2007)

55
Percentage of New Delhi's 15 million people who are connected to the city's sewage system

80
Percentage of Yamuna River's pollution resulting from raw sewage

500
Millions of dollars that the Indian government has spent trying to clean up the Yamuna River

855
Length in miles of the Yamuna from the Himalayas to the Ganges

1,815
Growth of India's population every hour

Sources: Daniel Pepper, "India's Rivers Are Drowning in Pollution," *Fortune*, June 4, 2007; Rakesh Jaiswal, "India in Peril," *Smithsonian*, October 31, 2007

Scarce Water and Drought

Water is growing scarce for the 1.5 billion people in South Asia. Because of increasing population and climate change, the region has had severe droughts recently. Elizabeth Kolbert wrote in *National Geographic* magazine:

> Since higher temperatures lead to increased evaporation, even areas that continue to receive the same amount of overall precipitation will become more prone to drought. (April 2009)

Bangladesh has already experienced severe drought. One reason is that upstream dams in India have reduced water flow into Bangladesh's rivers. In addition, the country's exploding population has taken millions of gallons of water from **aquifers**. Less water means less food, resulting in widespread hunger.

Water scarcity is also caused by shrinking glaciers in the Himalayas, which supply water to rivers and lakes in Asia. The glaciers used to build up in winter and then melt in spring, supplying water to rivers and lakes in Asia. According to some scientists, climate change is causing the glaciers to melt faster and shrink in size. These scientists fear that the glaciers will supply less water to rivers. People will have less water for drinking, sanitation, and irrigation.

While some areas will have droughts, other areas may experience more flooding. According to Kolbert, climate change is increasing air temperatures, and warmer air holds more moisture. As a result, the monsoons that hit South Asia every year may grow more severe.

COMPARE ACROSS REGIONS

China and India's Solutions

South Asians are studying how other countries have worked to improve water quality. In China, for example, 70 percent of the rivers and lakes are polluted. In *National Geographic* magazine, Brook Larmer explained that the Chinese have formed thousands of organizations urging a reduction in water pollution. (May 2008)

India's government is also trying to clean up rivers and lakes. In 1986, it passed the Environmental Protection Act. The law allows the government to stop companies from polluting.

The people of South Asia are making significant contributions to their society by taking steps to clean up the region's rivers and lakes. Citizens are calling for companies to stop polluting and are pushing their governments to take action.

Before You Move On

Summarize What is causing water scarcity in South Asia?

ONGOING ASSESSMENT
READING LAB GeoJournal

1. **Analyze Data** Analyze the data provided in "Pollution by the Numbers" and the map in Section 1.5. Which factors explain why the Ganges is more polluted than the Brahmaputra?

2. **Make Inferences** Based on the rate of India's population growth, what do you think might happen to the other figures shown?

3. **Write Comparisons** How are China and India dealing with water pollution? Write a sentence comparing their tactics.

2.1 Early Civilizations

TECHTREK
myNGconnect.com For an online civilization map

Maps and Graphs

Main Idea South Asia's ancient civilizations developed around the region's river systems.

The physical geography of a place often influences its history. Good farmland and the geographic **isolation** (separation) of South Asia made the Indus and Ganges river valleys **cultural hearths**, or centers of civilization from which ideas spread. Mountains and oceans provided natural boundaries that limited invasions. Fertile soil along the rivers provided arable farmland. Two major civilizations that thrived here were the Harappans and the Aryans.

Harappan Civilization

The first urban civilization in South Asia was the **Harappan** (huh RA puhn) civilization. It developed along the Indus River in what is now Pakistan. The land was fertile and good for farming. As a result, people formed farming villages that grew into cities. The two greatest cities of this civilization were Mohenjo-Daro and Harappa. These cities were well planned and laid out in a grid pattern of straight streets. The cities also had brick houses, indoor plumbing, and a sewer system. These cities provide early examples of organized city planning that helped to support cultural growth.

The Harappan people developed advanced technologies and a system of measurement using weights and bricks that were a standard size. Among the ruins of these cities, archeologists have found stone seals with images of animals and script. Based on this evidence, scholars believe the Harappans may have developed a system of writing, but the scripts have not yet been translated.

After a period of prosperity, the Harappan civilization started to decline between 2000 and 1700 B.C. Historians believe some possible causes include the change of the Indus River's course, floods, earthquakes, and invaders.

Aryan Migration

Many historians believe a group of nomadic herders called the **Aryans** (AIR ee uhnz) migrated from Central Asia into the Indus Valley around 2000 B.C. From there, they moved into northern India, and eventually migrated further south. Their language, **Sanskrit**, became the basis of many modern languages in South Asia. The Aryans recorded religious teachings in Sanskrit in sacred texts called the Vedas. The early religion of the Aryans established the beginnings of Hinduism, the major religion of India today.

2600–2500 B.C. Emergence of the Harappan, or Indus Valley, civilization

2000–1700 B.C. Decline of the Harappan civilization

3000 B.C.   2500 B.C.   2000 B.C.

2500 B.C. A sculpture fragment of a priest-king from the site of Mohenjo-Daro

2000 B.C. Beginning of Aryan migration into the Indus Valley

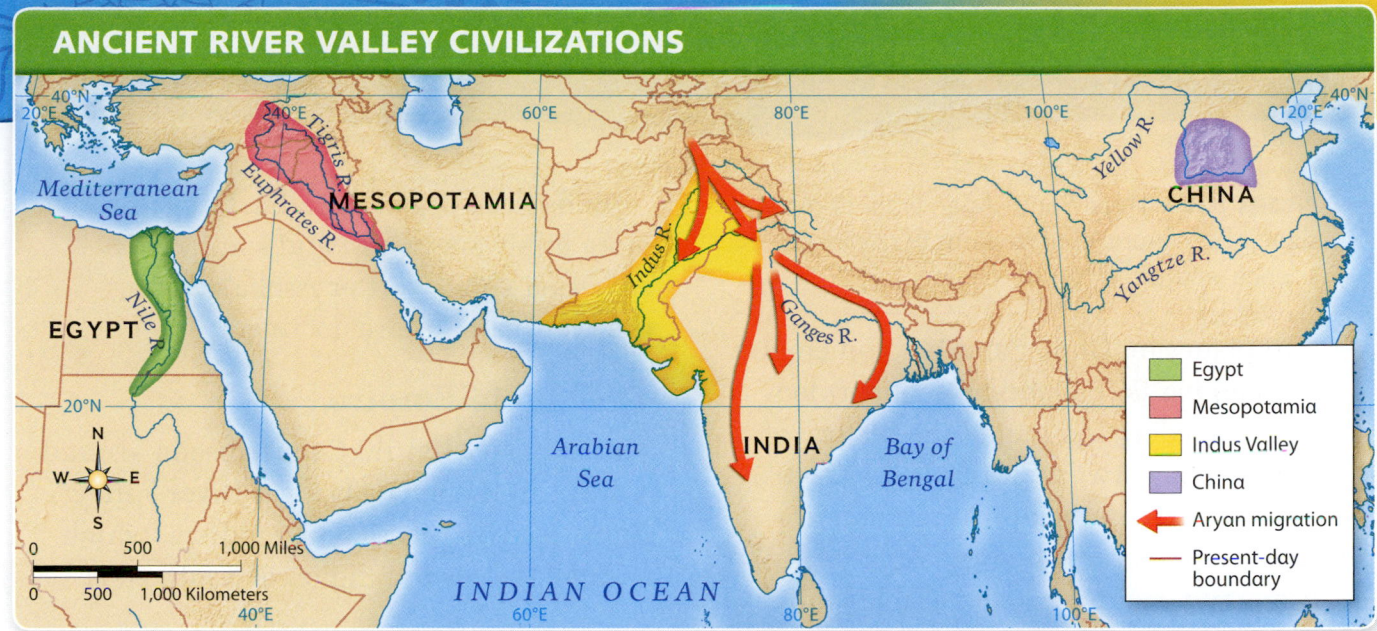

ANCIENT RIVER VALLEY CIVILIZATIONS

■	Egypt
■	Mesopotamia
■	Indus Valley
■	China
←	Aryan migration
—	Present-day boundary

Aryan society was organized into different social groups, or varnas, based on ancestries, family ties, and a person's occupation. Their social structure became known as the **caste system**. The caste system was made of four groups:

Brahmans: priests and scholars
Kshatriyas: rulers and warriors
Vaisyas: merchants and professionals
Sudras: artisans, laborers, and servants

Over the centuries, the caste system grew into thousands of subgroups. While its past influence made it an important part of South Asian culture, the caste system is slowly becoming less prominent as the region modernizes.

Before You Move On
Monitor Comprehension Where did South Asia's ancient civilizations develop?

ONGOING ASSESSMENT

LANGUAGE LAB  GeoJournal

1. **Compare** Read aloud the Sanskrit words in the chart. Compare them to the English words.

SELECTED INDO-EUROPEAN LANGUAGES			
Sanskrit	pitar	matar	dvi
English	father	mother	two
Greek	patéras	matros	dyo
Latin	pater	mater	duo
Spanish	padre	madre	dos

2. **Draw Conclusions** A language family is a group of languages that come from a common ancestor. After looking at the chart, what can you conclude about languages in the same language family?

3. **Interpret Time Lines** During what period of time were the Vedas written?

1500–1200 B.C.
Vedas written

1500 B.C.

1500 B.C.
Aryan migration into northern India

1500 B.C. Pages from the Vedas, sacred text in Sanskrit

500 B.C.

2.2 Historic Empires

TECHTREK
myNGconnect.com For an online map on
empires and photos of artifacts and architecture

Maps and Graphs

Digital Library

Main Idea Three South Asian empires made significant cultural contributions in religion, science, and the arts.

The Mauryan, Gupta, and Mughal empires (an **empire** is the land or people ruled by one leader) dominated the history of South Asia between 321 B.C. and A.D. 1858. Like the Harappan and Aryan empires, these later empires had the advantage of protective mountains and arable land. The Mauryans, Guptas, and Mughals also helped spread three major religions in the region—Buddhism, Hinduism, and Islam.

The Mauryan Empire

The Mauryan (MOWR yuhn) Empire was founded in the Ganges River valley in 321 B.C. The Mauryans built an efficient and organized government, which allowed them to run the empire. They also had a standing, or permanent, army.

The leader **Asoka** ruled for nearly 40 years and brought the empire to its height around 250 B.C. As the Mauryan empire grew and flourished, Asoka turned away from conquest in favor of more peaceful policies. He studied Buddhist nonviolent teachings and built many stupas, which are Buddhist religious structures.

Buddhism began in India, but gained more followers in east and southeast Asia as it spread. After Asoka's death, the Mauryan Empire declined.

The Gupta Empire

The Gupta (GUP tuh) Empire began around A.D. 321 in the fertile Ganges River valley. Gupta leaders practiced Hinduism, which became the major religion of South Asia.

Gupta artists and scientists created lasting cultural contributions. Advances in metal working, literature, mathematics (including the development of the decimal), and astronomy were part of this legacy. Eventually, invasions weakened the Guptas and by A.D. 540, their reign was over.

The Mughal Empire

One thousand years later, the Mughal (MOO guhl) Empire was established in 1526 by Babur. Mughal rulers practiced Islam and came from Central Asia. Islam became a unifying force in South Asia and grew into a large religious minority.

Akbar the Great came to power in 1556 and ruled for 49 years. Akbar expanded the empire and practiced religious **tolerance**, or respect for others' beliefs.

321 B.C.–185 B.C.
Mauryan Empire

Asoka's Pillar, c. 273 B.C., inscribed with his laws. The lion is a symbol of India.

A.D. 321–A.D. 540
Gupta Empire

Gold coin from Gupta Empire, c. A.D. 321

500 B.C.   A.D. 1   A.D. 500

c. 250 B.C.
Height of the Mauryan Empire

400
Height of the Gupta Empire

321 B.C.
Chandragupta Maurya founds the Mauryan Empire.

321
Chandra Gupta I founds the Gupta Empire.

EMPIRES OF SOUTH ASIA

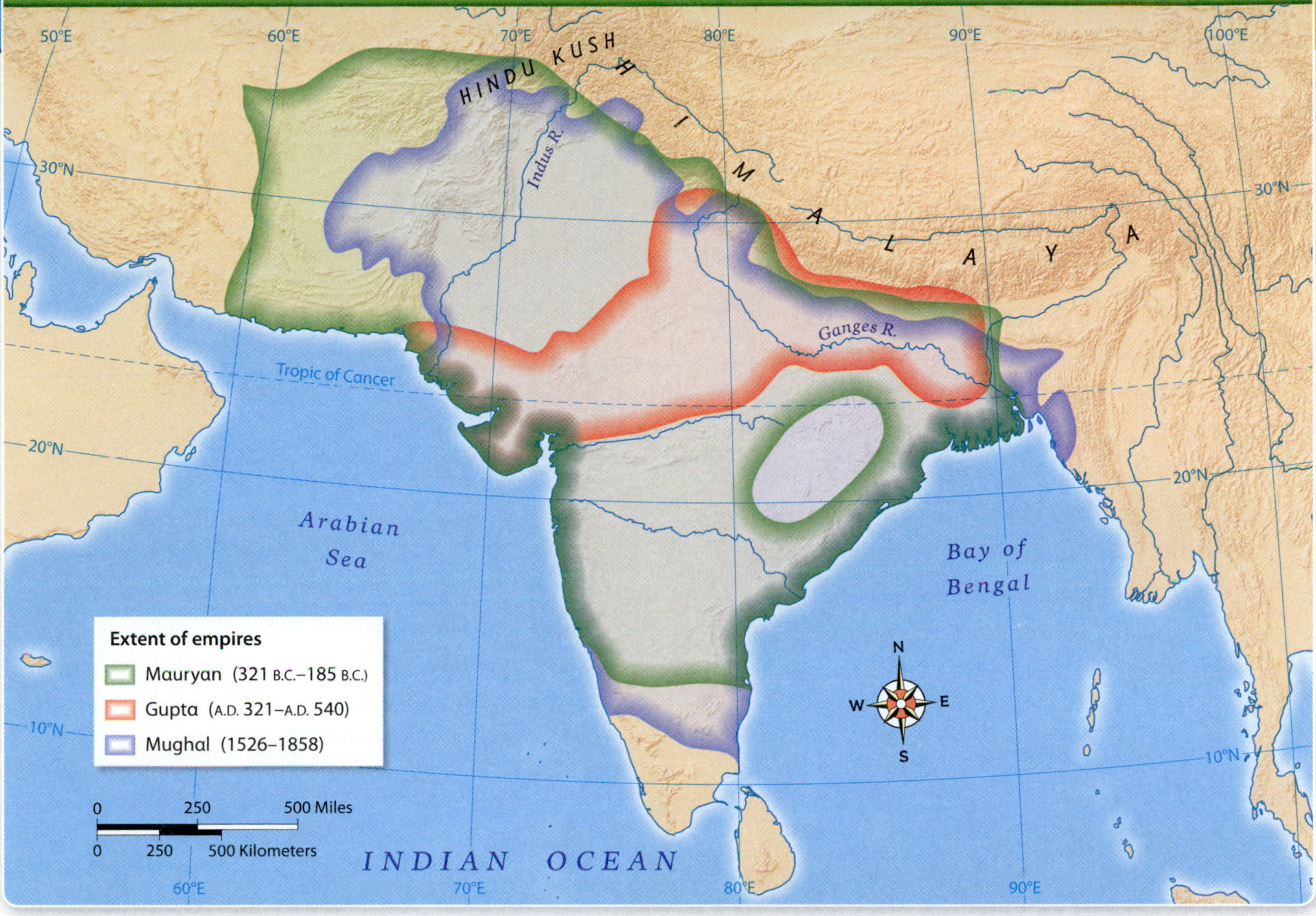

Extent of empires
- Mauryan (321 B.C.–185 B.C.)
- Gupta (A.D. 321–A.D. 540)
- Mughal (1526–1858)

HINDU KUSH

HIMALAYA

Indus R.

Ganges R.

Tropic of Cancer

Arabian Sea

Bay of Bengal

INDIAN OCEAN

0   250   500 Miles
0   250   500 Kilometers

Mughal leaders developed a large empire through military conquest. Cultural contributions of Mughal artists included the architecture of the **Taj Mahal** and detailed paintings called miniatures. The empire came to an end in 1858 when the British took control of the territory.

Before You Move On
Monitor Comprehension What cultural contributions did the three empires make?

ONGOING ASSESSMENT

MAP LAB

 GeoJournal

1. **Movement** What physical features influenced the empires' ability to control their territory?

2. **Make Generalizations** Based on the map and what you know about arable lands, what geographic features were vital for the success of these empires?

1526–1858
Mughal Empire

Taj Mahal, 1600s, built as a tomb for Shah Jahan's wife

1000

1500

PRESENT

1700
Height of the Mughal Empire

1526
Babur founds the Mughal Empire.

TECHTREK

myNGconnect.com For an online map and data on South Asian religions

 Maps and Graphs   Connect to NG

Main Idea Religion was an important part of South Asia's history. It is still important to the culture today.

Religion is an important part of South Asia's history because it shaped borders and cultures. Two of the world's main religions—**Hinduism** and **Buddhism**—were founded in India along with two other religions, **Jainism** and **Sikhism**. Buddhism spread into East Asia and Southeast Asia. Today, Buddhism has more followers in those regions than in South Asia. **Islam** was founded in Saudi Arabia, but it quickly spread to South Asia. Today, it is the second-largest religion in the region.

SIKHISM Sikhism began in India in the late A.D. 1400s. It was created to combine aspects of Hinduism and Islam. Sikhs believe in one god, truthful living, and equality of humankind. Today, most Sikhs live in the Indian state of Punjab.

Followers in South Asia: Approximately 1 percent of the population

JAINISM  Jainism developed in the 7th century B.C. Jains believe in *ahimsa*, or nonviolence towards all living things. Most Jains live in northwestern India.

Followers in South Asia: Approximately 1 percent of the population

BUDDHISM Buddhism was founded in India around 525 B.C. by a prince named Siddhartha Gautama. Gautama left his royal life to find a solution to human suffering. After six years, he discovered and began teaching the Four Noble Truths: 1) Suffering is a part of life; 2) Selfishness is a cause of suffering; 3) It is possible to move beyond suffering; and 4) There is a path that leads to this point.

People who followed Gautama's teachings called him the Buddha, or "Enlightened One." They recorded his teachings in a set of books called the *Tripitaka*, or "Three Baskets." During the next few centuries, Buddhism spread throughout Asia. In South Asia, it is the main religion of Bhutan and Sri Lanka. You can learn more about Buddhism in the chapters on East Asia.

Followers in South Asia: Approximately 2 percent of the population

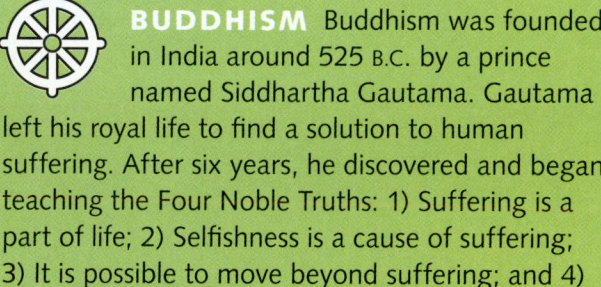

HINDUISM Hinduism began as a blend of native beliefs and the religion of the Aryan people. It has developed over thousands of years. Today it is the main religion in India and Nepal. Hindus worship many deities, or supreme beings, but they believe that every deity is a part of one universal spirit, called Brahman. Hindus also believe in reincarnation, or the rebirth of the soul. After death, a person's soul is reborn into another physical life. The kind of life is determined by the soul's *karma*, or actions during a previous life. If the soul has lived a good life, it is reborn into a better life. If the soul has lived an evil life, it is reborn into a worse life. This process continues until the soul lives a perfect life.

These beliefs and others are recorded in many different texts. The most important are the **Vedas**, the *Puranas*, the *Ramayana*, and the *Mahabharata*.

Followers in South Asia: Approximately 63 percent of the population

SOUTH ASIA'S RELIGIONS

Map Legend:
- Buddhism
- Christianity
- Hinduism
- Islam
- Jainism
- Sikhism

PAKISTAN

NEPAL

BHUTAN

BANGLADESH

Arabian Sea

I N D I A

Bay of Bengal

SRI LANKA

I N D I A N   O C E A N

MALDIVES

ISLAM Islam arrived with Muslim traders in the early A.D. 700s. Today it is the main religion of Pakistan and Bangladesh. It is also the largest minority religion in India. You can learn more about Islam in the chapters on Southwest Asia.

Followers in South Asia: Approximately 30 percent of the population

Before You Move On

Summarize Why is religion important to South Asia's history?

ONGOING ASSESSMENT
DATA LAB

 GeoJournal

1. **Create Graphs** Using the map and data in this section, make a circle graph that shows the percentages of followers of the religions in South Asia. How does the graph help you understand the distribution of religions across South Asia?

2. **Pose and Answer Questions** With a partner, ask and answer three questions about South Asia's religions.

3. **Location** In which two South Asian countries is Islam the dominant religion? What South Asian country separates these two Islamic countries from each other?

2.4 Colonialism to Partition

TECHTREK

myNGconnect.com For an online map, photos, and current news about Kashmir

 Maps and Graphs

 Digital Library

 Connect to NG

Main Idea Colonialism, independence, and conflict set new national borders in South Asia.

In 1600, during the Mughal Empire, the British established the East India Company, beginning an extensive trading relationship with countries in East and South Asia. As the Mughal Empire declined, the East India Company was able to take control over parts of India. Eventually, this power grew into **colonialism**, the control by one power over a dependent area or people.

Colonialism Limits Growth

In order to profit from India's rich natural resources, the British used trade practices that favored Britain. They shipped India's raw materials back to England and forced India to import British goods. The British also tried to prevent Indian manufacturers from producing certain goods that British manufacturers made, so that only British goods could be sold. This crippled India's economic growth for more than 100 years.

In 1857, Indians unsuccessfully rebelled against British control. The East India Company was disbanded in 1858, but the British established direct rule over India called the *raj*. British colonization had a lasting impact on the region's borders.

Seeking Independence

In 1885, Hindus formed the Indian National Congress (INC). Muslims formed the Muslim League in 1906. Both groups opposed British rule. India's movement for independence grew in the 1930s under the leadership of lawyer **Mohandas Gandhi** (GAHN dee). Gandhi campaigned for **civil disobedience**, or the nonviolent disobeying of laws, against the British.

India also faced conflict between Hindus and Muslims. Many Muslims called for their own land. In July 1947, the British Parliament passed the India Independence Act. British India would be divided into two countries: majority-Hindu India and majority-Muslim Pakistan, East and West. The part of Pakistan known as East Pakistan became Bangladesh in 1971.

Partition Creates Boundaries

Before **Partition**, or division, people scrambled to decide where to live. Many Muslims in India moved to Pakistan, and many Pakistani Hindus migrated to India. For hundreds of thousands of Indians, the movement was not peaceful. In August of 1947 they were **displaced** (forced to leave their homes) or even killed in the religious conflicts that followed.

Silk textiles are a thriving industry in South Asia.

1850

1900

1857
First Indian revolt for Independence

1885
Indian National Congress Party leads India's movement for independence.

1906
Muslim League leads India's Muslims in movement for their own nation.

In 1948, Gandhi was assassinated. The boundaries set by Partition have strained relations in South Asia to the present.

Conflict Over Kashmir

Partition established the present-day borders of South Asia. After Partition, India and Pakistan began a violent conflict over a region known as Kashmir. Both India and Pakistan believed the region belonged to them.

In 1949, a cease-fire was called, and Kashmir was divided. India maintains control over the portion called Jammu and Kashmir; Pakistan controls the portion known as Gilgit-Baltistan. However, Indian and Pakistani soldiers stand guard along the disputed border, and tension remains at a high level.

Part of the continuing conflict centers around the control of water for drinking and irrigation. Many rivers, including the Indus, begin in Kashmir, and maintaining control of the water is important to both countries. In fact, both countries have threatened to use nuclear weapons to resolve the dispute.

Before You Move On

Summarize What impact did British rule in India have on borders in South Asia?

CONFLICT OVER KASHMIR

ONGOING ASSESSMENT

MAP LAB

GeoJournal

1. **Interpret Maps and Time Lines** On this map, striped areas represent conflict between two countries. Which three countries disagree about their borders? Based on the time line, when was Kashmir divided?

2. **Location** How does the presence of rivers factor into the cause of the conflict in Kashmir?

3. **Turn and Talk** What physical features of this region might make it hard to control borders? Turn to your classmate and identify details from this map, or other physical maps in the chapter, that might help answer this question. Take notes to share with the class.

1947
India gains independence. Partition occurs.

1950

Mohandas Gandhi

PRESENT

1930
Gandhi continues his civil disobedience campaigns.

1949
Kashmir divided between India and Pakistan

Every religion has sacred text, and in Hinduism the most important may be the **Bhagavad Gita** (bah guh vahd GEE tah). It is a poem, part of the larger *Mahabharata* (muh hah BAH ruh tuh), which makes up much of Hindu mythology, or set of stories, traditions, and beliefs. It is an example of literature that has spread beyond Indian society and conveys religious themes. The Bhagavad Gita was written around the first or second century A.D. in Sanskrit, an ancient language of South Asia. The poem is still loved by millions of Hindus.

DOCUMENT 1

Gandhi on the Bhagavad Gita

> When disappointments stare me in the face, and when I see not one ray of light . . . I turn to the Bhagavad Gita and find a verse to comfort me, and I immediately begin to smile in the midst of overwhelming sorrow. My life has been full of external tragedies, and if they have not left any visible and indelible [permanent] effect on me, I owe it to the teaching of the Bhagavad Gita.

Gandhi with his granddaughters

CONSTRUCTED RESPONSE

1. Think about what you learned in Section 2.4. What specific "external tragedies" might Gandhi be referring to?

DOCUMENT 2

from the Bhagavad Gita
translated by Ranchor Prime

The Bhagavad Gita includes a conversation between the deity Krishna and a young warrior, Arjuna. Krishna explains to Arjuna his responsibilities as a warrior. Gandhi especially liked this passage:

> One without self-control cannot have a clear intelligence or a steady mind. An unsteady mind finds no peace, and without peace where is joy? As a strong wind sweeps away a boat on the water, so the mind dwelling on even one of the senses carries away the intelligence.

CONSTRUCTED RESPONSE

2. According to this passage, humans should not allow emotions to overcome logic. Think about what you have learned about Hinduism. What does this passage add to your understanding of how Hinduism unifies its followers?

DOCUMENT 3

Vishnu, Preserver of the Universe

Hindus believe Krishna, the narrator of the Bhagavad Gita, is a human form of Vishnu, a Hindu deity. Every element in this image of Vishnu is a **symbol** (something that stands for another thing) of an idea in Hinduism. For example, the crown symbolizes Vishnu's highest authority.

CONSTRUCTED RESPONSE

3. The lotus flower in his left hand shows that Vishnu is spiritually perfect. How does this symbol reflect the qualities described in Documents 1 and 2?

ONGOING ASSESSMENT

WRITING LAB  GeoJournal

DBQ Practice Think about the Bhagavad Gita and the picture of Vishnu. Why would these ideas and symbols be important to Gandhi and what he wanted for Hindus?

Step 1. Review Gandhi's influence on history in Section 2.4 and the ideas of Hinduism in Section 2.3.

Step 2. On your own paper, jot down notes about the main ideas expressed in each document.

Document 1: Quotation from Gandhi
Main Idea(s) _____

Document 2: Excerpt from the Bhagavad Gita
Main Idea(s) _____

Document 3: Image of Vishnu
Main Idea(s) _____

Step 3. Construct a topic sentence that answers this question: Which ideas and symbols from the Bhagavad Gita and the image of Vishnu would be important to Gandhi?

Step 4. Write a detailed paragraph that explains why each idea or symbol in your topic sentence was important to Gandhi and the history of India. Go to **Student Resources** for Guided Writing support.

VOCABULARY

For each pair of vocabulary words, write one sentence that explains the connection between the two words.

1. subcontinent; plate

> The Indian subcontinent is on a different plate than the rest of Asia.

2. evaporation; drought
3. arable; sustainable
4. conservation; ecosystem
5. deity; reincarnation
6. mythology; symbol

MAIN IDEAS

7. How were the Himalaya Mountains formed? (Section 1.1)

8. How do people living at high elevations in Bhutan feed themselves? (Section 1.2)

9. What are the two kinds of monsoons in India, and what are the effects of each? (Section 1.3)

10. What were two results of the Green Revolution in India during the 1960s? (Section 1.4)

11. What factors contributed to the pollution of the Ganges River? (Section 1.5)

12. What are the consequences of water pollution in South Asia? (Section 1.6)

13. How did Hinduism develop in India? (Sections 2.1, 2.3)

14. What events led to India's independence from Britain? (Section 2.4)

15. How did Islam spread to India? (Section 2.3)

GEOGRAPHY

ANALYZE THE ESSENTIAL QUESTION

How do South Asia's water systems affect how people in the region live?

Critical Thinking: Analyze Cause and Effect

16. What is the geographic connection between snowmelt in the Himalayas and South Asia's extensive river systems?

17. What are the effects of the wet and dry monsoon seasons on India's farmers?

18. What geographic factors make the Ganges Delta one of the primary agricultural regions of the Indian subcontinent?

INTERPRET CHARTS

THE GREEN REVOLUTION AND INDIAN VILLAGES (1955–1986)				
	1955	1965	1975	1986
Percentage of farmland irrigated	60	80	100	100
Use of farm machinery per village	none	none	4 tractors	9 tractors
Population per village	876	*N/A	N/A	1,076
Average size of farms	4.8 acres	N/A	5.7 acres	N/A

*Not available
Source: Brace, Steve. *India*. Des Plaines, IL: Heinemann Library, 1999, p. 29

19. **Make Generalizations** What happened to the amount of land irrigated in the average Indian village between 1955 and 1975?

20. **Summarize** What was the trend in the use of farm machinery? What do you think explains this trend?

ANALYZE THE ESSENTIAL QUESTION

How have physical features, religion, and empires shaped South Asia's borders?

Critical Thinking: Summarize

21. How did the Mauryan and Mughal empires bring religious diversity to India?

22. How did the Aryan migrations help to create cultural unity in the Indian subcontinent?

23. What religious differences led to boundaries between India and Pakistan?

INTERPRET MAPS

ARYAN MIGRATIONS
(1500–250 B.C.)

24. **Movement** What physical barrier did the Aryans have to cross to migrate into India?

25. **Make Inferences** Why might the Aryans' path of migration have avoided the western region of India?

ACTIVE OPTIONS

Synthesize the Essential Questions by completing the activities below.

26. **Write a Speech** Imagine that you live in a small village in South Asia today. You must persuade the village government to support sustainable agriculture. Write and deliver a one-minute speech to the village about its benefits. **Share your speech with the class.**

> **Writing Tips**
> - Make sure your introduction is lively enough to catch the reader's attention.
> - Present clear, strong evidence in support of your opinion.
> - End with a strong conclusion that summarizes your position.

TECHTREK **myNGconnect.com** For research links on South Asia today

27. **Create Charts** Make a three-column chart showing comparisons between India, Pakistan, and Bangladesh. Use the research links at **Connect to NG** and other online sources to gather the data. Then answer this question: What conclusions can you draw from the data recorded?

	India	Pakistan	Bangladesh
Year country was formed			
Population			
Square miles of land			
Main crops			
Form of Government			
Religion of majority of population			

South Asia
TODAY

PREVIEW THE CHAPTER

KEY VOCABULARY

- karma
- pilgrimage
- henna
- sari
- literacy rate
- shalwar-kameez
- cricket
- popular culture

ACADEMIC VOCABULARY
discrimination

TERMS & NAMES

- Krishna
- Bollywood

KEY VOCABULARY

- democracy
- federal republic
- infrastructure
- outsourcing
- developed nations
- developing nations
- microlending
- modernization

ACADEMIC VOCABULARY
accommodate, establish

TERMS & NAMES

- Parliament
- Golden Quadrilateral

KEY VOCABULARY

- slum
- push-pull factors
- military dictatorship
- overpopulation
- cyclone

ACADEMIC VOCABULARY
deny

TERMS & NAMES

- Mohammed Ali Jinnah
- Benazir Bhutto

TECHTREK

FOR THIS CHAPTER

Student eEdition

Maps and Graphs

Interactive Whiteboard GeoActivities

Digital Library

Connect to NG

Go to **myNGconnect.com** for more on South Asia.

A woman speaks on her mobile phone while riding on a bicycle rickshaw in Delhi, India.

1.1 Hinduism Today

TECHTREK
myNGconnect.com For photos of Hindu deities

Digital Library

Main Idea Hinduism unifies its followers through a wide variety of religious beliefs.

Approximately 63 percent of South Asians identify themselves as Hindus. In India, the majority is even larger. Hindus make up over 80 percent of the population, more than 930 million people. Hinduism in India is more than a religion—it is a way of life shared by a diverse population.

⌃ **Critical Viewing** This tenth-century painting shows Krishna as the protector of cows. What details in the painting tell you that cows are valued by Hindus?

Basic Beliefs

Hindus believe that everything in the universe is part of one spiritual force called Brahman. To become one with Brahman, every soul must go through reincarnation, in which it is reborn into different forms of life. A soul's **karma**, or actions during a life, determines the soul's form in the next life. If a soul has good karma, it will be reborn into a higher state, such as a human. If the soul has bad karma, it will be reborn into a lower state, such as a plant.

These beliefs are recorded in sacred texts called the Vedas, which date back to Aryan civilization. Hindu texts, such as the *Mahabharata* and the *Ramayana*, teach Hindu beliefs in the form of epic poems. As you've learned, the Bhagavad Gita is part of the *Mahabharata*.

Hindu Deities and Sacred Spaces

According to Hindus, there are many gods and goddesses, but all come from Brahman. The three most important deities are Brahma, the creator of the universe; Vishnu, the preserver of the universe; and Shiva, the destroyer of the universe. **Krishna**, an avatar of Vishnu, is a popular Hindu deity. He is often pictured playing his flute.

Hindus believe that the Ganges River is sacred. Millions of Hindus make **pilgrimages**, or religious journeys, to the city of Varanasi to worship on the *ghats*, or stone steps and platforms along the river. Hindus also believe that bathing in the water gives them better karma.

The Caste System

In addition to Hinduism, the Aryan migration into India also brought the caste system. This system divided society into different social groups, or varnas, based on a person's occupation. As you have learned, there were four original varnas in the caste system: the *Brahmans* (priests and scholars); the *Kshatriyas* (rulers and warriors); the *Vaisyas* (merchants and professionals); and the *Sudras* (artisans, laborers, and servants).

Eventually, an unofficial fifth group was created to include those outside the varna system. These "Untouchables" performed the lowest jobs in Indian society, such as tanning animal skins and collecting garbage. Today, this group prefers to call itself *Dalit*.

The rules of the caste system were very strict. People could not marry out of their caste or even eat with people of a higher caste. It was nearly impossible to move into a different caste. In the last 50 years, however, these rules have become less rigid. India's constitution now forbids **discrimination**, or unfair treatment, against members of any caste and many groups have formed to struggle against such discrimination.

Before You Move On

Monitor Comprehension What religious beliefs unite Hindus?

Critical Viewing How is this Hindu temple similar to other religious buildings you have seen?

ONGOING ASSESSMENT

READING LAB   GeoJournal

1. **Summarize** Name and describe the three most important Hindu deities.

2. **Compare and Contrast** Think about other world religions that you've learned about. How are they similar to or different from Hinduism? Copy the graphic organizer and use it to compare and contrast religions.

3. **Write a Response** Why is Hinduism so important to India? Think about the size of the Hindu population and Hinduism's part in Indian history. Write two to three sentences explaining Hinduism's importance.

1.2 **Changing Traditions**

TECHTREK

myNGconnect.com For photos of
South Asian traditions

 **Digital
Library**

 **Magazine
Maker**

Main Idea In South Asia, traditions blend with modern practices.

Many groups have helped shape South Asia's culture, from the Aryans to the British. Some aspects of the region's culture are very old and some are modern. These photographs show the ancient cultural traditions practiced by people in South Asia today.

Traditional Decoration

Since ancient times, Indian women and girls have decorated their hands with fancy patterns for special occasions, such as weddings or festivals. They use a reddish powder called **henna** to create the intricate designs (shown at right). Each design has a special meaning: A flower can mean happiness; a square might mean honesty; and a triangle might represent creativity.

Before You Move On

Make Inferences How do traditions blend with modern practices in South Asia?

> **Critical Viewing** What traditional and modern elements can you identify in this photo?

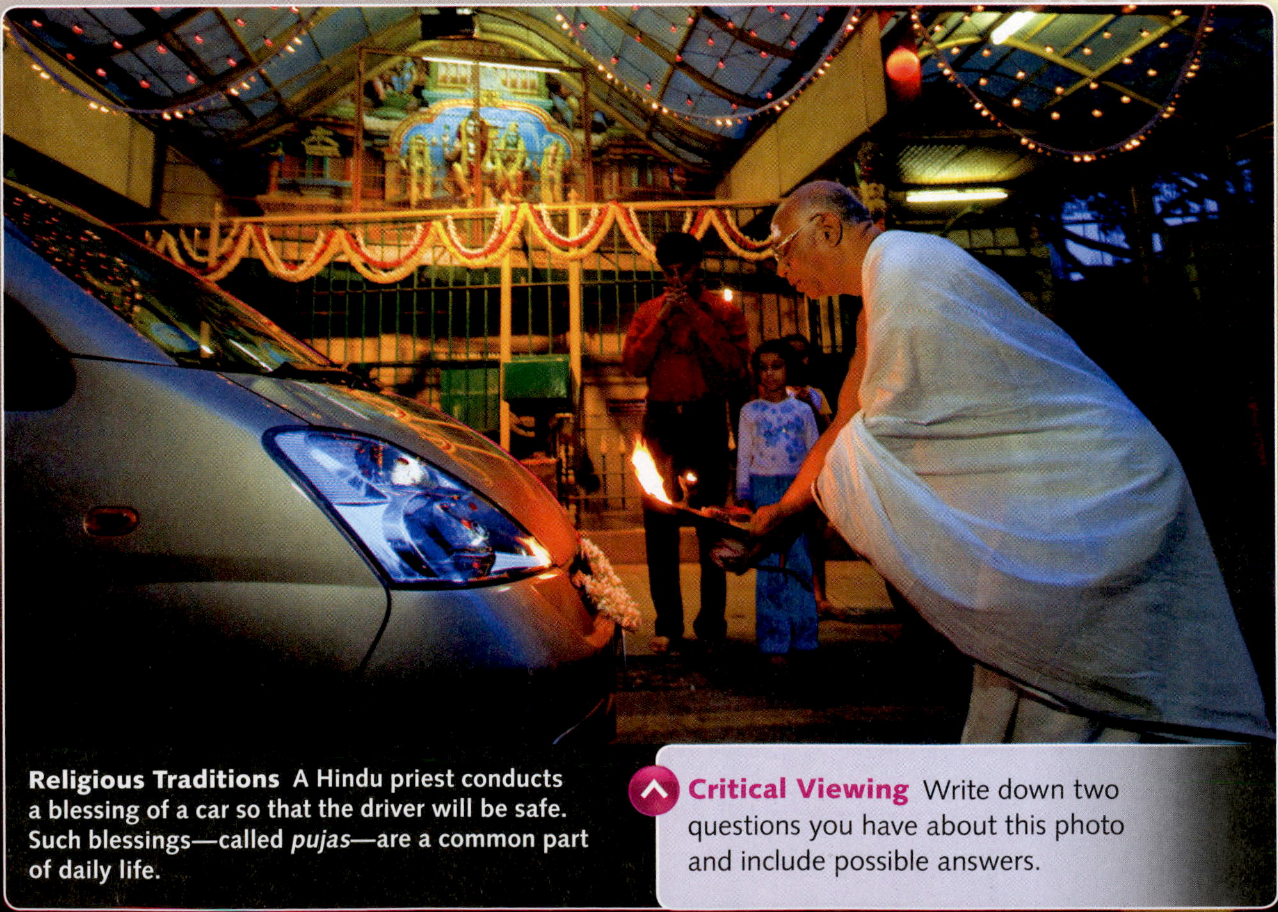

Religious Traditions A Hindu priest conducts a blessing of a car so that the driver will be safe. Such blessings—called *pujas*—are a common part of daily life.

∧ **Critical Viewing** Write down two questions you have about this photo and include possible answers.

Visual Vocabulary A sari is a traditional Indian garment for women worn wrapped around the body. Many saris are silk.

Visual Vocabulary Henna is a reddish powder used to create designs on skin. Henna is used for special occasions.

PHOTO LAB
GeoJournal

1. **Create a Photo Essay** Look at the two photos of traditions in South Asia. Now go to the **Digital Library** to find more photos of traditions. Choose several photos and print them out. Then paste them on paper or use the **Magazine Maker** to create a photo essay of traditions in South Asia. Write a caption for each photo that explains the tradition and how it is a part of modern life.

2. **Make Inferences** Many South Asians wear Western-style clothing. Why might wearing saris continue to be popular today?

TECHTREK

myNGconnect.com For photos of daily life in South Asia and Visual Vocabulary

Digital Library

Main Idea Some aspects of culture in South Asia show how tradition is important in everyday life.

Modern practices have spread to many parts of South Asia. However, people continue to observe early traditions.

Schools

Most countries in South Asia offer some level of free public education like the public schools in the United States. India's schools require attendance until age 14. In Bhutan, formal eduation has only been in place since 1950. However, education is definitely a shared value in the region. Schools are set up wherever there is space: mountain huts, river boats, desert tents, and even train platforms.

The **literacy rate** (the percentage of people who can read and write) for 15-to-24-year-olds ranges from 74 percent in Bangladesh to 96 percent in the Maldives. Rising literacy rates indicate that the region is developing.

Clothing

South Asia's climate extremes determine what people wear. For example, Bhutanese herders wear coats and yak-hair hats to protect themselves from the cold, while Sri Lankan fishermen dress lightly because of the heat.

Many men and women wear a **shalwar-kameez** (see photo opposite), a long shirt with loose-fitting pants. Once considered a traditional form of Muslim dress, the shalwar-kameez is now worn by Muslims and Hindus alike. Some women wear a sari, a single piece of cloth (often silk) wrapped to form a long dress. Western clothes are increasingly popular too.

< Critical Viewing A young boy lights tiny oil lamps to celebrate the festival of Diwali. This marks the victory of good over evil. Based on this photo, what holidays that you celebrate seem similar to Diwali?

Food

South Asia's cuisine—the food commonly served in a region—is richly spiced and often vegetarian. Cows are sacred to Hindus, so many Indians do not eat meat. Pakistan and India serve curry dishes with potatoes, eggplant, and okra. In Bhutan, yaks provide meat and milk, which is used to make butter and cheese. Each region prepares its own mixture of spices, or *masala*, to use in sauces. Rice is often the center of a meal, especially in Bangladesh and eastern and southern India. *Chapati*, a flat wheat bread, is common in northern India. *Dal*, a blended stew of peas, beans, or lentils, is a frequent side dish.

Visual Vocabulary shalwar-kameez

Visual Vocabulary cricket

hockey, golf, and soccer are also popular. India has professional cricket and soccer leagues that draw many enthusiastic fans.

Before You Move On

Summarize What are three ways in which tradition is part of everyday life in South Asia?

Rice, potatoes, and wheat are staples of South Asian cuisine.

Sports

Traditional sports like kabaddi (described as a combination of wrestling and rugby) have existed in India for nearly 4,000 years. Western sports such as cricket (a team game similar to baseball), field

---

ONGOING ASSESSMENT

WRITING LAB   GeoJournal

1. **Describe Culture** What is your culture like? Create a culture diagram to describe it

   Step 1 Cut a large paper circle. Write "My Culture" in the center, and divide it into four sections with the labels Schools, Clothing, Food, Sports.

   Step 2 Find photos from magazines or online that represent each category of U.S. culture. Paste photos in the appropriate sections.

   Step 3 Turn the circle over and write a detailed paragraph that describes your culture in each category.

2. **Compare** How is South Asia's daily life different from daily life in the United States? How is it the same? Use your diagram to compare. Discuss your ideas with a classmate.

1.4 Popular Culture

TECHTREK

myNGconnect.com
For samples of Indian music

 **Digital Library**

Main Idea Music and movies help unify the diverse peoples and cultures of South Asia.

Music and movies are a lively part of South Asia's popular culture, which consists of the arts, music, and other elements of everyday life in a region.

Unifying a Culture

South Asia has a long musical history, ranging from religious music to blockbuster movie soundtracks. Traditional music echoes in today's spectacles of Bollywood movies. With wide availability and appeal, music and movies foster the development of a common culture.

Music

Young people in South Asia listen to Bollywood music, Indian rock, and Western pop. Many also enjoy Bhangra, a form of folk dance and music from the Punjab region of India and Pakistan. Some modern Indian musicians are fusing Indian and Western styles to create a new sound that keeps Indian musical traditions alive.

Critical Viewing Young people in New Delhi dance to a wide variety of music. What influences of Western popular culture do you see in the photograph?

The musician on the left plays traditional music near the Ganges River at Varanasi. The DJ on the right plays music in Mumbai. Classical Indian music is based on religious themes and can be heard in both traditional and contemporary music.

Movies

India's film industry began in Bombay (Mumbai) and is called **"Bollywood"** after its American counterpart, Hollywood. Many Bollywood movies have themes based on Hindu stories, feature musical numbers, and use the Hindi language. India produces more feature films than any other country in the world, and their international audience is growing.

The southern city of Chennai has its own film industry that produces films in the Tamil language. This film center is called "Kollywood" after the Kodambakkam district of Chennai.

Cast members from the 2008 hit movie *Slumdog Millionaire* pose at the Academy Awards. Movies like *Slumdog Millionaire* showed that films about India are often popular in Western culture.

Before You Move On

Make Inferences How do music and movies unify people of South Asia?

🔼 **Critical Viewing** Actors perform a scene in a Bollywood movie. Based on details in this scene, how does this movie compare to movies you have seen?

ONGOING ASSESSMENT

SPEAKING LAB   GeoJournal

Turn and Talk Do music and movies bring people together in the United States? Think about your answer, and then work with a group to develop a response that can be presented orally to the class. Your first sentence will answer the question. Each person in the group adds one sentence for support. Be prepared to present your response to the class.

2.1 The Largest Democracy

TECHTREK

myNGconnect.com For current events in India

 Connect to NG

> **Main Idea** The democratic government of India faces many challenges in governing more than one billion citizens.

In 2009, 64-year-old Meira Kumar became the Speaker of the lower house of India's **Parliament**. She was the first woman elected to this position— a milestone for the world's largest democracy. India's Parliament is the legislative branch of the government. It has two houses like the U.S. Congress.

Governing One Billion People

India has passed many milestones since its independence. One of the first was crafting a new constitution in 1949 to create a **democracy**, or a government in which citizens make decisions either directly or indirectly through elected representatives. India's constitution established a **federal republic**, which is a democratic form of government in which voters elect representatives and the central government shares power with the states. The United States is also a federal republic.

India's government has three branches: the legislative, the executive, and the judicial. The legislative branch has two houses, the *Lok Sabha* (similar to the U.S. House of Representatives) and the *Rajya Sabha* (similar to the U.S. Senate). Members of the *Lok Sabha* are elected every five years, and those of the *Rajya Sabha* serve six-year terms.

Parliament and the states' legislatures elect the president. The president appoints justices to the Supreme Court. However, the leader with the most power is the prime minister, who is head of the political party with the most members in the *Lok Sabha*. The prime minister heads the Council of Ministers, who operate the government.

India's Politics Today

In 2009, India held its national elections. Voting took place in phases over a one-month period and brought Meira Kumar into the headlines. Kumar promised to work toward a "casteless" society in India.

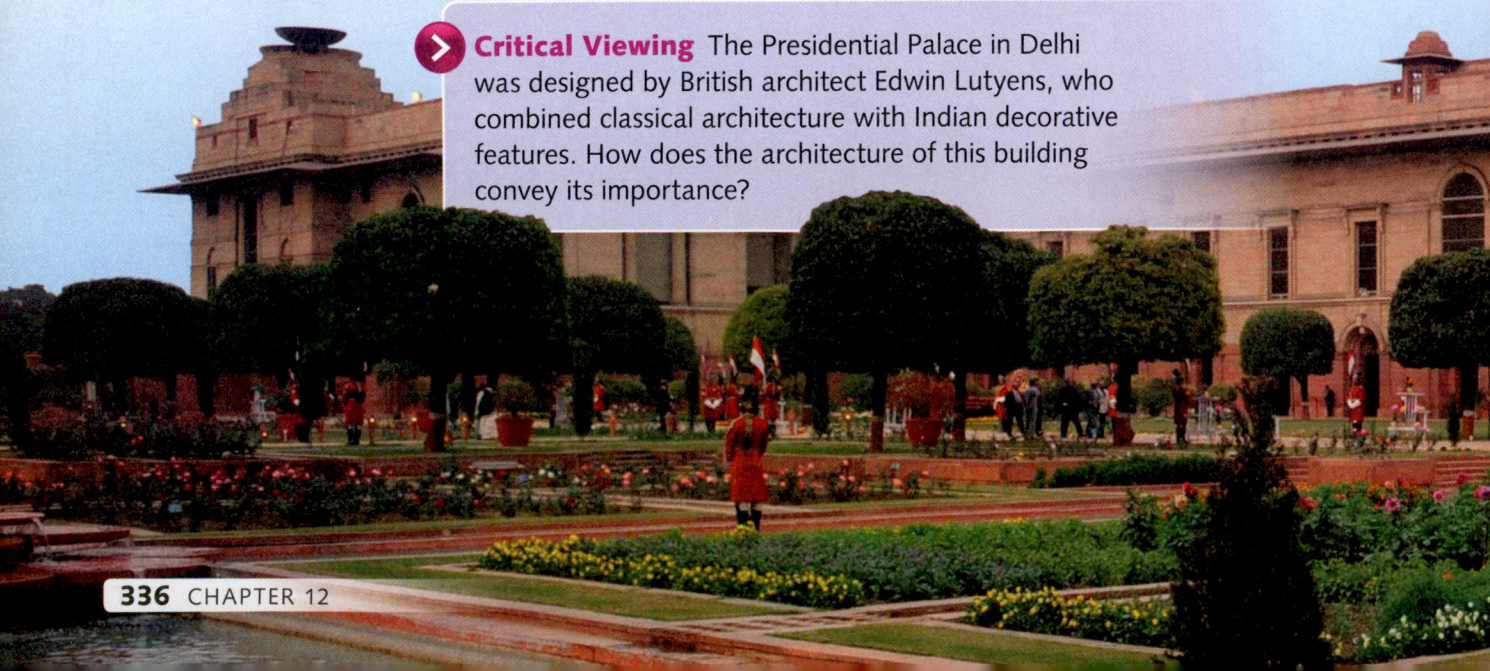

> **Critical Viewing** The Presidential Palace in Delhi was designed by British architect Edwin Lutyens, who combined classical architecture with Indian decorative features. How does the architecture of this building convey its importance?

INDIA'S GOVERNMENTAL STRUCTURE

LEGISLATIVE (Parliament)

Rajya Sabha
Upper House:
House of States
(250 members)

Lok Sabha
Lower House:
House of the People
(545 members)

EXECUTIVE

President

Prime Minister

Council of Ministers

JUDICIAL

Supreme Court
• Chief Justice
• 25 associate justices

The Indian government is tackling major challenges, such as India's fast-growing population. By 2030, India is expected to pass China as the world's most populous nation. The government is already working to improve its infrastructure in order to **accommodate,** or make room for, that growth. **Infrastructure** includes the basic systems that a society needs, such as roads, bridges, and sewers. You will read about one of India's massive road projects, the Golden Quadrilateral, later in this chapter. In addition, because of the shortage of fossil fuels in India, the government is increasing its investment in nuclear power to produce electricity.

Another challenge for the government is reaching an agreement with Pakistan over control of largely Muslim Kashmir. This is a goal most observers believe is important to the region's future.

Before You Move On

Make Inferences What is a major challenge facing India in governing its citizens?

ONGOING ASSESSMENT

READING LAB   GeoJournal

1. **Compare and Contrast** Name two ways in which the government of India is similar to the government of the United States. Name one way they are different.

2. **Analyze Visuals** How is India's government organized?

3. **Make Inferences** Why do you think the Prime Minister of India has the most power?

SECTION 2 FOCUS ON INDIA

2.2 Economic Growth

TECHTREK
myNGconnect.com For photos of economic growth and current events
Digital Library
Connect to NG

Main Idea Many factors have contributed to India's rapid economic growth.

If you have a computer problem and call for technical help, your call may be answered in the southern Indian city of Bangalore. Bangalore is the leading center of India's telecommunications industry. Companies there handle support services for U.S. computer and software companies. **Outsourcing**, the shifting of jobs to workers outside of a company, has been a big part of India's economic growth.

A Growing Economy

The value of the goods and services produced in a country divided by the number of people in that country is called the per capita Gross Domestic Product (GDP). Countries with a high per capita GDP are known as **developed nations**, and countries with a low per capita GDP are called **developing nations**. Developed nations have healthier, more educated people, consume more goods and services, use more energy, and employ more people in manufacturing and service industries than developing nations. Developing nations have a lower standard of living and less developed infrastructure.

Most South Asian countries are developing nations. But some economists see India as an emerging market, with a high growth rate and goods and services that compete in global trade. India's GDP has increased by an average of 7 percent every year for the last 20 years—about twice as fast as that of the United States. As a result, its estimated middle class population of more than 300 million people is nearly the same as the population of the entire United States.

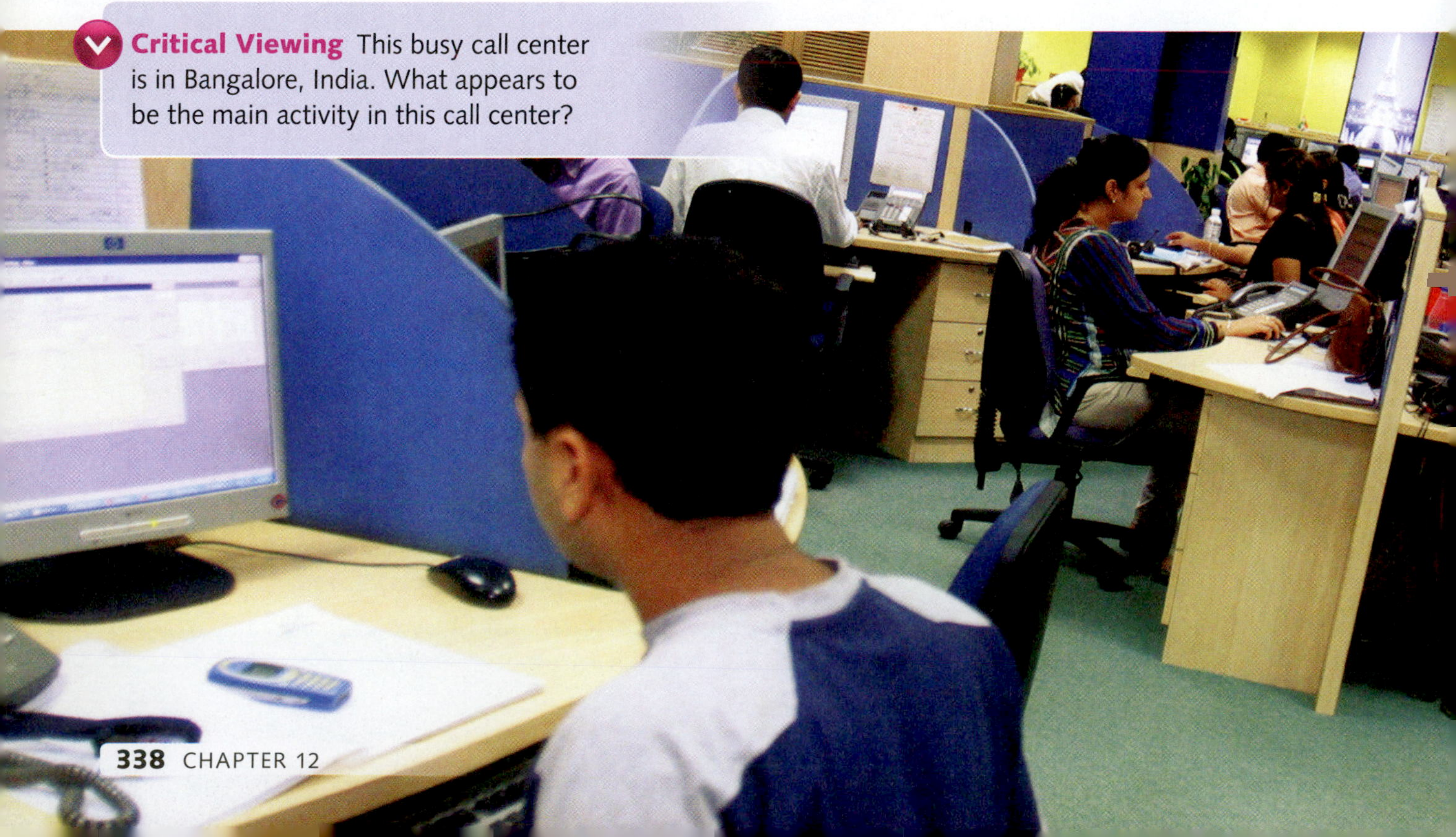

Critical Viewing This busy call center is in Bangalore, India. What appears to be the main activity in this call center?

Factors Affecting Growth

The teaching of English—the language of international commerce—is a legacy of British colonialism that has worked in India's favor. A common language, a stable democratic government, and improvements in infrastructure have encouraged foreign investment.

What does India's growth mean to the average person? With 75 percent of the people still trying to exist on only two dollars a day, it means slow but improving access to more consumer goods and more opportunities. **Microlending**, the practice of making small loans to people starting their own businesses, is one way that the country's growing prosperity reaches even its poorest citizens.

Before You Move On

Monitor Comprehension What economic factors contributed to rapid growth in India?

COMPARISON OF SELECTED CONSUMER GOODS, SOUTH ASIA AND THE UNITED STATES

	Televisions	Cell Phone Subscriptions	Personal Computers	Total Number of Automobiles
	(per thousand people)			
Bangladesh	85	217	12	185,000
Bhutan	33	172	16	10,574
India	78	44	12	8,619,000
Maldives	40	345	109	3,393
Nepal	7	7	4	66,395
Pakistan	131	32	4	1,559,284
Sri Lanka	111	115	28	293,747
U.S.A.	893	774	755	136,431,000

Between 2000 and 2007   **Source:** Encyclopedia Britannica

2.3 The Golden Quadrilateral

TECHTREK

my**NG**connect.com For online maps and photos of India's infrastructure

 Maps and Graphs

 Digital Library

Main Idea India is modernizing its roadways to support its economic growth.

In the mid-1990s, India's prime minister complained, "Our roads don't have a few potholes. Our potholes have a few roads." Since then, the country has invested large amounts of money to improve its roads.

A Road to the 21st Century

In 1998, the prime minister announced that India would build a 3,633-mile superhighway called the **Golden Quadrilateral** (GQ) to connect four major cities: Delhi, Mumbai, Chennai, and Kolkata. The GQ is an example of India's efforts to improve its infrastructure. **Modernization** efforts like the GQ bring countries up to present standards in technology and other areas.

The GQ will help spread economic growth from India's bustling cities to its thousands of poor villages.

When India gained independence in 1947, drivers shared unpaved roads with cattle and slow-moving tractors. In 1998, the government **established** (put into place) the National Highways Authority to build thousands of miles of new roads, including the Golden Quadrilateral, at a cost of more than $30 billion. It is the largest public works project in the country's history. With the GQ's advanced technology, road sensors will automatically notify crews when it needs repairing. Providing faster and more reliable transportation is one step toward modernizing India's infrastructure and preparing for a profitable future.

Critical Viewing Traffic backs up in Bangalore at a GQ construction site. Based on details in the photograph, what challenges do you think a company might face when building a road in a heavily populated area?

The Impact of the GQ

India's investment in the GQ has already had a tremendous impact on its economy—and on people's lives. For example, a farmer who could sell his crops only in nearby towns can now use the GQ to carry his products to Chennai and sell them for higher prices.

As India builds better roads, more Indians are purchasing automobiles and trucks—a clear sign of economic growth. In fact, recently India's economy has grown at an annual rate of 9 percent—a rate second only to China's growth. For strong growth to continue, India will need to keep investing in major projects like the Golden Quadrilateral.

Before You Move On

Monitor Comprehension How have improvements in infrastructure helped India's economic growth?

INDIA'S HIGHWAYS AND THE GOLDEN QUADRILATERAL

Golden Quadrilateral
— Completed
— Planned

North-South and East-West Corridors
— Completed
— Planned

⑦ Highway Route Number

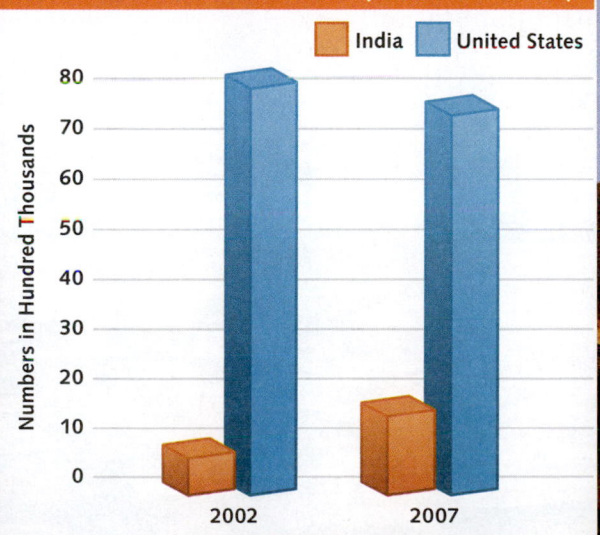

SALES OF PASSENGER VEHICLES IN INDIA AND THE UNITED STATES (2002 AND 2007)

India United States

Numbers in Hundred Thousands

Source: U.S. Bureau of Transportation Statistics; Society of Indian Automobile Manufacturers

3.1 The Impact of Urbanization

Main Idea South Asia's cities are facing the challenges of rapid growth.

Dharavi is a slum in Mumbai, India. A **slum** is an area in a city that has crowded, unclean housing, poor sanitation, and bad living conditions. Residents of Dharavi often walked a mile to get water, but now some have their own water taps. It's a small sign of progress in an area where improvement comes slowly.

Growing Cities

Since independence, cities in India have exploded in population. In 1951, about 62 million Indians lived in urban areas—cities and their surrounding communities. In 2001, nearly 300 million people lived in cities. Thirty-five cities have more than a million residents. People have flocked to India's cities because of a number of push-pull factors. **Push-pull factors** are the reasons why people migrate. "Push" factors cause people to leave a place, and "pull" factors draw them to another place.

One key factor pushing people to leave the countryside is poverty. Factors pulling people to cities include job opportunities and better education.

India's cities have gradually divided into small wealthy sections and vast slums. The slums have very poor infrastructure and lack clean water and electricity. Education and health-care services are inadequate. Permits to build new structures are hard to get. City centers are also jammed with growing numbers of automobiles, creating traffic congestion and adding to air pollution—which causes thousands of deaths each year. Despite these problems, the jobs and educational opportunities in the cities do help some people to improve their lives.

Pakistan faces similar issues. There, nearly 36 percent of Pakistan's population live in cities. The city of Karachi (kuh RAH chee) is approaching 13 million people—about 6 percent of Pakistan's entire population.

Critical Viewing Slums of Mumbai stand amidst new construction. What does this photo suggest about the differences between the way wealthy and poor people live?

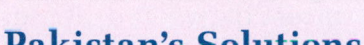

POPULATION GROWTH IN SOUTH ASIA (1971–2009)

● India   ● Bangladesh   ● Pakistan   ● Sri Lanka

Population in Millions

Years: 1971, 1981, 1991, 2001, 2009 (est)

Source: EncyclopaediaBritannica.com

^ Critical Viewing This crowded city street is in Karachi, Pakistan. What can you infer about life in a Pakistani city from this photo?

Pakistan's Solutions

As in India, people are leaving rural areas in Pakistan and moving to the cities. With urban populations increasing at a rate of 3 percent per year, Pakistan's infrastructure has trouble keeping up. This leads to overcrowding in slums.

One solution to overcrowding has been the development of "secondary" cities around major urban centers. For example, Gujranwala is a secondary city to Lahore.

Secondary cities provide rural migrants with an affordable place to settle and find employment. The secondary cities also relieve strain on big cities such as Lahore. They do this by slowing the flow of people into the main city. Other South Asian cities, particularly in Bangladesh and Sri Lanka, face similar pressures.

Before You Move On

Make Inferences What are some solutions to the challenges of rapid growth in South Asian cities?

ONGOING ASSESSMENT

DATA LAB

 GeoJournal

1. **Analyze Data** How much did Bangladesh's population grow from 1971 to 2009? How much did Pakistan's grow?

2. **Make Inferences** Think about the total land area of India, Pakistan, and Sri Lanka, their arable land, and the populations indicated on the graph. Based on what you know, explain the rates of population growth in each country.

3.2 Pakistan's Changing Government

Main Idea Pakistan's government has alternated between democracy and military dictatorships.

Pakistan was created in 1947 as an Islamic country. Since then, it has alternated between civilian governments and **military dictatorships**, in which the army runs the government and **denies,** or refuses to recognize, citizens' rights. The changes in Pakistan's government have limited social and economic progress in the country.

A Lack of National Unity

One cause of Pakistan's changes in government has been a lack of unity. The country is divided into four provinces: Punjab, Sindh, Balochistan, and the North-West Frontier Province. Each province has its own tribal groups and languages. No leader has been able to pull the provinces together to form a unified nation.

The lack of unity was a problem from the start. In 1947, the country had two parts. West Pakistan was west of India. East Pakistan was east of India, more than 1,000 miles away from West Pakistan.

Most leaders came from West Pakistan, which troubled the people of East Pakistan. In 1971, East Pakistan declared independence and became Bangladesh.

Military Dictatorships

Another cause of Pakistan's changes in government has been inconsistent leadership. **Mohammed Ali Jinnah**, who led the founding of Pakistan, believed in democracy. When he died in 1948, the country lacked a strong leader. Pakistanis grew angry about slow economic development, and unrest between different groups of Pakistanis increased. In 1958, the military took over the government and established a military dictatorship.

After a civil war in 1971, Zulfikar Bhutto was elected to lead Pakistan. He promised to take steps to end poverty. However, in 1977, the military took control of Pakistan again, and Bhutto was executed in 1979.

> **Critical Viewing** These women are getting ready to vote at a polling center in Pakistan. What can you infer about their attitude toward voting?

Benazir Bhutto's Election

Elections were held again in 1988 after the death of the military leader. Pakistanis elected Bhutto's daughter, **Benazir Bhutto**, as prime minister. Her opponents accused her of corruption. She denied the charges, but there were many threats on her life, and she fled Pakistan in 1999.

With the military in power again, Pakistan experienced moments of prosperity. Yet it also continued to deny people's rights. In 2007, the military promised elections, and Benazir Bhutto returned to Pakistan. However, she was assassinated later that year. Bhutto's husband, Asif Ali Zardari, was elected president in 2008.

Continuing Challenges

Pakistan's relationship with its larger neighbor, India, continues to be strained, particularly over the region of Kashmir. The two countries are frequently at a stand-off over the area. Violence regularly erupts, and war is a very real threat. Difficulty controlling the border with its eastern neighbor, Afghanistan, also continues to put a great deal of pressure on Pakistan's government.

Pakistan's internal problems show that successful countries need strong leaders who can bring different groups together. They also demonstrate the importance of having an effective government that makes sure the country's constitution is enforced, guaranteeing rights to all of its citizens.

Before You Move On

Make Inferences Explain why the military has taken power in Pakistan at various times.

PAKISTAN POLITICAL

MAP TIP The Federally Administered Tribal Areas are seven areas along the border with Afghanistan that are inhabited by mainly Pashtun tribes. These areas are loosely governed by regulations put in place by the British *raj*.

ONGOING ASSESSMENT
READING LAB GeoJournal

1. **Analyze Cause and Effect** What circumstances have caused Pakistan to alternate between democratic governments and military governments?

2. **Location** Find the five major cities in Pakistan on the map. How might geographic factors explain their location?

3. **Make Inferences** Why would it be difficult to govern the four provinces of Pakistan?

4. **Evaluate** Why has it been difficult for Pakistan to move forward as a country?

3.3 Fighting Poverty in Bangladesh

TECHTREK

myNGconnect.com For an online resources and industries map

Maps and Graphs

Main Idea Bangladesh is one of the poorest countries in the world, but it is making progress toward improving its economy.

The biggest problem in Bangladesh is poverty. In 2009, the gross domestic product (GDP) per capita was only $1,500. In contrast, U.S. GDP per capita that year was $46,000. More than eight out of ten Bangladeshis live on less than two dollars per day.

Poverty in Bangladesh

Economists identify three problems that help explain why Bangladesh is poor. One problem is **overpopulation**, or too many people living in one place. In 2010, over 156 million people in Bangladesh were living on only about 55,600 square miles of land. That is roughly equivalent to half of the U.S. population living in an area the size of the state of Illinois.

The second problem facing Bangladesh is natural disasters. Rivers like the Ganges surge over their banks during summer monsoons. Many **cyclones**—the name for hurricanes in South Asia—also hit Bangladesh. Floods and cyclones destroy crops, adding to the problems of an uneven food supply and poverty.

Lack of education is a third problem. Only 48 percent of Bangladeshis can read and write. In contrast, 99 percent of Americans can read and write. In addition, about 8 million Bangladeshi children work in factories and at other jobs. Because they work for wages, they are not able to go to school.

Toward a Brighter Future

When Bangladesh became an independent nation in 1971, it had a military government like Pakistan's. Since 1991, though, it has had a democratically elected government. The government has improved the economy by encouraging exports and building new industries.

It is not easy to find products to export. Bangladeshi farmers grow rice, wheat, tea, and sugarcane, but those crops must feed the country's large population, and there is usually nothing left over to export. However, the country does export huge amounts of jute. Jute is a strong fiber used in making ropes, carpets, types of paper, and other products.

Critical Viewing Garments are a major export in Bangladesh. What conditions shown in this photo of a factory help explain why the garment industry has been successful?

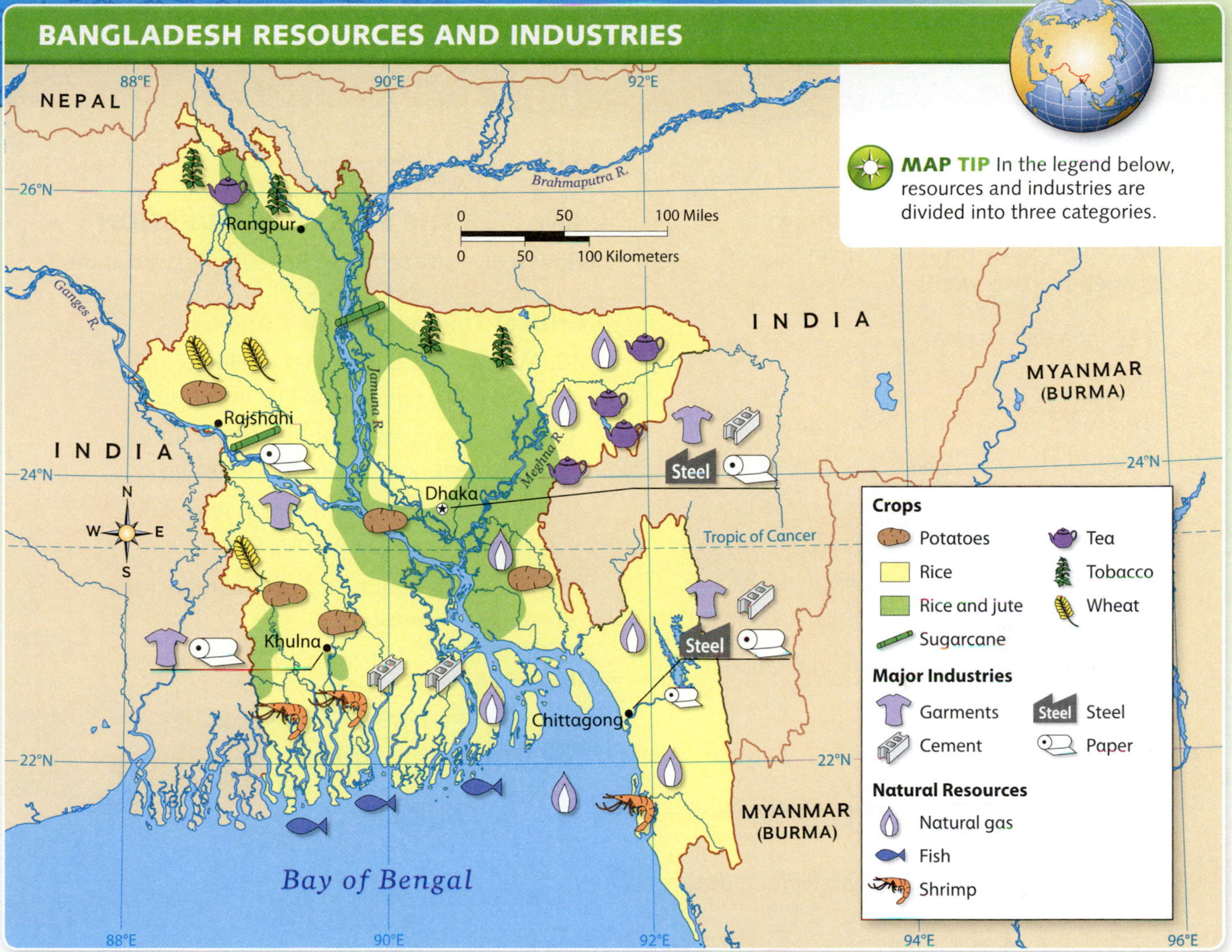

BANGLADESH RESOURCES AND INDUSTRIES

MAP TIP In the legend below, resources and industries are divided into three categories.

NEPAL

Brahmaputra R.

Rangpur

INDIA

Ganges R.

Rajshahi

Jamuna R.

INDIA

Dhaka

Meghna R.

Steel

MYANMAR (BURMA)

Tropic of Cancer

Khulna

Steel

Chittagong

MYANMAR (BURMA)

Bay of Bengal

Crops
	Potatoes		Tea
	Rice		Tobacco
	Rice and jute		Wheat
	Sugarcane		

Major Industries
	Garments	Steel	Steel
	Cement		Paper

Natural Resources
	Natural gas
	Fish
	Shrimp

Bangladesh has also built several industries. New factories produce paper, cement, and steel. (Locate them on the resources map above.)

Garment manufacturing—shown in the photo at left—has been particularly successful. In fact, garments are the country's largest export. Most of the employees in the garment industry are women, and today the clothing they produce is sold in countries around the world. This economic progress gives hope for a brighter future for Bangladesh.

Before You Move On
Summarize What are two strategies that Bangladesh has used to develop its economy?

1. **Human-Environment Interaction** What natural resource appears most frequently on the map? How would this natural resource contribute to building industries in Bangladesh?

2. **Write a Paragraph** Pick one resource on the map. Write a paragraph that explains where that resource is located in Bangladesh and why it might be important to Bangladesh's economy. Share your paragraph with classmates.

3. **Make Inferences** How has a lack of education interfered with economic development in Bangladesh?

4. **Make Generalizations** Why do you think that the garment industry has grown so quickly in Bangladesh?

VOCABULARY

For each pair of vocabulary words, write one sentence that explains the connection between the two words.

1. pilgrimage; karma

> *Hindus believe that a pilgrimage to the Ganges River will give them good karma.*

2. shalwar-kameez; sari
3. democracy; federal republic
4. infrastructure; modernization
5. developed nations; developing nations

MAIN IDEAS

6. In what ways does Hinduism unify South Asia? (Section 1.1)

7. For what purpose do Indian women use henna? (Section 1.2)

8. How do foods differ from region to region in South Asia? (Section 1.3)

9. How do music and movies foster a common culture in South Asia? (Section 1.4)

10. Name two challenges facing the government of India today. (Section 2.1)

11. How has outsourcing contributed to rapid growth? (Section 2.2)

12. What is the Golden Quadrilateral, and what cities will it connect? (Section 2.3)

13. What problems are caused by rapid growth in South Asia's cities? (Section 3.1)

14. What are two causes of Pakistan's many changes in government? (Section 3.2)

15. What three factors have contributed to poverty in Bangladesh? (Section 3.3)

CULTURE

ANALYZE THE ESSENTIAL QUESTION

How is diversity reflected in South Asia's cultures?

Critical Thinking: Draw Conclusions

16. What is one basic belief of Hinduism? How does it influence Indian culture?

17. Why do you think established traditions have continued in South Asia today?

18. In what ways is South Asian food similar to and different from American food?

19. How does South Asia's music reflect the region's diversity?

INTERPRET MAPS

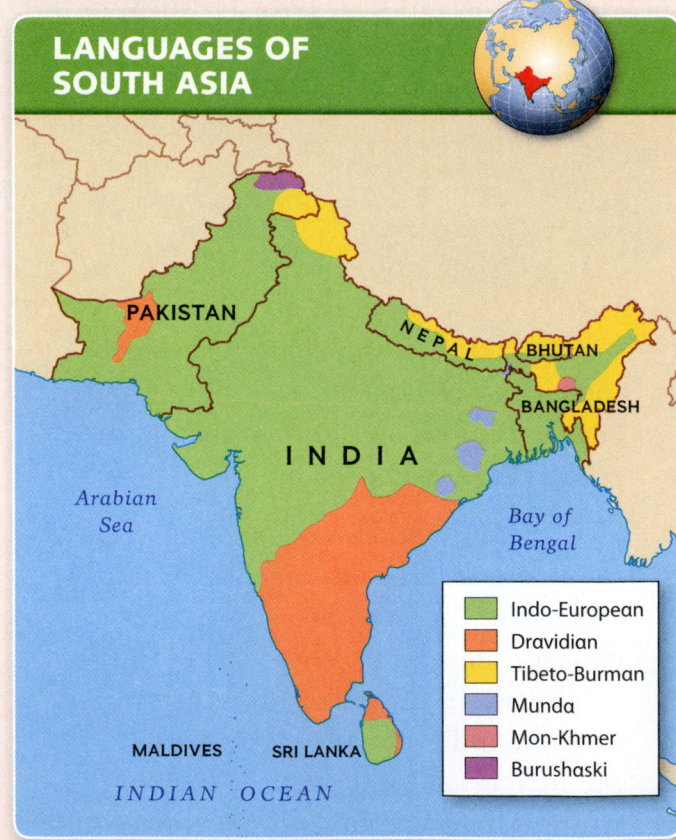

LANGUAGES OF SOUTH ASIA

PAKISTAN

NEPAL

BHUTAN

BANGLADESH

INDIA

Arabian Sea

Bay of Bengal

Legend:
- Indo-European
- Dravidian
- Tibeto-Burman
- Munda
- Mon-Khmer
- Burushaski

MALDIVES   SRI LANKA

INDIAN OCEAN

20. **Region** According to the map, what is the major language family in South Asia?

21. **Make Inferences** Why might there be so many different language families in South Asia?

FOCUS ON INDIA

ANALYZE THE ESSENTIAL QUESTION

Why has India experienced an economic boom?

Critical Thinking: Make Generalizations

22. In what ways is India's government similar to and different from the U.S. government?

23. If India's economy keeps expanding, what effect might that have on the other countries of South Asia?

24. How will the Golden Quadrilateral help to build India's economy?

GOVERNMENT & ECONOMICS

ANALYZE THE ESSENTIAL QUESTION

What are some effects of South Asia's rapid changes?

Critical Thinking: Analyze Cause and Effect

25. How do Pakistan's secondary cities help relieve strain on the major cities?

26. How has Pakistan's unstable government affected its citizens?

INTERPRET CHARTS

LITERACY RATE AND GDP PER CAPITA IN BANGLADESH (1981–2008)				
	1981	1991	2001	2008
Adult Literacy Rate (% people)	29%	35%	47%	55%
GDP Per Capita ($)	$319	$514	$836	$1,335

Source: The World Bank

27. **Analyze Data** What is the relationship between Bangladesh's adult literacy rate and GDP per capita?

28. **Evaluate** How would you explain the relationship between these two economic indicators? Why might GDP per capita affect the literacy rate?

ACTIVE OPTIONS

Synthesize the Essential Questions by completing the activities below.

29. **Write a Letter** Describe in a letter to a friend the different things that bring South Asians together. Also describe in your letter what you think the future holds for South Asia as a region and for specific countries in that region. Use the following tips to help you write your letter. **Read your letter to a classmate.**

> **Writing Tips**
> - Take notes on your ideas about what unifies South Asia.
> - Use an informal tone and style to describe those ideas in your letter.
> - Be sure to include colorful details to keep your friend's interest.

TECHTREK myNGconnect.com For photos of South Asia

30. **Create Visuals** Make a slide show of South Asia photos from the **Digital Library** or other online sources. Pick one photo for each section in the chapter and give it a title. Then write a couple of sentences explaining how the photo represents South Asia. Copy the chart below to help you organize your information.

Photo	How It Represents South Asia
Section 1	
Section 2	

Emerging Markets

 **TECHTREK**

myNGconnect.com For an online infographic and research links about emerging markets

Student Resources

Connect to NG

In this unit, you learned that India, the largest country in South Asia, is an emerging market. That means that India is a developing country experiencing rapid growth and industrialization.

Countries in other regions of the world are undergoing similar economic growth. For example, in the early 1990s, Hungary replaced the government-controlled economic system the country had used under communism with a free market economy. By 1995, Hungary's economy had developed into one of the most prosperous in Eastern Europe.

Compare

- India
- Hungary
- Vietnam
- South Africa

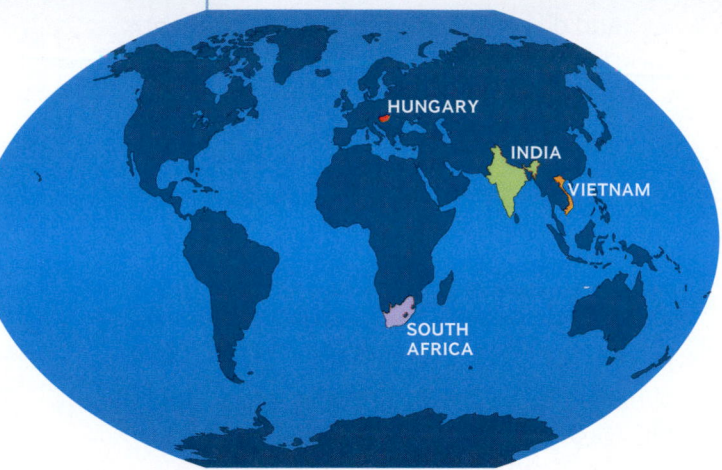

CHARACTERISTICS

Although some of the forces fueling emerging markets may differ from one region to another, they all share the following characteristics.

- They have fast-growing economies that contribute to growth in trade.

- They are sources of new markets for goods and services.

- They offer businesses and individuals in other countries opportunities to make money through investments.

- They inspire creative people to start new types of businesses.

- Their success or failure can affect the countries around them.

- They are changing economic policies that did not work, such as government-controlled industries.

KEY MEASUREMENTS

Investors use several key measurements to evaluate emerging markets. Here are some of them:

- size of the country's population

- standard of living: the financial health of the country, which is often measured by the Gross National Income (GNI) per capita (per person)

- adult literacy rate: a measure of knowledge and education

- life expectancy (at birth): a snapshot of health and expected length of life

The graphs on the next page use these measurements to show the growth in the economies of India and Hungary between 1990 and 2009. Compare the data in the graphs and answer the questions that follow about these emerging markets.

EMERGING MARKETS IN INDIA AND HUNGARY
(1990 AND 2009)

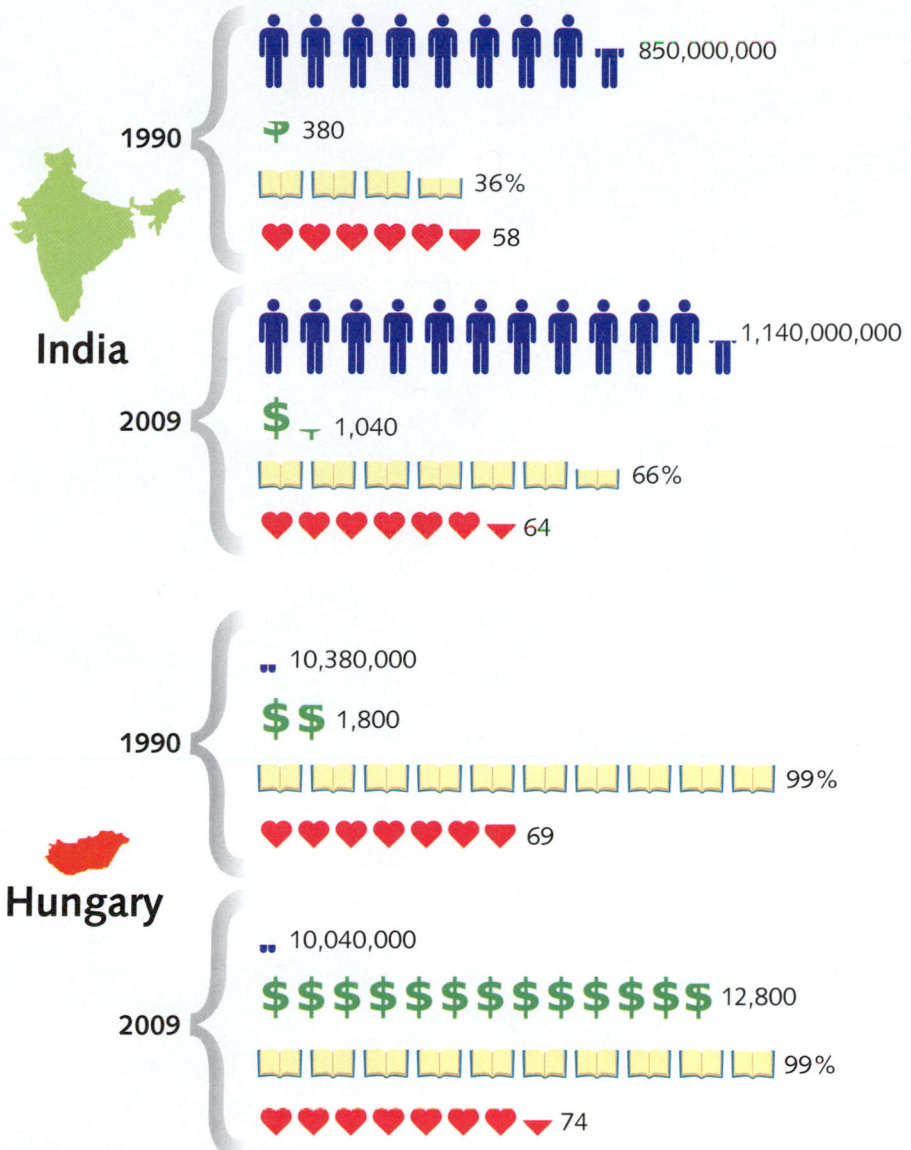

India

1990
- 850,000,000 (Population)
- $ 380 (GNI)
- 36% (Adult Literacy Rate)
- 58 (Life Expectancy)

2009
- 1,140,000,000 (Population)
- $ 1,040 (GNI)
- 66% (Adult Literacy Rate)
- 64 (Life Expectancy)

Hungary

1990
- 10,380,000 (Population)
- 1,800 (GNI)
- 99% (Adult Literacy Rate)
- 69 (Life Expectancy)

2009
- 10,040,000 (Population)
- 12,800 (GNI)
- 99% (Adult Literacy Rate)
- 74 (Life Expectancy)

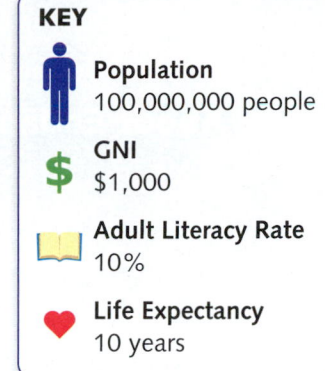

KEY

Population	100,000,000 people
GNI	$1,000
Adult Literacy Rate	10%
Life Expectancy	10 years

U.S. Benchmark
Population: 307,006,550
GNI: $47,240
Adult Literacy Rate: 89%
Life Expectancy: 74 years

Sources: World Bank, Encyclopedia Brittanica, CIA World Factbook

---

ONGOING ASSESSMENT
RESEARCH LAB GeoJournal

1. **Compare and Contrast** Which measurement increased the most in both countries between 1990 and 2009?

2. **Analyze Data** How might population growth affect economic growth in India and Hungary?

Research and Create Graphs Research the emerging markets of Vietnam in Southeast Asia and South Africa in Sub-Saharan Africa. Create a graph for each country, showing its most recent key emerging market measurements. Then compare the data for all four countries—India, Hungary, Vietnam, and South Africa. Based on the key measurements, which country's economy do you think might emerge as the strongest?

Active Options

TECHTREK

myNGconnect.com For photos of South Asia's culture and Guided Writing

 Digital Library

 Student Resources

 **Magazine Maker**

ACTIVITY 1

Goal: Extend your understanding of endangered animals.

Write About Endangered Species

National Geographic explorer Shafqat Hussain created Project Snow Leopard to protect the endangered snow leopard species in South Asia. With a partner, research one of the following endangered animals from the region:

- Bengal tiger
- Indian rhinoceros
- Ganges River dolphin
- Indus River dolphin
- Indian elephant
- Lion-tailed macaque

Create a presentation that tells about the animal, why the animal is endangered, and why it is important to protect the animal. Present your findings to the class. **Go to Student Resources** for Guided Writing support.

snow leopard

ACTIVITY 2

Goal: Research South Asia's culture.

Create *South Asia Today* Magazine

You and your classmates are on the staff of the magazine *South Asia Today*. Work together and use the **Magazine Maker** to create next month's issue. Go to the **Digital Library** to find images and information on:

- clothing
- festivals
- food
- movies
- music
- schools
- shopping
- sports

ACTIVITY 3

Goal: Review South Asia with a game.

Host a Geography Bee

Write a question about South Asia and its answer on a card. Your teacher will collect the questions and ask them to the class. Stand to answer the question. The last person standing wins!

NATIONAL GEOGRAPHIC
World Cultures and Geography

GEO

Explore
East Asia
with NATIONAL GEOGRAPHIC

MEET THE EXPLORER

NATIONAL GEOGRAPHIC

Using advanced tools such as satellite imagery and remote sensors, Emerging Explorer Albert Yu-Min Lin searches for the tomb of Genghis Khan. His search leaves the land undisturbed and respects Mongolian beliefs and tradition.

INVESTIGATE GEOGRAPHY

The upper levels of a temple overlook snowcapped Mount Fuji in central Japan. Mount Fuji is a volcano that has been dormant for over 300 years. A shrine on its peak (12,388 feet) attracts thousands of climbers every summer.

CONNECT WITH THE CULTURE

With a population of over 1.3 billion, China is the world's most populous country. A large and growing population creates challenges, such as polluting auto traffic and overcrowded public transportation. Bicycle use helps to ease the problem.

STEP INTO HISTORY

Construction on the Great Wall in China began in the 7th century. It was continually rebuilt through the 1500s. Its total length runs nearly 4,500 miles.

CHAPTER 13

East Asia
GEOGRAPHY & HISTORY

PREVIEW THE CHAPTER

KEY VOCABULARY

- mainland
- basin
- alluvium
- loess
- archipelago
- eruption
- typhoon
- demilitarized zone (DMZ)
- steppe
- semiarid
- ger
- animal-borne
- carapace

ACADEMIC VOCABULARY
erode, refuge

TERMS & NAMES

- Kunlun Mountains
- Tibetan Plateau
- Chang Jiang
- Huang He
- North China Plain
- Ring of Fire
- Kanto Plain
- Gobi desert

KEY VOCABULARY

- dynasty
- dynastic cycle
- empire
- terra cotta
- ethical system
- moral
- caravan
- maritime
- tribute
- expedition
- occupy
- collective farm
- elite

ACADEMIC VOCABULARY
ban, barter

TERMS & NAMES

- Shang
- Zhou
- Qin
- Shi Huangdi
- Great Wall of China
- Han
- Silk Roads
- Confucianism
- Zheng He
- Jiang Jieshi
- Mao Zedong
- Cultural Revolution

KEY VOCABULARY

- samurai
- shogun
- competitive
- zaibatsu
- rivalry
- celadon
- retreat
- armistice

ACADEMIC VOCABULARY
dissolve, seize

TERMS & NAMES

- Silla
- Koguryo
- Paekche
- Koryo
- Choson
- 38th parallel
- Cold War

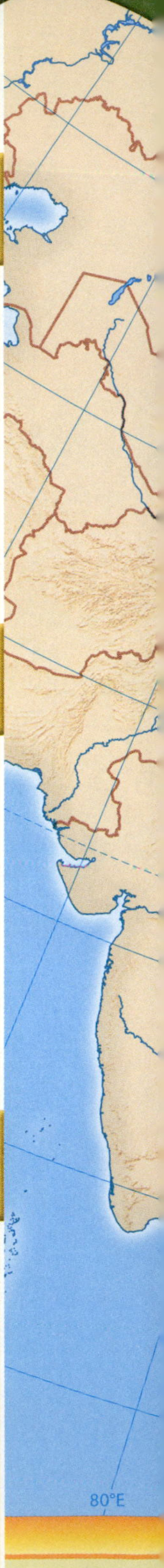

80°E

Chinese panda

60°E 70°E 80°E 90°E 100°E 110°E 120°E 130°E 140°E 150°E 160°E

Arctic Circle

A

Ertis R.

Sea of Okhotsk

50°N

B

MONGOLIA

Selenge R.

Ulaanbaatar

Amur R.

Songhua R.

JAPAN

40°N

C

Tarim R.

Liao R.

Beijing

NORTH KOREA

Pyongyang

Sea of Japan (East Sea)

Tokyo

D

Indus R.

CHINA

Huang He (Yellow R.)

Seoul

SOUTH KOREA

Yellow Sea

30°N

Ganges R.

Brahmaputra R.

Chengdu

Shanghai

East China Sea

PACIFIC OCEAN

E

Mt. Everest 29,035 ft. (8,850 m)

Chongqing

Chang Jiang (Yangtze R.)

Tropic of Cancer

20°N

Irrawaddy R.

Kunming

Mekong R.

Xi R.

Taipei

TAIWAN

F

Hong Kong

Philippine Sea

Bay of Bengal

South China Sea

N W E S

10°N

G

INDIAN OCEAN

0 400 800 Miles

0 400 800 Kilometers

0°

H

Equator

90°E 100°E 110°E 120°E 130°E

2 3 4 5 6

SECTION 1 GEOGRAPHY

1.1 Physical Geography

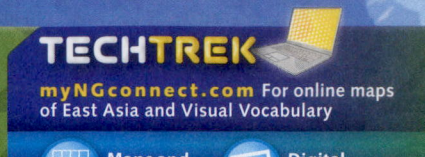

TECHTREK
myNGconnect.com For online maps of East Asia and Visual Vocabulary

Maps and Graphs | Digital Library

EAST ASIA PHYSICAL

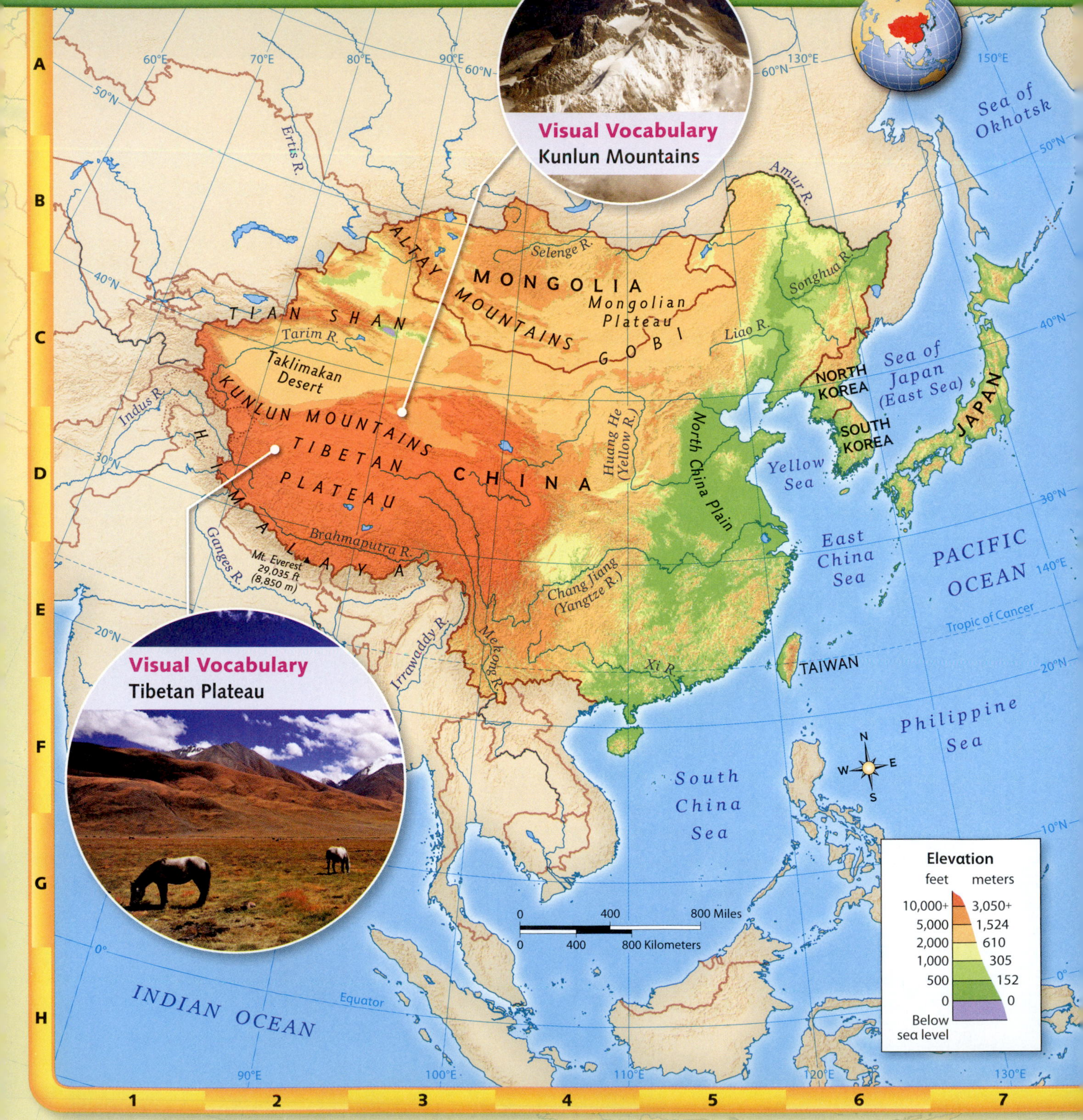

Visual Vocabulary
Kunlun Mountains
Visual Vocabulary
Tibetan Plateau
A
B
C
D
E
F
G
H
1
2
3
4
5
6
7
60°E
70°E
80°E
90°E
100°E
110°E
120°E
130°E
140°E
150°E
60°N
50°N
40°N
30°N
20°N
10°N
0°
Ertis R.
Amur R.
Sea of Okhotsk
Selenge R.
ALTAY MOUNTAINS
MONGOLIA
Mongolian Plateau
Songhua R.
TIAN SHAN
Tarim R.
GOBI
Liao R.
Taklimakan Desert
KUNLUN MOUNTAINS
TIBETAN PLATEAU
CHINA
Huang He (Yellow R.)
North China Plain
NORTH KOREA
SOUTH KOREA
Sea of Japan (East Sea)
JAPAN
Yellow Sea
Indus R.
HIMALAYA
Ganges R.
Brahmaputra R.
Mt. Everest 29,035 ft (8,850 m)
Chang Jiang (Yangtze R.)
East China Sea
PACIFIC OCEAN
Tropic of Cancer
Irrawaddy R.
Mekong R.
Xi R.
TAIWAN
Philippine Sea
South China Sea
N
S
E
W
0 400 800 Miles
0 400 800 Kilometers
Equator
INDIAN OCEAN
Elevation
feet meters
10,000+ 3,050+
5,000 1,524
2,000 610
1,000 305
500 152
0 0
Below sea level

CLIMATE

Climate Regions

- Humid Equatorial–Short dry season
- Humid Equatorial–Long dry season
- Dry–Semiarid
- Dry–Arid
- Humid Temperate–No dry season
- Dry Winter
- Humid Cold–No dry season
- Humid Cold–Dry winter
- Unclassified Highlands

POPULATION DENSITY

One dot represents 100,000 people

Main Idea East Asia's landforms, bodies of water, and climate influence where people live.

East Asia includes China and Mongolia, which stretch across the <mark>mainland</mark>, or continental landmass, of Asia. The region also contains the island countries of Japan and Taiwan as well as North Korea and South Korea on the Korean Peninsula.

Landforms and Water

High mountains and plateaus, such as the **Kunlun Mountains** and **Tibetan Plateau**, cover much of East Asia. The Gobi desert covers parts of Mongolia and China.

Major rivers, such as the **Chang Jiang** (chahng jyahng) in China, flow through East Asia. River <mark>basins</mark>, the low areas drained by these rivers, support large populations. Many East Asians also live near coastal areas and on low-lying plains near the Pacific Ocean. People throughout the region depend on the Pacific for fish, trade, and transportation.

Climate

East Asia's climate ranges from tropical, in parts of China, South Korea, and Japan, to desert, in much of Mongolia. Monsoons greatly influence the region's climate. These seasonal winds bring hot, rainy summers and cool, dry winters to some parts of East Asia.

Before You Move On

Summarize In what ways do landforms, bodies of water, and climate help determine where people live in East Asia?

ONGOING ASSESSMENT

MAP LAB  GeoJournal

1. **Interpret Maps** Use the physical map to locate the Tibetan Plateau on the population density and climate maps. Why do you think few people live on the plateau?

2. **Compare and Contrast** Compare the climate of inland East Asia with that of its coastal areas. What role might climate play in the population distribution of these areas?

1.2 China's Rivers and Plains

TECHTREK

myNGconnect.com For an online
map and photos of China's rivers and cities

 Maps and
Graphs

 Digital
Library

Main Idea China's main river systems support large populations.

China's two largest river systems begin in the melting snow and ice of the country's western mountains and plateaus. As the waters flow down, they **erode**, or wear away, the land and deposit alluvium in river basins. **Alluvium** is soil carried by flowing water and is ideal for farming.

River Systems

The Chang Jiang, or Yangtze, is China's longest river. It passes through the Sichuan (sheesh whan) Basin and the Chang Jiang Plain. China's second longest river, the **Huang He** (hwahng huh), flows through the **North China Plain**. The Huang He is often called the Yellow River because it flows through an arid region that contains **loess** (less), or fine yellowish soil. The wind carries the loess and deposits it in the river.

China's river valleys and fertile plains provide rich agricultural land for growing rice and wheat. They also support many cities. The capital, Beijing (bay ghing), is in the North China Plain. Shanghai, the country's largest city, is at the mouth of the Chang Jiang on the East China Sea.

Controlling the Rivers

Frequent flooding of the rivers leaves behind deposits of fertile soil. Over the centuries, however, the flooding has also caused millions of deaths. Chinese engineers have built barriers such as dams and levees to hold the water back.

To connect China's major rivers, which flow from west to east, the ancient Chinese built a structure called the Grand Canal, which flows from north to south.

> **Critical Viewing** Boats on a branch of the Chang Jiang sail past skyscrapers in Shanghai. Based on the photo, how do the people in the city use the river?

CHINA'S RIVERS AND CITIES

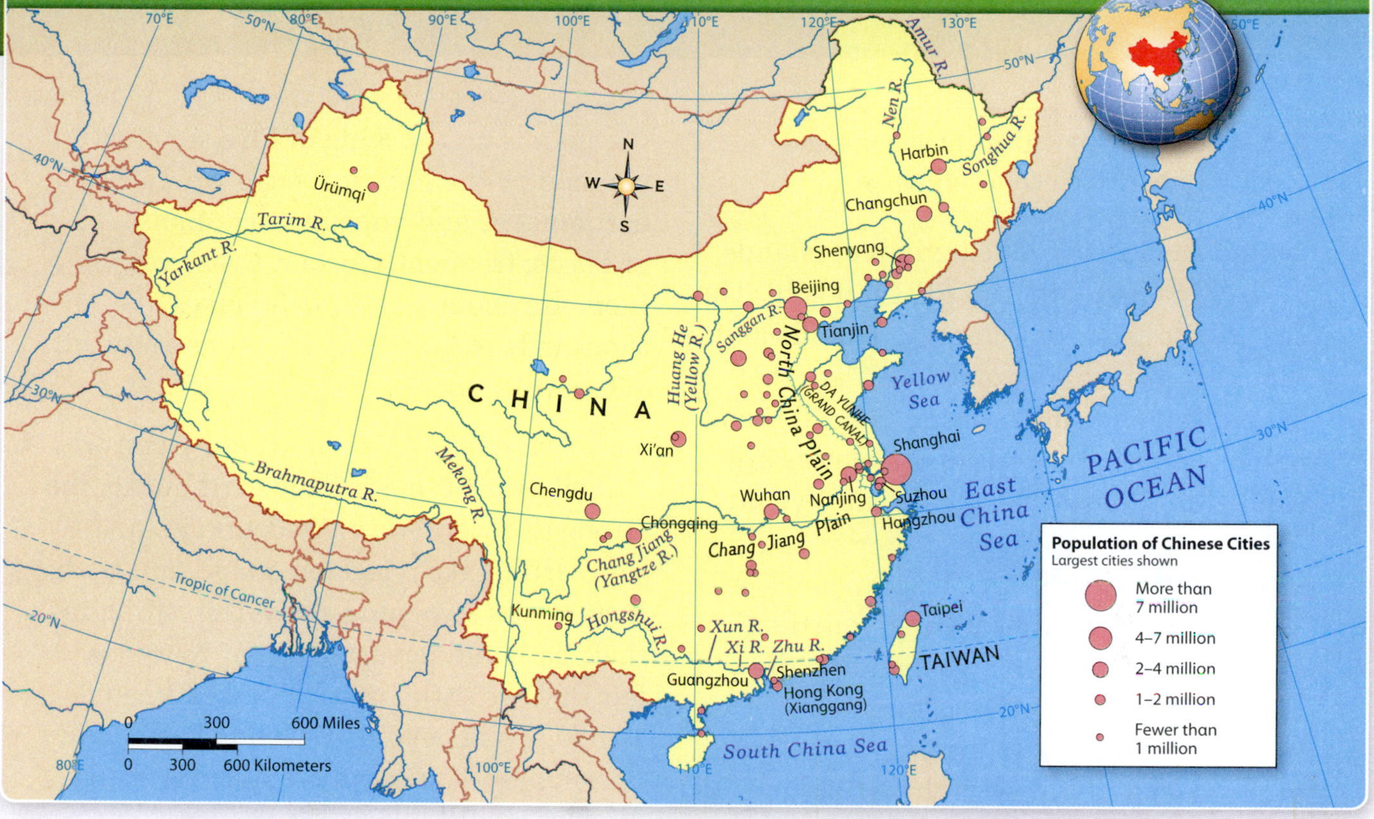

Population of Chinese Cities
Largest cities shown

- More than 7 million
- 4–7 million
- 2–4 million
- 1–2 million
- Fewer than 1 million

The roughly 1,100-mile-long canal—the longest in the world—links Beijing in the North China Plain to Hangzhou, just south of Shanghai. Barges travel along the canal, carrying bulk goods such as coal and gravel.

Before You Move On

Monitor Comprehension In what ways do China's rivers support large populations?

ONGOING ASSESSMENT

MAP LAB   GeoJournal

1. **Location** Study the map. How does the Grand Canal promote the transport of goods in China?

2. **Evaluate** Use the map above and the physical map in Section 1.1 to identify the landform where many of China's large cities are located. Why do you think the cities developed there?

1.3 The Island Arc of Japan

TECHTREK

myNGconnect.com For an online earthquake map and photos of Japan

 **Maps and Graphs**    **Digital Library**

Main Idea Japan is a mountainous island country with limited natural resources.

Japan is a closely related group of islands, or **archipelago** (ahr kuh PEH luh goh). The country consists of four main islands and thousands of smaller ones in an arc that extends about 1,400 miles in length. From north to south, the four main islands are Hokkaido, Honshu, Shikoku, and Kyushu.

A Mountainous Land

Mountains cover almost three-fourths of Japan and run like a spine down the middle of the four main islands. The Japanese islands were formed by the top part of a mountain range thrusting up from the bottom of the Pacific Ocean.

Japan is on the **Ring of Fire**, an area rimmed by the Pacific Ocean where earthquakes and volcanic **eruptions**, or blasts, frequently occur. Every year there are about 1,500 earthquakes and thousands of eruptions from the country's active volcanoes.

On March 11, 2011, northern Japan was rocked by a 9.0-magnitude earthquake, the strongest recorded in the country's history. The earthquake triggered a tsunami, with huge waves sweeping away everything in their path and devastating entire towns. A few days after the disaster, an estimated 12,000 people were believed to be dead or missing. Most experts predicted that the number would rise.

> **Critical Viewing** A tsunami wave washes over a street in the town of Miyako. What details in the photo help convey the size of the wave?

In addition to destroying countless homes and businesses, the earthquake and tsunami severely damaged several of Japan's nuclear power plants. Many people feared that high levels of radiation from these plants might be released into the environment and harm nearby residents.

A Crowded Country

Because so much of Japan is covered with mountains and forests, the people crowd on its plains. About 80 percent of Japan's nearly 130 million people live on the plains of Honshu.

The **Kanto Plain** lies to the east of the Japanese Alps, which cross the central part of Honshu. The flat land of the Kanto Plain is good for agriculture and industry. Tokyo, Japan's capital and largest city, is located there.

Limited Natural Resources

Japan's most important natural resource is fish, caught in the Pacific Ocean and Sea of Japan. The country's fishing industry is one of the largest in the world. The seas are also essential to trade, which is vital to Japan's economy.

Because Japan has only small quantities of mineral resources, the country must import raw materials such as iron ore and lead. Japan also imports most of its energy resources, including petroleum and coal. Japan uses these imports in its industries and then exports finished products, such as automobiles and electronics.

Before You Move On

Summarize How do Japan's mountains and limited natural resources affect life in Japan?

EARTHQUAKE 2011

MAP TIP The map shows the location, or epicenter, of Japan's 2011 earthquake and its larger aftershocks, a series of quakes that occur after the main one.

Sea of Okhotsk
Hokkaido
Sapporo
Akita · Miyako
Sea of Japan (East Sea)
JAPAN
Sendai
Honshu
Nagano
Kanto Plain
Japanese Alps
Tokyo
Kyoto
Hiroshima · Kōbe · Osaka
Mt. Fuji 12,388 ft (3,776 m)
Nagasaki
Kyushu
Shikoku
East China Sea
Ryukyu Islands
Okinawa
PACIFIC OCEAN
Tropic of Cancer

ASIA | NORTH AMERICA
JAPAN
PACIFIC OCEAN
AUSTRALIA | SOUTH AMERICA
Ring of Fire

0 — 200 — 400 Miles
0 — 200 — 400 Kilometers

Earthquake Magnitude
March 11–April 11, 2011
7.2–9.0
6.2–7.1
5.7–6.1
5.0–5.6
Epicenter

ONGOING ASSESSMENT

MAP LAB

 GeoJournal

1. **Location** Locate the area hit by the earthquake on the inset map. Why might this area be prone to earthquake activity?

2. **Make Inferences** Study the map of Japan's 2011 earthquake. Based on the map, between which two major cities do you think most of the devastation occurred?

1.4 The Korean Peninsula

TECHTREK

myNGconnect.com For an online map of Korea and Guided Writing

Maps and Graphs

Student Resources

Main Idea North Korea and South Korea have a similar geography, but these two countries on the Korean Peninsula remain divided.

The Korean Peninsula was divided into two countries after World War II. The two countries are the Democratic People's Republic of Korea, or North Korea, and the Republic of Korea, or South Korea. North Korea has a Communist government, while South Korea is democratic.

Geography and Climate

Like Japan, the Korean Peninsula is largely mountainous, with coastal plains and river valleys where most of the people live. With a population of about 50 million, South Korea has more than twice as many people as North Korea and is much more densely populated.

In South Korea, the southern and western plains are important areas for farming. The country's climate supports agriculture, with hot, humid summers and cold, dry winters. South Korea's two main industrial cities—Seoul, the capital, and Busan—are also located on the plains.

North Korea is more mountainous than South Korea. As a result, North Korea has less agricultural land than its neighbor. In general, North Korea's climate is colder and wetter than South Korea's, but **typhoons**, or hurricanes, can bring heavy rain and flooding to both countries.

Natural Resources and Industries

South Korea's natural resources include iron ore, copper, and coal. However, because these resources are limited, South Korea imports much of the raw materials it needs for industrial production. South Korea has a highly industrialized economy. Shipbuilding and steel are among the country's major industries, along with the manufacture of electronics and automobiles.

North Korea's natural resources include coal, magnesite, and iron ore. Unlike South Korea, many of North Korea's industries produce military equipment.

> **Critical Viewing** Steel is a major industry in South Korea. What details in the photo suggest that this plant in Busan is thriving?

A Divided Land

From 1950 to 1953, the two Koreas fought in a conflict known as the Korean War. North Korea initiated the war by invading South Korea with the goal of taking over the peninsula. At the end of the war, the two countries remained divided. You will learn more about the Korean War in Section 3.

After the war, a **demilitarized zone (DMZ)**, or neutral area, was created between North Korea and South Korea. The DMZ is about 150 miles long and 2.5 miles wide. Soldiers still patrol the land, which has become a wilderness.

Over time, the isolation of the DMZ has created a **refuge**, or safe place, for large numbers of wildlife. Animal species, such as rare cranes, tigers, and bears live there. Scientists once thought that these animals had died out on the Korean Peninsula.

Before You Move On

Monitor Comprehension In what ways are North Korea and South Korea alike and different?

LAND USE AND NATURAL RESOURCES

Forest
Woodland
Grassland
Mixed-use, including crops
Cropland
Intensive cropland
Coal
Copper
Gold
Iron ore
Magnesite
Demilitarized Zone

NORTH KOREA
Pyongyang
Korea Bay
Sea of Japan (East Sea)
Seoul
SOUTH KOREA
(Pusan) Busan
Korea Strait

N
W   E
S

0   50   100 Miles
0   50   100 Kilometers

40°N
35°N
125°E
130°E

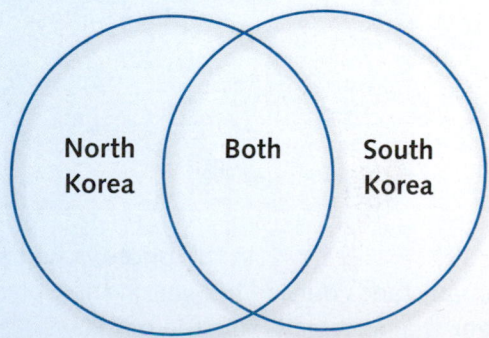

ONGOING ASSESSMENT

WRITING LAB   GeoJournal

Compare and Contrast Write a paragraph in which you compare and contrast North Korea and South Korea. Use a Venn diagram like the one below to organize information about each country's geography, climate, natural resources, and industries. Exchange your finished paragraph with a partner and discuss your ideas. Go to **Student Resources** for Guided Writing.

North Korea   Both   South Korea

1.5 Mongolia's Desert Landscape

TECHTREK
myNGconnect.com For an online physical map of Mongolia and photos of gers

Maps and Graphs   Digital Library

Main Idea Mongolians have adapted to life in a harsh, dry environment.

Mongolia is a landlocked country located between northern China and eastern Russia. The **Gobi desert**, the largest desert in Asia, covers much of southern Mongolia and extends into China.

High and Dry

Mountains, plateaus, **steppes**, or dry grassland, and desert make up the physical geography of Mongolia. All of the country's land is 1,700 feet or more above sea level.

Most of the **semiarid**, or somewhat dry, land gets fewer than 20 inches of rain, and much of it falls in July and August. In the Gobi desert, only about seven inches of rain fall every year. Some parts of the Gobi receive no rain at all. *Gobi* means "place without water." The extreme dryness and strong winds that sometimes blow through the desert result in blinding dust storms.

Summers in Mongolia are generally short and hot, while winters are long and cold. In the Gobi desert, temperatures can reach 113°F in the summer and -40°F in the winter. In addition, temperatures can rise and fall in the desert by as much as 60 degrees within the same day.

People and Environment

In Mongolia's harsh environment, not much livable space is available. Nearly half of the country's roughly three million people live in urban areas. About one-third of the population lives in the capital city of Ulaanbaatar (oo lahn BAH tawr).

Less than one percent of the land in Mongolia is arable. As a result, most of the people in rural areas are livestock herders. Many lead a nomadic life. They move from place to place with their sheep or goats, looking for good pastureland. Some herders live in **gers**. In Central Asia, these tents are called yurts.

> **Visual Vocabulary** In this photo, a nomadic Mongolian woman stands outside her ger on the Gobi desert. A **ger** is a portable tent made of felt.

MONGOLIA PHYSICAL

Preserving the Desert

As you have seen, the Mongolian people have learned to adapt to their desert environment. They also want to preserve it. The Gobi is expanding by a process called desertification, in which fertile land changes to desert. Much of this expansion is a result of human activities, including allowing livestock to overgraze pastureland. The Mongolians are trying to stop the process by reducing the number of livestock that graze the pastureland and by cutting down fewer trees.

Parts of the Gobi desert have also been turned into nature reserves and national parks. One of the parks in the southern part of the Gobi contains land of great interest to paleontologists, scientists who study prehistoric life. The area is one of the richest sources for dinosaur fossils in the world. In fact, the first dinosaur eggs were discovered here in the 1920s. The Gobi's harsh climate and remote location have helped protect and preserve the fossils for millions of centuries.

Before You Move On

Monitor Comprehension How have Mongolians adapted to their harsh, dry environment?

ONGOING ASSESSMENT

MAP LAB GeoJournal

1. **Region** Study the physical map of Mongolia. What does the map tell you about sources of water in the country?

2. **Synthesize** Compare the map on this page with the climate and population maps in Section 1.1. In what ways do climate and land features explain the population density of Mongolia?

3. **Draw Conclusions** Why do you think most of Mongolia's cities are located in the north?

1.6

SECTION **1** GEOGRAPHY

NATIONAL GEOGRAPHIC

TECHTREK

myNGconnect.com For photos of an
Explorer's work and an Explorer Video Clip

Digital
Library

Tracking Aquatic
Wildlife
with Katsufumi Sato

> **Main Idea** Animal-borne recorders help scientists study how
> animals behave and learn how to protect them.

Tracking Technology

Since the 1990s, scientists have been attaching electronic
data recorders to animals to gather information about
everything from how fast the animal moves to how much it
eats. These recorders are **animal-borne**, which means that
the devices are carried by the animals themselves. Many of
the recorders include cameras that take photos and videos of
what the animal sees under natural conditions.

Emerging Explorer Katsufumi Sato is continually looking
for ways to improve the technology, noting "We always
work to perfect the instruments . . . and find better ways
to attach and retrieve them." For example, when he first
started studying sea turtles, he used a harness to attach
the recorder. However, the harness slowed the turtle down
as it swam. Eventually, Sato figured out how to glue the
instrument to the turtle's **carapace**, or shell, to make it
lighter and less awkward for the animal to carry.

myNGconnect.com

For more on Katsufumi
Sato in the field today

Unexpected Findings

Sato tracks sea creatures of all kinds. One of his research subjects is the cormorant, a large seabird that dives underwater to catch fish in its bill. For years, Japanese fishermen have been convinced that cormorants steal some of the fish in their nets. As a result, the fishermen have hunted and killed the birds. By attaching tiny animal-borne recorders to the birds, Sato hopes to learn the exact amount and type of fish the birds actually eat. The cormorants may not be as destructive as the fishermen believe.

Sato has also studied the loggerhead sea turtle. Many scientists thought that these turtles were dying because they ate the plastic bags that litter the ocean, mistaking them for jellyfish. The scientists believed that the turtles swallowed the bags and died of suffocation or starvation. However, animal-borne cameras proved that the turtles actually swim away from the plastic bags, recognizing that they are not food. This finding told scientists that they need to continue researching to learn what is really causing the turtles' declining numbers.

In another surprising study, Sato discovered that king penguins regulate how much air they inhale before a dive. They inhale more air before a deep dive and less air before a shallower dive. This behavior suggests that the penguins may be capable of planning their actions.

Before You Move On

Make Inferences How might animal-borne recorders be used to save animals?

Critical Viewing King penguins, similar to those Sato studied, gather on a bay north of Antarctica. What data might one of Sato's devices record about the penguins in this environment?

TECHTREK

myNGconnect.com For photos of ancient Chinese artifacts

Digital Library

Main Idea Powerful families ruled and shaped ancient China for about 2,000 years.

In the 2100s B.C., farming settlements developed along the Huang He in eastern China. Over time, some of these settlements grew into cities and marked the beginning of an early civilization. Because Chinese culture developed from this early society, China is said to have the world's longest continuous civilization.

Shang and Zhou Dynasties

Around 1766 B.C., kings from the **Shang** family took over some of the cities along the Huang He. The Shang established a **dynasty**, a series of rulers from the same family. Eventually, the Shang ruled much of the area along the North China Plain.

Shang society was largely agricultural, but the leaders also built large walled cities. The society was structured, with nobles at the top and peasants at the bottom. The Shang used horse-drawn chariots to defend themselves against invaders. They also developed a system of writing, which helped unify their lands.

The **Zhou** (joh) people often fought with the Shang. Around 1050 B.C., they defeated the Shang and established their own dynasty. Throughout much of their long rule—longer than any other dynasty in Chinese history—the Zhou rulers waged war against invaders from the north and west. The Zhou also fought among themselves. Zhou kings had placed lords in charge of different parts of the region. These lords battled each other to acquire more land. The continual warfare led to great disorder in Chinese society.

Chinese rulers believed that the length of their dynasty was determined by the gods worshiped in ancient China. The pattern in the rise and fall of dynasties came to be known as the **dynastic cycle**, which is shown on the next page.

The Qin Dynasty

The **Qin** (chihn) Dynasty gained control of China in 221 B.C. *China* is thought to come from the name of this dynasty. Qin ruler **Shi Huangdi** (shee hwahng dee) strengthened the central government and expanded the lands under his control. He brought these lands together to form an **empire**, a group of states ruled by a single strong ruler, and became China's first emperor. Shi Huangdi unified his empire by building a system of roads and standardizing the Chinese currency.

1766 B.C.
The Shang establish a dynasty on the North China Plain.

1600 B.C.

Shang Dynasty oracle bone used to pose questions to the gods

Bronze vessel from the Zhou Dynasty

900 B.C.

1050 B.C.
The Zhou Dynasty begins its long rule.

THE DYNASTIC CYCLE

1. The people believe that the gods approve of the new dynasty.

2. The dynasty weakens.

3. Disasters occur.

4. The people believe that the gods no longer approve of the dynasty.

5. The dynasty is overthrown.

6. A new dynasty re-establishes order.

Shi Huangdi also began construction on the **Great Wall of China** to protect against invaders from the north. Thousands were forced to labor on the huge project. Later rulers continued to expand the wall until it stretched about 4,500 miles.

Qin rule ended four years after Shi Huangdi's death in 210 B.C. Archaeologists discovered thousands of terra cotta, or baked clay, warriors buried near his tomb in 1974. Experts believe these life-sized statues were created to guard his tomb.

The Han Dynasty

The **Han** Dynasty came to power in China in 206 B.C. and lasted until A.D. 220. Han leaders expanded the empire and established a strong central government. During this time, China began to trade with Europe and Central Asia along routes known as the **Silk Roads**. The goods and ideas traded helped create a prosperous civilization and an advanced culture in China. Many Chinese still call themselves "the people of the Han." You will learn more about the Silk Roads in Section 2.3.

Before You Move On

Monitor Comprehension How were the ancient Chinese dynasties alike, and how did they differ?

A terra cotta warrior found in Shi Huangdi's tomb

221 B.C.
The Qin Dynasty forms an empire.

200 B.C.

206 B.C.
The Han Dynasty expands the Chinese Empire.

ONGOING ASSESSMENT

READING LAB  GeoJournal

1. **Make Inferences** The Qin Dynasty ended shortly after the death of Shi Huangdi. What does that fact suggest about the ruler that followed him?

2. **Form and Support Opinions** Which dynasty do you think had the greatest impact on Chinese society? Explain your reasons.

3. **Evaluate** Why did the Chinese people probably accept the overthrow of an old dynasty and the rise of a new one?

NATIONAL GEOGRAPHIC

Photo Gallery • Terra Cotta Army

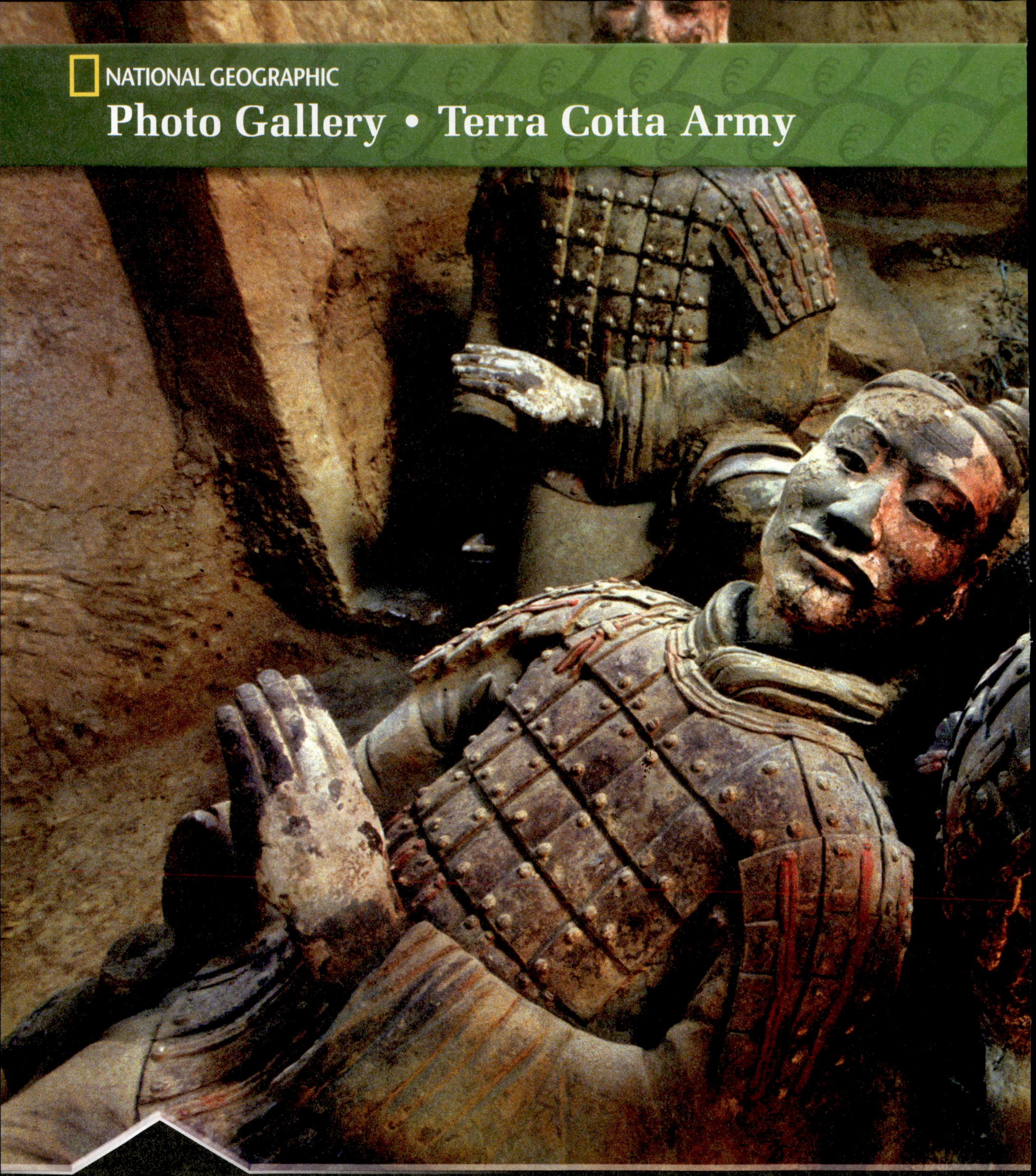

For more photos from the National Geographic Photo Gallery, go to the **Digital Library** at myNGconnect.com.

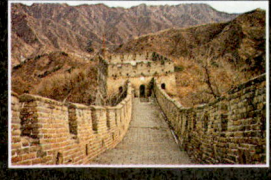

Great Wall of China

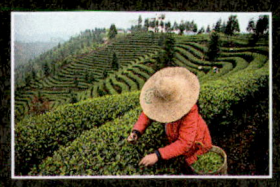

Tea farm, China

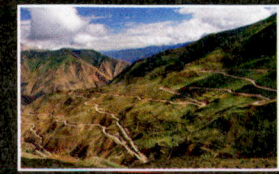

Tibet's Zar Gama Pass

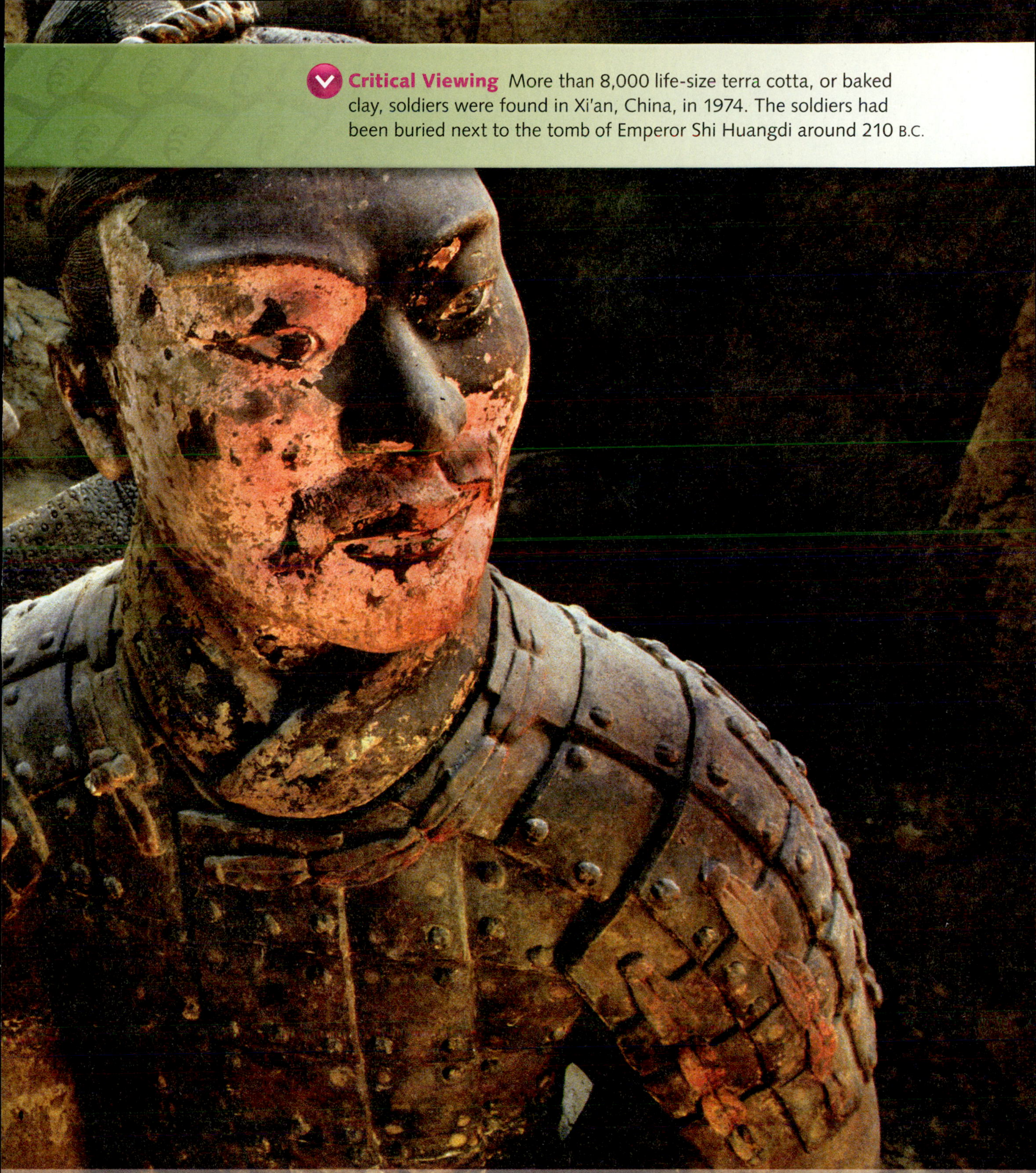

Critical Viewing More than 8,000 life-size terra cotta, or baked clay, soldiers were found in Xi'an, China, in 1974. The soldiers had been buried next to the tomb of Emperor Shi Huangdi around 210 B.C.

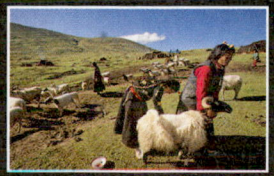

Nomads in Tibet

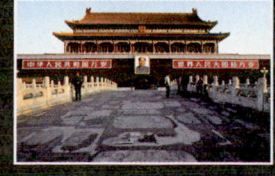

Forbidden City, Beijing

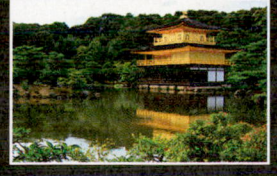

Kyoto's Golden Temple

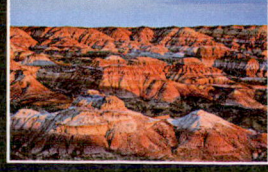

Gobi desert

2.2 Confucianism

TECHTREK

myNGconnect.com For photos of Confucian temples

 **Digital Library**

Main Idea The teachings of Confucius have influenced Chinese society for more than 2,000 years.

As you have read, the Chinese people endured long periods of conflict and disorder during the Zhou Dynasty. Confucius, a Chinese scholar and teacher born during that period, wanted to bring peace to his country. He developed ideas about the proper conduct of rulers and subjects and taught these ideas to others. His teachings form the basis of an ethical system called **Confucianism**. An **ethical system** teaches **moral**, or right, behavior. For more than 2,000 years, Confucius' teachings guided Chinese thought.

Teachings

Confucius' teachings are based on five relationships: father and son; elder brother and younger brother; husband and wife; friend and friend; and ruler and subject. Confucius taught that if these relationships were conducted respectfully, peace would be restored to society. The teachings below illustrate Confucius' emphasis on the importance of social relationships and study.

This Confucian temple in Qufu, China—Confucius' birthplace in 551 B.C.—was built shortly after the teacher's death.

Don't worry if people don't recognize your merits; worry that you may not recognize theirs. (Analects 1.16)

To study without thinking is futile [useless]. To think without studying is dangerous. (Analects 2.15)

What you do not wish for yourself, do not do to others. (Analects 15.24)

Legacy

Confucius died believing that he had been a failure. In his lifetime, his teachings did not change Chinese society. However, his students wrote down his teachings and gathered them into a book called the *Analects* for future generations to read. In time, Han Dynasty rulers adopted Confucius' teachings and used them to select government officials based on their accomplishments rather than wealth.

Confucianism continued to influence Chinese society until the Communist Party seized power in 1949. The Communists **banned**, or outlawed, Confucianism because they considered it a religion. The government ended its ban in 1977. Since then, Confucianism has regained much of its influence. It is taught in schools, and many leaders have rediscovered Confucius' wisdom. Some officials have begun to use the teachings to guide their work and behavior.

Before You Move On

Summarize What are the basic teachings of Confucianism?

ONGOING ASSESSMENT

PHOTO LAB

 GeoJournal

1. **Analyze Visuals** What mood do you think the temple in the photo is meant to inspire?

2. **Make Inferences** Study the photo. Why do you think so many people visit the temple?

3. **Analyze Primary Sources** Reread the third sample of Confucius' teachings. What Confucian values does this teaching reflect?

2.3 Silk Roads and Trade

TECHTREK

myNGconnect.com For an online map of the Silk Roads and photos of trade items

 Maps and Graphs

 Digital Library

Main Idea Goods were traded and ideas spread on the Silk Roads, which connected China with much of the world.

You have learned that the Silk Roads were a series of trade routes that began during the Han Dynasty. The routes connected China with Europe, India, Central Asia, and North Africa. The Silk Roads got their name from the trade of silk cloth, which was made only in China at that time.

Trade Routes

The main overland route began in the city of Chang'an, the capital of ancient China. The route split in two to go around the Taklimakan Desert. It divided again to avoid the highest peaks of the Hindu Kush. Trade routes stretched west to Central Asia and Africa and south to India. Most of the overland traders traveled with **caravans**, or groups, with camels—which were well suited to the difficult terrain and dry climate.

The Silk Roads also included **maritime**, or sea, routes. From Nanjing, traders carrying Chinese goods traveled to Japan. From the ancient cities of Antioch and Tyre, trade continued to Rome.

Goods and Ideas

The overland roads covered about 4,000 miles, but few traders traveled the routes from end to end. Most Chinese traders were forbidden to go beyond their country's borders. As a result, they mainly exchanged their silk, jade, and spices with Central Asian nomads and traders from India. These, in turn, traded with merchants from the Mediterranean. Market towns sprang up along the routes, and some became major cities.

> **Critical Viewing** Camels can travel for long periods without water. Why do you think that ability might have been useful on the Silk Roads?

TRADE ROUTES OF THE SILK ROADS

Silk Roads and Trading
c. 100 B.C.

— Silk road
— Maritime route
— Other major trade route
···· Great Wall

Present-day boundaries shown

0   500   1,000 Miles
0   500   1,000 Kilometers

Traders carried a variety of goods. Indian traders sold gems and sweet-smelling woods such as cedar. European goods included glass, pearls, and wool. Since there was not a single currency used by all the traders, they **bartered**, or exchanged goods without using money. For example, in exchange for their silk, Chinese traders might have received purple dyes from Mediterranean traders.

Government officials and missionaries also traveled on the routes. Their ideas spread along with the traders' goods. For instance, the Silk Roads helped spread Buddhism from India, where it began, to China and eventually to Korea and Japan.

Before You Move On

Monitor Comprehension How did the Silk Roads promote the exchange of goods and ideas?

ONGOING ASSESSMENT

MAP LAB

 GeoJournal

1. **Movement** Based on the map, what route might traders have taken from Dunhuang, China, to Delhi, India?

2. **Make Inferences** Study the map. Explain why Kashgar developed into a thriving city.

3. **Create Sketch Maps** Use the map and the information in the text to create a thematic map showing the activity on the Silk Roads. Use symbols to represent different goods and ideas.

2.4 Exploration and Isolation

TECHTREK

myNGconnect.com For an online map
and photos representing Zheng He's voyages

Maps and
Graphs

Digital
Library

Main Idea Zheng He made seven voyages
of exploration that were intended to impress
foreigners with China's wealth and power.

China dominated East Asia during the
Ming Dynasty, which began in 1368. Ming
emperor Yongle (yung lu) decided that he
wanted to show the world China's wealth
and power and take control of maritime
trade. He also wanted other countries to
pay him tribute, money demonstrating
their recognition of his power.

Zheng He

To carry out this mission, Yongle ordered
Admiral **Zheng He** (jung huh) to lead
seven expeditions, or voyages, between
1405 and 1433. Zheng He's fleet was much
larger than those of European explorers
such as Christopher Columbus, who would
set out almost 100 years later. Zheng He
commanded thousands of men and sailed
with hundreds of treasure ships designed
to carry huge stores of trade goods.

Shortly after Zheng He's last expedition,
China withdrew into isolation. This
isolation continued until the 1800s, when
European powers began controlling
China's economy.

Before You Move On
Summarize In what ways did Zheng He's
expeditions demonstrate China's wealth
and power?

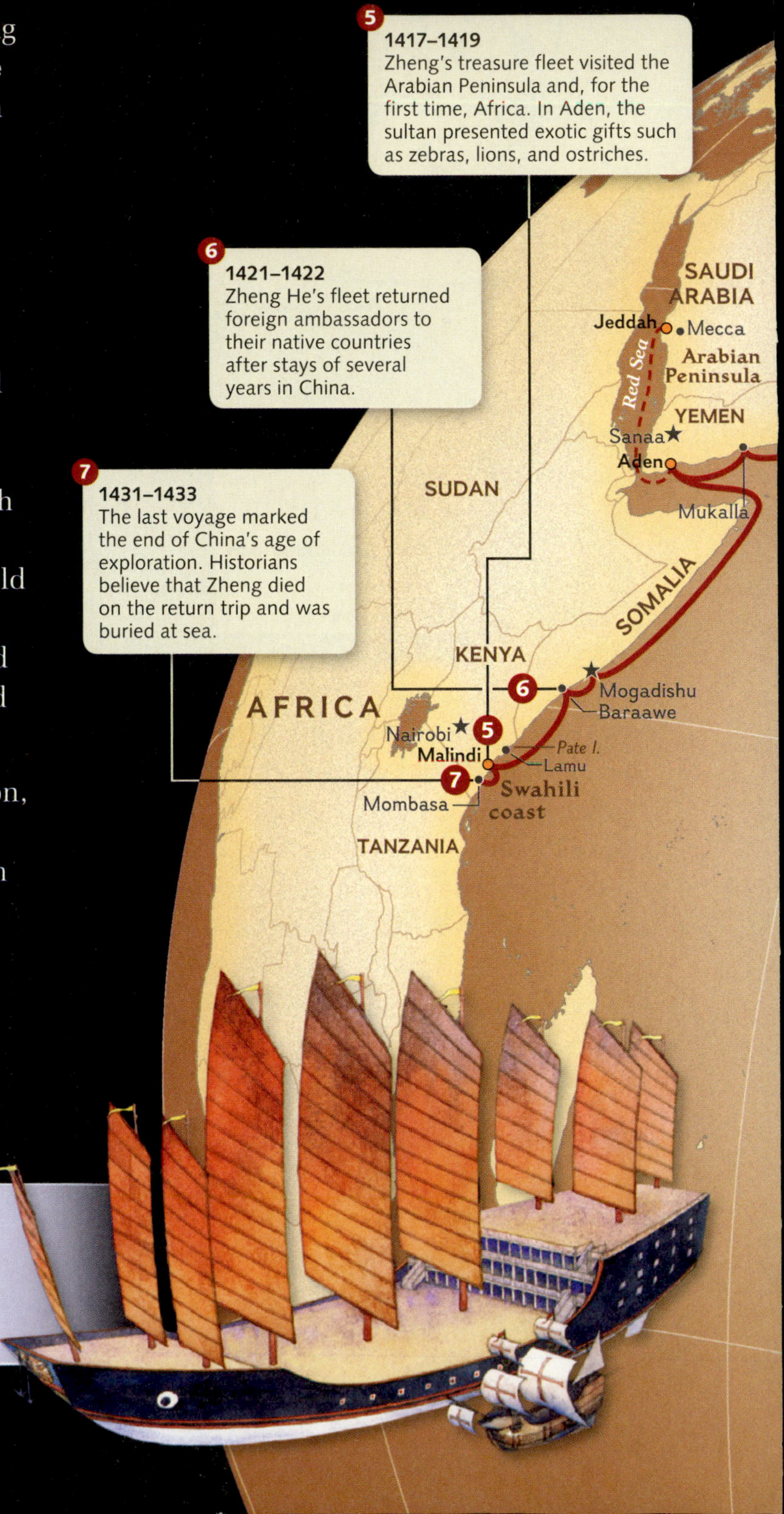

> **Critical Viewing** This illustration compares
Zheng He's large ship with that of a European
explorer. What impression might Zheng's ship
have made when it arrived in a foreign port?

5 1417–1419
Zheng's treasure fleet visited the
Arabian Peninsula and, for the
first time, Africa. In Aden, the
sultan presented exotic gifts such
as zebras, lions, and ostriches.

6 1421–1422
Zheng He's fleet returned
foreign ambassadors to
their native countries
after stays of several
years in China.

7 1431–1433
The last voyage marked
the end of China's age of
exploration. Historians
believe that Zheng died
on the return trip and was
buried at sea.

SAUDI
ARABIA

Jeddah ○ ● Mecca

Arabian
Peninsula

YEMEN

Sanaa ★

Aden ○

Mukalla

Red Sea

SUDAN

SOMALIA

KENYA

Mogadishu
Baraawe

AFRICA

Nairobi ★ **5**

Malindi

Pate I.
Lamu

7

Swahili
coast

Mombasa

TANZANIA

4 1413–1415
As a result of the voyage, an estimated 18 countries sent tribute and foreign ambassadors to China.

3 1409–1411
During this voyage, Zheng He conducted a land battle in Sri Lanka. The voyage was also marked by Zheng's offering of gifts to a Buddhist temple.

2 1407–1409
The fleet returned foreign ambassadors from Sumatra, India, and elsewhere who had traveled to China on the first voyage.

1 1405–1407
In July, the fleet with 317 ships and 27,870 men left Nanjing with silks, porcelain, and spices for trade.

ZHENG HE'S VOYAGES, 1405–1433

This map shows the main and subsidiary, or secondary, routes of Zheng He's seven expeditions. Note that the map labels include place names from the 1400s as well as present-day names.

— Main route
--- Subsidiary route
◦ Major trading center
4 Destination

Present-day boundaries shown
Scale varies in this perspective.

IRAN

Persian Gulf

4 Hormuz

OMAN

— Dhofar

Arabian Sea

INDIA

Origin of all 7 voyages

ASIA

CHINA

Yangtze

Great Wall — Beijing

MING EMPIRE

Grand Canal

• Dalian

JIANGSU

• Nanjing

YUNNAN
• Kunyang

BANGLADESH

• Chittagong

FUJIAN
Changle
Quanzhou •
Xiamen •

East China Sea

1
2
3

Kozhikode (Calicut)

Malabar Coast

Cochin
Quilon

• Jaffna

SRI LANKA
CEYLON

Colombo —
• Galle
— Dondra Head

MALDIVES

Bay of Bengal

Andaman Islands (India)

Nicobar Islands (India)

THAILAND
SIAM
• Ayutthaya

CAMBODIA

VIETNAM

Hainan

CHAMPA

— Qui Nhon

TAIWAN

South China Sea

INDIAN OCEAN

EQUATOR

Banda Aceh
Semudera

Strait of Malacca

Kelantan

MALAYSIA
• Pahang
• Malacca

Sumatra

Palembang •

INDONESIA

Java
• Surabaya

2.5 Communist Revolution

TECHTREK

myNGconnect.com For an online
map and photos of Communist China

Maps and
Graphs

Digital
Library

Main Idea The Chinese Communist Party under Mao Zedong significantly changed life in the People's Republic of China.

As you have read, China fell under the control of European powers in the 1800s. The Chinese Nationalist Party wanted to end this foreign control and modernize China. The party overthrew the ruling government and established the Republic of China in 1912. Over time, the Chinese people began to turn to the Communist Party, which arose in Shanghai in 1921. Many young people in particular believed that the Communists would be better able to modernize China. The Nationalists and Communists began a long struggle for power. By 1930, the two groups were locked in a bloody civil war.

Civil War

In 1933, Nationalist leader **Jiang Jieshi** (jee ahng jee shee) gathered a large army and attacked the Communists in their mountain base in Jiangxi (jee ahng shee) Province. Knowing they were outnumbered, the Communists fled to their base in Shaanxi (shahn shee) Province. Between 1934 and 1935, **Mao Zedong** (MOW dzuh dahng) led about 100,000 Communists over rough terrain on a series of marches that came to be known as the Long March. Tens of thousands of Communists died during the marches.

In 1937, the Japanese invaded China and **occupied**, or took possession of, parts of the country. The Nationalists and Communists set aside their differences to fight their common enemy. The two groups remained united against Japan throughout World War II. After the war, the Nationalists and Communists started fighting again. On October 1, 1949, the Communists declared victory and established the People's Republic of China on the mainland. The Nationalists fled and established a government in Taiwan, an island off the coast of the mainland.

Chairman Mao

As Chairman of the Communist Party, Mao became the leader of China and established a totalitarian dictatorship. The government seized farmland and forced the peasants to work on **collective farms**. Between 200 and 300 families lived and worked on each collective farm, which was strictly supervised by the government. The government also took over all businesses and industries. By 1957, industrial output had greatly increased.

1912
Nationalists establish the Republic of China.

1920

1921
Communist Party is founded in China.

1934–1935
Communists undertake the Long March.

1937–1945
Communists and Nationalists unite to fight against Japan.

1940

Statue celebrating heroes of the Communist revolution

Encouraged by this initial success, Mao set a plan in motion the following year called the Great Leap Forward to make China's economy grow even faster. The plan failed. The government's poor management of China's industries slowed economic growth, and crop failures severely reduced agricultural yields. After an estimated 20 million people had died of hunger, Mao ended the plan in 1961.

Five years later, Mao began a new plan called the **Cultural Revolution** to remove what he considered anti-Communist elements from China. To help carry out the plan, young people belonging to military groups known as the Red Guards attacked anyone whom they considered <mark>elite</mark>, or superior. They particularly targeted teachers and intellectuals. Thousands of people were killed in the violence.

After Mao died in 1976, the situation slowly began to change in China. Today, the country is still Communist, but reforms begun in the 1980s have resulted in great economic growth and improved the lives of many Chinese people.

Before You Move On
Summarize What happened in China as a result of Communist control under Mao Zedong?

THE LONG MARCH

China's Communist Revolution

- 🟩 China, 1949
- ⭕ Communist centers, 1934
- ➡️ Long March, Oct. 1934–Oct. 1935

ONGOING ASSESSMENT
MAP LAB
GeoJournal

1. **Movement** Trace the route on the map from Jiangxi to Shaanxi Province. What directions did the route follow?

2. **Describe** Use the map on this page and the physical map in Section 1.1 to describe the terrain encountered by those who marched from Jiangxi to Shaanxi Province.

3. **Interpret Time Lines** For how many years did Mao Zedong rule China?

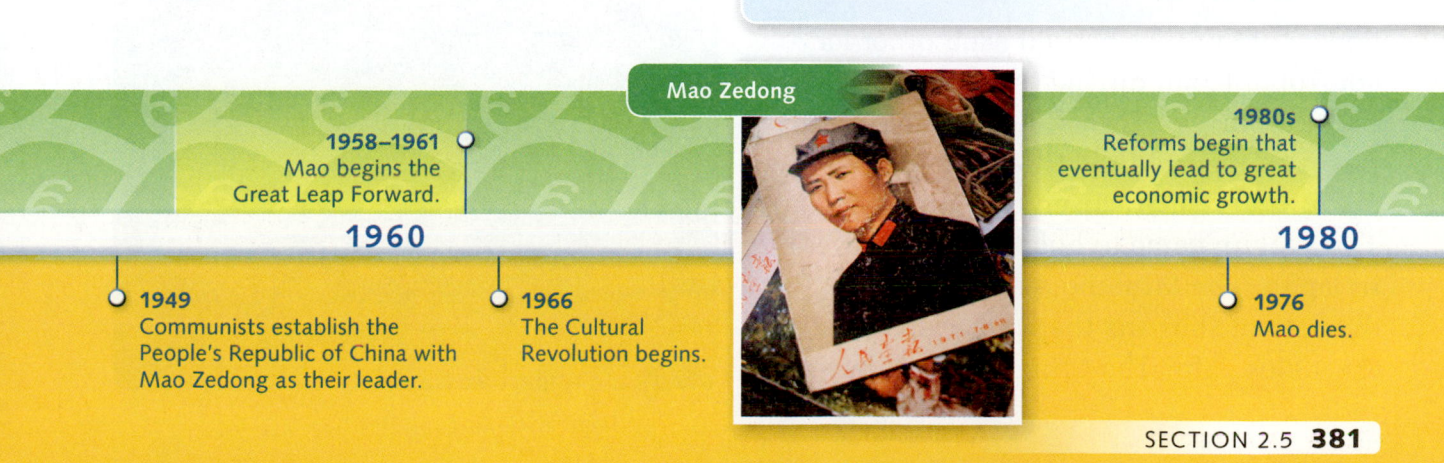

Mao Zedong

1958–1961
Mao begins the Great Leap Forward.

1960

1980s
Reforms begin that eventually lead to great economic growth.

1980

1949
Communists establish the People's Republic of China with Mao Zedong as their leader.

1966
The Cultural Revolution begins.

1976
Mao dies.

3.1 Japanese Samurai

Main Idea Samurai were part of a feudal system that continued in Japan for about 700 years.

Emperors ruled Japan for hundreds of years, but in the 1000s, they began to lose authority. Landowning lords who belonged to wealthy families expanded their estates and grew increasingly powerful. They formed private armies and hired warriors known as **samurai** (SAM uh ry) to protect their estates.

Samurai Code and Culture

Samurai means "one who serves." The relationship between lords and samurai was part of a Japanese feudal system similar to the one that arose in Europe during the Middle Ages. Like the knights of the Middle Ages, the samurai developed a code of behavior based on the values of honor, bravery, and loyalty to their lords.

The samurai were not only skilled warriors but also developed their own culture. They wrote poetry and created fine ink paintings. The samurai values of loyalty and hard work and the warriors' artistic interests influenced Japanese society, art, and literature.

Shogun Rule

Powerful families battled each other for control of Japan during the 1100s. In 1192, a lord from the powerful Minamoto family won and received the title of **shogun**, or "military ruler," from the emperor. The shogun came to hold the true power in Japan. The Minamoto established a dynasty of shoguns that ruled Japan until the 1300s. After the Minamoto lost power,

rival families fought for control. Japan became divided into many warring states and experienced a long period of disorder.

Finally, in 1603, a shogun named Tokugawa Ieyasu (ee yeh yah soo) defeated his rivals. He united Japan and moved the capital from Kyoto to Edo, which is present-day Tokyo. The Tokugawa shoguns restored peace. However, they also isolated Japan from the rest of the world. In 1868, this isolation ended when the Japanese people persuaded the shogun to resign. Imperial rule was restored, and the new emperor put an end to the samurai class.

Before You Move On

Monitor Comprehension What role did the samurai play in Japan's feudal system?

Samurai armor

Critical Viewing This painting shows samurai on horseback during a battle. What details in the painting illustrate the samurai's skill and bravery?

PHOTO LAB

GeoJournal

1. **Analyze Visuals** Study the photo of samurai armor. What words would you use to describe the warrior's equipment?

2. **Compare and Contrast** Compare the samurai in the photo and in the painting. Which details are similar? Which are different?

3. **Write Reports** Think about another group of warriors that you know about from history, comics, or movies such as *Star Wars*. In what way does their code compare with that of the samurai? Write a report comparing the two.

3.2 Japan Industrializes

> **Main Idea** Japan transformed itself into an
> industrialized economy beginning in 1868.

As you have learned, the emperor was
restored to power in Japan in 1868. The
restoration came about because many
of the Japanese people wanted to end
Japan's isolation. They wanted to make
the country ==competitive== with, or able
to challenge, the newly industrialized
Western nations. After the last of the
Tokugawa shoguns resigned, a new
era called *Meiji* (may gee), meaning
"enlightened rule," began.

Building on Western Models

For hundreds of years, Japan's economy
had been based on agriculture and fishing.
The Meiji government looked to Western
nations as models for a new industrial
economy. Many foreign experts came to
Japan to instruct the Japanese in English,
engineering, and science. British advisers
helped the Japanese government design
railroad and communication systems and
set up new industries.

At first, the government invested
directly in coal and zinc mines and
large-scale industries, especially those
needed to build a modern military force.
These industries included shipyards and
weapons factories. The government also
developed factories to produce textiles and
silk. The government planned to export
these goods so the country could acquire
raw materials. Japan turned to China and
Korea as sources for these raw materials
and as potential markets for the finished
products. By the 1880s, however, the
government could not afford to continue

This woodblock print by the Meiji-era artist
known as Hiroshige III illustrates a Tokyo railway
station and train built in 1872 with the help of
Western advisers and engineers.

this level of investment in the country's
industries and sold the industries to
private investors.

Large Businesses Develop

Over time, business in Japan became
concentrated in several large, family-
controlled organizations called ==zaibatsu==
(zeye BAHT sue). Each zaibatsu owned a
number of different businesses, such as
manufacturing, transportation, banking,
insurance, trade, and real estate.

Some of the zaibatsu were rooted in the shogun era. For example, the Mitsui family had been successful textile merchants under the Tokugawa shoguns. This family expanded first into banking and, in time, owned more than 270 companies.

Other zaibatsu began during the Meiji era. Mitsubishi, for instance, started as a large shipping firm. The family eventually established financial services and real estate businesses and industries such as oil, steel, and shipbuilding.

The zaibatsu were largely **dissolved**, or broken up, after World War II ended in 1945. Businesses owned by the zaibatsu were **seized**, or taken control of, and reorganized into smaller holdings. However, the value the zaibatsu placed on hard work and thoroughness continues to influence Japanese industry today.

ONGOING ASSESSMENT

READING LAB

GeoJournal

1. **Make Inferences** Why do you think Japan turned to Western nations for help in becoming industrialized? In what way was this a reversal of Japanese policy during the Tokugawa period?
2. **Analyze Cause and Effect** What impact do you suppose industrialization had on Japan's economy and people?
3. **Synthesize** Some former samurai were able to adapt to and find roles in the newly industrialized society. Based on what you have learned about these warriors, what qualities might have made them good employees?

Before You Move On

Summarize By what means did Japan transform into an industrialized economy beginning in 1868?

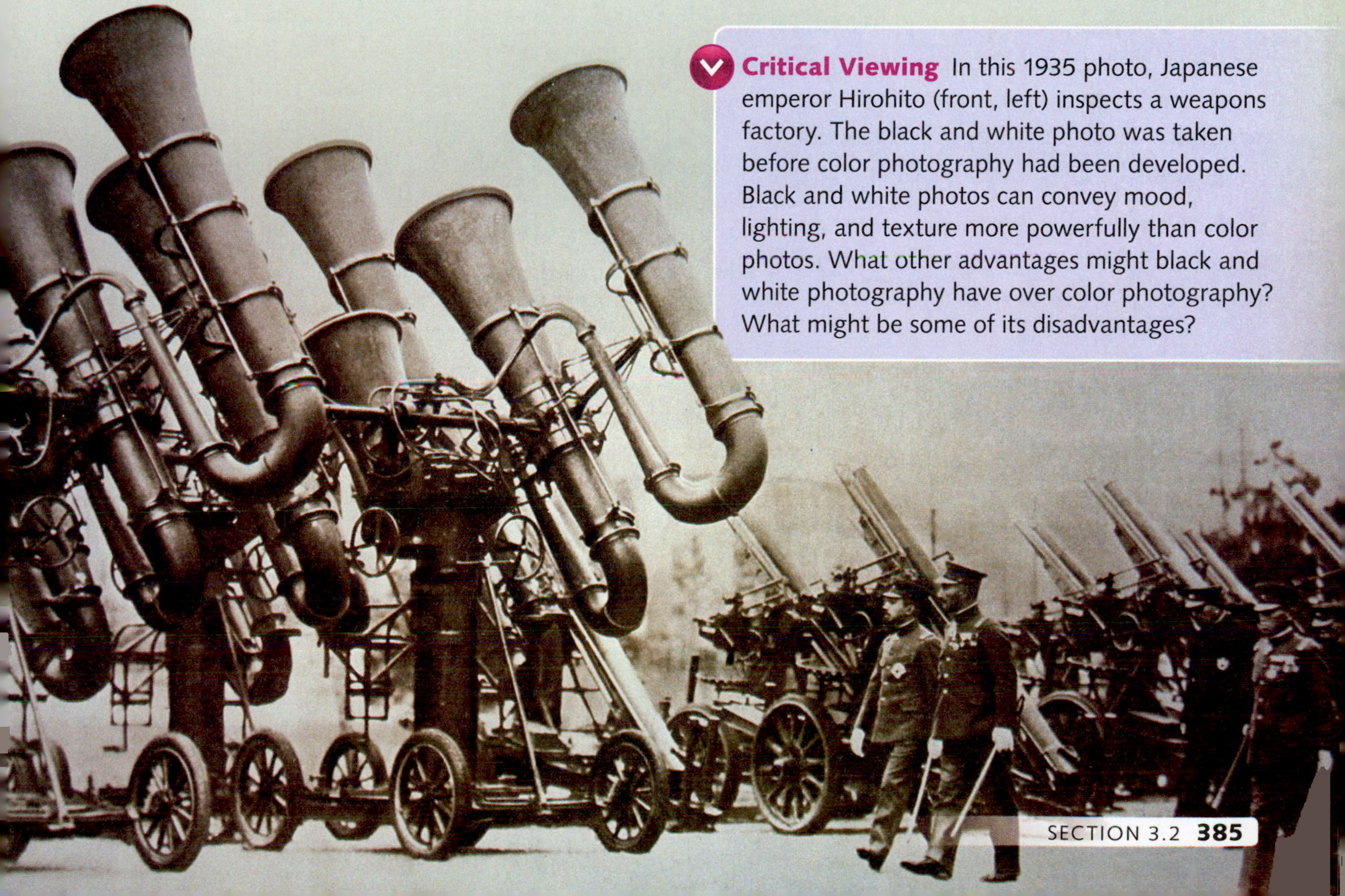

Critical Viewing In this 1935 photo, Japanese emperor Hirohito (front, left) inspects a weapons factory. The black and white photo was taken before color photography had been developed. Black and white photos can convey mood, lighting, and texture more powerfully than color photos. What other advantages might black and white photography have over color photography? What might be some of its disadvantages?

3.3 Korea's Early History

TECH TREK
myNGconnect.com For an online map and images of Korea's early history

Maps and Graphs

Digital Library

Main Idea A series of kingdoms and dynasties arose in Korea, developing a distinct culture.

In Section 2, you learned about the early dynasties that came to power in China. In 108 B.C., China's Han Dynasty expanded into Korea and took control of an area in the northwest. At that time, the Korean people belonged to scattered tribes. Eventually, groups of tribes united and formed three kingdoms on the peninsula: **Silla** (SIHL uh), **Koguryo** (koh gur YOO), and **Paekche** (PAHK CHAY).

The Three Kingdoms

Silla, the first kingdom to form, arose around 57 B.C. in the southeastern part of the peninsula. Koguryo emerged about 20 years later when several tribes united in the northern part of the peninsula and parts of eastern Manchuria in China. Paekche formed around 18 B.C. in the southwest. At first, Koguryo was the largest and most powerful of the three.

The three kingdoms were rivals and often invaded each other's territory. In spite of their <mark>rivalry</mark>, or opposition, the kingdoms had similar cultures, and their people spoke the same language. The kingdoms produced items such as leather goods, tools, and woolen clothing, which they exported to China. In return, the kingdoms received porcelain, paper, silk, and weapons. The kingdoms also absorbed some ideas from China, including its writing system, Confucianism, and Buddhism. Eventually, Buddhism spread from the Korean kingdoms to Japan.

In A.D. 660, the Tang Dynasty in China joined with Silla to defeat the other two kingdoms. Shortly after, Silla succeeded in driving out the Chinese. By 668, the Silla kingdom ruled the entire peninsula.

Koryo and Choson Dynasties

By 935, Silla had weakened, and its kingdom was overthrown. The Korean Peninsula came under the control of the **Koryo** Dynasty, which ruled for more than 450 years. The Koryo Dynasty modeled its government after that of China. Like the early kingdoms, Koryo was strongly influenced by Confucianism.

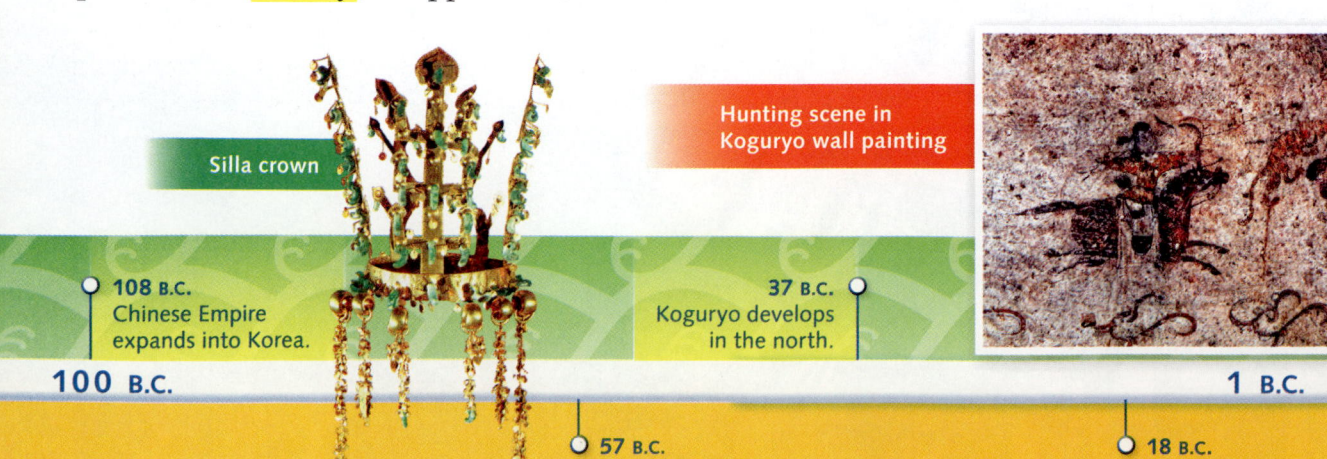

Silla crown

Hunting scene in Koguryo wall painting

108 B.C.
Chinese Empire expands into Korea.

37 B.C.
Koguryo develops in the north.

100 B.C.

1 B.C.

57 B.C.
Silla arises in the southeastern part of the peninsula.

18 B.C.
Paekche forms in the southwest.

A distinctly Korean culture developed during the Koryo period. Artists created **celadon** pottery, with its characteristic green glaze. They also carved all of the Buddhist scriptures onto more than 80,000 wooden blocks for printing. The blocks are called the Tripitaka Koreana. Invading Mongols destroyed the blocks in 1232, but a new set was re-created within 20 years. The temple of Haeinsa in South Korea is home to the set today and has been named a UNESCO World Heritage Site.

In 1392, the Koryo Dynasty was defeated and replaced by the **Choson** Dynasty, which lasted for 518 years. During this time, Korea continued to adopt elements of Chinese culture. Korean culture also flourished at this time, with developments in architecture, science, and technology. The Choson Dynasty ended in 1910 when Korea came under the control of Japan. Japanese occupation continued until 1945, when World War II ended.

Before You Move On

Summarize How were the Korean dynasties similar and different?

A monk holds a block of the Tripitaka Koreana.

EARLY KINGDOMS, A.D. 500s

0 50 100 Miles
0 50 100 Kilometers

RUSSIA

CHINA

KOGURYO

Nong'an

Tonggou

Sea of Japan
(East Sea)

NORTH
KOREA

Korea
Bay

Pyongyang

	Silla
	Koguryo
	Paekche
	Koryo Dynasty
	Choson Dynasty

Hanseong
(Seoul)
Kwangjiu
SOUTH
KOREA
Puyo   SILLA
Kyongjiu

Kimhae

JAPAN

PAEKCHE

Yellow Sea

Present-day boundaries shown

45°N
40°N
35°N
125°E   130°E

ONGOING ASSESSMENT

MAP LAB

GeoJournal

1. **Interpret Maps** On the map, use your finger to trace the outline of the Koguryo Kingdom. What parts of present-day countries did the kingdom consist of?

2. **Draw Conclusions** Remember that Silla conquered Koguryo and Paekche in 668. How would the size of Silla at its greatest extent have compared with that of the Koryo and Choson dynasties, as shown on the map?

A.D. 500

A.D. 668
Silla controls the
entire peninsula.

A.D. 935
Koryo Dynasty
overthrows Silla.

A.D. 1392
Choson Dynasty
begins.

1500

Celadon pillow

3.4 The Korean War

TECHTREK
myNGconnect.com For an online
map and photos of the Korean War

 Maps and Graphs
 Digital Library

Main Idea After three years of war, North Korea and South Korea remained divided.

When the Japanese occupation ended in 1945, the Korean Peninsula was divided along the 38° North latitude line—usually referred to as the **38th parallel**. The United States held the land south of that line, and the Soviet Union occupied the land to the north. The United States and the Soviet Union came to be locked in a **Cold War**. This means that they did not fight one another directly in battle. Instead, they supported opposing groups in wars that took place in other parts of the world. One such war occurred in Korea.

Fighting Begins

In 1947, the United Nations called for free elections that would create one government for Korea. The elections were held, but the Soviet Union stepped in and established a Communist government in the north. The following year, Communist North Korea and democratic South Korea were created, with the 38th parallel serving as the border.

On June 25, 1950, North Korea attacked South Korea. The United Nations called for an international force to come to South Korea's aid. The United States supplied most of the troops and placed them under the command of General Douglas MacArthur, who had been an important military leader during World War II.

The fighting wore on for several years. At one point, the North Koreans pushed deep into South Korea and captured most of the peninsula. Then the UN forces made a surprise landing at Inchon and took back the South Korean capital of Seoul. As the North Koreans **retreated**, or drew back, the UN forces pushed north and captured the North Korean capital of Pyongyang. The course of the war shifted from one side to the other through 1952, but in the end, little territory was gained or lost.

Impact of the War

As many as four million soldiers and civilians were killed in the war. Much of the Korean Peninsula was damaged by bombs dropped by jet aircraft.

General MacArthur lands at Inchon.

JUNE 1950
North Korea invades South Korea.

OCTOBER 1950
UN forces capture Pyongyang.

1950

SEPTEMBER 1950
North Korea captures most of the peninsula; MacArthur's forces land at Inchon and take back Seoul.

Korean civilians particularly suffered during the war. Half of all industries and a third of all homes were destroyed. In addition, many people died of starvation.

The War Ends

In 1953, the UN forces and North Korea signed an <mark>armistice</mark>, or agreement to stop fighting, but a treaty was never signed. An area set near the 38th parallel, called the demilitarized zone (DMZ), still divides the two countries today. Since 1953, North Korean troops have guarded one side of the zone, while South Korean and American troops have guarded the other.

Today, Communist North Korea is largely isolated from the rest of the world, while South Korea is a democracy with a global, market economy. Over the years, the two countries have discussed the possibility of uniting. However, political differences and the threat of North Korea's nuclear arms program have made this unification increasingly unlikely.

Before You Move On

Make Inferences What, if anything, did the Korean War accomplish?

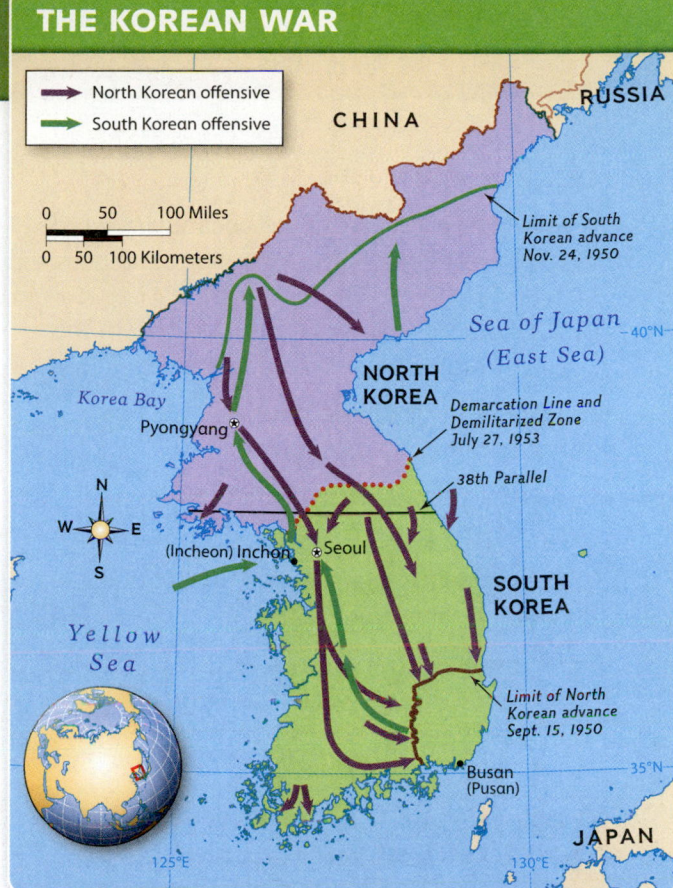

THE KOREAN WAR

→ North Korean offensive
→ South Korean offensive

0 50 100 Miles
0 50 100 Kilometers

RUSSIA
CHINA
Sea of Japan (East Sea) 40°N
Korea Bay
Pyongyang ✛
NORTH KOREA
Limit of South Korean advance Nov. 24, 1950
Demarcation Line and Demilitarized Zone July 27, 1953
38th Parallel
(Incheon) Inchon
✛ Seoul
SOUTH KOREA
Yellow Sea
Limit of North Korean advance Sept. 15, 1950
Busan (Pusan) 35°N
JAPAN
125°E   130°E

ONGOING ASSESSMENT

MAP LAB

GeoJournal

1. **Movement** Study the map. Which country's forces moved further into enemy territory?
2. **Interpret Maps** Based on the map, why do you think it was easy for Chinese forces to rapidly come to North Korea's aid?
3. **Analyze Time Lines** Based on the time line captions, how would you characterize the course of the war?

Statues of American soldiers in the Korean War Veterans War Memorial in Washington, D.C.

JULY 1953
An armistice is signed, but North and South Korea remain divided.

1952   1954

JANUARY 1951
Chinese forces occupy Seoul.

MARCH 1951
UN forces retake Seoul.

South Korean soldiers patrol the DMZ.

VOCABULARY

On your own paper, write the vocabulary word that completes each of the following sentences.

1. _____ is yellow silt that blows from the Gobi desert onto the Huang He.

2. Tropical hurricanes called _____ bring heavy rains to North Korea.

3. The rise and fall of dynasties in China followed a pattern known as the _____.

4. Confucianism is considered an _____, which teaches right behavior.

5. Japanese warriors known as _____ were similar to the knights of the European Middle Ages.

MAIN IDEAS

6. Where do most East Asians live? Why? (Section 1.1)

7. What purpose does the Grand Canal in China serve? (Section 1.2)

8. What are the main natural resources of North Korea and South Korea? (Section 1.4)

9. What steps has Mongolia taken toward preserving the Gobi desert? (Section 1.5)

10. Who was Shi Huangdi and what were his accomplishments? (Section 2.1)

11. Why did the Chinese people embrace Confucianism? (Section 2.2)

12. What were the Silk Roads? (Section 2.3)

13. In what ways did Mao Zedong control life in China? (Section 2.5)

14. What qualities were valued under the samurai code of behavior? (Section 3.1)

15. Why did Japan begin to industrialize after 1868? (Section 3.2)

16. What happened at the conclusion of the Korean War? (Section 3.4)

GEOGRAPHY

ANALYZE THE ESSENTIAL QUESTION

How did geographic factors affect population distribution?

Critical Thinking: Analyze Cause and Effect

17. Why is eastern China more densely populated than western China?

18. Why do many of Japan's people live on the island of Honshu?

19. How has Mongolia's harsh environment affected the ways people live and work?

INTERPRET TABLES

RIVERS IN CHINA	
River	Length (miles)
Chang Jiang	3,900
Huang He	3,395
Xi Jiang	1,250
Yalu	490

Source: CIA World Factbook

20. **Analyze Data** About how many times bigger is the Chang Jiang than the Yalu?

21. **Draw Conclusions** Based on the table, what conclusions can you draw about rivers in China?

CHINA'S HISTORY

ANALYZE THE ESSENTIAL QUESTION

What influences, beliefs, and encounters helped shape China?

Critical Thinking: Make Generalizations

22. What impact has Confucianism had on Chinese government and society?

23. What did China gain from its trade on the Silk Roads and its voyages of exploration?

INTERPRET MAPS

SPREAD OF BUDDHISM

● Origin of Buddhism
— Present-day boundaries shown

JAPAN
NORTH KOREA
SOUTH KOREA
CHINA
Huang He (Yellow R.)
Chang Jiang (Yangtze R.)
East China Sea
TAIWAN
PACIFIC OCEAN
Indus R.
Brahmaputra R.
Ganges R.
INDIA
MYANMAR (BURMA)
Mekong R.
Khmer
Bay of Bengal
South China Sea

0   400   800 Miles
0   400   800 Kilometers

24. Interpret Maps Study the map. Why do you think Buddhism spread to China before it came to Korea and Japan?

25. Make Predictions What other region shown on the map was most likely influenced by Buddhism?

HISTORY OF JAPAN & KOREA

ANALYZE THE ESSENTIAL QUESTION

What factors had an impact on the histories of Japan and Korea?

Critical Thinking: Make Inferences

26. Why do you think Japan's rulers chose to isolate the country once they had restored peace in the early 1600s?

27. In what ways has China influenced Korea throughout its history?

ACTIVE OPTIONS

Synthesize the Essential Questions by completing the activities below.

28. Write a Speech Choose a historical figure from the chapter that particularly interests you and write a speech from the figure's point of view. To get information for your speech, do some research online on the historical figure. Then use the tips below to help you write the speech. **Once you have finished writing the speech, deliver it to your class.**

> **Writing Tips**
> - Describe some of the historical figure's beliefs and accomplishments.
> - Include brief stories from the person's life to hold your audience's interest.
> - Use a tone that will help convey the person's personality. For example, for Shi Huangdi, you might use a boastful tone. For Confucius, you would probably use a more modest tone.

TECHTREK   myNGconnect.com For research links on East Asia

29. Create a Poster Choose two countries of East Asia and compare and contrast their geography, history, and situation today. Use the research links at **Connect to NG** and other online sources to help you gather your data. Then organize your ideas in a Venn diagram like the one below. Finally, create your poster, using photos to illustrate your ideas.

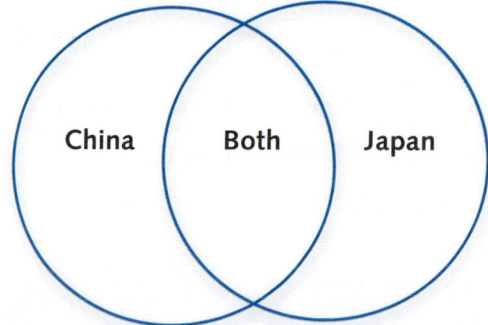

China   Both   Japan

CHAPTER 14

East Asia TODAY

PREVIEW THE CHAPTER

Essential Question How do traditions and modernization create a unique way of life in East Asia?

SECTION 1 • CULTURE

KEY VOCABULARY

- meditation
- animism
- polytheism
- monotheism
- porcelain
- movable type
- multinational corporation
- economic globalization
- anime
- manga
- bullet train
- magnetic levitation

ACADEMIC VOCABULARY
aerodynamic

TERMS & NAMES

- Daoism
- Shinto
- Special Economic Zone

Essential Question What problems does East Asia face today, and what are its opportunities?

SECTION 2 • GOVERNMENT & ECONOMICS

KEY VOCABULARY

- entrepreneur
- gross domestic product (GDP)
- one-child policy
- fertility rate
- gorge
- reservoir
- martial law
- pagoda
- capital
- free trade
- drought
- famine

ACADEMIC VOCABULARY
comply, controversy

TERMS & NAMES

- Three Gorges Dam
- Lost Decade

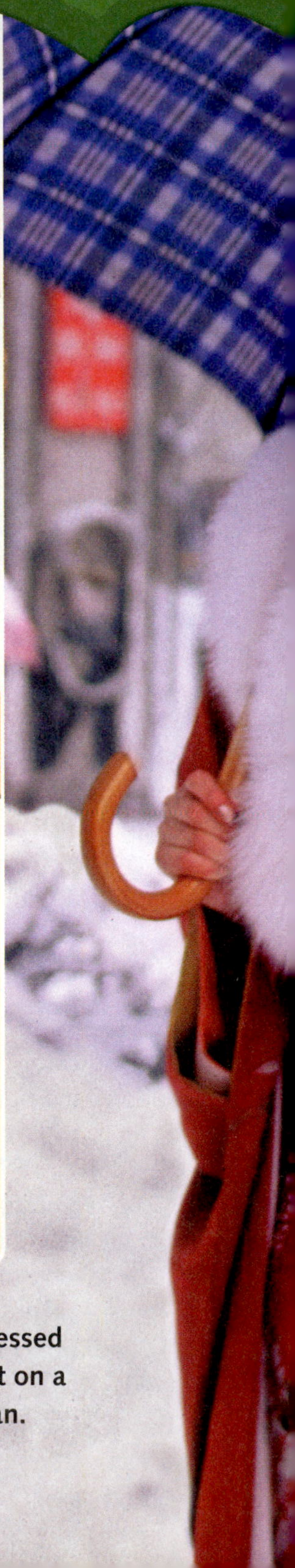

Young Japanese women dressed in traditional kimonos meet on a snowy day in Nagano, Japan.

TECHTREK

FOR THIS CHAPTER

Student eEdition

Maps and Graphs

Interactive Whiteboard GeoActivities

Digital Library

Connect to NG

Go to **myNGconnect.com** for more on East Asia.

1.1 Religious Traditions

TECHTREK

myNGconnect.com For online graphs of religions in East Asia and photos of Buddhism

 **Maps and Graphs**

 Digital Library

Main Idea Buddhism and other religions spread throughout East Asia.

As you have learned, Buddhism began in India, but it came to have a far greater influence in East Asia. Missionaries traveling on the Silk Roads during the Han Dynasty first brought Buddhist teachings to China. Over time, Buddhism blended with other religious traditions throughout East Asia.

Blending Beliefs

Before Buddhism spread to East Asia, many people in the region practiced Confucianism and **Daoism**. Like Confucianism, Daoism is an ethical system. It stresses the harmony between people and nature. As Buddhism gained popularity, many East Asians combined elements of all three traditions.

Buddhism focuses on helping people end their physical and mental suffering by teaching them to give up worldly possessions. According to Buddhist teachings, one way to achieve this goal is through meditation. **Meditation** is the practice of using concentration to quiet and control one's thoughts.

In Mongolia, Buddhism mixed with **animism**, a belief that everything has a soul, including objects in nature. The native religion of Japan, **Shinto**, is similar to animism. The followers of this religion worship the spirits of their ancestors. They believe that these spirits exist in natural forces, including trees, rocks, and rivers.

> **Critical Viewing**
> Prayer flags flutter in the wind at a Buddhist temple in China. How might this temple's setting aid meditation?

Absorbing New Beliefs

People who practice animism and Shinto worship many different gods. Belief in more than one god is called **polytheism** (PAHL ee thee ihz uhm). Eventually, religions that teach belief in one god, such as Christianity and Islam, came to East Asia. Belief in one god is called **monotheism** (MAHN uh thee ihz uhm). Muslim traders in northwestern China helped spread Islam in the 700s. Today, it is the dominant religion in that part of China. Christian missionaries began to come to East Asia in the 1600s to spread their religion. Today, about 30 percent of South Koreans are Christians.

Throughout East Asia's history, governments supported different religions and philosophies. Han rulers, for example, promoted the ideas of Confucianism as a model for government and society. When Communist leaders came to power in China and North Korea, they banned the practice of religion, emphasizing Communist philosophy instead. Since the 1970s, however, there has been more religious tolerance in China.

Before You Move On

Monitor Comprehension How were religions spread and absorbed in East Asia?

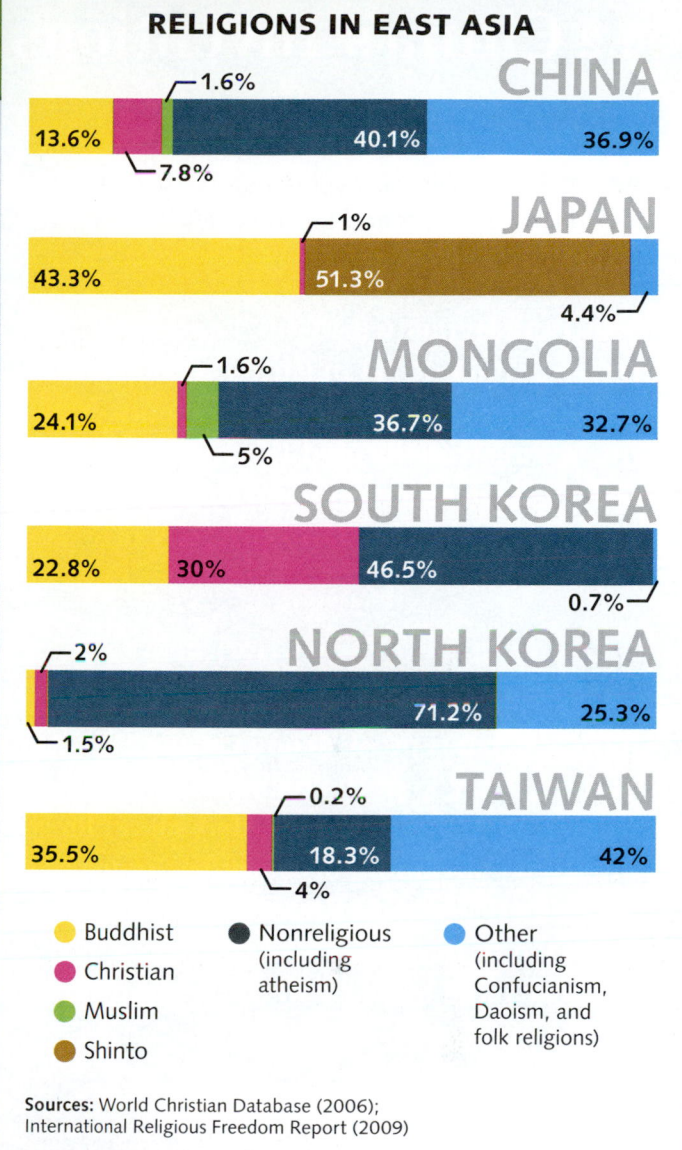

RELIGIONS IN EAST ASIA

CHINA
13.6% | 7.8% | 1.6% | 40.1% | 36.9%

JAPAN
43.3% | 51.3% | 1% | 4.4%

MONGOLIA
24.1% | 5% | 1.6% | 36.7% | 32.7%

SOUTH KOREA
22.8% | 30% | 46.5% | 0.7%

NORTH KOREA
2% | 1.5% | 71.2% | 25.3%

TAIWAN
35.5% | 4% | 0.2% | 18.3% | 42%

- ● Buddhist
- ● Christian
- ● Muslim
- ● Shinto
- ● Nonreligious (including atheism)
- ● Other (including Confucianism, Daoism, and folk religions)

Sources: World Christian Database (2006); International Religious Freedom Report (2009)

ONGOING ASSESSMENT

DATA LAB

 GeoJournal

1. **Interpret Graphs** On the graph, notice the high percentages of "Other" in China, Mongolia, and Taiwan. What probably accounts for the numbers in this category?

2. **Analyze Data** Why do so many people in China consider themselves nonreligious?

3. **Draw Conclusions** As you have learned, many East Asians combine and practice more than one religion. What does this suggest about the representations in the graphs?

1.2 China's Inventions

Main Idea Chinese inventions introduced products and technologies that continue to be used today.

The Tang Dynasty (618–907) and Song Dynasty (960–1279) in China were periods of great technological advancements. New inventions transformed life in China and spread throughout much of the world. Many of these inventions are part of modern life.

Porcelain and Gunpowder

In the 700s, the Chinese developed porcelain, a type of strong ceramic pottery. They kept the technology used in making it a closely guarded secret for hundreds of years. Porcelain became a valuable trade good that was—and is—exported to many parts of the world. Because of its close tie to Chinese culture porcelain is often referred to as "china."

Critical Viewing This porcelain teapot was made during the Song Dynasty. Why do you think porcelain became so desirable?

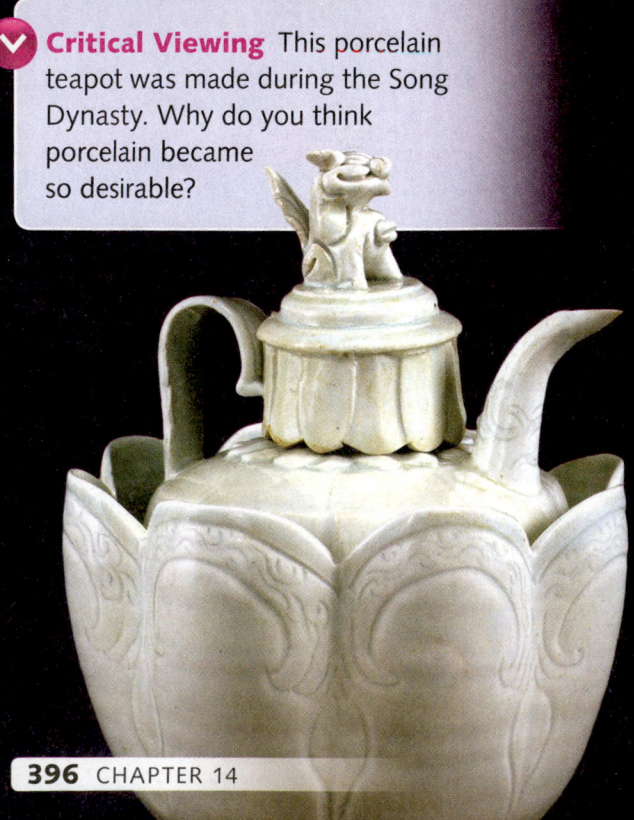

During the 800s, the Chinese invented a very different product: gunpowder. Gunpowder was first used for fireworks. Within 300 years, however, the Chinese were using it to shoot weapons. Fireworks, meanwhile, became a popular feature of Chinese celebrations because they were believed to frighten away evil spirits. Today, China is the world's largest manufacturer and exporter of fireworks.

Printing and Paper Money

The Chinese invented paper around A.D. 100. By the 700s, they began printing books using wooden blocks upon which text had been carved. Before block printing, people had reproduced books by copying each one by hand.

In the 1040s, the Chinese advanced printing further with the invention of movable type. Using this technique, individual characters carved on clay or metal blocks were placed in a frame to form a page of text. The blocks could be moved around to create many different pages. The Chinese used their printed books to spread the teachings of Confucius and Buddha.

The first paper money appeared in China in the 800s, but it did not become widely used until the 1020s. During the Song Dynasty, paper currency replaced the heavy metal coins that merchants and buyers previously carried. Paper money made trade easier and greatly expanded China's economy. However, people did not trust the new currency at first. After all, unlike the coins formerly used in trade, the paper itself had no value.

Magnetic Compass

Like paper, the compass was another early Chinese invention, dating from the 200s B.C. The compass from this period was used to position buildings and furniture in ways believed to bring good luck to the owners. By the 1100s, the Chinese had developed a magnetic compass for navigation by floating a magnetized needle in a bowl of water. They had learned that the needle always pointed in the north-south direction. With this floating compass, the Chinese explorer Zheng He traveled throughout Asia and Africa, demonstrating China's power and spreading its inventions.

Before You Move On
Make Inferences In what ways have Chinese inventions influenced today's society?

Critical Viewing Fireworks mark the opening of the 2008 Summer Olympics in Beijing. What aspects of this photo help explain why fireworks are so popular today?

1.3 Shanghai's Rapid Growth

TECHTREK
myNGconnect.com For photos reflecting Shanghai's growth
Digital Library

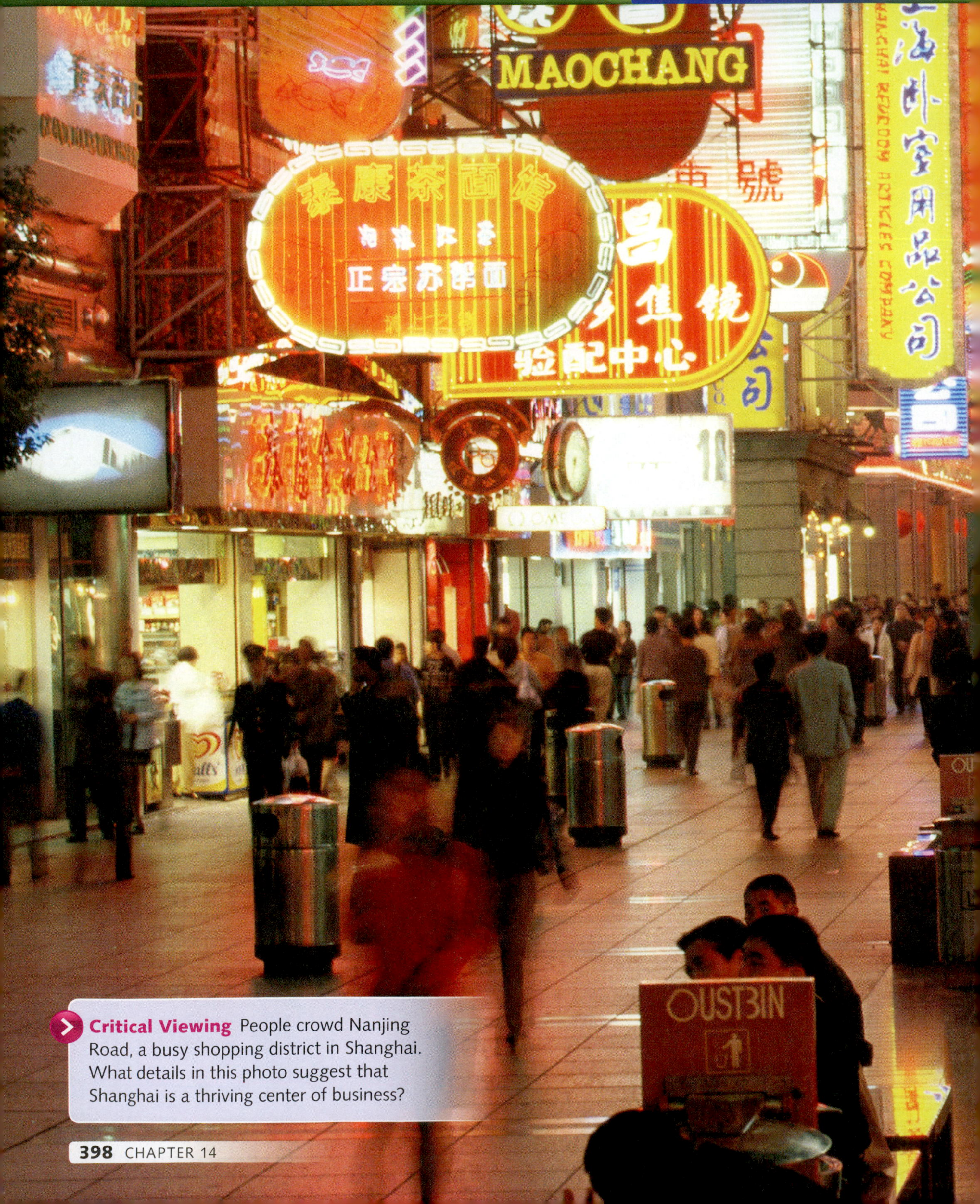

Critical Viewing People crowd Nanjing Road, a busy shopping district in Shanghai. What details in this photo suggest that Shanghai is a thriving center of business?

> **Main Idea** Shanghai's growth has resulted in a building and population boom.

Shanghai's location near the mouth of the Chang Jiang has attracted settlers since ancient times. Over the centuries, it became an important port city and busy trade center. Today, with a population of around 20 million, Shanghai has grown into China's largest city. It is also the country's center for finance and business.

Factors Fueling Growth

Shanghai's rapid growth began in the 1990s, when economic reforms led to new investments in the city. The economy of much of China is controlled by the central government. In 1990, however, the government allowed an area in Shanghai called Pudong to become a **Special Economic Zone**. This is an area that has the ability to develop a market economy with no government control of business.

As a result, Shanghai began to attract multinational corporations. A **multinational corporation** is a large company that is based in one country but establishes offices, branches, or plants in several others. Shanghai now hosts hundreds of multinational corporations. They have contributed to bringing economic globalization to China. **Economic globalization** occurs when economic activities are conducted across national borders.

Building and Population Boom

To make room for these corporations, the city witnessed a building boom in the 2000s. Thousands of new skyscrapers were constructed. One of the most important is the Shanghai World Financial Center, which opened in Pudong in 2008. It is the tallest building in China and home to the Shanghai Stock Exchange.

As Shanghai's economy has grown, so has its population. Since the 1990s, about four million people from China's rural areas have come to Shanghai seeking jobs. They now make up about 25 percent of the workforce. The city center became so crowded that about a half million households were relocated to the suburbs. The exploding growth in population also resulted in higher levels of pollution.

To address these problems, city planners are building several new towns around Shanghai's urban center. Each one will house about 500,000 people. To protect its environment, Shanghai has also funded efforts to use buses and taxis that run on cleaner fuel and has moved most of its factories out of the city center.

Before You Move On

Monitor Comprehension What factors fueled Shanghai's growth, and what happened in the city as a result?

NATIONAL GEOGRAPHIC

Photo Gallery • Shanghai's Skyscrapers

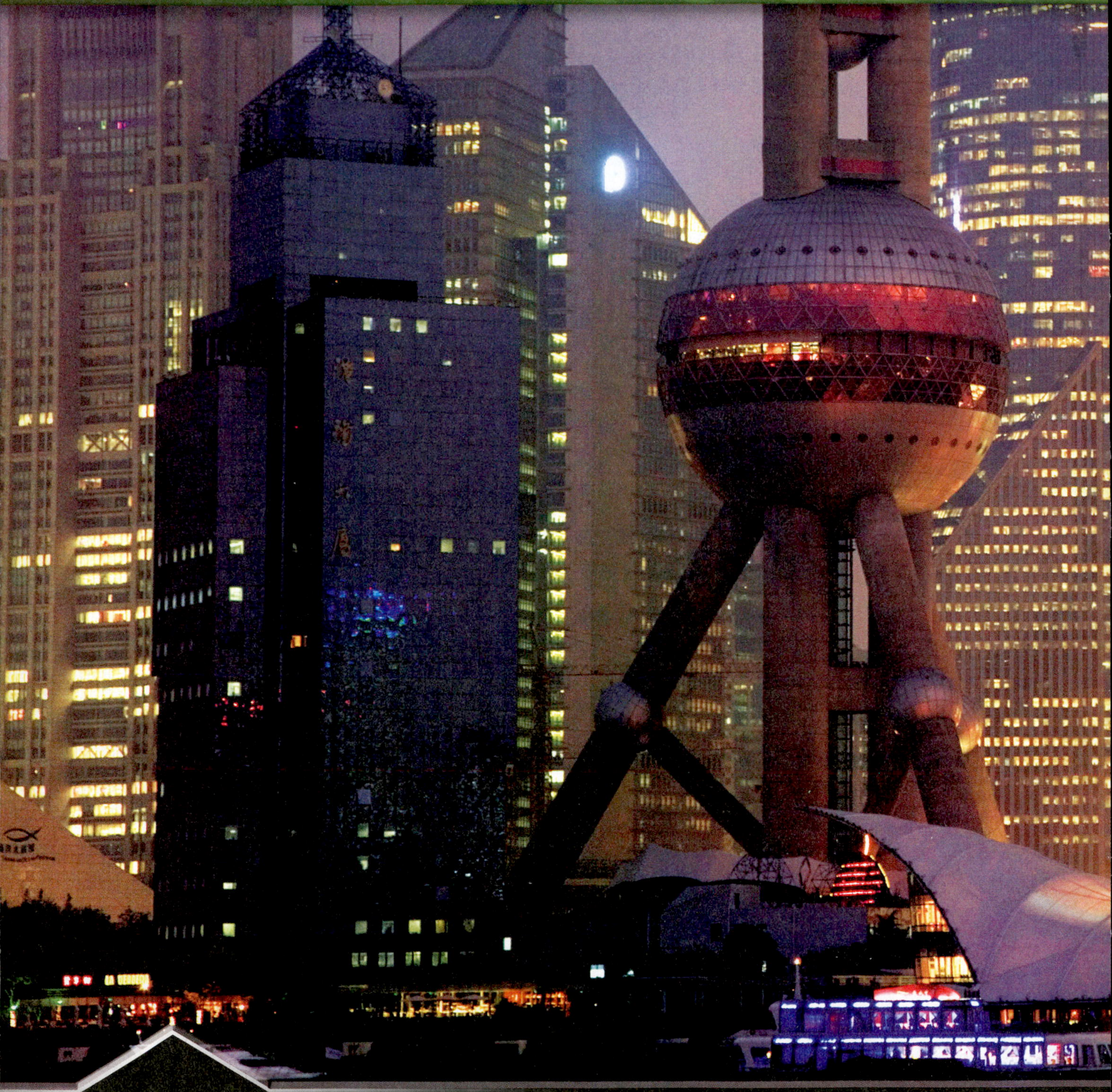

For more photos from the National Geographic Photo Gallery, go to the **Digital Library** at myNGconnect.com.

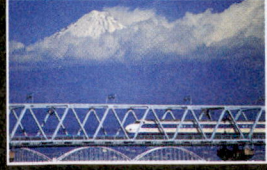

Bullet train passes Mt. Fuji

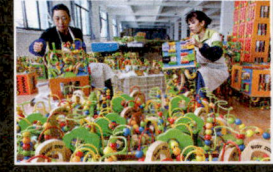

Toy factory, China

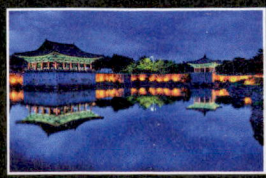

Silla palace, South Korea

Critical Viewing The Oriental Pearl Tower and other skyscrapers line the banks of the Huangpu River in Shanghai. The Oriental Pearl, which is a television tower, features the revolving restaurant shown in the photo.

Japanese Kabuki drama

Buddhist monastery, Tibet

Japanese tea ceremony

Celebration in Taiwan

1.4 Japanese Anime

Main Idea Anime reflects both Western and Japanese cultural influences and has become popular around the world.

Anime is a style of animation, or cartoon, developed in Japan. The characters and settings in this art form are either drawn by hand or computer-generated. Anime is closely related to **manga**, or Japanese comic books. Manga stories are told using a series of panels. Many anime films are based on the art and stories in popular manga. In fact, an anime film may even use manga panels to introduce a story.

⌃ Critical Viewing The artist in the top photo is drawing a panel, like that shown at bottom, for an anime film. What skills are probably needed to create anime?

Beginnings

Although anime films were produced in the early 1900s, the style and techniques used today didn't develop until the 1960s. Around that time, manga artist and animator Osamu Tezuka began to adapt some of the techniques used by Walt Disney in his own films. For example, Disney cartoons such as *Bambi* inspired Tezuka to draw characters with large eyes, which became a typical anime feature.

Anime uses elements from Western-style cartoons, but it also draws on Japanese culture. Ancient Japanese myths, Shinto, and Buddhism have all influenced many of the cartoons. For example, Shinto nature spirits populate the world depicted in *Spirited Away*, an anime film made in 2001 by director Hayao Miyazaki.

Popularity

Anime was a popular art form in Japan from the beginning, but Western audiences did not embrace it until the 1980s. Since then, anime has continued to gain fans— and respect—around the world.

One reason for anime's popularity is its broad range of subjects. Anime is used to tell stories in many different forms, including fantasy and science fiction. As a result, it appeals to all ages and both genders. Anime is also big business. In Japan, the art form and products related to anime earn more than $5 billion a year.

Before You Move On

Summarize In what ways does anime reflect Western and Japanese cultural influences, and why is it so popular?

This image shows a scene from Miyazaki's *Castle in the Sky*, in which a young boy and girl search for a floating castle.

Lighting and colors convey an exciting mood.

The boy and girl have round, expressive eyes like many anime characters.

The camera angle in this scene shows perspective and depth.

ONGOING ASSESSMENT

PHOTO LAB
GeoJournal

1. **Analyze Primary Sources** Examine the anime image above. Who might be the intended audience for the film? What details in the image support your ideas?

2. **Analyze Visuals** What personality traits do you detect in the characters shown above?

SECTION **1** CULTURE

TECHTREK
myNGconnect.com For an online map
and photos of high-speed trains in East Asia

Maps and Graphs   Digital Library

1.5 Bullet Trains

Main Idea Bullet trains and other high-speed rail services have transformed travel in much of East Asia.

Japan pioneered the use of high-speed rail in 1964, with trains that traveled at speeds of more than 125 miles per hour (mph). They earned the nickname **bullet trains** because of their appearance and speed. Each train has a rounded nose like that of an airplane and a sleek, **aerodynamic** design. This means that the train moves with little resistance from the wind. The technology has spread, and now bullet trains connect cities in countries throughout East Asia.

> **Critical Viewing** A Japanese bullet train arrives in a station in Tokyo. Based on the photo, how would you describe the location of the station?

High-Speed Rail Spreads

The first bullet trains linked Tokyo with Osaka and allowed people to travel the 320-mile route in four hours instead of six. Today, the system connects Tokyo with all major cities on the main island of Honshu. Additional lines are being built on Kyushu as well. The trains on these lines travel at speeds up to 186 mph. The newer trains are also quieter and more energy efficient.

By the 2000s, bullet trains were running throughout much of East Asia. In 2004, South Korea established a high-speed rail system, connecting Seoul with major industrial cities and ports. In 2007, a high-speed rail system began in Taiwan and connected Taipei with cities in the southwestern part of the country.

EAST ASIA HIGH-SPEED RAIL ROUTES, 2010

Legend:
— Current high-speed rail lines
— Planned high-speed rail routes

MONGOLIA
Ulaanbaatar

CHINA

Lanzhou
Baoji
Chengdu
Chongqing
Dali
Kunming
Guangzhou
Hong Kong
Hainan
Shijiazhuang
Taiyuan
Zhengzhou
Wuhan
Hangzhou
Beijing
Shenyang
Harbin
Shanghai
Ningbo
Fuzhou
Taipei
Kaohsiung
TAIWAN
Dalian
Qingdao
Yellow Sea
Gwangju
NORTH KOREA
Pyongyang
SOUTH KOREA
Seoul
Busan (Pusan)
Sea of Japan (East Sea)
Honshu
Hakodate
Tokyo
Osaka
JAPAN
Kyushu
Kagoshima
East China Sea
Philippine Sea
PACIFIC OCEAN
South China Sea
Tropic of Cancer

Scale: 0 300 600 Miles / 0 300 600 Kilometers

The Chinese began developing high-speed rail in the 1990s. By 2009, China operated the fastest trains in the world, with speeds up to 245 mph. China has also taken advantage of another high-speed train technology: the magnetic levitation train, or Maglev for short. A **magnetic levitation** train rides on a cushion of air over tracks laid with many powerful magnets. Today, Maglev trains run between the Shanghai airport and Pudong, while achieving a top speed of 268 mph.

Benefits

The greatest benefit of high-speed trains may be the time passengers save by riding them. However, the trains have also given an economic boost to once-remote rural locations that are now linked to cities and towns. In addition, bullet trains produce less pollution than traditional trains. In fact, the Maglev is almost pollution-free.

Before You Move On

Summarize In what ways have high-speed rail services transformed travel in East Asia?

ONGOING ASSESSMENT

MAP LAB
GeoJournal

1. **Movement** On the map, study the rail routes in Japan. What generalization can you make about the routes?

2. **Region** According to the map, in what part of the country are most of China's routes concentrated? Why do you think that is so?

3. **Make Inferences** Why might people prefer to travel from one city to another on a bullet train rather than on a plane or by car?

2.1 China's Economy Today

TECHTREK

myNGconnect.com For online graphs of China's GDP and photos of Chinese industry

 Maps and Graphs

Digital Library

Main Idea Since 1979, China has become a leader in the global economy.

As you have learned, the Chinese government established a Special Economic Zone, or SEZ, in Shanghai in 1990. You may remember that an SEZ is an area in which businesses are free to develop without government control. China began to allow the creation of SEZs in the early 1980s. As a result, the Chinese people today enjoy a higher standard of living, and China has the fastest-growing economy in the world.

New Strategy for Growth

China limited its SEZs to cities and provinces along the Pacific Coast because its leaders believed that this location would promote and facilitate overseas trade. The first SEZ was established in 1980 in Shenzhen, which at the time was a small fishing village just north of Hong Kong. Since then, Shenzhen has become a thriving city with about ten million people. Other important SEZs are in Xiamen and the province of Hainan.

Chinese leaders encouraged trade and foreign investment in these SEZs by offering lower taxes and fewer regulations on imports and exports. The lower costs also attracted **entrepreneurs**, people who start up new businesses. In addition, the lower wages paid to Chinese workers persuaded regional companies and multinational corporations to open branches in the country.

> **Critical Viewing** Chinese workers build electric bikes for domestic use and for export. What words would you use to describe the bikes in this photo?

Overall, China's strategy has been extremely successful. In 2010, China had the second largest economy in the world after the United States. Economists measure the size of a country's economy based on **gross domestic product (GDP)**, the total value of all the goods and services produced in a country in a given year. China's GDP increases about ten percent per year.

A Diverse Economy

China's growing economy is powered by industry, services, and agriculture. The country's main industries include mining and iron and steel manufacturing. Globally, China is most known for its exports of clothing and textiles, electronics, and toys. China exports more than any other single country in the world. About 20 percent of its exports go to the United States.

Inside China's borders, service industries provide activities such as banking, insurance, trade, communication, education, health care, recreation, and transportation. As China welcomes more and more visitors, tourism is also becoming a large part of its economy.

Agriculture contributes little to China's GDP, but it employs 40 percent of the country's workers. Only about 15 percent of Chinese land is arable, but the country produces most of the food it needs for its large population.

Before You Move On

Summarize What steps has China taken to grow and diversify its economy?

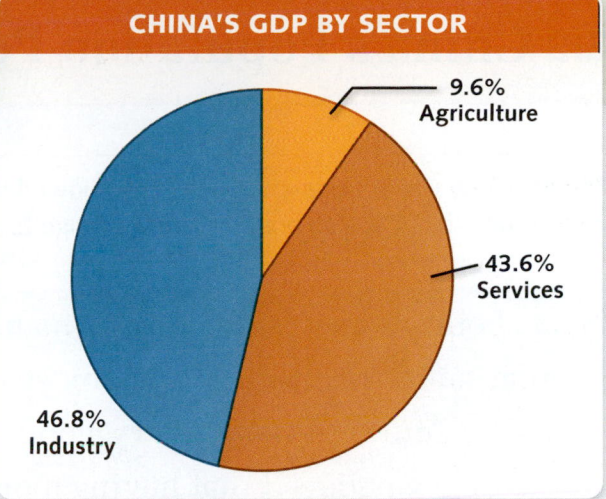

CHINA'S GDP BY SECTOR

9.6% Agriculture

43.6% Services

46.8% Industry

GDP OF SELECTED COUNTRIES

in Trillions of U.S. Dollars

China $9.872
India $4.046
United States $14.720

Source: CIA World Factbook 2010

ONGOING ASSESSMENT

DATA LAB  GeoJournal

1. **Draw Conclusions** According to the pie chart, how much of China's GDP is supplied by agriculture? Since 40 percent of the Chinese people work in agriculture, what conclusions can you draw about their earnings?

2. **Analyze Data** Study the bar graph and compare China's GDP with that of India. What do these numbers suggest about the relative rate of each country's GDP growth?

3. **Make Inferences** Based on what you have read, why are Chinese-made consumer goods sold at low prices in countries such as the United States?

TECHTREK

myNGconnect.com For an online population pyramid of China

Student Resources

Global Issues

2.2 China's Population Policy

Main Idea China's one-child policy has slowed the population growth rate, but it is also bringing great change to Chinese society.

With about 1.3 billion people, China has the largest population in the world. For many years, Communist leaders encouraged the country's population to grow because they believed that having more people—and workers—would strengthen China's economy. In time, however, it became difficult to provide food, housing, education, and jobs for the country's citizens.

KEY VOCABULARY

one-child policy, n., law that restricts urban Chinese families to one child

fertility rate, n., average number of children born per woman

ACADEMIC VOCABULARY

comply, v., follow or obey

Slowing Population Growth

To curb their rapidly growing population, the government introduced the **one-child policy** in 1979. This law limited families living in urban areas to one child. Those who **complied**, or followed, the policy were rewarded with more food, improved housing, and better education and job opportunities for that child. Those who did not comply had to pay large fines. As a result, the rate of population growth declined to about 0.65 percent per year by 2010, which is about half of what it was in 1979. There are some exceptions to the policy. For example, minority populations and rural families may have two or more children. The Chinese government also planned to loosen the policy in 2011 for people in provinces with low birth rates.

Before You Move On

Monitor Comprehension What policies did China put in place to decrease population growth?

CHINESE POPULATION, 2010

Source: U.S. Census Bureau, 2010

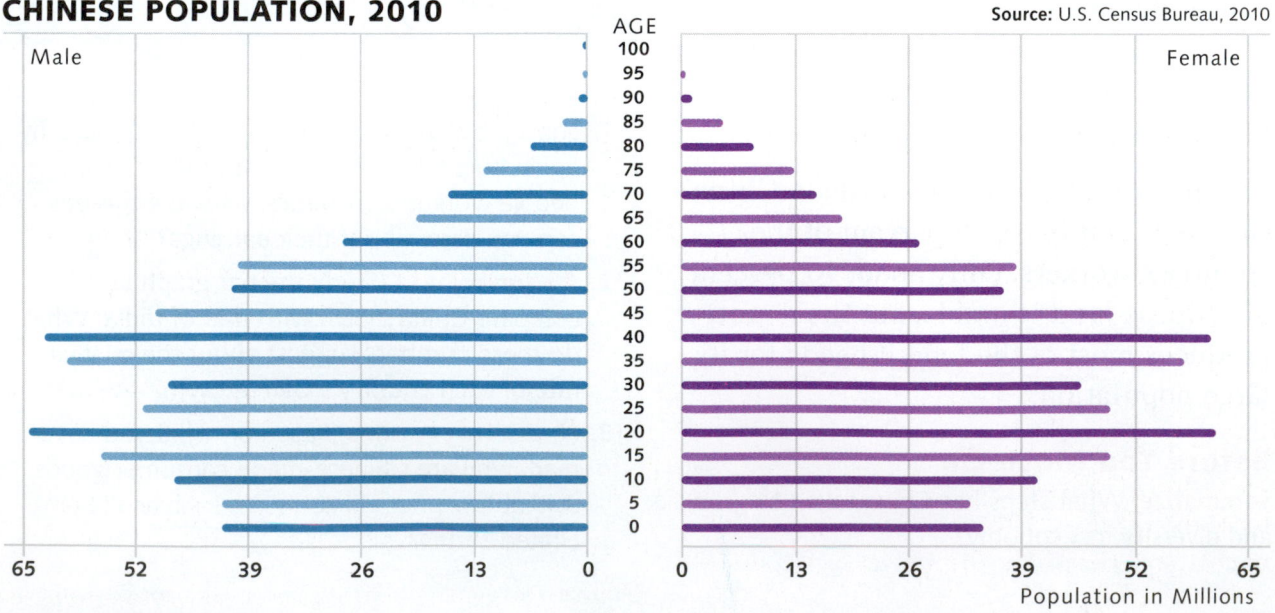

Effects of China's Policy

In general, China's one-child policy has worked. The **fertility rate**, or the average number of children born per woman, dropped from about six in the 1950s to about two by 1995. The Chinese government claimed that, since it was put in place, the policy has helped reduce the number of births in China by 400 million. According to the government, the policy has also improved the standard of living for many Chinese people.

The policy's biggest long-term impact will be on the age distribution of China's population. With fewer children being born, the elderly will soon outnumber young people in China. This aging population will place a serious burden on those born under the one-child policy, who will have to care for retiring seniors. Traditionally, daughters and daughters-in-law have been the primary caretakers of elderly parents. However, if a single adult child cannot take care of his or her parents, the seniors may have to depend on retirement funds or charity for support. In response, the Chinese government has begun to create community services to help the elderly who live alone.

The policy may also have an unintended impact. With no brothers and sisters, the "only" children produced by China's population policy are often seen as spoiled and selfish. Chinese society has traditionally emphasized children's respect for their parents. Now some parents complain that they are neglected by their adult children, who are focused mainly on their own lives and careers.

India's Population Growth

Like China, India has a huge population. With about 1.2 billion people, India is second only to China in population—for now. Scientists predict that India's population will exceed China's by 2030. In contrast with China, about 50 percent of India's population is under the age of 25. While this statistic guarantees that India will have a large workforce, the young population will also put a strain on the country's schools and other resources.

In spite of this growth, India does not enforce strict population policies. Instead, the country encourages the advancement and education of its women. India's leaders believe that educated women are more likely to have smaller families. The government also offers cash bonuses to women in rural areas who wait to start a family and agree to have fewer children.

Before You Move On

Summarize What have been the effects of China's one-child policy?

ONGOING ASSESSMENT

READING LAB GeoJournal

1. **Make Inferences** In what way might China's one-child policy lead to resentment between the younger and older generations?

2. **Interpret Graphs** Study the population pyramid. If these numbers hold, which groups will be the largest in 20 years? What impact might this situation have on China's economy?

3. **Form and Support Opinions** In your opinion, which country's policies have a better chance of working, China's or India's? Why?

TECHTREK
myNGconnect.com For photos of the
Three Gorges Dam and Guided Writing

Student Resources   Digital Library

Main Idea The Three Gorges Dam in China is the largest dam in the world and has both benefits and drawbacks.

The **Three Gorges Dam** on the Chang Jiang River is nearly 1.3 miles wide and 600 feet high. The dam was built to prevent flooding along the eastern part of the river between the cities of Chongqing and Wuhan. It was also designed to generate hydroelectric power, promote economic development in the region, and ease navigation along the Chang Jiang.

Building a Giant Dam

Geographic factors determined where the dam would be built. The Three Gorges is an area where the Chang Jiang is very narrow—only about 350 feet wide as the river flows between steep cliffs. The area is named for this geographic feature. A **gorge** is a deep, narrow passage surrounded by steep cliffs. When snows melt in the spring or heavy rains fall during summer monsoons, the river can rise very quickly in the narrow area. Over the years, this seasonal rise has caused major flooding and millions of deaths.

As work on the dam began in 1994, a large artificial lake called a **reservoir** formed behind it to store the water being held back. This reservoir is about 500 feet deep, and it is longer than Lake Superior in the United States, the largest natural body of fresh water in the world. During dry periods, the dam will slowly release water from the reservoir. This raises the level of the river downstream to allow large ships to travel to China's interior.

Work on the dam was completed in 2006, but it did not become fully operational until 2011. The dam is a major engineering achievement. However, it has stirred **controversy**, or debate, both within China and abroad.

Critical Viewing The Three Gorges Dam has been called China's Great Wall across the Chang Jiang. Based on this photo, what feelings might the dam inspire in the Chinese people?

Benefits and Drawbacks

The Chinese government expects to reap great benefits from the dam, which has already prevented serious flooding along the Chang Jiang. Improving conditions for commercial shipping along the river should increase economic development in the region. In addition, the government predicts that the hydroelectric power created by the dam will provide about 10 percent of China's total energy.

Critics, however, claim that the drawbacks of the dam, which cost an estimated $25 billion, outweigh the benefits. To create the reservoir, a large expanse of land had to be cleared. About 1.5 million people were displaced from their cities and villages, and more than 1,000 historic and cultural sites are now under water. The critics also fear that the dam may cause landslides and destroy the habitat of endangered animals. Many argue that building several smaller dams and using newer technology for producing energy would have been a better solution.

Before You Move On

Summarize Why did China build the Three Gorges Dam?

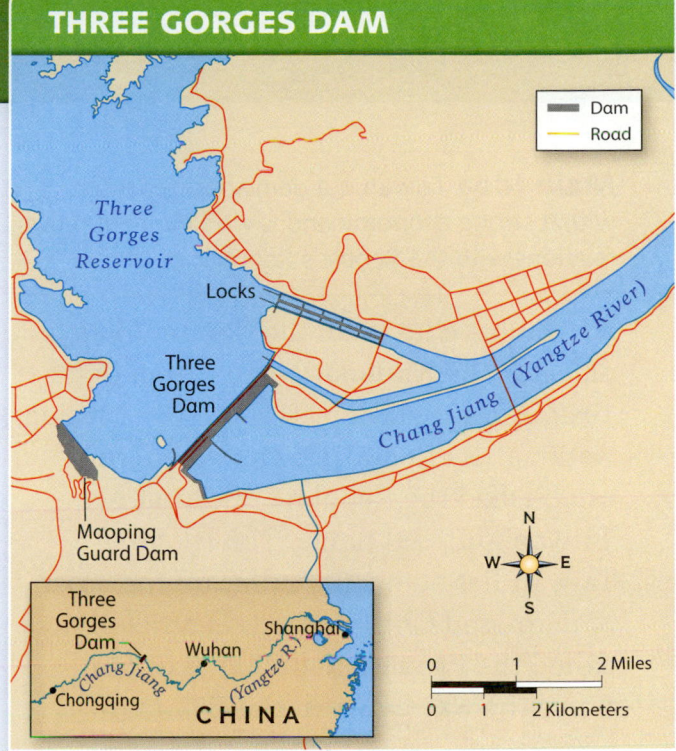

THREE GORGES DAM

Legend:
- Dam
- Road

Three Gorges Reservoir
Locks
Three Gorges Dam
Chang Jiang (Yangtze River)
Maoping Guard Dam

Three Gorges Dam
Wuhan
Shanghai
Chang Jiang (Yangtze R.)
Chongqing
CHINA

0 1 2 Miles
0 1 2 Kilometers

 MAP TIP
The smaller map shows where the Three Gorges Dam is located on the Chang Jiang. Notice that the dam is west of Shanghai in eastern China.

ONGOING ASSESSMENT

WRITING LAB   GeoJournal

Form and Support Opinions Do you think that building the Three Gorges Dam was a good idea? Use a chart like the one below to list the dam's benefits and drawbacks. Then write a brief paragraph in which you state and support your opinion with evidence. You may want to do additional research to learn more about the dam. Go to **Student Resources** for Guided Writing Support.

BENEFITS	DRAWBACKS
1.	
2.	

2.4 Republic of China (Taiwan)

TECHTREK

myNGconnect.com For an
online map and photos of Taiwan

Maps and
Graphs

Digital
Library

Main Idea Taiwan is a democratic country with a strong economy and is working to improve relations with the People's Republic of China.

As you have learned, the Communists defeated the Nationalists in China in 1949. After their defeat, about two million Nationalists fled to Taiwan where they established the Republic of China, Taiwan's official name. The island of Taiwan lies about 100 miles off the coast of mainland China, which is officially called the People's Republic of China. Since the war between the Communists and Nationalists never formally ended, there are still tensions between the two neighboring countries.

A Democratic Government

In 1949, the Nationalists placed Taiwan under **martial law**—a government maintained by military power—which continued into the 1980s. During their long, one-party rule, Nationalist leaders claimed that they had authority over all of China, including the mainland.

Beginning in 1987, more political parties were allowed to take part in elections, and martial law was lifted. As a result, people on the island experienced more democratic freedom. However, the country's political status, or legal position, has remained uncertain.

> **Visual Vocabulary** A large pagoda rises on a mountain in Taiwan. Most often found in Asia, a **pagoda** is a multistoried structure that is usually used for religious purposes.

Critical Viewing Motorcyclists wait at an intersection during rush hour in Taipei. What conclusions can you draw about Taiwan's capital based on this photo?

During the Cold War, many countries regarded Taiwan as the legitimate government of China. Today, these countries recognize the government of the People's Republic of China in Beijing but unofficially treat Taiwan as an independent state. Mainland China considers Taiwan one of its provinces. It has chosen not to challenge Taiwan, however, as long as the island country does not declare its independence.

A Strong Economy

During the 1960s and 1970s, Taiwan began to develop a market economy. Today, it has the 20th largest economy in the world with a gross domestic product (GDP) of about $800 billion.

Exports have fueled the growth of Taiwan's GDP. Important exports include electronics and machinery. Because Taiwan depends so much on exports, its economy declines when world demand for its goods decreases. Farming in Taiwan is limited because about two-thirds of the country is covered in mountains.

Taiwan wants to strengthen economic ties with mainland China, its leading export and import partner. Since 2010, Chinese investors have been able to invest directly in businesses in Taiwan, and Taiwanese financial firms have opened on mainland China. This improved economic cooperation may also lead to better diplomatic relations between the two.

Before You Move On

Monitor Comprehension What steps has Taiwan taken to improve its government, economy, and relations with China?

TAIWAN PHYSICAL

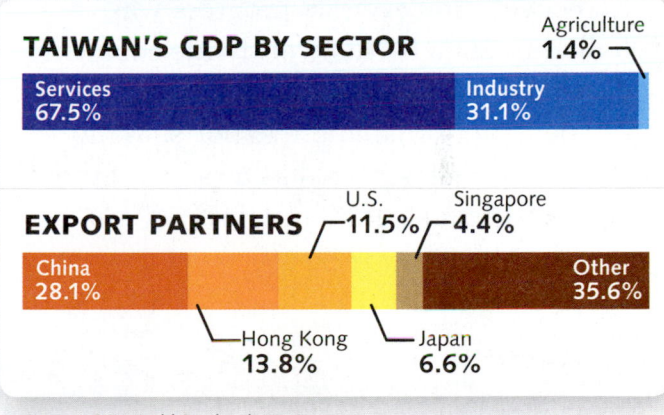

TAIWAN'S GDP BY SECTOR

| Services 67.5% | Industry 31.1% | Agriculture 1.4% |

EXPORT PARTNERS

China 28.1% — Hong Kong 13.8% — U.S. 11.5% — Singapore 4.4% — Japan 6.6% — Other 35.6%

Source: CIA World Factbook 2010

2.5 Japan's Economic Future

TECHTREK

myNGconnect.com For a population graph of Japan, photos of its industries, and current events

 Maps and Graphs   Digital Library   Connect to NG

Main Idea Japan's economy has declined sharply since the 1990s, and the country faces many challenges.

From the 1960s until the late 1980s, Japan had the strongest economy in East Asia. It was second only to the United States in gross domestic product (GDP) worldwide. However, by 1990, the country's economic growth had begun to slow down. Many businesses had borrowed huge amounts of money and were deeply in debt. With little **capital**, or money for investment, production declined. The situation continued throughout the 1990s, a period sometimes called Japan's **Lost Decade**.

In the 2000s, Japan's economy began to recover. However, its progress was halted in 2008 by the worldwide economic downturn, which led to a decrease in the demand for Japanese exports. Many industries also suffered as a result of the 2011 earthquake and tsunami.

Economy Today

Today, Japan's GDP is the third largest in the world, behind the United States and China. About 77 percent of Japan's economy is invested in service industries and about 22 percent in manufacturing.

Because Japan has few natural resources, the country must import most of the raw materials needed for its many industries. Japan's economy depends heavily on exported goods produced by these industries, including cars, computers, and other electronics.

Japan is still among the world's top five exporters. However, it has been less aggressive than many other Asian countries in promoting free trade with its neighbors. **Free trade** is trade that does not impose tariffs, or taxes on imports. These taxes can make Japan's exports more expensive than those of other Asian countries with free-trade agreements.

> **Critical Viewing** A technician inspects flat-screen televisions in a factory in Osaka. How do the televisions in this photo compare with those made in America?

Challenges Ahead

In addition to competition from other countries, Japan's aging population may have a harmful effect on the country's economy. In 2050, Japan's population is expected to be about 20 percent smaller than it was in 2009. Furthermore, about one-third of the people in 2050 are expected to be over age 65. These statistics suggest that Japan will have a shortage of workers as well as of entrepreneurs, since these risk takers are generally younger.

Currently, Japanese workers are required to retire at age 60. In addition, men hold most of the higher paying jobs in most industries. Japan could increase its workforce by encouraging older workers to keep working and by providing more economic opportunities for women. The country could also relax the limits it places on foreign investments in Japanese companies. Foreign investors could bring in new capital and ideas that might help make Japan become more competitive.

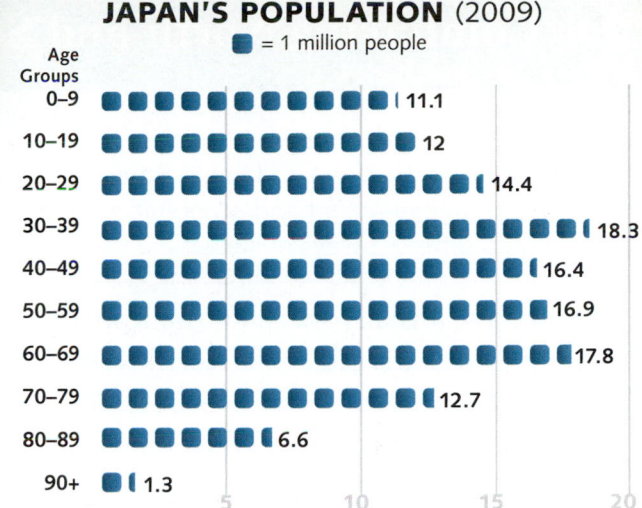

JAPAN'S POPULATION (2009)

■ = 1 million people

Age Groups	Value
0–9	11.1
10–19	12
20–29	14.4
30–39	18.3
40–49	16.4
50–59	16.9
60–69	17.8
70–79	12.7
80–89	6.6
90+	1.3

Source: Population Census, Statistics Bureau of Japan

Japan's economy also faced challenges following the 2011 earthquake and tsunami. After the disaster, manufacturing fell sharply and power supplies decreased when several of Japan's nuclear power plants became permanently disabled. The cost of repairing the damage caused by the natural disaster was estimated at about $300 billion. Some analysts feared that the disaster could put a halt to Japan's economic recovery.

Before You Move On

Summarize What were the causes and effects of Japan's economic decline?

ONGOING ASSESSMENT

DATA LAB

 GeoJournal

1. **Make Predictions** According to the graph, how many Japanese people were in the 0–9 age group in 2009? What will this statistic mean for Japan in 20 years?

2. **Interpret Graphs** Study the graph. What were Japan's four largest age groups in 2009? Why might these statistics have a negative impact on Japan's economy in 20 years?

3. **Draw Conclusions** What burdens might Japan's aging population place on young Japanese workers?

SECTION 2 GOVERNMENT & ECONOMICS

2.6 Comparing North and South Korea

TECHTREK

myNGconnect.com For current
events and photos of the two Koreas

Digital Library   Connect to NG

Main Idea North Korea and South Korea have different government and economic systems and different ways of life.

As you have already learned, the Korean Peninsula was divided into North Korea and South Korea after World War II. Since then, these neighbors have developed in very different ways.

North Korea

North Korea, which is officially called the Democratic People's Republic of Korea, has a Communist government that follows a policy of self-reliance. This means that the country depends mostly on its own efforts and abilities. The policy has kept North Korea isolated from most other countries.

The North Korean government controls most of the country's economy. The country's isolation policy has restricted trade, but it does conduct a limited amount of commerce, particularly with China. Much of its industry is devoted to the production of military equipment.

Only about 18 percent of North Korea's land is arable, and the climate does not support a long growing season. In the 1990s, floods and **drought**, a long period of extremely dry weather, greatly decreased food production. The resulting **famine**, or extreme food shortage, is believed to have caused as many as three million deaths.

South Korea

South Korea, which is officially called the Republic of Korea, was ruled largely by repressive military leaders until 1987.

Since then, the country has developed into a successful democracy, with free elections and multiple political parties.

Unlike North Korea, South Korea did not isolate itself from the rest of the world. It accepted grants and loans from countries such as the United States and Japan. As a result, South Korea's market economy grew quickly beginning in the 1960s. Today, it is the 13th largest economy in the world.

South Korea's economy depends heavily on manufacturing. Shipbuilding has long been a major industry. In fact, South Korea is the world's leading builder of ships. The country also produces automobiles and electronics. Like Japan, South Korea has few natural resources and relies largely on the goods it exports to fuel its economy.

Reunification

North Korea and South Korea have had a tense relationship since the end of the Korean War in 1953. In the early 2000s, meetings between the countries' leaders suggested that relations were improving. Attempts were made to encourage interaction between the two countries. However, North Korea's development of nuclear weapons and its military attack on a South Korean island in 2010 stirred up hostilities once again. Since then, there has been little hope that the two countries would reunite soon.

Before You Move On

Make Inferences What impact have the different governments and economic systems in North Korea and South Korea had on their people?

⌄ Critical Viewing This photo shows a street in Seoul, South Korea's capital. What does the photo suggest about the city?

This photo shows an empty street in the North Korean city of Kaesong. For the most part, only government officials own cars, which are considered symbols of status and power.

ONGOING ASSESSMENT

PHOTO LAB

GeoJournal

Turn and Talk Study the two photos of North Korea and South Korea. What do the photos suggest to you about what life might be like in each country? Discuss your thoughts with a partner and use details from the photos to support your ideas. Be ready to share your thoughts with the class.

NATIONAL GEOGRAPHIC

Photo Gallery • Buddhas in South Korea

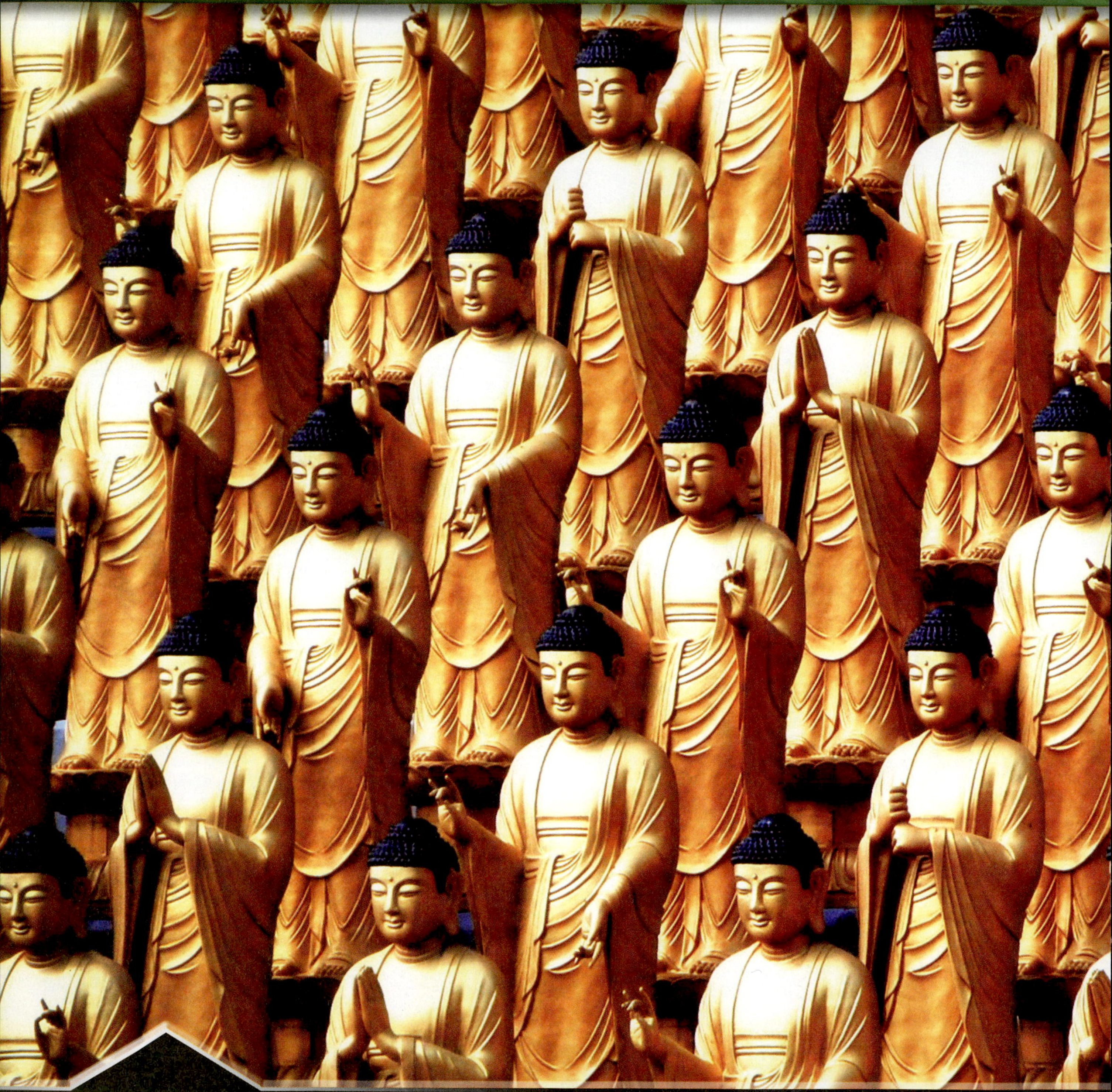

For more photos from the National Geographic Photo Gallery, go to the **Digital Library** at myNGconnect.com.

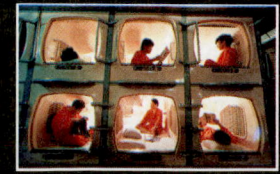

Capsule hotel, Japan

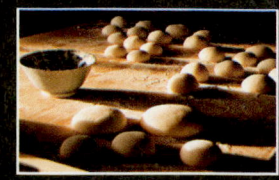

Japanese rice cakes

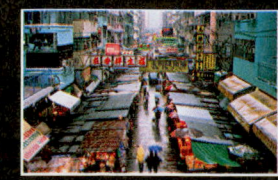

Hong Kong street market

Critical Viewing Rows of golden Buddhas welcome visitors to a South Korean temple. Buddhism spread from India to China in the A.D. 100s. Chinese monks eventually introduced the religion to the Korean Peninsula.

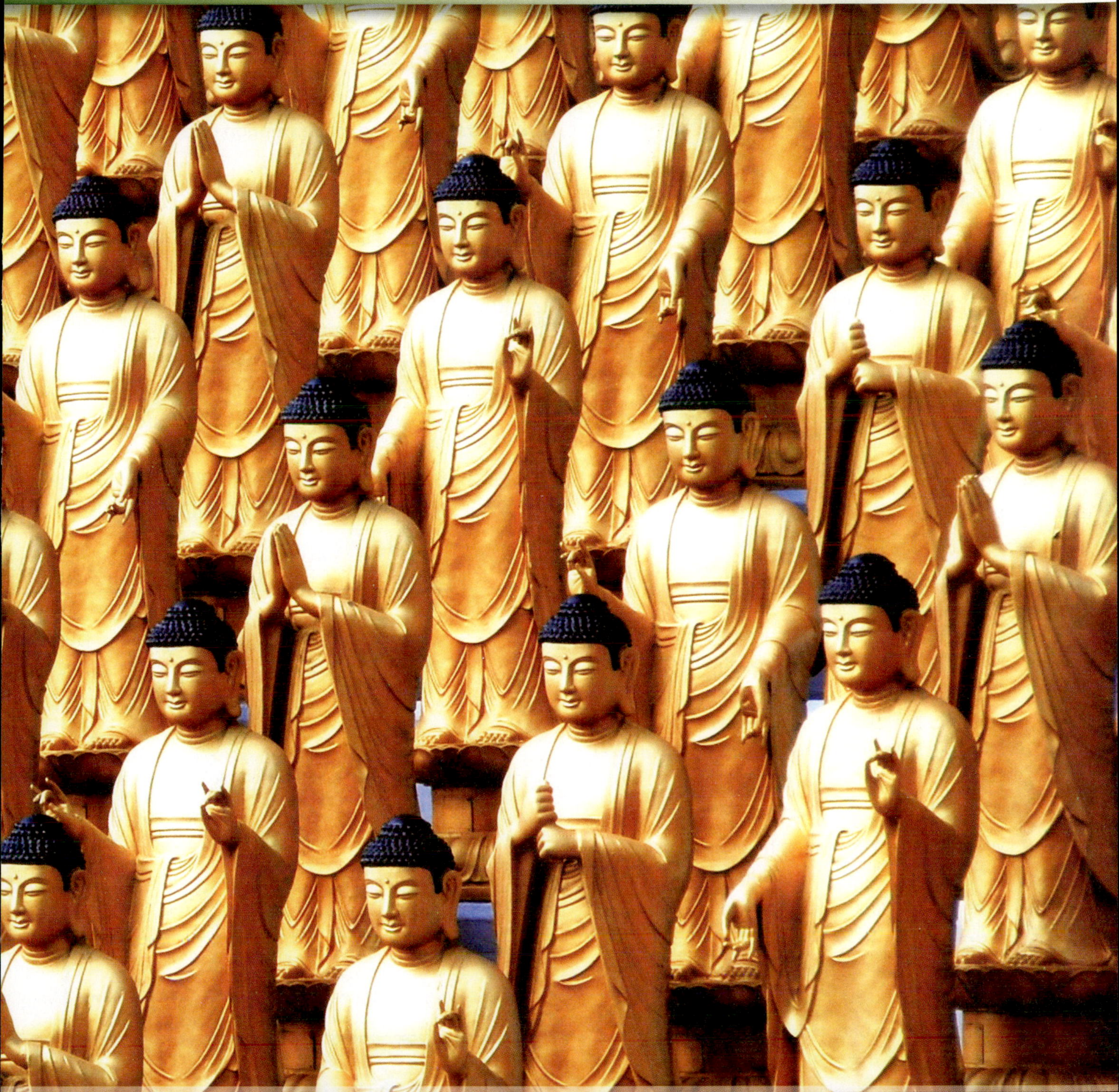

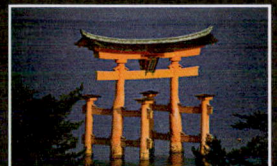

Shinto torii gate

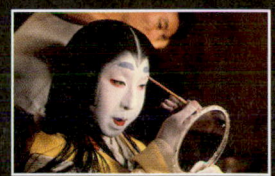

Kabuki theater actor

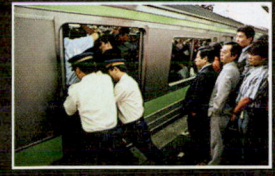

Crowded subway, Japan

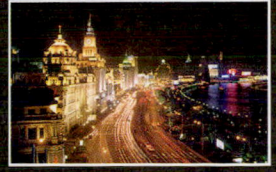

The Bund, Shanghai

VOCABULARY

For each pair of vocabulary words, write one sentence that explains the connection between the two words.

1. monotheism; polytheism

> *Belief in one god is called monotheism, while belief in more than one god is called polytheism.*

2. anime; manga

3. gorge; reservoir

4. free trade; capital

5. famine; drought

MAIN IDEAS

6. What are East Asia's religious traditions and in what manner did they arise in the region? (Section 1.1)

7. How did the Chinese use the inventions of movable type and paper money? (Section 1.2)

8. What factors helped Shanghai become China's largest city? (Section 1.3)

9. Why did Japan develop a high-speed rail system in the 1960s? (Section 1.5)

10. Where did China encourage most economic growth after 1979? (Section 2.1)

11. What did China hope to accomplish with its one-child population policy? (Section 2.2)

12. Why are some people critical of the Three Gorges Dam? (Section 2.3)

13. What is Taiwan's relationship with China? (Section 2.4)

14. In what way is Japan's population related to its economic future? (Section 2.5)

15. How do the governments and economic systems of North Korea and South Korea differ? (Section 2.6)

CULTURE

ANALYZE THE ESSENTIAL QUESTION

How do traditions and modernization create a unique way of life in East Asia?

Critical Thinking: Draw Conclusions

16. Why did many East Asians blend new religious ideas with traditional beliefs?

17. Why did China allow an area of Shanghai to become a Special Economic Zone?

18. In what ways does high-speed rail travel promote economic development in East Asia?

INTERPRET MAPS

LANGUAGES OF EAST ASIA

Color	Language	Color	Language	Color	Language
	Danzhou		Kalmyk-Oirat		Tibeto-Burman
	Darkhat		Kazakh		Tuvin
	Gan		Korean		Wu
	Hakka		Mandarin		Xiang
	Halt Mongolian		Min		Yue
	Hui		Mongolian Buriat		
	Japanese		Pinghua		Sparsely populated
	Jin		Southern Altai		

19. **Region** Which language is spoken in much of China? What might be the disadvantages for those who don't speak the language?

20. **Compare and Contrast** Which East Asian countries are most united by a common language? What does this fact suggest about the people of these countries?

GOVERNMENT & ECONOMICS

ANALYZE THE ESSENTIAL QUESTION

What problems does East Asia face today, and what are its opportunities?

Critical Thinking: Compare and Contrast

21. Compare the economic opportunities in China's coastal cities and rural areas. Which area provides greater opportunities? Why?

22. In what ways are the economies of Japan, South Korea, and Taiwan similar?

23. What do North Korea and South Korea have in common? Do you think these similarities might help the two countries unite some day? Explain why or why not.

INTERPRET CHARTS

LIFE IN THE TWO KOREAS, 2010		
	North Korea	South Korea
GDP (valued at U.S. prices)	$40 billion	$1.364 trillion
agriculture	23%	3%
industry	43%	39%
services	34%	58%
GDP per person	$1,900	$28,000
Life expectancy at birth	63.81 years	78.72 years
Infant mortality	51.3 deaths/ 1,000 live births	4.26 deaths/ 1,000 live births
Literacy	99%	97.9%

Sources: CIA World Factbook 2010

24. **Draw Conclusions** Based on the statistics in the chart under GDP per person, what conclusions can you draw about the standard of living in North Korea and South Korea?

25. **Make Inferences** Study the statistics on infant mortality. What do these numbers suggest about health care in each country?

ACTIVE OPTIONS

Synthesize the Essential Questions by completing the activities below.

26. **Create a Web Page** Make a home page for a Web site about East Asia today. Write a short introduction to the region and include links to each country. Use visuals, including photos, maps, charts, and graphs, that illustrate East Asia's culture and economic endeavors. Use the tips below to help you prepare your Web page. **Display your Web page and invite your classmates to "click" on the links. Be prepared to summarize what your viewers would find on each link.**

Writing Tips
- Use bullet points to summarize cultural and economic topics in East Asia today.
- Provide labels for your links that will capture your audience's attention.
- Include captions with your visuals that are brief and informative.

 **TECHTREK** myNGconnect.com For research links on East Asia today

27. **Design an Anime Storyboard** Get together in a small group to design a storyboard for a scene in an anime film. First, brainstorm to come up with ideas for the scene's characters and plot. You can use the research links at **Connect to NG** to learn more about anime. Then sketch and write dialogue for each panel of the scene. Use a chart like the one below to organize your ideas.

PANEL 1	PANEL 2	PANEL 3
Characters:	Characters:	Characters:
Action:	Action:	Action:
Dialogue:	Dialogue:	Dialogue:

Cell Phone Usage

  TECHTREK

myNGconnect.com For an online graph and research links on cell phone use

Maps and Graphs

Connect to NG

In this unit, you learned about the strong economies in much of East Asia. One of the factors that could further stimulate economic growth in the region is cell phone use. Some studies have suggested that cell phone usage can increase gross domestic product (GDP) by about 0.5 percent. For a country the size of China—with more than 1.3 billion people—this would mean an increase of about $12 billion. China might already be benefiting from such an increase because the country leads the world in the cell phone market. In 2007, over 500 million Chinese used cell phones. By contrast, subscribers in the United States, which is fourth in the market, numbered only about half as many.

Compare

- Argentina
- Bangladesh
- Brazil
- China
- India
- Mongolia
- Nepal
- South Korea
- United States
- Venezuela

CELLS VERSUS LANDLINES

Before the development of cell phones, people used landline telephones. This type of phone requires a system of wires or cables to be functional. Landline phones are common in developed countries where these systems were established decades ago at a relatively low price.

In developing countries, by contrast, landline systems were often only installed in large cities and at great expense. Only workers in large urban areas could afford a telephone. People in rural locations— more than half the population in many developing countries—were left with no communications network.

Cell phones, which have low set-up fees, changed communications in developing countries. Today, many people in these countries are connected.

ECONOMIC BENEFITS

In some developing countries, the cost of a cell phone may equal two or three months of a worker's income, However, many people believe that the benefits of a cell phone outweigh its purchase price. The following are some of the economic benefits of cell phone use:

- Small business owners can use the phones to keep informed of prices and other current market news.
- Government benefits with the revenue it receives from licenses and taxes.
- Mobile networks attract more foreign investment in the country.

Cell phone use is on the rise in both East Asia and South Asia. The graphs on the next page show cell phone usage in selected countries within these two regions. Compare the data in the graphs and use it to answer the questions.

CELL PHONE USAGE*

NUMBER OF PEOPLE USING CELL PHONES

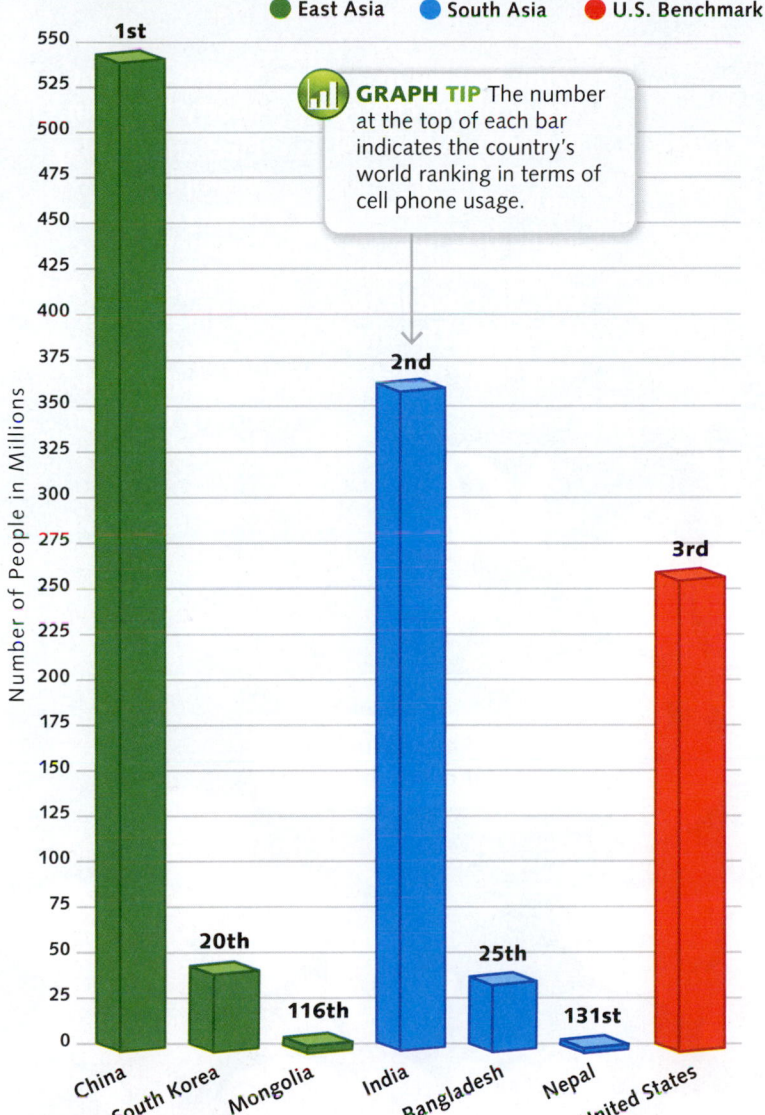

- ● East Asia
- ● South Asia
- ● U.S. Benchmark

GRAPH TIP The number at the top of each bar indicates the country's world ranking in terms of cell phone usage.

Number of People in Millions

1st — China
2nd — India
3rd — United States
20th — South Korea
25th — Bangladesh
116th — Mongolia
131st — Nepal

*Sources: UN Database, 2007

PERCENTAGE OF POPULATION

These graphs show the percentage of the country's population that uses cell phones.

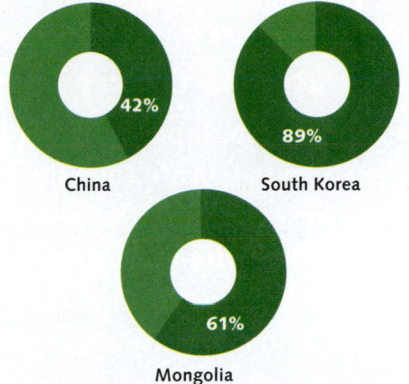

China 42%
South Korea 89%

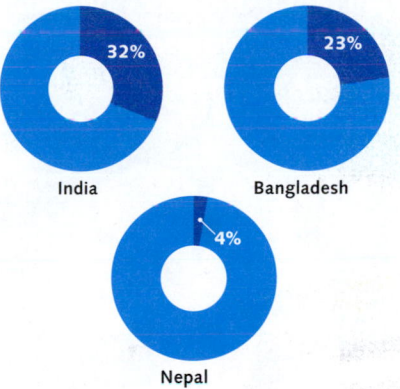

Mongolia 61%

India 32%
Bangladesh 23%

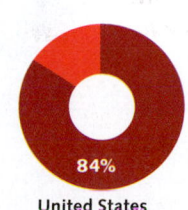

Nepal 4%

United States 84%

RESEARCH LAB   GeoJournal

1. **Analyze Data** Study the numbers for China and India. Why do you suppose these countries have the most cell phone subscribers?

2. **Draw Conclusions** Note that Mongolia has more than 2.5 million people while Nepal has more than 29 million. What conclusions can you draw about economic development in Mongolia and Nepal based on their cell phone usage?

Research and Compare Research cell phone usage in the South American countries of Brazil, Argentina, and Venezuela. Use graphs similar to those above to compare the numbers of cell phone subscribers in those countries. Then compare your findings with the statistics shown here. Which South American country has the most cell phone use? What might this statistic suggest about economic development in the country?

Active Options

TECHTREK

myNGconnect.com For photos of cherry blossoms in Japan and Korean kimchi

 Digital Library   **Connect to NG**   **Magazine Maker**

ACTIVITY 1

Goal: Learn about the importance of cherry blossoms in Japan.

Celebrate the Cherry Blossom

Every year, the Japanese people hold festivals to celebrate the cherry blossom. In Japanese culture, the beauty of the blossoms and the brief duration of their bloom symbolize the fleeting nature of life. The blossoms are often used in manga and anime. Research to learn more about the importance of cherry blossoms in Japan. Then choose one of the following ways to celebrate the flower:

- Write a haiku in honor of the cherry blossom. A haiku is a three-line poem with five syllables in the first line, seven in the second line, and five in the third line. The poem does not rhyme.

- Sketch or paint a cherry blossom.
- Find and display examples of cherry blossoms in Japanese art.
- Research and report on three other places outside Japan that hold cherry blossom festivals.

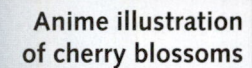
Anime illustration of cherry blossoms

ACTIVITY 2

Goal: Extend your knowledge of China's culture.

Create a Chinese Culture Magazine

Learn more about China's rich cultural heritage. Get together in a small group to design a magazine page on one part of Chinese culture. You might focus on the country's art, music, food, or martial arts. Research your topic and use the **Magazine Maker** to create your page. Combine the pages created by different groups into a magazine.

ACTIVITY 3

Goal: Find out about Korean cuisine.

Learn about Kimchi

A dish called kimchi is Korea's best-known food. Kimchi is made of pickled vegetables and eaten at almost every meal. Find out more about kimchi, including how it is prepared and stored. Share your findings with the class.

explore
Southeast Asia
with **NATIONAL GEOGRAPHIC**

MEET THE EXPLORER

NATIONAL GEOGRAPHIC

Emerging Explorer Jenny Daltry searches for unknown species of snakes, frogs, and crocodiles in unexplored corners of South Asia and Southeast Asia. Her work helps conserve these animals' habitats. Here, she inspects the fangs of a snake.

INVESTIGATE GEOGRAPHY

Mount Merapi is a volcanic mountain peak located near the center of the densely populated island of Java, Indonesia. It is the most active of the country's volcanoes. Its ash creates fertile soil, luring farmers in spite of the dangers.

STEP INTO HISTORY

A Buddhist monk prays at a statue in the Angkor Wat temple complex, Cambodia. Angkor Wat was originally built as a Hindu worship center in the 12th century. It remains the largest religious structure in the world.

CONNECT WITH THE CULTURE

The busy floating market in Damnern Saduak, Thailand has been attracting buyers since 1872. Hats, shown here, along with fruits, vegetables, flowers, and other food, are available.

427

CHAPTER 15

Southeast Asia

GEOGRAPHY & HISTORY

PREVIEW THE CHAPTER

Essential Question What are the geographic conditions that divide Southeast Asia into many different parts?

SECTION 1 • GEOGRAPHY

KEY VOCABULARY

- land bridge
- landlocked
- typhoon
- tsunami
- subsistence fishing
- ecologist
- bauxite
- biodiversity
- dynamic
- dormant
- zoologist
- wallaby

ACADEMIC VOCABULARY

enhance

TERMS & NAMES

- Ring of Fire
- Mekong River
- Chao Phraya River
- Irrawaddy River
- Malay Peninsula
- Foja Mountains

Essential Question How have physical barriers in Southeast Asia influenced its history?

SECTION 2 • HISTORY

KEY VOCABULARY

- complex
- bas-relief
- monopoly
- colonialism
- fossil
- commerce
- launch
- resistance

ACADEMIC VOCABULARY

transform

TERMS & NAMES

- Khmer Empire
- Angkor Wat
- Borobudur
- Dutch East India Company
- Manila
- Emilio Aguinaldo
- Ho Chi Minh

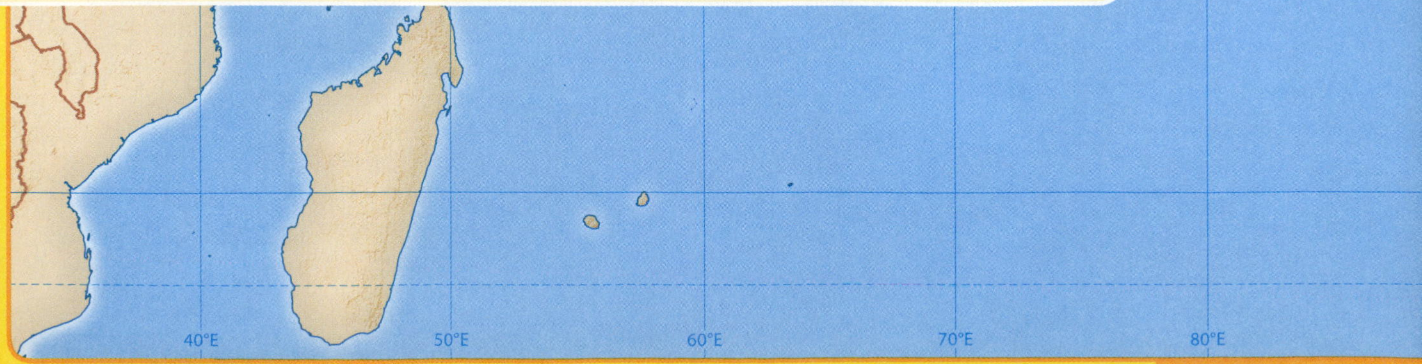

Komodo dragon

TECHTREK FOR THIS CHAPTER

Student eEdition

Maps and Graphs

Interactive Whiteboard GeoActivities

Digital Library

Connect to NG

Go to **myNGconnect.com** for more on Southeast Asia.

MYANMAR (BURMA)

Nay Pyi Taw

Yangon (Rangoon)

Irrawaddy R.

Salween R.

Red R.

Black R.

Hanoi

LAOS

Vientiane

Ping R.

Mekong R.

VIETNAM

THAILAND

Krung Thep (Bangkok)

CAMBODIA

Phnom Penh

Ho Chi Minh City (Saigon)

Gulf of Thailand

Andaman Sea

Strait of Malacca

SUMATRA

Kuala Lumpur

SINGAPORE

MALAYSIA

Bandar Seri Begawan

BRUNEI

Kapuas R.

BORNEO

CELEBES

South China Sea

Sulu Sea

Celebes Sea

Manila

PHILIPPINES

Philippine Sea

PACIFIC OCEAN

Tropic of Cancer

Equator

INDONESIA

Java Sea

Jakarta

JAVA

Banda Sea

Dili

TIMOR-LESTE (EAST TIMOR)

Timor Sea

Arafura Sea

NEW GUINEA

INDIAN OCEAN

N
W E
S

0   300   600 Miles
0   300   600 Kilometers

Tropic of Capricorn

A B C D E F G H

2 3 4 5 6

100°E 110°E 120°E 130°E 140°E

30°N 20°N 10°N 0° 10°S 20°S

SOUTHEAST ASIA PHYSICAL

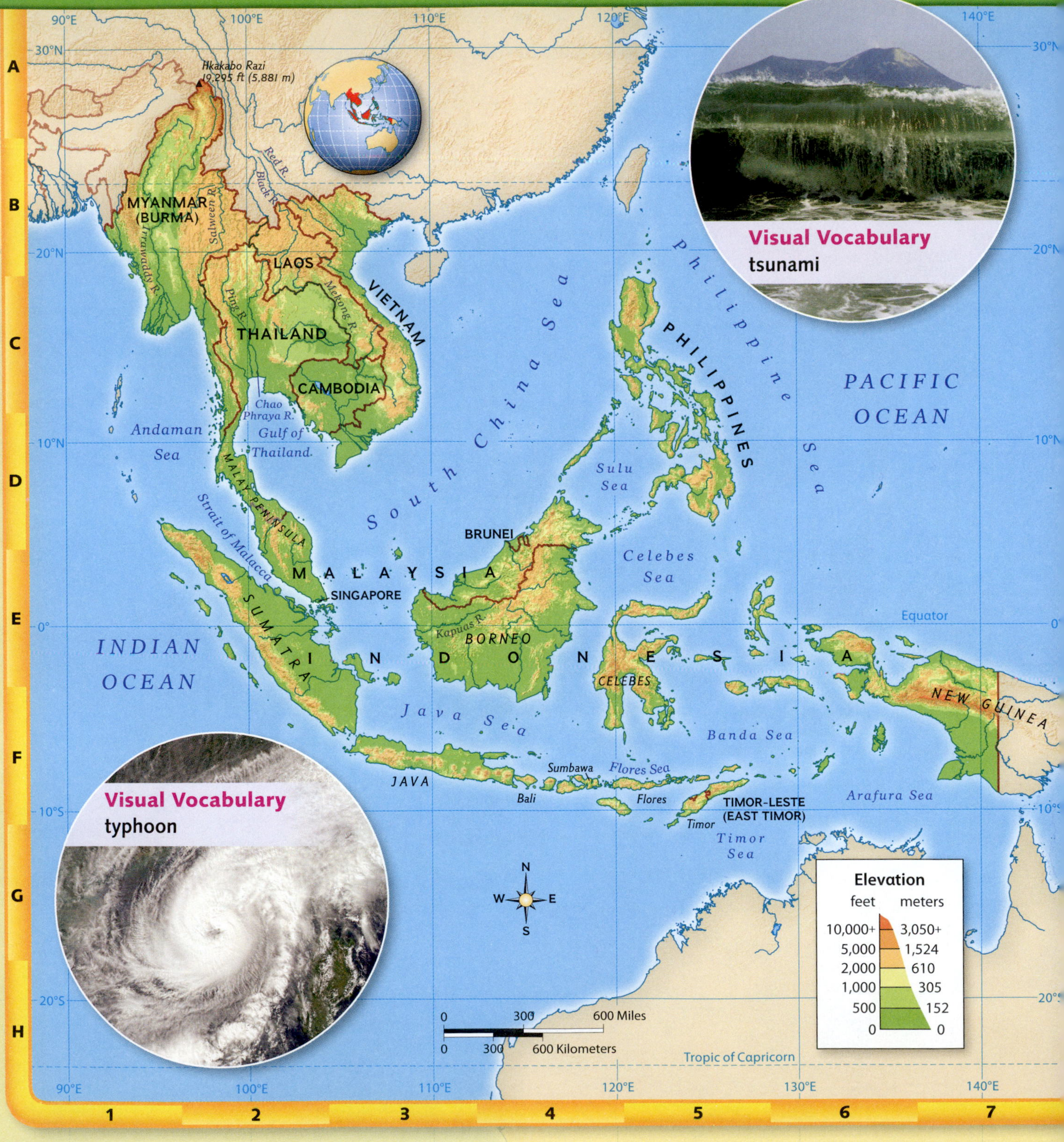

Hkakabo Razi
(19,295 ft (5,881 m)

MYANMAR
(BURMA)

LAOS

VIETNAM

THAILAND

CAMBODIA

Andaman Sea

Chao Phraya R.

Gulf of Thailand

MALAY PENINSULA

Strait of Malacca

SUMATRA

INDIAN OCEAN

MALAYSIA

SINGAPORE

Kapuas R.

BORNEO

I   N   D   O   N   E   S   I   A

Java Sea

JAVA

Bali

Sumbawa

Flores Sea

Flores

Timor

South China Sea

BRUNEI

Sulu Sea

Celebes Sea

CELEBES

PHILIPPINES

Philippine Sea

PACIFIC OCEAN

Equator

NEW GUINEA

Banda Sea

Arafura Sea

TIMOR-LESTE
(EAST TIMOR)

Timor Sea

Visual Vocabulary
tsunami

Visual Vocabulary
typhoon

N
W · E
S

Tropic of Capricorn

Elevation	
feet	meters
10,000+	3,050+
5,000	1,524
2,000	610
1,000	305
500	152
0	0

0   300   600 Miles
0   300   600 Kilometers

Main Idea Southeast Asia is a mountainous region with both mainland and island countries.

Southeast Asia has two kinds of countries: mainland and island. Indonesia and the Philippines are islands that were once connected by **land bridges**, strips of land connecting two land masses. Glaciers that melted over 6,000 years ago caused the sea level to rise, which separated these land masses. Malaysia is unique in that it includes land on the Asian continent as well as the island of Borneo.

Mainland Countries

The region's mainland countries are part of the Asian continent and include Myanmar, Thailand, Cambodia, Vietnam, and Laos. This cluster is linked by a long coastline. Only Laos is **landlocked**, or surrounded by land on all sides. Elevation is generally higher in the northern and eastern coast of mainland Southeast Asia. Mountains hold the source of several of the region's major rivers, which people rely on for transportation, food, drinking water, and irrigation.

The people of Myanmar, Laos, Vietnam, and Cambodia live mostly in small villages in the mountains or near waterways. However, many of the region's river deltas are densely populated. Bangkok, the country's most developed and densely populated city, is in the delta of the Chao Phraya River. Sediment deposits from this and other rivers created the fertile soil of Thailand's Central Plain, which is ideal for growing rice.

Southeast Asia generally has a tropical climate, although temperatures vary based on elevation and distance from the ocean. Mainland countries receive rainfall from May to September, the wet monsoon season. **Typhoons**, fierce tropical storms with heavy rains and high winds, often strike during this time. The rest of the year is the dry monsoon season.

Island Countries

Southeast Asia's island countries include Indonesia and the Philippines. They sit on the **Ring of Fire**, a volcanic zone around the Pacific Ocean where the plates that make up the earth's crust meet. Both countries have many active volcanoes. Undersea earthquakes can cause a **tsunami** (soo NAH mee), a giant ocean wave with enormous power. In 2004, an earthquake-caused tsunami near Sumatra killed at least 225,000 people. Its effects were felt as far away as East Africa.

Before You Move On

Monitor Comprehension How did islands form in the region?

ONGOING ASSESSMENT

MAP LAB  GeoJournal

1. **Categorize** Use the map and text to explain the differences between the two types of countries that make up Southeast Asia.

2. **Make Inferences** Find Laos on the map. What difficulties might it face as a result of being landlocked? What might be a benefit?

3. **Human-Enviroment Interaction** Why are the region's river deltas densely populated?

1.2 Parallel Rivers

TECHTREK

myNGconnect.com For a map and photos of Southeast Asia's rivers

Maps and Graphs   Digital Library

> **Main Idea** River systems in Southeast Asia support life in many ways.

Three parallel rivers run through mainland Southeast Asia: the Mekong (may KONG), the Chao Phraya (chow PRY uh), and the Irrawaddy. They begin in the highlands and flow south through valleys between mountains. As they approach the sea, they divide into a triangular shape made up of smaller streams. These river deltas are composed of silt, or fertile soil the rivers carried from upstream.

The Mekong River

At 2,600 miles, the **Mekong River** is the longest in Southeast Asia. It runs through the middle of the mainland and forms part of the borders of Myanmar, Laos, and Thailand. The mouth of the river, where it empties into the South China Sea, is in Vietnam near Ho Chi Minh City.

The Mekong Delta covers nearly 25,000 square miles, about the size of West Virginia. The densely populated delta is a rich rice-growing region. Some countries in the region are working to harness the river's power to produce hydroelectricity.

The Chao Phraya River

The **Chao Phraya River** is the most important river in Thailand. It is used to irrigate rice fields and serves as a major transportation route through the country. The capital city of Bangkok is located along its banks.

The Irrawaddy River

The **Irrawaddy River** 3 is about half as long as the Mekong. It also supports rice farming and is used as a transportation network. As a result of the soil carried by the river and dumped at its mouth, the delta is growing by about 165 feet a year.

Visual Vocabulary An ecologist is a scientist who studies the relationship between organisms and their environments. National Geographic Fellow Zeb Hogan is an aquatic ecologist working here in the waters of the Mekong River.

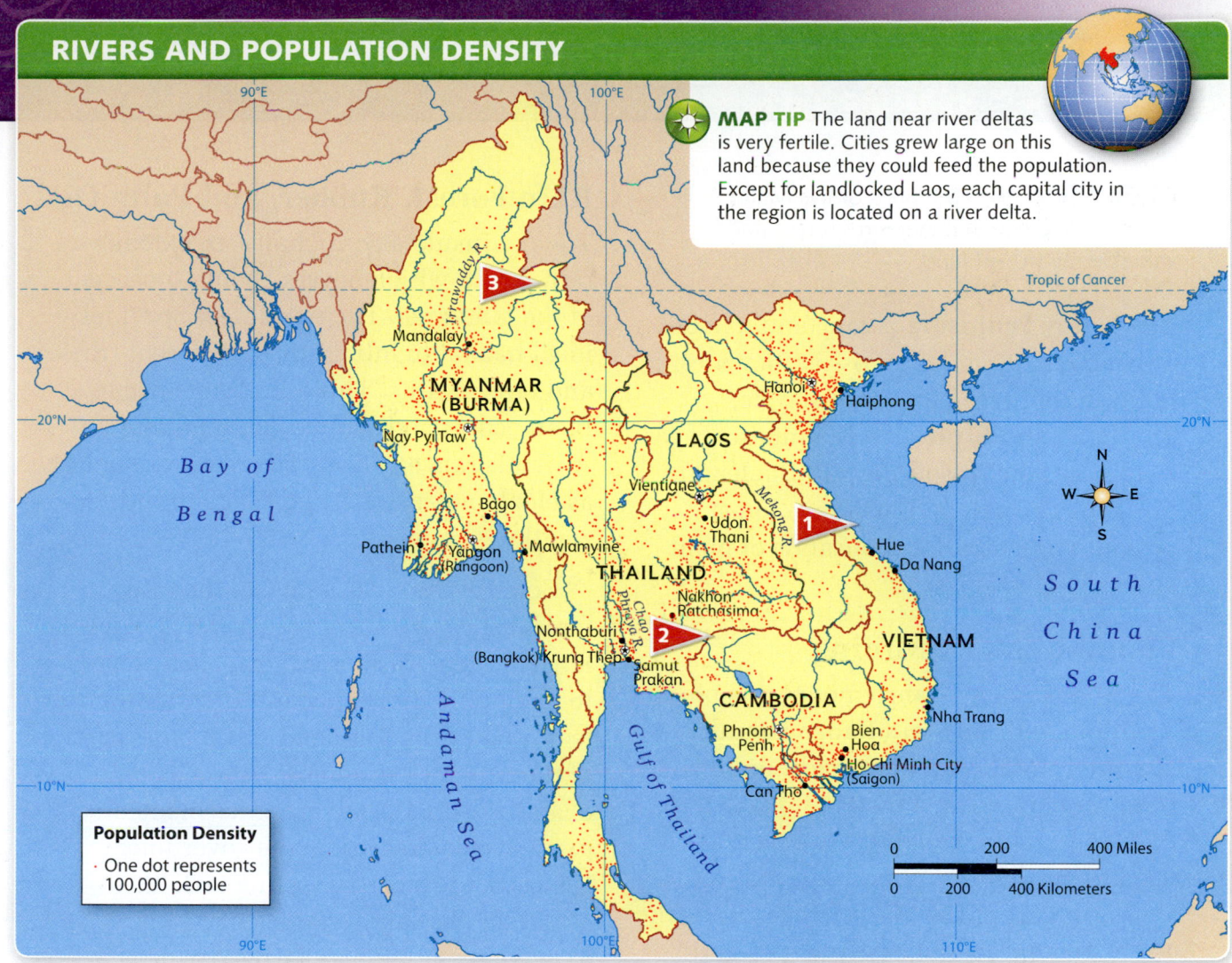

MAP TIP The land near river deltas is very fertile. Cities grew large on this land because they could feed the population. Except for landlocked Laos, each capital city in the region is located on a river delta.

Population Density
· One dot represents 100,000 people

In the rainy season, the Irrawaddy can rise more than 30 feet. Ports must have two areas for docking, one for each season. Farmers have adapted by storing water during the rainy season and releasing it onto their fields in the dry season.

Challenges of the Rivers

The rivers of Southeast Asia are used by many people for **subsistence fishing**, catching just enough fish to live on. Countries in the region must work together to control the threat of overfishing. **Ecologist** Zeb Hogan is part of a program that buys live fish from local fishermen for study. Efforts like this can protect endangered fish while still allowing local people to make a living.

Dams built along the rivers help control water levels, but they also sometimes interfere with transportation and disrupt river environments.

Before You Move On

Summarize What do river systems provide to support life in the region?

ONGOING ASSESSMENT
READING LAB 📓 GeoJournal

1. **Monitor Comprehension** How are dams a threat to the region's rivers?

2. **Make Inferences** Why is water level a challenge for people who depend on the rivers?

3. **Region** Trace the course of the Mekong River on the map. What countries share this river?

1.3 The Malay Peninsula

TECHTREK

myNGconnect.com For maps and photos of the Malay Peninsula

Maps and Graphs

Digital Library

Main Idea The mountains of the Malay Peninsula are rich in mineral resources and valuable rain forest land.

The **Malay Peninsula** is long and narrow, only about 200 miles across at its widest point. It includes parts of Malaysia, Thailand, and Myanmar. Mountain ranges rich in mineral resources run the length of the peninsula, and lush rain forest provides a habitat for thousands of plant and animal species.

Mountains and Mining

The Bilaukataung Range mountains of Thailand and the Main Range mountains of western Malaysia have traditionally been mined for tin, a metal often used in food containers. **Bauxite**, the raw material used to make aluminum, is mined in the southern part of the Main Range. Since the 1970s, the number of easily accessed tin and bauxite deposits has been shrinking, causing a steady decline in mining.

Rain Forest, Rubber, and Palm Oil

The peninsula also includes an extensive rain forest, which covers about 40 percent of the land area. The rain forest provides ideal habitat to hundreds of different trees and other plants. This variety of species in an ocosytem is called **biodiversity**. Animals range from large creatures such as elephants, rhinos, and tigers to the very small deer mouse.

Some of the trees native to the rain forest have significant value, and as a result large areas of rain forest have been cleared to plant only those species. At one time, teak wood from Thailand was a large part of the country's economy. However, after a landslide in 1989, which was blamed on excessive deforestation, the government imposed a ban on harvesting teak. In Malaysia, the rain forest is cut down to make room for large farms of rubber and palm oil trees. Palm oil is used with machinery, to make soap, and for cooking.

^ **Critical Viewing** This land in Malaysia is being prepared for a palm oil farm. What can you infer was present before the land was cleared?

Villlage on a hillside in Cameron Highlands, Malaysia

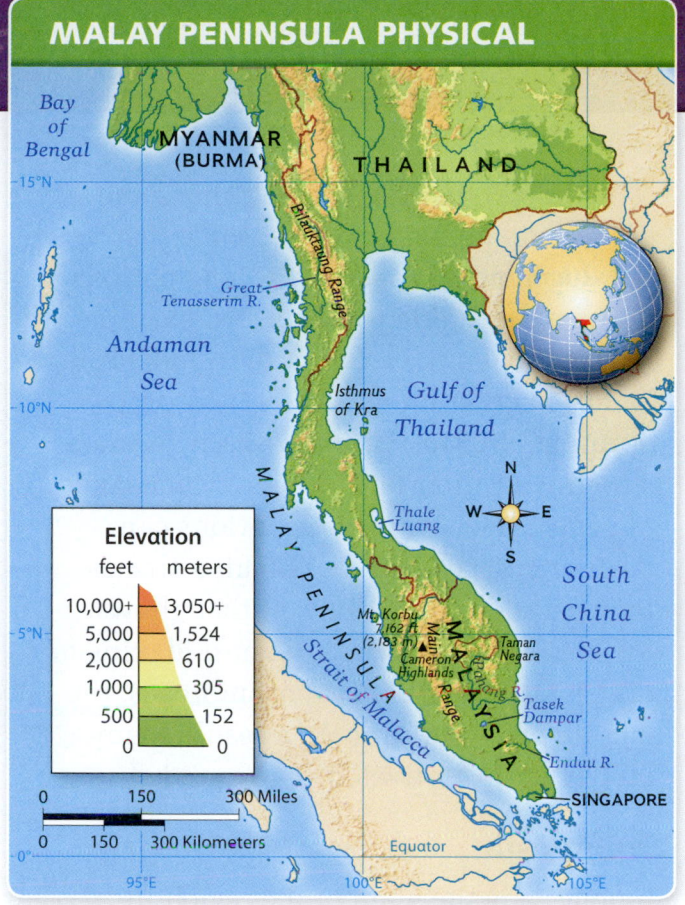

MALAY PENINSULA PHYSICAL

Bay of Bengal

MYANMAR (BURMA)

THAILAND

Bilauktaung Range

Great Tenasserim R.

Andaman Sea

Isthmus of Kra

Gulf of Thailand

MALAY PENINSULA

Thale Luang

South China Sea

Elevation

feet	meters
10,000+	3,050+
5,000	1,524
2,000	610
1,000	305
500	152
0	0

Mt. Korbu 7,162 ft (2,183 m)

Maur R.

Taman Negara

Cameron Highlands

Strait of Malacca

Pahang Range

Tasek Dampar

Endau R.

SINGAPORE

Equator

0   150   300 Miles
0   150   300 Kilometers

95°E   100°E   105°E

CLIMATE

Yangon (Rangoon)

Bay of Bengal

MYANMAR (BURMA)

THAILAND

Krung Thep (Bangkok)

Andaman Sea

Gulf of Thailand

Humid Equatorial
- No dry season
- Short dry season
- Long dry season

Humid Temperate
- Dry winter

South China Sea

MALAYSIA

Kuala Lumpur

Strait of Malacca

SINGAPORE

Equator

0   150   300 Miles
0   150   300 Kilometers

95°E   100°E   105°E

Critics claim that after depleting its mineral resources, Malaysia is now destroying its forests. To clear the land, farmers first cut the trees and then burn anything leftover. Fertilizers are then applied, permanently changing the soil and making it impossible for the rain forest to regrow quickly.

The environmental damage has an economic impact. Beginning in the early 2000s, rain forest tourism became an important part of the peninsula's economy. As a result governments are working to find a balance between land development and rain forest conservation.

Before You Move On

Summarize How do environmental concerns conflict with the production of palm oil?

ONGOING ASSESSMENT

MAP LAB

 GeoJournal

1. **Region** According to the map, which country has all four of the climates shown?

2. **Make Generalizations** Look at the physical map. Which type of land makes up most of the Malay Peninsula?

3. **Make Inferences** What effect might tourism have on the clearing of rain forest land?

SECTION **1** GEOGRAPHY

TECHTREK

myNGconnect.com For maps and photos
of the island nations of Southeast Asia

 Maps and Graphs

 Digital Library

1.4 Island Nations

Main Idea Geographic conditions on the islands affect settlement in Southeast Asia.

Five countries of Southeast Asia are islands or groups of islands: Indonesia, Singapore, Brunei, East Timor, and the Philippines. A part of Malyasia is on the island of Borneo. Mountains and water barriers among the islands have given rise to isolated cultures with distinct features.

Volcanic Activity

Located where the Eurasian, Indian, Philippine, and Australian plates come together, the islands of Southeast Asia are in a **dynamic**, or continuously changing, geographic zone. These plates are constantly—yet slowly—moving into and over each other. Most islands in the region were formed by the crashing together of these plates. The collisions gradually formed small land masses with high volcanic mountains that slope downward toward coastal plains.

Although many of the volcanoes above ground are no longer active, some that are **dormant**, or inactive for long periods of time, can suddenly erupt. For example, Mount Sinabung on Sumatra in Indonesia had been quiet for 400 years before erupting in 2010. Although many farmers stayed with their farms, tens of thousands of people fled. Volcanic eruptions can destroy villages, but the ash also creates the fertile, nutrient-rich soil that allows for successful farming. The humid climate of the islands **enhances**, or improves, the quality of agriculture. As a result, many crops can be cultivated year-round.

Indonesia

Indonesia is the giant of the region in both land area and population. It is nearly three times the size of Myanmar, the next largest. In population, Indonesia is more than three times as large as the Philippines, and it has more volcanoes than any other country in the world.

> **Critical Viewing** This village sits less than two miles from Mount Batur, an active volcano in Indonesia. What might be some advantages and disadvantages of living in this location?

EURASIAN
PLATE

Tropic of Cancer

Mt. Pinatubo,
1991

South
China
Sea

PHILIPPINE
PLATE

Andaman Sea

Gulf of
Thailand

Strait
of Malacca

PACIFIC
PLATE

Equator

SUMATRA

BORNEO

SULAWESI

JAVA

NEW GUINEA

INDIAN
OCEAN

AUSTRALIAN
PLATE

Tropic of Capricorn

— Plate boundary
▲ Volcanic eruption

0   400   800 Miles
0   400   800 Kilometers

100°E   110°E   120°E   130°E   140°E

Tropic of Cancer

Manila

PACIFIC
OCEAN

South
China
Sea

PHILIPPINES

Davao

Strait of
Malacca

Kuala
Lumpur

Medan

BRUNEI Bandar Seri
Begawan

MALAYSIA

SINGAPORE

Equator

I N D O N E S I A

Palembang

Jakarta
Bandung

Semarang
Surabaya

Dili

TIMOR-LESTE
(EAST TIMOR)

INDIAN
OCEAN

Tropic of Capricorn

Population Density
· One dot represents
   100,000 people

0   400   800 Miles
0   400   800 Kilometers

100°E   110°E   120°E   130°E   140°E

Indonesia is made up of thousands of islands. The five largest islands in size are Sumatra, Java, Borneo, Sulawesi, and New Guinea. Though Java is the smallest of the five, it is also the most populous, with more than half of Indonesia's 240 million people. Four of Indonesia's five largest cities are on Java, including the capital, Jakarta. With the exception of the largest cities, most urban areas are more like large towns, each with its own local culture.

The Philippines

Settlement patterns in the Philippines are similar to those in Indonesia. The greatest concentration of people is on lowland plains, which provide soil made fertile by volcanic eruptions. Like Indonesia, the Philippines has an extremely large capital city. This city, Manila, has more than ten million people.

The Philippines also has many small rural settlements that subsist on fishing or rice farming. Houses near the ocean are built on columns made of timber to allow for changing tides and boat traffic.

Before You Move On

Summarize In what ways has geography of the island nations affected life in Southeast Asia?

ONGOING ASSESSMENT

MAP LAB

 GeoJournal

1. **Place** Based on the map of tectonic plates, which large island seems to be unaffected by active volcanoes or earthquakes? Why might this be?

2. **Compare** Based on the Population Density map, how does the population density of Borneo compare to that of Java? What might be the cause of this difference in population?

TECHTREK

myNGconnect.com For a map of New Guinea, photos, and an Explorer Video Clip

Maps and Graphs

Digital Library

Discovering ● New Species

with Kristofer Helgen

Main Idea The unexplored Foja Mountains of Indonesia may be home to unidentified plant and animal species.

Undiscovered Species

Today knowledge of animal and plant life is extensive and well documented. However, National Geographic Emerging Explorer Kristofer Helgen is a ==zoologist==, a scientist who studies animals, and he knows there are hundreds of species that have yet to be discovered.

The Foja Mountains

The **Foja Mountains** of Indonesia have wildlife not found in any other region. The area is also largely unexplored. Helgen described the Foja (FOY ya) Mountains as "one of the few places in the world with no villages, no roads, no human population at all." Human presence is not part of life in this rain forest. "The animals there just don't know people," he said. "That's a very, very rare thing in this day and age."

In 2005, Helgen took part in a research trip to this remote area. The team included Helgen, who studies mammals, and other experts on plants, butterflies, reptiles, and birds.

The researchers found 20 new frog species, 5 new kinds of butterfly, and several new plant species. Helgen himself found new rat and mouse species. He also discovered a unique type of ==wallaby==, a small relative of the kangaroo, which he hopes to name as a new species.

myNGconnect.com

For more on Kristofer Helgen in the field today

New species of blossom bat

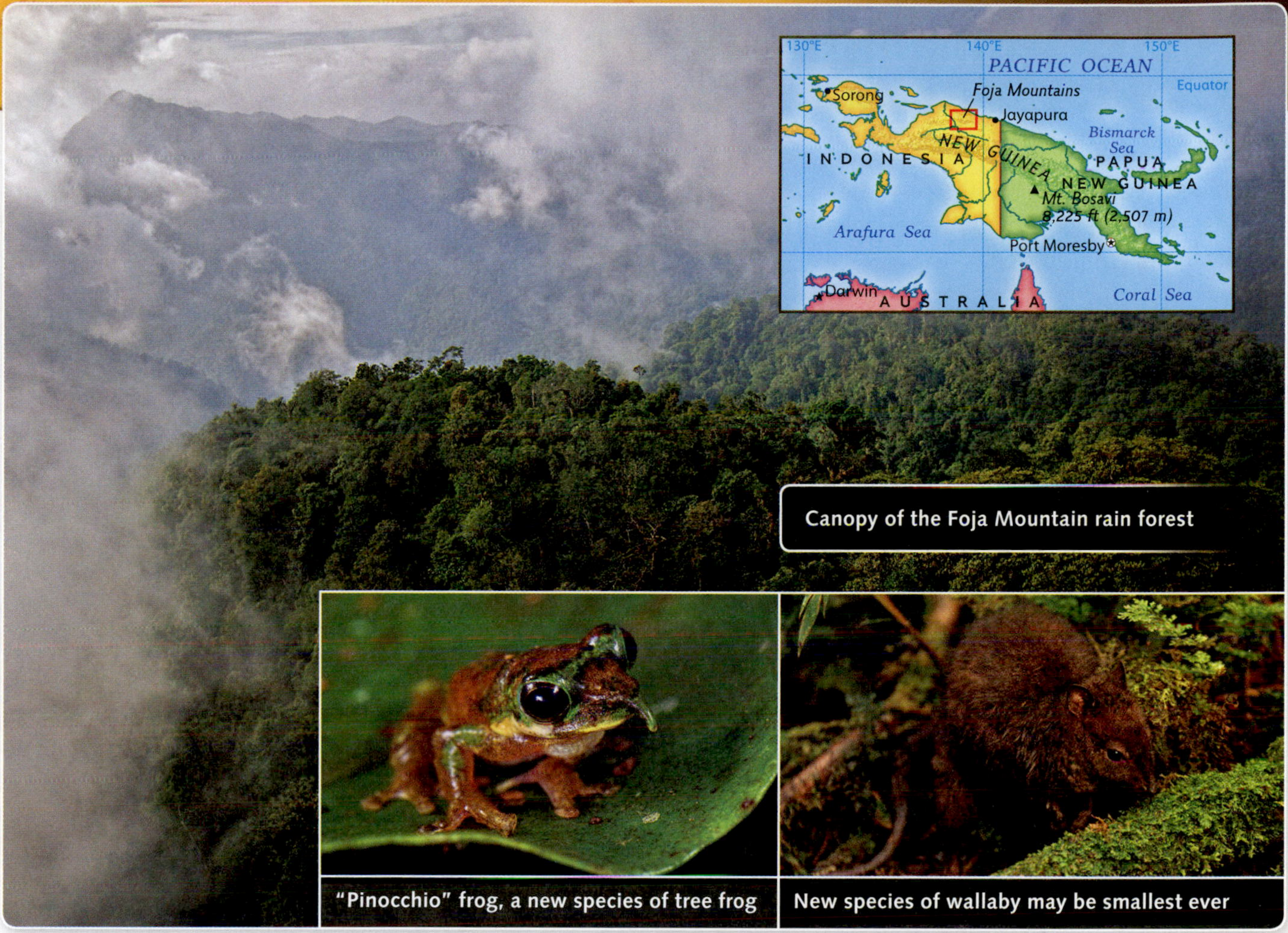

Canopy of the Foja Mountain rain forest

"Pinocchio" frog, a new species of tree frog

New species of wallaby may be smallest ever

Scientific Opportunity

The research conducted in the Foja Mountains provides a chance for scientists to learn more about the diversity of animal and plant life. Because they have not been explored, the mountains also offer new scientists an opportunity to become trained in the discovery of new species. Students can join experienced scientists in future trips to the region. They will learn about working in the field even as they help make new discoveries.

Before You Move On

Make Inferences Why might there be many undiscovered species of plants and animals in the Foja Mountains?

ONGOING ASSESSMENT

VIEWING LAB

GeoJournal

1. **Analyze Visuals** Go to the **Digital Library** and view the Explorer Video Clip about Kristofer Helgen. How would you describe the land?
2. **Place** What in the video explains why scientists are excited about this area?
3. **Make Inferences** How do you think animals that have never encountered humans would behave the first time they see a person? Why?
4. **Create Charts** Make a chart listing the kinds of new species that might be found in a place like the Foja Mountains. Remember that the region is home to a rain forest.

PLANTS	ANIMALS

2.1 Ancient Valley Kingdoms

TECHTREK
myNGconnect.com For photos of ancient Southeast Asian kingdoms
Digital Library

Main Idea The development of Southeast Asia was influenced by nearby powers because of its important location for trade.

The location of Southeast Asia between the Pacific and Indian Oceans meant its surrounding waterways were on important trade routes. Two powerful civilizations, China to the north and India to the west, heavily infuenced the cultural direction of the region. The impact came through military force and invasion, as well as through trade.

Mainland Empires

Chinese culture first came to Southeast Asia in 111 B.C., when the Chinese invaded and conquered part of what today is Vietnam. India had already established trade before the Chinese arrived. This trade had a strong influence on the area's religious practices.

By the A.D. 700s, Buddhist and Hindu empires were competing for influence and power in other parts of the region. The largest and longest lasting was the **Khmer Empire** of Cambodia. Centered along the Mekong River valley, the Khmer (kuh MAYR) Empire covered much of Southeast Asia and lasted from the A.D. 800s to the 1430s.

At the peak of the empire's success in the 1100s, its ruler, King Suryavarman II, built the massive Hindu temple **Angkor Wat** in the capital city.

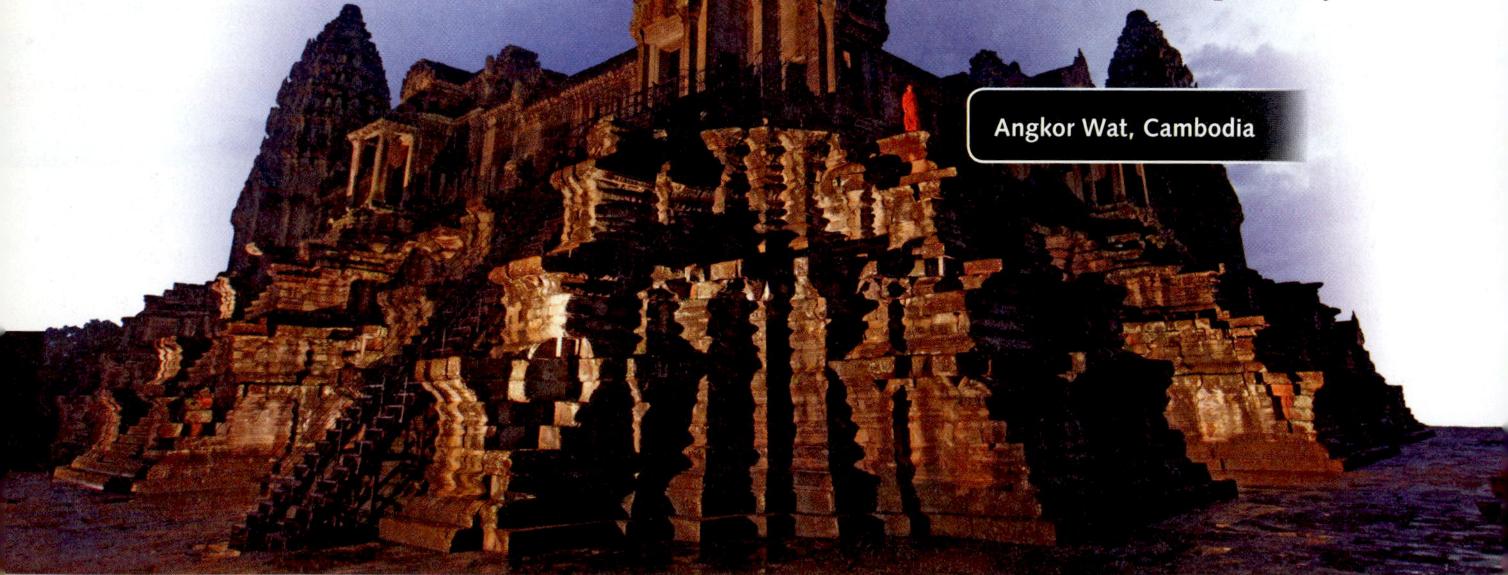

Angkor Wat, Cambodia

111 B.C.
China conquers Vietnam; Buddhism reaches Southeast Asia.

c. A.D. 780–850
Sailendra rulers of Java build Borobudur temple.

c. A.D. 890
Khmer Empire sets capital at Angkor.

100 B.C. | A.D. 600 | 800

A.D. 600s–c. 1100
Srivijaya Empire of Sumatra

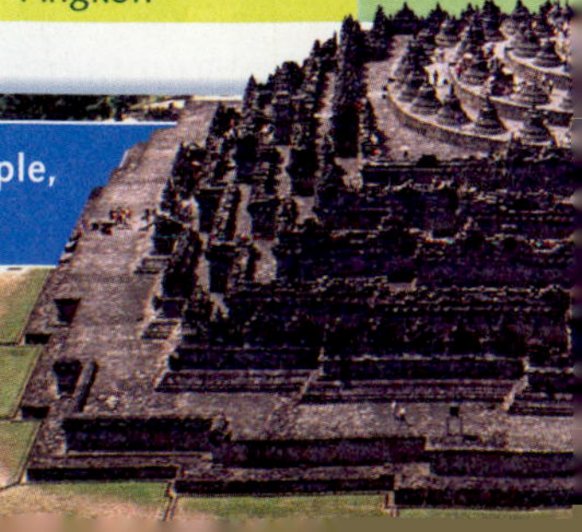

Borobudur temple, Java, Indonesia

This religious **complex**, or set of interconnected buildings, was dedicated to the Hindu god Vishnu and served as a tomb for the king. Beautiful **bas-reliefs**, or sculptures that slightly project from a flat background, cover the walls with scenes from Hindu stories. Eventually forces from modern-day Thailand conquered the city and the complex fell into ruin.

In A.D. 939 the people of Vietnam broke from China and established the independent kingdom of Dai Viet. Though influenced by Chinese culture, Vietnam had its own cultural traits. For instance, women in Vietnam had higher social standing than women in China. Eventually Dai Viet grew weak and was reconquered by China in 1407.

Island Empires

Modern Indonesia was also home to powerful empires. The earliest was Srivijaya (sree vi JY ah), which arose on southern Sumatra in the A.D. 600s. This kingdom controlled the Strait of Malacca and was therefore able to control trade from South Asia to China. It was known throughout Asia as a center of trade and also as a center of Buddhist study. The empire declined around A.D. 1100.

A second Indonesian power was the Sailendra dynasty, which arose in Java and flourished from about A.D. 780 to 850. Sailendran rulers built another famous temple complex, **Borobudur.** Each of the temple's three levels symbolizes a step toward enlightenment, the ultimate spiritual goal of Buddhism.

Another trading kingdom arose on Java around A.D. 1300. It is named for its capital city of Majapahit (mah jah PAH heet) in eastern Java. It gained power through the control of trade. However, by the 1500s, other powers had replaced it.

Before You Move On

Summarize How did Chinese and Indian empires influence life in the region?

ONGOING ASSESSMENT

VIEWING LAB   GeoJournal

1. **Interpret Time Lines** Based on the time line, how long was Vietnam able to maintain its independence from China?

2. **Make Generalizations** Look at the photo of Angkor Wat. How would you describe the building skill of the Khmer Empire? Support your answer with details from the photo.

3. **Compare and Contrast** Based on the photos, how are the Borobudur temple and Angkor Wat similar and different?

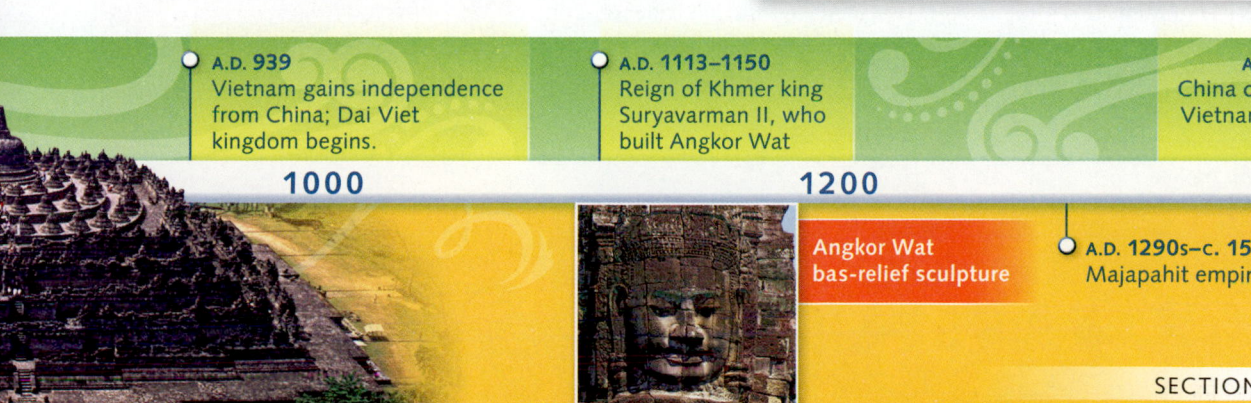

A.D. 939
Vietnam gains independence from China; Dai Viet kingdom begins.

A.D. 1113–1150
Reign of Khmer king Suryavarman II, who built Angkor Wat

A.D. 1407
China conquers Vietnam again.

1000   **1200**   **1400**

Angkor Wat bas-relief sculpture

A.D. 1290s–c. 1500
Majapahit empire in Java

Main Idea The development of the spice trade in Southeast Asia led to colonization of the region.

As you have learned, trade with India and China brought the influence of these cultures to Southeast Asia. European influence arrived in the 1500s, as merchants hoped to establish a spice trade **monopoly**, or complete control of the market. Spices found in the region, such as cinnamon, nutmeg, and black pepper, could be sold for a high profit in Europe. Traders from Spain and Portugal came first, but the Netherlands' **Dutch East India Company** dominated the region for many years. This success established a strong Dutch influence in Indonesia.

European Control

From the 1600s to the 1800s, Europeans tried to gain an economic hold on Southeast Asia. By 1850, through a combination of alliances, favorable trade agreements, and even military force, the majority of the region was ruled by European powers. (See the time line below.) Only Thailand and parts of the Philippines were independent. Britain, France, Spain, and the Netherlands controlled the rest.

The economic motives that led European powers to Southeast Asia changed the region. Increased production and an ongoing demand for goods strengthened the region's economy. However, trade and wealth, once in the hands of indigenous kingdoms, were now held by distant economic powers. **Colonialism**—one country ruling and developing trade in another country for its own benefit—continued in Southeast Asia well into the 20th century.

Before You Move On

Monitor Comprehension How did development of the spice trade lead to colonization?

SPAIN

1521
Ferdinand Magellan lands at the Philippines and claims them for Spain.

1565
Spain makes first settlement on the Philippine islands.

1571
Spain captures site of Manila.

1830s
Spain opens Manila to trade.

1892
Filipinos begin movement aimed at independence from Spain.

1898
United States defeats Spain in the Spanish-American War and wins control of Philippines.

NETHERLANDS

1619
Dutch East India Company makes base at Batavia (modern Jakarta).

1641
The Netherlands captures Malacca from Portugal and becomes a major power in spice trade.

1824
The Netherlands and Britain reach agreement on control of Java and Sumatra (to Dutch) and Singapore and Malacca (to British).

1825–1839
The Netherlands fights to put down revolts on Java.

1860
The Netherlands and Portugal sign treaty to divide Timor between them.

GREAT BRITAIN

1781
Britain captures Sumatra from Dutch.

1786–1809
Britain gains control of Malaya trade.

1819
Britain founds Singapore, which becomes a major port city.

1824–1826
Britain controls western Burma.

1886
Britain completes control of Burma in Third Burmese War.

1888
Britain wins southern Burma.

1888
Britain gains control of northern Borneo.

SOUTHEAST ASIA UNDER COLONIAL RULE, c. 1895

Legend:
- British possession
- French possession
- Portuguese possession
- Dutch possession
- Spanish possession
- Independent

Map labels: Tropic of Cancer, BRITISH BURMA, Rangoon, SIAM, Bangkok, FRENCH INDO-CHINA, Phnom Penh, Saigon, Andaman Sea, South China Sea, Manila, PHILIPPINE ISLANDS, Philippine Sea, Penang, BRITISH MALAYA, Malacca, Singapore, Strait of Malacca, BRUNEI, BRITISH NORTH BORNEO, SARAWAK, Sarawak, Celebes Sea, Palembang, DUTCH EAST INDIES, Batavia, Java Sea, Banda Sea, Hollandia, INDIAN OCEAN, Equator, PORTUGUESE TIMOR, Timor Sea

Scale: 0 400 800 Miles / 0 400 800 Kilometers

FRANCE

1644
France forms French East India Company.

1789
French East India Company disbanded during French Revolution.

1858
France captures Saigon, Vietnam.

1863
France seizes Cambodia.

1887
France creates Indo-Chinese Union (Cambodia, Vietnam).

1896
France and Britain agree to allow Siam to remain independent to separate their colonies.

ONGOING ASSESSMENT

MAP LAB

 GeoJournal

1. **Region** Based on the map, which European countries held the most territory in the region around 1895?

2. **Movement** Think about the importance of waterways in influencing trade in this region. Which European country was in the best position to control trade? Support your answer with evidence from the map.

3. **Interpret Time Lines** Based on the dates and events, what do you think was the attitude of the people of Southeast Asia toward European control? Why do you think so?

2.3 Indonesia and the Philippines

Main Idea Indonesia and the Philippines are island countries that have faced similar challenges in becoming independent.

As you know, Southeast Asia has a long history of diversity because of its geographic location and unique resources. The island nations of Indonesia and the Philippines have been influenced and even controlled by other cultures throughout history. However, after long struggles for independence, each one has become its own nation.

Indonesia

Indonesia may have been home to the earliest species of humans. **Fossils**, or preserved remains, found on Java suggest that human life existed there as early as 1.7 million years ago. Evidence shows that ancient societies there used hand tools, made implements out of metals, and wove cloth. They also traveled the sea as early as 2500 B.C., possibly to trade with other areas of Asia and beyond.

As its civilization matured, Indonesia became an intersection for trade in the East. Part of the country became known to Europe as the Spice Islands, for the many exotic spices that were a strong attraction for explorers and traders. For example, trade in nutmeg, a spice native to Indonesia, became extremely profitable for the Dutch who had settled there.

Throughout the 1800s, the Dutch expanded their control. Some revolts occurred, but the Dutch were able to maintain power. In 1830, they began a system that required all villages to give part of their crops to the government for export. As the Dutch gained wealth, Indonesians suffered. During the 1900s their resistance efforts became more organized. The Japanese seized control from the Dutch during World War II. As these two countries fought for control, Indonesians continued to resist. At the same time, a strong sense of national identity was developing. The country finally won independence in 1949.

nutmeg

mace

Mace (shown here) is a spice that covers nutmeg in its raw form. Both nutmeg and mace are highly valued for their distinct flavor.

Critical Viewing Emilio Aguinaldo (front, center) and members of the assembly of the First Philippine Republic, 1899. In early photography, subjects were required to sit for long periods to capture an image. How might this explain the expressions on these men's faces?

The Philippines

When the Spanish seized control of **Manila** in 1571, they made it the capital of their new colony. Manila was and still is the economic, political, and cultural center of the Philippines. Trade with China led many Chinese people to settle in Manila. They became the major force in <mark>commerce</mark>, or the business of trading goods and services.

In the 1800s, Spain's economic power began to fade and Manila became open to trade with more countries. This allowed some Filipinos to gain wealth and influence they had not known before.

Emilio Aguinaldo was a key leader in the movement for Filipino independence. He fought alongside the United States when it defeated Spain in the Spanish-American War in 1898, and thought the islands would become independent. However, after the war, the United States kept the islands as its own colony.

In late 1941, during World War II, Japan attacked the United States at Pearl Harbor, then attacked the Philippines. The outnumbered U.S. forces stationed on the islands were forced to leave. Later in the war, the United States took back control and remained in power until 1946, when the Philippines were granted independence. Today, it is a stable democracy.

Before You Move On

Summarize What challenges did Indonesia and the Philippines face in gaining independence?

ONGOING ASSESSMENT

READING LAB   GeoJournal

1. **Location** Why was Southeast Asia so important to Europeans?
2. **Make Inferences** Why were Indonesians able to develop a strong national identity?
3. **Draw Conclusions** Was trade or conquest more important in shaping these countries? Why?

2.4 The Vietnam War

TECHTREK

myNGconnect.com For photos of the war in Vietnam and Guided Wrting

 Digital Library

 Student Resources

In 1954 Vietnam was divided into two parts, North and South. President **Ho Chi Minh** of North Vietnam established a Communist government. War erupted in 1959 when he sent aid to overthrow the government in South Vietnam and create one country. The United States supported South Vietnam—it feared that defeat by the North would spread communism to other countries. U.S. forces fought there from 1964 to 1973. In 1975, South Vietnam surrendered. The reunited country became the Socialist Republic of Vietnam.

DOCUMENT 1

President Lyndon Johnson on U.S. Policy

In 1965, Johnson explained reasons for the involvement of the United States in the fighting in Vietnam:

> We are . . . there to strengthen world order. Around the globe . . . are people whose well-being rests, in part, on the belief that they can count on us if they are attacked. To leave Viet-Nam . . . would shake the confidence of all these people in the value of an American commitment. . . . Let no one think for a moment that retreat from Viet-Nam would bring an end to conflict. The battle would be renewed in one country and then another. The central lesson of our time is that the appetite of aggression is never satisfied.

CONSTRUCTED RESPONSE

1. How are Johnson's two reasons for fighting the Vietnam War related?

DOCUMENT 2

Letter from Ho Chi Minh

In 1967, Johnson sent Ho Chi Minh a letter suggesting that the two sides begin peace talks. Here is Ho Chi Minh's response:

> The Vietnamese people have never done any harm to the United States. But . . . the United States Government has constantly intervened in Viet-Nam, it has launched [started] and intensified the war of aggression in South Viet-Nam for the purpose of prolonging the division of Viet-Nam and of transforming [remaking] South Viet-Nam into an American neo-colony and an American military base. . . .
>
> The Vietnamese people . . . are determined to continue their resistance [opposition] until they have won real independence.

CONSTRUCTED RESPONSE

2. How does Ho Chi Minh respond to the suggestion of peace talks?

THE VIETNAMESE PEOPLE . . . ARE DETERMINED TO CONTINUE THEIR RESISTANCE UNTIL THEY HAVE WON REAL INDEPENDENCE.

— HO CHI MINH

DOCUMENT 3

Photo of Warfare

The physical geography and climate of Vietnam presented challenges to soldiers. These challenges might be especially great for soldiers who were not native to the region and were not accustomed to heavy rains and dense plant life of the jungle.

CONSTRUCTED RESPONSE

3. What does the photo suggest about the conditions the soldiers faced in the war in Vietnam?

ONGOING ASSESSMENT

WRITING LAB — GeoJournal

DBQ Practice Think about how North Vietnam compares in size and power to the United States. What factors might have allowed North Vietnam a chance to win the war?

Step 1. Think about Johnson's determination to fight in Vietnam, Ho Chi Minh's statements, and what the photo shows about the war in Vietnam.

Step 2. On your own paper, jot down notes about the main idea expressed in each document.

> Document 1: Excerpt: Johnson's Speech
> Main Idea(s) _____
> Document 2: Excerpt: Ho Chi Minh's Letter
> Main Idea(s) _____
> Document 3: Photo of Warfare
> Main Idea(s) _____

Step 3. Construct a topic sentence that answers this question: What factors might have affected North Vietnam's ability to fight and win a war against South Vietnam and the United States?

Step 4. Write a detailed paragraph that answers the question above, using evidence from the documents. Go to **Student Resources** for Guided Writing support.

VOCABULARY

Use the following vocabulary words in a sentence that shows understanding of each term's meaning.

1. landlocked

> All countries in Southeast Asia have some coastline except Laos, making it the region's only landlocked country.

2. subsistence fishing

3. dynamic

4. dormant

5. commerce

6. launch

MAIN IDEAS

7. How does the climate of the two parts of Southeast Asia differ? (Section 1.1)

8. Why are the upper reaches of the region's rivers less populated? (Section 1.2)

9. Where do you think most people live on mainland Malaysia? Why? (Section 1.3)

10. How do Manila and Java show that population in the Philippines and Indonesia tends to cluster? (Section 1.4)

11. What has allowed some species in the region to have gone undiscovered? (Section 1.5)

12. Why did trade play such a great role in the development of island kingdoms? (Section 2.1)

13. Which region of the world practiced colonialism in Southeast Asia and why? (Section 2.2)

14. What struggles did Indonesia and the Philippines face in developing their own culture? (Section 2.3)

15. What were North Vietnam and South Vietnam fighting about? (Section 2.4)

GEOGRAPHY

ANALYZE THE ESSENTIAL QUESTION

What are the geographic conditions that divide Southeast Asia into many different parts?

Focus Skill: Analyze Cause and Effect

16. Why are rivers and seas so important in this region?

17. Why are island countries like Indonesia and the Philippines good for farming?

18. In what way has Laos's lack of a coastline impacted its ability to engage in international trade?

INTERPRET MAPS

19. **Summarize** What is the advantage of the location of the mainland portion of Malaysia?

20. **Make Inferences** Look at the locator map. In what ways has Malaysia's location helped it establish trading partners?

HISTORY

ANALYZE THE ESSENTIAL QUESTION

How have physical barriers in Southeast Asia influenced its history?

Focus Skill: Draw Conclusions

21. Why do you think one great empire never arose to unite all of Southeast Asia?

22. Which other cultural regions do you think had the most influence on Southeast Asia? Explain your reasons.

23. What made it difficult for Europeans to establish complete control of the countries in this region?

INTERPRET MAPS

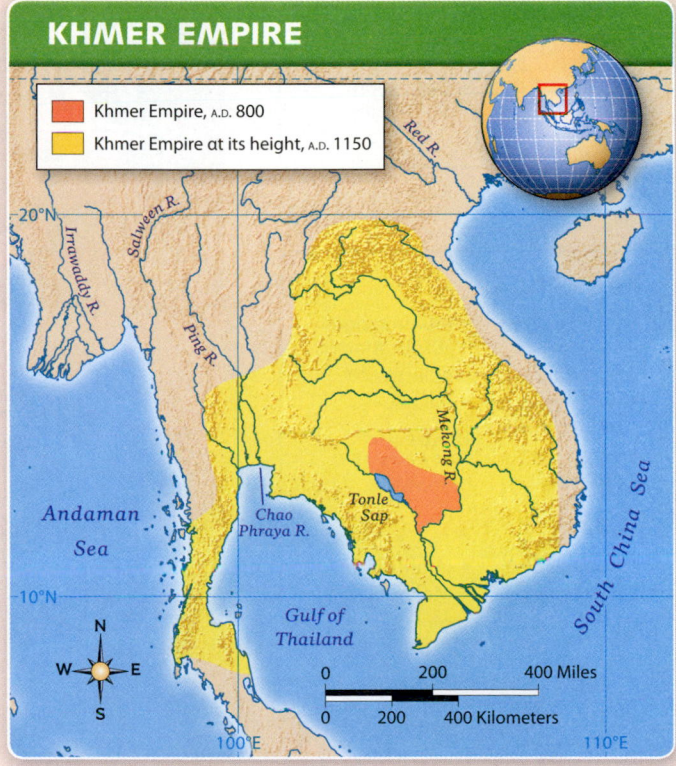

KHMER EMPIRE

Khmer Empire, A.D. 800
Khmer Empire at its height, A.D. 1150

Red R.
20°N
Irrawaddy R.
Salween R.
Ping R.
Mekong R.
Tonle Sap
Andaman Sea
Chao Phraya R.
South China Sea
10°N
Gulf of Thailand

N
W E
S

0   200   400 Miles
0   200   400 Kilometers

100°E   110°E

24. **Region** How much of Southeast Asia did the Khmer Empire control at its greatest point? In what year was that the case?

25. **Make Inferences** The kingdom of Dai Viet existed along the South China Sea. What factors would have allowed it to remain independent of the Khmer Empire?

ACTIVE OPTIONS

Synthesize the Essential Questions by completing the activities below.

26. **Write a Press Release** You work for an art museum that is staging an exhibition of art and objects from Angkor Wat. Write a 3- or 4-paragraph press release to announce the exhibition. Include information about how long the exhibition will run and the museum's schedule. Be sure to word the press release in a way that would attract visitors to the exhibition. Use these tips to prepare your press release. **Read your press release aloud with a partner to evaluate each other's ideas.**

> **Writing Tips**
> - Make sure you include details that will attract visitors.
> - Clearly explain what kinds of objects the exhibition will include.
> - Remember to use vivid, appealing language to describe the objects.
> - Be sure to give the dates and times of the exhibition.

TECHTREK   myNGconnect.com For research links on Southeast Asia

27. **Create a Time Line** Use information in the lessons and research links at **Connect to NG** to gather your facts about five key dates in the history of two countries in the region. Then construct a time line showing the dates for both countries and why they are significant. Illustrate the time line with photos connected to some of the events.

	COUNTRY 1	COUNTRY 2
Event 1		
Event 2		
Event 3		
Event 4		
Event 5		

CHAPTER 16

Southeast Asia TODAY

PREVIEW THE CHAPTER

Essential Question How have local traditions and outside influences shaped cultures in Southeast Asia?

SECTION 1 • CULTURE

KEY VOCABULARY

- prehistoric
- ritual
- attribute
- wat
- monk
- metropolitan area
- dialect
- adapt
- language diffusion
- poach
- restore
- domesticate

ACADEMIC VOCABULARY

predominant

TERMS & NAMES

- Java
- Sumatra
- Bali
- Cardamom Mountains

Essential Question How are Southeast Asia's governments trying to unify their countries?

SECTION 2 • GOVERNMENT & ECONOMICS

KEY VOCABULARY

- fragmented country
- remittance
- relocate
- trend
- port
- industrialize
- multinational corporation
- emergence
- reliable

ACADEMIC VOCABULARY

potential

TERMS & NAMES

- Madura
- inner islands
- outer islands
- Malaysia

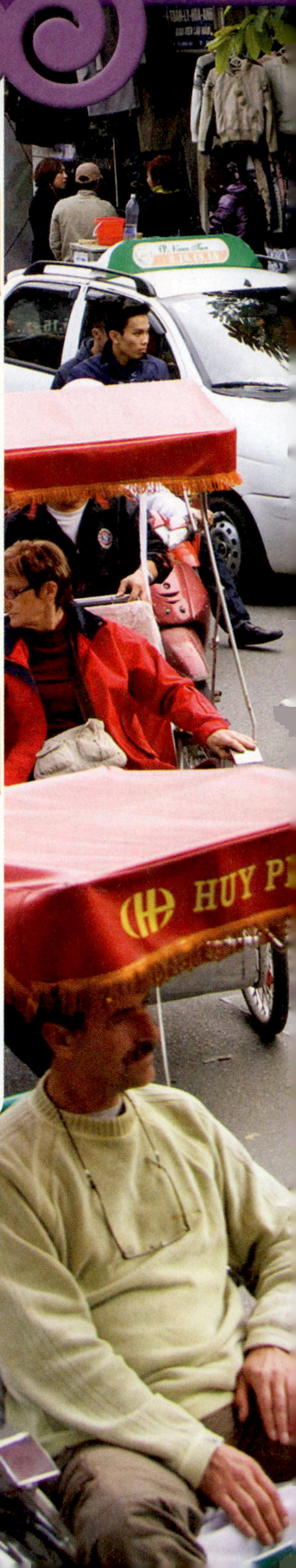

TECHTREK FOR THIS CHAPTER

Student eEdition

Maps and Graphs

Interactive Whiteboard GeoActivities

Digital Library

Connect to NG

Go to **myNGconnect.com** for more on Southeast Asia.

Tourists and locals use rickshaws and motorbikes to get around in Hanoi, Vietnam.

SECTION 1 CULTURE

1.1 Religious Traditions

TECHTREK
myNGconnect.com For photos of religious traditions in Southeast Asia
Digital Library

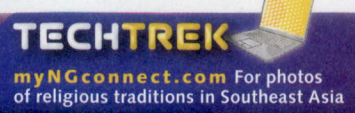

Main Idea Religions in Southeast Asia have been shaped by both local traditions and outside influences.

Religious practices in Southeast Asia today blend many influences from several centuries. The **predominant**, or most common, belief systems have shifted in many parts of the region.

Traditional Religion

The traditional religion in Southeast Asia is animism, the belief that spirits exist in animals, plants, objects, and places. These spirits are believed to influence people's lives. Many historians think that animism began in **prehistoric** times before there were written histories. Animists perform **rituals**—formal regularly repeated actions—to please spirits so they bring good fortune to their human families or villages. Many small, tribal groups in the region still practice forms of animism.

Outside Influences

Other cultures entering Southeast Asia through trade or conquest brought their traditions with them. The Chinese brought Buddhism when they conquered Vietnam in 111 B.C. They also introduced a philosophy, or system of thought, called Confucianism, which became an important part of Vietnam's local religions. Beginning in the A.D. 100s, traders from India spread Buddhism and Hinduism across the region. By the 400s, Buddhism had taken hold on **Java** and later spread to **Sumatra**. In the 1100s, Cambodians built Angkor Wat to worship a Hindu god.

Arab traders carried Islam to Southeast Asia during the 1300s, where it spread from Malaysia to parts of Indonesia. Later, Europeans spread Christianity. Spain brought Roman Catholicism to the Philippines in the 1500s. The French introduced it to the mainland in the 1700s.

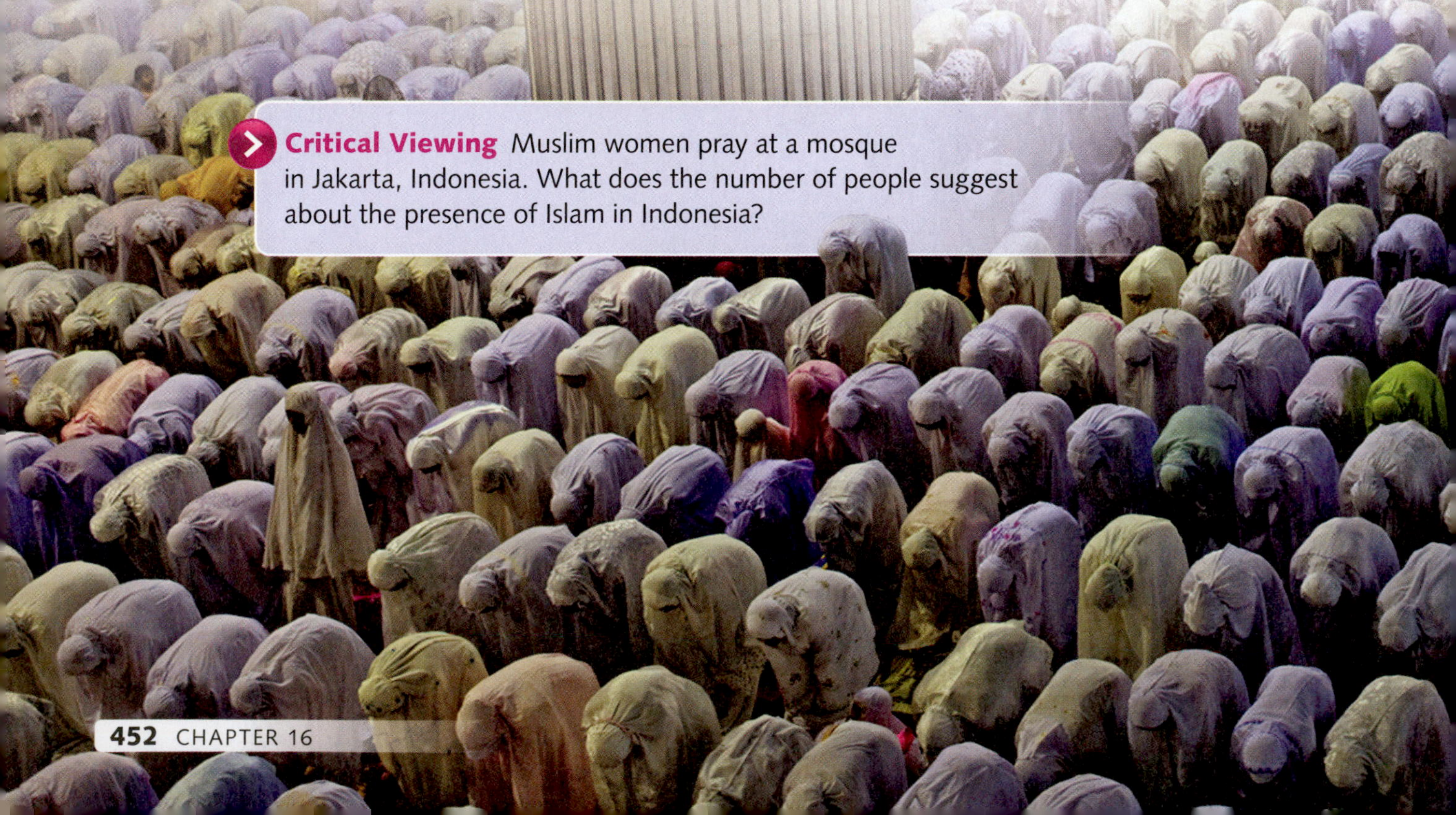

Critical Viewing Muslim women pray at a mosque in Jakarta, Indonesia. What does the number of people suggest about the presence of Islam in Indonesia?

Children attend a prayer service in a Catholic school in Makassar, Indonesia.

Forest roots devour ruins at Ta Prohm, a Buddhist temple at Angkor, Cambodia.

Religion Today

Over the centuries, the predominant religion of a country sometimes shifted depending on the beliefs of the ruling power. Today, Southeast Asia has a mix of religions. Buddhism is most prominent in mainland countries. Around 95 percent of the people in Thailand are Buddhists, and many Buddhist holy days are national holidays. Buddhism is also predominant in Myanmar. Islam is the main religion of Indonesia, which is the most populous Muslim country in the world. In Malaysia, about three out of five people are Muslim. Most people living in the Philippines and East Timor continue to practice Roman Catholicism, introduced by Europeans.

While some religions dominate each country, the region as a whole has religious diversity. For example, **Bali**, an island in Indonesia, is largely Hindu.

Islam has many followers in the southern Philippines. Diversity is also found in many of the ancient, local traditions that are still practiced in each country today.

Before You Move On

Monitor Comprehension Which religions were brought to the region by outside cultures?

1.2 Thailand Today

TECHTREK

myNGconnect.com For photos of modern Thailand

Digital Library

Main Idea Thai culture today reflects traditional foundations and modern influences.

Modern Thai culture includes regional traditions and other global infuences that blend into a unique Thai identity.

Classical Architecture

One attribute, or specific quality, of Thai culture is its remarkable architecture. Traditional buildings have steeply slanted roofs designed to shed the heavy monsoon rains. Many are built on legs to keep them high off the ground during the monsoon floods. The most important buildings in Thai architecture are wats, or Buddhist temples, influenced by designs from India, the Khmer empire, and China.

Buddhist Monks

As you have learned, Buddhism is the dominant religion in Thailand. Almost every village has a wat with a community of monks, men who devote themselves to religious work. Buddhist monks wear orange or yellow robes, live simply, and focus on religious practices such as meditation and other rituals.

Most young men traditionally became monks for at least three months during one rainy season. However, as more and more young people migrate away from rural communities and attend non-religious schools, young men are making shorter commitments to religious life, or sometimes none at all.

Visual Vocabulary A **wat** is a Buddhist temple. Building on Marble Wat (above) started in 1900, during a period of rapid growth in the capital city of Bangkok.

The Chao Phraya River cuts through downtown Bangkok at dusk.

Modern Influences

About four out of five young men and women now work in cities, especially Bangkok's large metropolitan area, the populated location that includes the city limits and surrounding area. Many still identify with their villages even though they mostly live and work in cities.

Urban life has also changed people's clothing, food, and entertainment. Most people in Thailand now wear Western-style clothing. Many urban women buy prepared food at local stores on their way home from work instead of cooking. Most homes now have televisions and other modern conveniences.

Young people in urban areas also turn to the Internet as a source of news, entertainment, and communication.

Before You Move On
Monitor Comprehension What traditional and modern influences can be seen in Thai culture today?

Critical Viewing Suvarnabhumi Airport in Bangkok opened in 2006. Based on the photo, what can you say about the design for the airport?

PHOTO LAB

 GeoJournal

1. **Make Inferences** What does the large photo show about life in modern Thailand?

2. **Analyze Visuals** What does each photo reveal about Thai culture? Use details from the photos to explain your answer.

3. **Human-Environment Interaction** How was traditional Thai architecture well suited to its environment?

1.3 Regional Languages

Main Idea Geographically isolated cultures and large historic migrations have created a diversity of languages in the region.

As you have learned, language is an important part of any culture. People use language to express their ideas, values, and history. Like religion, language can bring people together or it can divide them. The people of Southeast Asia speak hundreds of different languages.

Native Languages

Each country in the region has a dominant native language. Generally the name of this language reflects the name of the country or its largest ethnic group. The dominant language is the official language used by government, business, education, and the media. In countries such as Indonesia and Malaysia, a common language has helped unify geographically fragmented areas. In Myanmar, the diversity of minority languages has made unifying the country more difficult.

Many people in Southeast Asia speak a **dialect**, a regional variation of a main language. Speakers of different dialects often belong to small groups that live in isolated communities. Some of these dialects exist only in oral form, and many dialects and ethnic languages are in danger of disappearing. People learn the official language as a way to **adapt**, or adjust to common practices, especially as cultures become more globally connected. When older speakers of an isolated language die, that language may disappear, resulting in a loss of traditional culture.

Critical Viewing Many people in rural villages in Vietnam speak a native language or a dialect of a main language. How might speaking a dialect influence this woman's business?

Language Migration

The region's location has historically attracted diverse people to trade. This movement led to **language diffusion**, or the spread of languages from their original homes. Traders had to find a common language in order to communicate. Malay served this purpose for early traders from Arab countries and different parts of China. Today, English is often used as the common language.

Immigrants also brought their languages to Southeast Asia. For example, Chinese is the dominant language in Singapore. Many speakers of Chinese dialects live in Malaysia and Brunei and in major cities throughout the region. Immigrants from India brought various Indian languages to Malaysia, Singapore, and Myanmar.

SPOKEN LANGUAGES IN SELECTED SOUTHEAST ASIAN COUNTRIES

CAMBODIA
Official language: Khmer

TOTAL LIVING LANGUAGES: 23

Other Asian Languages
• Chinese
• Vietnamese
Western Languages
• English
• French

PHILIPPINES
Official language: Filipino and English

TOTAL LIVING LANGUAGES: 171

Other Asian Languages
• Chinese
• Languages of India and Southeast Asia
Western Languages
• English

INDONESIA
Official language: Bahasa Indonesia

TOTAL LIVING LANGUAGES: 719

Other Asian Languages
• Chinese
• Vietnamese
Western Languages
• English
• Dutch

SINGAPORE
Official languages: Malay, Mandarin Chinese, and English

TOTAL LIVING LANGUAGES: 21

Other Asian Languages
• Mandarin Chinese
• Tamil
Western Languages
• English

LAOS
Official language: Lao

TOTAL LIVING LANGUAGES: 84

Other Asian Languages
• Chinese
• Vietnamese
Western Languages
• English
• French

THAILAND
Official language: Thai

TOTAL LIVING LANGUAGES: 74

Other Asian Languages
• Mandarin Chinese
• Tamil
Western Languages
• English

MALAYSIA
Official language: Bahasa Malaysia

TOTAL LIVING LANGUAGES: 137

Other Asian Languages
• Chinese
• Languages of India and Southeast Asia
Western Languages
• English

VIETNAM
Official language: Vietnamese

TOTAL LIVING LANGUAGES: 106

Other Asian Languages
• Khmer
• Chinese
Western Languages
• English
• French

Sources: CIA World Factbook, www.ethnologue.com

European countries began establishing colonies in Southeast Asia in the 1500s. As the British, Dutch, French, and Spanish gained control of various countries, they used their languages to rule and do business. The United States controlled the Philippines for a time after Spanish rule and made English common there. As the countries of Southeast Asia gained independence in the 20th century, their governments chose dominant native languages to be the official languages. However, many people still speak a Western language as a second language.

Before You Move On

Summarize In what ways is language use changing as cultures become more connected?

ONGOING ASSESSMENT

LANGUAGE LAB GeoJournal

1. **Movement** According to the chart, what East Asian language is most widespread in these Southeast Asian countries?

2. **Draw Conclusions** Based on the chart, what Western language is a Southeast Asian business person most likely to learn? Explain.

3. **Identify Problems and Solutions** What might a country do to save endangered languages spoken by its people?

1.4 Saving the Elephant

Main Idea Some countries in Southeast Asia are working to protect the endangered Asian elephant.

Although smaller than their African cousins, Asian elephants are huge and awe-inspiring creatures. Some have been trained to use their enormous strength for the benefit of people. However, most elephants live in the wild. As the human populations of the region increase, their growing numbers increasingly threaten the Asian elephant population.

Asian Elephants

As recently as 1900, it is estimated 80,000 Asian elephants may have been living in the wild. Today, their population is thought to range from 30,000 to 50,000.

Human behavior, such as **poaching**, or illegal hunting of a wild animal, is one reason for that population loss. People kill male Asian elephants for their ivory tusks. Ivory is highly valued for its beauty and hard texture. An international agreement banned trade in ivory in 1989, but it still continues illegally.

A bigger problem for Asian elephants is their loss of habitat. These huge animals need large areas of rain forest to find food, but humans have cleared much of the land for alternative uses, such as logging and iron mining. Land is also cleared for growing crops such as coffee. Once crops are planted, some elephants wander in from the remaining forests to eat them, and farmers trying to protect their crops sometimes kill those raiding elephants.

Protecting Elephants

Many countries in Southeast Asia have tried to **restore**, or bring back, the wild Asian elephant population. In Cambodia's **Cardamom Mountains**, for example, many conservationists have begun to use modern technology, such as electric fences that run off solar power, to keep elephants confined to protected places. Other methods are more basic. Hammocks hung near crops make the elephants think humans are in the fields, so they stay away.

> **Critical Viewing** Asian elephants look for food in Sumatra. Based on the photo, how would you describe their habitat?

Visual Vocabulary Many elephants are **domesticated**, or trained to work with humans. Domesticated elephants can provide service or entertainment.

ASIAN ELEPHANTS BY THE NUMBERS

11+
Height in feet

12,000+
Weight in pounds

300
Pounds of food (plants, grain) consumed in one day

80,000
Population estimate, wild, beginning of 20th century

30,000
Population estimate, wild, 2008

15,000
Population in captivity (protected), 2006

Sources: World Wildlife Federation, Fauna & Flora International, U.S. Fish & Wildlife Service

The results have been dramatic. The elephant-human interaction that leads to population loss was reduced so much that from 2005 to 2010, no elephants were killed anywhere in Cambodia. These efforts can lead to long-term protection of this endangered species.

Before You Move On

Monitor Comprehension What efforts have been made to protect the wild Asian elephant in Southeast Asia?

ONGOING ASSESSMENT
READING LAB GeoJournal

1. **Region** Which country has successfully reduced elephant killings?

2. **Interpret Charts** How has the population of Asian elephants changed over time?

3. **Human-Environment Interaction** What human activities threaten Asian elephants? Why are they a threat?

2.1 Governing Fragmented Countries

TECHTREK
myNGconnect.com For photos from fragmented countries
Digital Library

Main Idea Geographic and ethnic divisions make it difficult for some countries in Southeast Asia to become unified.

In Southeast Asia, Indonesia, Malaysia, and the Philippines face challenges in forming unified countries. All three are **fragmented countries**, or countries that are physically divided into separate parts, such as a chain of islands. The three countries also have diverse ethnic groups.

Indonesia

Indonesia's 17,000 islands span across 3,200 miles and are more different than they are similar. Java, for example, is densely populated and urbanized. Sumatra, on the other hand, is rural and contains large plantations. The country includes more than 300 ethnic groups and more than 700 languages are spoken.

To meet the complex challenges of fragmentation, Indonesia's government has tried to create a sense of nationhood. The country's motto, a saying that guides them, is "Diversity in unity." However, unity is not always easy to achieve. For example, there is conflict between the majority Malays and minority Chinese, and groups in northern Sumatra and Borneo have recently tried to gain independence. The government has focused on improving people's standard of living so that these groups will see the advantages of staying part of Indonesia.

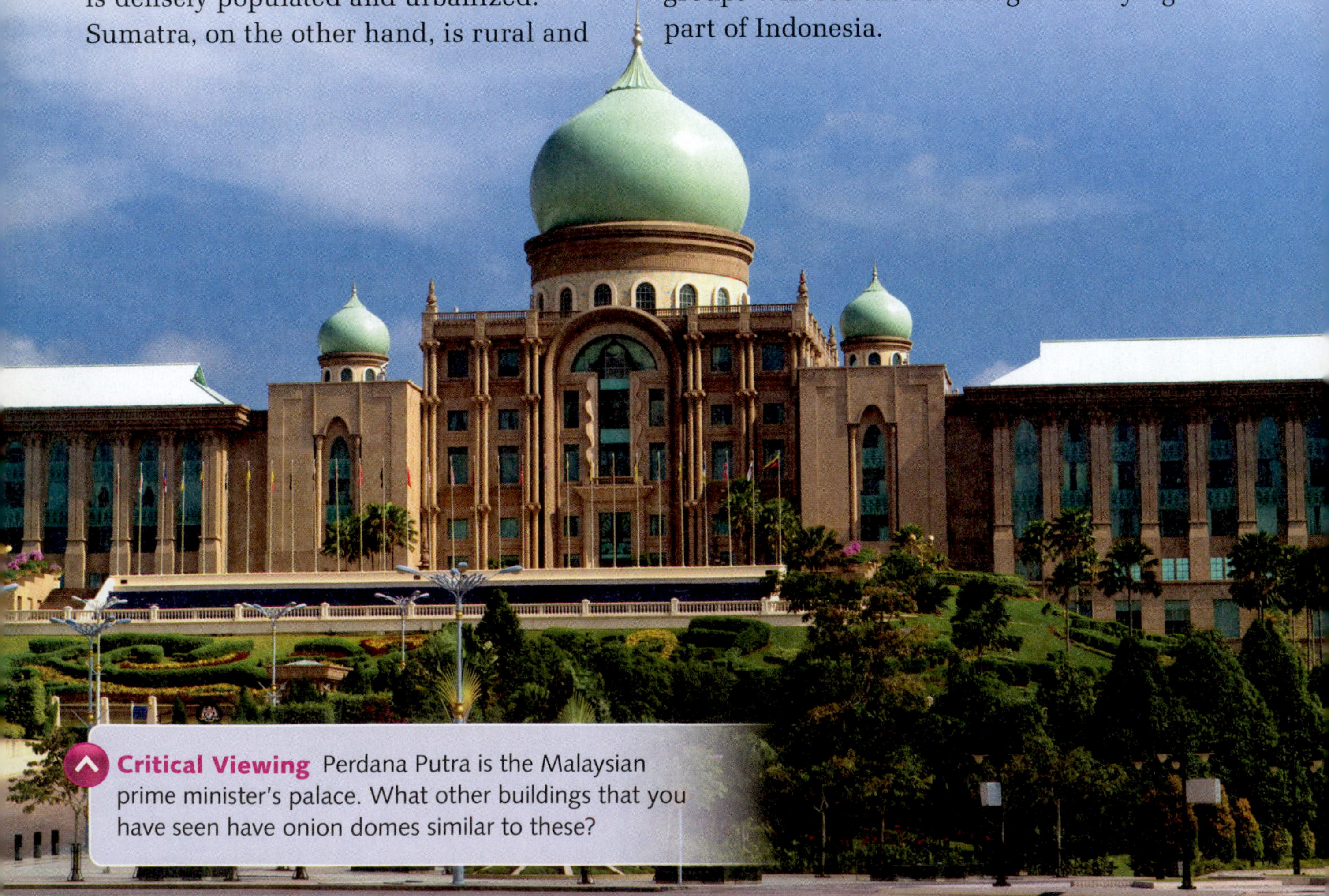

Critical Viewing Perdana Putra is the Malaysian prime minister's palace. What other buildings that you have seen have onion domes similar to these?

Malaysia

Malaysia includes both mainland and island areas. The mainland section is on the Malay Peninsula, and the island section is part of the island of Borneo. The challenge to Malaysia's government is to unify two parts of a country that are separated by several hundred miles of ocean. About half the total population is Malay, and their numbers dominate mainland Malaysia. The country also has sizable Chinese and Indian minorities, and these groups have generally achieved economic success. However, government policies have typically favored Malays, which has created tension between the two groups.

The Malaysian government has tried to achieve unity in several ways. Foremost is emphasis on economic growth. The country has made strong progress toward becoming a developed nation. This economic growth has helped ease some of the tensions among ethnic groups.

The Philippines

Like Indonesia, the Philippines consists of thousands of islands, most of which are less than a square mile in size. Its wide variety of ethnicities includes Malays, Chinese, Japanese, Arab, and Spanish. Many Americans have immigrated to the country as well. The country has blended these groups into its own Filipino culture. The widespread use of Filipino, one of the country's official languages, helps to form the national identity. The other official language, English, is also widely spoken.

Filipino teacher, Leonora Jusay, gives a lesson to the 59 students in her class. Education is underfunded but school is widely attended.

Although the Philippine economy has grown, nearly a third of all people are poor. Due to a lack of jobs, a few million people have left the Philippines to find work in other countries. They send a share of their earnings back to the Philippines as **remittances** to help their families at home.

Before You Move On

Make Inferences Why would geographic and ethnic divisions make it difficult for a country to come together?

2.2 Migration Within Indonesia

TECHTREK
myNGconnect.com For photos and a graph of Indonesia's population
Student Resources
Digital Library

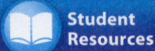

Main Idea Efforts to bring unity to Indonesia's islands through relocation have had mixed results.

Indonesia is the fourth most populous country in the world. However, the majority of the people live on just a few of Indonesia's many islands. Living on the remote islands isolates citizens from the greater population, and the country remains fragmented.

Relocation Policy

The Dutch, who had colonized the area, recognized the problem of unifying the vast chain of islands. In the 1800s, they began **relocating**, or moving, individuals and families from Java, **Madura**, and Bali—called the **inner islands**—to the surrounding and less-central islands, the **outer islands**. After Indonesia won independence in 1949, the Indonesian government continued relocating people.

Currently, more than half of Indonesia's people live on the island of Java—an island with a small percentage of the country's total land area. The government hopes that continuing to spread out the Javanese people will help to unify the country. Indonesians native to Java speak the official language, and their presence on the outer islands can help spread the official language to those places where it is infrequently heard. A common language can help unify a fragmented country.

Effects of Internal Migration

So far, however, the practice of moving people among the islands has had unintended results. New arrivals came into conflict with native people, altering their way of life. Modern farming practices clashed with traditional land use and sometimes damaged the environment.

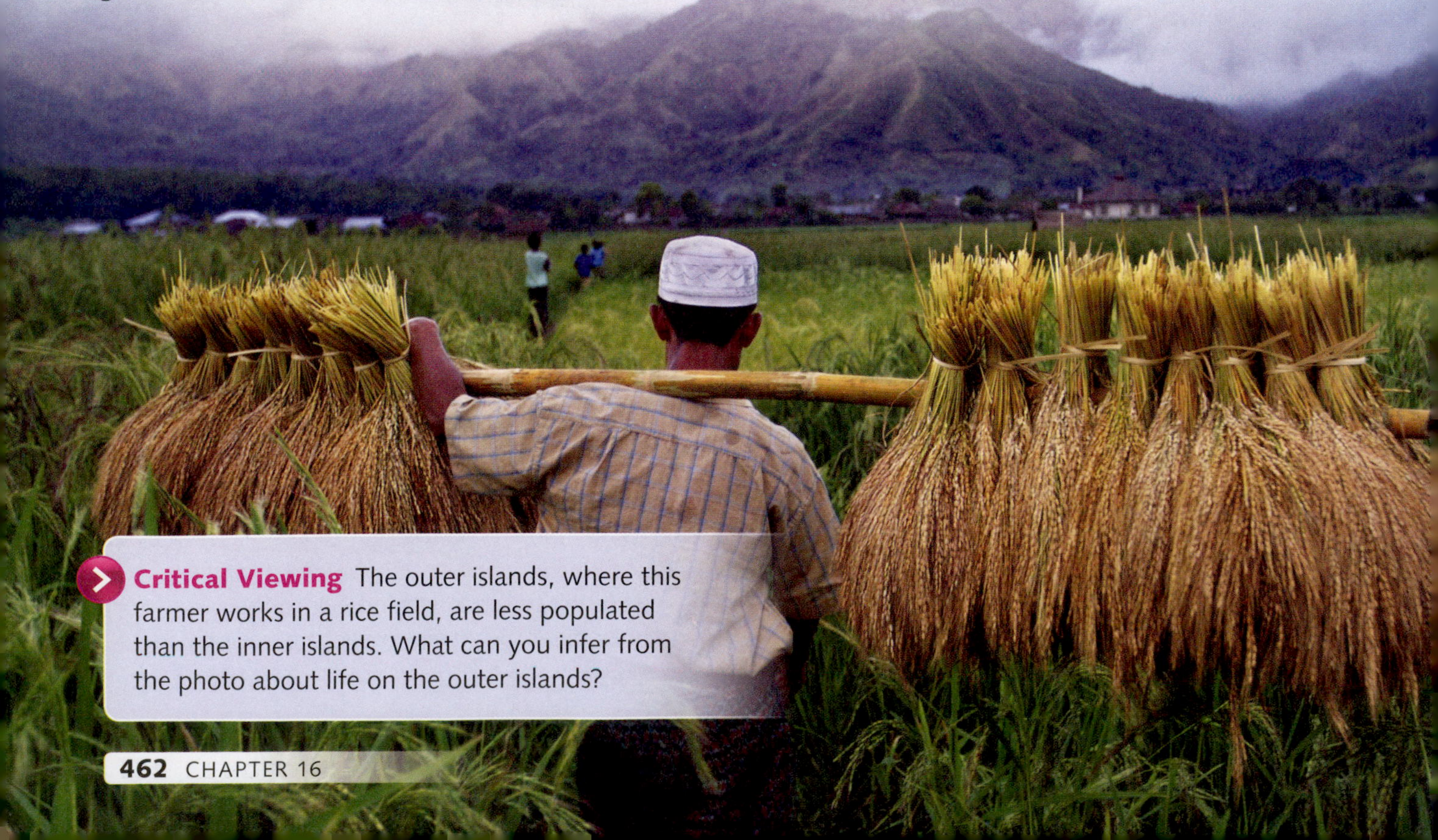

Critical Viewing The outer islands, where this farmer works in a rice field, are less populated than the inner islands. What can you infer from the photo about life on the outer islands?

Critical Viewing Indonesians return to Jakarta after visiting their homes on other islands. What does the photo suggest about the movement from rural to urban areas like Jakarta?

INDONESIA: ISLAND POPULATIONS

INNER ISLANDS:
Java + Madura + Bali

124.6 million

OUTER ISLANDS:
Sumatra

42.4 million

Sulawesi

14.9 million

Borneo

11.3 million

Lombok + Sumbawa

4.0 million

Flores + Sumba + Timor

4.0 million

New Guinea

2.2 million

Moluccas

2.0 million

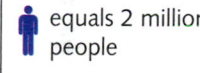

= equals 2 million people

Bangka + Belitung

0.9 million

Source: Indonesian Census 2000

The government hoped the new settlers could build successful farms, but many had trouble supporting themselves. Some ended up abandoning their new homes to return to the inner islands. As a result, Java and Bali remain crowded, making the government's relocation program ineffective. Crowding on Java and Bali grew even worse due to other **trends**, or changes over time. Indonesians living on the outer islands migrated there to flee rural poverty and find work on busier islands. Decades after the program began, Java and Bali are far more densely populated than Indonesia's other islands.

Before You Move On

Summarize How have government policies and economic factors determined migration within Indonesia?

1. **Interpret Graphs** According to the graph, which is the most populous outer island? Which is the least populous?

2. **Synthesize** When measuring population density, what does higher and lower population density indicate?

3. **Place** Look at the map of Indonesia in Section 1.1 of the previous chapter. Compare the sizes of the islands to the population figures here. Which of the outer islands do you think would have the highest population density? Why?

2.3 Singapore's Growth

Main Idea Singapore has grown economically due to its geographic location and economic policies.

The British established the modern port of Singapore in the early 1800s in an effort to compete with the Dutch in trade. Located just off the southern tip of the Malay Peninsula, the island has a perfect location near the shipping routes that link the Indian and Pacific oceans. This location gave it the <mark>potential</mark>, or possibility, of becoming a great <mark>port</mark>, a place where ships can exchange cargo. Today, the tiny island country is one of the world's busiest ports—and a strong economic power.

Building Success

In 1963, Malaysia gained independence from Britain. Singapore was part of this new country. However, conflict arose between the majority Chinese population of Singapore and the Malays of the rest of Malaysia. In 1965, to reduce the tension, the government offered Singapore its independence, and Singapore took it.

Singapore thrived because of its prime location. It served as the main transit point for sending raw materials such as timber, rubber, rice, and petroleum from Southeast Asia to other parts of the world. Manufactured goods from the United States and Europe came into the port and were shipped to other ports in Southeast Asia. Cars and machinery were shipped into the city from the west to be distributed around the region.

Prime Minister Lee Kuan Yew led Singapore from 1959 to 1990. He emphasized the country's role as an important port and led the drive to <mark>industrialize</mark>, or develop manufacturing. However, the government strictly controlled life in Singapore. Streets were kept clean, and there was very little crime.

Before You Move On

Monitor Comprehension What geographic assets have helped Singapore be part of the global economy?

KEY VOCABULARY

port, n., a place where ships can exchange cargo

industrialize, v., to develop manufacturing

multinational corporation, n., large business that has operations in many different countries

ACADEMIC VOCABULARY

potential, n., possibility or promise

SINGAPORE'S ANNUAL GDP
(1960–2010)

In billions of U.S. dollars

225
200
175
150
125
100
75
50
25
0

1960 1970 1980 1990 2000 2010

Source: CIA World Factbook

Keeping Pace

From 1965 to 2003, the economic output per person in Singapore grew to more than $24,150. That was more than twice the output of Malaysia. Incomes rose, and unemployment was low. Modern new buildings replaced slums. Singapore became the regional home to many **multinational corporations**.

Singapore's leaders have set goals based largely on economic success. They invest heavily in infrastructure improvements to gain countrywide access to the most current technologies available. To attract foreign investment, Singapore offers low tax rates and other economic incentives. Economic policy focuses on key growth industries such as telecommunications and other technologies. Because these industries rely on educated, highly skilled workers, the country emphasizes improving the level of education in the workforce.

COMPARE ACROSS REGIONS

As in Singapore, a large sector of the workforce in India is highly educated and trained in the newest technologies. Economic policy offers incentives to foreign investors, and many multinational corporations have set up operations there. However, India continues to face many challenges as it works to manage its economic growth. The poverty level is higher and the education level is lower than those of Singapore. India also struggles to improve its infrastructure to keep up with a global economy.

Critical Viewing This late-night street scene shows the business district in Singapore. Which details in the photo are typical of a prosperous city?

Greater stability has allowed Singapore to take advantage of changes happening elsewhere. When the British handed Hong Kong back to China in 1999, some business owners were concerned that Chinese rule would limit their freedom and prosperity. Singapore welcomed business owners who chose to move their companies there, which boosted economic growth.

Before You Move On

Summarize How have Singapore and India's participation in the global economy differed?

ONGOING ASSESSMENT
WRITING LAB GeoJournal

1. **Make Inferences** Is Singapore's location still important to its economy? Why or why not?
2. **Write Analyses** How has Singapore been affected by globalization? How has its economy been affected by other countries? Go to **Student Resources** for Guided Writing support.

2.4 Malaysia and New Media

TECHTREK

myNGconnect.com For graphs and photos of communication in Malaysia

 Maps and Graphs

 Digital Library

Main Idea Access to new media sources is changing the strict control on information formerly held by the government in Malaysia.

Since gaining independence in 1963, **Malaysia** has enjoyed prosperity and calm. The government has concentrated on building the economy. However, as in nearby Singapore, Malaysia has limited the freedoms its people can enjoy.

Controlling Information

One of these limits is government control of access to information. Freedom of the press is limited. Newspapers must obtain licenses from the government to operate, and the government can cancel those licenses at any time. Similar laws place restrictions on companies that want to run radio or television stations. Also, the government withholds information from media.

Tough laws punish media outlets that criticize the government. Members of the ruling party own many major newspapers. Even independently owned newspapers usually do not criticize the government.

New Media

The restrictions on the flow of information may be starting to loosen. Nearly two-thirds of Malaysians can now connect to the Internet. New media, like the Internet, give people access to new sources of information that the government may have more difficulty controlling.

Officially the government promises limitless Internet use. However, interference is not uncommon. Many Web site operators focusing on Malaysian news have been arrested multiple times for government criticism. Other journalists have been targeted as well.

> **Critical Viewing** A young woman uses her laptop in a public space in Malaysia. In what ways does the photo show both government control and freedom?

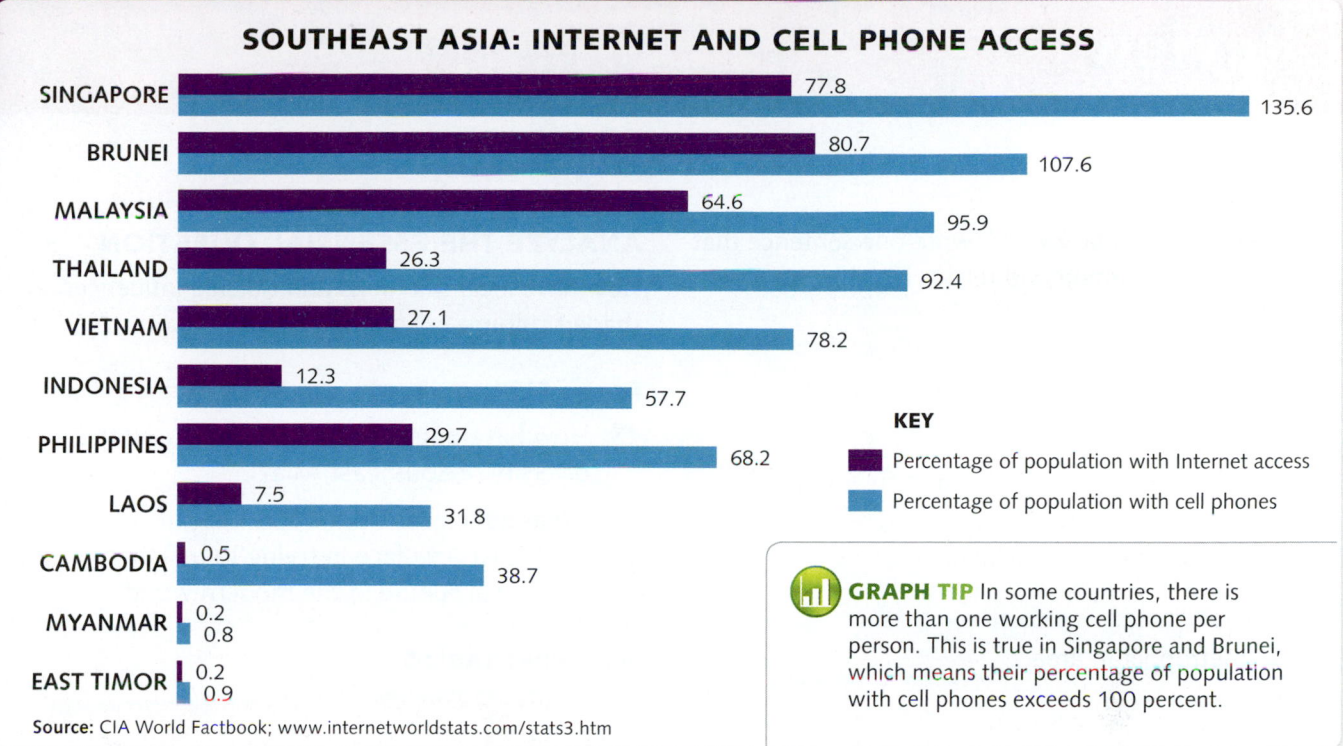

SOUTHEAST ASIA: INTERNET AND CELL PHONE ACCESS

Country	Internet access	Cell phones
SINGAPORE	77.8	135.6
BRUNEI	80.7	107.6
MALAYSIA	64.6	95.9
THAILAND	26.3	92.4
VIETNAM	27.1	78.2
INDONESIA	12.3	57.7
PHILIPPINES	29.7	68.2
LAOS	7.5	31.8
CAMBODIA	0.5	38.7
MYANMAR	0.2	0.8
EAST TIMOR	0.2	0.9

KEY
■ Percentage of population with Internet access
■ Percentage of population with cell phones

GRAPH TIP In some countries, there is more than one working cell phone per person. This is true in Singapore and Brunei, which means their percentage of population with cell phones exceeds 100 percent.

Source: CIA World Factbook; www.internetworldstats.com/stats3.htm

In other instances, police raided Web site headquarters and seized computers in order to find specific people who wrote articles critical of the government.

Many Malaysians believed that until the **emergence**, or arrival, of the Internet the government controlled and manipulated information available to citizens. During elections in 2008, Web sites became the leading source of news for people in the country.

Web sites were outside of government control. They also were completely open to opinions and ideas from anyone in the country. Any citizen could file stories or post videos to the sites. Because of such openness, Web sites may have influenced the 2008 elections. As a result of that vote, the ruling party lost 58 seats in the national legislature. It was the worst result for the party in more than 40 years.

New media technologies give people access to information that is difficult for people in power to control. Poll results found that older citizens trust newspapers and television news and are less likely to embrace additional sources of media. However, among voters in their twenties and thirties, only a small minority trusted traditional media, while more than 60 percent said online news sources were **reliable**, or trustworthy.

Before You Move On

Summarize How has access to new media changed the strict government control of information in Malaysia?

ONGOING ASSESSMENT

DATA LAB

 GeoJournal

1. **Analyze Data** How do percentages for Internet access and cell phone use in Malaysia compare to those of other countries in the region?
2. **Make Inferences** Why do you think the government tries to keep such tight control of the news in Malaysia?
3. **Make Predictions** What do the poll results suggest about the way traditional media will be accepted in the future? Why?

VOCABULARY

For each vocabulary word, write one sentence that explains its meaning and relates it to the content of the chapter.

1. prehistoric

> Many religions in Southeast Asia are prehistoric, meaning they existed before history was written down.

2. metropolitan area
3. dialect
4. domesticate
5. relocate
6. multinational corporation

MAIN IDEAS

7. How has colonial history shaped religion in the region? (Section 1.1)
8. How does modern Thailand reflect both tradition and external influences? (Section 1.2)
9. In what way has globalization changed language use in Southeast Asia? (Section 1.3)
10. What methods can help protect the wild Asian elephant population in Southeast Asia? (Section 1.4)
11. How has fragmentation been a problem for Indonesia, Malaysia, and the Philippines? (Section 2.1)
12. What have been the results of Indonesia's policy of moving people to new areas? (Section 2.2)
13. How has Singapore built its economy? (Sections 2.3)
14. How have new media changed politics in Malaysia? (Section 2.4)

CULTURE

ANALYZE THE ESSENTIAL QUESTION

How have local traditions and outside influences shaped cultures in Southeast Asia?

Focus Skill: Make Generalizations

15. How have outside influences increased the diversity of Southeast Asia?
16. What difficulties do the countries of Southeast Asia face in trying to maintain traditional culture in the modern world?

INTERPRET TABLES

PERCENTAGES OF ETHNIC GROUPS IN SELECTED SOUTHEAST ASIAN COUNTRIES		
Indonesia	Javanese: 41% Sundanese: 15% Madurese: 3%	Minangkabau: 3% Other: 38.4%
Laos	Lao: 55% Khmou: 11%	Hmong: 8% Other: 26%
Malaysia	Malay: 50% Chinese: 24%	Indian: 7% Other: 19%
Philippines	Tagalog: 28% Cebuano: 13% Ilocano: 9%	Bisaya/Binisaya: 8% Hiligaynon Ilonggo: 8% Other: 35%
Singapore	Chinese: 77% Malay: 14%	Indian: 8% Other: 1%
Thailand	Thai: 75% Chinese: 14%	Other: 11%
Vietnam	Kinh (Viet): 86%	Other : 14%

Source: CIA World Factbook

17. **Analyze Data** Which three countries have a single dominant ethnic group? What is the group in each case?
18. **Make Generalizations** Would you expect countries with one dominant ethnic group to have a single official national language? Why or why not?

GOVERNMENT & ECONOMICS

ANALYZE THE ESSENTIAL QUESTION

How are Southeast Asia's governments trying to unify their countries?

Focus Skill: Summarize

19. What economic and political concerns led Indonesia to adopt a policy of moving people from the inner islands to the outer islands?

20. Why did Singapore split from Malaysia and become independent?

21. How is the policy of Malaysia's government to limit freedom of the press related to the issue of fragmentation?

INTERPRET MAPS

PHILIPPINES POLITICAL

22. **Interpret Maps** Why is it appropriate to call the Philippines a fragmented country?

23. **Draw Conclusions** What transportation improvements might the government invest in to help unify the country? Explain.

ACTIVE OPTIONS

Synthesize the Essential Questions by completing the activities below.

24. **Write a Feature Article** Introduce a friend to Southeast Asia today by writing a feature article that describes one country in the region. In your article, highlight the ways religious practices and regional languages shape culture in that country. Describe the modern developments and traditional practices that impact the daily lives of people who live there. **Share your article with your friend.**

> **Writing Tips**
> - Provide a clear, concise introduction of your country.
> - Keep your paragraphs tightly focused on one topic.
> - Include details about the country's religions, languages, and form of government.
> - Provide smooth transitions between paragraphs.
> - In your conclusion, include a paragraph that summarizes your article.

TECHTREK myNGconnect.com For research links on Southeast Asia today

25. **Create Graphs** Use the research links at **Connect to NG** to make a bar graph showing the per capita gross domestic product (GDP) of Cambodia, Malaysia, and the Philippines. Write a paragraph explaining which countries' economies might be affected by fragmentation. Use the example below as a guide for your graph.

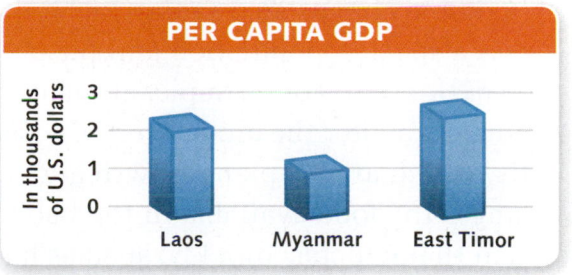

PER CAPITA GDP

Endangered Species

TECHTREK

myNGconnect.com For an online graph and research links on endangered species

Maps and Graphs

Connect to NG

Endangered species of both plants and animals can be found in every region of the world. A species is endangered when it runs the risk of becoming extinct, or disappearing completely from the world. Over time, various species have not survived. In fact, historically there have been five mass exinctions, in which a large number of existing species died out.

Today, the rate of extinction for plant and animal species has become hundreds of times faster than what scientists have observed through the fossil record. Many scientists believe the earth is currently in the midst of a sixth mass extinction. Those same scientists believe the main cause to be the destruction of habitat.

Compare Ranges of Endangered Big Cats

- Asiatic Lion
- Cheetah
- Iberian Lynx
- Jaguar
- Tiger

CAUSES OF EXTINCTION

Mining, logging, and clearing of forests for grazing cattle and growing crops all greatly change the natural landscape, or habitat, on which most species depend. Other development, such as the building of dams, highways, and housing, can increasingly divide animal populations into smaller, less diverse pockets. Some groups, like the big cats shown at right, face additional threats from hunters fearful of the animals' ability to harm people and livestock.

Climate change also can seriously stress a species' population and push it to extinction. For example, climate change can alter the amount of rain that falls, which affects plant growth and changes the food available in the habitat. When that happens quickly, species have difficulty adapting.

CONSERVATION

Species and their ecosystems contribute much to the health and well being of humans. A diversity of plants and animals, and the habitats in which they are found, provide fertile soils, medicines, clean air and water, fibers, building materials, and food.

The International Union for Conservation of Animals (IUCN) is one organization trying to help countries around the world find solutions for balancing human needs and environmental challenges. To help target the habitats and animals in need of support, the IUCN maintains a database called the "Red List of Endangered Species." The list identifies animal species as near threatened, vulnerable, endangered, and critically endangered, as shown on the diagram at right.

ENDANGERED BIG CATS

Source: The IUCN Red List of Threatened Species™

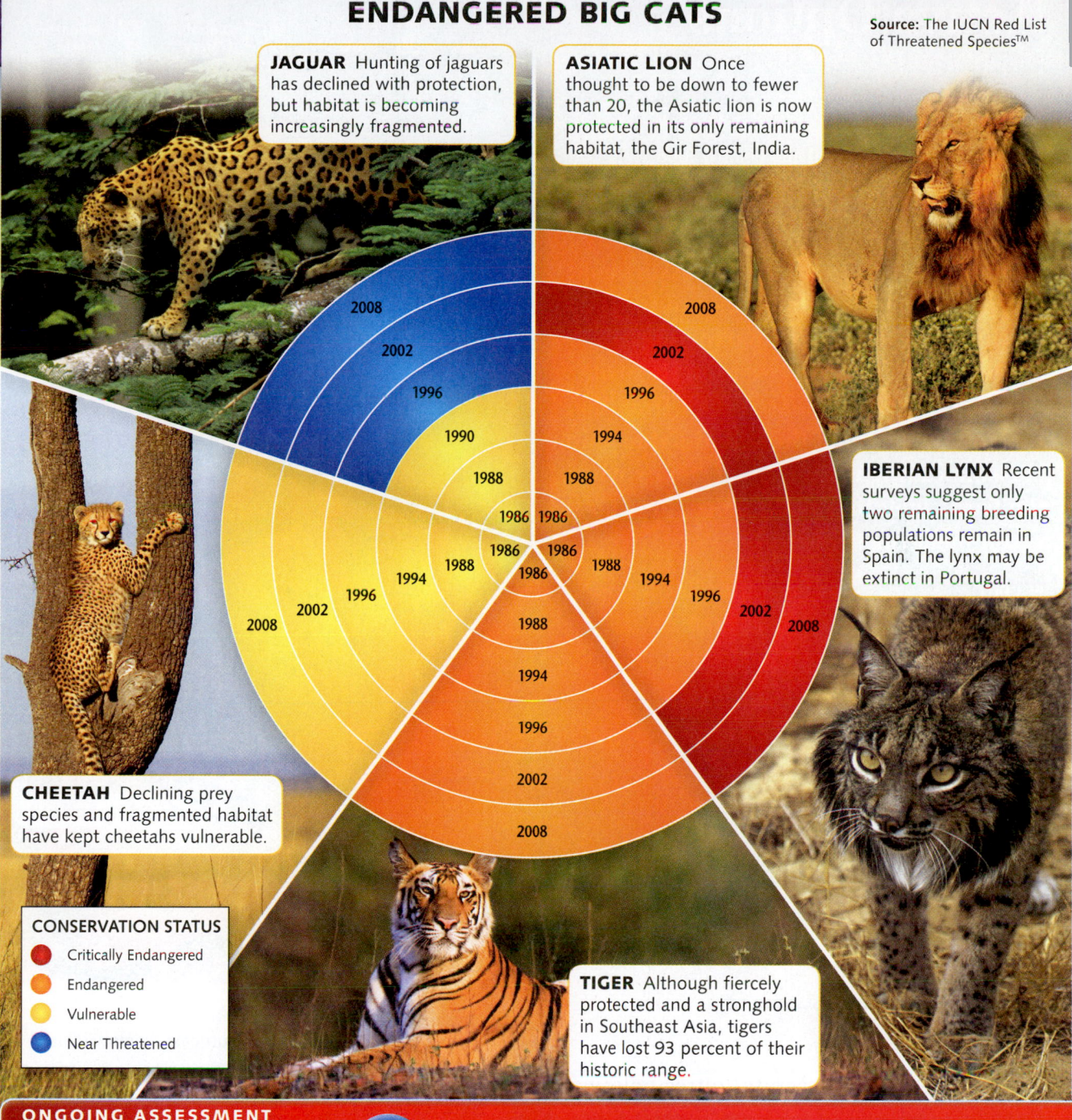

JAGUAR Hunting of jaguars has declined with protection, but habitat is becoming increasingly fragmented.

ASIATIC LION Once thought to be down to fewer than 20, the Asiatic lion is now protected in its only remaining habitat, the Gir Forest, India.

IBERIAN LYNX Recent surveys suggest only two remaining breeding populations remain in Spain. The lynx may be extinct in Portugal.

CHEETAH Declining prey species and fragmented habitat have kept cheetahs vulnerable.

TIGER Although fiercely protected and a stronghold in Southeast Asia, tigers have lost 93 percent of their historic range.

CONSERVATION STATUS
- 🔴 Critically Endangered
- 🟠 Endangered
- 🟡 Vulnerable
- 🔵 Near Threatened

ONGOING ASSESSMENT

RESEARCH LAB GeoJournal

1. **Analyze Data** Look at the diagram for the Asiatic lion. What happened between 2002 and 2008? How can you explain this?

2. **Draw Conclusions** Look at the diagram for the jaguar and the Iberian lynx. Think about where these cats live and the population density in those ranges. Why might this help explain why their levels of endangerment have changed?

Research and Compare Use the IUCN Red List of Threatened Species™ to research and compare the levels of endangerment within an animal group such as antelopes, seals, bats, bears, big apes, dolphins or otters. Identify the region and how humans are affecting the populations. What is being done to conserve the species? Create a table to describe your findings.

Active Options

TECHTREK

myNGconnect.com For research links and photos of fruit

Connect to NG

Digital Library

Magazine Maker

ACTIVITY 1

Goal: Learn about unusual and healthy food.

Make a Recommendation

Southeast Asia is home to a rich variety of plants that includes a wide assortment of fruits. Several of the more exotic fruits are listed below. With a partner, use the list below and the research links at Connect to NG to research fruit from the region. Recommend three fruits that your classmates might want to try. Show what the fruits look like and give reasons for recommending those three.

- ciku
- dragon fruit
- durian
- jackfruit
- langsat
- longan
- mangosteen
- rambutan
- salak
- sapodilla
- soursap
- star apple

dragon fruit

ACTIVITY 2

Goal: Research a Southeast Asian city.

Promote a City

Research one of the capital cities below to receive an award for Best Tourist Site of Southeast Asia. Use the **Magazine Maker** to present and describe the site that causes the city to earn the award. Include things to do and places to visit that would delight tourists.

- Bangkok, Thailand
- Kuala Lumpur, Malaysia
- Phnom Penh, Cambodia
- Singapore, Singapore

ACTIVITY 3

Goal: Extend knowledge of Southeast Asian wildlife through research and drama.

Set the Stage

Some of the world's oldest rain forests are in Southeast Asia. With a group, compose a one-act play set deep in the rain forest. Focus the drama on the region's rain forest issues, and include in your play one of the animals found there, such as the Komodo dragon, flying snake, or orangutan. Perform the play for the class.

EXPLORE
AUSTRALIA,
THE PACIFIC REALM & ANTARCTICA
WITH NATIONAL GEOGRAPHIC

MEET THE EXPLORER

NATIONAL GEOGRAPHIC

National Geographic Fellow Elizabeth Lindsey helps conserve knowledge and traditions of Polynesian cultures. Her work focuses on such elements as Micronesian chants and traditional navigation methods that do not use any instruments.

STEP INTO HISTORY

Easter Island, a UNESCO World Heritage site, is located 2,200 miles west of Chile. It is known for its enormous stone statues, standing from 10 to 40 feet tall. More than 600 statues remain on the island.

CONNECT WITH THE CULTURE

The opera house in Sydney, New South Wales, Australia, is one of the most famous buildings in the world. Sydney, now one of the largest ports in the South Pacific, began as a convict settlement in the 18th century.

Washington, D.C.

9,762 miles

Canberra, Australia

Go to **myNGconnect.com** for maps of Australia, the Pacific Realm, and Antarctica.

INVESTIGATE GEOGRAPHY

Adélie penguins jump off of a rock ledge in Armstrong Reef, western Antarctica. They will feed heavily in the food-rich waters of the southern Pacific Ocean.

CHAPTER 17

AUSTRALIA, THE PACIFIC REALM & ANTARCTICA
GEOGRAPHY & HISTORY

PREVIEW THE CHAPTER

Essential Question How did geographic isolation influence the development of this region?

KEY VOCABULARY

- coral reef
- atoll
- indigenous
- marsupial
- invasive species
- extinct
- feral
- hotspot
- coral island
- marine
- exoskeleton

ACADEMIC VOCABULARY

preserve

TERMS & NAMES

- Southern Hemisphere
- Southern Alps
- South Pole
- New Guinea
- Kingman Reef
- Great Barrier Reef

Essential Question How did geographic isolation shape the history of Australia and the Pacific Realm?

KEY VOCABULARY

- land bridge
- clan
- pictograph
- seafarer
- outrigger canoe
- navigation
- convict
- assisted migration
- generation
- linguist

ACADEMIC VOCABULARY

transport, convey

TERMS & NAMES

- Aborigine
- James Cook
- New South Wales
- Commonwealth of Australia
- Enduring Voices Project

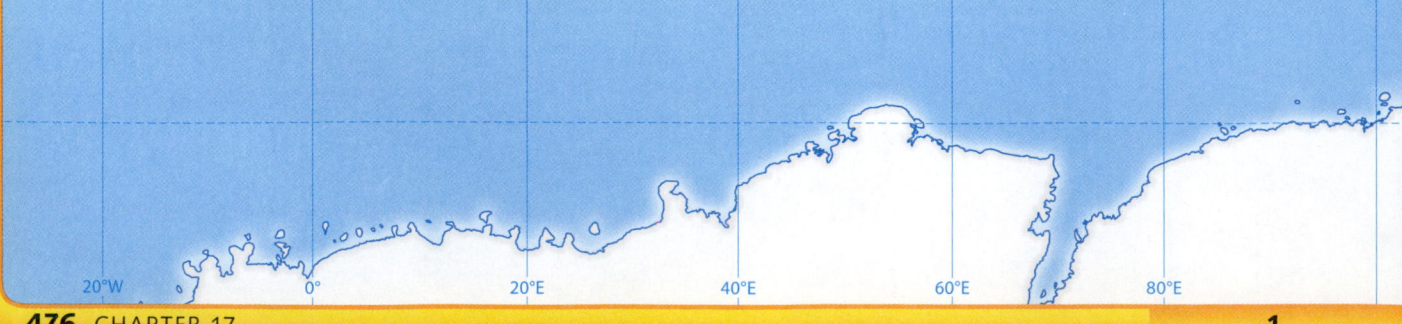

TECHTREK FOR THIS CHAPTER

- **Student eEdition**
- **Maps and Graphs**
- **Interactive Whiteboard GeoActivities**
- **Digital Library**
- **Connect to NG**

Go to **myNGconnect.com** for more on Australia, the Pacific Realm, and Antarctica.

koalas

Map labels

NORTH PACIFIC OCEAN

Tropic of Cancer

HAWAI'I (United States)

NORTHERN MARIANA ISLANDS (U.S.)

Saipan
Capital Hill
(Agana) Hagåtña
GUAM (U.S.)

Wake Island (U.S.)

Johnston Atoll (U.S.)

20°N

MARSHALL ISLANDS

Melekeok

PALAU

FEDERATED STATES OF MICRONESIA

Palikir

Majuro

Monday / Sunday

Palmyra Atoll (U.S.)

Equator 0°

NAURU
Yaren

Tarawa (Bairiki)

Howland Island (U.S.)
Baker Island (U.S.)

LINE ISLANDS

PAPUA NEW GUINEA

Arafura Sea

Port Moresby

SOLOMON ISLANDS

Honiara

TUVALU
Funafuti

KIRIBATI

TOKELAU (N.Z.)

Marquesas Islands

Timor Sea

Darwin

Great Barrier Reef

Coral Sea

VANUATU

Îles Wallis (France)

Port-Vila

SAMOA
Apia

Îles de Horne (France)

AMERICAN SAMOA (U.S.)
Pago Pago

COOK ISLANDS (N.Z.)

Niue (N.Z.)

TUAMOTU ARCHIPELAGO

Tahiti Papeete

NEW CALEDONIA (France)

Suva

FIJI

Nouméa

NORTHERN TERRITORY

Tropic of Capricorn

QUEENSLAND

AUSTRALIA

Nuku'alofa

TONGA

20°S

FRENCH POLYNESIA (France)

WESTERN AUSTRALIA

SOUTH AUSTRALIA

Darling R.

Brisbane

Norfolk Island (Australia)

Raoul Island (N.Z.)

PITCAIRN ISLANDS (U.K.)

Perth

Great Australian Bight

Adelaide

Murray R.

NEW SOUTH WALES

VICTORIA

Melbourne

Sydney
Canberra, AUSTRALIAN CAPITAL TERRITORY

Mt. Kosciuszko 7,310 ft (2,228 m)

Lord Howe Island (Australia)

Date Line

SOUTH PACIFIC OCEAN

Auckland

NEW ZEALAND

40°S

TASMANIA
Tasmania
Hobart

Tasman Sea

(Mt. Cook) Aoraki 12,316 ft (3,754 m)

Wellington

Christchurch

Chatham Islands (N.Z.)

INDIAN OCEAN

N
W E
S

Auckland Islands (N.Z.)

Bounty Islands (N.Z.)

Antipodes Islands (N.Z.)

Campbell Island (N.Z.)

Macquarie Island (Australia)

0   500   1,000 Miles
0   500   1,000 Kilometers

Monday / Sunday

60°S

Antarctic Circle

WILKES LAND

ANTARCTICA

120°E   140°E   160°E   180°

2   3   4   5   6

Inset map

ATLANTIC OCEAN

Antarctic Circle

Weddell Sea

Antarctic Peninsula

QUEEN MAUD LAND

INDIAN OCEAN

30°W   0°   30°E

75°S

60°S

PACIFIC OCEAN

Vinson Massif 16,067 ft (4,897 m)

South Pole

ANTARCTICA

WILKES LAND

Ross Sea

INDIAN OCEAN

90°E

0   500   1,000 Miles
0   500   1,000 Kilometers

150°W   180°   150°E

TECHTREK

myNGconnect.com For maps of Australia, the Pacific Realm, and Antarctica and Visual Vocabulary

Maps and Graphs

Digital Library

AUSTRALIA, THE PACIFIC REALM, AND ANTARCTICA PHYSICAL

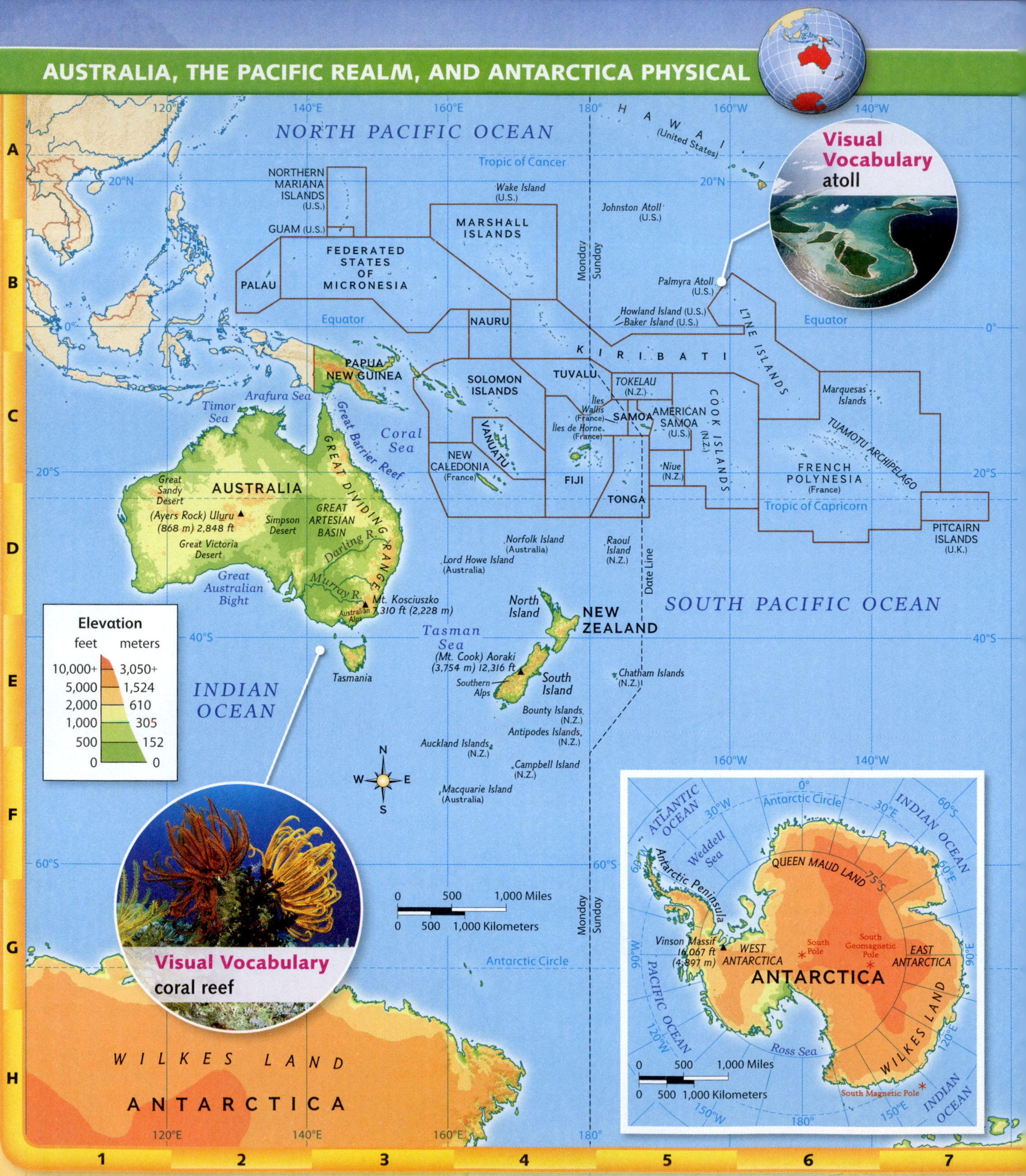

NORTH PACIFIC OCEAN

Tropic of Cancer

HAWAII (United States)

NORTHERN MARIANA ISLANDS (U.S.)

GUAM (U.S.)

Wake Island (U.S.)

Johnston Atoll (U.S.)

Visual Vocabulary
atoll

MARSHALL ISLANDS

PALAU

FEDERATED STATES OF MICRONESIA

Monday Sunday

Palmyra Atoll (U.S.)

Equator

NAURU

Howland Island (U.S.)
Baker Island (U.S.)

KIRIBATI

LINE ISLANDS

Equator

0°

PAPUA NEW GUINEA

SOLOMON ISLANDS

TUVALU

TOKELAU (N.Z.)

COOK ISLANDS

Marquesas Islands

Timor Sea

Arafura Sea

Great Barrier Reef

Coral Sea

VANUATU

Îles Wallis (France)
Îles de Horne (France)

SAMOA

AMERICAN SAMOA (U.S.)

COOK ISLANDS (N.Z.)

TUAMOTU ARCHIPELAGO

AUSTRALIA

Great Sandy Desert

(Ayers Rock) Uluru ▲
(868 m) 2,848 ft

Simpson Desert

GREAT ARTESIAN BASIN

GREAT DIVIDING RANGE

NEW CALEDONIA (France)

FIJI

*Niue (N.Z.)

FRENCH POLYNESIA (France)

Tropic of Capricorn

Great Victoria Desert

Darling R.

TONGA

PITCAIRN ISLANDS (U.K.)

Great Australian Bight

Murray R.

Mt. Kosciuszko
7,310 ft (2,228 m)
Australian Alps

Norfolk Island (Australia)

Lord Howe Island (Australia)

Raoul Island (N.Z.)

Date Line

SOUTH PACIFIC OCEAN

Elevation

feet	meters
10,000+	3,050+
5,000	1,524
2,000	610
1,000	305
500	152
0	0

INDIAN OCEAN

Tasmania

North Island

Tasman Sea
(Mt. Cook) Aoraki
(3,754 m) 12,316 ft

NEW ZEALAND

Chatham Islands (N.Z.)

Southern Alps

South Island

Bounty Islands (N.Z.)

Antipodes Islands (N.Z.)

Auckland Islands (N.Z.)

Campbell Island (N.Z.)

Macquarie Island (Australia)

N W E S

0   500   1,000 Miles
0   500 1,000 Kilometers

Visual Vocabulary
coral reef

Antarctic Circle

Monday Sunday

ATLANTIC OCEAN

Antarctic Circle

INDIAN OCEAN

Weddell Sea

Antarctic Peninsula

QUEEN MAUD LAND

Vinson Massif
16,067 ft
(4,897 m)

WEST ANTARCTICA

South Pole

South Geomagnetic Pole

EAST ANTARCTICA

PACIFIC OCEAN

ANTARCTICA

WILKES LAND

Ross Sea

WILKES LAND

ANTARCTICA

South Magnetic Pole

0   500   1,000 Miles
0   500 1,000 Kilometers

Main Idea This region lies in the Pacific Ocean and is geographically isolated from other parts of the world.

Australia, much of the Pacific Realm, and Antarctica are located in the **Southern Hemisphere**. The region is spread across a vast ocean area.

Australia

Australia is the only country in the world that is also a continent. Australia is sometimes called the "island continent" because it is surrounded by water. However, the climate is mostly dry. In fact, nearly 20 percent of the land mass of Australia is classified as desert. Few people live in the inland parts of the continent. Instead, most people live along the coasts where rainfall is plentiful.

The Pacific Realm and Antarctica

The Pacific Realm is a large area of the Pacific Ocean made up of thousands of small islands and coral reefs. <mark>Coral reefs</mark> are rock-like structures built by layers of coral organisms. Coral reefs thrive in warm ocean waters between latitudes of 30°N and 30°S. <mark>Atolls</mark>—ring-shaped reefs, islands, or chains of islands made of coral—also dot this area of the Pacific.

New Zealand, located southeast of Australia, is made up of two major islands, North Island and South Island. New Zealand's mountain range, the **Southern Alps,** includes peaks that rise more than 12,000 feet.

Antarctica is centered on the **South Pole,** the southernmost point of Earth's axis. A thick layer of ice covers almost all

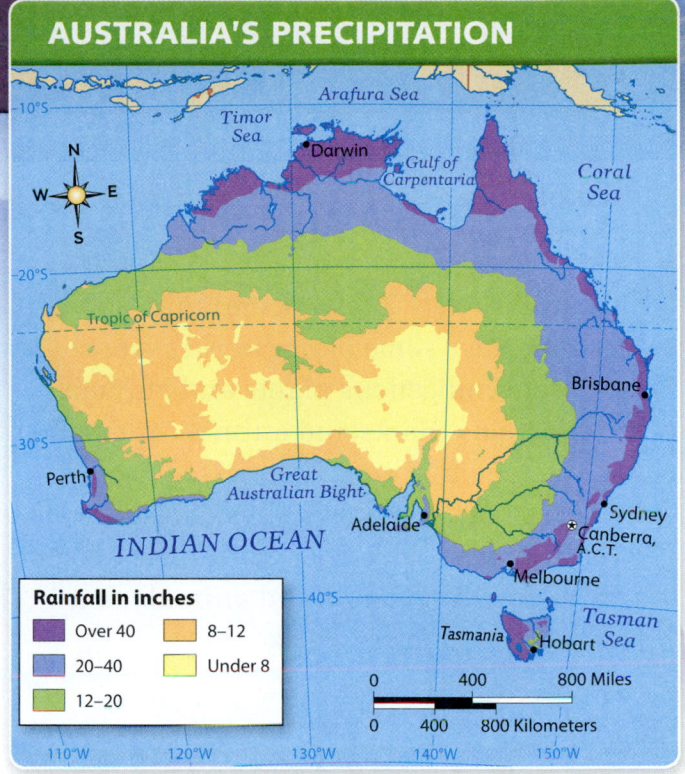

AUSTRALIA'S PRECIPITATION

Rainfall in inches

Over 40	8–12
20–40	Under 8
12–20	

the continent's mountains, valleys, and islands. Antarctica has the coldest climate on Earth and is the continent with the highest average elevation.

Before You Move On

Monitor Comprehension How is this region isolated from other parts of the world?

ONGOING ASSESSMENT

MAP LAB

GeoJournal

1. **Human-Environment Interaction** Based on the precipitation map, which part of Australia might produce few crops and why?

2. **Interpret Maps** Between what latitudes do the Pacific islands lie? Explain how their location is ideal for the formation of coral reefs.

3. **Compare and Contrast** Examine the physical map of the region. In what ways are Australia and Antarctica similar and different? Use a chart like this one to answer the question.

SIMILARITIES	DIFFERENCES

1.2 Indigenous Plants and Animals

TECHTREK

myNGconnect.com For photos of the region's indigenous plants and animals

Digital Library

Main Idea Many plants and animals are found exclusively in this region.

Australia, New Zealand, and the Pacific islands have some of the world's most unusual indigenous plants and animals. **Indigenous** plants and animals are native to the region in which they are found. Because of the region's physical isolation, these species developed with little outside influence. The plants and animals that developed here are unlike those found elsewhere.

Plant Species

Thousands of wildflowers, shrubs, and trees are native to the region. Among them are eucalyptus (YOO kuh LIHP tuhs) trees, whose leaves produce a fragrant oil used in medicines. These leaves are the only food that Australia's much-loved koalas eat. Another common plant is the acacia, with more than 700 varieties found in Australia. Most acacia have brightly colored flowers.

Native Animals

Animals native to this region include **marsupials** (mar SOO pee uhlz), which are mammals whose females carry their babies in a pouch. Four types of marsupials live in this region. Kangaroos are the biggest marsupials. Other marsupials include koalas, wallabies, wombats, and Tasmanian devils.

Other indigenous mammals of Australia and **New Guinea** include the platypus and the echidna (ih KID neh). These mammals are the only mammals that lay eggs. The platypus and the echidna have beaks, another bird-like feature not typically found in mammals.

The region is home to many kinds of birds, including New Zealand's kiwi and kakapo, which do not fly. The cassowary (KAS eh ware ee) and emu, also flightless, stand five to six feet tall. Australia's kookaburra, a large kingfisher, is famous for its call, which sounds like laughter.

Before You Move On

Summarize What plant and animal species are indigenous to the region?

Cassowaries grow as high as five feet tall and can be aggressive.

Critical Viewing Eucalyptus trees grow rapidly and some even reach 300 feet in height. Based on the photo, how would you describe a koala's habitat?

PHOTO LAB
GeoJournal

1. **Analyze Visuals** Look at the photos. Write two sentences that describe the physical features of the cassowary and the koala. What are some other birds and animals with similar physical features?

2. **Make Inferences** Based on the photos and text, why might protecting the health of eucalyptus trees and other unique plant species be important?

1.3 Biological Hitchhikers

TECHTREK

myNGconnect.com For a map and photos of the region's invasive species

Maps and Graphs

Digital Library

Main Idea The introduction of non-native plant and animal species has changed the natural habitats of Australia and New Zealand.

Plants and animals, and their habitats, developed as they did because of the physical isolation of the region. The introduction of new plants and animals to Australia and New Zealand has had negative consequences.

Invasive Species

Many non-native species were introduced to these countries on purpose—before people understood what problems might occur. New species also have arrived by accident. Planes and ships delivering produce and packages bring in hidden plant and animal "hitchhikers" that then invade natural habitats. Non-native plants and animals that disturb the habitats

 Visual Vocabulary Invasive species are non-native plants and animals that disturb the habitats of native life forms. These rabbits in Australia are an invasive species.

of native life forms are called invasive species. Because of invasive species, some native plants and animals have become extinct, meaning no more living members of that group exist. In fact, 19 different species of small mammals have become extinct in Australia since the introduction of non-native mammals such as cats, foxes, and rabbits.

Dingoes, Rabbits, and Other Pests

Dingoes have lived in Australia for many years, but they were not indigenous to the island continent. Sea travelers from Asia brought dingoes with them about 3,000 years ago. Some dingoes escaped, became

Critical Viewing Dingoes are hunters. Based on the photo, what other animals do dingoes look like?

EUROPEAN RABBITS IN AUSTRALIA, 1870–2000

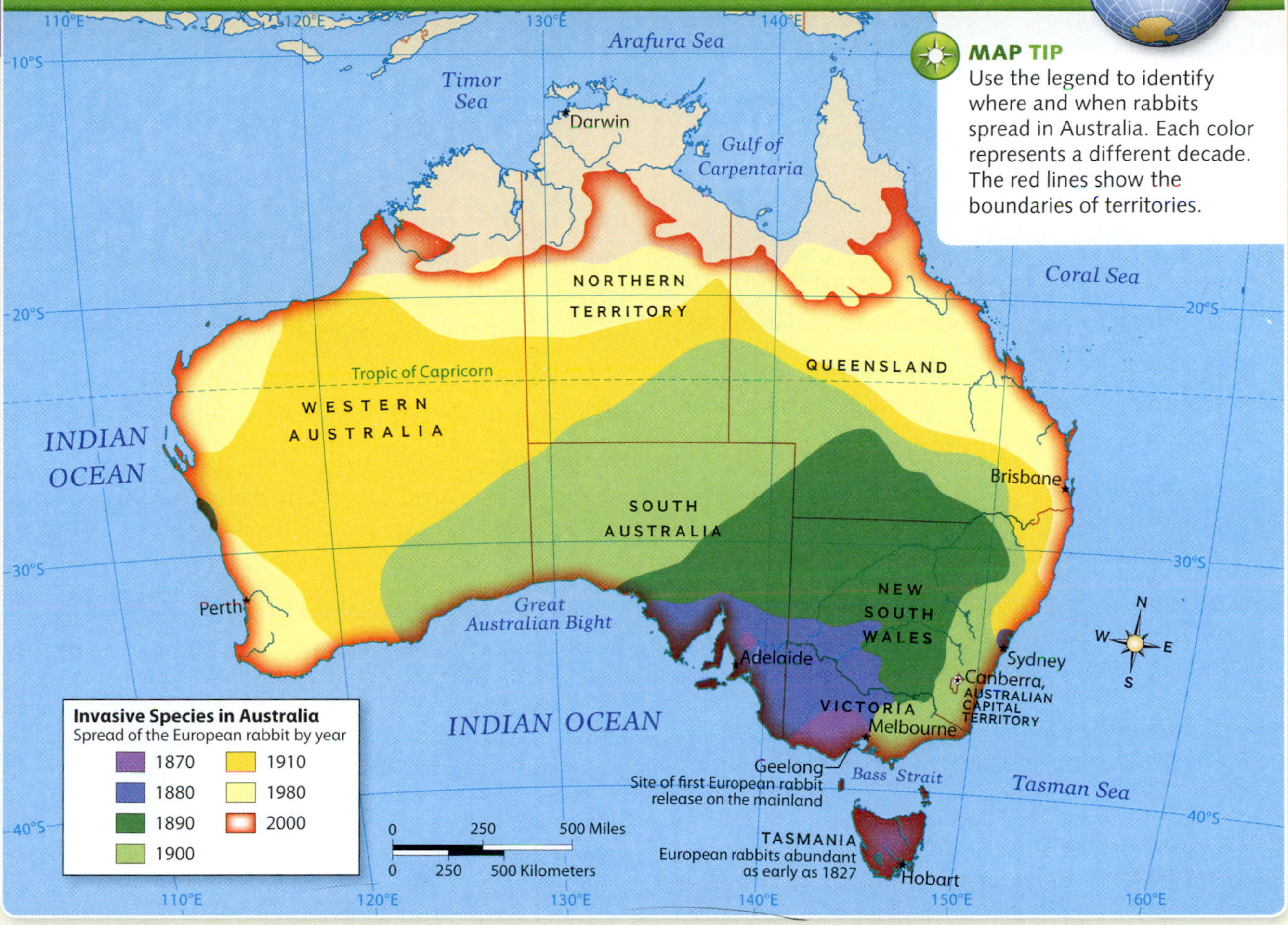

MAP TIP
Use the legend to identify where and when rabbits spread in Australia. Each color represents a different decade. The red lines show the boundaries of territories.

Arafura Sea

Timor Sea

Darwin

Gulf of Carpentaria

Coral Sea

NORTHERN TERRITORY

QUEENSLAND

Tropic of Capricorn

WESTERN AUSTRALIA

INDIAN OCEAN

SOUTH AUSTRALIA

Brisbane

Perth

Great Australian Bight

NEW SOUTH WALES

Adelaide

Sydney

Canberra, AUSTRALIAN CAPITAL TERRITORY

INDIAN OCEAN

VICTORIA
Melbourne

Geelong
Site of first European rabbit release on the mainland

Bass Strait

Tasman Sea

Invasive Species in Australia
Spread of the European rabbit by year

- 1870
- 1880
- 1890
- 1900
- 1910
- 1980
- 2000

| 0 | 250 | 500 Miles |
| 0 | 250 | 500 Kilometers |

TASMANIA
European rabbits abundant as early as 1827

Hobart

wild, and preyed on native kangaroos and wallabies. Today, dingoes continue to threaten native animals and livestock.

When Europeans began to settle in Australia in the 1800s, they brought along rabbits, which quickly became **feral**, or wild, and spread across the continent. Feral rabbits have caused serious damage to crops and grazing lands. In the 1950s, the Australian government tried to wipe out feral rabbits but was not successful.

Non-native plants have also done damage. In fact, in New Zealand, more types of invasive plants thrive in the wild than do native plants. One climbing plant called "old man's beard" has spread widely since the 1940s. It smothers native foliage, and can even bring down large trees.

Before You Move On

Summarize In what ways have invasive species impacted Australia and New Zealand?

ONGOING ASSESSMENT

MAP LAB

GeoJournal

1. **Interpret Maps** Look at the map. In what direction have rabbits moved across Australia? By what decade was the island of Tasmania overrun with rabbits?

2. **Interpret Maps** Based on the map, where were rabbits first released in Australia? Using the map scale, determine approximately how far rabbits spread between the years 1870 and 2000.

3. **Movement** What are invasive species? In what ways do human migrations encourage their spread?

SECTION 1 GEOGRAPHY

1.4 The Pacific Islands

TECHTREK
myNGconnect.com For photos of the Pacific islands
Digital Library

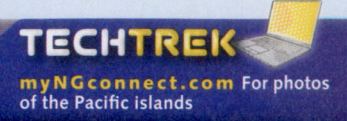

Main Idea Physical processes helped form the Pacific islands over a long period of time.

Northeast of Australia and New Zealand, the Pacific islands are a group of 20,000 to 30,000 islands scattered across millions of square miles of the Pacific Ocean. These islands were formed by geologic changes over thousands of years.

High Islands, Low Islands

Geographers divide the Pacific islands into two main categories: high islands and low islands. High islands are formed by volcanic activity. As tectonic plates move over **hotspots**, or unusually hot parts of Earth's mantle, magma rises up through the plates and produces a volcanic eruption. Cooled molten material forms underwater volcanic cones. This process repeats, and over time, these volcanic cones emerge above water as islands.

Low islands also form over long periods of time. They tend to be smaller than high islands and sit only a few feet above sea level. Low islands are **coral islands**, which are islands created by a gradual buildup of the skeletons of corals and other tiny marine animals. Coral islands often sit on top of coral reefs. Many coral reefs form around the base of high volcanic islands. Over time these reefs become atolls.

Climate, Agriculture, and Fishing

The climate in the Pacific islands is tropical and warm throughout the year. Most islands have a wet season and a dry season, but precipitation on the islands varies. Warm ocean air cools at higher elevations on high, mountainous islands and produces rain. High islands receive plentiful rainfall. Pineapples, sugarcane, and mangoes grow well in their rich

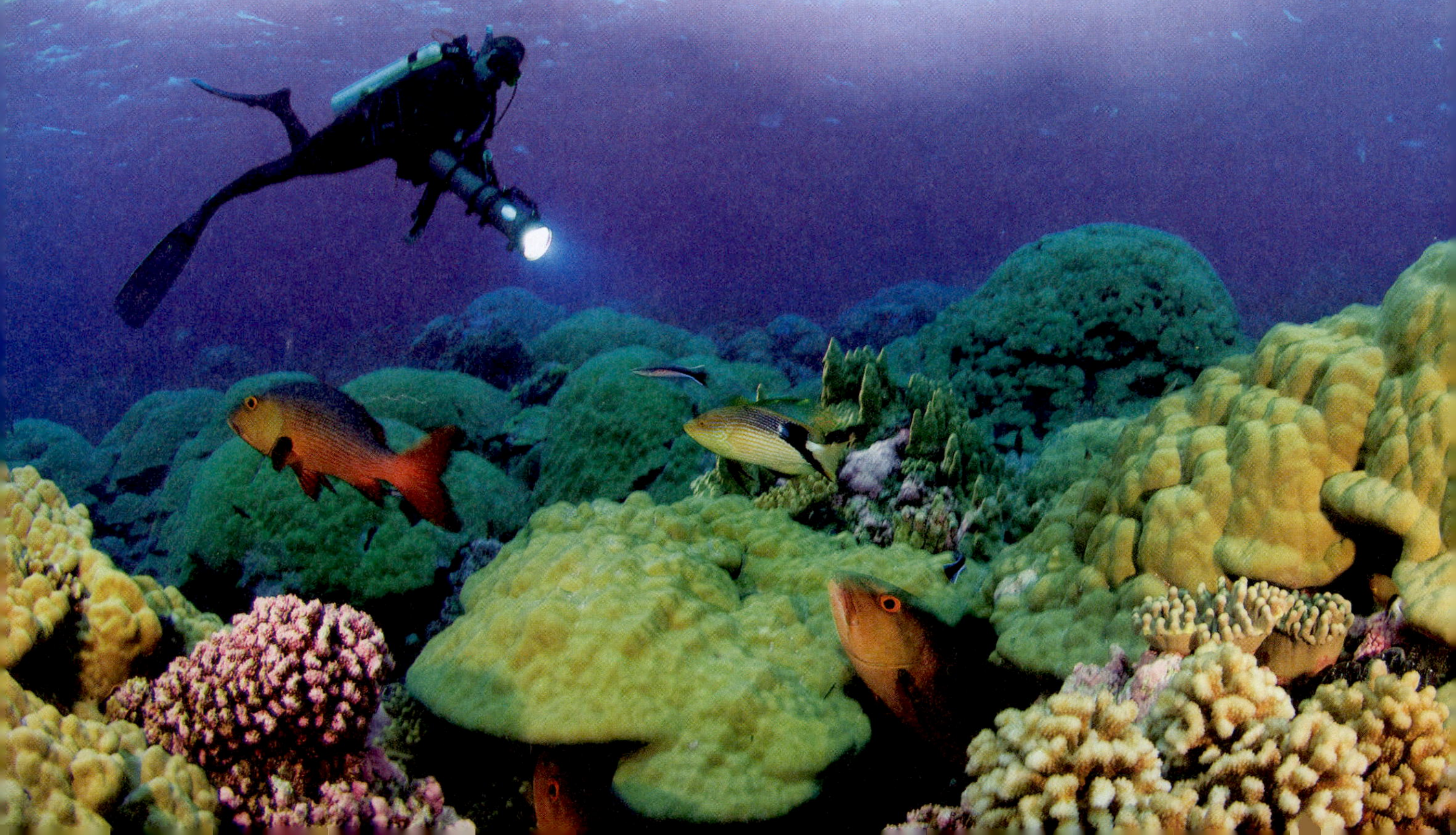

HIGH ISLAND FORMATION

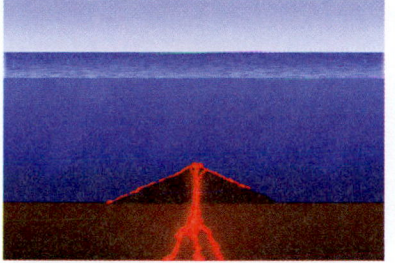

1 Magma rises from deep inside Earth's mantle.

2 Cooled molten material forms underwater volcanic cones.

3 Volcanic cones emerge from the water as high islands.

LOW ISLAND FORMATION

1 Coral reefs form on the outer edges of high islands.

2 High islands appear to sink as erosion wears them away.

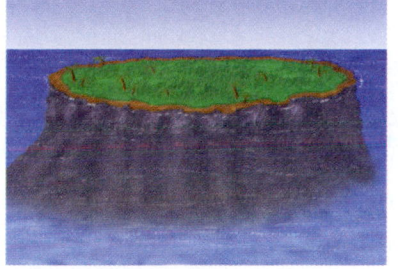

3 Over time, only the surrounding coral reef remains.

volcanic soil. Low islands, with no landforms to "catch" moisture, receive less rainfall. Though they are surrounded by ocean, drought is common on low islands. As a result, people on low islands depend less on agriculture and more on fishing.

Before You Move On

Monitor Comprehension How did the two types of Pacific islands form?

VIEWING LAB  GeoJournal

1. **Sequence Events** Based on the illustration above, describe in your own words the sequence of events in high and low island formation. Include the words *atoll* and *coral reef* in your answer.

2. **Analyze Visuals** Based on the illustration and the text, in what way is molten material involved in island formation?

3. **Human-Environment Interaction** Why might farming on high islands be more successful than on low islands?

Critical Viewing Kingman Reef lies under an atoll in the Line Islands and is one of the few unspoiled coral reefs remaining in the world. What details do you notice about life on this reef?

1.5 **Saving the Reefs**

TECHTREK

myNGconnect.com For photos
of Australia's Great Barrier Reef

Digital Library
Global Issues

Main Idea The Great Barrier Reef is a marine ecosystem that is threatened by human activities.

Each year, many people visit Australia's **Great Barrier Reef**. In order to protect this ecosystem and others like it around the world, Australia and other countries are balancing tourism with preservation.

The Great Barrier Reef

Located off the northeastern coast of Australia, the Great Barrier Reef is the world's largest coral reef. As you know, a coral reef is a complex **marine**, or sea-based, ecosystem. Small marine animals called corals secrete, or produce and discharge, calcium carbonate. This hardens into an **exoskeleton**, or a hard, external covering that provides protection. Over time, colonies of these corals form coral reefs that support marine life, from small fish to predators such as sharks.

The Great Barrier Reef stretches 1,250 miles and consists of around 2,900 separate reefs. It is home to one of the most diverse collections of plants and animals on Earth. Around 2,000 species of fish, 350 corals, and 4,000 mollusks, or soft-bodied marine animals with exterior shells, live here. In recent years, scientists have discovered hundreds of new animal species. Because of the diversity and extent of life on the reef, it is sometimes called "the rain forest of the sea."

Visitors to the Great Barrier Reef can scuba dive, snorkel, view coral reefs from glass-bottomed boats, and swim with dolphins. Since the 1970s, reef tourism has steadily increased. The Australian government made the Great Barrier Reef a national park in 1975. By doing so, it reduced the potential negative impact of tourism through controls on tourist activities in the park.

Before You Move On

Summarize What kinds of marine life does the Great Barrier Reef support?

GREAT BARRIER REEF BY THE NUMBERS

2,076,831
Number of tourists, June 2009–June 2010

72°–84°F
Ideal water temperature range for corals

$6 billion
Average number of tourist dollars per year

10,000
Average age of corals in years

2,900
Number of coral reefs that make up the Great Barrier Reef system

600
Species of starfish and sea urchins on the Great Barrier Reef

30
Species of whales and dolphins observed on the Great Barrier Reef

Sources: Great Barrier Reef Marine Park Authority, 2008–2010

Science at Work

Today, the Great Barrier Reef's survival is threatened. The reef's ecosystem depends on a specific temperature range for reef-building corals to thrive. These corals depend on algae, small oxygen-producing organisms, for nourishment. Even small increases in water temperatures of 1°–2°F can prompt the algae to leave the corals, resulting in the corals' death. This process is called "coral bleaching" because it causes the corals to turn white.

In addition to increased water temperatures, pollution and overfishing have also caused problems for corals and other marine life on the reef. In fact, if steps toward preservation are not taken, scientists estimate that corals on the Great Barrier Reef could be extinct by 2050.

Experts are working to **preserve**, or protect, the reef. One group of scientists is trying to identify all the different species living in the reefs. Another group is focused on the effects of rising water temperatures on corals and reef life. A deeper understanding of the reef could lead to new methods of saving it.

^ Critical Viewing White tip sharks like this one live on the Great Barrier Reef. How might their survival depend on the protection of the reef?

ocean floor have also been disturbed, causing the loss of sea life. As tourism has grown, the reef has also been overfished.

As in Australia, local communities in Belize are actively preserving their reef. For the reef to be saved, the Belize government must work more closely with the tourism industry to control human activities in this fragile part of the world.

Before You Move On

Monitor Comprehension How do increasing water temperatures threaten coral reefs?

COMPARE ACROSS REGIONS

Reef Preservation

Like the Great Barrier Reef, the Belize Barrier Reef—the second largest reef system in the world—attracts thousands of tourists each year. Belize is a small country in Central America that borders the Caribbean Sea. Mangrove trees, an important part of the reef's habitat, have been cut down. Sand and coral from the

ONGOING ASSESSMENT
READING LAB GeoJournal

1. **Summarize** In what ways have human beings contributed to the problems facing coral reefs?

2. **Make Inferences** What might Belize learn from Australia about reef preservation?

3. **Draw Conclusions** The Great Barrier Reef Marine Park keeps track of numbers of tourists to the reef. Why might it be doing this?

2.1 Indigenous Populations

TECHTREK
myNGconnect.com For photos of indigenous people and Australian rock art
Digital Library

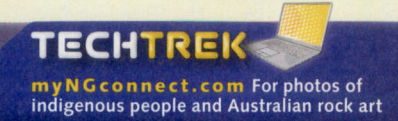

Main Idea The languages and culture of Australia's indigenous people developed in isolation on the continent.

About 50,000 years ago, humans migrated to present-day Australia from Asia. They may have arrived by sea, or by **land bridge**, a narrow and temporary link between two land masses. The **Aborigines** (AB uh RIHJ uh neez) were Australia's first developed human culture. Most scientists believe this culture dates from 30,000 B.C.

Populating the Continent

Australia's Aborigines are considered the oldest continuous human culture in the world. In fact, the phrase *ab origine* means "from the beginning." Because of Australia's isolated location, Aboriginal culture and languages developed apart from other human groups. Over time, more than 200 different languages developed. These languages were unlike any others in the world.

Like other prehistoric peoples, Aborigines were nomadic hunters and gatherers. To adapt to the continent's dry climate, they conserved water and carried it with them as they moved. They lived in distinct groups of larger **clans**, or family units. Each clan had a strong attachment to the land on which it hunted and gathered food. In addition to providing a source of food, the land held spiritual meaning for the Aborigines.

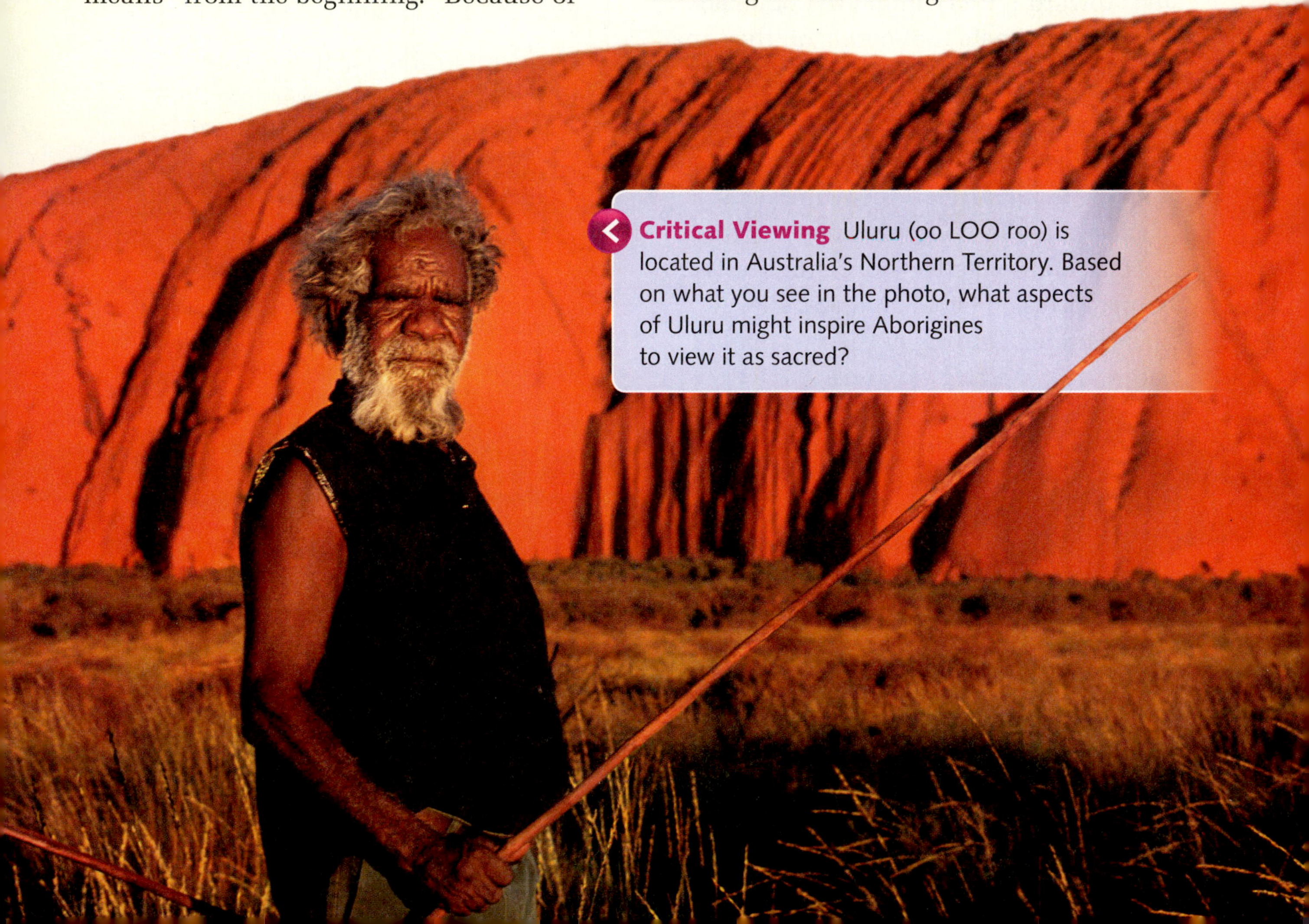

Critical Viewing Uluru (oo LOO roo) is located in Australia's Northern Territory. Based on what you see in the photo, what aspects of Uluru might inspire Aborigines to view it as sacred?

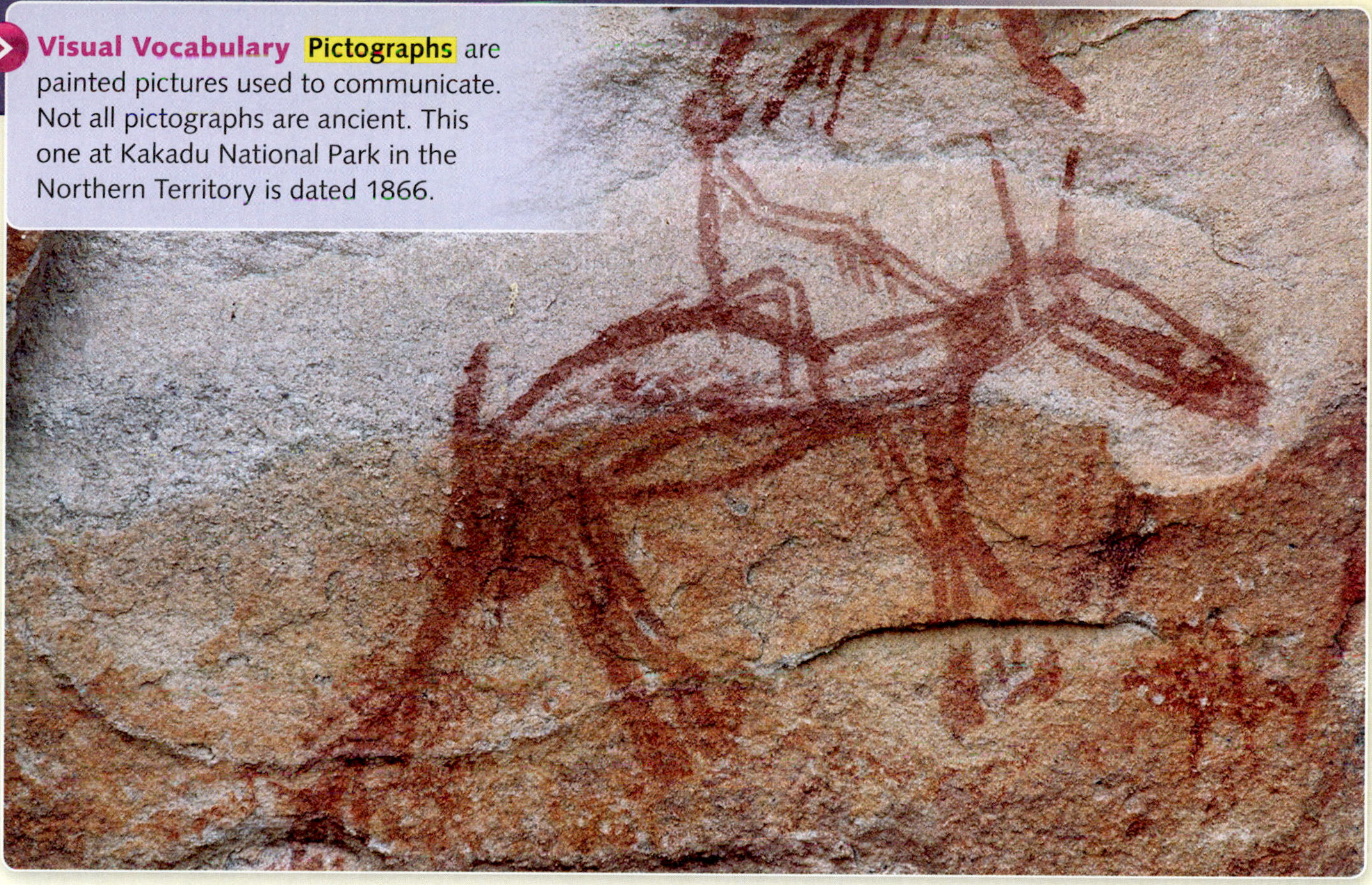

Aborigines have considered the huge, red rock called Uluru, pictured at left, a sacred site for thousands of years. Rock art painted on cave walls at Uluru and other sites reveals much about Aborigines' beliefs. Some ==pictographs== depict elements from the natural world while others portray religious ceremonies and important historical events.

Conflicts with Europeans

By 1788, when Europeans arrived in Australia, between 500,000 and one million Aborigines lived on the continent. Though Aborigines had established complex societies, most European settlers viewed their way of life as uncivilized.

Conflicts arose when Europeans seized Aborigines' lands. Separated from their land, Aborigines were not able to hunt, and many fell into poverty. Many died of disease because they did not have immunity or resistance to European illnesses. By 1921, the Aborigine population totaled only about 62,000. Today, Aborigines represent 2.5 percent of Australia's population, or 517,200 citizens.

Before You Move On

Summarize In what ways did Australia's location determine the way Aboriginal languages and culture developed?

ONGOING ASSESSMENT

PHOTO LAB  GeoJournal

1. **Analyze Visuals** Based on the photo of rock art, what can you infer about how Aborigines used their surroundings to record events?

2. **Make Inferences** Uluru is a popular tourist site in Australia. Based on the photo and text, what might you infer about how Aborigines would feel about Uluru as a tourist site?

3. **Movement** In what ways did conflicts over land cause problems for Aborigines?

2.2 Seagoing Societies

TECHTREK

myNGconnect.com For a map and photos of the Pacific islands

 Maps and Graphs   Digital Library

Main Idea The first people to inhabit the Pacific Realm developed specific skills to adapt to life on the islands.

Several thousand years ago, ancient **seafarers**, or sea travelers, set sail from Southeast Asia, southern China, and Taiwan. They traveled over wide expanses of ocean and settled on islands in the Pacific Ocean.

Living on the Ocean

The islanders built canoes with attached floats, called **outrigger canoes**. These early settlers were experts at **navigation**, or the process of determining location and routes, as they traveled between islands. Seafarers used the patterns of stars to map their course. They also studied the flight patterns of birds, which indicated where land was located. Seafarers created stick charts, an early form of maps, from sticks, shells, and twine. Shells on the charts indicated islands and different shapes of sticks represented ocean currents.

∧ **Critical Viewing** This stick chart is a type of map used by Pacific seafarers. How does this map differ from maps you use?

Over time, different cultures developed. On larger islands, inhabitants practiced advanced farming methods. On smaller islands, people supported themselves through fishing. Mountainous land on some islands kept groups of people isolated from one another and cultures developed independently. However, on

∧ **Visual Vocabulary** An **outrigger canoe** is a boat with an attached float that helps it balance. These boys use paddles to guide their outrigger canoe in Papua New Guinea.

PACIFIC ISLANDS

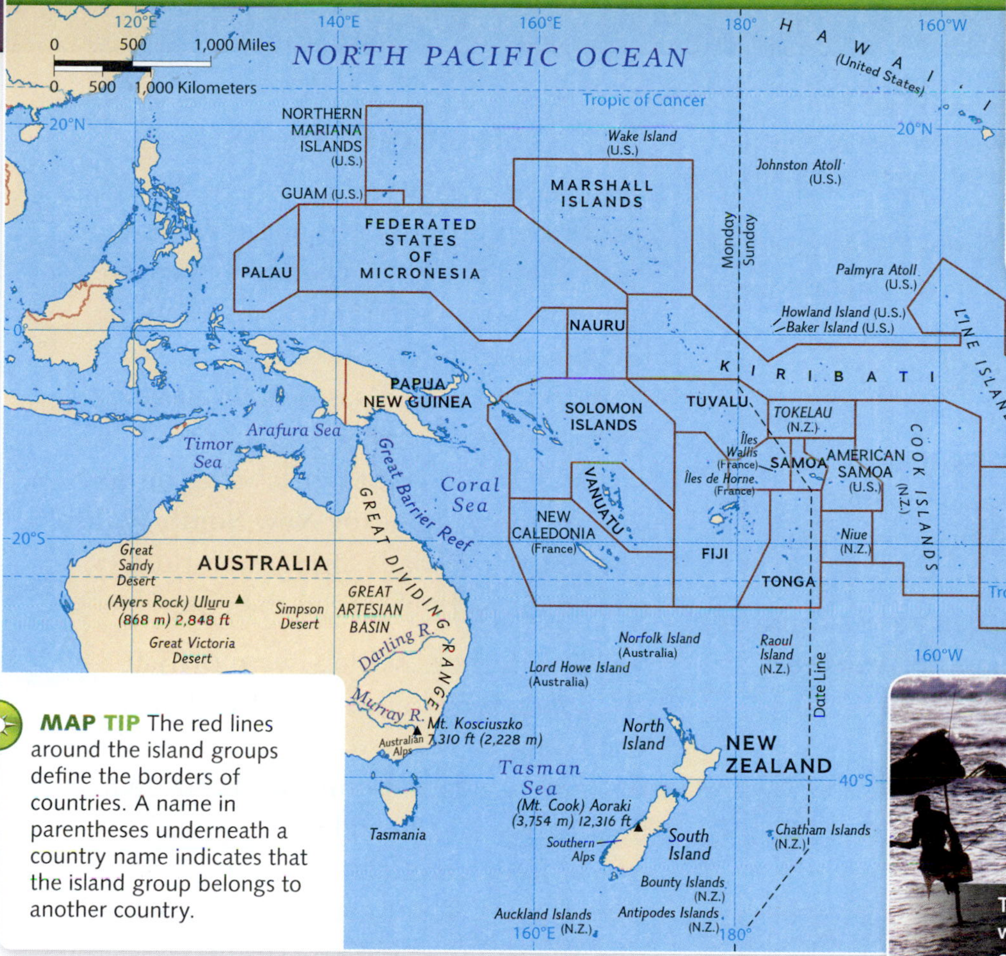

NORTH PACIFIC OCEAN

0   500   1,000 Miles
0   500   1,000 Kilometers

120°E   140°E   160°E   180°   160°W

H A W A I I (United States)

Tropic of Cancer

20°N   20°N

NORTHERN MARIANA ISLANDS (U.S.)

Wake Island (U.S.)

Johnston Atoll (U.S.)

GUAM (U.S.)

MARSHALL ISLANDS

FEDERATED STATES OF MICRONESIA

PALAU

Monday Sunday

Palmyra Atoll (U.S.)

NAURU

Howland Island (U.S.)
Baker Island (U.S.)

140°W

Equator   0°

K I R I B A T I

PAPUA NEW GUINEA

SOLOMON ISLANDS

TUVALU

TOKELAU (N.Z.)

Îles Wallis (France)
Îles de Horne (France)

SAMOA

AMERICAN SAMOA (U.S.)

Marquesas Islands

L I N E   I S L A N D S

Arafura Sea

Timor Sea

Coral Sea

VANUATU

NEW CALEDONIA (France)

FIJI

Niue (N.Z.)

TUAMOTU ARCHIPELAGO

C O O K   I S L A N D S   (N.Z.)

FRENCH POLYNESIA (France)

20°S   20°S

Great Barrier Reef

TONGA

Tropic of Capricorn

140°W

PITCAIRN ISLANDS (U.K.)

AUSTRALIA

Great Sandy Desert

(Ayers Rock) Uluru ▲ (868 m) 2,848 ft

Simpson Desert

Great Victoria Desert

GREAT ARTESIAN BASIN

G R E A T   D I V I D I N G   R A N G E

Darling R.

Murray R.

Mt. Kosciuszko 7,310 ft (2,228 m)
Australian Alps

Norfolk Island (Australia)

Lord Howe Island (Australia)

Raoul Island (N.Z.)

160°W

160°E

Date Line

Tasman Sea

North Island

NEW ZEALAND

40°S

Tasmania

(Mt. Cook) Aoraki (3,754 m) 12,316 ft

Southern Alps

South Island

Chatham Islands (N.Z.)

Bounty Islands (N.Z.)

Auckland Islands (N.Z.)

Antipodes Islands (N.Z.)

160°E   180°

N
W   E
S

MAP TIP The red lines around the island groups define the borders of countries. A name in parentheses underneath a country name indicates that the island group belongs to another country.

Nuka Hiva is the largest of the Marquesas Islands in French Polynesia.

These fishermen sit above the water as they fish off the coast of New Zealand.

islands with few land barriers, people shared a similar language and culture. Sometimes groups of people moved from one island to another. In the 1100s, for example, the Maori people migrated from eastern Pacific islands to the North Island of New Zealand.

Western Influence

For the most part, seafaring societies remained isolated from Western contact. A few Europeans explored the Pacific islands as early as the 1500s, but most islanders did not encounter Europeans until the 1700s. Some Europeans traveled to this region in search of wealth or to spread Christianity. Other Europeans, such as British navigator **James Cook**, charted new territories.

During multiple voyages in the 1770s, Cook mapped islands in the Pacific, including New Zealand. In fact, the Cook Islands, which lie northeast of New Zealand, are named for James Cook.

Before You Move On

Summarize In what ways did the first people in the Pacific Realm adapt to the islands?

ONGOING ASSESSMENT

MAP LAB   GeoJournal

1. **Interpret Maps** The Pacific islands cover a broad expanse of ocean. In what way are the Pacific islands represented on this map?

2. **Identify** In the 1800s, most Pacific islands became colonies of Western countries. Based on the map, which islands remain possessions of other countries? How can you tell?

2.3 From Convicts to Colonists

Main Idea Convicts and other immigrants to Australia led the development of a new country.

Seventeen years after James Cook explored Australia's coast, British ships sailed for Australia. From 1788 to 1868, the British government sent **convicts**, or prisoners, to overseas colonies in order to decrease overcrowding in British prisons. When the American colonies established independence in 1776, the government needed a new place to send convicts. British colonization began in Australia as a way for the government to **transport**, or ship, its convicts to prisons overseas.

Photograph of a replica of James Cook's ship *Endeavour*

Settling the Colonies

Most transported convicts were young, healthy, and unmarried. Only 20 percent were women. Those who were married sometimes brought their families with them. The majority of convicts came from the working classes in England and Ireland. The government used convicts' skills and relied on their labor to build the colonies, including **New South Wales**. Convicts built roads, bridges, buildings, and farms as they served their sentences.

Though these settlers were British prisoners, they had certain freedoms. For example, many were allowed to live in their own homes and run their own businesses. At the end of their prison terms, freed convicts often remained in the colonies. Their contributions helped the Australian colonies to grow.

In 1831, the British set up a program for **assisted migration**. This program encouraged people to move to Australia by giving them financial help. Twenty years later, gold was discovered in Australia. This discovery drew many settlers to the continent including immigrants from China and the surrounding Pacific islands. An increased population and more available workers led to the growth

1788
Convict transportation begins; British establish colony at New South Wales.

1750

1800

1770
James Cook explores Australian coast.

1787
First Fleet sets sail for Australia.

1831
British policy of assisted migration begins.

of industries, such as manufacturing and mining. As more of the continent was explored, some settlers began sheep farming. Wool became an important product and export.

Australia's population quickly grew from 400,000 in 1850 to more than one million by 1860. Population growth and European settlement pushed Aborigines off lands that had belonged to their ancestors for many years.

Becoming A Country

In the 1890s, many Australians began to think that the colonies would benefit by becoming a single country. Colonists wrote a constitution and submitted it to the British government. In 1901, Great Britain approved the constitution, and the **Commonwealth of Australia** came into being. Though Australians now had their own government, they were still part of the British Empire. Today, Australia remains part of the British Commonwealth. Australia recognizes the British monarch as its head of state, but it is an independent country.

Before You Move On

Make Inferences In what ways did convicts lead the development of Australia?

Critical Viewing Thomas Gosse painted *Founding of the Settlement of Port Jackson at Botany Bay in New South Wales* in 1799. What does his painting lead you to think about the settlement of Port Jackson?

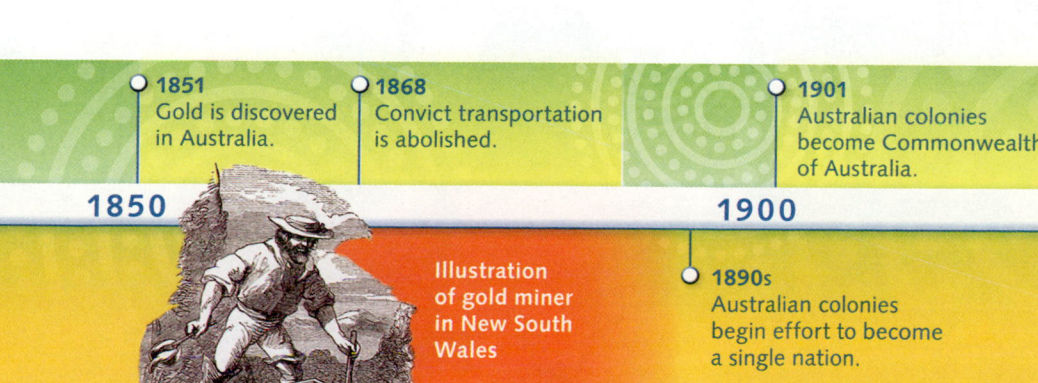

Australian flag, designed in 1901

1851
Gold is discovered in Australia.

1868
Convict transportation is abolished.

1901
Australian colonies become Commonwealth of Australia.

1850

1900

Illustration of gold miner in New South Wales

1890s
Australian colonies begin effort to become a single nation.

2.4

SECTION **2** HISTORY

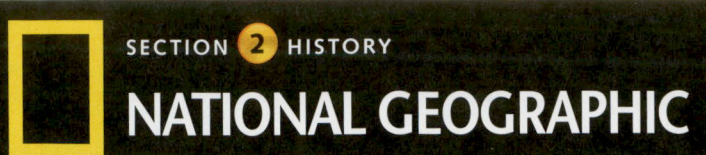
NATIONAL GEOGRAPHIC

TECHTREK

myNGconnect.com For photos from Enduring Voices and an Explorer Video Clip

 Digital Library

Exploring Vanishing Languages

with David Harrison and Greg Anderson

> **Main Idea** Many languages spoken by indigenous people are endangered but efforts are underway to revitalize them.

The Importance of Language

Nearly 7,000 languages are spoken in the world today. Language is a vital part of a people's identity. Each one tells a great deal about the people who speak it. Language **conveys**, or communicates, what is important to a culture and helps pass along history, stories, and songs from one generation to the next. A **generation** is a group of individuals who are born and live about the same time. Older generations share knowledge, like language, with younger generations.

Today, many languages are disappearing. Languages spoken by indigenous people are especially threatened. Many languages have no written form and are spoken only by a small group of people. As the children of these people are exposed to other languages, they often stop speaking their native language. Older members of the group are then the only ones left who remember the language. With no written record, the indigenous language disappears.

myNGconnect.com

For more on David Harrison and Greg Anderson in the field today

David Harrison and Greg Anderson work with Charlie Mangulda in Australia.

By 2100, more than half the world's languages may no longer exist. Some experts believe that on average a language is lost every two weeks. As these languages vanish, a wealth of information about history and culture goes with them. Because indigenous people often use language to share knowledge about native plants and animals, when their languages vanish, information about the natural world also disappears.

 Critical Viewing Harrison and Anderson interview Felix Andi in Papua New Guinea. Based on what you see here, what kinds of technology help linguists document endangered languages?

The Enduring Voices Project

National Geographic Fellows David Harrison and Greg Anderson are **linguists**, or language scientists. They founded National Geographic's **Enduring Voices Project** to study and document languages at risk. They work in many regions of the world. In 2010, in the foothills of the Himalayas in India, Harrison and Anderson documented a language previously unknown to researchers. Australia and Papua New Guinea are also among the places they identify as language hotspots because so many languages there are endangered. Over 800 languages are spoken in Papua New Guinea alone.

Harrison and Anderson use a variety of methods in their work. One is to interview the remaining speakers of the languages and make video and sound recordings. In addition to documenting the sounds and meanings of words, Harrison and Anderson gather valuable information about each group's history and culture. By studying these languages, they are also recording knowledge about the cultures.

Another method these linguists use is language revitalization. This involves supporting community-led efforts to bring vanishing languages back to life by teaching them to younger generations. Harrison and Anderson believe that educating young people is key to the survival of their languages.

Before You Move On

Summarize Why are many languages spoken by indigenous people endangered?

ONGOING ASSESSMENT

LANGUAGE LAB
GeoJournal

1. **Explain** Why is it important to revitalize indigenous languages?

2. **Make Inferences** Some words from indigenous languages have been adopted into English, such as *dingo* and *kangaroo*. Why might this be? Think about the special features of these animals before you answer.

3. **Analyze Visuals** Watch the Explorer Video Clip. Name at least two things you learned about the Enduring Voices Project. Then pose two questions you might ask David Harrison and Greg Anderson.

Review

VOCABULARY

For each pair of vocabulary words, write one sentence that explains the connection between the two words.

1. indigenous; marsupial

> The kangaroo, which is indigenous to Australia, is a marsupial.

2. invasive species; extinct
3. coral reef; atoll
4. seafarer; outrigger canoe
5. convict; assisted migration
6. linguist; generation

MAIN IDEAS

7. In what ways are Australia, the Pacific Realm, and Antarctica similar? (Section 1.1)
8. Why does the region have so many plants and animals unlike anywhere else in the world? (Section 1.2)
9. How were invasive species introduced in the region? (Section 1.3)
10. What two geologic processes created the Pacific islands? (Section 1.4)
11. In what ways are coral reefs around the world endangered? (Section 1.5)
12. How did migration and location shape the history of the Aborigines? (Section 2.1)
13. Why might Western influences be limited in the Pacific islands? (Section 2.2)
14. What groups helped develop the continent of Australia? (Section 2.3)
15. What steps are linguists taking to preserve vanishing languages? (Section 2.4)

GEOGRAPHY

ANALYZE THE ESSENTIAL QUESTION

How did geographic isolation influence the development of this region?

Critical Thinking: Summarize

16. Why is Australia called an "island continent"?
17. How did indigenous plants and animals develop in this region?
18. In what ways do invasive species threaten the indigenous plants and animals in the region?
19. What physical features make some Pacific islands better suited for farming than others?

INTERPRET MAPS

20. **Identify** What are the names of Australia's deserts? In what part of the continent do they lie?
21. **Location** In what sea is the Great Barrier Reef located? How is the name of the sea related to the reef?

HISTORY

ANALYZE THE ESSENTIAL QUESTION

How did geographic isolation shape the history of Australia and the Pacific Realm?

Critical Thinking: Draw Conclusions

22. In what ways did conflicts over land impact Aborigines in Australia?

23. What methods did ancient seafarers use to deal with the challenges of life on isolated islands in the Pacific Ocean?

24. Why might the British government have chosen Australia as a place to transport convicts?

25. On mountainous Pacific islands, how did physical features play a role in the establishment of multiple languages?

INTERPRET TABLES

COMPARING LANGUAGE USE IN AUSTRALIA AND PAPUA NEW GUINEA		
	Australia	Papua New Guinea
Percentage of people speaking English	78.5	2.0
Number of indigenous languages spoken	145	860

Sources: CIA World Factbook, 2010; Australian Government of Foreign Affairs and Trade 2010; BBC, 2010

26. **Compare and Contrast** Based on this table, in which country does the majority of the population speak English?

27. **Make Generalizations** Based on this table and what you have read, in which country—Australia or Papua New Guinea—did European settlers have a larger influence on indigenous populations?

ACTIVE OPTIONS

Synthesize the Essential Questions by completing the activites below.

28. **Write a Letter** Imagine that you are a settler in 1851 who has arrived in Australia to search for gold. Write a letter to a family member back in England, describing what you observed when you arrived. Explain whether you think this person should join you. **Send your letter to a friend and ask if it painted a clear picture of Australia.**

> **Writing Tips**
> • Address your letter to a particular person.
> • Draw on what you have read about Australia to describe what the country is like.

Go to **Student Resources** for Guided Writing support.

TECHTREK   myNGconnect.com
For photos of the region

29. **Create a Visual Presentation** Use the **Digital Library** or other online sources to create a visual presentation of this region. Choose a photo to go with each of the following statements. Describe your photos with additional research you conduct.

• Australia is home to many unusual plants and animals.

• Thousands of living creatures inhabit the Great Barrier Reef.

• Some Pacific islanders still practice a traditional way of life.

• Researchers work to revitalize endangered indigenous languages.

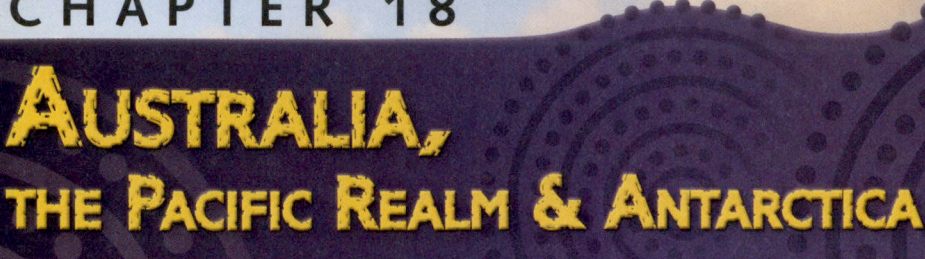

CHAPTER 18

AUSTRALIA, THE PACIFIC REALM & ANTARCTICA TODAY

PREVIEW THE CHAPTER

Essential Question How are Australia, the Pacific Realm, and Antarctica becoming connected to the rest of the world?

SECTION 1 • CULTURE

KEY VOCABULARY

- urban
- salinization
- immigration
- labor force
- oral tradition
- emigrate
- ice shelf
- scientific station

ACADEMIC VOCABULARY

deplete, maintain

TERMS & NAMES

- Sydney
- Polynesian Triangle
- Cook Islands
- Charles Wilkes
- Richard Byrd
- Antarctic Treaty

Essential Question What new economic patterns are emerging in the region?

SECTION 2 • GOVERNMENT & ECONOMICS

KEY VOCABULARY

- alliance
- free trade agreement
- reserve
- assimilation
- adventure tourism
- glacier
- crevasse
- renewable energy
- geothermal energy

ACADEMIC VOCABULARY

reclaim, lucrative

TERMS & NAMES

- Papua New Guinea
- Samoa
- Asia-Pacific Economic Cooperation (APEC)

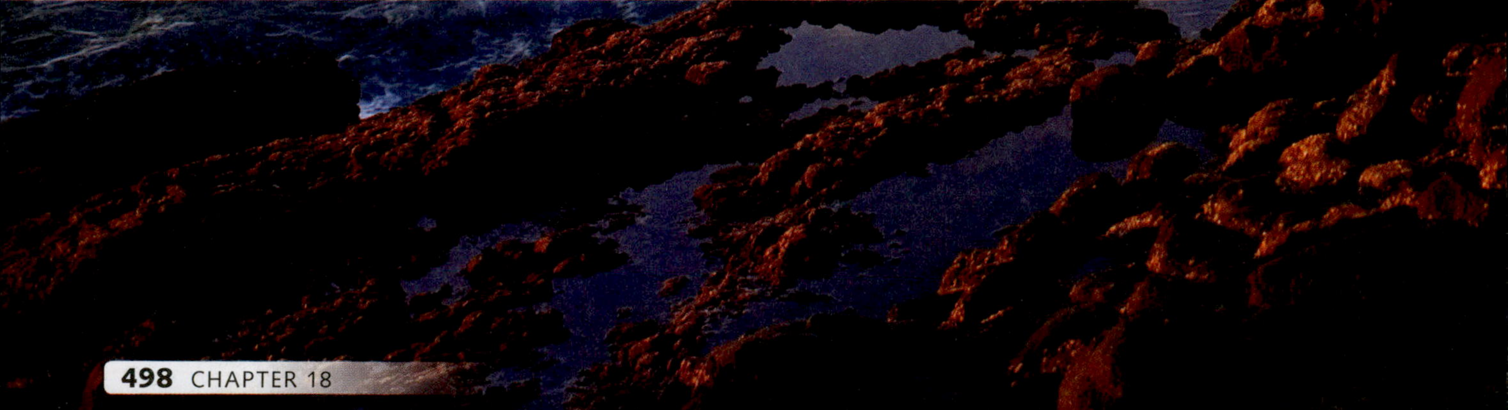

TECHTREK

FOR THIS CHAPTER

Student eEdition

Maps and Graphs

Interactive Whiteboard GeoActivities

Digital Library

Connect to NG

Go to **myNGconnect.com** for more on Australia, the Pacific Realm, and Antarctica.

Surfers study the waves at Piha, on the western coast of North Island, New Zealand.

1.1 From Ranch to City

TECHTREK
myNGconnect.com For a map and photos of ranches and cities in Australia, and current events

Maps and Graphs | Digital Library | Connect to NG

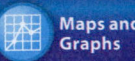

Main Idea Most Australians live in coastal cities, and many have migrated from rural areas.

Today, Australia is one of the world's most **urban** countries—most people live in cities or nearby suburbs. Lately, even farmers and ranchers whose families have lived in rural areas for generations are choosing to relocate. Economic pressures in farming and ranching communities are pushing people toward cities.

A Dry Climate Turns Drier

Much of Australia's climate is arid. About 60 percent of the continent receives fewer than 10 inches of rain per year. The dry climate of Australia's interior makes farming and ranching difficult. Farmers have mainly depended on irrigation, rather than rainfall, to water their crops. Sheep and cattle ranchers, whose livestock graze in dry areas, also depend on irrigation.

However, irrigation comes with problems of its own. Years of continuous irrigation have helped cause **salinization**, which occurs when salt accumulates in soil. Too much irrigation into normally dry land raises the natural water table and causes salts to rise to the surface. Crops cannot grow in this salty soil. Low rainfall and high evaporation rates make soil in the interior of Australia prone to salinization.

In 2002, the longest drought in Australia's history began. Increased demands **depleted**, or drained, major irrigation sources such as the Murray and Darling rivers in New South Wales. The drought caused severe water shortages and the amount of water allowed for irrigation was greatly reduced. Farmers could not water their crops and ranchers sold entire herds of sheep and cattle at auction.

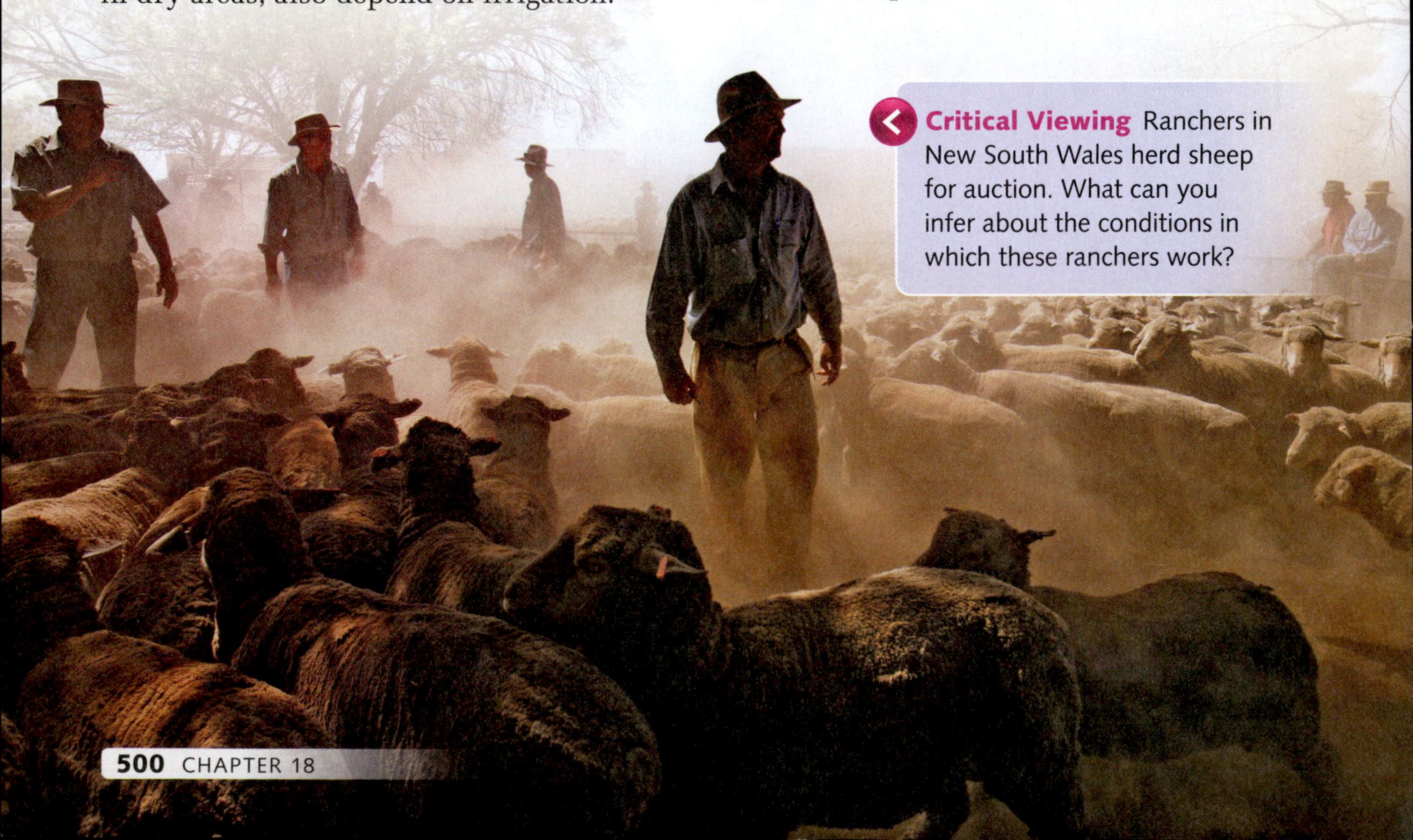

Critical Viewing Ranchers in New South Wales herd sheep for auction. What can you infer about the conditions in which these ranchers work?

The Pull to the Cities

In Australia's interior, few alternatives to farming and ranching exist. Faced with the combination of drought and poor soil, farmers and ranchers have had to make hard choices. Many are moving to the cities on the coasts.

Families are also drawn to urban areas because of more plentiful goods and services, including better schools for their children. Today, nearly 75 percent of the population lives in one of five coastal cities: **Sydney,** Melbourne, Brisbane, Perth, and Adelaide.

Before You Move On

Summarize Why do most Australians live in or near cities?

Critical Viewing Sydney is located on Australia's southeastern coast. Based on what you see in the photo, what might Sydney offer that might not be available in Australia's interior?

POPULATION DENSITY

Population Density

· One dot represents 25,000 people

ONGOING ASSESSMENT

MAP LAB

GeoJournal

1. **Location** Use the map to identify the locations of Australia's five largest cities. Which of those cities is most isolated?

2. **Interpret Maps** Locate South Australia and New South Wales on the map. In which state do more people live? What advantage might New South Wales have over South Australia, in terms of access to fresh water?

1.2 Immigration to Australia

TECH TREK

myNGconnect.com For photos of Australia's immigrant communities

Digital Library

Main Idea Australia includes people who have come from a wide variety of countries and cultures.

Since the end of World War II, the population of Australia has changed dramatically. A marked increase in **immigration**, or moving permanently from one country to another, has diversified the country.

Postwar Immigration

After World War II, refugees from war-torn Europe sought new lands to call home. At the same time, the Australian government wanted to increase population and expand its available **labor force**, or the number of people available to work. The government loosened immigration restrictions, increasing the number of immigrants permitted entry. Both Australia and its new settlers benefited from the decision.

Between 1950 and 1960, Australia's labor force increased by 60 percent due to immigration. New arrivals found jobs in postwar construction projects and other industries. Immigrants from Poland, Germany, Hungary, Greece, Italy, and other European countries changed the ethnic makeup of Australian cities. Some immigrants did not speak English and introduced their native languages to Australia. Many immigrants opened businesses that sold food and goods from their native countries.

▶ **Critical Viewing** Young men carry a colorful dragon in this traditional Chinese New Year's parade in Sydney, Australia. From what you can see in the photo, in what ways does this parade maintain cultural traditions?

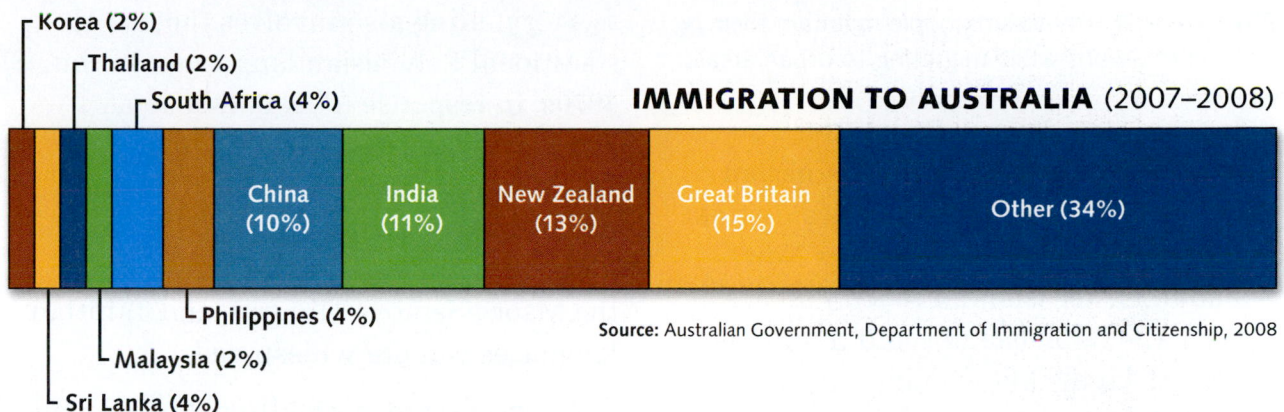

Korea (2%)
Thailand (2%)
South Africa (4%)

IMMIGRATION TO AUSTRALIA (2007–2008)

China (10%)
India (11%)
New Zealand (13%)
Great Britain (15%)
Other (34%)

Philippines (4%)
Malaysia (2%)
Sri Lanka (4%)

Source: Australian Government, Department of Immigration and Citizenship, 2008

Australia's Diversity

In the 1970s, new groups began to immigrate to Australia. For example, many refugees from the war in Vietnam fled by boat to the continent. Over the next few decades, immigrants from other Asian countries, including Japan, China, and India, relocated to Australia in search of job opportunities. Currently, about 10 percent of Australia's population is of Asian descent.

Many immigrants from New Zealand, the Pacific islands, Canada, and the United States have also moved to Australia since the 1970s. Unlike the early years of settlement, only about 15 percent of those who resettle in Australia each year come from Great Britain. Today, nearly one-quarter of Australians were born in other countries. English remains the official language, but more than 100 languages are spoken, including Chinese, Greek, and Italian.

Diversity is especially reflected in Australia's cities. Nearly 40 percent of new immigrants now settle in Sydney.

As Australia's largest city, Sydney is an attractive point of entry into the country. Melbourne, Australia's second largest city, boasts the largest Greek population in the world outside Greece. Brisbane and Perth receive large numbers of immigrants from Asia and Pacific island countries.

Before You Move On

Monitor Comprehension In what ways has immigration to Australia changed over the years?

ONGOING ASSESSMENT

DATA LAB

GeoJournal

1. **Movement** What percentage of immigrants to Australia between 2007–2008 arrived from the Asian countries included on the graph? Which two Asian countries represent the largest percentage of immigrants to Australia and why?

2. **Make Generalizations** Look at the data above. In what way does Australia's colonial past continue to be reflected in the number of people who immigrate to the country?

3. **Explain** What world events in the 20th century prompted different waves of immigration to Australia? What were different groups of immigrants seeking?

1.3 Polynesian Cultures

TECHTREK

myNGconnect.com For photos
of Polynesian culture today

 Digital
Library

> **Main Idea** Polynesian people maintain their cultural traditions while migrating to urban areas.

Polynesia spans a vast area in the South Pacific known as the **Polynesian Triangle**. This triangle captures a distinct Polynesian culture that includes New Zealand, Easter Island, French Polynesia, Samoa, the Cook Islands, and hundreds of other islands. Many Polynesians try to **maintain**, or preserve and carry on, their traditional cultures.

Traditional Polynesia

Historically, maintaining cultural practices was one function of the Polynesian family. Several generations of a single family lived together. Older members of the family taught younger members the arts of canoe-building, fishing, and navigation.

Polynesians have maintained traditional cultural practices and beliefs through the **oral tradition**, or passing stories by word of mouth from generation to generation. Polynesian stories explain how the universe was created and celebrate the connection between humans and the natural world.

Storytelling also involves the use of traditional Polynesian languages. In the 1970s, in response to fears that some languages were disappearing, many Polynesian schools established language programs that taught only traditional languages. Today, knowledge and use of the Maori, Samoan, Tongan, and Tahitian languages is more widespread.

In addition to storytelling, Polynesians have rich traditions in dance, music, and visual art forms such as woodcarving and basket-weaving. These art forms have provided ways for Polynesians to maintain and pass on their cultures. Dance, music, and visual arts are important elements of traditional Polynesian festivals, such as the Festival of Pacific Arts, held every four years in a different Polynesian country.

Polynesia Today

Like Australia, Polynesia has become increasingly urban since the 1960s. A lack of economic opportunities has pushed some Polynesians to migrate from rural villages to urban areas. Cities that have grown because of rural migration include Apia in Samoa, Pago Pago in American Samoa, and Nuku'alofa in Tonga.

Many Polynesians have **emigrated**, or left their home country, to live in other countries, including New Zealand and the United States. Between 1970 and 2000, emigration from Samoa to New Zealand doubled. Samoans emigrated to seek jobs and educational opportunities.

The **Cook Islands** provide examples of both rural to urban migration and emigration from Polynesia. Because of economic pressures, Cook Islanders have migrated from the countryside to the cities. Today, nearly 75 percent of Cook Islanders live in cities. Many others have emigrated to New Zealand to seek better economic futures.

Before You Move On

Summarize In what ways have Polynesian people maintained cultural traditions?

1. **Analyze Visuals** How would you describe the vegetation in the photo? How might you describe the climate?

2. **Make Inferences** Inferring from the photo, what physical features of this island do you notice? Why might those features make this work difficult?

3. **Explain** What are different ways in which Polynesians migrate? What are different reasons for migration? Use a chart like the one below to organize your ideas.

POLYNESIAN CULTURES	
Form of Migration	Reasons Why

Men on Olosega Island in American Samoa harvest leaves of the panandus plant, which will be dried and woven into mats.

1.4 Human Footprint in Antarctica

TECHTREK

myNGconnect.com For an online map and photos of Antarctica

Maps and Graphs

Digital Library

Main Idea People all over the world have been interested in Antarctica for hundreds of years.

Antarctica lies on the South Pole and has one of the most extreme climates on Earth. No humans live there permanently, but people have left their "footprint" through exploration and research.

Early Explorations

As early as 1773, Captain James Cook, a British naval captain, began looking for a southern continent. Ice blocked his southward journey. In 1820, British and Russian explorers were the first to report sighting land. American **Charles Wilkes** surveyed a 1,500-mile stretch of coastline in the late 1830s. His observations confirmed that Antarctica is a continent. Later explorers investigated Antarctica's thick sheets of ice and the ranges of the Transantarctic Mountains. They also explored the continent's coastal **ice shelves**, or floating sheets of ice that are attached to a landmass.

Twentieth century explorers made further discoveries. One dramatic event involved the race to the South Pole. In 1911, Robert Falcon Scott, from Great Britain, and Roald Amundsen, from Norway, each led expeditions to the South Pole. Amundsen's team reached the South Pole first. Though Scott's expedition

> **Critical Viewing** Researchers return to their station at the South Pole in Antarctica. Based on what you can see here, what preparations do people make in order to work in this extreme environment?

also reached the South Pole, none of its members survived the months-long journey back to base camp.

From 1929 through the 1940s, National Geographic supported American naval officer **Richard Byrd** as he explored the continent, both on land and by air. He studied its ice, rocks, and minerals and made observations about Earth's magnetism. Byrd's team also photographed Antarctica's coastline and discovered 26 new islands.

An International Treaty

By 1958, scientists from several different countries had built more than 50 <mark>scientific stations</mark>, or places to carry out research. Seven of these countries claimed sections of Antarctica as their own. However, some countries, including the United States, did not recognize the claims as legal. Then in 1959, 12 countries signed the **Antarctic Treaty.** According to the treaty, the continent can be used only for peaceful purposes. Any scientific discoveries must be shared internationally.

Today, Antarctica continues to serve as a center of research and global cooperation. One study, called the International Polar Year 2007–2008,

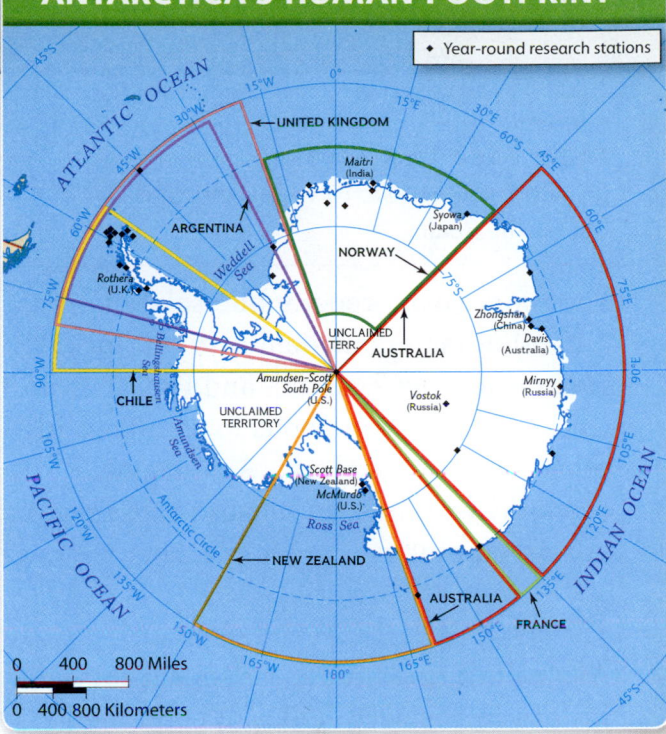

ANTARCTICA'S HUMAN FOOTPRINT

◆ Year-round research stations

involved scientists from more than 60 countries. Working together on more than 200 research projects, scientists collected data on thinning ice sheets in order to determine their effect on global sea levels. Studying changes in Antarctica helps scientists understand changes in other parts of the world.

Before You Move On

Make Inferences In what ways have people expressed interest in Antarctica?

ONGOING ASSESSMENT

MAP LAB

 GeoJournal

1. **Location** According to the map, which oceans surround Antarctica? From what central point do international claims originate?

2. **Interpret Maps** Based on the map, what country claims the largest area of Antarctica? What might explain the sizes of the South American and Australian claims?

3. **Explain** Why can no single country claim Antarctica as its own?

2.1 New Trade Patterns

TECHTREK

myNGconnect.com For a map and photos of exported goods from the region

 Maps and Graphs

 Digital Library

Main Idea Trade is a source of economic growth for Australia, New Zealand, and the Pacific Realm.

For many years, geographic isolation and other factors made it difficult for Australia, New Zealand, and Pacific island countries to export goods to countries in other parts of the world. As the global economy expands, countries in this region are developing new trade partners.

Trade Within the Region

Australia gained independence from Great Britain in the early 1900s. It continued to rely on Great Britain as its sole trading partner for several decades afterward. By the late 1900s, however, Australia began to increase trade with other countries.

As the largest and most populous country in this region, Australia serves as an economic and trade anchor, or hub, for surrounding countries. Australia, New Zealand, Papua New Guinea, and the other countries in the Pacific Realm have built trade partnerships with one another.

These trade partnerships benefit the entire region economically. For example, **Papua New Guinea** exports gold and coffee to Australia. **Samoa** exports cocoa beans to New Zealand. Regional economic development has strengthened ties among countries and has given the region a stronger global presence.

New Global Partnerships

This region's increased participation in the world economy is due in part to an organization called the **Asia-Pacific Economic Cooperation (APEC)**. Founded in 1989, APEC's goal is to strengthen economic ties and trade alliances, or partnerships, among its members. Three member countries—Australia, New Zealand, and Papua New Guinea—are located in this region. Other APEC member countries border the Pacific Ocean, including China, Japan, Mexico, the United States, Chile, and Peru. APEC alliances help these countries compete in the global economy.

> **Critical Viewing** Bales of wool in this New Zealand warehouse await export. From the photo, what can you infer this man is doing and why?

APEC MEMBER COUNTRIES

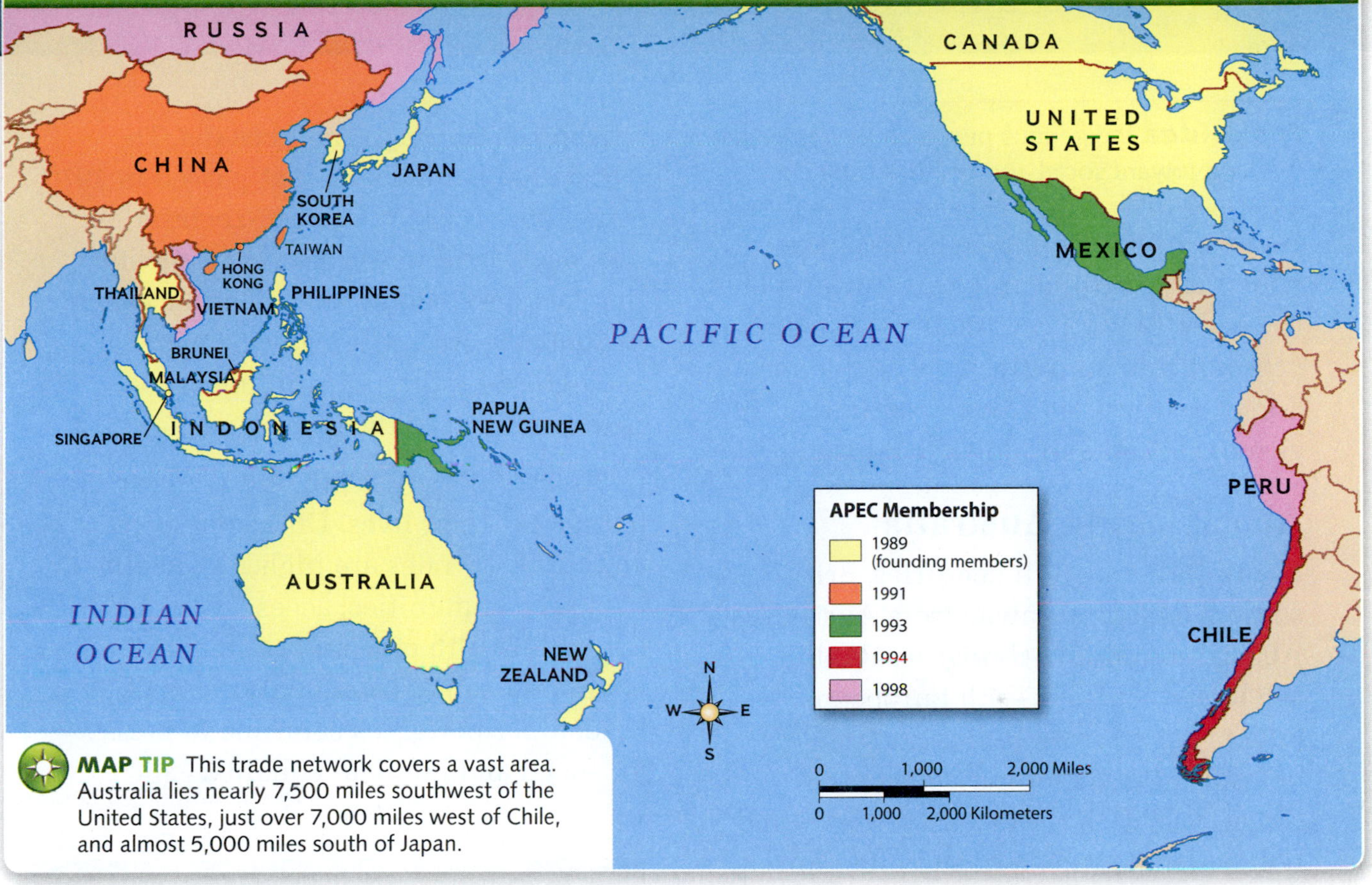

RUSSIA

CANADA

UNITED STATES

CHINA

JAPAN

SOUTH KOREA

TAIWAN

HONG KONG

THAILAND

PHILIPPINES

VIETNAM

MEXICO

BRUNEI

MALAYSIA

SINGAPORE

INDONESIA

PAPUA NEW GUINEA

PACIFIC OCEAN

AUSTRALIA

INDIAN OCEAN

NEW ZEALAND

PERU

CHILE

APEC Membership

- 1989 (founding members)
- 1991
- 1993
- 1994
- 1998

N W E S

MAP TIP This trade network covers a vast area. Australia lies nearly 7,500 miles southwest of the United States, just over 7,000 miles west of Chile, and almost 5,000 miles south of Japan.

0   1,000   2,000 Miles
0   1,000   2,000 Kilometers

One of APEC's main achievements has been to establish free trade agreements among member countries. **Free trade agreements** are treaties between two or more countries that encourage trade by limiting tariffs, or taxes, on that trade. Since 1989, APEC has helped form more than 30 free trade agreements.

Before You Move On

Monitor Comprehension In what ways have trade alliances helped this region become competitive in the global economy?

ONGOING ASSESSMENT

MAP LAB

GeoJournal

1. **Interpret Maps** Of the founding APEC member countries represented on the map, which are located in North America? Which APEC countries are located in Asia?

2. **Make Inferences** According to the map, when did Australia, New Zealand, and Papua New Guinea join APEC? Based on your reading, in what ways might APEC membership benefit their economies?

3. **Evaluate** In what ways might limiting tariffs on trade between countries help strengthen trade alliances?

2.2 Rights for Indigenous People

TECHTREK

myNGconnect.com For current events in the region

Maps and Graphs

Digital Library

> **Main Idea** Indigenous people in the region are working toward social and political equality.

As you have read, direct British rule ended in Australia and New Zealand in the early 1900s. However, the effects of colonial rule on indigenous people remain. Today, indigenous groups continue to address social and political problems.

Aborigines in Australia

In the 18th and 19th centuries, British settlers took lands away from Australia's Aborigines. Many Aborigines died in violent conflicts with Europeans. Others were forced to live on **reserves**, or land set aside for native people. Colonial British rulers enacted laws and policies that promoted, and often forced, assimilation. **Assimilation** is a process in which a minority group is pressured to give up its cultural practices and be absorbed into another society's culture.

One way in which the British tried to force Aborigines to assimilate was by separating Aboriginal children from their parents. Children were placed in mission schools and homes. This practice took place from the early 1900s until the 1960s. An estimated 100,000 children were separated from their families, most of them permanently. They are referred to as the "Stolen Generation."

In the 1960s, the Aborigines' struggle to gain basic civil rights intensified. By 1962, Aborigines were able to vote in national elections for the first time. Five years later, they were finally included as citizens in the Australian census. The Aboriginal Land Rights Act passed in 1976 gave Aborigines the right to **reclaim**, or take back, land in the Northern Territory that had once belonged to their ancestors. In 2008, the Australian government apologized for the years of unjust treatment toward the Aborigines.

Today, Aborigines are still working to improve their lives. Unemployment and illiteracy rates are high. Especially in rural communities, access to adequate health care and education is unpredictable. After many years of hardship, Aborigines continue to press for social and political equality.

Critical Viewing Lowitja O'Donoghue, a member of the Stolen Generation, and Australia's Prime Minister Kevin Rudd meet in 2008. What does the photo lead you to think about indigenous peoples' reaction to the government's apology?

Maori in New Zealand

The Maori (MOW ree) are the indigenous people of New Zealand. They arrived from Polynesia in the 1300s and were New Zealand's only inhabitants until the British arrived in the late 1700s. Like the Aborigines in Australia, the Maori struggled with the British for control of their land. Although the 1840 Treaty of Waitangi (WY tahng ee) granted the Maori legal and land rights, it was largely ignored for more than 100 years. The Maori people have organized politically, and, like Australia's Aborigines, have had some success achieving civil rights and reclaiming lands.

Before You Move On

Make Inferences In what ways are indigenous people in the region regaining social and political equality?

Critical Viewing Maori protesters gather in New Zealand to demand that the Treaty of Waitangi be followed. What can you infer about the strategy these protesters are using?

2.3 Adventure Tourism

TECHTREK
myNGconnect.com For photos of adventure tourism

Digital Library

Main Idea Tourism that involves physical activity and adventure is a major industry in the region.

Australia, the Pacific Realm, and Antarctica are frequent destinations for adventure tourism. **Adventure tourism** takes travelers to remote areas and involves active enjoyment of the physical environment.

Adventure Activities

New Zealand and Australia are popular destinations for adventure tourists. In New Zealand, rock climbing, mountain biking, and cave exploration are a few of the ways people enjoy the rugged terrain. Adventurers explore New Zealand's interior rivers and lakes by rafting, kayaking, and canoeing. Whale-watching expeditions provide opportunities to see the many different species that inhabit the waters off New Zealand's coast.

People seeking active enjoyment of unusual physical environments also travel to Australia—particularly to its Great Barrier Reef, which draws millions of tourists every year. In 2009, more than two million people visited this natural wonder. Activities on the Great Barrier Reef include snorkeling, scuba diving, fishing, and exploring reef waters in glass-bottomed boats.

In Antarctica, hikers trek across **glaciers**, which are large masses of ice and packed snow. Expert guides help hikers cross dramatic **crevasses**, or deep, open cracks in glaciers. Visitors to Antarctica can also observe emperor

penguins, leopard seals, and other Antarctic birds and animals in their natural habitat.

Economic Benefits

In recent years, adventure tourism has become a **lucrative**, or profitable, industry. People work as expert adventure guides and as drivers on tours. Others find jobs in hotels, stores, airports, and restaurants or on cruise ships that regularly carry thousands of people to the region.

In the Pacific islands, even those not directly involved with adventure tourism often benefit from it. Some local residents make a living creating and selling traditional baskets, mats, and masks. Other islanders earn income by entertaining guests with traditional feasts. Some critics worry that tourism as an industry might devalue, or trivialize, islanders' cultures and ways of life. However, those who support the industry focus on the opportunity to bring more income into their economies.

Before You Move On
Monitor Comprehension In what ways is this region an ideal destination for adventure tourism?

ONGOING ASSESSMENT

PHOTO LAB

 GeoJournal

1. **Human-Environment Interaction** Review each photo. In what ways do adventure tourists interact with the environment in this region?

2. **Make Generalizations** From what you can infer from the photos, what kinds of personal traits might adventure tourists share?

Visual Vocabulary A crevasse is a deep, open crack in a glacier. With the help of a guide, this group is crossing a crevasse on Franz Josef Glacier, South Island, New Zealand.

Critical Viewing Paramotoring, a form of motorized flying, is another popular activity on South Island. What kinds of scenery might this person be able to enjoy while flying?

Scuba divers use an underwater platform to work on diving skills in the waters of Australia's Great Barrier Reef.

2.4 New Zealand Today

TECHTREK

myNGconnect.com For a map and photos of New Zealand today

 Maps and Graphs

 Digital Library

Main Idea New Zealand protects its culture and environment while developing new uses for natural resources.

For some countries, the desire to improve their economies sometimes means choosing between commercial uses of the land and protecting their cultures and natural environments. New Zealand has been able to balance these concerns.

Preserving Maori Culture

As you have read, the Maori were New Zealand's first inhabitants. During the British colonial period, many Maori groups lost ancestral lands. Reclaiming lands that once belonged to their ancestors continues to be a key issue for the Maori. In addition to protecting land for future generations, the Maori people are embracing traditional customs in order to preserve their culture. They have renewed a focus on traditional arts and ceremonies as well as their native language. Maori-speaking elders and parents worked with the New Zealand government to establish language immersion schools for Maori children. These schools use the Maori language exclusively throughout the school day. In 1987, the government made the Maori language one of New Zealand's official languages.

Energy for the Future

New Zealand has also made a special effort to preserve its natural heritage. The government has established many national parks and wildlife preserves. In fact, nearly 30 percent of the land is protected by the government.

Another way New Zealand protects its environment is by producing energy from solar, water, and wind sources. This kind of energy is not depleted by its use and is called renewable energy. One-third of energy consumed in New Zealand comes from renewable sources and that percentage is expected to increase.

One source of renewable energy in New Zealand is geothermal energy. Geothermal energy is heat energy from within the earth that can be turned into electricity. Geothermal reservoirs of hot water and steam are common in New Zealand

⌃ Critical Viewing This sculpture in Auckland City is a modern representation of a traditional Maori entry gate. Based on the photo, in what ways does New Zealand balance tradition with modernity?

because the country lies in an active volcanic zone. Geothermal energy is low-cost and non-polluting and provides at least 10 percent of New Zealand's electricity needs.

New Zealand is considered an environment-friendly country because of its foward-thinking development of renewable energy sources. Through the establishment of wind farms, geothermal plants, and hydroelectric power stations, New Zealand is forging a new path for sustainable energy production.

Before You Move On

Summarize In what ways is New Zealand preserving its culture and its environment?

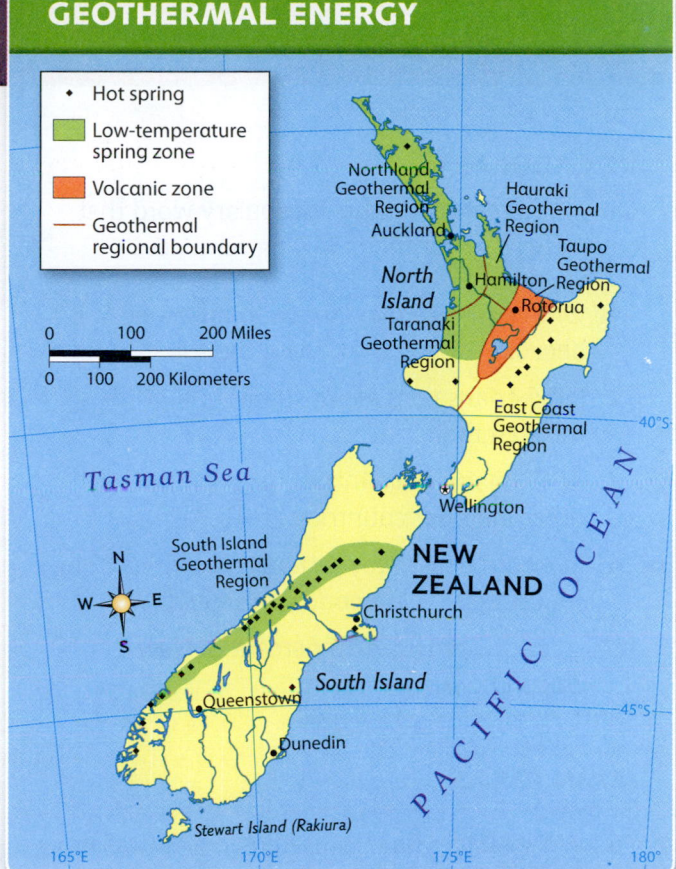

GEOTHERMAL ENERGY

- • Hot spring
- ▨ Low-temperature spring zone
- ▨ Volcanic zone
- — Geothermal regional boundary

ONGOING ASSESSMENT

MAP LAB   GeoJournal

1. **Interpret Maps** Geothermal energy zones are categorized by temperature. In what zone is the Taupo Geothermal Region? Based on the map, how might geothermal sites on South Island be described?

2. **Make Predictions** Most people live on New Zealand's North Island. What cities might benefit most from future development of geothermal energy plants?

∧ **Critical Viewing** This geothermal pool is located near Rotorua. What can you infer from the photo about the temperature of the water?

VOCABULARY

On your paper, write the vocabulary word that completes each of the following sentences.

1. One reason farmers move to urban areas is because _____ is ruining the soil.

2. Polynesians pass along stories and history through their _____.

3. APEC member countries have _____ with other member countries.

4. In Antarctica, many different countries have set up _____ to conduct research.

5. _____ provides 10 percent of New Zealand's energy needs.

MAIN IDEAS

6. What factors have caused many Australian farmers and ranchers to move to urban areas? (Section 1.1)

7. In what ways have populations of immigrants to Australia changed over the years? (Section 1.2)

8. What cultural traditions are Polynesians preserving? (Section 1.3)

9. What does the Antarctic Treaty state about the use of Antarctica? (Section 1.4)

10. In what way have Australia and the Pacific islands strengthened their presence in the global economy? (Section 2.1)

11. What are many indigenous people trying to reclaim? (Section 2.2)

12. In what ways does adventure tourism highlight the region's diverse physical features? (Sections 2.3)

13. Name three things the New Zealand government has done to protect traditional culture and preserve the environment. (Section 2.4)

CULTURE

ANALYZE THE ESSENTIAL QUESTION

How are Australia, the Pacific Realm, and Antarctica becoming more connected to the rest of the world?

Critical Thinking: Make Inferences

14. In what ways might Australia's long drought impact trade with other countries?

15. What characteristics attract immigrants from around the world to Australia?

16. How might rural life change as a result of current migration patterns?

INTERPRET MAPS

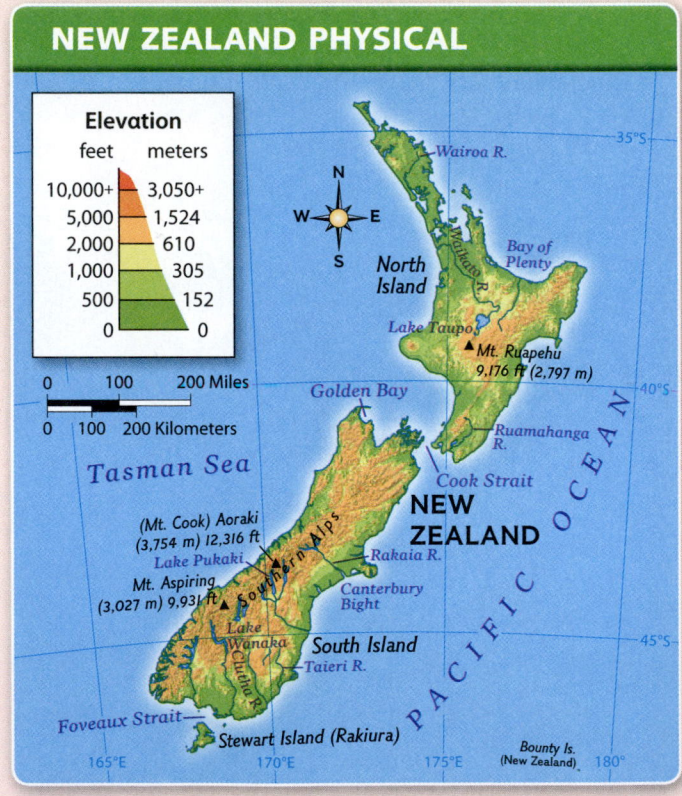

NEW ZEALAND PHYSICAL

17. **Place** What two main islands make up New Zealand? Which island has higher elevations?

18. **Region** Unlike Australia, New Zealand does not have extensive deserts. Instead, what physical feature can you infer from the map legend covers much of New Zealand?

GOVERNMENT & ECONOMICS

ANALYZE THE ESSENTIAL QUESTION

What new economic patterns are emerging in the region?

Critical Thinking: Draw Conclusions

19. How might new trade partnerships among different countries benefit the entire region?

20. What energy sources in New Zealand might represent the energy of the future?

21. In what ways might adventure tourism benefit countries in the region economically?

INTERPRET GRAPHS

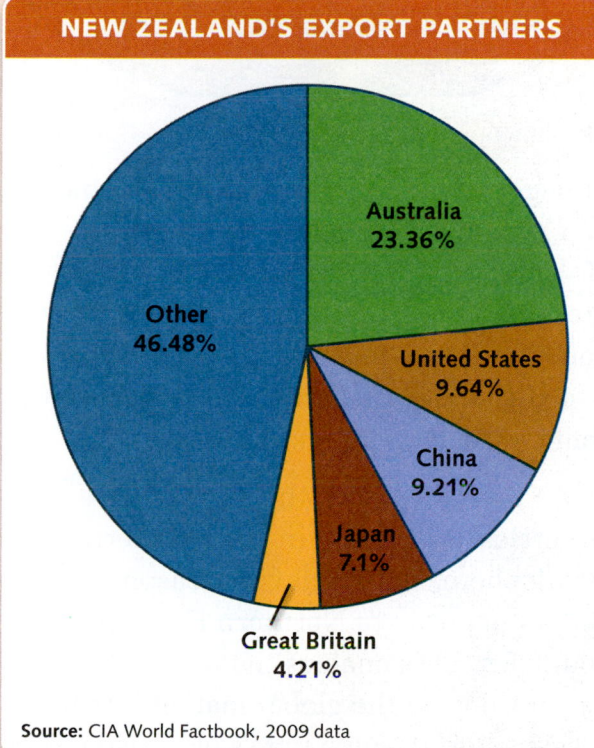

NEW ZEALAND'S EXPORT PARTNERS

Australia 23.36%

Other 46.48%

United States 9.64%

China 9.21%

Japan 7.1%

Great Britain 4.21%

Source: CIA World Factbook, 2009 data

22. **Analyze Data** What percentage of New Zealand's exports go to Australia? How does that number compare with Japan?

23. **Identify** Of the export partners shown on the pie chart, which are APEC members? Why might Australia represent nearly one-quarter of New Zealand's exports?

ACTIVE OPTIONS

Synthesize the Essential Questions by completing the activities below.

24. **Write a Travelogue** Imagine that you are a magazine writer trying to persuade your readers to visit one of the places you learned about in this chapter. Write two paragraphs that describe the features of the place and why it is a good choice for a vacation. **Share your paragraphs with a partner or a friend.**

> **Writing Tips**
> * Describe how your readers might travel to the location, and how long the trip might take.
> * Include specific details about important places travelers should make a point to visit.
> * Highlight at least one lesser-known attraction.

Go to **Student Resources** for Guided Writing support.

TECHTREK myNGconnect.com For research links on this region today

25. **Create Charts** Make a chart like the one below to compare information about three Pacific islands. Use the research links at **Connect to NG** and other online sources to gather the data. Then, with a partner, create three questions to ask other students.

	PAPUA NEW GUINEA	SAMOA	FRENCH POLYNESIA
Population			
Urban Population			
Square miles of land			
Exports			
Imports			

Pastures and Crops

TECHTREK

myNGconnect.com For research links about land use

Maps and Graphs

Student Resources

Australia's grasslands have been used for many years to support both ranching and agriculture. Countries with large, open grasslands use them for similar purposes, including grazing livestock and growing a variety of crops.

Decisions about land use are influenced by many factors, such as climate, crop yield, and sustainability. In most countries, mixed use of grasslands is more common than devoting land solely to ranching or agriculture. Comparing two countries, Australia and Argentina, reveals similarities in how these two countries use grasslands.

Compare

- Argentina
- Australia
- China
- India
- New Zealand
- Ukraine
- United States

GRAZING

Grasslands provide grazing, or feeding, grounds for ranch livestock. Cattle, horses, and sheep are the most common ranch livestock. Other livestock on ranches around the world include goats, alpacas, emus, ostriches, bison, deer, and elk.

Ranching industries have been in place in Australia and Argentina since the 1880s. Cattle were not native to either continent. European settlers imported cattle during the colonial periods.

Australia and Argentina are top beef producers today. Australia's beef industry is export-oriented, while much of Argentina's beef is consumed domestically. Each country faces challenges with its beef industry. Drought on pasture lands in Australia prompted ranchers to reduce herds. In Argentina, beef producers struggle with government policies that limit how much beef can be exported.

GROWING

Though their climates differ, Australia and Argentina are both top exporters of wheat. Most of Australia's wheat production is located in the southern and southwestern parts of the country. Argentina's fertile Pampas is a productive center of wheat cultivation.

As you can see on the map, both countries are located in the Southern Hemisphere. Their growing season occurs opposite of Northern Hemisphere countries. This enables them to make wheat sales on the global market during the Northern Hemisphere's off-season.

Around the world, demand for food production is increasing as the population climbs. In order to increase grain production, both Australia and Argentina are beginning to use marginal, or less desirable, lands to plant more wheat and other crops.

Grassland Use Around the World
Major Crops and Livestock of Selected Countries

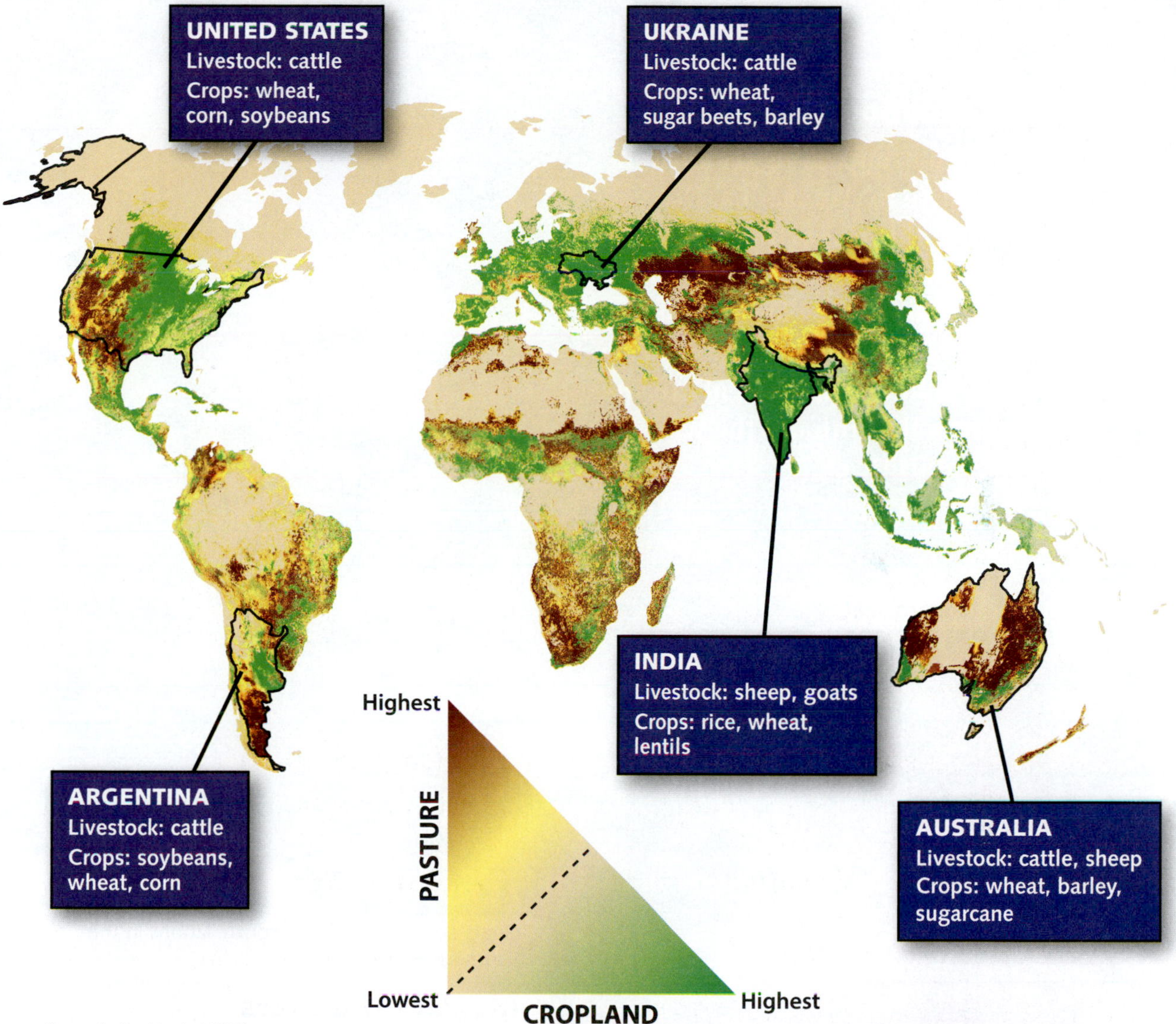

UNITED STATES
Livestock: cattle
Crops: wheat, corn, soybeans

UKRAINE
Livestock: cattle
Crops: wheat, sugar beets, barley

INDIA
Livestock: sheep, goats
Crops: rice, wheat, lentils

ARGENTINA
Livestock: cattle
Crops: soybeans, wheat, corn

AUSTRALIA
Livestock: cattle, sheep
Crops: wheat, barley, sugarcane

Highest

Lowest

PASTURE

CROPLAND

Highest

Source: Remankutty, N. et al 2007

ONGOING ASSESSMENT

RESEARCH LAB GeoJournal

1. **Identify** Look at the map. What crop do all five countries shown above grow?

2. **Compare and Contrast** Find India and Australia on the map. Which country devotes more land use to crops, and which country devotes more to grazing? What climate factors might account for these choices?

3. **Make Inferences** Which country do you think produces more beef, Australia or Ukraine? Why do you think so?

Research and Write a Report Consider the similarities of Australia and Argentina in their use of grasslands. Research other countries including New Zealand and China to compare their use of grasslands. Write a report to present your findings. Consider these topics to guide your research:

- climate
- location of grasslands
- typical livestock
- crops grown for export

Active Options

TECHTREK

myNGconnect.com For photos of animal life in the region and research links

 Connect to NG

 Digital Library

 Magazine Maker

ACTIVITY 1

Goal: Extend your knowledge about Antarctica.

Prepare a Guidebook

You are leading a group going on a seven-day summer cruise to Antarctica. Do research and use the Magazine Maker to prepare a guidebook for your group that gives some facts about the continent and describes what visitors might expect to see on the cruise. Include examples of marine life and sea birds that inhabit the waters surrounding Antarctica.

leopard seal

ACTIVITY 2

Goal: Research indigenous animal species in Australia.

Write a Report

The Tasmanian devil and the Australian dingo are animals found only in Australia. Research one of these animals using research links at **Connect to NG.** Write a report that details its habitat, what it eats, and what threats it might face. Include photos of your animal from the **Digital Library** with your written report.

ACTIVITY 3

Goal: Learn more about constellations in the Southern Hemisphere.

Identify the Stars

National Geographic Fellow Elizabeth Kapu'uwailani Lindsey studies the traditions of Pacific island seafarers who relied on their knowledge of the stars, waves, and birds to guide them. Many of the stars they observed in the Southern Hemisphere are not visible in the night skies in North America. Conduct research and identify five constellations of stars visible only in the Southern Hemisphere. Explain what shape each constellation forms and how that shape relates to its name.

NATIONAL GEOGRAPHIC
World Cultures and Geography

REFERENCE
HANDBOOK

GEO

Skills Handbook

Find Main Idea and Details

Every book, paragraph, or passage has a **main idea**, which is the sentence or sentences that state the subject of the text. A main idea can sometimes be unstated, or *implied*. In that case, the **details** in the text provide clues about the main idea. To find a main idea and details, follow the steps at right.

Step 1 Look for a stated main idea in the first and last sentences of the paragraph. If no main idea is clear, look for details that offer clues about what the implied main idea is.

Step 2 Next, find details that support and clarify the main idea. If the main idea is in the first sentence, supporting details follow it. If the main idea is in the last sentence, the details come before it.

GUIDED MODEL

Habitat Loss and Restoration

A The loss of habitats can destroy an entire ecosystem. An ecosystem is a community of plants and animals and their habitat. **B** <u>Earth has many different ecosystems that interact with each other</u>. **B** <u>The destruction of one affects all the other ecosystems</u>. **B** <u>For example, many scientists believe the destruction of rain forest habitats has led to global climate change</u>.

TIP When an author doesn't directly state the main idea of a passage, it's up to you to figure out the **implied main idea**. To do this, ask yourself, "What do the details of the passage have in common?" Find the connection between the details, put it into your own words, and you have the implied main idea.

Step 1 Look for a stated main idea.

Read the first and last sentences in the paragraph. Do they state the main idea? If the main idea is not obvious, look for details that offer clues about what the main idea is. Some details explain the main idea. Other details give examples of the main idea.

MAIN IDEA A The loss of habitats can destroy an entire ecosystem. In this example, the stated main idea is the first sentence of the paragraph.

Step 2 Once you have figured out the main idea, find details that support it.

If the main idea is in the first sentence, it is usually followed by details that support and explain it. If the main idea is in the last sentence, the supporting details come before it. In the example at left, the details follow the main idea.

DETAIL B Earth has ecosystems that interact with each other.

DETAIL B The destruction of one affects all the other ecosystems.

DETAIL B Many scientists believe the destruction of rain forest habitats has led to global climate change.

APPLY THE SKILL

Turn to *Europe Geography & History*, Section 2.2, "Classical Greece." Read the "Golden Age of Greece" passage. Identify the main idea and details. Record them in a web like the one at right.

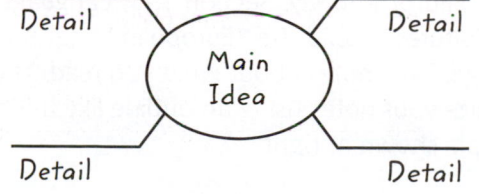

Take Notes and Outline

Taking notes while you read helps you understand and remember important information, including facts, ideas, and details. Many writers organize their notes in an **outline**. To take notes, follow the steps shown at right.

Step 1 Read the title to learn the topic of the passage.

Step 2 Write down the main ideas and important details of the passage.

Step 3 Summarize the main ideas and details in your own words.

Step 4 Search for key words and write down the words and their definitions.

Step 5 To save time and space, use abbreviations.

GUIDED MODEL

A Economic Indicators

B A country's economic strength can be measured by several indicators. One is **C gross domestic product** (GDP), or the total value of the goods and services that a country produces. **C** Other indicators include **GDP** per capita (or per person), income, basic literacy rate, and life expectancy.

B Economies are also placed in one of three categories. Countries with high GDPs, such as the United States, are **C more developed countries**. Countries with lower GDPs are **C less developed countries**.

TIP Outlines list the main ideas and supporting details in the order in which they best make sense. You can use time order, order of importance, or the order in which they appear in the text. Once the outline is complete, you can use it as a guide for your writing.

Step 1 Read the title to learn the topic of the passage.
TITLE A Economic Indicators

Step 2 Write down the main ideas and important details of the passage.
MAIN IDEAS B A country's economic strength can be measured by several indicators. Economies are also placed in one of three categories.

Step 3 Summarize the main ideas and details in your own words.
You can measure a country's economic strength using indicators, such as GDP, GDP per capita, income, basic literacy rate, and life expectancy. Countries can be placed in categories, such as as more developed, less developed, and newly industrialized.

Step 4 Search for key words and write down the words and their definitions.
SAMPLE KEY WORD: C Gross domestic product is the total value of the goods and services a country produces.

Step 5 To save time and space, use abbreviations. Here, for example, use "N" for *north* and GDP for *gross domestic product*.

APPLY THE SKILL

Turn to *Europe Today*, Section 1.1, "Languages and Cultures." Read the "European Languages" passage. Take notes about what you read. Then organize your notes using an outline like the example shown at right.

I. Main Language Groups in Europe
  A. Romance
    1. French
    2. Spanish
    3.
  B. Germanic

Summarize

When you **summarize**, you restate text in your own words and shorten it. A summary includes only the most important information and details. You can summarize a paragraph, a chapter, or a whole book. To summarize, follow the steps shown at right.

Step 1 Read the text looking for the most important information. Watch for topic sentences that provide the main ideas.

Step 2 Restate each main idea in your own words.

Step 3 Write your summary of the text using your own words and including only the most important information.

GUIDED MODEL

Foods of Eastern Europe

A Eastern Europe's cold climate causes a shorter growing season than that in Western Europe. In Russia, root vegetables such as turnips and beets are well adapted to the country's climate. A soup called borscht made from beets is a traditional dish on winter nights.

B The fertile soil of Hungary allows Hungarian farmers to grow grains and potatoes. These crops are used to make a variety of breads. A meat stew called goulash is Hungary's national dish. It is made with beef, potatoes, and vegetables and seasoned with a spice called paprika.

TIP Try using a chart to record and organize the main idea and details you want to summarize. You can then use your notes to write your summary.

Step 1 Read the text looking for the most important information. Watch for topic sentences that provide the main ideas.

TOPIC SENTENCE A Eastern Europe's cold climate causes a shorter growing season than that in Western Europe.

TOPIC SENTENCE B The fertile soil of Hungary allows Hungarian farmers to grow grains and potatoes.

Step 2 Restate each main idea in your own words.

RESTATED A The climate of Eastern Europe is cold, so it has a shorter growing season than Western Europe.

RESTATED B Hungarian farmers grow grains and potatoes.

Step 3 Write your summary of the text using your own words and including only the most important information.

SUMMARY: Eastern Europe has a cold climate, so it has a shorter growing season than Western Europe. Certain crops grow well there, such as potatoes and grains in Hungary.

APPLY THE SKILL

Turn to *Europe Today*, Section 1.1, "Languages and Cultures." Read the "Cultural Traditions" passage. Identify the main ideas and important information, and restate them in your own words. Then write a summary of the passage using the topic sentence shown here.

Cultural Traditions

Europe's cultural traditions reflect the region's ethnic diversity. _____

_____

_____

Sequence Events

When you **sequence events**, you put them in order based on when they occurred in time. Thinking about events in time order helps you understand how they relate to each other. To sequence events, follow the steps shown at right.

Step 1 Look for time clue words and phrases such as names of months and days, or words such as *before, after, finally, a year later,* or *lasted* that help you sequence events.

Step 2 Look for dates in the text and match them to events.

GUIDED MODEL

Greek Culture Spreads

Greece's golden age ended around **B** 431 B.C., when war broke out between Athens and Sparta. The conflict, known as the Peloponnesian War, **A** lasted 27 years. The war weakened both Athens and Sparta.

Around **B** 340 B.C., King Philip II of Macedonia took advantage of their weakness and conquered Greece. In **B** 334 B.C., Philip's son, Alexander the Great, began to extend the Macedonian Empire. Alexander's love of Greece led him to spread its culture throughout his empire. Alexander died in **B** 323 B.C. at the age of 33.

TIP A **time line** is a visual tool that can be useful for sequencing events. Time lines often move from left to right, listing events from the earliest to the latest.

Step 1 Look for time clue words and phrases.
TIME CLUES **A** lasted 27 years

Step 2 Look for specific dates in the text.

Be sure to read the text carefully because the dates in a paragraph may not always be listed in time order. Always match an event with its date. A chart like the one below can be a useful way to sequence dates and events.

SAMPLE DATE **B** 431 B.C., when Greece's golden age ends and the Peloponnesian War begins

DATE IN TEXT	EVENT
431 B.C.	Greece's golden age ends; Peloponnesian War breaks out between Athens and Sparta.
340 B.C.	King Philip of Macedonia conquers Greece.
334 B.C.	Alexander the Great starts extending the Macedonian Empire.
323 B.C.	Alexander the Great dies.

APPLY THE SKILL

Turn to *Europe Geography & History*, Section 2.3, "The Republic of Rome." Read the "A Republic Forms" passage. Identify the dates and events that occurred on each date. Use a time line like the one below to sequence the events.

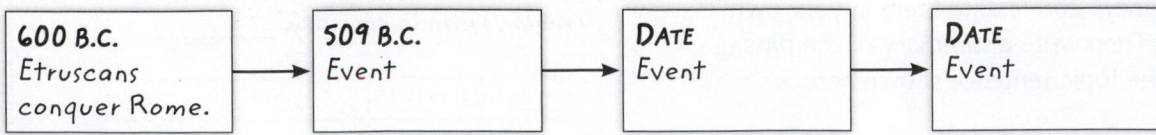

| 600 B.C. Etruscans conquer Rome. | → | 509 B.C. Event | → | DATE Event | → | DATE Event |

Categorize

When you **categorize**, you sort things into groups, or categories. Almost everything can be categorized, including objects, ideas, people, and information. Categorizing is important because it helps you recognize data patterns and trends. To categorize, follow the steps shown at right.

Step 1 Read the title and text and ask yourself what the passage is about to determine how the information can be categorized.

Step 2 Look for clue words to help you categorize information.

Step 3 Decide what the categories will be.

Step 4 Sort the information from the passage into the categories.

GUIDED MODEL

A Political and Physical Maps

Cartographers, or mapmakers, create B different kinds of maps for many B different purposes.

A A political map shows features that humans have created, such as countries, states, and cities. These features are labeled, and lines show boundaries, such as country borders.

A A physical map shows features of physical geography. It includes landforms, such as mountains, plains, deserts, and bodies of water. A physical map also shows elevation and relief. Elevation is the height of a physical feature above sea level. Relief is the change in elevation from one place to another.

TIP Every day, we put information into categories, such as healthy food versus junk food or fiction versus nonfiction. Knowing how information is categorized helps you understand how things are related and organize ideas.

Step 1 Read the title and text and ask yourself what the passage is about to determine how the information can be categorized.

THE PASSAGE IS ABOUT A the features of two types of maps.

Step 2 Look for clue words to help you categorize information.

CLUE WORDS B *different kinds of maps, different purposes*

Step 3 Decide what the categories will be.

THE CATEGORIES political maps and physical maps

Step 4 Sort the information into the categories.

Use a chart like the one below to help you organize the information.

TYPE OF MAP	PURPOSES
Political	Shows man-made features, such as countries, states, cities, and boundaries
Physical	Shows features of physical geography, such as mountains, plains, deserts, and bodies of water; also shows elevation and relief

APPLY THE SKILL

Turn to *Europe Today*, Section 1.2, "Art and Music." Read the section carefully. Determine how categorizing can help you understand and organize the information about European art. Then create a chart like the one at right and categorize the information.

EUROPEAN ART	
PERIOD	**CHARACTERISTICS**
Middle Ages	religious subjects; two-dimensional
Romantic Period	landscapes, nature, conveys emotion

Describe Geographic Information

When you read text or study a chart or graph about a geographic subject, you take in information. One way to enhance your understanding of **geographic information** is to describe it. You can describe geographic information by following the steps shown at right.

Step 1 Read the title of a passage, chart, or other visual to find out what geographic information it contains.

Step 2 Read the passage or study the visual and identify its main ideas or topics.

Step 3 Describe the information by summarizing it or responding to questions.

GUIDED MODEL

A ### The Geography of Europe

B Europe forms the western peninsula of Eurasia, the landmass that includes Europe and Asia. In addition, Europe contains several peninsulas and significant islands.

Four land regions form Europe. The Western Uplands are hills, mountains, and plateaus that stretch from the Scandinavian Peninsula to Spain and Portugal. The Northern European Plain is made up of lowlands that reach across northern Europe. The Central Uplands are at the center of Europe. The Alpine region consists of the Alps and several other mountain ranges.

TIP You can use the steps above to describe the geographic information in a chart. Just imagine that the column headings are the main ideas and the information in the rows are the supporting details.

Step 1 Read the title of a passage, chart, or other visual to find out what geographic information it contains.

A *From its title, I see that this passage is about the geography of Europe.*

Step 2 Read the passage or study the visual and identify its main ideas or topics.

MAIN IDEAS B *Europe is a peninsula on Eurasia. It is formed by four land regions.*

Step 3 Describe the information by summarizing it or responding to questions.

SUMMARY *Europe is a peninsula and contains several peninsulas and islands. It is formed by four land regions: the Western Uplands, the Northern European Plain, the Central Uplands, and the Alpine region.*

QUESTIONS

- *What landforms does Europe contain?*
- *What land regions make up Europe?*

APPLY THE SKILL

Turn to *Europe Geography & History*, Section 1.3, "Mountain Chains." Use the information in this section to describe the location and natural resources of Europe's mountains. You might use questions like those shown at right to help you describe the information.

- What mountain chains are part of the Alpine region?
- What natural resources do Europe's mountain chains provide?

Express Ideas Through Speech

When you **express ideas through speech**, you say what you are thinking using effective language. The purpose of your speech may be to narrate events, explain information, or convince your audience of your point of view. You can express ideas through speech by following the steps at right.

Step 1 Determine the topic you will be presenting and the purpose of your speech. Do research to gather information using a wide variety of reliable sources.

Step 2 Organize your ideas and create speaking notes and any visual aids you may need.

Step 3 Practice speaking honestly, respectfully, clearly, and appropriately.

GUIDED MODEL

A <u>European Literature</u>

B Ancient Greeks and Romans (focus on journeys, heroes, and adventures)
- Iliad and Odyssey (Homer)
- Aeneid (Virgil)

Middle Ages (focus on religious beliefs and politics of the time)
- Divine Comedy (Dante)

Renaissance (focus on human experiences)
- Hamlet (Shakespeare)
- Don Quixote (Cervantes)
—first modern novel

TIP When you express ideas through speech, it is important to hold your audience's attention. Vary your volume, speaking rate, and sentence structure to add interest to your presentation and emphasize meaning. Your body language is also important. Gestures, posture, facial expressions, and eye contact will engage your audience and enhance your speech.

Step 1 Determine the topic and purpose of your speech and do research to gather the information you need.

A The topic of this student's speech is the history of European literature. The purpose is to explain the main periods of European literature.

Step 2 Organize your ideas and create speaking notes and any visual aids you may need.

Notes and visual aids will help you stay on topic, remember your points, and emphasize your message.

B Well-organized notes will help you if you lose your place while speaking or forget a point you wanted to make. Keep your notes simple— just list headings and key points and important people and events.

Step 3 Practice speaking honestly, respectfully, clearly, and appropriately.

Consider your audience and subject matter. Speak honestly and respectfully and use language that is appropriate to your audience.

APPLY THE SKILL

Select a topic from the book, research it, and prepare speaking notes for a presentation about it like those shown at right. Then express your ideas through speech to an audience.

<u>World War I</u>
Began 1914 (explain causes)
- nationalism
- alliances
- Archduke Ferdinand

Write Outlines for Reports and Comparisons

Writing an outline is a useful way to synthesize, organize, and summarize information before you write a report or comparison. A good outline lists the most important ideas and supporting details in order, either chronologically or by level of importance. To write an outline, follow the steps shown at right.

Step 1 Give your outline a title.

Step 2 Determine and record the main ideas in logical order using Roman numerals.

Step 3 Determine and record the ideas that support and explain the main ideas.

Step 4 Determine and record specific details below the appropriate supporting ideas.

GUIDED MODEL

(A) <u>Earth's Water</u>
(B) I. Fresh Water
  (C) A. Sources
    (D) 1. Rivers, water flowing downward
    2. Lakes, large inland bodies of water
    3. Streams, brooks, and creeks
  B. Uses
    1. Drinking and cooking
    2. Irrigating crops
II. Salt Water
  A. Oceans
    1. Large bodies of salt water
    2. Atlantic, Pacific, Indian, Arctic
    3. Ocean currents and their effects
  B. Seas
    1. Small bodies of salt water enclosed partly or completely by land
    2. Red, Caspian, Arabian

Step 1 Give your outline a title.
(A) If you are writing an outline to organize information for a report or comparison, the outline title will probably become the title of your paper.

Step 2 Determine and record the topics in logical order using Roman numerals.
(B) Each topic will probably make up at least one paragraph in your paper.

Step 3 Determine and record the main ideas that support and explain each topic.
(C) Label these using indented capital letters below the topics. Not all topics need the same number of main ideas.

Step 4 Determine and record specific details below the appropriate main ideas.
(D) Label these details using more deeply indented numerals below the main ideas. Not every main idea will have the same number of details.

TIP When you write an outline for a comparison, you can list all of the features of one subject under one topic and all of the features of the other subject under another topic. You can also use the features as topics and list facts about both subjects under each one.

APPLY THE SKILL

Turn to *Europe Geography and History,* Section 3.1, "Exploration and Colonization." Then organize the information from the passage in an outline like the one shown at right. Use your outline to write a report.

<u>European Exploration</u>
I. Portuguese explorations
  A. Reasons
    1. Find gold
    2. Establish trade

Write Journal Entries

When you **write journal entries** in response to something you have read, you record your reactions, make notes, ask and answer questions, and respond to new information in an informal written format. To use written journal entries to respond to a question or writing prompt, follow these steps.

Step 1 Use what you have learned and read about the topic to answer the question.

Step 2 Respond and react in a personal way to the question.

Step 3 Record any related questions, comments, or notes you might have.

GUIDED MODEL

QUESTION:

What challenges might immigrants moving to Europe face today?

JOURNAL ENTRY:

A Some immigrants may feel pressure to assimilate into European society and fear they will lose their identity and culture. Those belonging to non-Christian religions may feel discriminated against simply for having different religious beliefs.

B When my family moved, I hated that students at my new school had to wear a uniform. I refused to wear the jacket for the first week and got in trouble. I felt angry that I wasn't allowed to dress the way I wanted. I think I know a little bit about how the immigrants living in Europe feel.

C NOTE: Look up countries in Europe with large immigrant populations on the CIA World Factbook Web site.

Step 1 Use what you have learned and read about the topic to answer the question.

A The journal writer answers the question using facts from the book and his or her prior knowledge.

Step 2 Respond and react in a personal way to the question.

B The journal writer relates to the topic on a personal level and provides insight into the question based on personal experience.

Step 3 Record any related questions, comments, or notes you might have.

C The journal writer writes himself or herself a note to look up more information about the topic online.

TIP Journals can be used to record the steps in a research project (such as your observations during an experiment). You can also use a journal to write your thoughts and questions as you read a book or carry out a project.

APPLY THE SKILL

Turn to *Europe Today*, Section 1.1, "Languages and Cultures." Then write a journal entry to respond to the following question: What is life like for Europeans who live in cities today? The beginning of a sample entry is shown at right.

Most Europeans today live in cities that reflect the cultures and ethnic origins of immigrant populations.

Pose and Answer Questions

Pose questions about what you have read to check your understanding, keep track of details, and identify points you may have missed. To **answer questions**, review what you have read, search for answers in the text, and use your prior knowledge. To pose and answer questions, follow these steps.

Step 1 Read the passage and identify information that is important or confusing.

Step 2 Pose questions about key concepts or confusing information using question words, such as *who, what, when, where, why,* and *how*.

Step 3 Answer the questions using the text and what you already know.

GUIDED MODEL

Establishing Colonies

A European exploration and colonization resulted in an exchange of goods and ideas known as the Columbian Exchange. From the Americas, Europeans obtained new foods, such as potatoes, corn, and tomatoes. Europeans introduced wheat and barley to the Americas. They also introduced diseases like smallpox. The diseases killed millions of native peoples.

TIP A two-column chart is a helpful way to record your questions and answers. Be sure to include the page number for each answer you found in the text in case you need to review or reread a passage or section.

Step 1 Read the passage and determine which information is important or confusing.

A The most important topic seems to be the result of European exploration in the colonies. I'm confused about what the Columbian Exchange is.

Step 2 Pose questions about key concepts or confusing information.

My Questions

1. **What** is the Columbian Exchange?
2. **Who** were killed by European diseases?
3. **Which** foods did Europeans obtain from the Americas? **Which** foods did Europeans introduce to the Americas?

Step 3 Answer your questions using the text and what you already know.

Don't forget to examine charts, graphs, maps, and photographs as you answer.

Answers

1. The Columbian Exchange is the exchange of goods and ideas between Europeans and the areas they explored and colonized.
2. European diseases killed native peoples in the Americas.
3. Europeans obtained potatoes, corn, and tomatoes; the Americas obtained wheat and barley.

APPLY THE SKILL

Turn to *Human & Physical Geography,* Section 1.1, "Earth's Rotation and Revolution." Read "Revolution and Rotation" and pose questions about information in the passage that confuses you. Record your questions in a chart like the one at right. Use the text and diagram to answer your questions.

MY QUESTIONS	ANSWERS
What process creates the four seasons?	the rotation and tilt of Earth

Make Predictions

When you **make predictions,** you think about the events described in a passage or selection, use your prior knowledge, and guess or predict what will happen next. Making predictions as you read can help you understand and remember what you have read. To make predictions, follow these steps.

Step 1 Preview the passage or selection to anticipate what it is about.

Step 2 Use your personal knowledge. Ask yourself what you know about the topic.

Step 3 As you read, make predictions about what will happen next.

Step 4 Confirm or revise your predictions as you continue to read.

GUIDED MODEL

Ⓐ European Music

European music began in ancient Greece and Rome. Musicians played on Ⓒ simple instruments. During the Ⓑ Middle Ages, music was used in religious ceremonies. Also, singers called troubadours sang about knights and love. These songs influenced Ⓑ Renaissance music, when the Ⓒ violin was introduced.

The new instruments helped inspire the complex music of the Ⓑ Baroque period (1600–1750). Opera, which tells a story through words and music, was born then. The Ⓑ Classical and Romantic periods followed the Baroque and continued until about 1910.

TIP Use a prediction chart to take notes as you read. A prediction chart allows you to record predictions, state whether they were correct, and explain why.

Step 1 Preview the passage or selection to anticipate what it is about.

Ⓐ *The title tells me that this passage is about European music. By skimming ahead, I see the names of many different types of music,* Ⓑ *musical periods, and* Ⓒ *words relating to instruments.*

Step 2 Use your personal knowledge.

I know that the Romantic period in art took place in the early 1800s. I have listened to Baroque music before.

Step 3 As you read, make predictions about what will happen next.

I predict that this text will describe how music evolved in Europe.

Step 4 Confirm or revise your predictions as you continue to read.

My prediction was correct. The passage describes how music evolved from the Middle Ages to the present.

PREDICTION	CORRECT?	EVIDENCE
This passage will explain music I am familiar with.	Yes	It describes what opera is and when it developed.

APPLY THE SKILL

Turn to *Human & Physical Geography,* Section 1.2, "Earth's Complex Structure." Read the "Earth's Layers" text. Then use a prediction chart like the one shown at right to record and analyze your predictions.

PREDICTION	CORRECT?	EVIDENCE
This passage will describe the high temperatures deep within the earth.		

R13

Compare and Contrast

When you **compare** two or more things, you examine the similarities and differences between them. When you **contrast** things, you focus on only their differences. To compare and contrast, follow the steps shown at right.

Step 1 Determine what the subject of the passage or paragraph is.

Step 2 Identify two or more ideas, examples, or features relating to the subject that can be compared and contrasted.

Step 3 Search for clue words that indicate similarities (comparing) and differences (contrasting).

GUIDED MODEL

Ⓐ Varied Climates

Most of Europe lies within the humid temperate climate region. The North Atlantic Drift, an ocean current of warm water, keeps temperatures relatively warm. Winds also affect climate. The sirocco sometimes blows over the Mediterranean Sea and brings humid weather to southern Europe at different seasons. The mistral is a cold wind in winter that blows through France, bringing cold, dry weather.

Ⓒ In general, Ⓑ a Mediterranean climate brings cool, wet winters and hot, dry summers. This climate supports a long growing season. Ⓒ In contrast, Ⓑ Eastern Europe has long, cold winters. Greenland, northern Scandinavia, and Iceland have polar climates and a limited growing season.

Step 1 Determine what the subject of the passage or paragraph is.
SUBJECT Ⓐ the climates of Europe

Step 2 Identify features of the subjects that can be compared and contrasted.
FEATURE Ⓑ climates of southern and eastern European countries

Step 3 Search the passage for clue words that indicate similarities and differences.
Those sentences will help you compare. Then look for clue words that indicate how the two aspects are different. Those sentences will help you contrast.
CLUE WORDS Ⓒ in general (comparing); in contrast (contrasting)

TIP A Y-Chart and a Venn diagram, shown below, are useful graphic organizers for comparing and contrasting two topics. In a Y-Chart, list unique information on the branches and shared characteristics in the straight section. In a Venn diagram, list unique characteristics in the left and right sides and common characteristics in the overlapping area.

APPLY THE SKILL

Turn to *Human & Physical Geography*, Section 2.4, "Natural Resources." Read the "Categories of Resources" passage. Compare and contrast renewable resources and nonrenewable resources using a Y-Chart or Venn diagram.

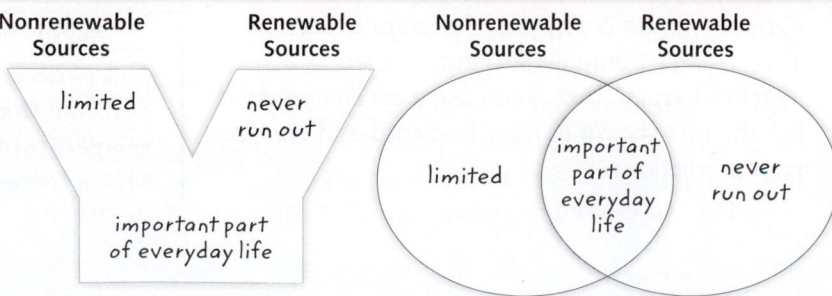

Analyze Cause and Effect

A **cause** is an event or action that makes something else happen. The **effect** is an event that happens as a result of the cause. Analyzing cause-and-effect relationships can help you understand how events are related. To analyze cause and effect, follow the steps shown at right.

Step 1 Determine the cause of an event. Look for signal words that show cause, such as *because, due to, since,* **and** *therefore*.

Step 2 Determine the effect that results from the cause. Look for signal words such as *led to, consequently,* **and** *as a result*.

Step 3 Look for a chain of causes and effects. An effect may be the cause of another action or event.

GUIDED MODEL

Rome's Decline

For about 500 years, the Roman Empire was the most powerful in the world and extended over three continents. **Ⓐ** <u>Beginning around A.D. 235, however, Rome had a series of poor rulers.</u> **Ⓐ** <u>In addition, German tribes began invading the empire.</u>

Ⓑ <u>As a result, in 312, Emperor Constantine moved the capital from Rome to Byzantium, in present-day Turkey, and renamed the city Constantinople.</u> In 395, the empire was divided into an Eastern Empire and a Western Empire with two different emperors. **Ⓒ** <u>Because Rome had been severely weakened, invaders overthrew the last Roman emperor and ended the Western Empire in 476.</u>

Step 1 Determine the cause.

Ask yourself why the event took place. Notice that an event may have more than one cause.

CAUSE(S)

Ⓐ Beginning around A.D. 235, Rome had a series of poor rulers.

Ⓐ German tribes began invading the empire.

Step 2 Determine the effect.

Ask yourself what happened as a result of the event or events.

EFFECT Ⓑ As a result, in 312, Emperor Constantine moved the capital from Rome to Byzantium.

Step 3 Look for a chain of causes and effects.

An effect may cause another event.

CAUSE/EFFECT Ⓒ Because Rome had been severely weakened, invaders overthrew the last Roman emperor and ended the Western Empire in 476.

TIP Test whether events have a cause-and-effect relationship by using this construction: "Because [insert cause], [insert effect] happened." If the construction does not work, one event did not lead to the other.

APPLY THE SKILL

Turn to *Human & Physical Geography*, Section 3.5, "Protecting Human Rights." Read "The Impact of Human Rights" passage. Identify a cause-and effect-relationship in the passage. Record the cause(s) and effect(s) in a graphic organizer like the one at right.

CAUSE	EFFECT
Declaration of Human Rights was developed	World pressured South Africa to grant human rights to non-whites

Make Inferences

Inferences are conclusions or interpretations a reader makes from information that a writer does not state directly. When you make inferences, you use common sense and your own experiences to figure out what the writer means. To make inferences, follow the steps shown at right.

Step 1 Read the text looking for facts and ideas.

Step 2 Think about what the writer does not say but wants you to understand.

Step 3 Reread the text and use what you know to make an inference.

GUIDED MODEL

Aryan Migration

Many historians believe a group of nomadic herders called the Aryans migrated from Central Asia into the Indus Valley around 2000 B.C. From there, they moved into northern India. **A** Their language, Sanskrit, became the basis of many modern languages in South Asia. The Aryans recorded religious teachings in Sanskrit in sacred texts called the Vedas. **B** The Aryans established the beginnings of Hinduism, the major religion of India today.

TIP Use a two-column chart to keep track of inferences. In the left column, write details, quotations, examples, statistics, and other facts. In the right column, write the inference that you draw from each fact. Note that an inference can be based on one fact or several facts.

Step 1 Read the text looking for facts and ideas.

You must know the facts before you can make inferences. Ask yourself, "What facts does the writer state directly in this text?"

FACT A The Aryan language, Sanskrit, became the basis of many modern languages in South Asia.

FACT B The early religion of the Aryans established the beginnings of Hinduism, the major religion of India today.

Step 2 Think about what the writer does not say but wants you to understand.

Ask yourself:
- *How do these facts connect with what I already know?*
- *How does this information help me better understand Hinduism?*

Step 3 Reread the text and use what you know to make an inference.

INFERENCE A Many modern languages have ancient origins.

INFERENCE B Modern religions are influenced by ancient religions.

APPLY THE SKILL

Turn to *Europe Geography & History*, Section 1.2, "A Long Coastline." Read the "Exploration and Settlement" text. Then make two inferences about how Europe's location near large bodies of water affected its people. Use a two-column chart like the one at right to record the facts you found and the inferences you made.

FACTS	INFERENCES

Draw Conclusions

When you **draw conclusions**, you make a judgment based on what you have read. You analyze the facts, make inferences, and use your own experiences to form your judgment. To draw a conclusion, follow the steps shown at right.

Step 1 Read the passage to identify the facts.

Step 2 Make inferences based on the facts.

Step 3 Use the inferences you have made and your own experiences and common sense to draw a conclusion.

GUIDED MODEL

Ⓐ The euro was launched in 1999, but paper money and coins of the currency were not issued until 2002. Ⓐ As of early 2011, 17 of the 27 European Union (EU) nations had adopted the euro. The countries using the euro are known as the eurozone. Ⓐ Some EU countries, including Romania and Bulgaria, hope to join the eurozone soon. Other countries, including Great Britain and Sweden, have not adopted the euro. They believe that giving up their own currency might result in a loss of control over their economies.

TIP Use a diagram to organize the facts you have identified, the inferences you have made, and your conclusion. A diagram can help you clarify your thinking.

Step 1 Read the passage to identify the facts.

FACTS Ⓐ The euro was introduced to Europe in 1999. By 2011, 17 out of 27 EU nations used it. Some countries hope to join the eurozone. Others, including Great Britain and Sweden, have not adopted the euro because they think it would result in a loss of control over their economies.

Step 2 Make inferences based on the facts.

INFERENCES *I see that most of the EU countries have adopted the euro, and at least two more want to adopt it. However, some countries have chosen not to adopt the currency. That must mean that there are both advantages and disadvantages to using the euro.*

Step 3 Use the inferences you have made and your own experience and common sense to draw a conclusion.

CONCLUSION: *In the future, more European countries will probably adopt the euro and become part of the eurozone. However, other countries will not adopt the euro because there are disadvantages to doing so.*

APPLY THE SKILL

Turn to *Human & Physical Geography*, Section 2.3, "Extreme Weather." Read the "Scientific Solutions" passage and draw conclusions about the solutions scientists have come up with to combat the dangers of extreme weather. Write a few sentences summarizing your conclusion. A sample sentence is shown at right.

Scientists are making great progress in predicting extreme weather _____
_____
_____

Make Generalizations

Generalizations are broad statements that apply to a set of ideas, a group of people, or a series of events. You make generalizations based on information you have read or heard. You can also draw on personal experiences. To make generalizations, follow the steps shown at right.

Step 1 Look for the overall theme or message of the selection.

Step 2 Find information in the passage that supports the theme.

Step 3 Draw on your personal knowledge.

Step 4 Make a generalization about the topic and put it into sentence form.

GUIDED MODEL

Life in the Ocean

Dr. Sylvia Earle has led more than 60 diving expeditions to explore marine life. During these dives, she has seen the incredible variety of ocean life— **B** "More than 30 major divisions of animals are known, from sponges and jellies to many kinds of beautiful worms and mollusks."

A Dr. Earle has also seen how people have harmed the oceans. **B** "Taking too much wildlife out of the sea is one way," she claims. "Putting garbage, toxic chemicals, and other wastes in is another." Dr. Earle has witnessed a huge drop in the number of fish in the sea. She has also noted pollution's impact on coral reefs.

TIP You can also make generalizations about information from multiple sources. Determine what information the sources have in common. Then form a generalization that all of the sources would support.

Step 1 Look for the overall theme.
> **COMMON THEME A** Humans harm the diverse marine life in the oceans.

Step 2 Find information that supports the theme.
> **SUPPORTING INFORMATION B** Dr. Sylvia Earle, an ocean expert, states that people harm ocean life by taking too much wildlife out of the sea and putting garbage, toxic chemicals, and other wastes into it. Dr. Earle has witnessed a huge drop in the number of fish in the sea and pollution's impact on coral reefs.

Step 3 Draw on your personal knowledge.
> *I know that living in polluted water is bad for living things. I watched a television program on how toxic waste in the oceans has caused the extinction of some species.*

Step 4 Make a generalization about the topic and put it into sentence form.
> **GENERALIZATION** *If we don't stop polluting the waters and taking too much wildlife from the oceans, many species could become extinct.*

APPLY THE SKILL

Turn to *Europe Geography & History*, Section 2.5, "Middle Ages and Christianity." Read "The Feudal System" passage. Then make generalizations about the feudal system's role in Europe, like the one at right, using the information in the passage and what you may already know.

> *The feudal system provided security for Europeans during the Middle Ages.*

Form and Support Opinions

When you **form an opinion**, you determine and assess the importance and significance of something. While an opinion is a person's own judgment, not a fact, sound opinions must be supported with examples and facts.

Step 1 Read the passage. Look for reliable information about the subject, including facts, statistics, and quotations.

Step 2 Form your own opinion about the subject.

Step 3 Find facts to support your opinion.

GUIDED MODEL

Germany in World War II

A Hitler became the leader of the National Socialist German Workers' Party, or the Nazis. He forged alliances with Italy and Japan.

A Germany's invasion of Poland in 1939 started World War II. Great Britain and France, the Allies, declared war on Germany soon after, but Germany conquered Poland and quickly defeated most of Europe. In 1941, the United States entered the war on the side of the Allies.

A Finally, in May 1945, Germany surrendered. Near the end of the war, Allied troops were stunned to find Nazi concentration camps where six million Jews and other victims had been murdered. This mass slaughter was called the Holocaust.

TIP To distinguish facts from opinions, ask yourself, "Can this sentence be proven?" If evidence can prove a statement, it is a fact. If the sentence cannot be proven because it expresses a thought or emotion, it is an opinion.

Step 1 Read the passage. Look for reliable information about the subject.

SUBJECT Germany in World War II

IMPORTANT INFORMATION ABOUT THE SUBJECT A Germany, Italy, and Japan formed an alliance. Germany's invasion of Poland in 1939 started World War II. Germany soon defeated most of Europe. In 1945, Germany surrendered and Allied troops discovered evidence of the Holocaust.

Step 2 Form your own opinion.

Decide the subject's importance or wider meaning. What do you believe about the subject?

MY OPINION *Hitler and the Nazi Party's actions in the 1930s and 1940s dramatically violated human rights.*

Step 3 Find facts to support your opinion.

If you cannot find facts to support your opinion, you must revise it.

FACTS THAT SUPPORT MY OPINION *Hitler was a powerful leader of the Nazis. They murdered six million Jews and other victims during the Holocaust.*

APPLY THE SKILL

Turn to *Europe Geography & History*, Section 2.2, "Classical Greece." Read the "Greek Achievements" passage. Provide an opinion about what you have read and list facts to support your opinion. Use a chart like the one at right to organize your ideas.

OPINION	EVIDENCE
Greece's golden age was a period of extraordinary achievements.	

Identify Problems and Solutions

Throughout history, people have faced problems and learned how to solve them. When you **identify problems,** you find the difficulties people faced. When you **identify solutions,** you learn how people tried to fix their problems. To identify problems and solutions, follow the steps shown at right.

Step 1 Read the text and determine what problem or problems people faced.

Step 2 Determine what caused the problems. There may be multiple causes.

Step 3 Identify the solutions people used to resolve, or fix, the problems.

Step 4 Determine if the solutions succeeded. Ask yourself, "Was the problem solved?"

GUIDED MODEL

Impact of the Revolution

The Industrial Revolution had a tremendous impact on how people worked and lived. Cities grew rapidly because people migrated to them for factory jobs. Standards of living rose, and a prosperous middle class grew.

A However, factory workers often faced harsh conditions. **B** Laborers worked as many as 14 hours a day. Child labor was common. Children as young as ten years old worked in factories and mines.

B Many workers lived in small, crowded houses in neighborhoods where open sewers were common. Diseases spread quickly in these buildings. **C** Over time, the workers' quality of life improved as public health acts were passed to provide sewage systems and larger buildings.

Step 1 Read the passage and determine what problem people faced.

PROBLEM A Factory workers faced harsh conditions during the Industrial Revolution.

Step 2 Determine what caused the problem.

CAUSES B long work hours, child labor, crowded lodgings, open sewers

Step 3 Identify the solutions people used to resolve the problem.

SOLUTIONS C Larger buildings for workers to live in were built, and sewage systems for improved sanitation were installed.

Step 4 Determine if the solutions succeeded.

SUCCESS? Yes. Laws regulated working and living conditions, which are now much cleaner and safer. Sewers reduced disease. Children are not permitted to work in factories.

TIP A chart like the one below is a useful tool for recording information about a problem and solution. Organize your chart to have separate rows for a description of the problem, its cause, the solutions, and the success of the solution.

APPLY THE SKILL

Turn to *Human & Physical Geography,* Section 2.5, "Habitat Preservation." Read the "Natural Habitats" passage and the "Habitat Loss and Restoration" passage. Identify the problems and solutions using a chart like the one at right.

What was the problem?	
What were the causes?	
What was the solution?	
Did the solution work?	

Analyze Data

Information, or data, can be collected in charts, databases, graphs, diagrams, models, and maps. Regardless of its format, you can **analyze data** to draw conclusions, make comparisons, identify trends, and improve your understanding of the information. To analyze data, follow the steps at right.

Step 1 Identify the data source, which could be a chart, graph, database, model, diagram, or map. Read its title to determine what information it contains.

Step 2 Read any headings or subheadings to see how the information is organized.

Step 3 Study the data to answer questions, draw conclusions, make comparisons, and identify trends.

GUIDED MODEL

Ⓐ ECONOMIC INDICATORS OF SELECTED COUNTRIES*

Ⓑ COUNTRY	GDP PER CAPITA (U.S. $)	LITERACY RATE (percent)
Ⓑ Afghanistan	366	28.0
Brazil	8,536	90.0
China	3,422	93.3
Ethiopia	321	35.9
Germany	44,525	99.0
Mexico	10,249	92.8
United States	47,210	99.0

Source: The World Bank and the United Nations

* GDP per capita figures are for 2008, while literacy rate is for 2007

TIP Data can be represented in ways other than charts and graphs. When analyzing data from a model, diagram, or map, examine all callouts, symbols, illustrations, and text boxes. Important data may be found where you least expect it.

Step 1 Identify the data source and determine what information it contains.

Ⓐ *This data source is a chart. Its title is "Economic Indicators of Selected Countries." That tells me it includes economic data on various countries.*

Step 2 Read any headings or subheadings to see how the information is organized.

In a chart, columns go up and down, and rows go from left to right. Other data sources, such as graphs, models, and diagrams, organize information in different ways.

This chart gives two types of economic indicator data recorded for each country. Ⓑ *The columns record the economic indicator data.* Ⓑ *The rows list information for each country.*

Step 3 Study the data to answer questions, draw conclusions, make comparisons, and identify trends.

Using the data in this chart, I can compare the data in the literacy rate column to the data in the GDP per capita column. I can draw the conclusion that a high literacy rate is tied to a high GDP per capita.

APPLY THE SKILL

Turn to *Human & Physical Geography,* Section 1.2, "Earth's Complex Structure." Examine the "Tectonic Plate Movements" diagram. Then analyze the data in the diagram to answer the following questions.

1. What type of data does this diagram provide?

2. What type of plates are shown?

3. What are the four types of plate movements?

4. What is subduction?

5. Compare divergence and convergence.

6. Based on your analysis of the data in the diagram, what conclusions can you draw about the movement of Earth's plates?

Distinguish Fact and Opinion

It is important to **distinguish fact from opinion** to separate someone's personal beliefs from concepts or events that are known to be true. To distinguish facts from opinions, follow the steps shown at right.

Step 1 Read the passage and identify facts: information that can be proven to be true.

Step 2 Identify opinions: claims or feelings about a topic or statements of personal beliefs.

Step 3 Ask yourself, "Can this statement be checked to see if it is true?" If the answer is yes, you should decide how or where you can verify it.

GUIDED MODEL

Changes in Eastern Europe

A After communism fell in Yugoslavia in 1991, civil war broke out among its ethnic groups. The war ended in 1995. Over time, Yugoslavia was broken up into several new democratic countries.

Ukraine has also had setbacks. In 2004, the Ukrainian people staged the Orange Revolution and peacefully removed their prime minister, Viktor Yanukovych. **B** Many believed that he was corrupt and being controlled by Russia. However, their new leader disappointed the Ukrainians. **A** In 2010, they re-elected Yanukovych.

TIP When you identify personal beliefs in a historical text, you should also check for bias or a writer's prejudiced point of view. Look for words, phrases, or statements that reflect the positive or negative opinions of a particular group, social class, or political party.

Step 1 Read the passage and identify information that can be proven to be true.

FACTS A Communism in Yugoslavia fell in 1991. Civil war broke out among ethnic groups. The war ended in 1995. Yugoslavia was broken into several new countries. In 2004, the Ukrainian people staged the Orange Revolution and peacefully removed their prime minister, Viktor Yanukovych. In 2010, Ukrainians re-elected Yanukovych.

Step 2 Identify opinions.

Look for personal beliefs, judgments, and feelings. Key words such as "could be" and "believed" are clues that a statement is an opinion.

OPINIONS B Many believed that Viktor Yanukovych was corrupt and being controlled by Russia.

Step 3 Ask yourself, "Can this statement be checked to see if it is true?" If the answer is yes, decide how and where to verify it.

STATEMENT After communism fell in Yugoslavia in 1991, civil war broke out among its ethnic groups.

WHERE TO CHECK encyclopedias, history books, reliable online sources

FACT OR OPINION fact

APPLY THE SKILL

Turn to *Europe Geography & History*, Section 1.4, "Protecting the Mediterranean." Read the "Marine Reserves" passage. Use a chart like the one at right to record statements of fact and opinion and decide where you could verify it. After you have checked the statement, record whether it is a fact or opinion.

Statement: Scandola is a marine reserve that prohibits people from fishing, swimming, or anchoring their boats.

Where to check:

Fact or opinion:

Evaluate

When you read an informational text, you must **evaluate**, or assess what you have read. Sometimes you evaluate a passage to determine if its claims are believable. You may also evaluate to understand someone's actions. To evaluate something that took place, follow the steps shown at right.

Step 1 Identify the action or event you want to evaluate.

Step 2 Gather evidence about the positive impact of the action or event.

Step 3 Gather evidence about the negative impact of the action or event.

Step 4 Decide if the evidence is adequate.

Step 5 Form your evaluation of the action.

GUIDED MODEL

Ⓐ Eastern European countries have had mixed results since changing to a market economy. Ⓑ Poland has had the greatest success. It has a fast-growing economy and exports goods throughout Europe. Ⓒ Other countries have been slower to establish new businesses and become competitive. They have also experienced rises in prices and unemployment.

Ⓑ The leaders of many eastern European countries wish to integrate with the rest of Europe. They want to join the European Union and NATO, a military alliance of democratic states in Europe and North America.

TIP Making a list of the positive and negative outcomes of a decision, event, or action can help you evaluate. Read the passage and list the positives and negatives. Then review your list and form your evaluation.

Step 1 Identify the action or event.

Ⓐ the impact of Eastern Europe's adoption of a market economy

Step 2 Gather evidence about the positive impact of the action or event.

Ⓑ Poland's economy has grown quickly. Poland now exports goods throughout Europe. It has caused many eastern European leaders to reach out to the rest of Europe.

Step 3 Gather evidence about the negative impact of the action or event.

Ⓒ Some countries have been slow to establish new businesses and become competitive. They have experienced rises in prices and unemployment.

Step 4 Decide if the evidence is adequate.

There is evidence to support both the positive and negative effects of the action. The evidence seems factual.

Step 5 Form your evaluation of the action.

Changing to a market economy has had negative effects on some eastern European countries, However, it has been successful in others and has encouraged Eastern Europe to deepen connections with the rest of the world.

APPLY THE SKILL

Turn to *Europe Geography & History,* Section 2.5, "Middle Ages and Christianity." Read "The Growth of Towns" passage. Record evidence from the passage and determine whether it is positive or negative using a chart like the one at right. Then write a brief evaluation of the impact of the growth of towns on Europe.

EVIDENCE	POSITIVE/NEGATIVE
Trade and business developed.	

Synthesize

When you read, you take in information, details, clues, and concepts. When you **synthesize**, you combine all of that data to form an overall understanding of what you have read. To synthesize, follow the steps shown at right.

Step 1 Look for solid, factual evidence.

Step 2 Look for explanations that connect facts.

Step 3 Think about what you have experienced or already know about the topic.

Step 4 Use evidence, explanations, and your prior knowledge to form a general understanding of what you have read.

GUIDED MODEL

Immigrants in Australia

Ⓐ Like Europe, Australia has an aging population. Ⓑ The Australian government believes that immigration can help change that trend. As a result, each year the government identifies gaps in the country's workforce. It then determines the number of skilled immigrants that can come to Australia. Between 2008 and 2009, more than 110,000 immigrants came on this skilled migration program. Most of Australia's immigrants came from Great Britain, India, and China.

TIP To synthesize, you must be able to determine what information is the most important. Once you have identified the most important facts, organize them, find explanations for them, and fit them in with what you already know. Synthesis occurs as you extract the most important information from a passage and give it personal meaning.

Step 1 Look for factual evidence.

Identifying the facts will help you base your synthesis on reliable evidence.

FACT Ⓐ Like Europe, Australia has an aging population.

Step 2 Look for explanations that connect facts.

In this passage, the facts have a problem-solution connection.

EXPLANATION Ⓑ Australia believes that immigration can help change its aging population trend.

Step 3 Think about what you already know.

I know that immigration made the United States diverse. I assume that the cultural diversity of Australia has grown due to that country's skilled migration program.

Step 4 Use evidence, explanations, and your prior knowledge to form a general understanding of what you have read.

Because of its aging population, Australia has a government-regulated immigration policy to draw workers from other countries. Through this program, Australia is both solving the problem of its aging population and increasing its cultural diversity.

APPLY THE SKILL

Turn to *Europe Geography & History,* Section 2.3, "The Republic of Rome." Read "The Roman Way" passage. Use a chart like the one at right to organize the evidence. Then write a brief synthesis statement about the importance of Roman values in the development of Rome.

Evidence: Romans applied the values of self-control, working hard, doing one's duty, and pledging loyalty to Rome.
Supporting explanation:
Synthesis:

Analyze Primary and Secondary Sources

Primary sources are materials written or provided by people who have had personal experience with an event. **Secondary sources** are materials written by people who did not witness or experience an event directly. To analyze primary and secondary sources, follow the steps at right.

Step 1 Identify whether the material is a primary or secondary source.

Step 2 Determine the quality and credibility of the primary or secondary source.

Step 3 Determine the main idea of the secondary source material.

Step 4 Identify the author of the primary source and his or her main point.

GUIDED MODEL

Mandela's Inspiration

(A) Enlightenment thinkers asserted that **(B)** people have natural rights, such as life, liberty, and property. They inspired Nelson Mandela in his struggle to end apartheid in South Africa. For his efforts, Mandela received the 1993 Nobel Peace Prize. The following is from his speech.

(C) *The value of our shared reward will and must be measured by the joyful peace which will triumph.*

Thus shall we live, because we will have created a society which recognizes that all people are . . . entitled . . . to life, liberty, prosperity, human rights, and good governance.

(D) —Nelson Mandela, 1993

TIP When you analyze a primary or secondary source, separate facts from opinions. These opinions may reflect the author's bias and the beliefs and ideas of his or her time.

Step 1 Identify whether the material is a primary or secondary source.

Most textbook passages are secondary sources. An observation from an expert or eyewitness is likely to be a primary source. The secondary source material in this textbook passage is marked with **(A)**.

Step 2 Determine the main idea of the secondary source material.

MAIN IDEA (B) People have natural rights that all humans should enjoy.

Step 3 Determine the quality and credibility of the primary or secondary source.

The primary source in this passage is marked with **(C)**. Ask yourself, "Is the author a reliable source of information? *Nelson Mandela is a political activist and Nobel Peace Prize winner. He is a credible source. The secondary source is my textbook, which is also a reliable source.*

Step 4 Identify the author of the primary source and his or her main point.

AUTHOR (D) Nelson Mandela

MAIN IDEA (B) People have natural rights that all humans should enjoy.

APPLY THE SKILL

Turn to *Europe Geography & History*, Section 1.4, "Protecting the Mediterranean." Read the "Under the Sea" passage. Determine what parts of the passage are primary and secondary sources. Then use a chart like the one at right to analyze the passage.

Quality of the source:
Main idea of secondary source material:
Author of primary source material:
Information provided by primary source:

Analyze Visuals

Visuals, such as charts, graphs, maps, photos, artwork, and diagrams, illustrate ideas within a text. When you **analyze visuals**, you determine what information is being presented and how it relates to other information about that topic. To analyze visuals, follow the steps shown at right.

Step 1 Study the visual and determine what information it provides.

Step 2 Determine how the information in the visual relates to other information provided in a text.

Step 3 Analyze the information presented in the visual to enhance your understanding.

GUIDED MODEL

Solstices and Equinoxes

The moment at which summer and winter start is called a **B** solstice. June 20 or 21 is the summer solstice in the Northern Hemisphere. On December 21 or 22, the Northern Hemisphere has its winter solstice.

The beginning of spring and autumn is called an **B** equinox. In the Northern Hemisphere, the spring equinox occurs around March 21, and the autumn equinox occurs around September 23.

A EARTH'S FOUR SEASONS: NORTHERN HEMISPHERE

B SPRING EQUINOX (March 21)
B WINTER SOLSTICE (December 21 or 22)
SUN
North Pole
24 hours
Northern Hemisphere
B SUMMER SOLSTICE (June 20 or 21)
365 days
B AUTUMN EQUINOX (September 23)
Southern Hemisphere
South Pole

Step 1 Study the visual and determine what information it provides.

Examine any titles, labels, or captions for clues.

A *This diagram is titled "Earth's Four Seasons: Northern Hemisphere." The dates on the visual tell me the position of Earth in relationship to the sun throughout the year.*

Step 2 Determine how the information in the visual relates to other information provided in a text.

B *The summer and winter solstices and spring and autumn equinoxes are described in the text. The visual illustrates the information from the text.*

Step 3 Analyze the information presented in the visual to enhance your understanding.

Ask yourself: What information does the visual show? Why was this visual used? Does the visual represent information accurately? What does the visual show that is not explained in the text?

TIP Look for patterns and connections between items and information in a visual and the surrounding text. Study colors and symbols to understand what they represent.

APPLY THE SKILL

Turn to *Human & Physical Geography*, Section 1.3, "Earth's Landforms." Analyze the visual and answer the following questions.

1. What type of information does this visual provide?

2. Which text passage does this visual enhance?

3. Use the information in the related text passage and the visual to explain what the continental shelf is and where it is located.

4. What does the visual show you about continental rise that is not explained in the passage?

Interpret Physical Maps

Physical Maps provide information about the earth's physical features such as lakes, rivers, and mountains. You can learn about elevation, or relief, and absolute and relative location by studying physical maps. To read a physical map, follow the steps at right.

Step 1 Read the title of the map.

Step 2 Use the map legend.

Step 3 Use the map scale to measure distance.

Step 4 Use the compass rose or directional pointer to determine direction.

Step 5 Use the latitude and longitude gridlines to determine the region's location on Earth.

GUIDED MODEL

A SOUTH ASIA PHYSICAL

Step 1 Read the title of the map.

Ⓐ Read the title of the map to find out what type of map it is and what kind of information the map presents.

Step 2 Use the map legend.

Ⓑ A map legend explains the symbols used on the map. On a physical map, the legend usually provides information about physical features, such as mountains. The map also has a color-coded elevation scale to show how far above sea level each area is.

Step 3 Use the map scale.

Ⓒ Use the map scale to help you determine the distances between points on the map.

Step 4 Determine direction.

Ⓓ Use the compass rose or directional pointer to help you determine direction on the map.

Step 5 Determine latitude and longitude.

Ⓔ Examine the numbered gridlines on the map. The horizontal lines represent latitude. The vertical lines represent longitude.

TIP Making a chart is a good way to record important information from a physical map. Use a chart to record the map's title, legend information, scale information, latitude and longitude, and location.

APPLY THE SKILL

Turn to *Europe Geography & History*, Section 1.1, "Physical Geography." Interpret the map to answer the following questions.

1. What is the map about?

2. What type of map is this and what region does it represent?

3. What is the highest elevation in the country of Romania?

4. What is the approximate distance between the easternmost point in France and the westernmost point in Italy, in both miles and kilometers?

Interpret Political Maps

Political maps provide information about human-made features such as cities, capitals, and borders between countries. Unlike physical maps, political maps do not focus on the physical features of a country or region. To read a political map, follow the steps shown at right.

Step 1 Read the title of the map.

Step 2 Use the map legend.

Step 3 Use the map scale to measure distance.

Step 4 Use the compass rose or directional pointer to determine direction.

Step 5 Use the latitude and longitude gridlines to determine the region's location on Earth.

GUIDED MODEL

TIP On many political maps, countries or states are shown in different colors. This makes it easy to distinguish their borders. Cities are often designated by a dot of varying sizes based on the city's population.

Step 1 Read the title of the map.

A Read the title of the map to find out what type of map it is and what information the map represents.

Step 2 Use the map legend.

B A map legend explains the symbols used on the map. Symbols for capital cities and major cities are usually found in the legend of a political map. On this map, New Delhi, India's capital, is represented by a star.

Step 3 Read the map scale.

C Use the map scale to help you determine the distances between cities or countries on the map.

Step 4 Determine direction.

D Use the compass rose or directional pointer to determine direction on the map.

Step 5 Determine latitude and longitude.

E Examine the numbered gridlines that intersect over the map. The horizontal lines represent latitude, and the vertical lines represent longitude. Latitude and longitude lines can help you establish the location of the countries or major cities shown.

APPLY THE SKILL

Turn to the chapter introduction for *Europe Geography & History*. Locate the Europe Political map next to the Preview the Chapter page. Interpret the map and answer the following questions.

1. What type of map is this and what region does it represent?

2. What is the capital city of Spain?

3. What is the approximate distance between the easternmost and westernmost coasts of Iceland?

4. Where is Berlin located in relation to Warsaw?

Create Sketch Maps

In social studies textbooks, maps are a common visual aid. However, sometimes it is helpful to draw your own map to help you understand a place better or to visualize features described in the text. To **create a sketch map** of a country, follow the steps shown at right.

Step 1 Determine which map you are going to draw and give it a title.

Step 2 Sketch the outline of the location.

Step 3 Add important political and physical features to your map.

Step 4 Add a compass rose to your map.

Step 5 If appropriate, sketch and label surrounding countries or regions.

GUIDED MODEL

Step 1 Determine which map you are going to draw and give it a title.

Ⓐ Study existing maps and text describing that geographical location.

Step 2 Sketch the outline of the location.

Ask yourself, "What shape is this location? What borders it?" Remember that sketch maps do not have to be perfect. Everyone's sketch map will look different.

Step 3 Add important political and physical features to your map.

Ⓑ This student chose to draw and label the Ebro River in Spain.

Step 4 Add a compass rose to your map. Ⓒ

Step 5 If appropriate, sketch and label surrounding countries or regions.

Ⓓ On this sketch map, the country of Portugal has been identified.

TIP Sketch your map on graph paper so that you can draw your location and its surrounding area to scale. Also, use a pencil when you sketch so that you can easily erase any mistakes.

APPLY THE SKILL

Examine the Europe Political map next to the Preview the Chapter page in *Europe Geography & History*. Then examine the Europe Physical map and read the passage "A Peninsula of Peninsulas" in Section 1.1, "Physical Geography." Use a chart like the one at right to help you create a sketch map of Ireland.

Map title:
Capital:
Major cities:
Surrounding countries:

Create Charts and Graphs

To organize information, it is useful to **create charts** and **graphs.** Charts simplify and summarize information. Graphs present numerical information. To create charts and graphs, follow the steps shown at right.

Step 1 Determine whether you should use a chart or graph to represent your data.

Step 2 Give your chart or graph a title to tell what kind of information it shows.

Step 3 Create your chart or graph using appropriate labels for the data.

GUIDED MODEL

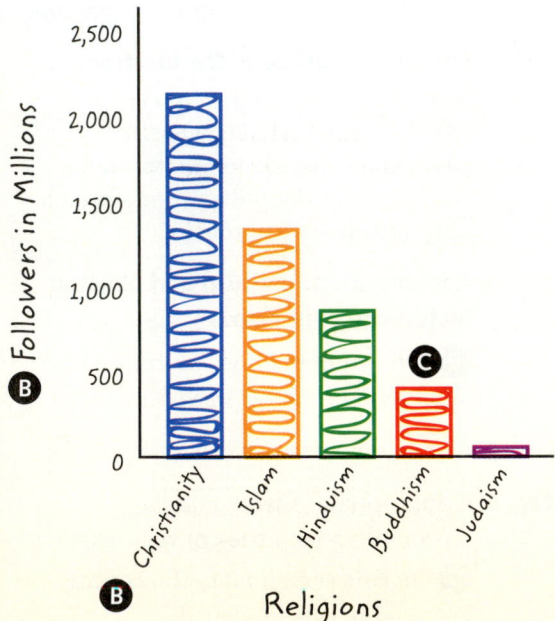

A Major World Religions

Followers in Millions (vertical axis: 0, 500, 1,000, 1,500, 2,000, 2,500)

Religions (horizontal axis: Christianity, Islam, Hinduism, Buddhism, Judaism)

TIP Different visuals are used to represent different types of data. Line graphs are useful to compare changes over time. Bar graphs compare quantities. Pie graphs show percentages of a whole. Charts can be structured in many different ways, but they always simplify and organize information.

Step 1 Determine whether you should use a chart or graph to represent your data.

I want to represent the number of people who practice the major world religions: Christianity, Islam, Hinduism, Buddhism, and Judaism. The data is numerical, so I will create a graph. I will use a bar graph to compare the data and show which religion has the most followers and which religion has the least.

Step 2 Give your chart or graph a title to tell what kind of information it shows.

A *I will call the bar graph "Major World Religions."*

Step 3 Create your chart or graph using appropriate labels for the data.

B *My horizontal axis will represent the five major world religions. My vertical axis will represent the number of followers in millions. I will label the axes accordingly.*

C *I will record my data on the bar graph using a different color to represent each world religion.*

APPLY THE SKILL

Turn to *Europe Geography & History,* and locate the Chapter Review. Use the "Miles of Railway Track in Selected European Countries" chart to create a line graph. To simplify the graph, include only the four countries with the highest totals. A line graph representing 1840 is shown at right.

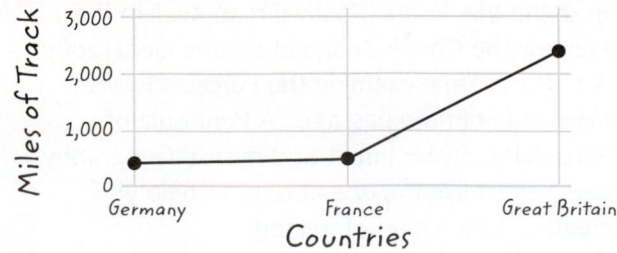

Miles of Track (vertical axis: 1,000, 2,000, 3,000)

Countries (horizontal axis: Germany, France, Great Britain)

Interpret Charts

A chart is a way to represent information or data visually. In a chart, information is organized, simplified, and summarized. To **interpret charts,** follow the steps shown at right.

Step 1 Read the title of the chart to find out what type of information the chart presents.

Step 2 Check the source of the data in the chart for reliability.

Step 3 Read any headings, subtitles, or labels to understand how the chart is organized.

Step 4 Examine the data in the chart.

Step 5 Summarize the information.

GUIDED MODEL

A **POPULATION OF MEDITERRANEAN CITIES**
(IN MILLIONS)

C City	1960	2015 (projected)
C Athens, Greece	2.2	3.1
Barcelona, Spain	1.9	2.73
Istanbul, Turkey	1.74	11.72
Marseille, France	0.8	1.36
Rome, Italy	2.33	2.65

B Source: UN, 2002

TIP When you interpret a chart, compare the data and draw conclusions from the information. For example, in the chart above, you might conclude that Istanbul is the fastest-growing city on the Mediterranean. You might also conclude that, as a result, the city may face challenges in housing and employing its large population.

Step 1 Read the title of the chart to find out what type of information it represents.

TITLE **A** Population of Mediterranean Cities (in millions)

Step 2 Check the source of the data in the chart for reliability.

SOURCE **B** The UN, or United Nations, is a known and reliable source.

Step 3 Read any headings, subtitles, or labels to understand how the chart is organized.

C This chart is organized in columns and rows. The rows feature major Mediterranean cities, and the columns give data for 1960 and 2015.

Step 4 Examine the data in the chart.

The data tells how certain cities are predicted to grow over a period of 55 years. Its purpose is to report past data and project future data.

Step 5 Summarize the information.

The data in this chart predicts that Istanbul, Turkey, will have the greatest growth and Rome, Italy, the least.

APPLY THE SKILL

Turn to *Human & Physical Geography*, Section 1.5, "Waters of the Earth." Interpret the "Longest World Rivers" chart to answer the following questions.

1. What information does the chart present?

2. How is the chart organized?

3. Where is the Amazon River located?

4. What is the length of the Mississippi-Missouri River?

5. What is the longest river in the world?

6. Of the two Asian rivers on the chart, which is longer?

Interpret Graphs

A graph is another way to represent information or data in picture form rather than written text. In a graph, data can be represented using numbers, symbols, or pictures. Common graphs include pie graphs, line graphs, and bar graphs. To **interpret graphs,** follow the steps shown at right.

Step 1 Read the title of the graph and identify what type of graph it is.

Step 2 Check the source of the data in the graph for reliability.

Step 3 Read the labels in the graph.

Step 4 Examine the data in the graph and look for patterns.

Step 5 Summarize the information in the graph.

GUIDED MODEL

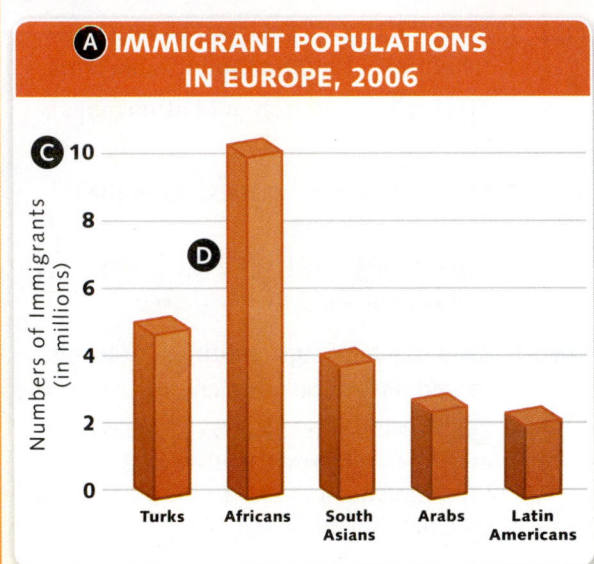

B Source: Council of Europe, 2006

> **TIP** A pie graph is used to compare parts of a whole, and each "slice" represents a percentage. Remember that the percentages represented by the slices in a pie graph always add up to 100.

Step 1 Read the title of the graph and identify what type of graph it is.

TITLE **A** Immigrant Populations in Europe, 2006; This is a bar graph, which is often used to compare quantities.

Step 2 Check the source of the data in the graph for reliability.

SOURCE **B** Council of Europe, which is a reliable source on Europe.

Step 3 Read the labels in the graph.

C The vertical axis is labeled "Numbers of Immigrants." Its scale goes to ten million. The horizontal axis is labeled with five immigrant groups.

Step 4 Examine the data and look for patterns.

D The graph compares the numbers of immigrants to Europe from various regions in a single year.

Step 5 Summarize the information in the graph.

The graph shows that Africans made up the largest group of immigrants in Europe in 2006. The smallest group came from Latin America.

APPLY THE SKILL

Turn to "Compare Across Regions: World Languages" at the end of the Europe unit. Study the Number of Languages Spoken by Continent graph and use it to answer the following questions.

1. What do the colors of the bars represent?

2. What do the words written on the bars represent?

3. On which continent are the most languages spoken?

4. On which continent are the fewest languages spoken?

Create and Interpret Time Lines

Another visual way to organize, represent, and review information is by creating a time line. When you **create a time line,** you record events and dates chronologically along an axis from left to right or from top to bottom. To create a time line, follow the steps at right.

Step 1 Decide what your time line will show and give your time line a title.

Step 2 Determine time line dates and place them in chronological order.

Step 3 Plot major events on the time line in the appropriate locations.

Step 4 Note the patterns that emerge on your time line.

GUIDED MODEL

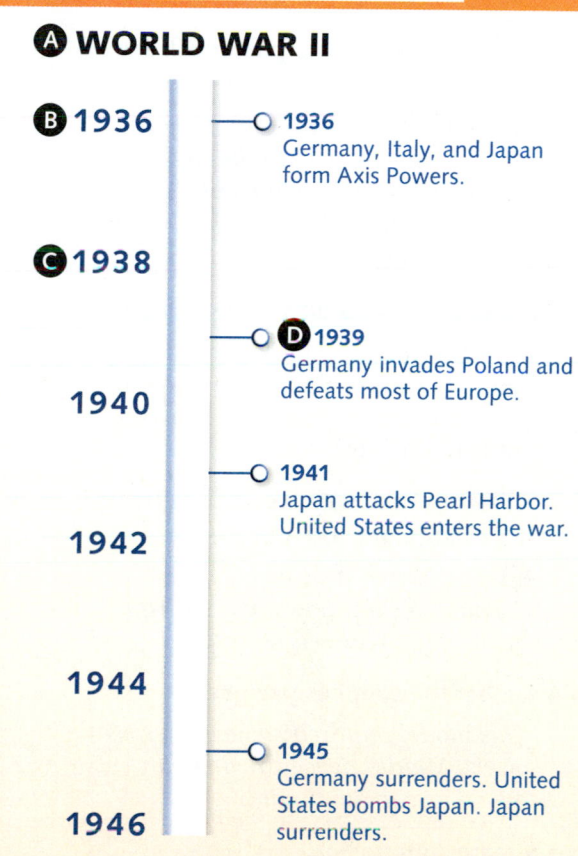

A WORLD WAR II

B 1936
○ 1936
Germany, Italy, and Japan form Axis Powers.

C 1938

○ **D** 1939
Germany invades Poland and defeats most of Europe.

1940

○ 1941
Japan attacks Pearl Harbor. United States enters the war.

1942

1944

○ 1945
Germany surrenders. United States bombs Japan. Japan surrenders.

1946

TIP When you record B.C. dates on a time line, be careful to place them in the correct order. The higher the number, the further back in time it occurred. For example, 850 B.C. took place 350 years before 500 B.C. The year A.D. 1 took place immediately after 1 B.C. No zero year exists.

Step 1 Decide what your time line will show and give your time line a title.

Time lines often feature events that are thematically related.

I will create a time line to record the major events that led up to and took place during World War II.

TITLE **A** World War II

Step 2 Determine time line dates and place them in chronological order.

Determine your start and end dates. Make sure the interval dates in between are regular.

B *The events occur from 1936 to 1945. My start date will be 1936, and my end date will be 1946.* **C** *I will put an interval date every two years.*

Step 3 Plot major events on the time line in the appropriate locations.

D *For an event that occurs on a date that isn't labeled, I will draw a line from the time line and label the date.*

Step 4 Note the patterns that emerge.

On my time line, I see that Germany was powerful in the late 1930s and that the United States played an important role in the war.

APPLY THE SKILL

Turn to *Europe Geography & History,* Section 2.2, "Classical Greece." Study the time line and use it to answer the following questions.

1. What title might you give the time line?

2. What is the time span of the time line?

3. What are the intervals between major dates on the time line?

4. When did the Greeks defeat Persia?

5. What patterns do you see on the time line?

Build Models

You can **build models** to show information about a place, an event, or a concept in a visual way. Models can be any visual representation of information, but the most common examples are posters, diagrams, mobiles, and dioramas. To create a poster, follow the steps shown at right.

Step 1 Determine what your poster will show and research the topic.

Step 2 Brainstorm ideas for your poster and sketch them.

Step 3 Think of visual ways you can represent information on your poster.

Step 4 Gather the supplies you need.

Step 5 Create your poster.

GUIDED MODEL

MANOR IN THE MIDDLE AGES (A)

The lord lived in relative safety and ease in his castle.

The church dominated everyone's life.

Serfs lived in tiny huts with dirt floors.

Guards protected the manor from rival lords.

TIP Dioramas are models that show a three-dimensional place, event, or scene, such as a battle or extreme type of weather. A diorama is built inside a small box, such as a shoe box. The inside of the box is painted with a background, and the scene is constructed using craft supplies and common household objects.

Step 1 Determine what your poster will show and research the topic.

This poster shows the structure of a manor during the Middle Ages. To create this model, the author researched the Middle Ages and the feudal system.

Step 2 Brainstorm ideas for your poster and sketch them.

A rough draft or sketch is always a good idea before you begin drawing the actual poster.

Step 3 Think of visual ways you can represent information on your poster.

(A) *The author uses pictures and callouts to show how a manor from the Middle Ages was structured.*

Step 4 Gather the supplies you need.

This model required some resources on the Middle Ages, drawing paper, a pencil or pen, and paint.

Step 5 Create your poster.

This model was drawn from a bird's eye view, which shows how the parts of the manor relate to each other.

APPLY THE SKILL

Turn to *Human & Physical Geography*, Section 2.4, "Natural Resources." Read the "Earth's Resources" and "Categories of Resources" passages. Create a poster or other type of model to represent how Earth's natural resources are categorized and show examples of each type. A sample poster is shown at right.

Earth's Natural Resources

 Coal

 Sugarcane

 Wheat

 Copper

Another type of model is a diagram, which can help observers visualize something that may not be visible in real life. Diagrams are also often used to explain how things work or to show how something is divided into parts. To build a diagram, follow these steps.

Step 1 Determine what your diagram will show and research the topic.

Step 2 Brainstorm ideas for your diagram and sketch them.

Step 3 Think of visual ways you can represent information on your diagram.

Step 4 Gather the supplies you will need.

Step 5 Create your diagram.

GUIDED MODEL

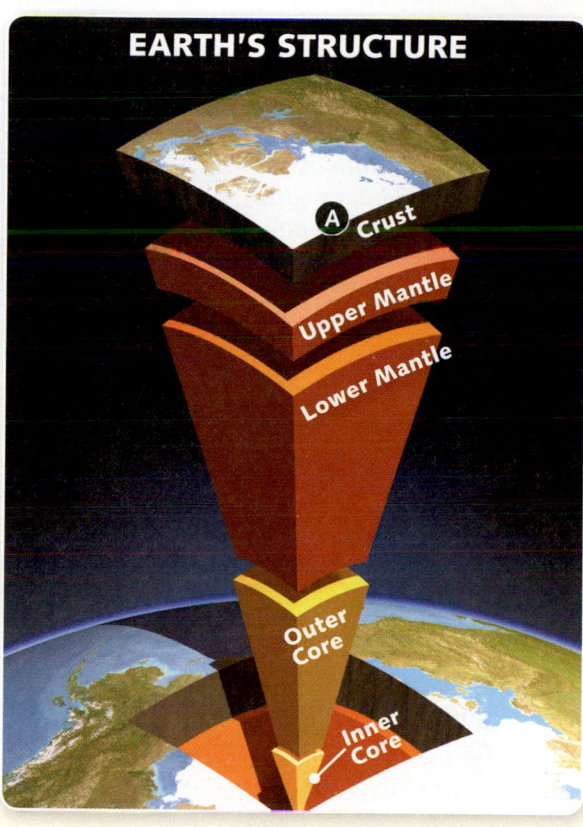

EARTH'S STRUCTURE

A Crust
Upper Mantle
Lower Mantle
Outer Core
Inner Core

TIP A mobile is a three-dimensional model. Mobiles are especially useful for showing how objects relate spatially to each other, such as how the planets circle the sun. Mobiles can be created by hanging objects from a simple wire hanger using yarn or thread.

Step 1 Determine what your diagram will show and research the topic.

This model shows the structure of Earth's layers. It is helpful because we cannot normally see them.

Step 2 Brainstorm ideas for your diagram and sketch them.

It is a good idea to determine and make a rough sketch of the approximate thicknesses of each layer ahead of time.

Step 3 Think of visual ways you can represent information on your diagram.

A *This diagram shows the layers of the earth, using a different color to represent each layer. The names of the layers are labeled.*

Step 4 Gather the supplies you will need.

This diagram required paints, markers, or colored pencils, as well as paper and a pen or pencil.

Step 5 Create your diagram.

This model clearly shows Earth's five layers, their approximate thicknesses, and their correct order.

APPLY THE SKILL

Turn to *Europe Geography & History*, Section 1.2, "A Long Coastline." Read the "Exploration and Settlement" passage. Draw a diagram like the one at right to represent the structure of a European fishing village. Use visual elements rather than text to represent information.

Create Databases

A database is a collection of information, or data, organized in a chart format so it can easily be viewed, used, and updated. You can use a computer program to **create databases** that allow you to search through the data to find only what you need. To create a database, follow these steps.

Step 1 Determine the information your database will contain and give it a title.

Step 2 Enter the column headings and/or row headings for your database.

Step 3 Enter the data in the appropriate columns and rows.

GUIDED MODEL

Ⓐ FAMOUS EUROPEAN EXPLORERS

Ⓑ Explorer's Name	Explorer's Country	Date of Exploration	Place Explored
Ⓒ Bartolomeu Dias	Portugal	1488	Southern tip of Africa
Vasco de Gama	Portugal	1498	Southern tip of Africa; also reached India and opened up trade with Asia
Christopher Columbus	Italy (but he was working for Spain)	1492	Islands in the Caribbean; continents of North and South America
Jacques Cartier	France	1530s	Northern part of North America
Sir Francis Drake	England	1577	Sailed around the world

TIP You might want to use index cards to write and organize the information you want to use in your database. Once you have gathered all of your information, you can put it in a written chart format or use database software to organize it.

Step 1 Determine the information your database will contain and give it a title.

Ⓐ *This database includes data from a passage on European exploration. It records data about which explorers went to which regions. An appropriate title is "Famous European Explorers."*

Step 2 Enter the column headings and/or row headings for your database.

Remember, columns go up and down, and rows go from left to right. Database headings most commonly appear across the top or down the left side of the database, or in both places.

Ⓑ *I have four categories of data, so I will use four columns, one per category. I will label each column with a heading that describes the data below it.*

Step 3 Enter the data in the appropriate columns and rows.

Ⓒ *I will extract information from the text I have read and insert it in the appropriate column in the database. I will use a separate row to provide information on each explorer.*

APPLY THE SKILL

Turn to *Human & Physical Geography*, Section 2.2, "World Climate Regions." Read about the five climate regions featured in this section. Then create a database like the one at right to record the names of each climate region, a description of its weather and plant life, and a list of places that have that climate.

WORLD CLIMATE REGIONS

CLIMATE REGION	WEATHER	PLANT LIFE	LOCATIONS

Create Graphic Organizers

While you read, you can **create graphic organizers** such as charts, diagrams, and time lines to take notes and explore how the information is related. To create a graphic organizer, follow the steps shown at right.

Step 1 Determine your needs.

Step 2 Determine what type of graphic organizer you need.

Step 3 Draw your graphic organizer.

Step 4 Use information from the passage, plus any relevant maps or other features, to fill in your graphic organizer.

GUIDED MODEL

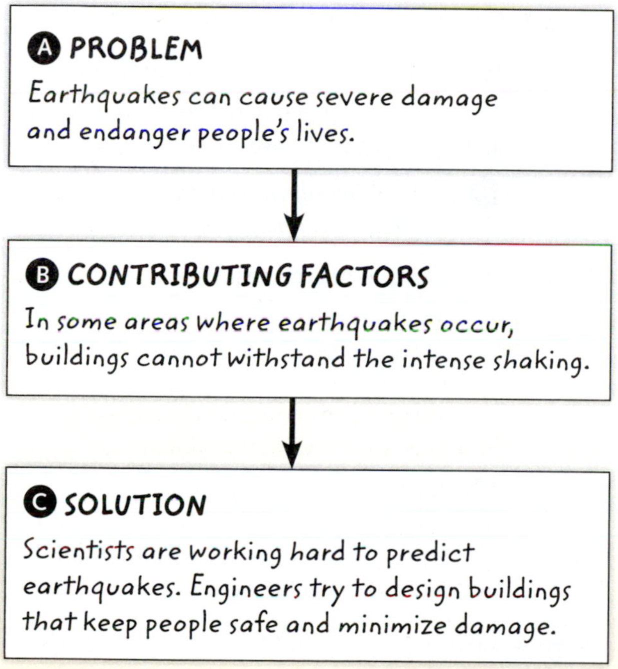

A PROBLEM

Earthquakes can cause severe damage and endanger people's lives.

B CONTRIBUTING FACTORS

In some areas where earthquakes occur, buildings cannot withstand the intense shaking.

C SOLUTION

Scientists are working hard to predict earthquakes. Engineers try to design buildings that keep people safe and minimize damage.

TIP You can use a cause-and-effect chain to help you understand the results of a particular event. Note that the chain can contain a single cause and multiple effects, or it can have many causes and only one effect.

Step 1 Determine your needs.

First, think about the type of information you are recording. Then determine what you will be doing with the information. Are you gathering facts, comparing two things, or recording a series of events?

Step 2 Determine what type of graphic organizer you need.

A chart can be used to record facts. A Venn diagram helps compare and contrast two things. A time line records a series of events over time. For this example, a problem-solution chart was created to identify a problem and understand how people are trying to solve it.

Step 2 Draw your graphic organizer.

Step 4 Use information from the passage, plus any relevant maps or other features to fill in your graphic organizer.

This student identified:

A the problem, in this case, the dangers posed by earthquakes;

B contributing factors, the situations that add to the problem;

C a possible solution to the problem.

APPLY THE SKILL

Turn to *Europe Geography & History*, Section 3.2, "The Industrial Revolution." Read "The Revolution Begins" passage. Then create a main idea and details graphic organizer like the one at right and use it to record important information from the text.

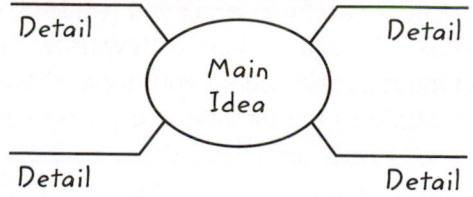

Conduct Internet Research

When you **conduct Internet research**, you search and access source material from all over the world via the Internet. As you probably know, the World Wide Web (or "www") is part of the Internet that allows you access to online information and data. To conduct Internet research, follow the steps shown at right.

Step 1 Choose and access a search engine.

Step 2 Type key words or phrases into the search field.

Step 3 Examine the search results.

Step 4 Visit the suggested Web sites and perform more specific searches for information.

Step 5 Use other reliable sources to verify any information you gather.

GUIDED MODEL

(A)

| elephant habitat | Search |

(B)

ELEPHANTS- Habitat & Distribution
ELEPHANTS- Habitat & Distribution...Discover animal, environmental, and zoological career facts as you explore in-depth topic coverage via SeaWorld, ...
www.seaworld.org/.../elephants- habitat & distribution.htm -

Elephant Facts - Defenders of Wildlife
Get the facts on **elephant**. **Elephants** are the largest land-dwelling mammals on earth. ... of **elephants' habitat** will become significantly hotter and drier, ...
www.defenders.org › Wildlife and Habitat -

African Elephants, African Elephant Pic ...
African **elephants** are the largest of Earth's **land** mammals. Their enormous ears help them to keep cool in the hot African climate. ...
animals.nationalgeographic.com/animals/... /african-**elephant**/ -

Elephants Habitat | Animal Habitats
A quick look into the **elephant's habitat**. In studying an **elephant's habitat**, the first thing one encounters is the fact that there are several species of ...
www.animalhabitats.org/elephants_habitat /elephants_habitat.htm -

TIP When you use the Internet to do research for a report or project, print the source material you find online. This will provide you with a record of the Web sites you have used and allow you to review the source materials as you write your report without having to go back online.

Step 1 Choose and access a search engine.

Step 2 Type key words into the search field.

Be as specific as possible to narrow your search and yield better results.

(A) *I typed "elephant habitat" into the search field and clicked "search."*

Step 3 Examine the search results.

Remember that Web addresses that end with ".edu," ".gov," and ".org" are often more reliable than ".com" sites. Review the addresses and summaries of each Web site and decide which ones to use.

(B) *This Web site's domain is ".org", which is usually a reliable source, and the information is relevant to my topic, so I will go to this Web site.*

Step 4 Visit the suggested Web sites and perform more specific searches for information.

Once I'm on the site, I will search "elephant habitat preservation" to find more specific information.

Step 5 Use other reliable sources to verify any information you gather.

I found the same information in an online encyclopedia and on the National Geographic Web site. This tells me the information is reliable.

APPLY THE SKILL

Think of a topic from this book you would like to research. Brainstorm key words to use on a search engine. Then, with your teacher's permission, use a Web browser to conduct Internet research on your topic. If possible, print copies of the information you find. Write a brief paragraph explaining your search process and results.

| Ancient Greece | Search |

Evaluate Internet Sources

It is important to use only the most reliable and credible information as a resource. To do so, you must **evaluate Internet sources** to be sure they are accurate and reputable. To evaluate Internet sources, follow the steps shown at right.

Step 1 Examine the Web site's Internet address.

Step 2 Identify the Web site's author.

Step 3 Identify when the Web site was created or last updated.

Step 4 Verify the information from the Web site using other reliable sources.

Step 5 Evaluate the Internet source.

GUIDED MODEL

TIP Another way to evaluate an Internet source is to determine its intended audience. Ask yourself to whom the author is writing, and study the writing style, vocabulary, and tone. Determine if the author's objective appears to be to inform, to explain, or to persuade, and whether he or she provides sufficient evidence.

Step 1 Examine the Web site's Internet address.

A *This Web site's address is www.cia.gov. A government agency created the site, which makes it reliable.*

Step 2 Identify the Web site's author.

Sites that clearly name the author are more reliable than anonymous Web sites.

B *The author of this Web site is the Central Intelligence Agency, or CIA. It is expert at collecting data.*

Step 3 Identify when the Web site was created or last updated.

C *This Web site is updated every week, so the information is current.*

Step 4 Verify the information from the Web site using other reliable sources.

I found the information in an encyclopedia and on the National Geographic Web site.

Step 5 Evaluate the Internet source.

I believe this is a reliable source because it is created by the government, updated frequently, and has the same information as other reliable sources.

APPLY THE SKILL

With your teacher's permission, visit the following Web address: www.nationalgeographic.com. Use the steps above to evaluate this Internet site as a reliable source of information. Write a brief paragraph explaining your evaluation of the site.

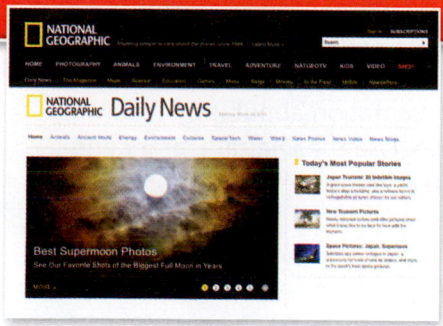

Create Multimedia Presentations

You can **create multimedia presentations** on any topic using media such as photographs, video clips, and audio recordings. To create a multimedia presentation, follow the steps shown at right.

Step 1 Determine the topic you will be presenting and the types of media you will use.

Step 2 Research your topic.

Step 3 Put your presentation together.

GUIDED MODEL

Unlike many species of crab, the one shown here can swim as well as walk.

TIP The **Magazine Maker** allows you to search for images and information on a wide variety of topics and create your own magazine by importing pictures and text. Use this program as you create your multimedia presentation.

Step 1 Determine the topic you will be presenting and the types of media you will use.

Certain types of media enhance topics more effectively than others. Once you have determined your topic, decide which media work best with it. Would your topic be enhanced by images, or would audio clips or music be more relevant? Would a video clip add meaning to your presentation? What about a map or graph? You decide.

Ⓐ This student used a photo from the **Digital Library** and wrote her own caption for it.

Step 2 Research your topic.

Use reliable library and online sources to research your topic. Write a brief script that provides information on your topic and ties your media together.

Step 3 Put your presentation together.

Combine your informational written script with the media you have selected. Make sure everything flows together well. You might use presentation software to display your information. Rehearse your presentation a number of times to identify and correct any problems.

APPLY THE SKILL

Select a topic from this text and create a multimedia presentation about it by following the steps above. Use the **Digital Library** and **Magazine Maker** to create your presentation.

HANDBOOK
ECONOMICS & GOVERNMENT

PART I ECONOMICS

agriculture *n., the development of plants and animals to provide food.* Agricultural products include crops such as wheat, corn, and barley. Agriculture also includes animals that have been domesticated, or tamed. These animals are often called livestock, and they include cattle, sheep, pigs, and horses.

business cycle *n., a period during which a country's economic activities increase and then decrease in a relatively predictable pattern.* A business cycle has four phases. During expansion, businesses do well. At the peak of the cycle, economic activity begins to slow. During a contraction, economic activity continues to decrease. A contraction is also called a recession. The trough is the lowest level of economic activity. Then business starts to improve, and a new business cycle begins.

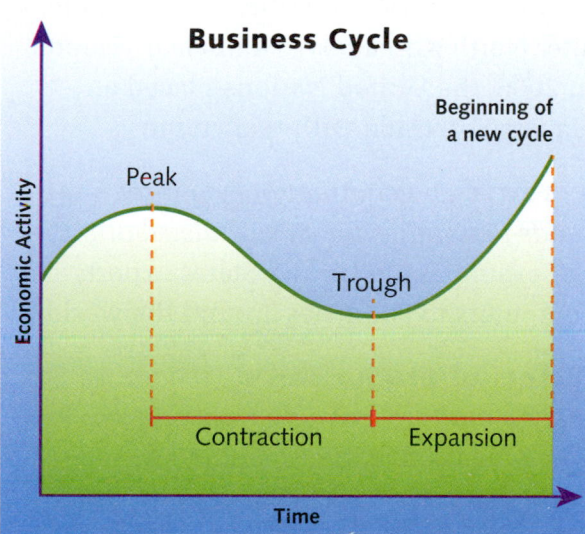

Business Cycle

capitalism *n., an economic system in which private individuals or groups own the resources and produce goods for a profit.* In capitalism, private individuals or groups decide to produce goods or offer services. They offer those goods and services for sale in markets, which are places where people buy and sell. Businesses raise capital, or money, to create new products and hire workers. Capitalism is also called free enterprise because people are free to start businesses.

Bookstore, Los Angeles, California

command economy *n., an economic system in which the government controls a country's economic activities.* The government owns and controls the factories, farms, and stores in a country.

communism *n., an economic and political system in which the government owns and controls economic activities.* Communism is a type of command economy. In a Communist system, the government owns and operates factories, farms, and other types of economic activities. For example, in the steel industry, government officials decide what kind of steel and how much steel will be produced. North Korea, Vietnam, and Cuba have Communist systems. Communist economies have been much less efficient than free enterprise economies in producing goods and services. See **Part II Government** for the definition of communism as a political system.

corporation *n., a company in which people own shares, or parts of the company.* A corporation sells stock, or shares of the company, to raise money to create products and services. Shareholders often receive a dividend, which is a share of the profits. Multinational corporations are corporations that operate in several countries. People purchase shares of a corporation on a stock exchange.

depression *n., a deep and long-lasting contraction of economic activities.* During a depression, business activity falls dramatically. Businesses hire few workers. The unemployment rate, or percent of people without work, rises. In the United States, the Great Depression lasted from 1929 until the early 1940s. A recession is also a contraction in business. It is less severe and shorter than a depression.

A church in New York City distributes food in a breadline during the Great Depression.

developed nation *n., a country with highly-developed industries, a high standard of living, and private ownership of most businesses.* Developed nations have industries that make products like automobiles and computers. Less developed countries have fewer industries and rely on agriculture. They also have lower standards of living. The United States is a developed nation. Many countries in Africa, Asia, and Latin America are less developed countries.

economy *n., a country's system for producing and exchanging goods and services.* A country's economic development is based on its level of economic activity. If a country has a great deal of industry, it has a high level of economic development.

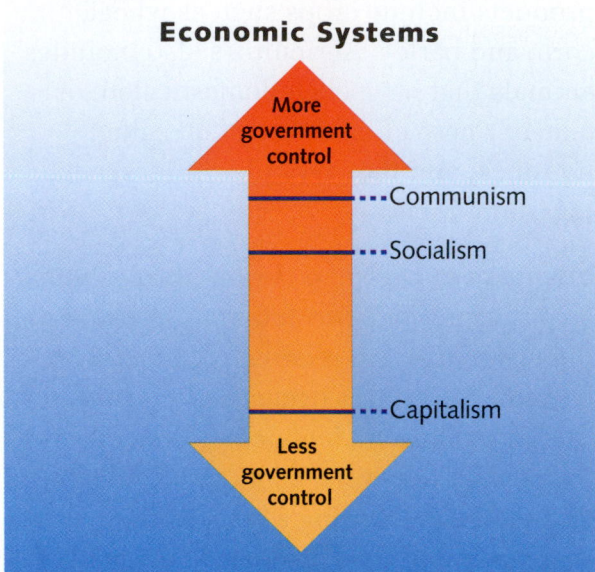

Economic Systems

More government control

- - - Communism

- - - Socialism

- - - Capitalism

Less government control

embargo *n., a government ban on trade with another country.* Countries set embargoes to show their disapproval of another country's activities. For example, after North Korea tested a nuclear weapon in 2006, the United Nations placed an embargo on trade with that country.

export *n., a good that one country sends to another for sale or distribution.* For example, the United States exports computers to countries around the world.

factors of production *n., the things that go into producing a good or a service.* Economists have identified four factors of production: land, labor, capital, and entrepreneurs. Land includes all natural resources, such as oil and silver. Labor is the work that people do. Capital is the machinery and other tools that are used to create a good or service. Entrepreneurs are people who start businesses.

free enterprise *n., an economic system in which businesses are privately owned, people buy and sell goods in free markets, and individuals freely make decisions whether to buy or sell.* The system is also called capitalism. The United States has a free enterprise system.

In a free enterprise system, producers are motivated by self-interest. To meet consumers' demand, producers decide to manufacture a good, such as an automobile, or offer a service. By doing so, they hope to make a profit. Consumers are also motivated by self-interest. They try to buy the best goods and services they can at the lowest price offered in a market.

Government makes sure that businesses compete fairly. It also ensures that food, medicines, and other products are safe. It provides services, such as defense, that are important to a country. The government also builds infrastructure, such as transportation facilities and roads.

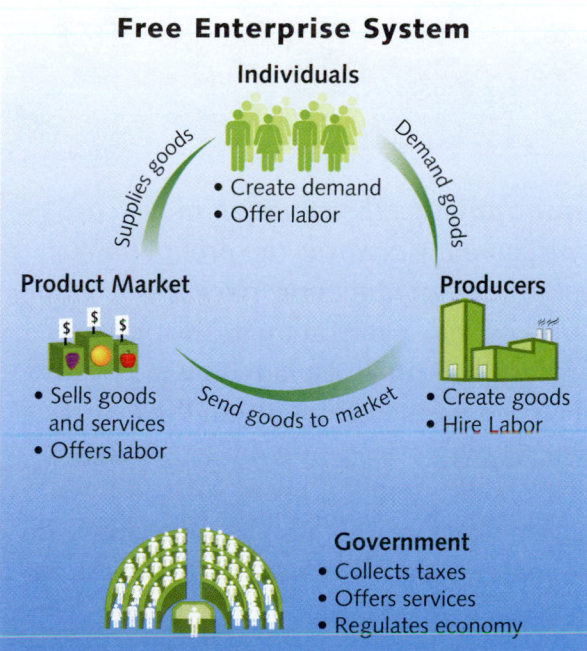

Free Enterprise System

gross domestic product (GDP) *n., the total value of all the goods and services that a country produces in a specific time period, such as a year.* The GDP is an important measure of an economy's

strength. Economists measure the GDP by adding together four kinds of goods and services. One is the goods and services that consumers buy. Another is the machines and other items that companies buy for their businesses. A third is goods and services that government buys. Fourth is the goods and services that a country exports to other countries.

GDP per capita is a country's GDP divided by the country's population. It shows how much the country produces per person.

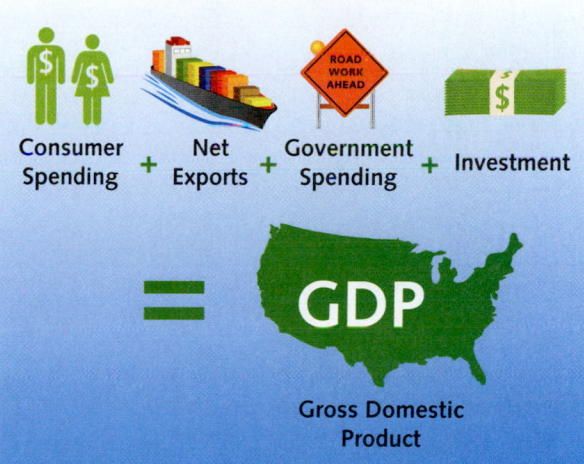

Gross Domestic Product

import *n., a good that one country receives from another for sale or distribution.* For example, the United States imports many cars and trucks from automobile manufacturers in Japan. Japan in turn, imports oil from Saudi Arabia.

industry *n., a group of businesses that produce a similar product or service.* For example, the film industry produces feature films. Common industries in the United States include construction, computers, pharmaceuticals, and electronics. In many industries, businesses take a raw material and turn it into a finished product. For example, in the clothing industry, businesses turn cotton, wool, and other materials into clothes.

inflation *n., an increase in the price of goods and services in a country.* In any given year, the average price of goods and services may go up. This is called an increase in the price level. The rate of increase is called the inflation rate. In 1980, for example, the United States had an inflation rate of about 14 percent. This means that the average prices that year were about 14 percent higher than in 1979.

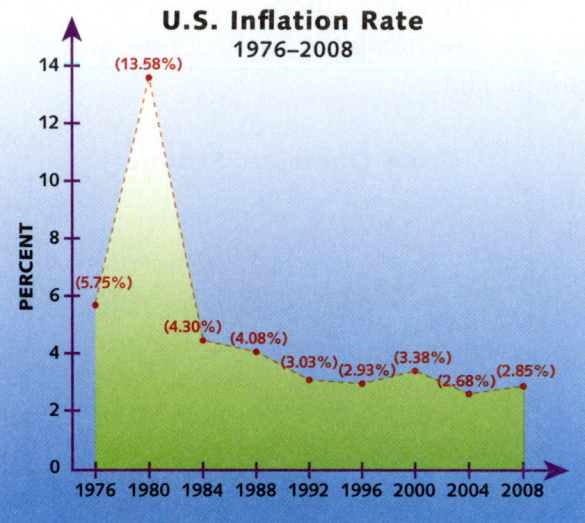

U.S. Inflation Rate
1976–2008

Source: U.S. Bureau of Labor Statistics

manufacturing *n., the production of physical products to be sold.* Manufacturing creates a wide variety of products, such as automobiles, steamships, airplanes, computers, and furniture. The term refers to the creation of items by machine and by hand. Light manufacturing means the creation of relatively small things, such as the circuits in computers. Heavy manufacturing means the creation of large objects, such as diesel engines for railroads. In 2010, manufacturing employed about 17 million workers in the United States and Canada.

market economy *n., an economic system in which people and businesses choose freely to buy and sell in markets.* A market is a place where people buy and sell goods and services. In a market economy, individuals make choices to buy and sell.

For example, an individual might decide to earn a living by making and selling T-shirts. After making them, the person would try to interest stores in selling the T-shirts. A store is a common type of market. The government establishes rules, such as the rule that stores should be safe and clean.

mechanization *n., the use of machines instead of humans or animals to perform tasks.* For example, in 1764, the Englishman James Hargreaves invented the spinning jenny. It was a machine that could spin yarn from wool. The machine allowed operators to create yarn much faster than they could by hand.

Spinning jenny

monopoly *n., the situation in which one company controls the production or selling of a product or service.* For example, suppose that a person owns the only grocery store in an isolated town. Because the store has no competition, the owner can set high prices on food. If someone else opens another grocery store in town, competition results. Because the stores are competing, they will have to lower prices and improve quality.

national debt *n., the total amount of money that the federal government owes.* If a government spends more money than it receives in taxes, it uses deficit spending. It must borrow money from individuals, companies, or other governments. As a government uses deficit spending, its national debt grows.

The national debt of the United States has grown rapidly since 1980. Critics say that the government should raise taxes, cut spending, or do a combination of the two to balance its budget.

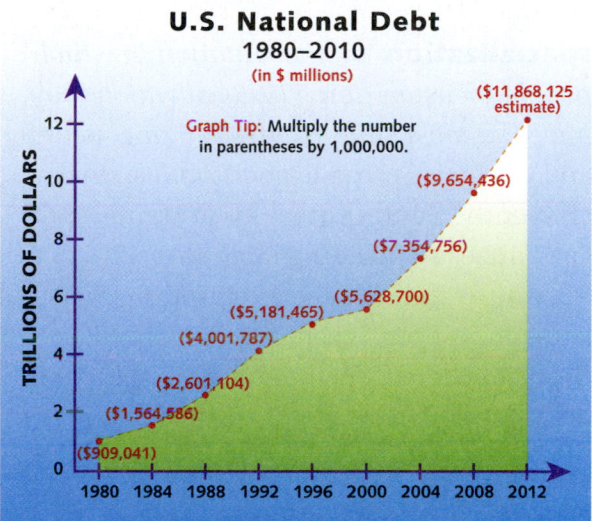

U.S. National Debt
1980–2010
(in $ millions)

Graph Tip: Multiply the number in parentheses by 1,000,000.

($11,868,125 estimate)
($9,654,436)
($7,354,756)
($5,628,700)
($5,181,465)
($4,001,787)
($2,601,104)
($1,564,586)
($909,041)

TRILLIONS OF DOLLARS

12
10
8
6
4
2
0

1980 1984 1988 1992 1996 2000 2004 2008 2012

Source: U.S. Office of Management and Budget

natural resources *n., resources such as oil, coal, timber, and water that exist naturally in a place.* Having ample natural resources can help a country to develop its economy. For example, the United States has numerous natural resources, including timber and fresh water. On the other hand, Japan has become wealthy even though it has few energy resources like oil. It has developed its wealth by inventing new technologies and building quality products.

Oil pump

opportunity cost *n., the opportunity that a person gives up when he or she chooses to buy one item instead of another.* For example, suppose that Benita must decide between buying a new flat-screen television or taking a vacation. If she chooses the television, her opportunity cost is the vacation she did not take.

poverty *n., the lack of enough money to buy necessary things like food, clothing, and shelter.* Poverty is a worldwide problem. It causes disease and hunger. Poor people often do not have adequate shelter and suffer greatly during cold or hot weather. Economists say that more than one billion people in the world are poor. The U.S. government measures the percentage of its population that is poor. The graph below shows how the rate has changed over the past several years.

U.S. Poverty Rate
1960–2009

Percent of Population

25
20
15
10
5
0

22.5%
15%
12.5%
13%
14%
13%
14%
12%
13%
14%

1960 1965 1970 1975 1980 1985 1990 1995 2000 2005 2009

Source: U.S. Census Bureaus

raw materials *n., the materials that are used in manufacturing a final product.* Raw materials often come from natural resources, such as wood, oil, and iron ore. For example, the basic raw material in creating a table is wood. Cotton and wool are raw materials used in clothing.

retail goods *n., goods that are sold directly to consumers.* When you go to a store and purchase a DVD, you are buying a retail good. Today, merchants sell retail goods in a variety of ways. They sell them in stores, through vending machines, over the telephone, or on the Internet.

service industries *n., companies, nonprofit organizations, and government organizations that provide services rather than products.* Services include many activities, such as health care, education, financial operations, retail selling, and legal advice. In the United States, more than 75 percent of people work in service industries. Less developed countries, on the other hand, have fewer people working in service industries.

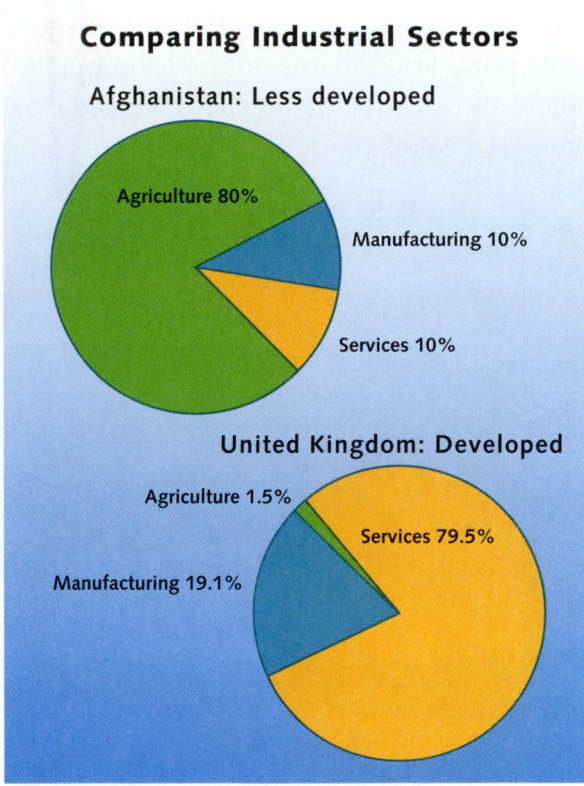

Comparing Industrial Sectors

Afghanistan: Less developed

Agriculture 80%
Manufacturing 10%
Services 10%

United Kingdom: Developed

Agriculture 1.5%
Services 79.5%
Manufacturing 19.1%

Source: www.NationMaster.com

socialism *n., an economic system in which the government owns and operates most businesses.* Socialism first developed in Europe in the early 1800s. Socialists wanted to eliminate poverty and improve working conditions. During the 1840s, a German socialist named Karl Marx said that the government should own all businesses. His ideas gave rise to communism.

Today, Sweden and some other European countries are often called democratic socialist countries. This means the government owns fewer businesses than in Communist countries, but it does provide many services, such as health care. In addition, taxes are high. Many economists claim that socialism is less efficient than free enterprise.

specialization *n., the situation in which people focus on doing tasks at which they have the most skill.* By specializing, people and countries provide goods and services efficiently. South Korea, for example, specializes in building large ships. It exchanges the revenue from selling the ships for oil produced in Saudi Arabia. Both countries benefit by doing what they do best.

standard of living *n., the level of economic well-being that people in a country have.* In a country with a high standard of living, people have the income to purchase such items as automobiles. In a country with a low standard of living, most of the population struggles to have enough food, clothing, and shelter.

stock market or stock exchange *n., a market where people buy and sell stocks and bonds.* Companies are able to raise money in one of two ways. They can sell stocks, which are shares of ownership in the company, or they can issue bonds. A bond is a written agreement to borrow money, and the company pays the money back with interest.

People may want to sell stocks or bonds to raise money or make a profit. The sale of stocks and bonds takes place on a stock exchange or a stock market. Brokers handle both the buying and selling of stocks and bonds. The largest stock market in the United States is the New York Stock Exchange on Wall Street in Manhattan.

strike *n., a situation in which workers stop working in order to win higher wages or benefits.* Strikes are most often called by unions. A union is an organization that represents workers in a certain industry. If a union's workers believe that wages are not high enough, they may go on strike. The strike brings the company's operations to a halt. A wildcat strike occurs when workers go on strike without the union's support.

Auto workers on strike, Naperville, Illinois, 2007

supply and demand *n., the economic forces that decide the price and amount of a product or service.* The supply of a good or service is the amount that a business is willing and able to offer for sale. If the company can get a higher price, it will create more of the product or service. The demand is the amount of a good or service that consumers are willing and able to buy at a given price. Usually, the demand for a good or service goes down as the price rises. The price at which supply equals demand is called the equilibrium price.

The graph shows how supply and demand work. The gray line stands for demand. The green line stands for supply. The lines meet at equilibrium, when supply equals demand. At prices above equilibrium, demand for the product goes down. At prices below equilibrium, consumer demand rises.

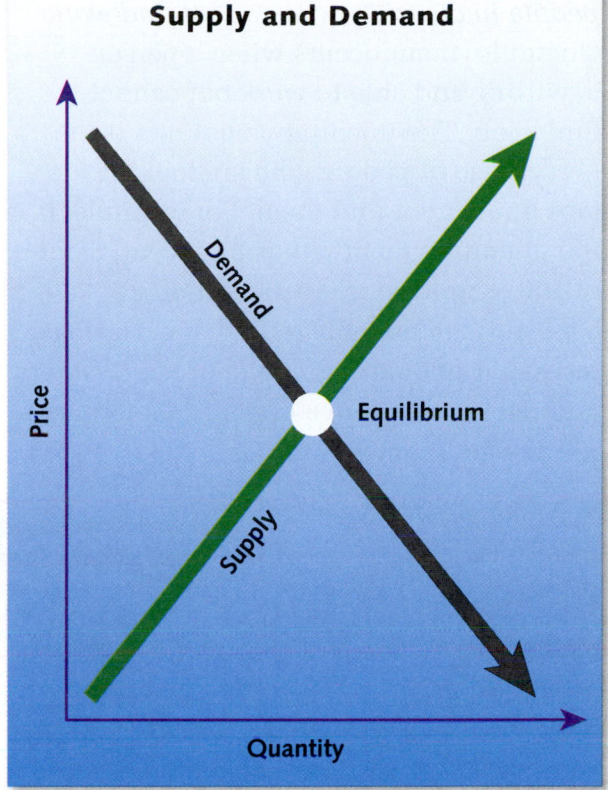
Supply and Demand

Price

Demand

Supply

Equilibrium

Quantity

tariff *n., a tax that a country places on goods imported from another country.* Countries use tariffs to protect their own companies from competition with other countries. For example, to protect auto manufacturers, a country might place tariffs on cars from other countries. The country's own manufacturers would thrive. However, consumers might pay more for their cars because there is less price competition.

trade *n., the exchange of services and goods.* Trade occurs when people cannot make things themselves but can get them from other people. For example, Chile grows asparagus. It exports the vegetable to countries around the world. With the money it earns, it imports automobiles from the United States, oil from Saudi Arabia, and so forth. This is an example of international trade. The trade within a country is called domestic trade.

unemployment rate *n., the percentage of people in a society who cannot find work.* Unemployment occurs when a person is willing and able to work but cannot find a job. The unemployment rate is the percentage of people who are looking for jobs but cannot find them. For example, if the unemployment rate is 8 percent, 8 out of 100 people could work, but do not have a job and are not able to find one. During a recession or depression, the unemployment rate rises. Unemployed people sometimes receive government aid.

wholesale goods *n., goods that producers sell to other business firms, such as stores.* Wholesalers buy goods in large quantities. They then sell those goods to retailers, who sell them to customers. For example, a factory in China might produce thousands of toys for children. A wholesaler buys the toys from the factory and ships the toys to retail stores around the world. Then the retailer sells them to you, the consumer. Without wholesalers, the modern economy would not work efficiently.

PART II GOVERNMENT

citizenship *n., membership in a state or a nation, with full rights and responsibilities.* A citizen of a country owes loyalty to that country. He or she is expected to perform certain duties, such as obeying the laws and paying taxes. In return, a citizen has certain rights, such as the right to vote. Civic participation is the way in which citizens participate in their government and society. Citizens let their representatives know what they think about important issues and participate in other ways.

Responsibilities of U.S. Citizens

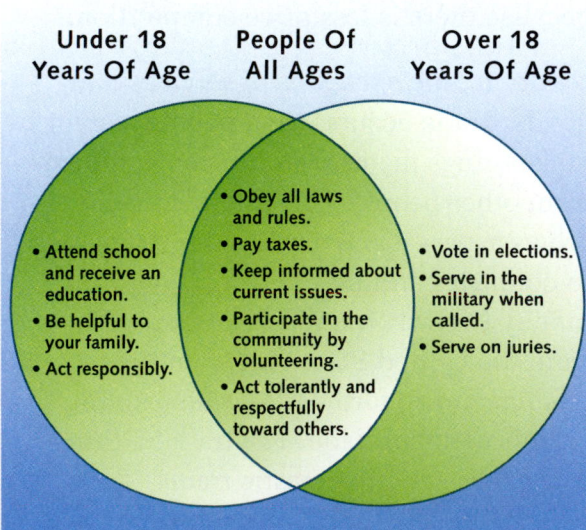

Under 18 Years Of Age	People Of All Ages	Over 18 Years Of Age
• Attend school and receive an education. • Be helpful to your family. • Act responsibly.	• Obey all laws and rules. • Pay taxes. • Keep informed about current issues. • Participate in the community by volunteering. • Act tolerantly and respectfully toward others.	• Vote in elections. • Serve in the military when called. • Serve on juries.

communism *n., an economic and political system in which the government owns and controls economic activities.* A communist government exerts great power over its people. This type of government is called totalitarian because it totally controls its people's lives. In the Soviet Union (1917–1991), the Communist Party outlawed all other parties, and the population had no freedom of speech, religion, or other rights.

constitution *n., a statement that explains the basic principles and rules of an organization.* The constitution of a government explains how leaders are selected, how laws are passed, and how laws are interpreted and enforced. The U.S. Constitution was ratified, or agreed to, by the states in 1789. In 1791, the states approved of the first 10 amendments, which are known as the Bill of Rights. These amendments guarantee rights, such as the right to speak freely. The U.S. Constitution can be changed, but doing so requires significant majorities to propose amendments, and an even greater majority of states to ratifiy the change.

democracy *n., a form of government in which the citizens of a nation or state hold the power to pass laws and select leaders.* The United States is a representative democracy in which people elect fellow citizens to serve in a legislature and pass laws. They also elect their leaders, such as the President of the United States. Representative democracies are also known as democratic republics or federal republics (see definition on this page). Democracy had its origins in Greek city-states.

dictatorship *n., a form of government in which a ruler holds total power.* A dictator usually comes to power through force. One of the most brutal dictators was Adolf Hitler of Nazi Germany, who ruled from 1933 to 1945. He created the policy that led to the Holocaust, or the murder of 6 million Jews and other people during World War II.

Adolf Hitler

executive branch *n., the branch of the federal government that implements and enforces laws.* In the United States, the president is the leader of the executive branch. He or she is the commander-in-chief of the armed forces. The president signs bills from Congress into law. The executive branch then makes sure that the laws are applied. To help enforce the laws, the president has a cabinet, and each member of the cabinet is responsible for a specific area of the federal government.

Executive Branch

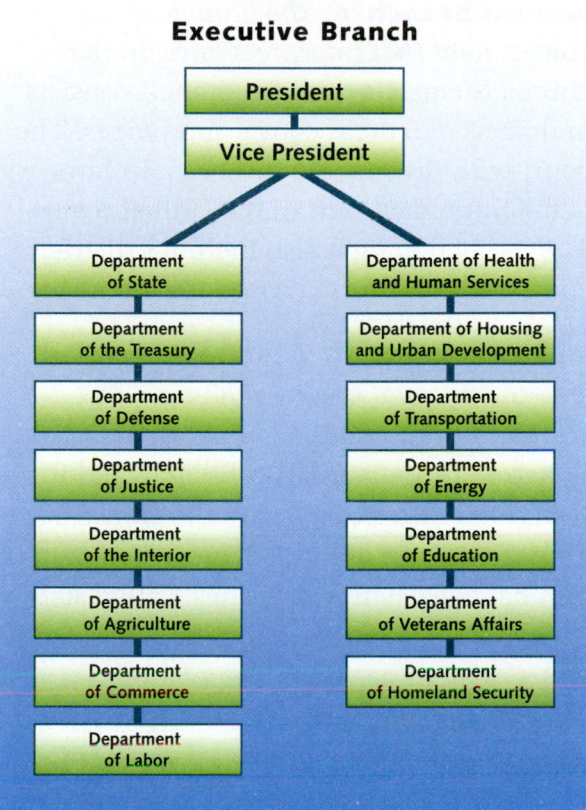

President

Vice President

Department of State	Department of Health and Human Services
Department of the Treasury	Department of Housing and Urban Development
Department of Defense	Department of Transportation
Department of Justice	Department of Energy
Department of the Interior	Department of Education
Department of Agriculture	Department of Veterans Affairs
Department of Commerce	Department of Homeland Security
Department of Labor	

fascism *n., a form of government that has a dictator and controls political, economic, and other activities.* The two most well-known examples of fascism in the 20th century were Italy under Benito Mussolini (1922–1943) and Germany under Adolf Hitler (1933–1945). Mussolini came to power in Italy in 1922. He promised to solve the economic crisis that Italy faced after World War I. He soon became the country's dictator. Fascism in Italy ended in 1943 when the country surrendered during World War II.

federal republic *n., a form of government in which the population elects representatives to pass and carry out laws.* Many Western countries, including the United States and Canada, are federal republics. In federalism, a national, or federal, government shares power with states or provinces. The federal government has certain powers, such as national defense. The states or provinces control other activities, such as education.

judicial branch *n., the branch of government that interprets laws.* In the United States, the judicial branch consists of dozens of federal courts and judges. The court with the highest authority on laws is the Supreme Court of the United States. The judicial branch also includes district courts, which oversee cases at local levels.

legislative branch *n., the branch of government that creates, changes, or eliminates laws.* In the United States, Congress is the legislative branch, and it is divided into two Houses: the House of Representatives and the Senate. The states also have legislative branches, called state legislatures. In addition to passing laws, the U.S. Congress has other responsibilities. The Senate, for example, must approve of treaties with other countries. Democracies around the world have legislative branches. In the United Kingdom, for example, the Parliament passes laws.

Parliament, United Kingdom

limited government; unlimited government *n., a limited government places limits on governmental powers. An unlimited government does not place limits on governmental powers.* The United States has a limited government. The Constitution places limits on its powers. For example, the First Amendment states that the government does not have the power to limit people's free speech.

In unlimited government, the power of the government is not limited by a constitution or other document. Absolute monarchies and dictatorships are unlimited governments. From 1917 until 1991, the Soviet Union had an unlimited government.

monarchy *n., a form of government in which a king or a queen rules a country.* A monarch inherits the throne, or position, from his or her parent. Until the 1700s, monarchy was a common form of government. Britain, France, and Russia all had monarchs. In Russia, kings were known as czars. Then, in 1789, the French Revolution limited the powers of the French king, King Louis XVI. In 1793, the revolutionary government executed Louis and formed a republic. Through the 1800s and early 1900s, governments around the world became more democratic. They either abolished the monarchies completely or limited the powers of kings and queens.

oligarchy *n., a kind of government in which a small group of people holds power.* In an oligarchy, the rulers usually are either the military or the wealthy classes. Oligarchies usually rule for their own interests, not for the interests of people in lower classes. Myanmar (also known as Burma) is an oligarchy. It is ruled by a small group of military officers.

totalitarian *adj., a form of government in which a dictator or small group holds total control over the lives of the people in a country.* In a totalitarian country, the party in power is usually intolerant of other viewpoints and ideas. For example, in the 1930s, the Nazi Party rose to power in Germany by promising to bring back prosperity and return Germany to glory. The Nazi Party shut down all other parties, stating they were obstacles to these goals.

From the beginning of human history, people have asked questions about what life means. What will happen after I die? What is the right way to act? Religion helps people find answers to questions like those.

World Religions OVERVIEW

A religion is an organized system of beliefs and practices. Thousands of religions exist in the world. Most teach that one or more gods, or supreme powers, exist. To help people relate to the divine power, most religions teach a set of beliefs and a moral or ethical code. These codes of conduct also teach people how to treat each other. Another way that religion affects human relationships is in the building of community. Groups of people who share beliefs often gather together to worship and celebrate important events. For a map showing the distribution of the world's religions, see Chapter 2.

WORLD POPULATION'S RELIGIOUS AFFILIATIONS

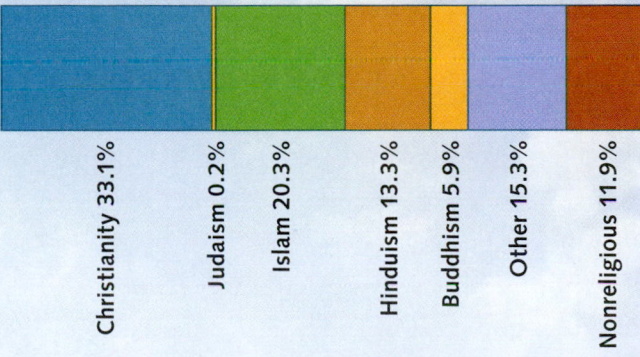

Christianity 33.1% Judaism 0.2% Islam 20.3% Hinduism 13.3% Buddhism 5.9% Other 15.3% Nonreligious 11.9%

Source: Encyclopedia Brittanica

✝CHRISTIANITY

Historical Origins

Christianity is based on the life and teachings of Jesus of Nazareth, also called Jesus Christ by Christians. Christians believe he was the son of God who died to save humanity from sin. He was a Jew who lived in the first century near Jerusalem, which was then part of the Roman Empire. Roman rulers put him to death fearing he might lead a revolt. The life of Jesus is recorded in the New Testament of the Christian Bible, which also contains stories of his followers and letters outlining Christian beliefs.

Detail of Jesus in Leonardo DaVinci's *The Last Supper*

Central Beliefs

Most Christians believe there is only one God, who exists in three forms: Father, Son, and Holy Spirit. The New Testament teaches that after Jesus was executed, he rose from the dead and ascended to heaven. Christians believe they gain salvation by believing in Jesus and following his teachings. Christians gather at places of worship called churches. Their religious leaders are called either priests or ministers.

Spread of Christianity

Jesus' followers, called disciples, carried their faith around the Mediterranean world. In the A.D. 300s, Christianity became the official religion of the Roman Empire. Later, during the period of colonization, Europeans spread Christianity around the globe. It is now the largest and most widespread religion.

Christians wave palm fronds in Managua, Nicaragua, as they celebrate Palm Sunday, the first day of Holy Week.

☪ ISLAM

Historical Origins

Islam teaches that in the year 610, an Arab trader named Muhammad was visited by the angel Gabriel. The angel told Muhammad that he was God's messenger. Muslims believe that through a series of these visits, Muhammad received the words of the Qur'an, or sacred book. According to Islam, Muhammad was the last prophet that Allah, the Muslim name for God, sent to humanity. Muslims believe he is a direct descendant of Abraham, who is also the founder of Judaism.

A 17th-century Turkish ceramic tile from a mosque

Central Beliefs

Muslims, or followers of Islam, believe there is only one God, the same God worshipped by Jews and Christians. The word *Islam* means surrender, and the goal of Islam is to surrender to the will of Allah. Muslims do this by practicing the Five Pillars of Islam. These are professing faith, praying five times a day, giving to charity, fasting, and making a journey to Mecca to reenact Abraham's dedication. Muslims worship in mosques.

Spread of Islam

In the centuries after Muhammad's death, Muslims spread their religion by conquest. Islamic rulers took control of Southwest Asia, Central Asia, North Africa, and parts of India and Spain. Today Islam continues to spread around the world through migration and conversion. It is the world's second largest religion.

Muslims engage in prayer at a mosque in Delhi, India, during Ramadan, a holy month of fasting.

✡ JUDAISM

Historical Origins

Judaism, the religion of the Jewish people, dates back more than 4,000 years. Its founder was Abraham, who lived in Mesopotamia. According to the Hebrew Bible, God told Abraham to move to Canaan in present-day Israel and Lebanon. God made an agreement with Abraham to bless his descendants. They later became known as Hebrews or Israelites. The Hebrew Bible contains books of law, history, and prophecy. Another important work is the Talmud, a collection of scholarly writings.

Moses, a descendant of Abraham, is rescued from the Nile.

Central Beliefs

Judaism was the first major religion to teach monotheism, or the belief in one god. Jews believe God is the creator of the whole universe and has given them special responsibilities. Jews are to live holy lives by treating others well and pursuing justice. Today, Jews worship in synagogues, and their leaders are called rabbis.

Spread of Judaism

For centuries, Judaism was practiced primarily in what is present-day Israel. Several times in history, empires conquered the region and drove many Jews from the area. The last major event occurred in A.D. 135 when Rome punished Jewish rebels attempting to regain independence. As Jews spread out around the world, Judaism spread with them.

A worshipper holds up a Torah scroll during a Passover blessing at the Western Wall in Jerusalem, Israel.

☸ BUDDHISM

Historical Origins

Buddhism is based on the teachings of Siddhartha Gautama, known as the Buddha or "enlightened one." He was born a prince in India in the 400s or 500s B.C. Siddhartha was protected from seeing sickness, death, poverty, or old age until he was 29. However, after he learned about suffering, he left his palace to lead a religious life. While he was meditating several years later, Buddhists believe he received enlightenment about the meaning of life.

A 14th-century painting of Buddha

Central Beliefs

Buddhists believe that a law of cause and effect called karma controls the universe. Buddhists teach that suffering occurs because people desire what they do not have. A person who gives up desire and other negative emotions will achieve a state called nirvana, or the end of suffering. The basic beliefs of Buddhism are summarized in the Four Noble Truths. The actions that help people achieve nirvana are called the Eightfold Path.

Spread of Buddhism

During its first century, Buddhism spread across northern India. Over time, missionaries and travelers carried it to the Himalayas, Central Asia, and China. China spread Buddhism to Japan and Korea. In the 1800s, immigrants introduced Buddhism to the United States. In the late 20th century, the religion gained popularity in the United States and other Western countries.

FOUR NOBLE TRUTHS

- Suffering is a part of life.
- Selfishness is the cause of suffering.
- It is possible to move beyond suffering.
- There is a path that leads to the end of suffering.

Buddhist monks in Siem Reap, Cambodia, celebrate the birthday of Buddha.

ॐ HINDUISM

Historical Origins

Hinduism, one of the world's oldest religions, originated in India in about 1500 B.C. Scholars believe that it developed from the beliefs of a group of Indo-European people who spoke Sanskrit. The sacred writings of Hindus include the Vedas, which are poems and hymns, and the Puranas, which are sacred stories. Other Hindu texts such as the *Mahabharata*, of which the Bhagavad Gita is a part, teach Hindu beliefs in the form of epic poems.

Scene from the Bhagavad Gita, a sacred Hindu text

Central Beliefs

In general, Hindus believe in one eternal force called Brahman. This divine spirit takes the form of many gods and goddesses. The most important deities are Brahma, the creator; Vishnu, the preserver; and Shiva, the destroyer. Hindus believe that souls are constantly being reborn. Karma, the negative or positive effect of one's actions, determines if the soul moves to a higher or lower state of being. Hindu religious practice includes worship, study, and rituals such as bathing in the Ganges River.

Spread of Hinduism

Hinduism spread from India through parts of Southeast Asia, but now is practiced by few people in that region. In general, Hinduism has remained mainly a religion of the Indian people. Nearly 80 percent of Indians are Hindus. Indian immigrants have brought Hinduism to the United States.

Women light candles in observance of Diwali, the Hindu Festival of Lights, in Jaipur, India.

☬ SIKHISM

Historical Origins

Followers of Sikhism are called Sikhs, which means "learner." A teacher named Guru Nanak established the religion in India in the late 1400s. After his death, a line of nine other teachers, or gurus, followed him. Sikhs believe that all ten gurus were inspired by a single spirit. They also believe that after the tenth guru died, this spirit inhabited the Sikhs' sacred scripture, the Guru Granth Sahib or Adi Granth.

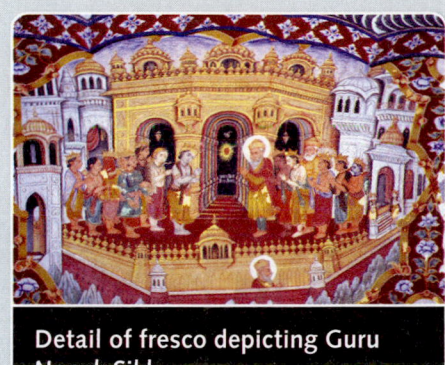

Detail of fresco depicting Guru Nanak Sikh

Central Beliefs

Sikhs believe in one god who does not take physical form. Like Hindus, they believe in reincarnation—the rebirth of the soul. The goal of Sikhism is to form a close, loving relationship with God. Sikh practices include prayer several times a day, worship, and meditation. Sikhs do not use tobacco or alcohol, and they often follow a strict dress code, which includes never cutting their hair.

Spread of Sikhism

Sikhism is practiced by nearly 25 million people, most of whom live in the Punjab region of northwest India. However, immigrants have introduced the religion to Western countries, including the United States, Canada, and the United Kingdom.

A Sikh pilgrim visits the Golden Temple—Sikhism's holiest site—in Amritsar, India.

☯ CONFUCIANISM

Historical Origins

Confucianism is an ethical system and philosophy based on the teachings of a Chinese public official and teacher named Kongfuzi. He is called Confucius in Western countries. He lived from 551 to 479 B.C. His goals were to revive traditional values and establish education as a way to improve society. The sayings and writings of Confucius were collected in a work called the *Analects*.

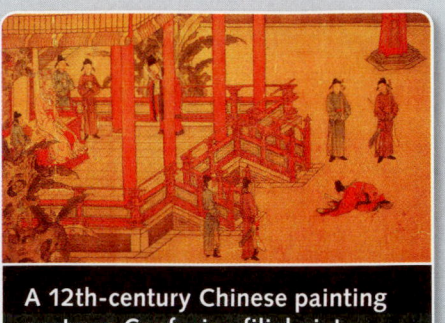

A 12th-century Chinese painting portrays Confucian filial piety.

Central Beliefs

One of the most important things in Confucianism is the concept of filial piety, in which children obey and honor their parents. Confucius applied this idea to other areas. For example, subjects should obey their rulers. Confucius taught that education, right relationships, and moral behavior would create an orderly society.

Spread of Confucianism

Confucianism became a way of life in China. Because China had a major influence on East Asia and Southeast Asia, Confucianism spread throughout these regions. It continues to be a strong cultural force in China and other Asian countries such as South Korea, Japan, and Singapore.

Celebrants participate in a festival in Qufu City, Shandong Province, China, the birthplace of Confucius.

ONGOING ASSESSMENT

RESEARCH LAB GeoJournal

Express Ideas Through Speech Get together in a group and prepare a panel discussion in which you analyze what the major world religions have in common.

Step 1 Appoint each person in your group to become an expert on a different religion. Each person should review the pages in the text that cover their assigned religion.

Step 2 Each student should summarize the religion they studied. As a group, identify characteristics that all or most of the religions and ethical systems have in common.

Step 3 Hold a panel discussion before the class and present your conclusions. After you have finished, allow time for questions.

GLOSSARY

A

Aborigine (AB uh RIHJ uh nee) *n.*, Australia's first developed culture

absolute location *n.*, the exact point where a place is located, identified by latitude and longitude coordinates

abstract *adj.*, an artistic style that stresses form and color over realism

accommodate *v.*, to make room for

Acropolis *n.*, a rocky hill in Athens, Greece, that served as a fortress for the ancient city's most important buildings

adapt *v.*, to adjust or modify to fit

adherent *n.*, a follower of a religion, cause, or person

adventure tourism *n.*, a type of tourism in which travelers engage in physical activities such as mountain climbing, water sports, or hiking

aerodynamic *adj.*, designed to move with little wind resistance

African National Congress (ANC) *n.*, founded in the early 20th century, an organization of black South Africans, including Nelson Mandela, who protested discriminatory treatment

African Union *n.*, an organization of African countries that work toward economic progress

aging population *n.*, a demographic trend that occurs as the average age of a population rises

agricultural revolution *n.*, a period in which humans began to grow crops instead of gathering plants

Aguinaldo, Emilio *n.*, a leader in the movement for Filipino independence in the 1890s

al-Qaeda *n.*, a terrorist group based in Southwest Asia

Alexander the Great *n.*, the conqueror who extended the Macedonian Empire and spread Greek culture throughout Eurasia from 334–323 B.C.

alliance *n.*, a partnership between countries

alluvial (a LOO vee ahl) *adj.*, describing the sediment deposited by a river

alluvial plain *n.*, a flat area of land next to a stream or river that floods

alluvium *n.*, the soil, or silt, carried by flowing water, which is ideal for farming

Alps *n.*, a European mountain chain

Angkor Wat *n.*, a large temple built in Cambodia in the 1100s dedicated to the Hindu deity Vishnu

animal-borne *adj.*, carried by animals

anime *n.*, a style of animation or cartoon developed in Japan

animism *n.*, a belief that everything, including objects in nature, has a soul

Antarctic Treaty *n.*, a 1959 agreement by 12 countries to use Antarctica for peaceful purposes and to share scientific discoveries

apartheid (uh PAHRT hyt) *n.*, the legal separation of the races; a system that denied black South Africans their rights

aqueduct *n.*, a transport system for carrying water long distances, sometimes raised on a bridge

aquifer *n.*, a layer of rock beneath Earth that contains water

arable *adj.*, fertile; suitable for farming

Aral Sea *n.*, a saltwater lake in Central Asia that has been greatly reduced in size due to diversion for irrigation of the rivers that flow into it

archipelago (ahr kuh PEH lug goh) *n.*, a chain of islands

arid *adj.*, very dry, having almost no rainfall

aristocrat *n.*, a member of the upper class

armistice *n.*, an agreement to stop fighting

artifact *n.*, an object made by humans from a past culture

Aryans (AIR ee uhnz) *n.*, nomads who migrated from Central Asia into the Indus Valley around 2000 B.C.

Asia-Pacific Economic Cooperation (APEC) *n.*, a global partnership founded in 1989 to strengthen economic ties among countries in the Pacific

Asoka *n.*, the leader at the height of the Mauryan Empire, around 250 B.C.

assimilate *v.*, to be absorbed into a society's culture

assimilation *n.*, a process in which a minority group adopts the culture of the majority

assisted migration *n.*, a program set up in 1831 in which the British gave people money to move to Australia

Aswan High Dam *n.*, a dam on the Nile River completed in 1970

atoll *n.*, a ring-shaped reef, island, or chain of islands made of coral

attribute *n.*, a specific quality

Augustus *n.*, the first emperor of Rome in 27 B.C.

autonomy *n.*, a country or people's self-governance

B

ban *v.*, to outlaw or forbid

Bantu *n.*, an African group who moved from west Central Africa south and east across sub-Saharan Africa from 2000–1000 B.C.

barbarian *n.*, a soldier or warrior considered to be culturally less developed than those being fought; the term originally comes from the German tribes who invaded the Roman Empire in A.D. 235

Baroque period *n.*, a period from 1600–1750 in which music had complicated patterns and themes

barter *v.*, to trade or exchange goods without using money

bas-relief *n.*, a sculpture that is slightly raised from a flat background

basin *n.*, a region drained by a river system

bauxite *n.*, the raw material used to make aluminum

bay *n.*, a body of water surrounded on three sides by land

Bedouin (BEHD u ihn) *n.*, a nomadic Arabic-speaking people of Southwest Asia

benefit *v.*, to be useful to

Berlin Conference *n.*, a meeting of European nations in 1884 to settle disputes over their colonial claims in Africa

Berlin Wall *n.*, wall that divided Communist East Berlin from democratic West Berlin; it was torn down in 1989

Bhutto, Benazir *n.*, the female prime minister of Pakistan, 1988–1999

Biko, Stephen *n.*, the president of a student protest organization in South Africa who was arrested and killed in 1977

biodiversity *n.*, the variety of species in an ecosystem

Black Sea *n.*, an inland sea bounded by Europe, Russia, Georgia, and Turkey

Blue Mosque *n.*, an historic mosque built in Istanbul by Sultan Ahmed I, completed in 1617

Bollywood *n.*, India's film industry in Mumbai

Bolshevik *n.*, a political party in Russia led by Lenin that overthrew the czar in 1917

Borobudur *n.*, a Buddhist temple complex in central Java

Bosporus Strait *n.*, the waterway that passes through Istanbul and connects the Black Sea and the Sea of Marmara

breakwater *n.*, a barrier built to protect a harbor

Buddhism *n.*, a religion founded in 525 B.C. in India by Siddhartha Gautama

bullet train *n.*, a train that travels more than 125 miles per hour, and is shaped like a bullet

butte *n.*, a hill or mountain with steep sides and a flat top

Byrd, Richard *n.*, an American who explored Antarctica by land and air in the 1940s

Byzantine Empire *n.*, the eastern part of the Roman Empire after A.D. 395 that lasted about 1,000 years

C

Caesar, Julius *n.*, a general who became the ruler of Rome, 46–44 B.C.

canal *n.*, a human-made waterway through land for boats and ships

capital *n.*, a country's wealth and infrastructure

carapace *n.*, the hard shell of an animal such as a turtle

caravan *n.*, a group of merchants traveling together for safety

Cardamom Mountains *n.*, mountain range in Cambodia

cartographer *n.*, a mapmaker

Caspian Sea *n.*, the largest enclosed body of water on Earth, bounded by Russia, Kazakhstan, Turkmenistan, Azerbaijan, and Iran

caste system *n.*, a social structure in India composed of four main levels, which was started by the Aryans

categorize *v.*, to group information; to classify

Catherine the Great *n.*, the empress of Russia from 1762 to 1796

celadon *adj.*, green-glazed, describing pottery made by Korean artists

Chang Jiang (chahng jyahng) *n.*, the longest river in Asia, which flows through China; also known as the Yangtze

Chang Jiang Plain *n.*, the plains along the Chang Jiang River

Chao Phraya River *n.*, a river that flows through Thailand

Chernobyl *n.*, a city in the Ukraine in which a nuclear reactor exploded in 1986

Choson *n.*, a dynasty in Korea that lasted from 1392 to 1910

Christian Bible *n.*, the sacred book of Christianity

Christianity *n.*, a religion based on the life and teachings of Jesus of Nazareth

citizen *n.*, a person living within a territory who has rights and responsibilities granted by the government

city-state *n.*, an independent state made up of a city and the territories depending on it

civil disobedience *n.*, the nonviolent disobeying of laws

civilization *n.*, a society with a highly developed culture, politics, and technology

clan *n.*, a large family-based unit with loyalty to the group

Classical period *n.*, a period from 1750 to 1900 in which music followed the standard rules of form and complexity, such as sonatas or symphonies

climate *n.*, the average condition of the atmosphere of an area over a long period of time, including temperature, precipitation, and seasonal changes

climograph *n.*, a graph showing a region's climate through average precipitation and temperature

coalition *n.*, an alliance

Cold War *n.*, a long period of political tension without fighting between the United States and the Soviet Union, roughly from 1948 to 1991

collective farm *n.*, in the Soviet Union, a large farm where workers grew food to be distributed to the entire population

collide *v.*, to crash together

colonialism *n.*, the practice of one country directly ruling and developing trade in a foreign territory for its own benefit

colony *n.*, an area controlled by a distant country

commerce *n.*, the business of trading goods and services

commodity *n.*, a material or good that can be bought, sold, or traded

Common Market *n.*, the European Economic Community formed in 1957

Commonwealth of Australia *n.*, the country of Australia that became part of the British Commonwealth in 1901

communal *adj.*, shared

communism *n.*, a system of government in which a single political party controls the government and the economy

competitive *adj.*, able to try to win a contest or race

complex *n.*, a connected set of buildings

comply *v.*, to follow a rule or an order

concentrated *adj.*, gathered in one central area

concentration camp *n.*, an area where Jews and others were held during World War II and murdered by the Nazis

condensation *n.*, the process of water vapor turning into liquid droplets due to cooling in the hydrologic cycle

Confucianism *n.*, an ethical system based on the teachings of Confucius

Congo River *n.*, a major river in Central Africa that ends in the Atlantic Ocean

conservation *n.*, the protection of the environment

Constantine *n.*, Roman Emperor, 306–337 B.C., who made Christianity the official religion of the empire

consumer *n.*, a person who buys goods

contaminated *adj.*, infected; unfit for use because of the presence of unsafe elements

continent *n.*, a large landmass on Earth's surface; Earth has seven continents

continental drift *n.*, the slow movement of continents on tectonic plates

continental shelf *n.*, the edge of a continent that extends under water into the sea

controversy *n.*, a debate or quarrel

convert *v.*, to persuade someone to change religious beliefs

convey *v.*, to communicate

convict *n.*, a person sentenced to prison for committing a crime

Cook Islands *n.*, a country near New Zealand in the South Pacific made up of 15 small islands

Cook, James *n.*, a British navigator who explored and mapped the Pacific Islands in the 1770s

coral island *n.*, an island created by a gradual buildup of coral skeletons

coral reef *n.*, a rock-like structure built by layers of coral organisms

cosmopolitan *adj.*, bringing together many different cultures and influences

Counter-Reformation *n.*, a movement within the Roman Catholic Church to reform its own practices

coup (KOO) *n.*, a sudden, illegal takeover of government by force

crevasse *n.*, a deep, open crack in a glacier

cricket *n.*, a team game similar to baseball

Crusades *n.*, military expeditions of the Roman Catholic Church to take back holy lands in the Middle East from Muslim control, 1096–1291

cuisine *n.*, the food and cooking traditions common to a certain region

cultural hearth *n.*, center of civilization from which ideas and technology spread

Cultural Revolution *n.*, social plan in China that lasted from 1966 to 1969 in which the government attempted to remove any capitalistic and anti-Communist elements from society

culture *n.*, a group's way of life, including types of food, shelter, clothing, language, behavior, and ideas

culture region *n.*, an area that is unified by common cultural traits

cuneiform (kyoo NEE uh form) *n.*, the earliest known form of writing, from Sumeria

currency *n.*, a form of money

cyclone *n.*, a storm with rotating winds, called a typhoon in the Eastern Hemisphere; a hurricane in other parts of the world

czar *n.*, term used for an emperor in Russia

D

Danube River *n.*, a river that starts in Germany and ends at the Black Sea

Daoism *n.*, an ethical system that emphasizes the harmony between people and nature; also called Taoism

deforestation *n.*, the practice of cutting down forests for crops or urban use

deity *n.*, a god or goddess

delta *n.*, an area where a river deposits sediment as it empties into a larger body of water

demilitarized zone (DMZ) *n.*, a neutral area between enemy countries, specifically between North Korea and South Korea

democracy *n.*, a form of government by the people, in which citizens often elect representatives to govern them

democratization *n.*, the process of becoming a democracy

demographics *n.*, the characteristics of a human population, such as age, income, and education

deny *v.*, to refuse to recognize

deplete *v.*, to drain or diminish a resource so as to make it no longer available or able to function as planned

desertification *n.*, a gradual transition from fertile to less productive land

developed nations *n.*, countries with a high per capita gross domestic product

developing nations *n.*, countries with a low per capita gross domestic product

dialect *n.*, a regional variation of a main language

Diaspora *n.*, the movement of Jews away from Israel

diffusion *n.*, a spreading out

diplomacy *n.*, discussion between groups or countries to resolve disputes or disagreements

discrimination *n.*, unfair treatment of an individual or group based on factors other than ability

displace *v.*, to force a people to leave their homes

disregard *v.*, to ignore

dissolve *v.*, to break up

distort *v.*, to change the usual shape or appearance; twist out of shape

distribute *v.*, to spread an asset or resource amongst a larger group

domestic policy *n.*, a government's plan for running affairs within its own borders

domesticate *v.*, to keep and use animals as a source of labor and food

dormant *adj.*, inactive, referring to a volcano

drought *n.*, a long period with little or no precipitation

Dubai *n.*, a young and wealthy city and state in United Arab Emirates on the Persian Gulf

Dutch East India Company *n.*, a Netherlands company that dominated the spice trade in Southeast Asia

dynastic cycle *n.*, a pattern in the rise and fall of dynasties in Chinese history

dynasty *n.*, a series of rulers from the same family

E

earthquake *n.*, a shaking of Earth's crust generally caused by the collision or sliding of tectonic plates

Eastern Hemisphere *n.*, the half of Earth east of the prime meridian

ecologist *n.*, a scientist who studies the relationships between living things and their environment

economic globalization *n.*, the practice of economic activities being conducted across national borders

economy *n.*, a system in which people produce, sell, and buy things

ecosystem *n.*, a community of living organisms and their natural environment, or habitat

ecotourism *n.*, a way of visiting natural areas that conserves the natural resources of the region

elevation *n.*, the height of a physical feature above sea level

elite *adj.*, superior

emergence *n.*, the development and widespread use of something new

emigrate *v.*, to leave one's home country to live in another country

emirate *n.*, a state in the country of the United Arab Emirates

empire *n.*, a group of peoples or states ruled by a strong single ruler

Enduring Voices Project *n.*, a National Geographic project to study and preserve languages at risk of dying out

enhance *v.*, to improve the quality of something

Enlightenment *n.*, a social movement of the 1700s that worked for the education and rights of the individual

enlist *v.*, to volunteer for military service

entrepreneur *n.*, a person who starts up a new business

entrepreneurship *n.*, the characteristics of creativity and risk existing in a person or society

epic poem *n.*, a long poem that tells of a hero's adventures

epidemic *n.*, an outbreak of a disease affecting a great number of the population in a particular community

equator *n.*, an imaginary circle around Earth that is the same distance from the North and South Poles and divides Earth in half; the center or 0° line of latitude

equinox *n.*, when day and night are of equal length; occurs twice each year on March 21 or 22 and September 21 or 22

erode *v.*, to wear away

erosion *n.*, the process by which rocks and soil slowly break apart and are worn away

eruption *n.*, a blast or explosion

escarpment *n.*, a steep slope

essential *adj.*, necessary

establish *v.*, to put into place

ethical system *n.*, a belief system that teaches moral behavior

ethnic group *n.*, a group of people who share a common culture, language, and sometimes racial heritage

ethnobotanist *n.*, a scientist who studies the relationship between cultures and plants

euro *n.*, the common currency of the European Union

European Union (EU) *n.*, an economic organization composed of 27 European member countries (2011)

eurozone *n.*, the countries that have adopted the euro as money

evaporation *n.*, the process of water turning into vapor and rising into the atmosphere due to the sun's heating; part of the hydrologic cycle

evident *adj.*, clearly present or observable

exchange *v.*, to convert money into another currency

exile *n.*, a state of absence from one's home country

exoskeleton *n.*, a hard external covering that protects corals and other sea animals

expand *v.*, to make larger

expedition *n.*, a journey or voyage of some length, usually to other lands

extent *n.*, the distance or degree to which something spreads

extinct *adj.*, completely gone, such as a species of animal or plant

extremist *n.*, a person with religious or political views that are outside the range of majority opinion

F

factory system *n.*, a way of working in which each person works on only one part of a product

failed state *n.*, a country in which government, economic institutions, and order have broken down

fallout *n.*, radioactive particles from a nuclear explosion that fall through the atmosphere

famine *n.*, a long period of food shortage

fault *n.*, a fracture in Earth's crust

federal republic *n.*, a democratic form of government in which voters elect representatives and the central government shares power with the states

federal system *n.*, a system of government with a strong central government and local government units

feral *adj.*, wild, having gone wild after it was once domesticated

Fertile Crescent *n.*, the area around the Tigris and Euphrates rivers

fertility rate *n.*, the average number of children born per woman at a certain period of time

feudal system *n.*, a social structure during the Middle Ages consisting of a king, lords, vassals, and serfs

first language *n.*, the language learned by the children of a people

fjord (fee ORD) *n.*, a deep, narrow bay

floodplain *n.*, the low-lying land next to rivers, formed by sediment left by flooding

Foja Mountains *n.*, a mountain range on the island of New Guinea in eastern Indonesia

fossil *n.*, the preserved remains of ancient animals and plants

fragmented country *n.*, a country that is physically divided into separate parts, such as a chain of islands, and/or politically or culturally divided

free enterprise economy *n.*, a system in which privately-owned businesses create goods and services; also called a market economy or capitalism

free trade agreement *n.*, a treaty between countries that improves trade by limiting taxes on that trade

free trade *n.*, trade that does not impose taxes on imports

French Indochina *n.*, France's combined colonies of Vietnam, Cambodia, Laos, and others in Southeast Asia from 1887 through 1955

fuel cell *n.*, a small unit that, like a battery, combines chemicals to make energy

G

Gandhi (GAHN dee), **Mohandas** *n.*, Indian leader who pushed for India's independence from the British through civil disobedience in the 1930s; considered the father of modern India

Ganges Delta *n.*, fertile area where the Ganges River flows into the Bay of Bengal

gaucho *n.*, a cowboy of South America

gauge *n.*, the measurement of the width of railroad tracks

generation *n.*, a group of individuals who are born and live about the same time

Genghis Khan (JEHNG-gihs KAHN) *n.*, the Mongol ruler who established an empire in Central Asia in the early 1200s

genre *n.*, a form of literature such as poem, play, or novel

Geographic Information Systems (GIS) *n.*, computer-based devices that show data about specific locations

geographic pattern *n.*, a similarity among places

geothermal energy *n.*, heat energy from within the earth that can be turned into electricity

ger *n.*, a portable tent made of felt

Giza *n.*, a city on the Nile where ten pyramids were built in ancient Egypt

glacier *n.*, a large mass of ice and packed snow

glasnost (GLAHS-nuhst) *n.*, the Soviet Union's policy of openness that encouraged people to speak openly about government, introduced by Mikhail Gorbachev in the 1980s

Global Positioning System (GPS) *n.*, a satellite system based in space that finds absolute location and time anywhere on Earth

globe *n.*, a three-dimensional, or spherical, model of Earth

Gobi desert *n.*, the largest desert in Asia, covering much of southern Mongolia and extending into China

golden age *n.*, a period of great wealth, culture, and democracy in Greece

Golden Quadrilateral (GQ) *n.*, a superhighway that connects four major cities in India

Gorbachev, Mikhail (mih KYL GAWR buh chawf) *n.*, a leader in the Soviet Union 1985–1991; President from 1990 until it was dissolved in 1991

gorge *n.*, a deep, narrow passage surrounded by steep cliffs

government *n.*, an organization that keeps order, sets rules, and provides services for a society

Grand Bazaar *n.*, a center of shops and trading in Istanbul, Turkey

Great Barrier Reef *n.*, a huge ecosystem off the coast of Australia made up of coral reefs

Great Depression *n.*, worldwide economic downturn in the 1930s, marked by poverty and high unemployment

Great Escarpment *n.*, the steep slope from the plateau of Southern Africa to the coastal plains

Great Leap Forward *n.*, Mao Zedong's plan to make China's economy grow faster, 1958–1961

Great Pyramid of Khufu *n.*, the oldest and tallest pyramid in Egypt, built at Giza around 2550 B.C.

Great Rift Valley *n.*, a wide valley in East Africa, part of a chain of valleys formed when tectonic plates separated

Great Wall of China *n.*, a stone wall in northern China over 4,500 miles long, built to repel invaders from the north

Great Zimbabwe (zim BAH bwe) *n.*, a walled city of stone that the Shona people built in Southern Africa between 1200 and 1450

greenhouse gas *n.*, a gas that traps the sun's heat over Earth

griot (GREE oh) *n.*, a traditional African storyteller

gross domestic product (GDP) *n.*, the total value of all goods and services produced in a country in a given year

guest worker *n.*, a temporary laborer who migrates to work in another country

guillotine (GHEE uh teen) *n.*, a machine used to execute people during the French Revolution

Gutenberg, Johannes *n.*, a German printer who invented the printing press in 1450

H

habitat *n.*, the natural environment of a living plant or animal

Hagia Sophia *n.*, a museum in Istanbul, Turkey, that was originally built as a church during the Roman Empire and later served as a mosque.

half-life *n.*, the time needed for half the atoms in a radioactive substance to decay

Hammurabi *n.*, the king who developed a code of law in Babylonia in the 1700s B.C.

Han *n.*, a Chinese dynasty that lasted from 206 B.C. to A.D. 220

Harappan *adj.*, related to the first urban civilization in South Asia along the Indus River around 2600 B.C.

Hatshepsut (hat SHEHP soot) *n.*, a female pharaoh in Egypt around 1500 B.C.

Hebrew Bible *n.*, the sacred book of Judaism

hemisphere *n.*, one-half of Earth

henna *n.*, a reddish powder used to create designs on skin

hereditary *adj.*, passed on through the family

heritage *n.*, a tradition passed down from ancestors

Hermitage Museum *n.*, a museum of art and culture in St. Petersburg, Russia

hieroglyphics (HY ruh GLIHF ihks) *n.*, an ancient system of writing that uses pictures and symbols

highlands *n.*, areas of high mountainous land

Himalaya Mountains *n.*, the highest mountain range in the world, located in South Asia

Hinduism *n.*, religion practiced by more than 60 percent of the population of South Asia

Hitler, Adolf *n.*, German head of state from 1933 to 1945

Ho Chi Minh *n.*, the leader of North Vietnam who wanted to unify North and South Vietnam under communism; these efforts started the Vietnam War

Holocaust *n.*, the mass slaughter by the Nazis of six million Jews and others during World War II

homelands *n.*, during Apartheid, separate areas within South Africa where black South Africans were forced to live

hotspot *n.*, an unusually hot part of Earth's mantle

Huang He (hwahng huh) *n.*, China's second largest river, also called the Yellow River

human rights *n.*, the political, economic, and cultural rights that all people should have

hurricane *n.*, a strong storm with swirling winds and heavy rainfall

Hussein, Saddam *n.*, the president of Iraq from 1979 to 2003

hydroelectric power *n.*, a source of energy that uses flowing water to produce electricity

I

ice shelf *n.*, a floating sheet of ice attached to a landmass

immigration *n.*, the permanent movement of a person to a different country

impact *n.*, an effect that produces change

imperialism *n.*, the practice of extending a nation's influence by controlling other territories

Impressionism *n.*, an artistic style in which artists used light and color in short strokes to capture a moment in time

incentive *n.*, the reason or motive to do something

incorporate *v.*, to include; to combine with something already formed

indigenous *adj.*, native to the area in which something is found

indulgence *n.*, a fee paid to the church to relax the penalty for a sin during the Middle Ages

Indus River *n.*, a river in the western part of South Asia

Industrial Revolution *n.*, a period in the 1700s and 1800s in which workers in factories began to use machines and power tools for large-scale industry

industrialize *v.*, to develop manufacturing

infectious *adj.*, capable of spreading rapidly to others

infrastructure *n.*, the basic systems of a society such as roads, bridges, sewers, and electricity

inner islands *n.*, islands in Indonesia that include Java, Madura, and Bali

interact *v.*, to affect other people and be affected by them

interior *n.*, the land that is away from a seacoast

Internet *n.*, a communications network

intersection *n.*, a place where people meet or paths cross

intifada *n.*, an uprising or rebellion, usually meant to reference the Palestinian revolt against Israel

invader *n.*, an enemy who enters a country by force

invasive species *n.*, non-native plants or animals introduced to a new area, intentionally or unintentionally, that disturb habitats of native life forms

Iron Curtain *n.*, an imaginary boundary that separated Communist and non-Communist countries in Europe during the Cold War

Irrawaddy River *n.*, a river that flows through Myanmar

irrigation *n.*, the process of redirecting water to crops through channels and ditches

Isis *n.*, an important Egyptian goddess

Islam *n.*, religion founded in Saudi Arabia in the early A.D. 600s

isolated *adj.*, cut off from others

isolation *n.*, separation or being set apart from others

J

Jainism *n.*, a religion in northwest India started in the late A.D. 500s

Java *n.*, one of the inner islands located in Southeast Asia

Jerusalem *n.*, the capital city of Israel

Jiang Jieshi (jee ahng jee shee) *n.*, a Chinese nationalist leader who fought Communists

Jinnah, Mohammed Ali *n.*, a leader of the All-India-Muslim League who helped found Pakistan and was Pakistan's first Governor-General from 1947–1948

K

Kalahari *n.*, a major desert in southern Africa

Kanto Plain *n.*, flat area good for agriculture and industry located east of the Japanese Alps

karma *n.*, in Hinduism, the negative or positive effect one receives as a result of his or her actions

Kenyatta, Jomo (JOH moh ken YAA taa) *n.*, an African leader who helped gain Kenya's independence and became its first elected leader in 1963

Khmer Empire *n.*, the largest and longest-lasting empire in Cambodia from the 800s to 1430

Kievan Rus *n.*, the state established by the Varangian Russes in 882 that became part of modern Russia

Kilimanjaro *n.*, an inactive volcano in Tanzania that is 19,340 feet tall

kimono *n.*, traditional Japanese women's clothing

Kingman Reef *n.*, an unspoiled coral reef in the Line Islands between Hawaii and American Samoa

Knesset (kuh NEH set) *n.*, the Israeli parliament

Koguryo (koh gur YOO) *n.*, a kingdom in northern Korea in 37 B.C.

Kongo *n.*, a state founded in 1390 in Central Africa

Koryo *n.*, a dynasty that ruled Korea from 935 to 1392

Kremlin *n.*, a historic complex of palaces, armories, and churches in Moscow and the seat of Russian government

Krishna *n.*, in Hinduism, a deity in the bodily form of Vishnu

Kunlun Mountains *n.*, a mountain range in East Asia

Kurd *n.*, a member of a non-Arab ethnic group in Southwest Asia

L

labor force *n.*, the number of people available to work

land bridge *n.*, a strip of land connecting two landmasses

landlocked *adj.*, surrounded by land on all sides, with no direct access to a seacoast

landmass *n.*, a very large area of land

language diffusion *n.*, the spread of languages from their original home

latitude *n.*, an imaginary line around Earth that runs east to west, showing location relative to the equator

launch *v.*, to start

legume *n.*, peas or beans

Lenin, V.I. *n.*, the Bolshevik leader who overthrew the czar in the Russian Revolution of 1917 and led the new government

lingua franca (LEEN gwa FRAWN kah) *n.*, the common language among several groups of people

linguist *n.*, a language scientist

literacy rate *n.*, the percentage of people who can read and write

literate *adj.*, able to read and write

Locke, John *n.*, an English philosopher of the late 1600s who helped inspire the American Revolution and the Enlightenment

loess (less) *n.*, a yellow silt or sediment that forms thick deposits

longitude *n.*, an imaginary line running north to south from the North Pole to the South Pole that shows location relative to the prime meridian

Lost Boys of Sudan *n.*, a group of young men of Sudan who were orphaned by civil war and stuck together to escape the violence

Lost Decade *n.*, the 1990s in Japan, when production declined because businesses were deeply in debt

lucrative *adj.*, making money, or profitable

Luther, Martin *n.*, a German monk whose actions in 1517 led to the Reformation to address corruption in the Roman Catholic Church

M

Madura *n.*, one of the inner islands located in Southeast Asia

magnetic levitation (Maglev) *adj.*, type of train that rides on a cushion of air over tracks laid with many powerful magnets

mainland *n.*, the land connected to a continent, usually in reference to countries with both continental land and islands.

maintain *v.*, to preserve and carry on

Malay Peninsula *n.*, a peninsula in Southeast Asia that includes parts of Thailand, Myanmar, and the mainland section of Malaysia

Malaysia *n.*, a country in Southeast Asia at the bottom of the Malay Peninsula and the island of Borneo

malnutrition *n.*, the lack of enough food or nourishment

Mandela, Nelson *n.*, a leader of the African National Congress who was jailed for fighting apartheid, but who became the president of South Africa in 1994

manga *n.*, a type of Japanese comic book

Manila *n.*, the capital of the Philippines

Mao Zedong (MOW dzuh dahng) *n.*, the chairman of the Communist Party who led China from 1949–1976

Maori (MOW ree) *n.*, the native people of New Zealand

map *n.*, a two-dimensional, flat representation of Earth

marine *adj.*, sea-based

marine life *n.*, the plants and animals living in the ocean

marine reserve *n.*, an ocean area set aside to protect ocean life from humans

maritime *adj.*, related to the sea

marsupial *n.*, a mammal whose females carry their babies in a pouch

martial law *n.*, government maintained by military power

medicinal plant *n.*, plant used to treat illness

meditation *n.*, the practice of using concentration to quiet and control thoughts

Mediterranean climate *n.*, a climate that has hot, dry summers and mild, rainy winters

Mekong River *n.*, the longest river in Southeast Asia; flows through Myanmar, Laos, and Thailand

messiah *n.*, a leader or savior

methane *n.*, a colorless, odorless natural gas released from carbon

metropolitan area *adj.*, populated place around a city that includes the city limits and the surrounding communities

microcredit *n.*, a small loan of money

microlending *n.*, the practice of making small loans to people starting their own businesses

Middle Ages *n.*, a period in Western Europe after the fall of the Roman Empire, from about 500 to 1500

Middle Passage *n.*, the months-long trip across the Atlantic Ocean in which enslaved Africans were brought to European colonies in the Americas

migration *n.*, the movement from one place to another

military dictatorship *n.*, a form of government in which the army runs the government

mineral *n.*, a solid natural substance found in rocks and Earth that is inorganic and has its own set of properties

missionary *n.*, a person sent by a religious organization to convert others to that religion

modernization *n.*, policies and actions designed to bring a country up to date in technology and other areas

modify *v.*, to change, or make less extreme

Mongol Empire *n.*, an empire established in Central Asia in the early 1200s by Genghis Khan

monk *n.*, a man who devotes himself to religious work

monopoly *n.*, the complete control of the market for a service or product

monotheism (MAHN uh thee ihz uhm) *n.*, a belief in one god

monotheistic *adj.*, related to a religious belief in one god

monotheistic religion *n.*, a system of belief in one god or deity

monsoon *n.*, a seasonal wind that brings intense rainfall during part of the year

Montezuma *n.*, an Aztec ruler who was killed by the Spanish conquistador Hernán Cortés

moral *adj.*, right and good, describing human behavior

mosque *n.*, a Muslim place of worship

mouth *n.*, the place where a river empties into a sea

movable type *n.*, a printing invention in which individual characters can be moved to create different pages of text

multinational corporation *n.*, a large company based in one country that establishes branches in several others

N

Nairobi (ny ROE bee) *n.*, Kenya's capital city

Napoleon *n.*, Napoleon Bonaparte, the leader of France who conquered other European countries to build an empire, 1804–1815

nationalism *n.*, a strong sense of loyalty to one's country

natural rights *n.*, rights such as life, liberty, and property that people possess at birth

navigable *adj.*, wide or deep enough to be traveled easily by boats and ships

navigation *n.*, the science of finding position and planning routes

Nazi Germany *n.*, Germany as led by the Nazi party from 1933–1945

N'Dour, Youssou *n.*, a famous griot from West Africa who plays Mbalax music

Nebuchadnezzar (nehb buh kuhd NEHZ uhr) *n.*, king of Babylonia, 605–562 B.C.

New Guinea *n.*, the world's second largest island, located in the southwest Pacific

New South Wales *n.*, in Australia, a colony built by convicts in 1788

Nile River *n.*, the longest river in the world; flows through Egypt and Africa for 4,000 miles

Nkrumah, Kwame (KWAA may en KROO mah) *n.*, an African leader in the 1950s and 1960s who helped Ghana gain independence

nocturnal *adj.*, active at night rather than during the day

nomad *n.*, a person who moves from place to place

nonrenewable *adj.*, cannot reproduce quickly enough to keep up with its use

nonrenewable fossil fuel *n.*, a source of energy such as oil, natural gas, or coal that is in limited supply

nonrenewable resource *n.*, a source of energy that is limited and cannot be replaced, such as oil

North Anatolian Fault *n.*, a fracture in Earth's crust that runs east and west, just south of the Black Sea

North Atlantic Drift *n.*, a warm ocean current that warms the waters around the northwest part of Russia

North China Plain *n.*, the plain along the Huang He River

North Pole *n.*, the northernmost point on Earth, opposite the South Pole, where all lines of longitude meet

Northern European Plain *n.*, a vast lowland that stretches from France to Russia

Northern Hemisphere *n.*, the half of Earth north of the equator

novel *n.*, a long work of fiction with complex characters and plot

O

oasis *n.*, a fertile place with water in a dry desert area

occupy *v.*, to take over

Okavango Delta (oh kuh VAANG oh) *n.*, an inland delta in Botswana, where the Okavango River empties into a swamp

one-child policy *n.*, the Chinese law of 1979 limiting families living in urban areas to one child

opera *n.*, a performance that tells a story through words and music

oral tradition *n.*, the passing of stories or histories by word of mouth from one generation to the next

Orange Revolution *n.*, the Ukraine's peaceful removal of its prime minister in 2004

Osman *n.*, the first leader of the Turks in the 1300s, for whom the Ottomans were named

Ottoman Empire *n.*, a large, wealthy empire from 1453 to 1923 centered in what is now Turkey

outer islands *n.*, islands in Indonesia that include Sumatra, Borneo, New Guinea, and others

outrigger canoe *n.*, a boat with an attached float that helps it balance

outsourcing *n.*, the shifting of jobs to workers outside of a company, often to a foreign country

overpopulation *n.*, the condition of too many people living in one place

GLOSSARY

P

Paekche (PAHK CHAY) *n.*, a kingdom in southwest Korea in 18 B.C.

pagoda *n.*, multistoried religious structure found in Asian countries, usually used for religious purposes

Palestine Liberation Organization (PLO) *n.*, an organization created by Palestinian leaders

Pan-Africanism *n.*, a movement in the early 1900s to unite African people

pandemic *n.*, an outbreak of a disease that spreads over a wide geographic area

Papua New Guinea *n.*, a country in the South Pacific that is the eastern half of the island of New Guinea and includes nearby islands

papyrus *n.*, a paper-like material invented in ancient Egypt

Parliament *n.*, the legislative branch of India's government

parliamentary democracy *n.*, a government system in which the chief executive is the prime minister, chosen by the party with the most seats in Parliament

Partition *n.*, refers to the division of South Asia into the independent countries of India and Pakistan

patrician *n.*, a wealthy landowner in ancient Rome

Pearl Harbor *n.*, a U.S. naval base in Hawaii that the Japanese bombed in 1941 and that brought about U.S. entry into World War II

peat *n.*, the material from very old decayed plants that burns like coal

peninsula *n.*, a body of land surrounded on three sides by water

perestroika (pehr ih STORY kuh) *n.*, reform in the structure of the economy introduced by Mikhail Gorbachev in the Soviet Union in 1985

permafrost *n.*, permanently frozen ground

perspective *n.* an artistic way of showing objects as they appear to people in terms of relative distance or depth, as if in three dimensions

pesticide *n.*, a chemical that kills harmful insects and weeds

Peter the Great *n.*, Peter Romanov, who ruled Russia as czar from 1682 to 1725

petrochemicals *n.*, products made from petroleum, or oil

petroleum *n.*, raw material used to produce oil

pharaoh *n.*, a king in ancient Egypt

philosopher *n.*, a person who examines questions about the universe and searches for the truth

pictograph *n.*, a painted picture used to communicate

pilgrimage *n.*, a religious journey

pipeline *n.*, a series of connected pipes used to transport liquids or gases

plain *n.*, a level area on Earth's surface

plate *n.*, a rigid section of Earth's crust that can move independently

plateau *n.*, a plain high above sea level that usually has a cliff on all sides

plebeian *n.*, a farmer or lower class person in ancient Rome

poach *v.*, to hunt or fish illegally

poaching *n.*, illegal hunting or fishing

polder *n.*, land in the Netherlands that has been reclaimed from the sea for farming

pollution *n.*, chemical or physical waste that creates an unclean or dirty environment

Polynesian Triangle *n.*, a large area in the South Pacific that includes many islands

polytheism (PAHL ee thee ihz uhm) *n.*, the belief in more than one god

polytheistic religion *n.*, a system of belief in many deities or gods

popular culture *n.*, the arts, music, and other elements of everyday life in a region

porcelain *n.*, a type of strong ceramic pottery

port *n.*, a harbor for ships where goods are exchanged

potential *n.*, possibility

precipitation *n.*, the process of water falling to Earth as rain, snow, or hail

predominant *adj.*, main, most common, superior to others

prehistoric *adj.*, before written history

preserve *v.*, to protect

pride *n.*, a group of lions that live together

prime meridian *n.*, the line of 0° longitude that runs from the North to the South Pole, and passes through Greenwich, England

privatization *n.*, the process of government-owned businesses becoming privately owned

projection *n.*, a way of showing Earth's curved surface on a flat map

promote *v.*, to encourage

propaganda *n.*, information made to influence people's opinions or advance an organization or party's ideas

proportional representation *n.*, a system in which a political party gets the same percentage of seats as its percentage of votes

push-pull factors *n.*, reasons why people migrate; "push" factors cause them to leave; "pull" factors make them come to a place

pyramid *n.*, a stone monument built as a tomb in ancient Egypt

Q

qanat (kuh NOT) *n.*, a human-made underground tunnel on the Iranian Plateau, used for carrying water from the mountains

Qin (chihn) *n.*, the dynasty that ruled China from 221–206 B.C.

Qur'an *n.*, the sacred book of Islam

R

radical *n.*, a person who wants an extreme change or holds an extreme political position

radioactive *adj.*, giving off energy caused by the breakdown of atoms

rain forest *n.*, a forest with warm temperatures, high humidity, and thick vegetation that receives more than 100 inches of rain per year

Ramses II *n.*, Egyptian pharaoh who reigned around 1185 B.C. and expanded Egypt's empire

raw materials *n.*, unfinished or natural materials such as minerals, oil, or coal used to make finished products

Re *n.*, the sun god of the ancient Egyptians

rebellion *n.*, a revolt or resistance to authority

reclaim *v.*, to take back

Reformation *n.*, the movement in the 1500s to reform Christianity

refuge *n.*, a safe place

refugee *n.*, a person who flees a place to find safety

region *n.*, a group of places with common traits

regulate *v.*, to control

reign *n.*, the period of rule for a king, queen, emperor, or empress

Reign of Terror *n.*, a movement in France led by Maximilien Robespierre in which 40,000 people were beheaded in 1793–94

reincarnation *n.*, the birth of the soul into another physical life

relative location *n.*, the position of a place in relation to other places

reliable *adj.*, dependable or trustworthy

relief *n.*, the change in elevation from one place to another

religious tolerance *n.*, the acceptance of different religions to be practiced at the same time, without prejudice

relocate *v.*, to move

remittance *n.*, money sent to a person in another place

remote *adj.*, hard to reach, isolated

Renaissance *n.*, meaning "rebirth," a period in the 1300–1500s in which culture and the arts flourished

renewable energy *n.*, energy from sources that do not run out, such as wind, sun, and water

renewable resource *n.*, a raw material or energy source that replaces itself over time

reparation *n.*, after a war, money paid as punishment, usually by the aggressors in the conflict

republic *n.*, a form of government in which officials are elected by the people to govern

reserve *n.*, land set aside for a special purpose such as farming, preserving habitats, or housing specific groups of people; a future supply (of oil)

reservoir *n.*, a large artificial lake that stores water

resistance *n.*, opposition

restore *v.*, to bring back

retreat *v.*, to go backward, not forward

revenue *n.*, income

reverse *v.*, to go in the opposite direction

Rhine River *n.*, a river that starts in Switzerland and ends in the North Sea

rift valley *n.*, a deep valley formed when Earth's crust separated, as in East Africa

Ring of Fire *n.*, an area along the rim of the Pacific Ocean where tectonic plates meet, causing many active volcanoes and earthquakes

ritual *n.*, a formal action that is regularly repeated

rivalry *n.*, competition or opposition between people

river basin *n.*, a low area drained by a river

Romantic Period *n.*, an artistic period in the early 1800s when artists painted landscapes and natural scenes to convey emotions

Rub al Khali *n.*, a large desert in southern Saudi Arabia

Russian Revolution *n.*, the revolution in Russia in 1917 that overthrew the czar and put the Bolsheviks in power

Russification *n.*, the policy of putting Russians in charge of Soviet republics during the 1970s and 1980s

GLOSSARY

S

Sahara Desert *n.*, the largest hot desert in the world, covering most of North Africa

Sahel (saa HEL) *n.*, in Sub-Saharan Africa, a semiarid grassland that separates the Sahara in the north from tropical grasslands in the south

salinization *n.*, the building up of salt in soil

Samoa *n.*, a country governing the western Samoan islands in the South Pacific Ocean

samurai (SAM uh ry) *n.*, a skilled Japanese warrior

sanitation *n.*, measures such as sewers to protect public health

Sanskrit *n.*, the language of the Aryans, which became the basis for many languages in South Asia

sarcophagus *n.*, a coffin

sari *n.*, a traditional Indian garment for women worn wrapped around the body

savanna *n.*, grassland, as in the south of Sub-Saharan Africa

scale *n.*, the part of a map that indicates how big an area of Earth is shown

scientific station *n.*, a place to carry out research

scorched earth policy *n.*, in 1812, practice in which Russian troops, as they retreated from Napoleon's army, burned crops and resources that could supply the enemy

seafarer *n.*, a sea traveler

secular *adj.*, worldly, not connected to a religion

segregation *n.*, separation by race

seize *v.*, to take control of

self-rule *n.*, the government of a country by its own people

semiarid *adj.*, somewhat dry, with very little rainfall

serf *n.*, from the 1500s to the 1800s, a poor Russian or European peasant farmer who rented land from a landlord and had few rights

shalwar-kameez *n.*, a long shirt with loose-fitting pants worn in India and Southwest Asia

Shang *n.*, a family whose dynasty ruled China from 1766–1050 B.C.

sheikh *n.*, an Arab leader

Shi Huangdi (shee hwahng dee) *n.*, the ruler of the Qin dynasty, who became China's first emperor in 221 B.C.

Shi'ite (SHEE eyt) *adj.*, a branch of Muslims who believe that religious leaders should be descendants of Muhammad

Shinto *n.*, a native religion of Japan that is similar to animism

shogun *n.*, a military governor of Japan

Siberia *n.*, a huge region in central and eastern Russia

significant *adj.*, important

Sikhism *n.*, a religion in India started in the late A.D. 1400s

Silk Roads *n.*, ancient trade routes that connected Southwest and Central Asia with China

Silla (SIHL uh) *n.*, a kingdom in southeastern Korea in 57 B.C.

silt *n.*, fine particles of soil deposited along riverbanks

Slavs *n.*, people who came from around the Black Sea or Poland and settled in the Ukraine and western Russia around A.D. 800

slum *n.*, an area in a city that is crowded, with poor housing and bad living conditions

socialism *n.*, a system of government in which the government controls economic resources

solstice *n.*, the point at which the sun is farthest north or farthest south of the Equator; the beginning of summer and winter

South Pole *n.*, the southernmost point on Earth, opposite the North Pole, where all lines of longitude meet

Southern Alps *n.*, a mountain range in New Zealand

Southern Hemisphere *n.*, the half of Earth south of the equator

sovereignty *n.*, a country's control over its own affairs

Soviet Union *n.*, the Union of Soviet Socialist Republics, a country made up of Russia and other Eurasian states, from 1922 to 1991

spatial thinking *n.*, a way of thinking about space on Earth's surface, including where places are located and why they are there

Special Economic Zone *n.*, an area in China that was allowed to develop a market economy with less government control of business

sprawl *v.*, to spread out

standard of living *n.*, the level of goods, services, and material comforts of people in a country

staple *n.*, a basic part of people's diets

state *n.*, a defined territory with its own government

steppe *n.*, a very large plain of dry grassland

strait *n.*, a narrow waterway that connects two bodies of water

strike *n.*, a work stoppage by employees who refuse to work

subcontinent *n.*, separate region of a continent

GLOSSARY

subsistence farmers *n.*, farmers who grow food for their families to eat, not to sell

subsistence fishing *n.*, fishing for enough food to live on, not for profit

suffrage *n.*, the right to vote

Suleyman I *n.*, the emperor of the Ottoman Empire in the mid-1500s

sultan *n.*, a leader or ruler of the Ottoman Empire

Sumatra *n.*, one of the outer islands in Southeast Asia

Sunni (SOO nee) *adj.*, a branch of Islam that believes the religious leaders should be chosen from those most qualified, as opposed to being descendants of Muhammad

sustainable *adj.*, capable of being continued without damaging the environment or using up resources permanently

Swahili *n.*, the Bantu language that is mainly spoken in East Africa, also knows as Kiswahwali

Sydney *n.*, the largest city in Australia and the state capital of New South Wales

symbol *n.*, an object or idea that can be used to represent another object or idea

T

taiga (TY guh) *n.*, the large forest area that stretches through northern Russia, Canada, and other northern countries

Taj Mahal *n.*, a famous building built as a tomb for Shah Jahan's wife in India in the 1600s; now a UNESCO World Heritage site

Taliban *n.*, a group of Pashtuns in Afghanistan who began ruling Afghanistan in 1996

Taman Negara National Park *n.*, a national park in Malaysia within one of the world's oldest rain forests

tariff *n.*, a tax on imports and exports

tectonic plate *n.*, a section of Earth's crust that floats on Earth's mantle

terra cotta *n.*, baked clay

terrace *n.*, a flat surface that is built into a hillside

terrain *n.*, the physical features of the land

terrorist *n.*, a person who uses violence to achieve political results

textile *adj.*, related to cloth or clothing

theme *n.*, topic

thirty-eighth parallel *n.*, the border along the 38° north latitude that separates North and South Korea, created in 1945

Three Gorges Dam *n.*, the largest dam in the world, located on the Chang Jiang River in China

Tibetan Plateau *n.*, a vast high plateau in Central Asia

Timbuktu *n.*, a city in West Africa that was a center of education in the 1200s

tolerance *n.*, acceptance of others' beliefs

tomb *n.*, a burial place

tornado *n.*, a storm with powerful winds that follows an unpredictable path

totalitarian *adj.*, relating to a government ruled by a dictator that requires complete obedience to the state

trans-Atlantic slave trade *n.*, the business of trading slaves taken from Africa across the Atlantic Ocean to the Americas starting in the 1500s

trans-Saharan *adj.*, across the Sahara desert

Trans-Siberian Railroad *n.*, the world's longest continuous railroad, linking Moscow with eastern Russia across Siberia

transform *v.*, to remake or change

transition zone *n.*, an area between two geographic regions that has characteristics of both

transport *v.*, to send from one place to another

transportation corridor *n.*, a land or water route to move people and goods from place to place easily

Treaty of Versailles *n.*, a peace treaty that ended World War I in 1919

trench *n.*, a long ditch that protects soldiers from enemy gunfire

trend *n.*, a change over time in a specific direction

tribute *n.*, fees paid to another ruler or country for protection or as a token of submission

troubadour *n.*, a singer during the Middle Ages who performed songs about knights and love

tsunami (soo NAH mee) *n.*, a large, powerful ocean wave

Tuareg (TWAH rehg) *n.*, a semi-nomadic people who travel across the Sahara in caravans, trading salt

tundra *n.*, flat treeless land found in arctic and subarctic regions

typhoon *n.*, a dangerous tropical storm with heavy rains and high winds, known as a hurricane in the Western Hemisphere

U

United Arab Emirates *n.,* a country located on the Arabian Peninsula at the Persian Gulf

United Nations (UN) *n.,* an organization of countries formed in 1945 to keep peace among countries and protect human rights

Universal Declaration of Human Rights *n.,* an agreement approved by the United Nations that defines the rights that people all over the world should have

uplands *n.,* hills, mountains, and plateaus

Ur *n.,* an important Sumerian city-state, 2800–1850 B.C.

Ural Mountains *n.,* a mountain range that separates the Northern European Plain from the West Siberian Plain in Russia

urban *adj.,* describing or related to the city or suburbs

V

vaccine *n.,* a treatment to increase immunity to a particular disease

vegetation *n.,* plant life

veto *v.,* to reject a decision made by another government body

vocational *adj.,* related to job skills

volcano *n.,* a mountain that erupts in an explosion of molten rock, gases, and ash

vulnerable *adj.,* open, able to be hurt by outside forces

W

wallaby *n.,* a marsupial animal that is smaller than a kangaroo

wat *n.,* a Buddhist temple in Southeast Asia

waterway *n.,* a navigable route for traveling and transport

weapon of mass destruction (WMD) *n.,* a weapon that causes great harm to large numbers of people

weather *n.,* the condition of the atmosphere at a particular time, including temperature, precipitation, and humidity for a particular day or week

Western Hemisphere *n.,* the half of Earth west of the prime meridian

Wilkes, Charles *n.,* an American who surveyed the coastline of Antarctica in the late 1830s

Y

yurt *n.,* a traditional felt tent of Central Asia

Z

zaibatsu (zeye BAHT sue) *n.,* family–run organizations in Japan that owned several businesses

Zambezi River *n.,* a river in Southern Africa that flows through many south-central countries to the Indian Ocean

Zheng He (jung huh) *n.,* an admiral in the Chinese navy who led seven naval voyages exploring lands outside China; his last voyage began in A.D. 1431

Zhou (joh) *n.,* the dynasty that ruled China from 1050–221 B.C.

zoologist *n.,* a scientist who studies animals

A

Aborigine [aborigen] *s.*, primera cultura desarrollada de Australia

absolute location [ubicación absoluta] *s.*, punto exacto donde está ubicado un lugar, identificado por medio de las coordenadas de latitud y longitud

abstract [abstracto] *adj.*, estilo artístico que enfatiza la forma y el color por sobre el realismo

accommodate [albergar] *v.*, alojar

Acropolis [Acrópolis] *s.*, colina rocosa ubicada en Atenas, Grecia, que servía de fortaleza para los edificios más importantes de la ciudad antigua

adapt [adaptar] *v.*, ajustar o modificar para que sea apropiado

adherent [partidario] *s.*, seguidor de una religión, causa o persona

adventure tourism [turismo aventura] *s.*, tipo de turismo en el cual los viajeros realizan actividades físicas, como montañismo, deportes acuáticos o senderismo

aerodynamic [aerodinámico] *adj.*, diseñado para moverse con poca resistencia del viento

African National Congress (ANC) [Congreso Nacional Africano (CNA)] *s.*, organización de sudafricanos negros (entre ellos, Nelson Mandela) que protestaron en contra del trato discriminatorio a principios del siglo XX

African Union [Unión Africana] *s.*, organización de países africanos que trabajan para lograr el progreso económico

aging population [envejecimiento de la población] *s.*, tendencia demográfica que ocurre cuando aumenta la edad media de una población

agricultural revolution [revolución agrícola] *n.*, período en el cual los seres humanos comenzaron a cultivar en lugar de recolectar plantas

Aguinaldo, Emilio [Aguinaldo, Emilio] *s.*, líder del movimiento por la independencia filipina en la década de 1890

Alexander the Great [Alejandro Magno] *s.*, conquistador que extendió el Imperio Macedónico y difundió la cultura griega por Eurasia desde 334–323 A.C.

alliance [alianza] *s.*, sociedad entre países

alluvial [aluvial] *adj.*, dicho del sedimento depositado por un río

alluvial plain [llanura aluvial] *s.*, área de tierra plana ubicada junto a un arroyo o río que se desborda

alluvium [aluvión] *s.*, suelo o cieno arrastrado por el agua que fluye, que es ideal para los cultivos

Alps [Alpes] *s.*, cadena montañosa europea

al-Qaeda [Al Qaeda] *s.*, grupo terrorista con sede en el suroeste de Asia

Amazon River Basin [cuenca del río Amazonas] *s.*, la cuenca fluvial más grande de la Tierra, ubicada en América del Sur

Angkor Wat [Angkor Wat] *s.*, templo grande construido en Camboya en el siglo XII, dedicado a la deidad hindú Vishnú

animal-borne [de tracción animal] *adj.*, transportado por animales

anime [anime] *s.*, estilo de animación o historieta desarrollado en Japón

animism [animismo] *s.*, creencia de que todas las cosas, incluidos los objetos naturales, tienen alma

Antarctic Treaty [Tratado Antártico] *s.*, acuerdo de 1959 entre 12 países para usar la Antártida con fines pacíficos y para compartir los descubrimientos científicos

apartheid [apartheid] *s.*, separación legal de las razas; sistema que privaba de sus derechos a los sudafricanos negros

aqueduct [acueducto] *s.*, sistema de transporte para llevar agua a grandes distancias, a veces elevado sobre un puente

aquifer [acuífero] *s.*, capa de roca subterránea que contiene agua

arable [cultivable] *adj.*, fértil; apropiado para la agricultura

Aral Sea [mar Aral] *s.*, lago de agua salada ubicado en el centro de Asia, cuyo tamaño se ha reducido mucho como consecuencia del desvío para irrigación de los ríos que desembocan en él

archipelago [archipiélago] *s.*, cadena de islas

arid [árido] *adj.*, muy seco, que casi no recibe lluvia

aristocrat [aristócrata] *s.*, miembro de la clase alta

armistice [armisticio] *s.*, acuerdo para dejar de luchar

artifact [artefacto] *s.*, objeto hecho por seres humanos de una cultura pasada

Aryans [arios] *s.*, nómadas que migraron desde el centro de Asia al valle del Indo alrededor del año 2000 A.C.

Asia-Pacific Economic Cooperation (APEC) [Foro de Cooperación Económica Asia-Pacífico] *s.*, asociación global fundada en 1989 para fortalecer los vínculos económicos entre los países del Pacífico

Asoka [Asoka] *s.*, líder durante el apogeo del Imperio Maurya, alrededor del año 250 A.C.

assimilate [asimilarse] *v.*, incorporarse en la cultura de una sociedad

assimilation [asimilación] *s.*, proceso en el cual un grupo minoritario adopta la cultura de la mayoría

assisted migration [migración asistida] *s.*, programa establecido en 1831 en el cual los británicos ofrecían dinero a la gente para que se mudara a Australia

Aswan High Dam [Presa Alta de Asuán] *s.*, presa ubicada en el río Nilo, terminada en 1970

atoll [atolón] *s.*, isla, arrecife o cadena de islas con forma de anillo hechas de coral

attribute [atributo] *s.*, cualidad específica

Augustus [Augusto] *s.*, primer emperador de Roma, en 27 A.C.

autonomy [autonomía] *s.*, auto-gobierno de un país o pueblo

B

ban [proscribir] *v.*, declarar fuera de la ley o prohibir

Bantu [bantúes] *s.*, grupo africano que se trasladó desde el oeste de África central hacia el sur y hacia el este, a través del África subsahariana, entre 2000 y 1000 A.C.

barbarian [bárbaro] *s.*, soldado o guerrero considerado culturalmente menos desarrollado que aquellos contra los que luchaba; el término proviene originalmente de las tribus germanas que invadieron el Imperio Romano en el año 235 D.C.

Baroque period [Barroco] *s.*, período entre 1600 y 1750 en el cual la música presentaba esquemas y temas complicados

barter [canjear] *v.*, comerciar o intercambiar bienes sin utilizar dinero

basin [cuenca] *s.*, región bañada por un sistema fluvial

bas-relief [bajorrelieve] *s.*, escultura que sobresale ligeramente de un fondo plano

bauxite [bauxita] *s.*, materia prima que se utiliza para fabricar aluminio

bay [bahía] *s.*, masa de agua rodeada por tierra en tres de sus lados

Bedouin [beduinos] *s.*, pueblo nómade de habla árabe del suroeste de Asia

benefit [beneficiar] *v.*, hacer bien

Berlin Conference [Conferencia de Berlín] *s.*, encuentro de las naciones europeas en 1884 para resolver los conflictos sobre sus reclamos coloniales en África

Berlin Wall [Muro de Berlín] *n.*, muro que dividía Berlín Este communista de Berlín Oeste demucrático

Bhutto, Benazir [Bhutto, Benazir] *s.*, primera ministra mujer de Pakistán, 1988–1999

Biko, Stephen [Biko, Stephen] *s.*, presidente de una organización sudafricana de protesta estudiantil que fue arrestado y asesinado en 1977

biodiversity [biodiversidad] *s.*, variedad de especies que viven en un ecosistema

Black Sea [mar Negro] *s.*, mar interior que limita con Europa, Rusia, Georgia y Turquía

Blue Mosque [Mezquita Azul] *s.*, mezquita histórica construida en Estambul por el Sultán Ahmed I, terminada en 1617

Bollywood [Bollywood] *s.*, industria cinematográfica de la India, ubicada en Bombay

Bolshevik [bolchevique] *s.*, partido político de Rusia liderado por Lenin que derrocó al zar en 1917

Borobudur [Borobudur] *s.*, complejo de templos budistas ubicado en el centro de Java

Bosporus Strait [estrecho de Bósforo] *s.*, paso que atraviesa Estambul y conecta el mar Negro y el mar de de Mármara

breakwater [rompeolas] *s.*, barrera construida para proteger un puerto

Buddhism [budismo] *s.*, religión fundada por Siddhartha Gautama en 525 A.C. en India

bullet train [tren bala] *s.*, tren que viaja a más de 125 millas por hora y cuya forma es similar a una bala

butte [cerro testigo] *s.*, colina o montaña con laderas escarpadas y cima plana

Byrd, Richard [Byrd, Richard] *s.*, estadounidense que exploró la Antártida por cielo y tierra durante la década de 1940

Byzantine Empire [Imperio Bizantino] *s.*, parte oriental del Imperio Romano después de 395 D.C. que duró aproximadamente 1,000 años

C

Caesar, Julius [César, Julio] *s.*, general que se convirtió en gobernante de Roma, 46–44 A.C.

canal [canal] *s.*, vía fluvial construida por el hombre a través de la tierra para botes y barcos

capital [capital] *s.*, riqueza e infraestructura de un país

carapace [caparazón] *s.*, cubierta dura de ciertos animales, como la tortuga

caravan [caravana] *s.*, grupo de mercaderes que viajan juntos por razones de seguridad

Cardamom Mountains [colinas Cardamom] *s.*, cadena de montañas ubicada en Camboya

cartographer [cartógrafo] *s.*, persona que hace mapas

Caspian Sea [mar Caspio] *s.*, la masa de agua endorreica más grande de la Tierra, que limita con Rusia, Kazajistán, Turkmenistán, Azerbaiyán e Irán

caste system [sistema de castas] *s.*, en India, estructura social compuesta por cuatro niveles principales, que fue instaurada por los arios

categorize [categorizar] *v.*, agrupar información; clasificar

Catherine the Great [Catalina la Grande] *s.*, emperadora de Rusia desde 1762 hasta 1796

celadon [celadón] *adj.*, con barniz vítreo de color verde, referido a la cerámica realizada por los artistas coreanos

Chang Jiang [Chang Jiang] *s.*, el río más largo de Asia, que fluye a través de China; también llamado Yangtsé

Chang Jiang Plain [llanura de Chang Jiang] *s.*, llanuras ubicadas junto al río Chang Jiang

Chao Phraya River [río Chao Phraya] *s.*, río que fluye a través de Tailandia

Chernobyl [Chernóbil] *s.*, ciudad ubicada en Ucrania, en la cual un reactor nuclear explotó en 1986

Choson [Choson] *s.*, dinastía coreana que duró desde 1392 hasta 1910

Christian Bible [Biblia cristiana] *s.*, libro sagrado del cristianismo

Christianity [cristianismo] *s.*, religión basada en la vida y en las enseñanzas de Jesús de Nazaret

citizen [ciudadano] *s.*, persona que vive dentro de un territorio y tiene derechos y responsabilidades garantizadas por el gobierno

city-state [ciudad estado] *s.*, estado independiente compuesto por una ciudad y los territorios que dependen de ella

civil disobedience [desobediencia civil] *s.*, desobediencia no violenta de las leyes

civilization [civilización] *s.*, sociedad con cultura, política y tecnología altamente desarrolladas

clan [clan] *s.*, unidad grande basada en la familia con lealtad hacia el grupo

Classical period [Clasicismo] *s.*, período comprendido entre 1750 y 1900 en el cual la música seguía las reglas establecidas para la forma y la complejidad, como en el caso de las sonatas o las sinfonías

climate [clima] *s.*, el promedio de las condiciones de la atmósfera de un área durante un largo período de tiempo, incluidas la temperatura, la precipitación y los cambios estacionales

climograph [gráfica del clima] *s.*, gráfica que muestra el clima de una región a través de la precipitación y la temperatura medias

coalition [coalición] *s.*, alianza

Cold War [Guerra Fría] *s.*, largo período de tensión política sin lucha armada entre los Estados Unidos y la Unión Soviética, desde aproximadamente 1948 hasta 1991

collective farm [granja colectiva] *s.*, en la Unión Soviética, una granja grande donde los trabajadores cultivaban alimentos para distribuirlos a toda la población

collide [colisionar] *v.*, chocar

colonialism [colonialismo] *s.*, práctica de un país que directamente gobierna y desarrolla el comercio en un territorio extranjero para su propio beneficio

colony [colonia] *s.*, área controlada por un país lejano

commerce [comercio] *s.*, negocio de comprar y vender bienes y servicios

commodity [mercancía] *s.*, material u objeto que se puede comprar, vender o comerciar

Common Market [Mercado Común] *s.*, Comunidad Económica Europea formada en 1957

commonwealth [mancomunidad] *s.*, nación que se gobierna a sí misma pero es parte de un país mayor

Commonwealth of Australia [Mancomunidad de Australia] *s.*, la nación de Australia, que entró a formar parte de la Mancomunidad Británica de Naciones en 1901

communal [comunal] *adj.*, compartido

communism [comunismo] *s.*, sistema de gobierno en el cual un único partido político controla el gobierno y la economía

competitive [competitivo] *adj.*, capaz de intentar ganar un concurso o una carrera

complex [complejo] *s.*, conjunto de edificios vinculados

comply [acatar] *v.*, seguir una regla o una orden

concentrated [concentrado] *adj.*, reunido en un área central

concentration camp [campo de concentración] *s.*, área donde los judíos y otras personas fueron detenidos durante la Segunda Guerra Mundial y asesinados por los nazis

condensation [condensación] *s.*, proceso por el cual el vapor de agua se convierte en gotitas de líquido debido al enfriamiento durante el ciclo hidrológico

Confucianism [confucianismo] *s.*, sistema ético basado en las enseñanzas de Confucio

Congo River [río Congo] *s.*, río principal ubicado en África Central que desemboca en el océano Atlántico

conservation [conservación] *s.*, protección del medio ambiente

Constantine [Constantino] *s.*, emperador romano, 306–337 A.C., que hizo del cristianismo la religión oficial del imperio

consumer [consumidor] *s.*, persona que compra bienes

contaminated [contaminado] *adj.*, infectado; inapropiado para el uso debido a la presencia de elementos peligrosos

continent [continente] *s.*, gran masa de tierra sobre la superficie de la Tierra; la Tierra tiene siete continentes

continental drift [deriva continental] *s.*, movimiento lento de los continentes sobre las placas tectónicas

continental shelf [plataforma continental] *s.*, borde de un continente que se extiende bajo el agua y se adentra en el mar

controversy [controversia] *s.*, debate o disputa

GLOSARIO

convert [convertir] *v.*, persuadir a alguien para que cambie sus creencias religiosas

convey [transmitir] *v.*, comunicar

convict [convicto] *s.*, persona sentenciada a prisión por cometer un crimen

Cook Islands [islas Cook] *s.*, país ubicado cerca de Nueva Zelanda, en el Pacífico Sur, formado por 15 islas pequeñas

Cook, James [Cook, James] *s.*, navegante británico que exploró y cartografió las Islas del Pacífico en la década de 1770

coral island [isla de coral] *s.*, isla creada por la acumulación gradual de esqueletos de coral

coral reef [arrecife de coral] *s.*, estructura parecida a una roca construida por capas de organismos de coral

cosmopolitan [cosmopolita] *adj.*, que reúne muchas culturas e influencias diferentes

Counter-Reformation [Contrarreforma] *s.*, movimiento dentro de la Iglesia Católica Romana para reformar sus propias prácticas

coup [golpe de estado] *s.*, toma repentina e ilegal del gobierno por medio de la fuerza

crevasse [grieta] *s.*, abertura profunda en un glaciar

cricket [cricket] *s.*, juego de equipo similar al béisbol

Crusades [Cruzadas] *s.*, expediciones militares de la Iglesia Católica Romana para recuperar las tierras santas ubicadas en Medio Oriente que estaban bajo control musulmán, 1096–1291

cuisine [cocina] *s.*, alimentos y tradiciones culinarias comunes a cierta región

cultural hearth [centro cultural] *s.*, centro de civilización desde el cual se difunden las ideas y la tecnología

Cultural Revolution [Revolución Cultural] *s.*, en China, plan social que duró desde 1966 hasta 1969, en el cual el gobierno intentó eliminar todos los elementos capitalistas y anticomunistas de la sociedad

culture [cultura] *s.*, modo de vida de un grupo, que incluye el tipo de alimentación, vivienda, vestido, idioma, comportamiento e ideas

culture region [región cultural] *s.*, área que está unificada por rasgos culturales comunes

cuneiform [cuneiforme] *s.*, primera forma de escritura conocida, proveniente de Sumeria

currency [moneda] *s.*, forma de dinero

cyclone [ciclón] *s.*, tormenta con vientos giratorios, llamada tifón en el Hemisferio Oriental; denominada huracán en otras partes del mundo

czar [zar] *s.*, término utilizado en Rusia para designar al emperador

D

Danube River [río Danubio] *s.*, río que nace en Alemania y desemboca en el mar Negro

Daoism [taoísmo] *s.*, sistema ético que enfatiza la armonía entre las personas y la naturaleza

deforestation [deforestación] *s.*, práctica de talar bosques para despejar la tierra y utilizarla para cultivos o uso urbano

deity [deidad] *s.*, dios o diosa

delta [delta] *s.*, área donde un río deposita sedimentos cuando desemboca en una masa de agua más grande

demilitarized zone (DMZ) [zona desmilitarizada] *s.*, área neutral entre países enemigos, específicamente entre Corea del Norte y Corea del Sur

democracy [democracia] *s.*, forma de gobierno del pueblo, en la cual los ciudadanos suelen elegir representantes para que los gobiernen

democratization [democratización] *s.*, proceso de convertirse en una democracia

demographics [demografía] *s.*, características de una población humana, tales como la edad, el ingreso y la educación

deny [negar] *v.*, no reconocer

deplete [agotar] *v.*, extraer o disminuir un recurso hasta que no esté disponible o no sea capaz de funcionar como planeado

desertification [desertificación] *s.*, transición gradual de la tierra fértil a tierra menos productiva

developed nations [naciones desarrolladas] *s.*, países con un alto producto interno bruto per cápita

developing nations [naciones en vías de desarrollo] *s.*, países con un bajo producto interno bruto per cápita

dialect [dialecto] *s.*, variante regional de un idioma principal

Diaspora [Diáspora] *s.*, dispersión de los judíos por todo el mundo

diffusion [difusión] *s.*, propagación

diplomacy [diplomacia] *s.*, discusión entre grupos o países para resolver disputas o desacuerdos

discrimination [discriminación] *s.*, trato injusto de un individuo o grupo en base a factores distintos de la habilidad

displace [desplazar] *v.*, obligar a un pueblo a dejar su hogar

disregard [desconocer] *v.*, ignorar

dissolve [disolver] *v.*, deshacer

distort [distorsionar] *v.*, modificar la forma o apariencia usual; deformar

distribute [distribuir] *v.*, repartir un bien o recurso entre un grupo mayor

domestic policy [política interna] *s.*, plan de un gobierno para manejar los asuntos dentro de sus propias fronteras

domesticate [domesticar] *v.*, criar y usar animales como fuente de mano de obra y alimento

dormant [inactivo] *adj.*, latente, referido a un volcán

drought [sequía] *s.*, largo período con escasa o ninguna precipitación

Dubai [Dubái] *s.*, ciudad y estado joven y rico de los Emiratos Árabes Unidos, en el golfo Pérsico

Dutch East India Company [Compañía Neerlandesa de las Indias Orientales] *s.*, compañía de los Países Bajos que dominaba el comercio de especias en el sureste de Asia

dynastic cycle [ciclo dinástico] *s.*, patrón de ascenso y caída de las dinastías en la historia china

dynasty [dinastía] *s.*, serie de gobernantes de la misma familia

E

earthquake [terremoto] *s.*, sacudida de la corteza de la Tierra generalmente causada por el choque o el deslizamiento de las placas tectónicas

Eastern Hemisphere [Hemisferio Oriental] *s.*, la mitad de la Tierra ubicada al este del primer meridiano

ecologist [ecologista] *s.*, científico que estudia las relaciones entre los seres vivos y su medio ambiente

economic globalization [globalización económica] *s.*, práctica de actividades económicas que se realizan a través de las fronteras nacionales

economy [economía] *s.*, sistema en el cual las personas producen, venden y compran cosas

ecosystem [ecosistema] *s.*, comunidad de organismos vivos y su medio ambiente o hábitat natural

ecotourism [ecoturismo] *s.*, manera de visitar las áreas naturales que conserva los recursos naturales de la región

elevation [elevación] *s.*, altura de un accidente geográfico sobre el nivel del mar

eliminate [eliminar] *v.*, deshacerse de

elite [elite] *s.*, clase superior

emergence [aparición] *s.*, desarrollo y uso extendido de algo nuevo

emigrate [emigrar] *v.*, dejar el país natal para vivir en otro país

emirate [emirato] *s.*, estado del país de los Emiratos Árabes Unidos

empire [imperio] *s.*, grupo de pueblos o estados gobernados por un único gobernante fuerte

Enduring Voices Project [Proyecto Voces Perdurables] *s.*, proyecto de National Geographic para estudiar y conservar las lenguas que están en riesgo de desaparición

enhance [realzar] *v.*, mejorar la calidad de algo

Enlightenment [Ilustración] *s.*, movimiento social del siglo XVIII que trabajó a favor de la educación y de los derechos del individuo

enlist [alistarse] *v.*, presentarse como voluntario para el servicio militar

entrepreneur [empresario] *s.*, persona que comienza un negocio nuevo

entrepreneurship [espíritu empresarial] *s.*, características de creatividad y riesgo que existen en una persona o sociedad

epic poem [poema épico] *s.*, poema largo que relata las aventuras de un héroe

epidemic [epidemia] *s.*, brote de una enfermedad que afecta a una gran parte de la población de una comunidad en particular

equator [ecuador] *s.*, círculo imaginario alrededor de la Tierra que está a la misma distancia del Polo Norte y del Polo Sur y divide la Tierra por la mitad; la línea central o de 0° de latitud

equinox [equinoccio] *s.*, momento en que el día y la noche tienen la misma duración; ocurre dos veces al año, el 21 o 22 de marzo y el 21 o 22 de septiembre

erode [erosionar] *v.*, desgastar

erosion [erosión] *s.*, proceso por el cual las rocas y el suelo se rompen lentamente y se desgastan

eruption [erupción] *s.*, estallido o explosión

escarpment [escarpe] *s.*, pendiente pronunciada

essential [esencial] *adj.*, necesario

establish [establecer] *v.*, instituir

ethical system [sistema ético] *s.*, sistema de creencias que enseña conductas morales

ethnic group [grupo étnico] *s.*, grupo de personas que tienen una cultura y una lengua en común y a veces comparten la misma herencia racial

ethnobotanist [etnobotánico] *s.*, científico que estudia la relación entre las culturas y las plantas

euro [euro] *s.*, moneda común de la Unión Europea

European Union (EU) [Unión Europea (UE)] *s.*, organización económica compuesta por 27 países miembros europeos (2011)

eurozone [eurozona] *s.*, países que han adoptado el euro como moneda

evaporation [evaporación] *s.*, proceso por el cual el agua se convierte en vapor y sube a la atmósfera debido al calentamiento del Sol; una parte del ciclo hidrológico

evident [evidente] *adj.*, claramente presente u observable

exchange [cambiar] *v.*, convertir el dinero en otra moneda

exile [exilio] *s.*, situación en la que se está ausente del propio país natal

exoskeleton [exoesqueleto] *s.*, cubierta externa dura que protege a los corales y a otros animales marinos

expand [ampliar] *v.*, hacer más grande

expedition [expedición] *s.*, travesía o viaje de cierta extensión, usualmente a otras tierras

extent [extensión] *s.*, distancia o grado hasta el cual se difunde algo

extinct [extinto] *adj.*, completamente desaparecido, tal como una especie animal o vegetal

extremist [extremista] *s.*, persona con opiniones religiosas o políticas que están fuera del rango de la opinión de la mayoría

F

factory system [sistema fabril] *s.*, modo de trabajar en el cual cada persona trabaja en solo una parte de un producto

failed state [estado fallido] *s.*, país en el cual el gobierno, las instituciones económicas y el orden han colapsado

fallout [lluvia radiactiva] *s.*, partículas radiactivas provenientes de una explosión nuclear que caen a través de la atmósfera

famine [hambruna] *s.*, largo período de escasez de alimentos

fault [falla] *s.*, fractura en la corteza de la Tierra

federal republic [república federal] *s.*, forma democrática de gobierno en la cual los votantes eligen representantes y el gobierno central comparte el poder con los estados

federal system [sistema federal] *s.*, sistema de gobierno con un gobierno central fuerte y unidades gubernamentales locales

feral [asilvestrado] *adj.*, salvaje, que se ha convertido en salvaje después de haber sido domesticado

Fertile Crescent [Creciente Fértil] *s.*, área alrededor de los ríos Tigris y Éufrates

fertility rate [tasa de fertilidad] *s.*, promedio de niños nacidos por cada mujer en un cierto período de tiempo

feudal system [sistema feudal] *s.*, durante la Edad Media, estructura social que consistía en un rey, señores, vasallos y siervos

first language [lengua materna] *s.*, idioma que aprenden los niños de un pueblo

fjord [fiordo] *s.*, bahía angosta y profunda ubicada

floodplain [planicie aluvial] *s.*, terreno bajo a la orilla de los ríos, formado por el sedimento dejado por la inundación

Foja Mountains [montañas Foja] *s.*, cadena de montañas ubicada en la isla de Nueva Guinea, en el este de Indonesia

fossil [fósil] *s.*, restos conservados de plantas y animales antiguos

fragmented country [país fragmentado] *s.*, país que está físicamente dividido en partes separadas, tales como una cadena de islas, y/o política o culturalmente dividido

free enterprise economy [economía de libre empresa] *s.*, sistema en el cual las empresas de propiedad privada producen bienes y servicios, también llamada economía de mercado o capitalismo

free trade [libre comercio] *s.*, comercio que no grava las importaciones con impuestos

free trade agreement [acuerdo de libre comercio] *s.*, tratado entre países que mejora el comercio al limitar los impuestos a dicho comercio

French Indochina [Indochina Francesa] *s.*, grupo de colonias pertenecientes a Francia, formado por Vietnam, Camboya, Laos y otras, ubicadas en el sureste de Asia, desde 1887 hasta 1955

fuel cell [celda de combustible] *s.*, unidad pequeña que, al igual que una pila o batería, combina sustancias químicas para producir energía

G

Gandhi, Mohandas [Gandhi, Mahatma] *s.*, líder indio que impulsó la independencia de la India respecto del Reino Unido mediante la desobediencia civil, en la década de 1930; considerado el padre de la India moderna

Ganges Delta [delta del Ganges] *s.*, área fértil donde el río Ganges desemboca en la bahía de Bengala

gaucho [gaucho] *s.*, vaquero de América del Sur

gauge [trocha] *s.*, medida del ancho de las vías del ferrocarril

generation [generación] *s.*, grupo de individuos que nacen y viven aproximadamente en la misma época

Genghis Khan [Gengis Kan] *s.*, gobernante mongol que estableció un imperio en el centro de Asia, a principios del siglo XIII

genre [género] *s.*, forma literaria, como un poema, una obra de teatro o una novela

Geographic Information Systems (GIS) [Sistemas de Información Geográfica (SIG)] *s.*, aparatos computarizados que presentan datos sobre lugares específicos

geographic pattern [patrón geográfico] *s.*, similitud entre lugares

geothermal energy [energía geotérmica] *s.*, energía térmica proveniente del interior de la Tierra que se puede convertir en electricidad

ger [ger] *s.*, tienda portátil hecha de fieltro

Giza [Giza] *s.*, ciudad ubicada junto al Nilo, donde se construyeron diez pirámides en el antiguo Egipto

Glacier [glaciar] *s.*, masa grande de hielo y nieve acumulada

glasnost [glásnost] *s.*, política de la Unión Soviética de apertura que animó al pueblo a hablar abiertamente acerca del gobierno, introducida por Mijaíl Gorbachov en la década de 1980

Global Positioning System (GPS) [Sistema de Posicionamiento Global (GPS)] *s.*, sistema de satélites con base en el espacio que encuentra la ubicación absoluta y la hora de cualquier lugar de la Tierra

globe [globo terráqueo] *s.*, modelo tridimensional, o esférico, de la Tierra

Gobi desert [desierto de Gobi] *s.*, el desierto más grande de Asia, que cubre gran parte del sur de Mongolia y se extiende hasta el interior de China

golden age [edad dorada] *s.*, período de gran riqueza, cultura y democracia en Grecia

Golden Quadrilateral (GQ) [Cuadrilátero de Oro] *s.*, superautopista que conecta cuatro ciudades importantes de la India

Gorbachev, Mikhail [Gorbachov, Mijaíl] *s.*, líder de la Unión Soviética, 1985–1991; presidente desde 1990 hasta que fue disuelta en 1991

gorge [desfiladero] *s.*, paso profundo y angosto rodeado por acantilados empinados

government [gobierno] *s.*, organización que mantiene el orden, establece reglas y proporciona servicios para una sociedad

Grand Bazaar [Gran Bazar] *s.*, centro de tiendas y comercios en Estambul, Turquía

Great Barrier Reef [Gran Barrera de Coral] *s.*, ecosistema inmenso, ubicado cerca de la costa de Australia, hecho de arrecifes de coral

Great Depression [Gran Depresión] *s.*, descenso económico mundial ocurrido en la década de 1930, marcado por la pobreza y una alta tasa de desempleo

Great Escarpment [Gran Escarpe] *s.*, pendiente pronunciada que se extiende desde la meseta del sur de África hasta las llanuras costeras

Great Leap Forward [Gran Salto Adelante] *s.*, plan de Mao Zedong para hacer que la economía de China creciera más rápidamente, 1958–1961

Great Plains [Grandes Llanuras] *s.*, área de tierra baja y llana ubicada al este de las montañas Rocosas

Great Pyramid of Khufu [Gran Pirámide de Keops] *s.*, la pirámide más antigua y más alta de Egipto, construida en Guiza alrededor de 2550 A.C.

Great Rift Valley [Gran Valle del Rift] *s.*, valle ancho ubicado en África Oriental, parte de una cadena de valles formados cuando las placas tectónicas se separaron

Great Wall of China [Gran Muralla China] *s.*, muro de piedra ubicado en el norte de China de más de 4,500 millas de largo, construido para rechazar a los invasores provenientes del norte

Great Zimbabwe [Gran Zimbabue] *s.*, ciudad amurallada de piedra que el pueblo shona construyó en el sur de África entre 1200 y 1450

greenhouse gas [gas invernadero] *s.*, gas que atrapa el calor del Sol sobre la Tierra

griot [griot] *s.*, narrador africano tradicional

gross domestic product (GDP) [producto interno bruto (PIB)] *s.*, valor total de todos los bienes y servicios producidos en un país en un año dado

guest worker [trabajador huésped] *s.*, trabajador temporal que migra para trabajar en otro país

Guillotine [guillotina] *s.*, máquina utilizada para ejecutar a las personas durante la Revolución Francesa

Gutenberg, Johannes [Gutenberg, Johannes] *s.*, impresor alemán que inventó la imprenta en 1450

H

habitat [hábitat] *s.*, medio ambiente natural de una planta o animal vivo

Hagia Sophia [Santa Sofía] *s.*, museo ubicado en Estambul, Turquía, que se construyó originalmente como iglesia, durante el Imperio Romano, y luego sirvió de mezquita

half-life [vida media] *s.*, tiempo necesario para que la mitad de los átomos de una sustancia radiactiva se desintegren

Hammurabi [Hammurabi] *s.*, rey que desarrolló un código de leyes en Babilonia, en el siglo XVIII A.C.

Han [Han] *s.*, dinastía china que duró desde 206 A.C. hasta 220 D.C.

Harappan [Harappa] *s.*, primera civilización urbana del sur de Asia, desarrollada junto al río Indo alrededor de 2600 A.C.

Hatshepsut [Hatshepsut] *s.*, mujer que fue faraón de Egipto alrededor de 1500 A.C.

Hebrew Bible [Biblia hebrea] *s.*, libro sagrado del judaísmo

hemisphere [hemisferio] *s.*, una mitad de la Tierra

henna [alheña] *s.*, polvo rojizo usado para crear diseños sobre la piel

hereditary [hereditario] *adj.*, que se transmite de padres a hijos

heritage [herencia] *s.*, tradición que se transmite de los ancestros a los descendientes

Hermitage Museum [museo Hermitage] *s.*, museo de arte y cultura ubicado en San Petersburgo, Rusia

hieroglyphics [jeroglíficos] *s.*, sistema antiguo de escritura que emplea imágenes y símbolos

highlands [tierras altas] *s.*, áreas de terreno montañoso alto

Himalaya Mountains [Himalaya] *s.*, la cordillera más alta del mundo, ubicada en el sur de Asia

Hinduism [hinduismo] *s.*, religión practicada por más del 60 por ciento de la población de Asia del Sur

Hitler, Adolf [Hitler, Adolf] *s.*, jefe de estado alemán desde 1933 hasta 1945

Ho Chi Minh [Ho Chi Minh] *s.*, líder de Vietnam del Norte que quería unificar Vietnam del Norte y Vietnam del Sur bajo el comunismo; estos intentos dieron comienzo a la Guerra de Vietnam

Holocaust [Holocausto] *s.*, asesinato masivo cometido por los nazis contra seis millones de judíos y otras personas durante la Segunda Guerra Mundial

homelands [terruños] *s.*, durante el apartheid, áreas separadas dentro de Sudáfrica donde se obligaba a vivir a los sudafricanos negros

hotspot [punto caliente] *s.*, parte inusualmente caliente del manto de la Tierra

Huang He [Huang He] *s.*, el segundo río más largo de China, también llamado río Amarillo

human rights [derechos humanos] *s.*, derechos políticos, económicos y culturales que todas las personas deben tener

hurricane [huracán] *s.*, tormenta fuerte con vientos giratorios y lluvia intensa

Hussein, Saddam [Hussein, Saddam] *s.*, presidente de Iraq desde 1979 hasta 2003

hydroelectric power [energía hidroeléctrica] *s.*, fuente de energía que emplea agua en movimiento para producir electricidad

I

ice shelf [capa de hielo] *s.*, lámina flotante de hielo adherida a una masa de tierra

immigration [inmigración] *s.*, mudanza permanente de una persona a otro país

impact [impacto] *s.*, efecto que produce un cambio

imperialism [imperialismo] *s.*, práctica de extender la influencia de una nación controlando otros territorios

Impressionism [Impresionismo] *s.*, estilo artístico en el cual los artistas usaban la luz y el color en pinceladas cortas para capturar un momento de tiempo

incentive [incentivo] *s.*, razón o motivo para hacer algo

incorporate [incorporar] *v.*, incluir; combinar con algo ya formado

indigenous [indígena] *adj.*, originario del área donde algo se encuentra

indulgence [indulgencia] *s.*, durante la Edad Media, tarifa pagada a la iglesia para disminuir el castigo por un pecado

Indus River [río Indo] *s.*, río ubicado en la parte occidental de Asia del Sur

Industrial Revolution [Revolución Industrial] *s.*, período de los siglos XVIII y XIX durante el cual los trabajadores de las fábricas comenzaron a usar máquinas y herramientas eléctricas para producir a gran escala

industrialize [industrializar] *v.*, desarrollar la manufactura

infectious [infeccioso] *adj.*, capaz de propagarse rápidamente a otros

infrastructure [infraestructura] *s.*, sistemas básicos de una sociedad, tales como las carreteras, los puentes, las cloacas y el tendido eléctrico

inner islands [islas interiores] *s.*, islas de Indonesia entre las que se incluyen Java, Madura y Bali

interact [interactuar] *v.*, afectar a otras personas y verse afectado por ellas

interior [interior] *s.*, tierra que está lejos de la costa del mar

intersection [encrucijada] *s.*, lugar donde se encuentran las personas o se cruzan los caminos

intifada [intifada] *s.*, levantamiento o rebelión, usualmente se utiliza para referirse a la revuelta palestina en contra de Israel

invader [invasor] *s.*, enemigo que entra a un país por la fuerza

invasive species [especies invasoras] *s.*, plantas o animales no nativos introducidos en un área nueva, intencionalmente o no, que alteran los hábitats de los seres vivos nativos

Iron Curtain [Cortina de Hierro] *s.*, en Europa, frontera imaginaria que separaba los países comunistas de los no comunistas durante la Guerra Fría

Irrawaddy River [río Irawadi] *s.*, río que fluye a través de Myanmar

irrigation [irrigación] *s.*, proceso de redirigir el agua hacia los cultivos a través de canales y zanjas

Isis [Isis] *s.*, diosa egipcia importante

Islam [islamismo] *s.*, religión fundada en Arabia Saudita a principios del siglo VII D.C.

isolated [aislado] *adj.*, separado de los demás

isolation [aislamiento] *s.*, separarse o ser colocado aparte de los demás

J

Jainism [jainismo] *s.*, religión del noroeste de India que se inició a fines del siglo VI D.C.

Java [Java] *s.*, una de las islas interiores del Sureste Asiático

Jerusalem [Jerusalén] *s.*, ciudad capital de Israel

Jiang Jieshi [Chiang Kai-shek] *s.*, líder nacionalista chino que combatió contra los comunistas

Jinnah, Mohammed Ali [Jinnah, Mohammed Ali] *s.*, líder de la Liga Musulmana Pan India que ayudó a fundar Pakistán y fue el primer gobernador general de Pakistán desde 1947–1948

K

Kalahari [Kalahari] *s.*, desierto importante del sur de África

Kanto Plain [llanura de Kanto] *s.*, área llana y propicia para la agricultura e industria ubicada en el este de los Alpes japoneses

karma [karma] *s.*, en el hinduismo, efecto positivo o negativo que una persona recibe como consecuencia de sus actos

Kenyatta, Jomo [Kenyatta, Jomo] *s.*, líder africano que ayudó a obtener la independencia de Kenia y en 1963 se convirtió en su primer gobernante electo

Khmer Empire [Imperio Jemer] *s.*, el imperio más vasto y duradero de Camboya, que se extendió desde inicios del siglo IX hasta 1430

Kievan Rus [Rus de Kiev] *s.*, estado fundado por los rusos Varegos en 882, que pasó a formar parte de la Rusia moderna

Kilimanjaro [Kilimanyaro] *s.*, volcán inactivo de Tanzania, de 19,340 pies de altura

kimono [kimono] *s.*, vestimenta femenina tradicional del Japón

Kingman Reef [arrecife Kingman] *s.*, arrecife de coral impoluto que forma parte de las Islas de la Línea, ubicadas entre Hawái y Samoa Americana

Knesset [Knéset] *s.*, parlamento del Estado de Israel

Koguryo [Goguryeo] *s.*, reino ubicado en el norte de Corea, en el año 37 A.C.

Kongo [Congo] *s.*, estado del centro de África, fundado en 1390

Koryo [Goryeo] *s.*, dinastía que gobernó Corea de 935 a 1392

Kremlin [Kremlin] *s.*, complejo histórico de palacios, arsenales e iglesias ubicado en Moscú, sede del gobierno ruso

Krishna [Krishna] *s.*, en el hinduismo, deidad encarnada en Visnú

Kunlun Mountains [montañas de Kunlun] *s.*, cordillera del este de Asia

Kurd [kurdo] *s.*, miembro de un grupo étnico no árabe del sudoeste de Asia

L

L'Ouverture, Toussaint [L'Ouverture, Toussaint] *s.*, ex esclavo que lideró la exitosa revuelta haitiana para lograr independizarse de Francia

labor force [fuerza laboral] *s.*, cantidad de personas disponibles para trabajar

land bridge [puente de tierra] *s.*, franja de tierra que conecta dos grandes masas de tierra

landlocked [sin litoral] *adj.*, rodeado de tierras, sin acceso directo al mar

landmass [masa de tierra] *s.*, área muy extensa de tierra

language diffusion [difusión de la lengua] *s.*, la expansión de una lengua desde su lugar de origen

latitude [latitud] *s.*, línea imaginaria que se extiende de este a oeste alrededor de la Tierra y que indica la ubicación en relación con el ecuador

launch [lanzamiento] *v.*, para empezar

legume [legumbre] *s.*, arvejas o frijoles

Lenin, V.I. [Lenin, V.I.] *s.*, líder bolchevique que destituyó al zar en la Revolución Rusa de 1917 y tomó el mando del gobierno nuevo

lingua franca [lengua franca] *s.*, lengua común a varios grupos de personas

linguist [lingüista] *s.*, científico del lenguaje

literacy rate [alfabetismo] *s.*, porcentaje de personas que saben leer y escribir

literate [alfabetizado] *adj.*, que puede leer y escribir

Locke, John [Locke, John] *s.*, filósofo inglés de fines del siglo XVII que contribuyó a inspirar la Revolución Norteamericana y la Ilustración

loess [loes] *s.*, sedimento o limo de color amarillo que forma depósitos de gran espesor

longitude [longitud] *s.*, línea imaginaria que corre del Polo Norte al Polo Sur y que indica la ubicación en relación con el primer meridiano

Lost Decade [Década Perdida] *s.*, en Japón, la década de 1990, en la que disminuyó la producción porque las empresas estaban fuertemente endeudadas

Lost Boys of Sudan [Niños Perdidos de Sudán] *s.*, grupo de jóvenes de Sudán que quedaron huérfanos como consecuencia de la guerra civil y se mantuvieron unidos para escapar de la violencia

lowland [tierras bajas] *s.*, área de poca altura

lucrative [lucrativo] *adj.*, que hace ganar dinero o produce ganancia

Luther, Martin [Lutero, Martín] *s.*, monje alemán cuyos actos condujeron, en 1517, a la Reforma para enfrentar la corrupción de la Iglesia Católica

M

Madura [Madura] *s.*, una de las islas interiores ubicadas en el sudeste de Asia

magnetic levitation (Maglev) [levitación magnética] *adj.*, tipo de tren que se desliza sobre un colchón de aire por encima de vías dotadas de muchos imanes potentes

mainland [masa territorial] *s.*, tierra unida a un continente; usualmente se utiliza en relación con los países que poseen tanto un sector continental como islas

maintain [mantener] *v.*, conservar y continuar

Malay Peninsula [Península Malaya] *s.*, península del Sureste Asiático que abarca partes de Tailandia, Myanmar y la parte continental de Malasia

Malaysia [Malasia] *s.*, país del Sureste Asiático, situado en el extremo inferior de la península malaya y en la isla de Borneo

malnutrition [desnutrición] *s.*, insuficiencia de alimentos o nutrientes

Mandela, Nelson [Mandela, Nelson] *s.*, líder del Congreso Nacional Africano, encarcelado por luchar contra el apartheid, elegido presidente de Sudáfrica en 1994

manga [manga] *s.*, tipo de libro de historietas japonés

Manila [Manila] *s.*, capital de las Filipinas

Mao Zedong [Mao Zedong] *s.*, presidente del Partido Comunista que gobernó China de 1949 a 1976

Maori [maoríes] *s.*, pueblo nativo de Nueva Zelanda

map [mapa] *s.*, representación plana, bidimensional, de la Tierra

marine [marino] *adj.*, perteneciente al mar

marine life [vida marina] *s.*, plantas y animales que viven en el océano

marine reserve [reserva marina] *s.*, área del océano resguardada para proteger a los animales marinos de los seres humanos

maritime [marítimo] *adj.*, relacionado con el mar

marsupial [marsupial] *s.*, mamífero cuyas hembras transportan a las crías en una bolsa abdominal

martial law [ley marcial] *s.*, gobierno que se mantiene por el poder militar

medicinal plant [planta medicinal] *s.*, planta que se usa para tratar enfermedades

meditation [meditación] *s.*, práctica de usar la concentración para calmar los pensamientos y controlarlos

Mediterranean climate [clima mediterráneo] *s.*, clima de veranos calurosos y secos e inviernos templados y lluviosos

Mekong River [río Mekong] *s.*, el río más largo del Sureste Asiático; fluye a través de Myanmar, Laos y Tailandia

messiah [mesías] *s.*, líder o salvador

methane [metano] *s.*, gas natural incoloro que se libera a partir del carbono

metropolitan area [área metropolitana] *adj.*, sitio poblado alrededor de una ciudad que incluye los límites de la ciudad y las comunidades que la rodean

microcredit [microcrédito] *s.*, préstamo de una suma pequeña de dinero

microlending [micropréstamo] *s.*, práctica de otorgar préstamos de sumas pequeñas de dinero a las personas que inician sus propios negocios

Middle Ages [Edad Media] *s.*, período en Europa occidental posterior a la caída del Imperio Romano, desde aproximadamente 500 a 1500

Middle Passage [Pasaje del Medio] *s.*, viaje para cruzar el océano Atlántico, que tomaba meses, en el cual los africanos esclavizados eran llevados a las colonias europeas en América

migration [migración] *s.*, traslado de un lugar a otro

military dictatorship [dictadura militar] *s.*, forma de gobierno en la que el ejército ejerce el control del gobierno

mineral [mineral] *s.*, sustancia sólida y natural que se encuentra en las rocas y en la Tierra; es inorgánica y tiene un conjunto propio de propiedades

missionary [misionero] *s.*, persona enviada por una iglesia a convertir a otras personas a esa religión

modernization [modernización] *s.*, políticas y acciones diseñadas para que un país se actualice tanto tecnológicamente como en otras áreas

modify [modificar] *v.*, cambiar o hacer menos extremo

Mongol Empire [Imperio Mongol] *s.*, imperio establecido en Asia Central por Gengis Kan a comienzos del siglo XIII

monk [monje] *s.*, persona que dedica su vida a tareas religiosas

monopoly [monopolio] *s.*, control total del mercado para un servicio o producto

monotheism [monoteísmo] *s.*, creencia en un solo dios

monotheistic [monoteísta] *adj.*, relativo a la creencia religiosa en un solo dios

monotheistic religion [religión monoteísta] *s.*, sistema de creencias basadas en un solo dios o deidad

monsoon [monzón] *s.*, viento estacional que trae lluvias intensas durante parte del año

Montezuma [Moctezuma] *s.*, líder azteca asesinado por el conquistador español Hernán Cortés

moral [moral] *adj.*, correcto y bueno, referido al comportamiento humano

mosque [mezquita] *s.*, templo musulmán

mouth [desembocadura] *s.*, lugar donde un río desemboca en el mar

movable type [tipo móvil] *s.*, invento de aplicación en la imprenta, que permite mover los carácteres individuales para crear distintas páginas de texto

multinational corporation [corporación multinacional] *s.*, empresa grande que tiene su base en un país y que abre sucursales en muchos otros países

N

N'Dour, Youssou [N'Dour, Youssou] *s.*, famoso griot de África occidental que interpreta música Mbalax

Nairobi [Nairobi] *s.*, capital de Kenia

Napoleon [Napoleón] *s.*, emperador de Francia que conquistó otros países europeos y formó un imperio, 1804–1815

nationalism [nacionalismo] *s.*, profundo sentimiento de lealtad al propio país

natural rights [derechos naturales] *s.*, derechos como la vida, la libertad y la propiedad, que las personas poseen desde su nacimiento

navigable [navegable] *adj.*, suficientemente ancho o profundo para que los barcos o botes puedan navegar sin inconvenientes

navigation [navegación] *s.*, ciencia de averiguar la posición y planear rutas marítimas

Nazi Germany [Alemania nazi] *s.*, Alemania bajo el régimen del partido nazi, de 1933 a 1945

Nebuchadnezzar [Nabucodonosor] *s.*, rey de Babilonia, 605–562 A.C.

New Guinea [Nueva Guinea] *s.*, la segunda isla más grande del mundo, ubicada en el sudoeste del océano Pacífico

New South Wales [Nueva Gales del Sur] *s.*, colonia construida por convictos en Australia, en 1788

Nile River [río Nilo] *s.*, el río más largo del mundo; fluye 4,000 millas a través de Egipto y África

Nkrumah, Kwame [Nkrumah, Kwame] *s.*, líder africano de las décadas de 1950 y 1960, que contribuyó a lograr la independencia de Ghana

nocturnal [nocturno] *adj.*, activo de noche en lugar de durante el día

nomad [nómada] *s.*, persona que se desplaza de un lugar a otro

nonrenewable [no renovable] *adj.*, que no se puede reproducir con la misma rapidez con que se lo usa

nonrenewable fossil fuel [combustible fósil no renovable] *s.*, fuente de energía, como el petróleo, el gas natural o el carbón, cuya provisión es limitada

nonrenewable resource [recurso no renovable] *s.*, fuente de energía que es limitada, y no se puede reemplazar, como el petróleo

North Anatolian Fault [Falla del Norte de Anatolia] *s.*, fractura de la corteza terrestre que se extiende al este y al oeste, justo al sur del mar Negro

North Atlantic Drift [Corriente del Atlántico Norte] *s.*, corriente marina cálida que calienta las aguas que bañan la parte noroeste de Rusia

North China Plain [Llanura del Norte de China] *s.*, llanura que se extiende a lo largo del río Huang He

North Pole [Polo Norte] *s.*, punto ubicado en el extremo norte de la Tierra, opuesto al Polo Sur, donde convergen todas las líneas de longitud

Northern European Plain [Llanura del Norte de Europa] *s.*, vastas tierras bajas que se extienden desde Francia hasta Rusia

Northern Hemisphere [Hemisferio Norte] *s.*, la mitad de la Tierra que se encuentra al norte del ecuador

novel [novela] *s.*, extensa obra de ficción, con trama y personajes complejos

O

oasis [oasis] *s.*, sitio fértil con agua ubicado en un área seca y desértica

occupy [ocupar] *v.*, apoderarse

Okavango Delta [delta del Okavango] *s.*, delta del interior de Botsuana, donde el río Okavango desemboca en un pantano

one-child policy [política de hijo único] *s.*, ley china de 1979 que limitaba a las familias que vivían en áreas urbanas a tener solamente un hijo

opera [ópera] *s.*, representación que cuenta una historia mediante música y palabras

oral tradition [tradición oral] *s.*, transmisión verbal de historias o relatos de una generación a la siguiente

Orange Revolution [Revolución Naranja] *s.*, destitución pacífica del primer ministro de Ucrania en 2004

Osman [Osman] *s.*, primer líder de los turcos en el siglo XIV, por cuyo nombre pasaron a ser llamados otomanos

Ottoman Empire [Imperio Otomano] *s.*, vasto y rico imperio que existió desde 1453 hasta 1923, centrado en el territorio actual de Turquía

outer islands [islas exteriores] *s.*, islas de Indonesia que incluyen a Sumatra, Borneo, Nueva Guinea y otras

outrigger canoe [canoa hawaiana] *s.*, bote que tiene adosado un flotador que le da estabilidad

outsourcing [subcontratar] *s.*, transferir empleos a trabajadores que no pertenecen a la compañía, que a menudo se encuentran en un país extranjero

overpopulation [superpoblación] *s.*, situación en la que demasiadas personas viven en un mismo lugar

P

Paekche [Baekje] *s.*, reino del sudoeste de Corea, en el año 18 A.C.

pagoda [pagoda] *s.*, estructura religiosa de varios pisos que se encuentra en los países asiáticos, a menudo utilizada con fines religiosos

Palestine Liberation Organization (PLO) [Organización para la Liberación de Palestina (OLP)] *s.*, organización creada por líderes palestinos

Pan-Africanism [Panafricanismo] *s.*, movimiento surgido en los inicios del siglo XX para unir a los pueblos africanos

pandemic [pandemia] *s.*, brote de una enfermedad que se propaga por una vasta área geográfica

Papua New Guinea [Papúa Nueva Guinea] *s.*, país del Pacífico Sur que abarca la mitad oriental de la isla de Nueva Guinea y las islas cercanas

papyrus [papiro] *s.*, material similar al papel inventado en el antiguo Egipto

Parliament [Parlamento] *s.*, poder legislativo del gobierno de la India

parliamentary democracy [democracia parlamentaria] *s.*, sistema de gobierno en el cual el poder ejecutivo está presidido por el primer ministro, que es elegido por el partido que posee la mayoría de los escaños del Parlamento

Partition [Partición] *s.*, se refiere a la división del sur de Asia en los países independientes de India y Paquistán

patrician [patricio] *s.*, rico terrateniente de la antigua Roma

Pearl Harbor [Pearl Harbor] *s.*, base naval de los EE.UU. ubicada en Hawái, bombardeada por los japoneses en 1941, lo cual provocó el ingreso de EE.UU. en la Segunda Guerra Mundial

peat [turba] *s.*, material que se forma a partir de la descomposición de plantas muy antiguas y que arde como el carbón

peninsula [península] *s.*, masa de tierra rodeada por agua en tres de sus lados

perestroika [perestroika] *s.*, reformas en la estructura económica de la Unión Soviética introducidas por Mijaíl Gorbachov en 1985

permafrost [permafrost] *s.*, suelo que está permanentemente congelado

perspective [perspectiva] *s. modo artístico de mostrar los objetos de la manera en que son vistos por las personas*, en términos de distancia o profundidad relativa, como si estuvieran en tres dimensiones

pesticide [pesticida] *s.*, sustancia química que mata insectos y malezas nocivas

Peter the Great [Pedro el Grande] *s.*, Pedro Romanov, zar que gobernó Rusia desde 1682 hasta 1725

petrochemicals [petroquímicos] *s.*, productos elaborados a partir del petróleo

petroleum [petróleo] *s.*, materia prima que se usa para producir combustibles

pharaoh [faraón] *s.*, rey en el antiguo Egipto

philosopher [filósofo] *s.*, persona que examina las preguntas sobre el universo y busca la verdad

pictograph [pictograma] *s.*, imagen pintada usada para comunicar

pilgrimage [peregrinaje] *s.*, viaje religioso

pipeline [tubería] *s.*, serie de tubos o caños conectados para transportar líquidos o gases

plain [llanura] *s.*, área plana de la superficie terrestre

plate [placa] *s.*, sección rígida de la corteza terrestre que se puede mover de manera independiente

plateau [meseta] *s.*, llanura ubicada a gran altura sobre el nivel del mar que a menudo tiene un precipicio en todos sus lados

plebeian [plebeyo] *s.*, agricultor o persona de clase baja de la antigua Roma

poach [caza o pesca furtiva] *v.*, cazar o pescar ilegalmente

poaching [cazar o pescar furtivamente] *s.*, caza o pesca ilegal

polder [pólder] *s.*, tierras de los Países Bajos ganadas al mar que se destinan a la agricultura

pollution [contaminación] *s.*, desechos químicos o físicos que generan un medio ambiente sucio o poco limpio

Polynesian Triangle [Triángulo Polinésico] *s.*, vasta área del Pacífico Sur que abarca muchas islas

polytheism [politeísmo] *s.*, creencia en más de un dios

polytheistic religion [religión politeísta] *s.*, sistema de creencias basadas en varios dioses o deidades

popular culture [cultura popular] *s.*, artes, música y otros elementos de la vida cotidiana de una región

porcelain [porcelana] *s.*, tipo de cerámica dura

port [puerto] *s.*, embarcadero para barcos donde se intercambian mercancías

potential [potencial] *s.*, posibilidad

precipitation [precipitación] *s.*, proceso que hace caer agua sobre la Tierra, en forma de lluvia, nieve o granizo

predominant [predominante] *adj.*, principal, más común, superior a los demás

prehistoric [prehistórico] *adj.*, anterior a la historia escrita

preserve [preservar] *v.*, proteger

pride [manada] *s.*, grupo de leones que viven en comunidad

prime meridian [primer meridiano] *s.*, línea de longitud de 0° que se extiende desde Polo Norte al Polo Sur y que pasa por Greenwich, Inglaterra

privatization [privatización] *s.*, proceso por el que las empresas que eran propiedad del gobierno pasan a manos privadas

projection [proyección] *s.*, modo de mostrar la superficie curva de la Tierra sobre un mapa plano

promote [fomentar] *v.*, animar, estimular

propaganda [propaganda] *s.*, información que se difunde para influir sobre la opinión de las personas o para promover las ideas de un partido u organización

proportional representation [representación proporcional] *s.*, sistema en el cual un partido político consigue un porcentaje de escaños igual al porcentaje de votos que obtuvo

push-pull factors [factores de atracción y repulsión] *s.*, motivos por los que las personas emigran; los factores de "repulsión" las hacen partir de un sitio, los factores de "atracción" las hacen ir hacia otro sitio

pyramid [pirámide] *s.*, monumento construido en roca que servía de tumba en el antiguo Egipto

Q

qanat [qanat] *s.*, túnel subterráneo construido por el hombre en la meseta de Irán, usado para transportar agua desde las montañas

Qin [Qin] *s.*, dinastía que gobernó China durante el período 221–206 a.c.

Qur'an [Corán] *s.*, libro sagrado del islamismo

R

radical [radical] *s.*, persona que busca un cambio extremo o sostiene una posición política extrema

radioactive [radiactivo] *adj.*, que emite energía producida por la ruptura de un átomo

rain forest [bosque tropical] *s.*, bosque de temperatura cálida, humedad elevada y vegetación espesa que recibe más de 100 pulgadas de lluvia al año

rain shadow [sombra orográfica] *s.*, región seca ubicada sobre uno de los lados de una cordillera

rainshadow effect [efecto de la sombra orográfica] *s.*, proceso en el cual el aire húmedo asciende por una ladera de la cordillera y luego se enfría y cae en forma de precipitación, dejando el otro lado de la cordillera mayormente seco

Ramses II [Ramsés II] *s.*, faraón egipcio que reinó alrededor del año 1185 A.C. y expandió los límites del imperio egipcio

raw materials [materias primas] *s.*, materiales naturales o sin terminar, como minerales, petróleo o carbón, que se usan para elaborar productos terminados

Re [Ra] *s.*, dios solar de los antiguos egipcios

rebellion [rebelión] *s.*, revuelta o resistencia a la autoridad

reclaim [reclamar] *v.*, volver a tomar

Reformation [Reforma] *s.*, movimiento que surgió en el siglo XVI para reformar el cristianismo

refuge [refugio] *s.*, lugar seguro

refugee [refugiado] *s.*, persona que huye de un lugar para estar a salvo

region [región] *s.*, conjunto de sitios con características comunes

regulate [regular] *v.*, controlar

reign [reinado] *s.*, período de mando de un rey, reina, emperador o emperadora

Reign of Terror [El Terror] *s.*, movimiento francés liderado por Maximilien Robespierre, en el cual fueron decapitadas 40,000 personas, durante el período 1793–94

reincarnation [reencarnación] *n.*, el nacimiento de un alma en otra vida

relative location [ubicación relativa] *s.*, la posición de un lugar en relación con otros

reliable [confiable] *adj.*, fiable o de confianza

relief [relieve] *s.*, cambio en la elevación de un lugar a otro

religious tolerance [tolerancia religiosa] *s.*, aceptación de distintas religiones para que sean profesadas al mismo tiempo, sin prejuicios

relocate [relocalizar] *v.*, trasladar

remittance [remesa] *s.*, dinero enviado a una persona que se encuentra en otro lugar

remote [remoto] *adj.*, difícil de llegar, aislado

Renaissance [Renacimiento] *s.*, período que se desarrolló entre los siglos XIV y XVI, donde florecieron el arte y la cultura

renewable energy [energía renovable] *s.*, energía obtenida a partir de fuentes que no se agotan, como el viento, el Sol y el agua

renewable resource [recurso renovable] *s.*, materia prima o fuente de energía que se reemplaza a sí misma con el paso del tiempo

reparation [reparación] *s.*, dinero que, después de una guerra, pagan como castigo normalmente los agresores que iniciaron el conflicto

republic [república] *s.*, forma de gobierno en la cual las personas eligen funcionarios para que gobiernen

reserve [reserva] *s.*, tierras destinadas a propósitos especiales, como la agricultura, la preservación de los hábitats o para ser usadas como vivienda por determinados grupos de personas; futuro suministro (de petróleo)

reservoir [embalse] *s.*, lago artificial de gran tamaño para almacenar agua

resistance [resistencia] *n.*, oposición

restore [restaurar] *v.*, recuperar

retreat [retirarse] *v.*, ir hacia atrás, no hacia adelante

revenue [rentas] *s.*, ingresos

reverse [retroceder] *v.*, ir en la dirección opuesta

Rhine River [río Rin] *s.*, río que nace en Suiza y desemboca en el mar del Norte

rift valley [valle de fisura] *s.*, valle profundo que se formó al separarse la corteza terrestre, como en África Oriental

Ring of Fire [Anillo de Fuego] *s.*, área que se extiende a lo largo de las riberas del océano Pacífico, donde chocan las placas tectónicas, lo que genera terremotos y una gran actividad volcánica

ritual [ritual] *s.*, acto formal que se repite regularmente

rivalry [rivalidad] *s.*, competencia u oposición entre personas

river basin [cuenca de un río] *s.*, área baja por la que fluye un río

Romantic Period [Romanticismo] *s.*, período artístico de comienzos del siglo XIX, en el cual los artistas pintaban paisajes y escenas de la naturaleza para transmitir emociones

Rub al Khali [Rub al-Jali] *s.*, vasto desierto ubicado al sur de Arabia Saudita

Russian Revolution [Revolución Rusa] *s.*, revolución que tuvo lugar en Rusia en 1917, en la cual el zar fue depuesto y los bolcheviques tomaron el poder

Russification [rusificación] *s.*, política de designar ciudadanos rusos a cargo de las repúblicas soviéticas durante las décadas de 1970 y 1980

S

Sahara Desert [desierto de Sahara] *s.*, el desierto más grande del mundo, que cubre la mayor parte del norte de África

Sahel [Sahel] *s.*, pradera semiárida del África subsahariana, limitada al norte por el Sahara y al sur por las praderas tropicales

salinization [salinización] *s.*, acumulación de sal en el suelo

Samoa [Samoa] *s.*, país que comprende las islas occidentales de Samoa, ubicadas al sur del océano Pacífico

samurai [samurái] *s.*, diestro guerrero japonés

sanitation [sanidad] *s.*, medidas tomadas para proteger la salud pública, como la red cloacal

Sanskrit [sánscrito] *s.*, idioma de los arios, que se convirtió en la base de numerosas lenguas del sur de Asia

sarcophagus [sarcófago] *s.*, ataúd

sari [sari] *s.*, vestido tradicional femenino de la India, que se enrolla alrededor del cuerpo

saturate [saturar] *v.*, remojar completamente

savanna [sabana] *s.*, pradera, como la del sur del África subsahariana

scale [escala] *s.*, parte de un mapa que indica el tamaño en el que se muestra un área de la Tierra

scarcity [escasez] *s.*, falta o carencia de algo

scientific station [estación científica] *s.*, sitio para desarrollar una investigación

scorched earth policy [táctica de tierra quemada] *s.*, práctica llevada adelante por las tropas rusas en 1812, quienes, a medida que retrocedían ante el avance del ejército de Napoleón, quemaban los cultivos y todos los recursos que pudieran servir al enemigo para abastecerse

seafarer [navegante] *s.*, persona que viaja por el mar

secular [secular] *adj.*, terrenal, sin vínculo con una religión

segregation [segregación] *s.*, separación por raza

seize [apoderarse] *v.*, tomar el control

self-rule [autonomía] *s.*, gobierno ejercido por los propios habitantes de un país

semiarid [semiárido] *adj.*, algo seco, con muy poca lluvia

serf [siervo] *s.*, campesino ruso o europeo, pobre y con pocos derechos, que alquilaba tierras a un terrateniente entre los siglos XVI y XIX

shalwar-kameez [shalwar-kameez] *s.*, camisa larga con pantalones holgados que se usa en la India y en el suroeste de Asia

Shang [Shang] *s.*, familia cuya dinastía gobernó China desde 1766 hasta 1050 A.C.

sheikh [jeque] *s.*, líder árabe

Shi Huangdi [Qin Shi Huang] *s.*, líder de la dinastía Qin, que se convirtió en el primer emperador de la China en 221 A.C.

Shi'ite [chiita] *adj.*, rama de musulmanes que considera que los líderes religiosos deben ser descendientes de Mahoma

Shinto [sintoísmo] *s.*, religión nativa del Japón, similar al animismo

shogun [sogún] *s.*, gobernador militar japonés

Siberia [Siberia] *s.*, enorme región del centro y este de Rusia

significant [significativo] *adj.*, importante

Sikhism [sijismo] *s.*, religión que surgió a fines del siglo XV en la India

Silk Roads [rutas de la seda] *s.*, antiguas rutas comerciales que unían el sudoeste y el centro de Asia con China

Silla [Silla] *s.*, reino ubicado en el sudeste de Corea en el año 57 A.C.

silt [limo] *s.*, partículas finas de suelo que se depositan a lo largo de las márgenes de los ríos

Slavs [eslavos] *s.*, pueblo originario de los alrededores del mar Negro o Polonia, que se instaló en Ucrania y el oeste de Rusia alrededor del año 800 D.C.

slum [barriada] *s.*, área densamente poblada de una ciudad, con viviendas precarias y malas condiciones de vida

socialism [socialismo] *s.*, sistema de gobierno en el que el gobierno controla los recursos económicos

solstice [solsticio] *s.*, punto en que el Sol se encuentra a la distancia máxima, al sur o al norte, del ecuador; inicio del invierno y del verano

South Pole [Polo Sur] *s.*, punto más austral de la Tierra, opuesto al Polo Norte, donde convergen todas las líneas de longitud

Southern Alps [Alpes del Sur] *s.*, cordillera situada en Nueva Zelanda

Southern Hemisphere [Hemisferio Sur] *s.*, la mitad de la Tierra que se encuentra al sur del ecuador

sovereignty [soberanía] *s.*, control de un país sobre sus propios asuntos

Soviet Union [Unión Soviética] *s.*, Unión de las Repúblicas Socialistas Soviéticas, país formado por Rusia y otros estados euroasiáticos, que existió desde 1922 hasta 1991

spatial thinking [pensamiento espacial] *s.*, manera de pensar en el espacio que está sobre la superficie de la Tierra, incluyendo la ubicación de los distintos lugares y por qué se encuentran allí

Special Economic Zone [Zona Económica Especial] *s.*, área de China en la que se permitió el desarrollo de una economía de mercado, con menor control de los negocios por parte del gobierno

sprawl [expandirse] *v.*, extenderse

standard of living [nivel de vida] *s.*, nivel de acceso de los habitantes de un país a bienes, servicios y comodidades materiales

staple [alimento básico] *s.*, constituyente básico de la dieta de las personas

state [estado] *s.*, territorio determinado que posee un gobierno propio

steppe [estepa] *s.*, llanura muy extensa de praderas secas

strait [estrecho] *s.*, vía fluvial angosta que conecta dos masas de agua

strike [huelga] *s.*, interrupción del trabajo por parte de empleados que se niegan a trabajar

subcontinent [subcontinente] *s.*, región separada de un continente

subsistence farmers [agricultores de subsistencia] *s.*, agricultores que producen cultivos para alimentar a sus familias, no para vender

subsistence fishing [pesca de subsistencia] *s.*, pescar para tener comida para poder vivir, no para obtener ganancias

suffrage [sufragio] *s.*, derecho al voto

Suleyman I [Suleyman I] *s.*, emperador del Imperio Otomano a mediados del siglo XVI

sultan [sultán] *s.*, líder o gobernante del Imperio Otomano

Sumatra [Sumatra] *s.*, una de las islas exteriores del sudeste de Asia

Sunni [sunita] *adj.*, rama del islamismo que sostiene que los líderes religiosos deben ser escogidos entre los candidatos más capacitados, a diferencia de hacerlo entre los descendientes de Mahoma

sustainable [sustentable] *adj.*, capaz de ser continuado sin dañar el medio ambiente o sin agotar los recursos de manera permanente

Swahili [suajili] *s.*, lengua bantú hablada mayormente en África Oriental, también llamada kiswahili

Sydney [Sídney] *s.*, la ciudad más grande de Australia, capital del estado de Nueva Gales del Sur

symbol [símbolo] *s.*, objeto o idea que se puede usar para representar otro objeto o idea

T

taiga [taiga] *s.*, extensa área de bosques que se extiende a través del norte de Rusia, Canadá y otros países del norte

Taj Mahal [Taj Mahal] *s.*, famoso edificio construido en la India en el siglo XVII, para servir como tumba de la esposa de Shah Jahan; actualmente reconocido por la UNESCO como patrimonio de la humanidad

Taliban [talibán] *s.*, grupo de pashtunes de Afganistán que comenzaron a gobernar el país en 1996

Taman Negara National Park [Parque Nacional Taman Negara] *s.*, parque nacional de Malasia situado dentro de uno de los bosques tropicales más antiguos del mundo

tariff [arancel] *s.*, impuesto sobre las importaciones y exportaciones

tectonic plate [placa tectónica] *s.*, sección de la corteza terrestre que flota sobre el manto terrestre

terra cotta [terracota] *s.*, arcilla endurecida al horno

terrace [terraza] *s.*, superficie llana construida sobre la ladera de un monte

terrain [terreno] *s.*, características físicas de la tierra

terrorist [terrorista] *s.*, persona que emplea la violencia para obtener resultados políticos

textile [textil] *adj.*, relativo a la tela o ropa

theme [tema] *s.*, tópico

thirty-eighth parallel [paralelo treinta y ocho] *s.*, frontera que separa Corea del Norte y Corea del Sur, establecida en 1945 a lo largo del paralelo de 38° de latitud norte

Three Gorges Dam [presa de las Tres Gargantas] *s.*, la presa más grande del mundo, emplazada sobre el río Chang Jiang, en China

Tibetan Plateau [meseta tibetana] *s.*, meseta vasta y de gran elevación situada en Asia Central

Timbuktu [Tombuctú] *s.*, ciudad de África Occidental que fue un centro educativo en el siglo XIII

tolerance [tolerancia] *s.*, aceptación de las creencias de los demás

tomb [tumba] *s.*, lugar donde se realiza un entierro

tornado [tornado] *s.*, tormenta de vientos muy fuertes que sigue una trayectoria impredecible

totalitarian [totalitario] *adj.*, relacionado con un gobierno dirigido por un dictador que exige obediencia absoluta al estado

trans-Atlantic slave trade [tráfico transatlántico de esclavos] *s.*, negocio de traficar a América, a través del océano Atlántico, esclavos originarios de África, que se inició en el siglo XVI

transform [transformar] *v.*, rehacer o cambiar

transition zone [zona de transición] *s.*, área situada entre dos regiones geográficas y que posee características de ambas

transport [transportar] *v.*, enviar de un lugar a otro

transportation corridor [corredor de transporte] *s.*, ruta terrestre o marítima para trasladar personas o mercancías de un lugar a otro con facilidad

trans-Saharan [transahariano] *adj.*, que atraviesa el desierto del Sahara

Trans-Siberian Railroad [ferrocarril transiberiano] *n.*, el ferrocarril de servicio continuo más largo del mundo, que une Moscú con el este de Rusia, atravesando Siberia

Treaty of Versailles [Tratado de Versalles] *s.*, tratado de paz que puso fin, en 1919, a la Primera Guerra Mundial

trench [trinchera] *s.*, zanja extensa que protege a los soldados del fuego enemigo

trend [tendencia] *s.*, cambio que se produce en determinada dirección con el paso del tiempo

tribute [tributo] *s.*, sumas pagadas a otro país o gobernante a cambio de protección o como muestra de sumisión

troubadour [trovador] *s.*, cantante de la Edad Media que interpretaba canciones sobre caballeros y el amor

tsunami [tsunami] *s.*, ola enorme y muy potente que se forma en el océano

Tuareg [tuareg] *s.*, pueblo semi nómada que se desplaza en caravanas a través del Sahara, comerciando sal

tundra [tundra] *s.*, tierras llanas y sin árboles que se encuentran en las regiones árticas y subárticas

typhoon [tifón] *s.*, tormenta tropical peligrosa, que trae lluvias copiosas y vientos muy fuertes, que en el Hemisferio Occidental se denomina huracán

U

United Arab Emirates [Emiratos Árabes Unidos] *s.*, país situado en la península arábiga, en el golfo pérsico

United Nations (UN) [Naciones Unidas (ONU)] *s.*, organización de países formada en 1945, con el objetivo de mantener la paz entre los países y proteger los derechos humanos

Universal Declaration of Human Rights [Declaración Universal de los Derechos Humanos] *s.*, acuerdo aprobado por las Naciones Unidas que define los derechos que deben tener todas las personas del mundo

uplands [tierras altas] *s.*, colinas, montañas y mesetas

Ur [Ur] *s.*, importante ciudad estado sumeria, 2800–1850 A.C.

Ural Mountains [montes Urales] *s.*, cordillera que separa la Llanura del Norte de Europa de la Llanura de Siberia Occidental, en Rusia

urban [urbano] *adj.*, que describe o está relacionado con la ciudad o los suburbios

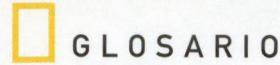

V

vaccine [vacuna] *s.*, tratamiento para incrementar la inmunidad a una enfermedad determinada

vegetation [vegetación] *s.*, formas de vida vegetal

veto [veto] *v.*, rechazar una decisión tomada por otro órgano de gobierno

vocational [profesional] *adj.*, relacionado con las destrezas laborales

volcano [volcán] *s.*, montaña que, al entrar en erupción, explota y lanza roca derretida, gases y ceniza

vulnerable [vulnerable] *adj.*, abierto, que puede ser lastimado por fuerzas externas

W

wallaby [walabí]*s.*, marsupial más pequeño que un canguro

wat [wat] *s.*, templo budista del Sureste Asiático

waterway [vía fluvial] *s.*, ruta navegable que se usa para los viajes y el transporte

weapon of mass destruction (WMD) [arma de destrucción masiva] *s.*, arma que produce un daño inmenso a grandes cantidades de personas

weather [tiempo atmosférico] *s.*, condiciones de la atmósfera en un momento determinado, incluyendo la temperatura, precipitación y humedad de un día o una semana determinada

Western Hemisphere [Hemisferio Occidental] *s.*, la mitad de la Tierra que se encuentra al oeste del primer meridiano

Wilkes, Charles [Wilkes, Charles] *s.*, estadounidense que exploró la costa de la Antártida a fines de la década de 1830

Y

yurt [yurta] *s.*, carpa tradicional de fieltro de Asia Central

Z

zaibatsu [zaibatsu] *s.*, organizaciones dirigidas por familias japonesas, propietarias de numerosos negocios

Zambezi River [río Zambeze] *s.*, río del sur de África, que fluye hacia el océano Índico atravesando varios países situados en el sur del centro de África

Zheng He [Zheng He] *s.*, almirante de la marina china que lideró siete expediciones navales para explorar las tierras situadas más allá de China; su último viaje se inició en 1431 D.C.

Zhou [Zhou] *s.*, dinastía que gobernó China en el período 1050–221 A.C.

zoologist [zoólogo] *s.*, científico que estudia los animales

INDEX

H

INDEX

◻ ACKNOWLEDGMENTS

Text Acknowledgments

322: Excerpts from *The Illustrated Bhagavad Gita*, translated by Ranchor Prime. Copyright © 2003 by Godsfield Press, text © by Ranchor Prime. Reprinted by permission of Godsfield Press.

374: Excerpts from *The Analects of Confucius*, translated by Simon Leys. Copyright © 1997 by Pierre Ryckmans. Used by permission of W. W. Norton & Company, Inc.

468: Data from the International Union for Conservation of Nature (IUCN) Red List of Threatened Species by IUCN. Data copyright © 2008 by the IUCN Red List of Threatened Species. Reprinted by kind permission of IUCN.

◻ National Geographic School Publishing

National Geographic School Publishing gratefully acknowledges the contributions of the following National Geographic Explorers to our program and to our planet:

Greg Anderson, National Geographic Fellow

Katey Walter Anthony, 2009 National Geographic Emerging Explorer

Ken Banks, 2010 National Geographic Emerging Explorer

Katy Croff Bell, 2006 National Geographic Emerging Explorer

Christina Conlee, National Geographic Grantee

Alexandra Cousteau, 2008 National Geographic Emerging Explorer

Thomas Taha Rassam (TH) Culhane, 2009 National Geographic Emerging Explorer

Jenny Daltry, 2005 National Geographic Emerging Explorer

Wade Davis, National Geographic Explorer-in-Residence

Sylvia Earle, National Geographic Explorer-in-Residence

Grace Gobbo, 2010 National Geographic Emerging Explorer

Beverly Goodman, 2009 National Geographic Emerging Explorer

David Harrison, National Geographic Fellow

Kristofer Helgen, 2009 National Geographic Emerging Explorer

Fredrik Hiebert, National Geographic Fellow

Zeb Hogan, National Geographic Fellow

Shafqat Hussain, 2009 National Geographic Emerging Explorer

Beverly and Dereck Joubert, National Geographic Explorers-in-Residence

Albert Lin, 2010 National Geographic Emerging Explorer

Elizabeth Kapu'uwailani Lindsey, National Geographic Fellow

Sam Meacham, National Geographic Grantee

Kakenya Ntaiya, 2010 National Geographic Emerging Explorer

Johan Reinhard, National Geographic Explorer-in-Residence

Enric Sala, National Geographic Explorer-in-Residence

Kira Salak, 2005 National Geographic Emerging Explorer

Katsufumi Sato, 2009 National Geographic Emerging Explorer

Cid Simoes and Paola Segura, 2008 National Geographic Emerging Explorers

Beth Shapiro, 2010 National Geographic Emerging Explorer

José Urteaga, 2010 National Geographic Emerging Explorer

Spencer Wells, National Geographic Explorer-in-Residence

Photographic Credits

vi (left column, top to bottom) ©Gemma Atwal, ©Ken Banks, ©Rebecca Hale/National Geographic Stock. (right column, top to bottom) ©Christina Conlee, ©Tyrone Turner/National Geographic Stock. (b) ©Kip Evans Photography. vii (left column, top to bottom) ©Rebecca Hale/National Geographic Stock, ©National Geographic Society, Explorer Programs and Strategic Initiatives. (right column, top to bottom) ©Mark Thiessen/National Geographic Stock, ©Rebecca Hale/National Geographic Stock, ©Victor Sanchez de Fuentes. (b) ©Mauricio Ramos. viii (left column, top to bottom) ©Beth Shapiro, ©Victor Sanchez de Fuentes. (right column, top to bottom) ©Rachel Etherington, ©David Evans/National Geographic Society. (b) ©Paul Hoekman/ViaNica.com (c) ©Brian Wallace. 8 (bkg)©Richard Barnes/National Geographic Stock (bl) ©Mitchell Funk/Photographer's Choice/Getty Images (cl) ©NASA Goddard Space Flight Center. (tl) ©David Evans/National Geographic Society. 10 ©Stephen Alvarez/National Geographic Stock. 11 ©Sunpix Travel/Alamy. 12 (bc) ©Mike Theiss/National Geographic Stock (cl) ©Blakeley/Alamy (cr) ©Michael S. Yamashita/National Geographic Stock (tl) ©John Wark/Wark Photography, Inc. 13 ©Mark Remaley /Precision Aerial Photo. 15 (b) ©Michael S. Yamashita/National Geographic Stock (t) ©Ma Wenxiao/Sinopictures/Photolibrary. 16 ©Susan Byrd/National Geographic My Shot/National Geographic Stock. 20 ©Stephen Alvarez/National Geographic Stock. 23 ©PictureLake/Alamy. 26 ©Brooks Kraft/Corbis. 28 (bc) ©James Forte/National Geographic Stock (bl) ©Kenneth Garrett/National Geographic Stock (br) ©Michael Poliza/National Geographic Stock (bkg)©National Geographic Maps. 29 (bcl) ©N.C. Wyeth/National Geographic Stock (bcr) ©Justin Guarlglia/National Geographic Stock (bl) ©Kenneth Garrett/National Geographic Stock (br) ©Abraham Nowitz/National Geographic Stock 33 (b) ©David Trood/Getty Images (tr) ©Peter Carsten/National Geographic Stock. 34 ©Andrew Hasson/Alamy. 38 ©George H.H. Huey/Corbis. 41 (tc) ©Images & Volcans/Photo Researchers, Inc. (tr) ©Chris Cheadle/Getty Images. 44 ©Bill Hatcher/National Geographic Stock. 46 (bc) ©Michael Doolittle/Alamy (cl) ©Daniel Dempster Photography/Alamy (tl) ©Tom Bean/Alamy. 47 (tl) ©Frank Krahmer/Corbis (tr) ©imagebroker/Alamy. 50 ©Panoramic Images/Getty Images. 52 ©Michael Nichols/National Geographic Stock. 53 ©Russ Bishop/Alamy. 54 ©Kip Evans Photography. 55 ©NASA Goddard Space Flight Center. 56 (bc) ©Gordon Wiltsie/National Geographic Stock (bl) ©Ralph Lee Hopkins/National Geographic Stock (bkg) ©George Grall/National Geographic Stock. (br) ©Paul Nicklen/National Geographic Stock. 57 (bcl) ©Norbert Rosing/National Geographic Stock (bcr) ©William Albert Allard/National Geographic Stock (bl) ©Stuart Franklin/National Geographic Stock (br) ©Priit Vesilind/National Geographic Stock. 58 (b) ©David R. Frazier Photolibrary, Inc./Alamy (cl) ©Olivier Asselin/Alamy. 59 ©Greg Elms/

Lonely Planet Images. 63 ©John Stanmeyer/National Geographic Stock. 64 ©Nic Bothma/epa/Corbis. 65 ©Alain Nogues/Corbis. 67 ©Michael Dunning/Photographer's Choice/Getty Images. 72 ©Romeo Gacad/AFP/Getty Images. 74 (b) ©Tetra Images/Corbis (bkg) ©John Burcham, National Geographic Stock (c) ©Walter Meayers Edwards, National Geographic Stock (t) ©Beth Shapiro. 77 ©Petra Engle/National Geographic Stock. 78 (b) ©Phil Schermeister/National Geographic Stock (bkg) ©Menno Boermans/Aurora Photos/Corbis (t) ©SeBuKi/Alamy. 80 ©Mike Grandmaison/Corbis. 82 ©Bill Hatcher/National Geographic Stock. 84 ©Daniel H. Bailey/Corbis. 86 (bkg) ©Jean-Pierre Lescourret/Corbis (cr) ©Mauricio Ramos. 88 ©Jon Arnold Images Ltd/Alamy. 89 ©American School Private Collection/Peter Newark American Pictures/The Bridgeman Art Library Nationality. 90 ©Thomas Sbampato/Photolibrary. ©91 ©The Granger Collection. 92 (l) ©The Granger Collection (r) ©photostock1/Alamy. 93 ©Visions LLC/Photolibrary. 95 ©William Manning/Corbis. 96 ©Bettmann/Corbis. 98 (l) ©Corbis (r) ©American School Private Collection/Courtesy of Swann Auction Galleries/The Bridgeman Art Library. 99 ©Corbis. 100 ©Bettmann/Corbis. 101 ©Lynn Johnson/National Geographic Stock. 102 ©The Art Archive/Museo Ciudad Mexico/Gianni Dagli Orti. 103 (l) ©Kenneth Garrett/National Geographic Stock (r) ©David R. Frazier Photolibrary, Inc./Alamy. 104 ©The Stapleton Collection/The Bridgeman Art Library. 105 ©The Stapleton Collection/The Bridgeman Art Library. 106 ©The Granger Collection. ©107 (b) ©The Granger Collection (t) ©Look and Learn Magazine Ltd/The Bridgeman Art Library. 108 (l) ©North Wind Picture Archives/Alamy (r) ©Randy Faris/Corbis. 109 (l) ©Corbis (r) ©Charles & Josette Lenars/Corbis. 110 ©North Wind Picture Archives/Alamy. 111 ©Hulton Archive/Getty Images. 114 ©Joe McNally/National Geographic Stock. 116 ©Jennifer Shaffer/National Geographic School Publishing. 117 ©Mike Theiss/National Geographic Society Image Sales. 118 ©Jennifer Shaffer/National Geographic School Publishing/Art Institute of Chicago. 120 ©Car Culture/Corbis. 123 (c) ©James Forte/National Geographic Stock (tr) ©Michael Dunning/Photographer's Choice/Getty Images. 124 ©Marjorie Kamys Cotera/Daemmrich Photography/The Image Works. 125 (b) ©Robb Kottmyer (t) ©Roger Meno. 126 ©Tono Labra/Photolibrary. 127 ©Keith Dannemiller/Alamy. 128 ©STR/Reuters/Corbis. 130 ©Alfredo Guerrero/epa/Corbis. 131 ©Blaine Harrington III/Alamy. 136 ©Corbis Premium RF/Alamy. 138 (b) ©Georgios Kollidas/Alamy (bkg) ©Raul Touzon, National Geographic Stock (c) ©Stephen Alvarez, National Geographic Stock. 141 ©Konrad Wothe/ Minden Pictures. 142 (b) ©Jon Arnold Images Ltd/Alamy (bkg) ©Menno Boermans/Aurora Photos/Corbis (t) © Danny Lehman/Corbis. 144 ©Dr. Richard Roscoe/Visuals Unlimited, Inc. 147 © Stuart Westmorland/Corbis. 148 (b) ©Paul Hoekman (t) ©Bryan Wallace. 150 (bc) ©Roy Toft/National Geographic Stock (bl) ©Michael Nichols/National Geographic Stock (bkg) ©Paul Sutherland/National Geographic Stock (br) ©Steve Winter/National Geographic Society Image Sales. 151 (bcl) ©Michael Melford/National Geographic Stock (bcr) ©Bobby Haas/National Geographic Stock (bl) ©Christian Ziegler/National Geographic Stock (br) ©Roy Toft/National Geographic Stock. 152 ©Steve Winter/National Geographic Stock. 155 ©The Bridgeman Art Library. 156 (br) ©Georgios Kollidas/Alamy (r) ©Hemis /Alamy. 157 ©Stuwdamdorp/Alamy. 158 (bl) ©Creativ Studio Heinemann/Westend61/Corbis (c) ©INTERFOTO /Alamy. 159 (br) ©Reuters/Corbis (tl) ©Bettmann/Corbis (tl) ©Creativ Studio Heinemann/Westend61/Corbis. 162 ©Walter Bibikow/JAI/Corbis. 164 ©Martin Gray/National Geographic Stock. 165 ©Danita Delimont/Alamy. 166 ©Nico Tondini/Photolibrary. 167 ©Rick Gerharter/Lonely Planet Images. 168 ©National Geographic Stock. 169 ©Frans Lanting/Corbis. 170 ©Photolibrary. 172 ©Yuan Man/Xinhua Press/Corbis. 173 ©Logan Abassi/UN Handout/Corbis. 174 ©JS Callahan/tropicalpix/Alamy. 175 ©Michael Dunning/Photographer's Choice/Getty Images. 176 ©Lonely Planet Images /Alamy. 178 (bkg) ©Christian Heeb/Aurora Photos (c) ©Roy Toft/National Geographic Stock (r) ©Micahel & Patricia Fogden/Minden Pictures/National Geographic Stock. 179 ©Arterra Picture Library/Alamy. 183 (l) ©Jacques Marais/Getty Images (r) ©Danny Lehman/Corbis. 184 ©Christian Zeigler/National Geographic Stock. 186 (b) ©Frans Lanting/Corbis (bkg) ©Rod Smith/National Geographic My Shot/National Geographic Stock (cl) ©David R. Frazier Photolibrary, Inc./Alamy (tl) ©Photograph by Victor Sanchez de Fuentes. 189 ©Nick Gordon/Oxford Scientific (OSF)/Photolibrary. 190 (bkg) ©Menno Boermans/Aurora Photos/Corbis (bl) ©Colin Monteath/Minden Pictures/National Geographic Stock (br) ©John Eastcott and Yva Momatiuk/National Geographic Stock. 192 ©Ivan Kashinsk/National Geographic Stock. 196 (bkg) ©Melissa Farlow/National Geographic Stock (br) ©Aldo Sessa/Tango Stock/Getty Images. 199 (br) ©Michael Nichols/National Geographic Stock (tr) ©Michael Dunning/Photographer's Choice/Getty Images. 200 ©Ethan Welty/ Aurora Photos/Alamy. 201 ©Gnter Wamser/F1online digitale Bildagentur GmbH/Alamy. 202 ©Christina Conlee. 203 (b) ©Christina Conlee (t) ©Robert Clark/National Geographic Stock. 206 (bc) ©Maria Stenzel/National Geographic Stock (c) ©Peruvian School/Museo Arqueologia, Lima, Peru/Boltin Picture Library/The Bridgeman Art Library International (tr) ©McConnell, James Edwin/Private Collection /Look and Learn/The Bridgeman Art Library International. 207 ©Cro Magnon/Alamy. 208 (bc) ©The Art Archive/Bibliothque des Arts Dcoratifs Paris/Gianni Dagli Orti (tr) ©The Art Archive/Bibliothque des Arts Dcoratifs Paris/Gianni Dagli Orti. 209 ©The Art Archive/Kharbine-Tapabor. 211 ©Luis Marden/National Geographic Stock. 215 ©Florian Kopp/imagebroker/Alamy. 216 ©Paolo Aguilar/epa/Corbis. 219 ©Corey Wise/Lonely Planet Images/Getty Images. 220 ©Mike Theiss/National Geographic Stock. 221 ©Richard Nowitz/National Geographic Stock. 223 ©Ivan Alvarado/Reuters/Corbis. 224 ©Jeremy Hoare/Alamy. 225 (b) ©James P. Blair/National Geographic Stock (c) ©Nicolas Misculin/Reuters (t) ©Kit Houghton/Corbis. 226 ©Jennifer Shaffer/National Geographic School Publisingg. 228 ©Keren Su/Corbis. 229 (t) ©Imaginechina/Corbis. 231 ©Robert Clark/National Geographic Stock. 232 ©Sebastiao Moreira/epa/Corbis. 234 (b) ©Mike Theiss/National Geographic Stock (bl) ©Charles Dharapak/Pool/Reuters. 240 ©Pete McBride/National Geographic Stock. 242 (b) ©Bob Krist, National Geographic Stock (bkg) ©Richard List, Corbis (c) ©Anne Keiser, National Geographic Stock. 245 ©Atlantide Phototravel/Corbis. 246 (b) ©Yann Arthus-Bertrand/Corbis (bkg) ©Menno Boermans/Aurora Photos/Corbis (t) ©Douglas Pearson/Corbis. 248 ©Owi-Diasign/Photolibrary. 250 ©National Geographic Stock. 252 ©Octavio Aburto. 254 (bc) ©Anne Keiser/National Geographic Stock (bl) ©Agnieszka Pruszek/National Geographic My Shot/National Geographic Stock (bkg) ©Jim Richardson/National Geographic Stock (br) ©Steve Raymer/National Geographic Stock. 255 (bcl) ©Greg Dale/National Geographic Stock (bcr) ©James P. Blair/National Geographic Stock (bl) ©Richard Nowitz/

National Geographic Stock (br) ©James L. Stanfield/ National Geographic Stock. 256 ©Panoramic Images/ National Geographic Stock. 258 (l) ©Richard Nowitz/ National Geographic Stock (r) ©The Art Gallery Collection/Alamy. 259 ©PoodlesRock/Corbis. 260 ©Jean-Pierre Lescourret/Corbis. 262 (l) ©Hoberman Collection/Corbis (r) ©The Bridgeman Art Library International. 263 ©North Wind Picture Archives/Alamy. 266 (bl) ©The Bridgeman Art Library (br) ©Underwood & Underwood/Corbis. 267 (bl) ©Doug Taylor/Alamy (br) ©The Bridgeman Art Library (t) ©The Bridgeman Art Library. 268 ©Richard Schlect/National Geographic Stock. 270 ©Paul Thompson/Corbis. 272 (l) ©The Bridgeman Art Library (r) ©The Gallery Collection/ Corbis. 273 (b) ©Peter Horree/Alamy (t) ©The Bridgeman Art Library. 274 ©The Bridgeman Art Library. 275 ©SCANFOTO/X00729/Reuters/Corbis. 276 (l) ©Stefano Bianchetti/Corbis (r) ©Clynt Garnham/Alamy. 277 (l) ©Michael Nicholson/Corbis (r) ©Michael Nicholson/ Corbis. 279 ©DC Premiumstock/Alamy. 282 ©Rudy Sulgan/Corbis. 284 ©MARKA /Alamy. ©286 (l) ©Leonardo da Vinci (1452-1519) Louvre, Paris, France/ Giraudon/ The Bridgeman Art Library (r) Claude Monet (1840-1926) Musee Marmottan, Paris, France/ Giraudon/ The Bridgeman Art Library Nationality. 287 ©Arnaud Chicurel/Hemis/Corbis. 288 ©The Gallery Collection/ Corbis. 289 ©Columbia/The Kobal Collection. 290 ©Jon Arnold/JAI/Corbis. 291 ©Sergiy Koshevarov/ StockPhotoPro. 292 ©Photolibrary. 294 ©Paul Seheult/ Eye Ubiquitous/Corbis. 295 ©Perutskyi Petro/ Shutterstock Photos. 296 ©Gregory Wrona/Alamy. 299 ©Michael Dunning/Photographer's Choice/Getty Images. 304 ©Grand Tour/Corbis. 306 (b) ©Gerd Ludwig/Corbis (bkg) ©Photolibrary (c) ©Gordon Wiltsie, National Geographic Stock (t) ©Rebecca Hale, National Geographic Stock. © 309 ©Klaus Nigge/National Geographic Stock. 310 (bkg) ©Menno Boermans/Aurora Photos/Corbis (br) ©Bruno Morandi/Robert Harding World Imagery/Corbis (tl) ©Maxim Toporskiy/Alamy. 312 ©Denis Sinyakov/Reuters/Corbis. 314 ©Cary Wolinsky/ National Geographic Stock. 316 ©National Geographic Stock. 318 ©Gerd Ludwig/National Geographic Stock. 319 (tl) ©U.S. Geological Survey (tr) ©NASA. 320 (l) ©Sisse Brimberg/National Geographic Society (r) ©James L. Stanfield/National Geographic Society. 321 (l) ©Massimo Pizzotti/Getty (r) ©Dallas and John Heaton/ Photolibrary. 322 ©Richard Klune/Corbis. 323 ©imagebroker/Alamy. 324 ©The Bridgeman Art Library (r) ©Cary Wolinsky/National Geographic Society. 325 (l) ©The Bridgeman Art Library. 326 ©North Wind Picture Archives/Alamy. 328 (l) ©Bettmann/Corbis (r) ©The Art Archive. 329 (b) ©Bettmann/Corbis (t) ©Thomas Johnson/Sygma/Corbis. 332 ©Paul Harris/JAI/Corbis. 334 ©Arne Hodalic/Corbis. 335 (tc) ©Michael Runkel/Robert Harding World Imagery/Corbis (tr) ©Maria Stenzel/ National Geographic Stock (tr) ©Sean Sprague/ Photolibrary. 336 ©Olaf Meinhardt/VISUM /Fotofinder. 339 ©Kristel Richard/Grand Tour/Corbis. 340 ©Shepard Sherbell/CORBIS SABA. 342 ©Imagesource/Photolibrary. 344 ©Oleg Nikishin/Stringer/Getty Images. 347 (c) ©iStockphoto ©Michael Dunning/Photographer's Choice/Getty Images. 352 ©iStockphoto. 354 (b) ©Ingo Arndt/Minden Pictures/National Geographic Stock (bkg) ©David Alan Harvey/National Geographic Stock (c) ©Mitsuaki Iwago/Minden Pictures/National Geographic Stock (t) ©Kakenya Ntaiya. 357 ©Top-Pics TBK/Alamy. 358 (bkg) ©Menno Boermans/Aurora Photos/Corbis (bl) ©tbkmedia/Alamy (tr) ©Michael Poliza/National Geographic Stock. 360 ©Michael Nichols/National Geographic Stock. 362 ©Philippe Bourseiller/Getty

Images. 364 ©Ian Nichols/National Geographic Stock. 366 ©Mike Hutchings/Reuters. 368 (bc) ©Beverly Joubert/National Geographic Stock (bl) ©Beverly Joubert/ National Geographic Stock. 370 ©Gerald Hoberman/ Hoberman Collection UK/Photolibrary. 372 (bl) ©The Trustees of the British Museum/Art Resource (br) ©ADB Travel/dbimages/Alamy. 373 ©HIP/Art Resource. 376 ©Private Collection/Look and Learn/The Bridgeman Art Library International . 377 (bl) ©Mary Evans Picture Library/The Image Works (br) ©Bruce Dale/National Geographic Stock. 378 (bc) ©James L. Stanfield/National Geographic Stock (bl) ©Tim Laman/National Geographic Stock (bkg) ©George Steinmetz/National Geographic Stock (br) ©Annie Griffiths/National Geographic Stock. 379 (bcl) ©Roy Toft/National Geographic Stock (bcr) ©Tino Soriano/National Geographic Stock (bl) ©Jodi Cobb/National Geographic Stock (br) ©Ed Kashi/ National Geographic Stock. 382 ©Ralph Lee Hopkins/ National Geographic Stock. 384 ©Vanessa Burger/Images of Africa Photobank Alamy. 386 ©Paul Gilham - FIFA/ FIFA via Getty Images. 387 ©David Alan Harvey/ National Geographic Stock. 388 (bc) ©Nigel Pavitt/John Warburton-Lee Photography/Alamy (bl) ©Sean Sprague/ Still Pictures/Photolibrary. 389 (cl) ©Michael Nichols/ National Geographic Stock (tr) ©Suzi Eszterhas/Minden Pictures/National Geographic Stock. 390 ©Jane Goodall Institute. 391 (bkg) ©Gerry Ellis/ Minden Pictures/ National Geographic Stock (br) ©Wade Davis/Ryan Hill. 392 ©Finbarr O'Reilly/Reuters. 394 ©George Steinmetz/ Corbis. 396 (b) ©Pascal Maitre /National Geographic Stock (tr) ©Joerg Boethling/Alamy. 398 ©Louise Gubb/ Corbis. 399 ©Michael Dunning/Photographer's Choice/ Getty Images. 400 ©Frederic Courbet/Still Pictures/ Photolibrary. 402 ©Ulrich Doering/Alamy. 403 ©Trinity Mirror/Mirrorpix/Alamy. 406 (bl) ©Chris Stenger/FN/ Minden Pictures/National Geographic Stock (br) ©Walker, Lewis W./National Geographic Stock. 407 (bkg) ©Clement Philippe/Arterra Picture Library/Alamy (bkg) ©Tim Fitzharris/Minden Pictures/National Geographic Stock (cf) ©Mattias Klum /National Geographic Stock (l) ©Tom Vezo/Minden Pictures/National Geographic Stock (rbkg) ©Ted Wood/Aurora Photos (rf) ©Thomas Lehne/ Alamy. 408 ©Anup Shah/Corbis. R39 ©Fresco J Linga/My Shot/National Geographic Stock. R40 ©George Grall/ National Geographic Stock. R41 ©SERDAR/Alamy. R42 ©Bettmann/Corbis. R44 ©Hulton Archive/Getty Images. R45 ©Svabo/Alamy. R47 ©Scott Olson/Getty Images. R49 ©Lordprice Collection/Alamy. R50 ©Parbul TV via Reuters TV/Reuters/Corbis. R51 ©Markus Altmann/ Corbis. R52 (b) ©Mario Lopez/epa/Corbis (t) ©Alinari Archives/Corbis. R53 (b) ©Anindito Mukherjee/epa/ Corbis (t) ©Werner Forman/Art Resource. R54 (b) ©Ronen Zvulun/Reuters/Corbis (t) ©The Art Archive/Museo del Prado Madrid. R55 (b) ©Mak Remissa/epa/Corbis (t) ©Rubin Museum of Art/Art Resource. R56 (b) ©Joe McNally/National Geographic Stock (t) ©Biju/Alamy. R57 (b) ©Robert Harding World Imagery/Corbis (t) ©Art Directors & TRIP/Alamy. R58 (b) ©Christian Kober/ Photolibrary (t) ©National Palace Museum Taiwan/The Art Archive.

Map Credits

Mapping Specialists, LTD., Madison, WI.
National Geographic Maps, National Geographic Society

Illustrator Credits

Precision Graphics